Structural Fire Engineering

Venkatesh K. R. Kodur, Ph.D., P.Eng.
M. Z. Naser, Ph.D., P.E.

Mc Graw Hill

New York Chicago San Francisco
Athens London Madrid
Mexico City Milan New Delhi
Singapore Sydney Toronto

Library of Congress Control Number: 2019957426

Structural Fire Engineering

1 2 3 4 5 6 7 8 9 CD 25 24 23 22 21 20

ISBN 978-1-260-12858-1
MHID 1-260-12858-X

Sponsoring Editor
Ania Levinson

Editorial Supervisor
Stephen M. Smith

Production Supervisor
Pamela A. Pelton

Acquisitions Coordinator
Elizabeth M. Houde

Project Manager
Poonam Bisht, MPS Limited

Copy Editor
Joanne Slike

Proofreader
Mohammad Taiyab Khan, MPS Limited

Indexer
Edwin Durbin

Art Director, Cover
Jeff Weeks

Composition
MPS Limited

Contents

About the Authors

Venkatesh Kumar R. Kodur, Ph.D., P.Eng., is University Distinguished Professor in the Department of Civil and Environmental Engineering at Michigan State University and Director of MSU's Center on Structural Fire Engineering and Diagnostics. He has published over 425 peer-reviewed papers in international journals and conference proceedings, and, per Google Scholar, he has more than 11,000 citations and an h-index of 58. The methodologies, techniques, and design guidelines resulting from Dr. Kodur's research have been incorporated in various codes and standards, as well as practical applications, in the United States and around the world and are instrumental in minimizing the destructive impact of fire in the built infrastructure. Widely recognized as an international expert on fire engineering, he was part of the FEMA/ASCE assessment team that studied the collapse of the World Trade Center after 9/11. In recognition of his scholarship and professional contributions, Dr. Kodur has been inducted as a Fellow of two academies (Canadian Academy of Engineering, Indian National Academy of Engineering) and of three institutes (American Society of Civil Engineers, Structural Engineering Institute, American Concrete Institute).

M. Z. Naser, Ph.D., P.E., is Assistant Professor in the Glenn Department of Civil Engineering at Clemson University. He specializes in fire engineering research, computational intelligence, and retrofitting of damaged/aging structures and is active in various engineering communities.

Preface

Fire represents one of the most severe environmental conditions to which structures may be subjected, and, hence, the provision of appropriate fire safety measures for structural members is an important aspect in the design of high-rise buildings and built infrastructure. These requirements can be attributed to the fact that when other measures for containing fire fail, structural integrity and stability are the last line of defense. In recent years, fires have become a growing concern around the world due to the increased number of high-rise buildings, higher fuel loads in buildings, the introduction of new construction materials and slender sectional shapes with lower fire resistance, and greater expectations from society in terms of safety and security.

The conventional approach for evaluating the fire resistance of structural members is through prescriptive-based methodologies. These methodologies have significant drawbacks and do not provide optimal and rational fire resistance designs. Therefore, the last three decades have witnessed important research endeavors devoted to developing better understanding of structural behavior under fire conditions, as well as rational calculation methodologies for evaluating the fire resistance of structures.

Further, a number of countries around the world are updating their codes and standards by introducing performance-based fire safety design provisions and associated rational methodologies for fire resistance evaluation. A performance-based design approach to fire safety often facilitates innovative, cost-effective, and rational engineering designs. European technical committees were in the forefront of implementing some of the latest research findings into Eurocodes and other standards to enable the application of rational fire engineering principles in the design of structures. More recently there have been growing efforts in a number of countries, including the United States and Canada, to introduce performance-based fire engineering provisions in codes and standards. However, to undertake such performance-based fire safety design, advanced numerical models, calculation methodologies, design manuals, textbooks, and trained personnel are needed.

While the performance-based design provisions and associated rational calculation approaches laid out in the codes and standards provide a needed framework for undertaking rational fire design, they are nevertheless far from being useful textbooks, lecture notes, or guidance documents, since there is no detailed commentary on or explanation for the various specifications or calculation methodologies. In addition, fire design is rarely taught as part of regular engineering curricula, and thus most engineers, architects, and regulators may not be fully versed in the necessary background to easily understand the relevant clauses, to make interpretations, or to recognize the limits of application of various rules. In other words, unless one has some level of expertise in fire safety engineering, it is not easy to apply the current provisions in codes and standards to most practical situations. Compounding this problem is that there are only a handful of textbooks in the area of structural fire engineering.

This book is aimed at filling the current gaps in structural fire engineering and presents fundamental behavioral aspects, as well as principles and concepts, of structural design under fire conditions in a systematic way. The concepts, methodologies, approaches, and guidelines for the design and analysis of civil engineering structures under fire conditions are explained in depth. The fire resistance design and analysis approaches presented in this book cover a wide range of structures made of different structural materials used in construction—namely concrete, steel (and composite), wood, and masonry. In addition, the book provides significant new information compared to currently available books in the field. Specifically, analysis approaches for evaluating the fire resistance of fiber-reinforced polymer (FRP)–strengthened concrete structures and transportation structures, such as bridges and tunnels, are presented in two separate chapters. Finally, a full chapter is dedicated to conditional assessment, as well as to residual capacity evaluation, of fire-damaged structures.

This book is organized into 13 chapters. Each chapter starts with introductory behavioral concepts to provide a fundamental understanding, and then presents advanced concepts, calculation methods, and the latest research findings in the field of structural fire engineering. In addition, strategies that can be developed for overcoming fire hazards in structures made of different material types are discussed in detail. Further, emerging trends and research needs in subtopics in each specific area are presented at the end of the chapter to assist researchers and Ph.D. students interested in undertaking further research to advance the state of the art of structural fire engineering.

Chapters 1 and 2 provide an overall introduction to fire safety design, highlight the magnitude of the fire problem in built infrastructure, and lay out basic design strategies for achieving fire safety in buildings. Chapter 3 discusses various aspects of fire growth in a compartment and lays out steps for fire load calculations for different building occupancies, together with basic heat transfer calculations. Chapter 4 presents the effects of elevated temperatures on thermomechanical properties of commonly used construction materials. Chapters 5 and 6 outline principles and procedures for evaluating fire resistance through testing and calculation methods. Then Chaps. 7, 8, 9, 10, and 11 address fire design of steel, concrete, composite (steel-concrete), wood, and FRP-strengthened concrete structures, respectively. In each of these chapters, behavioral aspects of structures are explained and then specific fire resistance design/analysis approaches—as per prescriptive-based methodologies and through advanced analysis procedures—are presented for the type of construction material. In addition, detailed strategies that can be developed to enhance fire resistance of a structure made of each material are presented. Chapter 12 discusses specific aspects and methodologies associated with fire resistance design and analysis of critical transportation structures, such as bridges and tunnels. Finally, Chap. 13 discusses steps for conditional assessment and residual capacity evaluation of fire-damaged structures. In each chapter the subject matter is explained with the aid of numerous illustrations, figures, tables, and step-by-step numerical examples to help readers grasp the topic with ease.

This book allows designers to go beyond current prescriptive approaches that generally do not yield a useful understanding of the actual performance during a fire and instead explore analyses that present a realistic evaluation of structural fire performance. The book is written to be a textbook for graduate and senior undergraduate students in civil, mechanical, and fire protection engineering programs, as well as for students in architectural disciplines. Chapters 7 to 12 are written as stand-alone chapters so that specific chapters can be adopted for introducing structural fire design in traditional civil engineering elective courses. Therefore, these chapters can be incorporated as additional lecture modules in senior undergraduate or graduate design courses relating to the design or analysis of steel, concrete, composite, wood, FRP-strengthened concrete, or transportation (bridge) structures, respectively. This book is also well suited for designers, engineers, practitioners in design/consulting firms, and regulatory officials in city/municipal agencies. The information in each chapter is presented such that those not familiar with fire safety engineering can make the relevant calculations for establishing the fire resistance of structures.

We hope that this book will enable researchers, architects, engineers, design professionals, and students to develop a greater insight into structural fire engineering so that safer structures can be designed for fire conditions.

If needed, an errata list will be placed on Dr. Kodur's website (www.egr.msu.edu/~kodur) and on Dr. Naser's website (www.mznaser.com).

Venkatesh K. R. Kodur, Ph.D., P.Eng.
M. Z. Naser, Ph.D., P.E.

Acknowledgments

From Dr. Kodur

Much of the information presented in this book is based on extensive research undertaken by various researchers in the field of structural fire engineering. I have had opportunities to interact and collaborate with a large number of these researchers from around the world and have benefited enormously from discussions with them. I am grateful to the researchers for such fruitful interactions. Also, this book draws on much of the expertise gained from my experiences at the National Research Council Canada and Michigan State University, and I am thankful for the support of these organizations, as well as numerous funding agencies and industries (from around the world), in my undertaking of numerous research projects relating to structural fire engineering. In particular, I am grateful to a number of M.S. and Ph.D. students, post-doctoral fellows, and visiting scholars who have worked with me over the last 25 years and have contributed to the development of numerous methodologies and approaches for structural fire engineering. I am also most indebted to my wife, son, daughter, and parents for their encouragement, support, and patience during the preparation of this book.

From Dr. Naser

I would like to express my gratitude for Dr. Kodur's continuous guidance and mentorship, and for giving me the opportunity to participate in this book. I would also like to thank my wife, family, friends, colleagues, former advisors, and students for their everlasting support and understanding.

Finally, we would like to thank our publisher, McGraw-Hill, and, in particular, original sponsoring editor Lauren Poplawski, for her patience, guidance, and support.

CHAPTER 1

Fire Hazard in Built Infrastructure

1.1 Overview

This chapter highlights the magnitude of fire problems in built infrastructure. Recent statistics related to fires in buildings is presented, and some major fire incidents are highlighted to illustrate the devastating effect fires can have on built infrastructure. In addition, current provisions in building codes and standards to overcome fire hazard in the built infrastructure are reviewed. Finally, the importance of provision of structural fire safety to overcome fire hazard in buildings is discussed.

1.2 Magnitude of Fire Hazard

1.2.1 Fire Problem

Fire represents one of the most severe hazards to which built infrastructure may be subjected during its service life (Kodur et al., 2010; Kodur and Naser, 2013). Fire can occur as a primary event, arising from accidents such as from cooking or electrical short circuit, or it can break out as a secondary event following other hazards such as earthquake, impact, blast, and so on. When grown to full size, such fires can have a devastating effect on people, property, and environment and can lead to significant losses, including human life and injury, damage to structure, temporary shutdown of the building or business operations, repair and rehabilitation costs, and environmental cleaning costs, among others. In most countries, the total monetary costs each year arising from fire hazard run in the order of 1% to 2% of the gross domestic product (GDP) of the country (Hall, 2014).

In recent years, there has been a surge in the number of fires occurring in built infrastructure around the world. This increase is attributed to a number of factors, such as an increased number of high-rise buildings, a high level of combustible materials (contents) in buildings, and the transportation of highly combustible chemicals across bridges and through tunnels (Gencer, 2013). The increase in intense fires in built infrastructure is leading to a growing number of losses from fire.

While it is impractical to prevent all fires from occurring in built infrastructure, the devastating effects of fire can be mitigated through the provision of appropriate fire safety measures. Provisions of such fire safety measures are critical not only for ensuring life safety but also to minimize property and environmental damage, as well as to minimize the interruption of business following a fire incident. The common fire protection features incorporated in buildings include measures to prevent ignition, to limit the growth of the fire, to resist the fire, and to fight the fire. These measures can be incorporated into the initial construction stage of a building and can also be implemented as an upgrade to an existing structure. Other fire safety measures include training occupants to perform basic fire safety-related skills like using fire extinguishers and performing orderly evacuations in emergency situations.

Although appropriate fire safety measures reduce the chances of fire and fire losses, such provisions incur high costs and involve a number of practical complexities with regard to installation and upkeep. Further, incorporating such fire safety features can reduce available usable space, and these measures therefore need to be optimally designed and installed to minimize needed space for operation and installation. For this reason, current fire safety features are mostly required in buildings where life safety is at risk. However, in other structures where endangerment to human life is very low, fire safety measures are not currently incorporated to the same extent as that in buildings. Examples of such structures include warehouses and transportation infrastructure such as bridges and tunnels.

It can be inferred from above discussion that fire is a complex and unique loading event with multidimensional scale which encompasses the domains of people (occupants and first responders), structure (property and associated damage costs), and the surrounding environment (air and water bodies). Due to this complex interaction of factors, rational design approaches based on engineering principles must be applied to achieve operational and cost-effective fire safety in built infrastructure in order to prevent or limit casualties and fire-induced damage. Applying such rational approaches to structures is much more critical, since these structures are to be designed for a number of loading events (wind, blast, earthquake) and fire being one of these loadings. The content presented in this book is aimed at providing design and analysis approaches through rational "structural fire engineering" principles in order to mitigate adverse effects of fire in built infrastructure.

1.2.2 Fire Statistics

To ensure safety and performance throughout their lifespan, structures must be designed to withstand loading effects arising from a number of external forces. Unlike other extreme events such as earthquakes, hurricanes, and floods, occurrence of fire is more often a local phenomenon and is much more frequent. Thus, while other hazards such as earthquake and hurricanes are more likely to hit specific areas that are prone to these events, fire hazard is a threat to the built environment in all regions of a country and in all countries in the world. However, while fire incidents are random and cannot be predicted, they do follow stochastic or probabilistic patterns and distributions that can be analyzed. The probabilistic nature of such events can be best described as a series of independent events that occur over an extended period of time, rather than to a geographic location.

In recent years, reliable statistics relating to fire incidents and associated losses are being compiled in many countries. The National Fire Protection Association (NFPA) produces annual reports on fire losses in the United States. The latest NFPA report reveals that U.S. fire departments responded to an estimated 1,342,000 fires in 2016. These fires caused 3390 civilian deaths, 90 firefighter deaths, 14,660 civilian injuries, and 68,085 nonfatal injuries to firefighters while on duty. Residential fires constituted the bulk of reported fires and contributed to 80.7% of total fire deaths and 60% of the total monetary losses. In 2016 in the United States, estimated property losses due to fire was $10.4 billion. Of this amount, $7.9 billion in property damage occurred due to structural fires and included $5.6 billion in property loss from residential fires.

While the fire losses in the United States seem astonishingly high, unfortunately, a review of similar fire statistics in Europe and around the world reveals similar trends. For example, recent reports from the United Kingdom indicate that 161,700 fires occurred in 2016 causing 261 deaths and 3133 injuries. Out of these fires, 714 occurred in high-rise buildings and 37,721 in dwellings. The total cost of fire in the United Kingdom was estimated at £7.7 billion, which is equivalent to approximately 0.9% of the gross value of the economy.

The magnitude of the fire problem is shaping up to be a serious problem to emerging economies as well. In India, although the total number of fires was not reported, fire-related losses in 2014 were estimated at 19,513 deaths and 1889 injuries. The root cause analysis of these fires reveals that 18.3% of reported fire accidents occurred in residential buildings. The overall monetary fire losses in India were estimated at more than $10 billion, which turns out to be 30% more than that estimated due to floods. In the case of China, a total of 189,000 fires were reported in 2010. These fires led to 11,419 deaths.[1] In the same year, 200,386 fires occurred in Russia, causing 10,400 deaths and 12,800 injuries. Note that the way fire incidents and related losses are counted varies from one country to other. Figure 1.1 illustrates fire-related deaths in some countries based on statistics compiled by The Geneva Association for the year 2010. These global fire statistics reveal that in most countries fires constitute a significant number of deaths and injuries as well as losses to economy.

The preceding discussion clearly indicates that fire hazard is a universal problem. Much of the fire losses occur in buildings, since the probability of a fire is much higher than in other built infrastructure, as ignition sources such as electrical wiring, cooking appliances, and heating/cooling equipment are commonly present. In addition, buildings contain significant fuel content and furnishings that constitute main combustibles.

The cause of fire in buildings varies because of differences in types of building layout (such as open-concept architecture) and construction materials (wood, steel, masonry, concrete, etc.), as well as local customs and environmental conditions. For example, until the 1980s, in the United States the main cause

[1] No reports are available on number of injuries due to fire incidents.

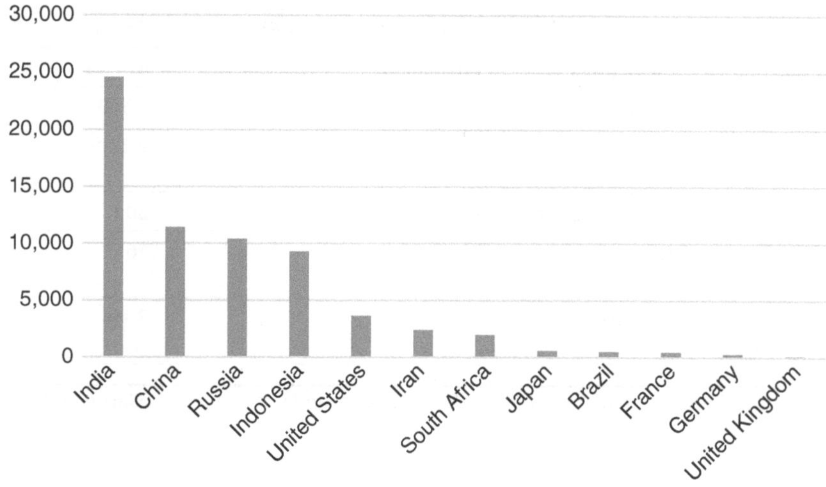

Figure 1.1 Fire-related deaths in countries around the world. (*Reproduced from the Geneva Association, 2010.*)

of ignition leading to fire was smoking. However, after the establishment and enforcement of laws banning smoking in many public buildings in recent years, the main source of ignition in the United States is now cooking. In India, electric short circuiting (possibly due to poor wiring) is the most prominent cause of fires. These aspects (ignition sources, building features, fuel load type) vary from one country to another because of such factors as construction methods, local customs materials, practices, design philosophy, degree of enforcement of fire safety regulations, and so on.

Fire statistics also indicate that not all fires grow to full size and not all fully developed fires result in major damage to structural systems. In order to quantify adverse effects of fire on structural systems, recent studies (Rahikainen and Keski-Rahkonen, 2004; Lin, 2005; Naser and Kodur, 2015) applied the Poisson distribution approach to evaluate the probability of a fire breaking out in a structure, as well as fires leading to the collapse of a structure. Poisson distribution principles assume that fire incidents are random events that follow a probabilistic approach, and thus are best described as a series of independent events that occur over time. This study reported that the annual probability of a fire breakout in a building is about 29.5%.

The above study also showed that the probability of fire-induced collapse of buildings can be deduced from data compiled by the National Institute of Standards and Technology (NIST) as well as a survey by Wardhana and Hadipriono (2003). In these two surveys, the total number of collapsed buildings from various catastrophic events, including fire, is 225, and those due to fire-induced collapse alone is 29. Utilizing this data in the Poisson distribution, the probability of fire-induced collapse of a building can be estimated to be 12.1% (Scheer, 2011). The same study also estimated the probability of fire breaking out and fire-induced collapse in other structures such as bridges, as shown in Table 1.1. The table shows that the probability of fire breaking out and fire-induced collapse in buildings (29.5%; about 1 in 3) is much higher than that on bridges. However, the probability of collapse of a building when fire occurs is 12.1% (about 1 in 8). This lower probability of fire-induced structural collapse in buildings is due to activation of fire safety systems.

1.2.3 Recent Fire Incidents

The preceding statistics clearly show that while fire breaking out in the built environment is still a rare event, only in limited cases can it grow into a large-scale fire that produces fatalities, property damage, and partial/complete collapse of structures. To further illustrate the vulnerability of buildings and infrastructure to fire hazard, the next sections discuss some of the notable recent fire incidents.

World Trade Center Twin Towers

The collapse of the Twin Towers of the World Trade Center (WTC) resulting from the September 11 terrorist strikes is one of the most destructive fire-induced collapses of tall buildings in history (see Fig. 1.2). This collapse mainly resulted from fires initiated by the crashing of a jet airliner into each of the towers on September 11, 2001 (McAllister and Corley, 2002). Each tower was 110 stories and was square in plan, with each face approximately 63 m wide. The North Tower (WTC 1) and the South Tower (WTC 2)

Category	Buildings (2012, 2006)	Bridges (2017)	Tunnels (2008)
Total number of structures	118,000,000	614,387	70,000
Reported fire incidents	480,500	4,500	177
Probability of a fire breaking out (yearly)	29.5%*	2.27%	1.5%
Number of collapsed structures	225	503	–
Number of collapsed structures due to fire	29	16	–
Probability of collapse due to fire (yearly)	12.1%**	3.1%	–

Table 1.1 Comparison of Probability of Fire Incidents and Fire-Induced Collapse in Bridges and Buildings

Figure 1.2 Moment of second jet impact on WTC 2 on September 11, 2011, New York City. (*Courtesy of Robert J. Fisch.*)

were of a total height of 417 m and 415 m, respectively. The twin towers, built in 1973, were designed as unique "tube-in-tube" structures, with 59 steel columns along each face of the building. The floors were composed of 100-mm lightweight concrete slabs, laid on a steel deck and connected to a network of trusses. These trusses spanned 18 m along the long-span area and 11 m along the short-span area. Overall, these monumental structures were designed and built with high levels of structural resiliency.

Results of the joint investigation conducted by the American Society of Civil Engineers (ASCE) and the Federal Emergency Management Agency (FEMA), as well as the National Institute of Standards and Technology (NIST), concluded that the structural systems of each tower were able to withstand the impact of the plane despite loss (fracture) of a number of columns. However, ensuing fires of very high intensity that started at multiple floors along the impacted zones rapidly spread to multiple floors. Due to failure or nonoperationality of various fire protection systems, including firefighting, ensuing fires caused further thermal damage to already-weakened steel trusses and columns, eventually leading to the total collapse of the towers. The collapse of the towers claimed the lives of 2753 people and caused at least $10 billion in property and infrastructure damage.

World Trade Center Seven

On September 11, 2001, another steel-framed building, the World Trade Center Seven (WTC 7), also collapsed entirely due to fire. This modern 47-story steel-framed office building, built in 1987, had a trapezoidal footprint of 100 m long and 43 m wide and was 190 m in height. The design of this building incorporated eight 14-m-long cantilever transfer girders that spanned the building core and the north side

FIGURE 1.3 Satellite top view and collapse of Faculty of Architecture Building at the Delft University of Technology, the Netherlands. *(With permission from ASCE.)*

of building at the seventh floor. The transfer girders, made of steel plate girders, varied in depth from 2.7 m at the north end and down to 1.37 m at the southern section closest to the core. Each transfer girder weighed approximately 52 tons and was connected to steel beams via single-plate shear connections through thin web. These transfer girders were designed to carry significant levels of concentrated loads coming from upper floors.

On September 11, a fire started on the seventh floor of WTC 7 as a result of debris and sparks from WTC 1 and WTC 2. The fire then spread to various floors and led to intermittent and simultaneous burning at multiple stories. About 5:00 p.m. there was intense burning at floors 5 and 7 in the vicinity of transfer girders. As a result, these girders were likely to have experienced a rapid rise in temperature and became overloaded with high levels of vertical shear forces. Failure of these transfer girders triggered the fire-induced progressive collapse of this building (Gann et al., 2008), leading to significant property damage.

Faculty of Architecture Building at the Delft University of Technology

On May 13, 2008, a serious fire occurred in the building of the faculty of architecture building at the Delft University of Technology in the Netherlands (Engelhardt et al., 2013). Completed in 1970, this building was a reinforced concrete structure composed of a 3-story structure, which effectively served as a podium, with a 13-story tower located on top (see Fig. 1.3). The building was 140 m long and 76 m wide. An electrical short circuit in a coffee machine on the sixth floor led to the start of the fire. This fire spread quickly in the building mainly due to wood finishings in the interior design. Fortunately, this fire did not cause any fatalities. This was attributed to the fast and efficient evacuation and also the withdrawal of firefighters from the building after about three hours of firefighting. Still, six stories of the concrete structure collapsed, leading to the dismantling of the entire building. The structural collapse was attributed to a 65% loss in flexural capacity in joists, as well as spalling of concrete cover in columns. The total estimated losses, including insurance payments, was put at more than 150 million euros. This incident exposed the vulnerability of concrete buildings in fire.

Plasco Building

On January 19, 2017, a fire broke out at the 17-story Plasco building in Tehran, Iran. This steel-framed building, dating from the early 1960s, included a shopping center and garment workshops, and was among the first high-rises built in the Iranian capital. The fire appeared to originate in the building's garment workshops on the upper floors, where tailors would cook food and use old kerosene heaters in winter (Naji and Rohani, 2017). The intense fire blazed through the upper stories for over three hours, sending out thick plumes of black smoke. More than 200 firefighters were at the scene to fight the fire. It was estimated that 20 to 30 firefighters were on the third floor of this building when the structure collapsed. Another 70 people were injured, 23 of them seriously (see Fig. 1.4). The post-fire investigation attributed the cause of fire to faulty electrical wiring and the ensuing progressive collapse to negligent enforcement of 22 of the national building fire regulations, including issues related to structural fire protection.

Kiss Nightclub

On January 27, 2013, a fire broke out at a college party at Kiss, a nightclub in Santa Maria, Rio Grande do Sul, Brazil. Although the capacity of this 615-square-meter club was under 700 people, it was packed with

FIGURE 1.4 Moment of collapse of the Plasco building, Tehran, Iran. (*Attribution: Tasnim News Agency.*)

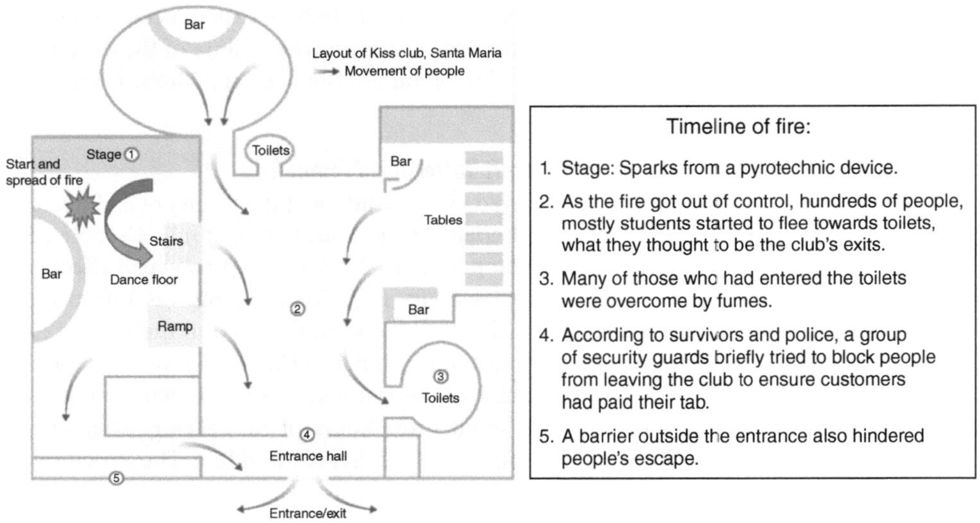

FIGURE 1.5 Layout of Kiss club, Santa Maria, Brazil. (*Courtesy of Fúlvio.*)

a crowd of an estimated 1200 to 1300. The fire started as the result of fireworks ignited by band members and then spread rapidly without giving much time for occupants to evacuate mainly due to overcrowding and a lack of proper exits. This tragic fire resulted in 242 deaths and 630 injuries. Most of the dead were college students 18 to 21 years old. This was one of the deadliest nightclub fires and the second most devastating fire disaster in the history of Brazil.

The fire started at around 2:30 a.m. when sparks from fireworks ignited flammable acoustic foam located above the stage. The blaze spread throughout the packed club, starting from the stage to the dance floor, as shown in Fig. 1.5, emitting thick and toxic smoke. Because the club lacked proper fire safety measures (sprinklers, exits, and proper ventilation), at least 90% of the victims died of smoke inhalation. Other contributors to fire spread were the use of cheap and combustible insulating foam material in the ceiling and faulty fire safety extinguishers. Another reason for the high death toll was that the club had only one exit. Firefighters had to open a hole in the outer wall to allow more people to escape. This incident highlighted the need for installing proper fire protection systems, including entry and exit paths.

Grenfell Tower

This 24-story tower block located in London and built in 1967 experienced a major fire incident on June 14, 2017. This tower housed more than 600 people in 129 apartments. Like many towers built in the UK at the time, Grenfell Tower had a single central staircase and was built mainly of concrete. Between 2012 and 2016, the Grenfell Tower underwent a major renovation that costed £8.7 million. As part of this renovation, the concrete structure received new windows and new aluminum composite rainscreen cladding. Two types of cladding were used: Arconic's Reynobond PE, which consists of two coil-coated aluminum sheets that are fusion-bonded to both sides of a polyethylene core, and Reynolux aluminum sheets. Beneath this cladding and fixed to the outside of the walls of the apartments was Celotex RS5000 PIR thermal insulation.

In the early hours of June 14, 2017, a fire started on the fourth floor, allegedly resulting from poor wiring in a faulty refrigerator. This fire quickly spread to upper floors on one side of the tower. The fire on the exterior of this tower moved upward, reentering the building as shown in Fig. 1.6. While residents started to alert neighbors and began to evacuate the building, it took only six minutes for the London Fire Brigade to arrive at the scene. A team of more than 250 firefighters from 70 fire engines attempted to control the blaze.

Figure 1.6 Development of Grenfell Tower fire, London. (*Courtesy of Natalie Oxford.*)

By sunrise, five hours after start of fire, firefighters were still battling the fire and trying to spray areas where people were seen trapped. The fire continued to burn on the tower's upper floors, and it was not brought under control until 13 hours later. Although fears were expressed on the structural stability and that the building could collapse, structural engineers determined that it was not in danger and that rescue teams could enter the damaged building to search for survivors and casualties. The fire brigade subsequently used an unmanned aerial vehicle (i.e., drone) to inspect the building and search for casualties. This fire caused 79 deaths, over 70 injuries, and significant property damage. Preliminary reports suggest that a number of fire protection features did not function effectively and the exterior cladding was one of the main issues that triggered the rapid spread of fire.

The fire incidents discussed in this section till now clearly highlight the adverse effects of fire on high-rise buildings, specifically vulnerability of structural system under combined effects of fire and structural loading. Similar to buildings, other infrastructure systems such as bridges, tunnels, and ports are vulnerable to the adverse effects of fire, including possible collapse. The next section highlights a few recent fire incidents in bridges and tunnels.

I-85 Bridge

The I-85 bridge, located on Route I-85 near Atlanta, Georgia, served 243,000 vehicles a day. The bridge is composed of 10 prestressed concrete girders resting on top of three reinforced concrete piers. This bridge was originally built in 1953 and reconstructed in 1985. In a recent inspection carried out in 2015, the bridge achieved a sufficiency rating of 94.6 on scale of 100, implying that the bridge was in excellent service condition. On March 30, 2017, a fire started underneath the bridge due to the burning of large polyvinyl chloride (PVC) pipes stored under the bridge.[2] The temperature of the fire rapidly grew to 900 to 1100°C.

During firefighting operations, first responders reported hearing explosive sounds, attributed to concrete spalling, and were advised to evacuate the fire scene. Thirty minutes into the intense heat, arising from burning of plastics and composite materials, one span 30.3 m long collapsed, while adjacent piers and spans underwent significant damage, as shown in Fig. 1.7. Post-fire investigation revealed major spalling in concrete piers and girders of the bridge. Losses due to this fire-induced collapse were estimated at $10 million. It took weeks to repair the fire-damaged bridge, resulting in significant traffic detours and indirect economic losses.

[2]Reports suggest that this was initiated from an act of vandalism.

Figure 1.7 Aftermath of fire incidents on I-85 bridge, near Atlanta, Georgia. (*Courtesy of GDoT.*)

MacArthur Maze Bridge

On April 29, 2007, a fire broke out at the two-span steel bridge of the MacArthur Maze interchange in Oakland, California, when a fuel tanker transporting 32,500 liters of fuel overturned under the bridge. The burning of highly combustible fuel led to intense heat, generating temperatures in the range of 1100°C in steel girders. The steel girders had no fire proofing, since as per current practice, no fire protection measures are required for structural members in bridges. Due to rapid rise in steel temperatures, the strength and modulus properties of steel rapidly degraded, which in turn amplified local instability (buckling) effects and led to large deflections and reduction in the capacity of the girders. The resulting large deflections caused significant fire-induced tensile forces that overstressed the bolted connections and caused shear failure of shear tab connections at the end of girders (Astaneh-Asl et al., 2009). As a result, steel girders, which are vulnerable to shear effects under fire conditions, collapsed in about 22 minutes. The fire-induced collapse of this bridge caused traffic detours and significant losses (estimated at $9 million), as it took weeks to repair the damaged bridge. Figure 1.8 shows the aftermath of the fire.

Mont Blanc Tunnel

On March 24, 1999, one of the worst recorded tunnel fires in Europe occurred in the 11.6-km-long Mont Blanc Tunnel connecting Chamonix, France, with Courmayeur, Italy. The fire was initiated when a Belgian truck with a refrigerated trailer carrying large quantities of margarine and flour (equivalent to a 6100 gallons of oil) caught fire due to an overheating refrigerator. The fire quickly spread to nearby vehicles, and the intense heat and smoke filled the entire tunnel section, preventing emergency evacuation and firefighting operations. The fire burned for 53 hours and temperatures reached in excess of 1832°F (1000°C). The intense heat and smoke led to 39 deaths, including a commanding firefighter officer, and injured 14 firefighters who were trapped in two of the emergency fire cubicles. In addition, 30 vehicles were burned. This tunnel fire caused losses of more than $390 million and significant damage to the tunnel structure, including severe spalling over 900 m of the concrete lining and roof, as shown in Fig. 1.9.

Shanxi Tunnel

A similar tunnel fire incident in China further illustrates the devastating effect of tunnel fires. On March 1, 2014, a methanol tanker rear-ended another tanker during a traffic jam in the 800-m-long highway tunnel near Jincheng in Shanxi province. As a result of this collision, methanol started to leak and both trucks involved in the accident caught on fire. The two drivers failed to alert police or nearby vehicles and fled the scene. The fire spread quickly as the leaked methanol flowed down the tunnel, lighting up other trucks loaded with coal and one tanker transporting natural gas. As the fire grew, thick smoke filled the entire tunnel, as the tunnel was not equipped with any ventilation facilities. About 100 minutes after the collision, a natural gas tanker exploded (see Fig. 1.10), sending one half of a truck flying 50 m inside the tunnel. A total of 42 vehicles were destroyed in this accident, which led to 31 deaths and significant damage to the tunnel.

Figure 1.8 Collapse of MacArthur Maze bridge due to fire, Oakland, California. (*Courtesy of Pete Klosterman.*)

Figure 1.9 Severe concrete spalling and damage of Mont Blanc Tunnel after fire, France and Italy. (*Courtesy of Elsevier.*)

Figure 1.10 Aftermath of recent fire incident in tunnel near Jincheng, Shanxi province, China. (*Courtesy of Elsevier.*)

The fire incidents discussed in this section involving bridges and tunnels highlight the fact that despite rare occurrence of fires in civil infrastructure, when such fires grow into intense fires, they can lead to fatalities, property damage, and partial/complete collapse of structure. In order to mitigate adverse effects of fire, a number of strategies need to be incorporated to minimize the effect of fire hazard in built infrastructure. These strategies are highlighted here and throughout the book.

1.3 Strategies to Overcome Fire Hazard

Since life safety is the primary consideration in building design, codes and standards require specific fire safety measures to be incorporated into buildings to minimize adverse effects from fire hazard. Fire safety in buildings is often achieved through the provision of active and passive fire protection systems. Active fire protection systems, which include water sprinklers, smoke and heat detectors, smoke control systems, exit paths, and firefighting, are intended to control the growth of fire through activation of a device such as a sensor or human action. On the other hand, passive protection systems are built into nonstructural members, such as doors and compartment walls, and structural members such as walls, floors, columns, and beams, and are intended to minimize the spread of fire and subsequent structural collapse.

In general, active fire protection systems are most effective during the early stages of fire breakout, when the fire is small enough to be extinguished or controlled. In the case of fully grown fires, especially when fire spreads throughout the structure, active systems such as sprinklers may not be effective in controlling the growing fire.[3] Thus, for controlling the spread of fully grown fires, passive protection systems are better suited, as structural and nonstructural systems are designed to withstand high levels of temperature rise with minimal strength degradation in structural members.

Passive fire protection features can be incorporated into structural and nonstructural members. These features are an inherent property of building members and do not require sensors to detect the fire and be activated. Although adding fire protection features in a building can significantly improve fire safety, these features can also increase construction costs, reduce usable space, and pose installation and maintenance challenges (Arablouei and Kodur, 2016). Thus, in order to develop optimum and economical fire protection solutions, appropriate strategies must be developed through proper design and analysis, as will be discussed throughout this book.

1.4 Structural Fire Safety

The primary goal of fire safety is to limit, to an acceptable level, the probability of death, injury, and property loss (including loss of functionality) during or in the aftermath of a fire. To achieve this goal, a set of fire protection measures must be incorporated into building design. One component of such measures is the provision of structural fire safety feature to structural and nonstructural members. Structural fire safety is measured in terms of fire resistance and is intended to minimize the spread of fire and to prevent collapse of the structure.

Fire resistance of a structural member (wall, beam, column, slab, or floor) is defined as the ability of a member to withstand the effects of fire so as to maintain load-bearing capacity and integrity function, and/or to act as an insulation barrier against the spread of fire during its entire duration. Fire resistance is measured in terms of time and is the duration from the start of fire exposure to the point at which the structural member ceases to perform insulation, integrity, and/or load-bearing functionality.

Building codes typically specify the required fire-resistance ratings for different structural members in a building. Fire-resistance ratings represent the duration a structural member has to withstand adverse effects of a representative standard fire through maintaining load-bearing capacity, insulation, and/or integrity criteria. The required fire-resistance ratings are dependent on the type of occupancy, height of the building, and area of compartment. Fire-resistance ratings are rounded off to the nearest 30 minutes or an hour period and are usually 30 min, 1 hr, $1^1/_2$ hr, 2 hr, 3 hr, or 4 hr. A structural member is said to satisfy codal requirements if the actual fire resistance of the member is higher than the required fire-resistance rating as per building code specifications.

1.5 Approaches to Evaluating Fire Resistance

Currently the primary method of evaluating fire resistance of a structural member is through prescriptive-based approaches. These approaches are mainly derived from data (fire ratings) generated through standard fire tests on typical structural members used in buildings. Such data are listed in tabulated form in directories of testing agencies or industry handbooks. These listings provide the minimum fire-resistance rating a structural member can have when subjected to similar conditions as that in a standard fire test. This simple "deemed-to-satisfy" approach has been embraced by the construction industry, engineering profession, and

[3]In some instances, active protection systems, such as sprinklers, may not properly operate when water pipes are damaged or have burst following an extreme event such as an earthquake.

regulation agencies for the last few decades. However, with their simplistic "select from a table/graph" nature, prescriptive approaches often do not account for various factors that influence fire resistance of structural members such as realistic fire scenario, load level, and restraint conditions present in an actual building during a fire event. As a result, these prescriptive approaches do not yield realistic fire performance in a structural member especially in extreme or unconventional fire loading, among other fire situations.

In order to overcome many of these limitations, engineering approaches based on rational design principles can be applied to evaluate the performance of structures under fire conditions. These rational approaches, often grouped under "performance-based fire safety design," can evaluate fire performance through simple engineering calculations or through advanced analysis approaches and can take into account actual dimensions of a structure, relevant material properties, realistic fire exposure, loading, and restraint conditions that are not considered under prescriptive approaches. Thus, rational approaches not only provide better assessment of fire performance but can also reflect actual performance of a representative structure. These approaches have been evolving over the last three decades, and their use in actual practice is still limited because these rational methods are not fully validated and require advanced expertise/training to carry out design and analysis.

Rational approaches also suffer from inadequate development of numerical platforms (algorithms and software, etc.), lack of reliable input parameters (i.e., material properties, fire scenarios), and limited range of applicability to modern construction (materials, fire exposure scenarios, etc.). Once these limitations are overcome, rational engineering approaches are expected to find increased applications in practice especially for unique and critical building or other critical infrastructure. For the sake of completion and in order to provide a wide range of approaches for fire design and analysis of structures, both prescriptive and performance-based approaches are discussed and treated equally within the subsequent chapters of this book.

1.6 Importance of Structural Fire Safety in Modern Construction

The role of structural fire safety to resist adverse effects of fire is much more critical in the context of modern construction. This is mainly due to a number of changes and innovations that are adopted in modern construction. Recently there has been a significant increase in number of high-rise buildings in urban areas due to growing population and high land costs. Modern construction is also evolving with respect to height, architectural and structural configuration, construction materials, and compartmental composition. Buildings are becoming larger in size and designs are made to accommodate multiple occupancy types with complex architecture. Advances in construction materials, analysis/design approaches, and sustainability aspects, combined with ever evolving architectural imagination and economic versatility, are leading to completely new types of buildings and structural shapes (Cowlard et al., 2013).

As indicted earlier, from a life safety consideration, fire resistance is one of the most important design considerations in high-rise buildings. Today's buildings are vastly different entities from those built few decades ago, and this propensity for change will be even greater in the immediate future. For example, new buildings are optimized to have leaner structural members and open-concept features to accommodate innovative architectural demands. In order to satisfy such demands, materials such as high-performance concrete, high-strength steel, and engineered wood that offer superior mechanical strength and stiffness properties, mostly at ambient conditions, are being developed and widely used. However, under fire conditions, these new construction materials exhibit poor fire performance. For instance, newer types of concrete are prone to a fire-induced spalling phenomenon, which can reduce the already optimized small structural member's cross-sectional area. In a similar notion, high-strength steel is vulnerable to fire-induced local buckling and engineering wood may experience delamination at relatively low temperatures.

Thus, structural members in modern buildings have inherently lower fire resistance, resulting from much lower thermal mass in the member and material-related problems of spalling, buckling, or delamination under fire conditions (Fletcher, 2009). Adding to the problem is the ever-increasing amount of fuel loads due to increasing contents from consumeristic lifestyles in building compartments and increased use of plastics in furnishings and building components in interiors, which create different and more severe type of fires than that experienced with the cellulosic materials that dominated building contents two decades ago. Further, the popular open-concept architecture in modern buildings, arising from modern lifestyle and the consumeristic nature of society, provides an optimum environment for fire growth and spread. Currently used prescriptive approaches that were developed based on conventional materials and standard fire loading scenarios do not take into consideration many of these parameters present in today's buildings. Therefore, application of rational engineering approaches is key for realistic fire-resistance assessment in modern construction.

1.7 Layout of the Book

Organized into 13 chapters, this book accommodates readers with different levels of background in structural engineering, fire protection engineering, and architectural fields. Each chapter starts with introductory behavioral concepts to provide a fundamental understanding and then presents advanced concepts, methods, and the latest research findings in structural fire engineering. Chapters 1 and 2 provide an introduction to fire safety design, highlight the magnitude of fire issues in built infrastructure, and lay out basic design strategies for achieving fire safety in buildings. Chapter 3 discusses various aspects of fire growth in a compartment and lays out steps for fire load calculations in different building occupancies, together with basic heat transfer calculations. Chapter 4 presents the effect of high temperature on thermomechanical properties of commonly used construction materials. Chapters 5 and 6 outline principles and procedures for evaluating fire resistance of structural members and systems through testing and calculation methods. Chapters 7, 8, 9, 10, and 11 present behavioral aspects, followed by analysis approaches. Each chapter begins with behavioral aspects of structures and then hones in on specific approaches for fire-resistance evaluation as per different codes of practice, presenting a thorough, advanced analysis of each construction material (steel, concrete and masonry, composite such as steel-concrete, wood, and FRP-strengthened concrete structures, respectively). Chapter 12 discusses steps for residual capacity assessment of fire-damaged structures. Finally, Chap. 13 discusses specific issues associated with fire-resistance design of critical transportation infrastructure such as bridges and tunnels. In each chapter the subject matter is explained with the aid of numerous illustrations, figures, and tables, and step-by-step numerical examples are presented to help readers grasp the subject matter with ease.

1.8 References

Arablouei, Amir, and Venkatesh Kodur. "Modeling Delamination of Fire Insulation from Steel Structures Subjected to Blast Loading." *Engineering Structures* 116 (2016): 56–69.

Astaneh-Asl, A., C. R. Noble, J. Son, A. P. Wemhoff, M. P. Thomas, and L. D. McMichael. "Fire Protection of Steel Bridges and the Case of the MacArthur Maze Fire Collapse." In *TCLEE 2009: Lifeline Earthquake Engineering in a Multihazard Environment*, edited by Alex K. K. Tang and Stu Werner 1–12, 2009.

Cowlard, Adam, Adam Bittern, Cecilia Abecassis-Empis, and José Torero. "Fire Safety Design for Tall Buildings." *Procedia Engineering* 62 (2013): 169–81.

Engelhardt, Michael D., Brian Meacham, Venkatesh Kodur, Adam Kirk, Haejun Park, IJsbrand van Straalen, Johan Maljaars, Kees van Weeren, Rene de Feijter, and Kees Both. "Observations from the Fire and Collapse of the Faculty of Architecture Building, Delft University of Technology." In *Structures Congress 2013: Bridging Your Passion with Your Profession*, 1138–49, Reston, VA: American Society of Civil Engineers, 2013.

Fletcher, I. A. (2009). "Tall Concrete Buildings Subject to Vertically Moving Fires: A Case Study Approach." PhD thesis, University of Edinburgh School of Engineering, May 2009.

Gann, Richard G., William L. Grosshandler, Hai S. Lew, Richard W. Bukowski, Fahim Sadek, Frank W. Gayle, John L. Gross et al. "Final Report on the Collapse of World Trade Center Building 7. Federal Building and Fire Safety Investigation of the World Trade Center Disaster (NIST NCSTAR 1A) 'DRAFT for Public Comments'." *National Construction Safety Team Act Reports (NIST NCSTAR)-1*(2008), National Institute of Standards and Technology, Gaithersburg, Maryland.

Gencer, Ebru A. "Natural Disasters, Urban Vulnerability, and Risk Management: A Theoretical Overview." In *The Interplay between Urban Development, Vulnerability, and Risk Management*, 7–43, Berlin/Heidelberg: Springer, 2013.

Hall, John R. "The Total Cost of Fire in the United States." Quincy, MA: National Fire Protection Association, 2014.

Kodur, V. K. R., and M. Z. Naser. "Importance Factor for Design of Bridges against Fire Hazard." *Engineering Structures* 54 (2013): 207–20.

Kodur, Venkatesh, Li Gu, and Maria Garlock. "Review and Assessment of Fire Hazard in Bridges." *Transportation Research Record* 2172 (2010): 23–29.

Lin, Yuan-Shang. "Estimations of the Probability of Fire Occurrences in Buildings." *Fire Safety Journal* 40, no. 8 (2005): 728–35.

McAllister, Therese, and Gene Corley. *World Trade Center Building Performance Study: Data Collection, Preliminary Observations, and Recommendations*. Washington, DC: Federal Emergency Management Agency, 2002.

Naji, Arash, and Mohamadreza Rohani. "Progressive Collapse Analysis of Reinforced Concrete Structures: A Simplified Procedure." *European Journal of Engineering Research and Science* 2, no. 10 (2017): 7–12.

Naser, M. Z., and V. K. R. Kodur. "A Probabilistic Assessment for Classification of Bridges against Fire Hazard." *Fire Safety Journal* 76 (2015): 65–73.

National Crime Records Bureau (NFPA). *The Geneva Association: World Fire Statistics 2011*. Avalon Consulting Analysis, 2010.

Rahikainen, J., and O. Keski-Rahkonen. "Statistical Determination of Ignition Frequency of Structural Fires in Different Premises in Finland." *Fire Technology* 40, no. 4 (2004): 335–53.

Scheer, Joachim. *Failed Bridges: Case Studies, Causes, and Consequences*. Wiley, 2011.

Wardhana, Kumalasari, and Fabian C. Hadipriono. "Analysis of Recent Bridge Failures in the United States." *Journal of Performance of Constructed Facilities* 17, no. 3 (2003): 144–50.

CHAPTER 2

Fire Safety in Buildings

2.1 Overview

This chapter presents general issues pertaining to fire safety in buildings. Provisions specified in building codes and standards for achieving fire safety in different building occupancies are reviewed. The strategies that can be developed to overcome fire hazard in buildings are laid out, and the role of structural fire safety within the context of overall fire safety is highlighted. Finally, different approaches that can be applied to evaluate fire resistance of structural members is introduced.

2.2 Design Principles for Structures

A structure in civil engineering terminology commonly refers to a constructed facility to provide functionality needs such as living, working, mobility, and storage. Typical civil structures are broadly grouped under buildings and nonbuilding structures. Both types of structures make up the civil infrastructure and are further subgrouped into a number of categories depending on functionality as high-rise buildings, single-family dwellings, bridges, tunnels, and so on.

A structure, depending on its functionality and geographic location, can be subjected to a wide variety of natural and manmade forces resulting from wind, flooding, earthquake, snow, landfill, and blasting, among others. Thus, for maintaining safety, stability, and functionality, a structure must be designed to withstand loads arising from external forces, often grouped under various loading types, such as dead, live, and environmental loads. For design purposes, estimates of such loading can be obtained by referring to various formulae and tables listed in design standards and civil engineering handbooks such as, ASCE 7 (ASCE, 2016).

One such loading effect that can occur in a structure during its lifetime is that resulting from fire. Fire can break out due to accidental situations such as electrical short circuit, cooking-related accidents, natural phenomenon such as lightning, or manmade causes, like blast, vandalism, and so on. If a fire breaks out and grows to full size, it will produce adverse conditions for occupants and the structure due to high temperatures developing in the structural members. Such high temperatures can produce significant degradation in strength and stiffness properties of constituent materials, which in turn can result in loss of load-carrying capacity in a structural member, leading to possible collapse of the structure. Thus, fires in civil infrastructure can lead to loss of life or injuries to occupants and first responders, as well as property damage to the contents in a building, the structure itself, and the surrounding environment.

Despite adverse consequences of fire in infrastructure, effects arising from fire to a structure still not given proper *engineering* consideration in design and analysis of structures. In fact, structures are often not designed for specific conditions present during a *fire* per se, rather structural members are simply provided with some level of structural fire protection to withstand a certain duration of typical fire exposure referred to as standard fire.[1] As will be illustrated in Chaps. 3 and 5, this standard fire exposure, as well as other conditions such as loading and restraint conditions considered in current fire protection approaches, does not represent realistic fire scenarios and loading conditions present in a structure during actual fire incidents. The standard fire, used for establishing certain fire protection level, is based on certain idealized fuel load and ventilation conditions, which often do not represent realistic compartment (fire) characteristics. In fact, a number of recent fire incidents in buildings and other infrastructure have shown that fire exposure and other conditions that are present during a fire incident in structures are entirely

[1] The standard fire exposure, also specified as temperature-time curve in standards, is a standard fire curve used internationally for fire resistance testing of components. This allows for a direct comparison between standard fire tests of components rather than giving an indication of how long the component will survive in the fire. Further discussion on standard fire exposure will be provided in Chap. 5.

different than that of a standard fire and other idealized conditions currently adopted to provide needed fire protection to structures (Khorasani-Zavareh and Shokouhi, 2018).

2.3 Fire Safety in Buildings

Buildings, among other structures, are much more vulnerable to adverse effects developed during a fire incident, since occupants can be trapped in buildings, specifically in the case of high-rise buildings. Therefore, from a life safety consideration, buildings are to be provided with adequate fire safety measures. As discussed in Chap. 1, the primary goal of providing fire safety in buildings is to limit probability of death and injury to occupants and also to minimize property damage (including loss of functionality) during or in the aftermath of a fire. To achieve this goal, current building codes specify various fire protection measures such as installation of sprinklers, smoke detectors, and exit paths. Such fire protection measures may not fully ensure building performance under severe fires, since reliability (and functionality) of fire safety measures cannot always be guaranteed. Further, firefighting may not always be fully effective due to varying response times, as well as rapid fire spread and growth. Therefore, structural members in a building must be provided with *structural fire protection* measures to control fire spread and to maintain stability and integrity of the members in the event of fire.

Designing structural members to withstand the effects of fire is quite different from that under normal (ambient) conditions, as behavior of construction materials and structures at elevated temperatures varies. For instance, strength and stiffness properties of most construction materials degrade with a rise in temperatures, which in turn can lead to rapid degradation in load-carrying capacity of structural members. If members are made of concrete and wood, exposure to elevated temperature can also lead to loss of cross section resulting from spalling or charring phenomena. A rise in cross-sectional temperatures in a member can also lead to development of additional fire-induced forces and associated deformations (distortions) not often experienced under ambient conditions. These forces can be substantial under certain conditions of fire and may trigger unconventional failure modes in a structure. Therefore, to evaluate realistic fire response of a structure, these effects, specific to the type of material, must be taken into consideration in the design of structures for fire safety. Undertaking such analysis and design requires the application of rational methodologies based on *structural fire engineering* principles.

2.3.1 Fire Scenarios in Buildings

Buildings typically contain large amounts of fuel in the form of furniture, furnishings, clothing, and lining materials. Much of the fuel in buildings are composed of cellulose-based fuels and organic materials derived from fabric, plastics, or wood. Oftentimes fire starts in a building due to ignition of fuel resulting from an accidental event or intended incident. The fuel gradually burns and can grow from local burning into full-size fire, provided there is adequate fuel and ventilation in the compartment and no measures are in place to extinguish the localized fire.

The development of fire within a typical compartment (room) can be represented by a time–temperature curve as shown in Fig. 2.1, where the horizontal axis represents the duration of fire and the vertical axis represents the progression of temperature in the compartment. The fire growth in the compartment can be grouped into four stages: initial combustion (*incipient*), pre-flashover (*growth*), post-flashover (*fully*

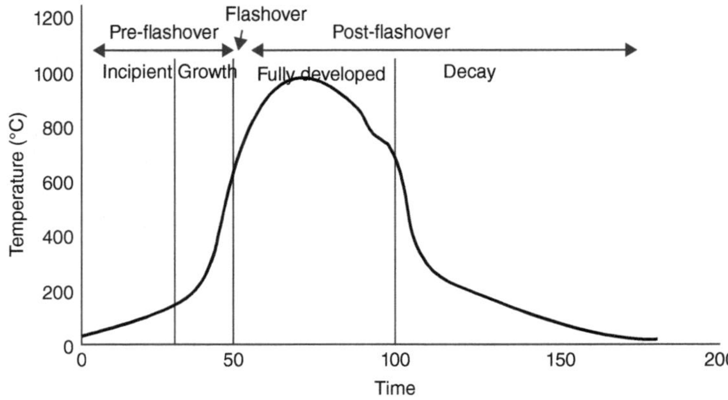

Figure 2.1 Time–temperature curve for the full fire development process inside a typical room with no suppression.

developed), and the decay (*cooling*) stage (Gosselin, 1987). Fire characteristics in each stage vary significantly in terms of peak temperature attained and duration of burning (or cooling) depending on the extent of fuel and air (oxygen) present, as well as confining boundaries of burning area (or objects). Despite grouping fire development into number of "stages," the actual fire growth process is continuous as fire transitions smoothly from one stage to the next (Drysdale, 2011).

The development of fire is a complex process that depends not only on the amount of available fuel and ventilation but also on the interaction of fire with compartment boundaries (walls/floors), fire protection systems, and firefighting activities. This is a good time to revisit the discussion of building codes requiring the installation of fire protection systems, whether in terms of active fire protection systems such as sprinklers and smoke detectors or passive protection measures such as fire protection to structural members. These measures are to be in place to control fire growth and flame spread beyond the compartment of fire origin to minimize adverse consequences.

The behavior of fire, together with the role of occupants and fire protection systems at different stages of fire development, is summarized in Table 2.1. In the early incipient stage of fire, heating of fuel (initial combustion) occurs locally around a few fuel items leading to ignition and breakout of fire. In this stage, the burning of fuel is in a small area and fire temperature is low (ranging from 50 to 150°C; see Fig. 2.1). During this stage, fire breakout can be detected visually or through smoke or heat detectors by residents and could possibly be extinguished by stopping ignition or put out through handheld fire extinguishers. If ignition spreads to other objects, these objects will start to burn quickly, causing the fire to rapidly grow within the compartment and leading to a much larger fire with temperature reaching to 200 to 300°C. At these temperatures heat detectors and sprinklers should be triggered to alert occupants to evacuate and to notify authorities for help in extinguishing the fire. This stage is referred to as growth period and is the time where evacuation of occupants needs to occur, in conjunction with alerting firefighters. Active fire protection systems such as sprinklers, heat detectors, and exit paths should be effective and fully functional during the growth period. This stage marks the transition of fire from medium-sized to "fully developed" fire.

During the burning of objects, hot gases and smoke rise up in the compartment and accumulate below the ceiling level. These gases facilitate further ignition of various objects in the compartment through radiative heat transfer. When the temperature from these hot gases reaches about 600°C at the ceiling level, the whole compartment ignites due to radiant feedback (heat transfer from trapped hot gases) to available unburnt fuel at the floor level. This transition point from local burning to whole compartment ignition is referred to as *flashover* and represents start of the *burning* (fully developed fire) stage. At this temperature level, windows can break and let fresh air flow into the burning compartment. Thus, fire in this stage is mainly controlled by the amount of ventilation available for burning. Since automated fire safety measures such as pipes used for sprinklers cannot survive extreme temperatures (600 to 800°C), their functionality cannot be taken for granted. Also, structural members and compartment boundaries play a critical role at this stage, since sectional temperatures in the members can exceed 500°C during the growth stage. Hence, the main mode of controlling and fighting the fire at this point is through professional firefighting or

Role	**Stage**			
Factor/Stage	**Incipient stage**	**Growth stage**	**Burning stage**	**Decay stage**
Fire development process	Heating of fuel and local burning	Portions of room burning—fuel controlled	Fully developed fire—ventilation controlled	Cooling off period—fuel controlled
Role of occupants	Preventing ignition	Use extinguishers to dose the fire, evacuate/escape to safety	Can result in death, if not evacuated	–
Role of detection systems	Triggering of smoke and heat detectors, etc.	Continued working of smoke detectors, heat detectors, etc.	Operation of smoke exhaust systems	–
Role of active systems	Evacuation routes to be fully functional	Activation of sprinklers and arrival of firefighters	Professional firefighting	Professional firefighting
Role of passive systems	–	Compartmentation boundaries (walls, floors) to be effective to limit fire spread	Compartmentation boundaries and structural members to be effective	Structural members to be fully effective

Table 2.1 Role of Occupants and Various Fire Protection Systems During Different Stages of Fire

through passive fire protection measures. Therefore, passive protection measures (such as fire insulation on compartment boundaries, i.e., walls and floors) must be fully effective in this stage of fire to minimize the spread of fire to other compartments/floors.

Once available fuel and/or ventilation (oxygen) gets depleted, burning in the compartment slows down and fire starts to decay, with fire temperature in the room gradually decreasing. This stage is referred to as the *decay* or *cooling* stage. During this stage, temperatures in structural members continue to rise due to the *lag effect* and continued heat transfer from surrounding air to members. During this stage, passive fire protection features such as fire resistance of structural members are critical to ensure the integrity and stability of the overall structural system.

In a nutshell, active fire protection systems are best suited to control and fight fire during incipient and growth periods, while passive systems are critical for controlling fire spread and minimizing structural collapse during the growth, burning, and decay stages of fire.

As a side note to fire development process, if the fire in the compartment is left uncontrolled, either through breakdown of compartment boundaries resulting from the failure of passive systems or lack of firefighting, then fire can spread to adjacent compartments and/or other stories of the building or even to neighboring buildings, depending on the layout and characteristics of the building. Fire spread to other compartments/floors can be through burning flames, radiant heat, and/or hot smoke traveling across multiple compartments/floors through open doors, windows, ceilings, and cavities or through flames penetrating through surrounding walls/floors. In such a scenario, windows can break and flames and/or radiant heat can ignite cladding on the exterior of the building, leading to the fire spreading into higher floors, as seen in a number of high-rise building fires in recent years, such as the Grenfell Tower fire in London (Guntrip, 2017) and the Windsor tower in Madrid, Spain (Fletcher et al., 2006). Further details on fire spread phenomenon in a compartment are discussed in Chap. 3.

2.3.2 Need for Fire Safety Measures

Fire, when grown to full size, can be one of the most severe conditions a structure has to endure during its lifetime. Since the 17th century, fire safety has been a major economic and societal concern, as conflagrations were common in major cities around the world. For example, the Great Fire of London was a major conflagration that swept through the central parts of the city in 1666, leading to burnout of numerous buildings for four days. The fire destroyed 13,200 houses, leaving at least 70,000 of the City of London's inhabitants homeless (out of a population of 80,000). Other examples of city conflagration include Baltimore in 1904, San Francisco in 1906, and Chelsea, Massachusetts in 1908, which led to the destruction of 80 city blocks, 28,000 buildings, and 3500 buildings, respectively (Fitzgerald and Meacham, 2017). In fact, on a single day on March 21, 1916, three separate city conflagrations broke out in three cities located in the states of Georgia, Texas, and Tennessee in the United States and resulted in the destruction of a total of 2770 buildings. Other examples of city conflagration include those in Copenhagen in Denmark (1728), Kiev in Ukraine (1811), Hamburg in Germany (1943), Tokyo in Japan (1945), and Brussels in Belgium (1967). The aftermath of these fires has highlighted the destructive nature of fire and led to a rise in public awareness of the need for fire safety measures in buildings to minimize the destructive effects of fire.

As life safety is a key consideration in design, and since buildings in the event of fire represent a kind of trapped environment for occupants, building codes and standards require appropriate fire safety features to be installed in buildings to withstand the adverse effects of fire. Fire safety measures are tailored at controlling the fire, as well as in protecting occupants, first responders, and the fabric of the structure, together with reducing the adverse consequences to the surrounding environment. Provisions of fire safety measures are also aimed at minimizing the loss of functionality of the structure for continued operations following the fire. In addition, such fire safety measures, through early control/extinguishment of a fully developed fire, can also limit environmental damage resulting from gaseous pollutants from smoke emissions, and liquid pollutants present in run-off water from firefighting.

2.4 Strategies to Overcome Fire Hazard

Fire is a stochastic event that can break out anytime in the life span of a structure. To mitigate the adverse consequences arising from fire, appropriate strategies must be put in place to control the effect of fire on people (occupants), structure, and surrounding environment. Building codes and standards recognize fire to be a significant hazard from life safety standpoint and as such require the compliance of specific fire safety goals and objectives through installation of active and passive fire protection measures in buildings.

2.4.1 Goals and Objectives

The primary goal of providing fire safety in buildings is to *reduce the burden of fire on the quality of life* by minimizing probability of death and injury to occupants and first responders, as well as to limit property damage and maintain continued functionality of the structure (mission/business) with minimum interruptions. This goal can be achieved through specifying a set of objectives in the design of buildings: life safety, property protection, functional continuity, and environmental protection.

Life safety is inarguably the most important objective for achieving overall fire safety goal in a building. The general public places a large expectation and trust in buildings to provide a certain level of safety for occupants and first responders. Unlike in open structures, such as bridges, people can get trapped in buildings, especially in high-rises, during a fire event. Thus, buildings should be able to withstand the adverse effects of fire for a certain duration of time to allow occupants to safely evacuate or remain safe in refuge areas. Buildings also need to maintain structural integrity to allow firefighters to safely douse the fire. Life safety not only is from the consideration of occupants of a burning building but also is intended to protect occupants of nearby buildings, bystanders, and of course first responders.

The second objective of fire safety is to ensure property protection in the building on fire, as well as any surrounding buildings. To achieve this objective, fire spread within the building or to the exterior, through windows or cladding, must be contained, and if possible, prevented. The third objective of incorporating continuing functionality as part of fire safety is analogous to that of property protection. However, instead of considering just losses occurring during a fire, mission continuity is more concerned with uninterrupted business operations following the fire incident. Often, large fires can lead to interruption to businesses and produce significant losses. These losses, described as *indirect losses*, result from relocation/closure of businesses. Such interruptions can be minimized through the provision of proper fire safety measures that include preventing or minimizing damage to the structure.

The objective of environment protection is to limit the impact of fire on the surrounding environment because burning arising in a fire can lead to the release of toxic fumes and hazardous gases, and the spread of pollutants in the form of smoke. Chemical agents used in the extinguishment of fire can also pollute the surrounding air or water bodies. Hence, to minimize the release of burning by-products (pollutants), and their effects on the environment, the magnitude of burning (i.e., the size of the fire) needs to remain small and firefighting activities have to be efficient.

2.4.2 Provisions for Fire Safety Features

The preceding set of fire safety objectives can be achieved through the provision of appropriate fire protection features in buildings. Such features include preventative measures to minimize ignition, extinguishing measures to control/douse fire, notification measures to alert and facilitate occupant evacuation, and collapse mitigation measures to prevent collapse of the structure. Fire protection features include those that are installed during the construction of a new building (e.g., sprinklers, egress paths) or those incorporated during the life of a structure, including training of occupants for orderly evacuation. Fire protection features integrated into a building are broadly grouped under two categories: namely *active systems* and *passive systems*. The various active and passive fire protection systems that are incorporated in high-rise buildings, together with their operational range, purpose, and design features, are presented in Table 2.2.

Active Fire Protection Systems

Active fire protection measures refer to devices and systems that require some level of action or motion to activate or trigger a device to control the growth of fire or to alert occupants to evacuate during a fire breakout. Active fire protection measures include smoke alarms, heat/smoke detectors, sprinklers, exit paths (signage), and fire extinguishers. These systems comprise a device which incorporates a detector (sensor) designed to sense heat, flame, smoke, or other products of combustion and then activate noise or signals to alert occupants, or release water to douse a growing fire. Performance and effectiveness of active protection systems depend on their location relative to the fire zone, movement of combustion products, detector sensitivity, and system reliability. Once a sign of fire is detected, an alarm (via sound, light, or other signals) breaks out to locally alert occupants in a fire emergency in the building. In some modern buildings, automated notification devices are also placed to notify control room within a building and/or local fire station to summon firefighters for assistance.

One of the most effective and commonly installed fire protection features in buildings is smoke detector. A smoke detector is a device that senses generation or development of smoke as an indicator of

Type – application	Effective operational range*	Purpose	Design features
Active systems			
Smoke/heat detectors—common in residential and high-rise buildings	Incipient stage	Sense smoke/heat/CO_2 and trigger high-decibel alarm (noise) to alert occupants	Sensors detect smoke/heat either optically (photoelectric) or through physical process (ionization)
Notification devices—in high-profile commercial buildings	Incipient stage	Sense fire/smoke in a compartment and alert building control room and firefighting department	Can be activated automatically from smoke and heat detectors or may also be activated via manual fire alarm activation devices such as manual call (emergency phone buttons) points or control rooms
Automatic sprinklers (dry, wet, foam agents etc.)—in most modern high-rise buildings	Incipient/ growth stage	Detect heat and trigger valves to release water to douse flames (fire)	Consist of a water supply system (tank) providing adequate pressure and flowrate to a water distribution piping system, and a sensor to detect and turn-on sprinklers
Special hazard or spot protection systems (smoke management)—in high-rise buildings	Growth/ burning/decay stage	Detect/monitor pressure (smoke) and keep stairwells clear of smoke and toxic fumes to facilitate safe evacuation and firefighting activates	Monitor and maintain positive pressure in stairwells through mechanical fans
Firefighting force—at strategic locations in cities/ towns/counties	Growth/ burning stage	Control fire spread and extinguish fire by professionally trained fire department personnel	Acceptable response time
Passive systems			
Fire-rated doors and barriers—in most modern high-rise buildings	Growth/ burning stage	To prevent spread of flame-heat or smoke-gas from one fire zone (space) to another zone (compartment)	Comprise of fire-rated doors and compartment barriers
Egress system—in all buildings	Throughout fire duration	Continuous and unobstructed multiple passage paths (stairs) for travel in and out of a building	Emergency staircases with fire-rated barriers and placed at strategic locations with a building
Refuge areas/floors—in some high-profile tall buildings	Throughout fire duration	Designed to house building occupants during a fire when evacuation to outside of building is not feasible/safe	Fire-safe compartment especially designed with additional life safety features; e.g.: supply of fresh or filtered outside air and made of high levels of fire-rated construction
Barrier/compartmentation—in most residential and commercial buildings	Growth/ burning and decay stages	To limit fire spread (through flames/gases/temperature) from one compartment to another compartment	Provisions of fire resistance (insulation and integrity features) in walls and slabs/floors
Structural fire resistance—all buildings except single family houses	Growth, burning, and decay stages	To maintain stability and load-bearing capacity in structural members	Provisions of appropriate fire resistance (insulation, integrity, and stability features) in structural members

*Refer to Fig. 2.1 for definition of different stages of fire.

Table 2.2 Typical Fire Protection Systems Incorporated in High-Rise Buildings

fire (or combustion) breakout. Once smoke is detected, the smoke alarm issues a signal to a fire alarm control panel as well as audible and/or visual notifications to alert firefighting department and occupants, respectively. Other active protection systems installed in modern high-rise buildings are sprinkler systems. A sprinkler system is composed of a water tank, control valves, a piping distribution system, nozzles to release water, and a monitoring system connected to detection devices to trigger nozzles. The main objective of such a sprinkler system is to extinguish or control the fire while the fire is still in an early stage, and of small size, such that a local fire does not grow to a full-size fire. Modern buildings are often fitted with additional measures to work in parallel with sprinklers and aid in fire suppression and facilitate egress. Such measures include a smoke management system to exhaust smoke from burning floors and pressurize stairwells and corridors to prevent smoke accumulation in stairwells (or escape paths). Other measures include detection devices to close fire doors held open by magnetic devices and the shutting of dampers in air ducts. To minimize adverse impact of a fire during its early stages, active fire protection measures[2] must be effective in extinguishing/controlling the fire. Table 2.2 lists various fire protection systems installed in high-rise buildings.

Building codes lay out specific provisions for providing *means of egress* to facilitate evacuation of occupants and to provide entry to firefighters in case of fire in a building. An egress path is a continuous and unobstructed passage from any location within a building (or structure) to an exit leading to outside of the building. Exit paths can consist of corridors, hallways, stairways, and/or aisles leading to exit doors. Egress systems are to be designed as per specified codal provisions. For example, International Building Code (IBC, 2015) specifies a minimum of two staircases to be installed in high-rise buildings of less than 20 stories high and requires three staircases for buildings higher than that. The same code also specifies a minimum corridor width of 1120 mm for residential and office buildings and about 2440 mm for high-functioning buildings (e.g., hospitals). These egress paths are critical for facilitating evacuation of occupants and entry of firefighters, since elevators or escalators cannot be operational during a fire breakout in a building.

Passive Fire Protection Systems

Passive fire protection refers to fire resistance features that are built into structural members. These are inherent or built-in properties of a structural member or system and do not require any sensing or activation to function during a fire. The fire resistance in a structural member is intended to achieve two specific objectives: (1) to limit/prevent the spread of hot gases and flames beyond the compartment of fire through satisfying insulation and integrity criterion and (2) to maintain the load-carrying capacity of main structural members through satisfying stability criterion.

Passive fire protection features can be grouped into two broad categories: compartmentation and structural integrity. *Compartmentation features* aim to limit the spread of fire through horizontal (slab or floor) and vertical (wall) partitioning boundaries of a given space (i.e., room) so as to prevent the spread of fire and hot gases, thereby reducing/delaying fire growth to other compartments in a building under fire or to any adjacent buildings. This is achieved through the provision of thermal fire resistance features in compartment barriers, namely *integrity* and *insulation* requirements. Typical barriers/elements include floors, roofs, fire doors, walls, windows, enclosed corridors, and shafts.

Structural integrity features ensure that structural members (columns, beams, etc.) maintain stability as well as load-bearing capacity during a fire incident. These members, when exposed to fire, experience a rise in cross-sectional temperatures, which lead to deterioration in mechanical properties of constituent materials the structural member is made of. This will result in loss of load-carrying capacity and could possibly lead to partial or full collapse of structural systems. To minimize such a scenario, structural members need to maintain stability and load-carrying capacity during a fire incident. This is referred to as *structural fire resistance*. Such fire resistance feature provides additional time for building occupants to evacuate and for carrying out firefighting operations, which minimize property damage. The ways and means to achieve fire safety can vary with materials and structural types used in construction. These are discussed in the next section.

Ways and Means of Achieving Structural Fire Safety

All construction materials lose strength and stiffness properties under elevated temperature exposure as encountered during a fire incident. However, different materials behave differently under fire conditions due to unique physio-chemical changes that occur in materials under elevated temperature exposure. As a result, techniques to be adopted for achieving fire resistance in structural members made of concrete, masonry, steel, wood, and so on are different. For example, in reinforced (and prestressed) concrete and masonry structures, the required fire resistance can be achieved through providing sufficient *cover* thickness to minimize temperature rise in steel reinforcing and/or prestressing bars. Alternatively, concrete members strengthened with fiber-reinforced polymer (FRP) composites require application of external fire

---

[2] It should be noted that firefighting is also considered to be part of active fire protection measures since it requires detection, notification, and action (either automatically or manually).

insulation to protect FRP from reaching its glass transition temperature and debonding. In the case of high-strength concrete columns, appropriate detailing of internal steel reinforcement such as providing stirrups with 135° hooks can significantly enhance fire resistance through mitigating fire-induced spalling (Khaliq et al., 2015).

Steel structures, due to high thermal conductivity and low specific heat of steel, experience a quick rise in cross-sectional temperatures, which can lead to rapid degradation in capacity and stiffness of members under fire conditions. Thus, an effective way to limit heat transmission to steel structural members, and thereby minimize degradation in load carrying capacity, is through the application of *external fire insulation*, such as spray-resistive fire resistance material (SFRM), or intumescent paints or concrete encasement.

In many instances, structural systems are composed of a combination of steel and concrete, referred to as *composite construction*. In such composite systems, superior fire resistance properties of concrete can be utilized to overcome weaker fire resistance properties of steel and potentially achieve high fire resistance in the composite member. Concrete not only can provide insulation effect to slow temperature rise in steel, but also can effectively carry a higher portion of load from the composite member during the fire incident. One such example is the case of concrete-filled steel columns, wherein concrete filling through *capacitive protection* can slow down the temperature rise in steel tubes through its high thermal capacity properties and also carry higher portion of load during a fire incident.

In the case of a wood member exposed to fire, the outer surface of wood chars due to incomplete combustion of wood (i.e., pyrolysis). Such charring, although it reduces the cross section of a wood member, acts as an insulation layer and protects the unburnt layers of wood. This slows down combustion and delays degradation in strength and stiffness properties of inner layers of wood. As a result, the member will experience a slower reduction of its load-bearing capacity. Therefore, one possible strategy to enhance fire resistance of a wood member is to use larger sectional dimensions for the structural members, so that the outer layers in the cross section act as a sacrificial layer and the required fire resistance is achieved.

Although the above discussed ways and means can be applied to achieve fire resistance in structural members, rational calculation procedures based on heat transfer and structural mechanics principles must be applied to realize effective fire resistance solutions in members made of different construction materials. Structural fire engineering concepts for different materials and structural systems to achieve required fire resistance are discussed in various chapters of this book.

2.5 Codes and Standards

The specifications and guidelines for the design, construction, and maintenance of civil infrastructure are laid out in codes and standards. This is intended to protect health and safety features, as well as the comfort of the public, and generally represent a compromise between optimum levels of safety and economic feasibility from a structural point of view. Specifications relating to fire safety—active and passive fire protection systems aimed at ensuring safety of occupants, first responders, and the structure itself—are specified in building codes.

2.5.1 General

Fire safety requirements and associated fire design specifications are contained in building codes and standards adopted by each country. Therefore, fire protection provisions in each country or region are heavily influenced by interpretation of specifications laid out in national building codes and standards as well as by local practice. Building codes are broad in scope, have legal applicability (mandatory regulations adopted through legislation or government approval), and include in their framework references to many construction-related standards. Fire design provisions in building codes specify minimum requirements for achieving required safety provisions relating to various fire protection features in buildings. As part of structural fire safety requirements, codes specify required fire ratings for various structural members in a building, and these ratings depend on occupancy type, building height, and maximum compartment areas.

Building codes reference a number of *standards* and *guidance documents* for further details on various specifications and provisions. Standards, developed by standards writing organizations, list out specific set of procedures, conditions, or requirements to be met by a material, product, process, or design methodology. Standards are developed by different groups or stakeholders through a consensus process among consumers, producers, designers, regulatory agencies, and independent agencies.

Construction-related standards, referenced in building codes, can be grouped under material standards, design (engineering practice) standards, and testing standards. Material standards (e.g., concrete, steel, wood, gypsum) establish minimum requirements of quality as measured by composition, material

Country	Applicable building/fire code	Fire resistance provisions
United States	• International Building Code (IBC) • National Fire Protection Association (NFPA) • Building and Construction Code 5000	Mostly prescriptive
Canada	• National Building Code (NBCC) • National Fire Code of Canada (NFCC)	Prescriptive- and objective-based provisions
Europe	Eurocodes	Both prescriptive- and performance-based fire design provisions
United Kingdom	British Building Regulations	Prescriptive- and performance-based fire design provisions
China	SAC/TC113 and GB50016-2014 Code of Design on Building Fire Protection and Prevention	Mostly prescriptive, but allows performance-based design in case of special structures
Hong Kong	Code of Practice for Fire Safety in Buildings	Mostly prescriptive, but allows performance-based design in case of special structures
New Zealand	Building Act	Prescriptive- and performance-based design features
Australia	Australian Building Codes Board	Prescriptive- and performance-based design features
Brazil	Brazilian Building Code	Prescriptive-based provisions
Japan	Building Standard Law (BSL) (1989)	Performance-based design with some prescriptive-based provisions
India	National Building Code (NBC)	Prescriptive-based provisions

Table 2.3 Building and Fire Codes Around the World

properties, dimensions, and uniformity of product. They may also include specifications for establishing methods of sampling and testing for verification of such quality. Design standards provide basic design procedures and engineering formulas, as well as special provisions to carry out engineering calculations (such as structural design) so as to provide a satisfactory level of performance during the lifetime of a structure. Some of these procedures include appropriate means to properly design structures to withstand effects of various loading types such as wind, earthquake, and fire.

Specifications for evaluating the *fire resistance rating* of a structural member through testing is contained in fire test standards such as ISO 834 (ISO, 2012), ASTM E119 (ASTM, 2016), and UL 263 (UL 263, 2011), among others. A fire resistance rating refers to the duration during which a structural member can withstand a standard fire exposure without reaching the failure limit state, as specified in the applicable fire test standard. Fire resistance ratings can also be evaluated by means other than fire testing. For example, codes and standards such as Eurocodes and ASCE 29 provide engineering approaches ranging from simplified formulae to advanced numerical analysis to evaluate fire resistance (rating) of structural members.

Codes are adopted by governmental bodies and hold authority over jurisdiction in a specific country or region. A brief list of building and fire codes adopted in different countries, together with the type of fire resistance provisions contained in these codes, are listed in Table 2.3 and further discussed below.

2.5.2 American Codes and Standards

In the United States, two main building codes, the *International Building Code (IBC)*[3] (2015) and the *National Fire Protection Association (NFPA) Building and Construction Code 5000* (NFPA, 2018), are currently used in building design and construction. These codes also lay out requirements for fire safety in

[3] The ICC's family of International Codes includes:

International Building Code (IBC): Applies to almost all types of new buildings.

International Residential Code (IRC): Applies to new one- and two-family dwellings and townhouses of not more than three stories in height.

International Existing Building Code (IEBC): Applies to the alteration, repair, addition, or change in occupancy of existing structures. The ICC publishes new editions of the International Codes every three years, and many states and localities have adopted them since the first editions were issued in 2000. In 2000, the three regionally based model code organizations (BOCA National Code, SBCCI Standard Code, and ICBO Uniform Code) combined together to form the ICC.

buildings and other structures to ensure public safety, health, and structural integrity. IBC and NFPA codes specify fire resistance ratings following a prescriptive philosophy, i.e., minimum required fire endurance times (fire ratings) for building elements, as well as prescriptive-based approaches (tables/graphs) for achieving fire ratings in structural members (beams, columns, slabs etc.). These codes do not contain specific provisions for undertaking performance-based fire design.

IBC refers designers to number of standards to evaluate fire resistance of different construction materials such as steel, reinforced concrete, masonry, and wood. The referenced standards include American Institute of Steel Construction standard (AISC, 2017) for steel and composite construction, American Concrete Institute standard for concrete and masonry construction (ACI 216.1), (ACI/TMS Standard, 2014), Precast/Prestressed Concrete Institute standard for prestressed concrete construction (PCI MNL-124), (Precast/Prestressed Concrete Institute, 2011), and American Wood Council standard for wood construction (NDS) (American Wood Council, 2018). In some cases, these standards further refer to specific guidance documents for evaluating fire resistance of elements. For example, the AISC design manual provides limited information on *explicit* fire design provisions for steel structures and hence refers designers to the use of simplified expressions, analytical approaches, and empirical relations listed in the American Society of Civil Engineers Standard Calculation Methods for Structural Fire Protection (ASCE/SFPE 29) (ASCE, 2007).

Other guidance documents used in the United States for fire design of structural members are published by the Society of Fire Protection Engineers (*SFPE*), such as SFPE Standard on Calculating Fire Exposures to Structures (SFPE S01, 2011) and SFPE Standard: Calculation Methods to Predict Thermal Performance of Structural and Fire Resistive Assemblies (SFPE S02, 2015). Detailed procedure for fire resistance calculations can also be found in the *SFPE Handbook of Fire Protection Engineering* (2016), specifically information on "high-temperature material properties" of various construction materials in Chapter 9, "fire exposure scenarios" in Chapter 38, "structural analysis under fire" in Chapter 52, and design of "steel, concrete, and timber structures under fire conditions" in Chapters 53, 54, and 55, respectively.

2.5.3 Canadian Codes and Standards

Specifications for buildings in Canada, including fire safety provisions, are contained in the *National Building Code of Canada (NBCC)* (2015). This code references standards such as *CAN/CSA-A23.3* (design of concrete structures), *S16-14* (design of steel structures), and *O86-14* (engineering design in wood) for structural design specifications of different construction materials. Part 3 of NBCC, the National Fire Code of Canada (NFCC), details technical provisions and regulations related to the construction, use, and/or demolition of buildings and facilities, the condition of specific elements of buildings and facilities, and needed fire protection measures in buildings through a combination of prescriptive- and objective-based design solutions. Many of the structural fire safety provisions adopted by the NBCC, such as those relating to prescriptive-based fire resistance ratings, are similar to the provisions contained in American codes and standards such as IBC, ACI 216.1, and ASCE29.

The objective-based code provisions for undertaking fire design is applicable for new construction or for upgrading an existing structure. These provisions are formulated to limit the probability of life loss and structural failure (including adjacent buildings) due to fire breakout.

The recent edition of the NFCC specifies new fire-related provisions to allow construction of six-story wood buildings (earlier this was limited to four story). The new specifications require unfinished timber building to install additional fire protection measures. Another notable change related to fire safety design is the introduction of the classification system used by Workplace Hazardous Materials Information System (WHMIS) to define dangerous and combustible products. Also, the recent edition of NBCC has provisions for retrospective application of fire protection features (such as installation of sprinklers) to existing buildings.

2.5.4 European Codes: Eurocodes

Provisions relating to the design and analysis of structures within the European Union (EU) are detailed in the *Eurocodes*. The Eurocodes were developed under the umbrella of European Committee for Standardization (CEN) and are currently adopted in most countries in Europe. The Eurocodes form a set of documents for determining various loadings on a structure and for undertaking design and analysis of the structure. Currently there are 10 Eurocodes, and each Eurocode is designated by a number in the CEN classification, starting from 1990 for the basis of design to 1999 for aluminum alloy structures, where the last digit of the number in the CEN designation is the same as the number of the corresponding Eurocode. Thus, the Eurocodes are numbered from Eurocode 0 (EC1990 or simply EC0) to Eurocode 9 (EC1999 or simply EC9), and these documents cover design rules and framework for various types of construction

Number	Scope	Ambient conditions	Fire conditions
EC0	Basis of design	EN 1990	–
EC1	Actions (loading)	EN 1991-1-1	EN 1991-1-2
EC2	Concrete structures	EN 1992-1-1	EN 1992-1-2
EC3	Steel structures	EN 1993-1-1	EN 1993-1-2
EC4	Composite steel-concrete structures	EN 1994-1-1	EN 1994-1-2
EC5	Timber structures	EN 1995-1-1	EN 1995-1-2
EC6	Masonry structures	EN 1996-1-1	EN 1996-1-2
EC7	Geotechnical design	EN 1997-1-1	–
EC8	Seismic design	EN 1998-1-1	–
EC9	Aluminum alloy structures	EN 1999-1-1	EN 1999-1-2

Table 2.4 Classification of the Eurocodes

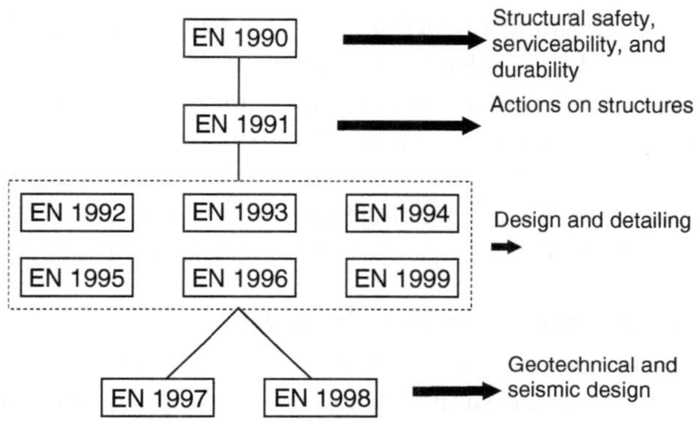

Figure 2.2 Synoptic view of different Eurocodes. (*Reproduced from CEN.*)

materials and loading situations. The different Eurocodes, together with their designation and scope, are listed in Table 2.4.

Eurocodes are the most advanced design codes, specifically with regard to design and analysis of structures under fire conditions. Most Eurocodes have two parts: Part 1-1 (such as EN1992 1-1), which covers the general rules for design at ambient temperature, and Part 1-2 (such as EN1992 1-2), which covers the design in a fire situation. Figure 2.2 gives a synoptic view of evolution of Eurocode documents and how these different codes relate to each other. Eurocode 0 (EN 1990) defines the general rules governing the ultimate limit state design, which is the basic philosophy of the Eurocodes. Eurocode 1 (EN 1991) gives the design values of the actions to various loadings and their load combinations to be considered in the design and analysis of civil engineering structures. Eurocodes 2 to 6 and Eurocode 9 provide rules for design and analysis of structures made of different materials: Eurocode 2 is for concrete structures, Eurocode 3 is for steel structures, Eurocode 4 is for steel-concrete composite structures, Eurocode 5 is for timber structures, Eurocode 6 is for masonry structures, and Eurocode 9 is for aluminum structures. Part 1-2 of Eurocodes 1 to 6 as well as Eurocode 9 are specifically devoted to provisions for design of structures subjected to fire loading. Eurocodes 7 and 8 deal with specifications relating to geotechnical and seismic design aspects.

2.5.5 British Codes

The *Building Regulations* in the United Kingdom contain detailed prescriptive- and performance-based fire safety provisions. It is generally accepted that the British codes have laid the foundation for development of regulation and design provisions in European codes and a number of other building codes. Until 1985, the building regulations for the United Kingdom were largely prescriptive and very restrictive. In an attempt to increase flexibility in design and to produce a more intelligent system, a reform of the building regulations was undertaken in the late 1970s and early 1980s. This resulted in major changes that shifted the prescriptive orientation nature of the fire regulations toward performance-based specifications in 1985.

With its publication in 1985, the *Building Regulations* had been reduced from 307 pages to 23 pages. This was made possible using functional or performance wording instead of prescriptive requirements.

Provisions relating to fire safety were revised again in 1991 to include approved documents (i.e., alternative methods based on fire safety engineering principles) in order to meet fire safety objectives of the stated regulations. However, the newly adopted methods lacked guidance on equivalency and recommended measures needed to demonstrate compliance with code provisions. To address this issue, the British Standards Institute (BSI) contracted a design team to develop a draft code of practice for the application of fire safety engineering principles in building design. This *Code of Practice* was published as a draft (DD240) for review by BSI in 1997. In 2001, BS7974 "Code of Practice on the Application of Fire Engineering Principles to the Design of Buildings" was published and supported by eight *Published Documents* containing technical guidance on different aspects of fire engineering and quantitative risk assessment (Charters, 2006).

In 2008, BS9999 "Code of Practice for Fire Safety in the Design, Management and Use of Buildings" came into effect. The concept behind the development of BS 7974 and BS 9999 is that technical guidance on fire safety is provided at three different levels: (1) general, (2) fire engineering (BS 9999), and (3) advanced analysis (BS 7974). This allows the proper selection of a design approach to correspond to the complexity of the building and to the degree of flexibility required.

Currently, the design, construction, and maintenance of high-rise residential buildings in the United Kingdom is governed by two legal frameworks: (1) *The Housing Act* and (2) *The Fire Safety Order*. These frameworks are enforced in high-rise buildings by the local authority and fire and rescue services, respectively. Specific guidance to achieve compliance for structural safety and fire safety in high-rise buildings and dwellings is contained in *Approved Documents* (*Part A–Structure* and *Part B–Fire Safety*, respectively). Similar to other building codes, the UK building regulations also refers to a number of standards such as *BS 5950* (for design of steel structures) as well as *BS 8110* (for design of concrete structures).

2.5.6 Codes and Standards in Other Countries

Fire safety provisions in other countries, specified in respective national buildings codes, are mostly prescriptive-based and are similar to (and in some cases and extension of) fire safety specifications contained in American, British, or European building codes. In China, for example, the National Technical Committee on Fire Protection of Standardization Administration (*SAC/TC113*) is responsible for enforcing fire-related technical standards. The Chinese fire code (*GB50016*: Code of Design on Building Fire Protection and Prevention) mainly lays out prescriptive-based fire safety requirements and is to be followed for typical buildings, such as low-rise residential buildings. However, for special types of structures, including high-rise buildings, large halls, and stadiums, this Chinese code allows designers to utilize *alternative* (rational) design methods based on performance-based provisions. The Chinese fire code provisions emphasize the need to provide appropriate fire mitigation measures aimed at minimizing breakout of fire. This code also requires building designs to be signed-off by a *charter engineer*, of "first or second class," depending on the complexity of the building and level of risk involved. Depending on the complexity of the design, these designs may undergo a peer review process by a qualified individual registered with the government or go before an expert review panel.

In Hong Kong, fire-related building regulations together with design provisions are contained in a series of four prescriptive codes of practice that cover the following areas: (1) means of escape, (2) access for firefighting and rescue, (3) fire-resistant construction, and (4) fire service installations and equipment. These provisions were developed, to a large degree, as an extension of provisions listed in British codes and standards during the time when Hong Kong was under British administration. Alternative fire engineering solutions, outside of the code provisions, can be arrived at with the approval by the Fire Safety Committee set up under the buildings department of the Hong Kong government. This committee is made up of representatives from the building department and the fire services department, as well as academics, engineering specialists, and practitioners.

New Zealand was one of the first countries to introduce well-articulated performance-based fire safety provisions, as early as 1992. The New Zealand Building Code requires developing performance-based solutions to occupancies (in buildings) with fire loads exceeding 1500 MJ/m. There are six Building Code clauses related to protecting people in and around buildings, limiting fire spread, and helping firefighting and rescue. These clauses comprise two verification methods and seven acceptable solutions. The three major clauses in the New Zealand fire code are as follows: (1) Change of use (occupancy) and alterations of an existing building can trigger the need for upgrades to the fire safety systems, (2) Specified systems and compliance schedules have information for building owners about their responsibility to maintain active and passive systems according to their compliance schedule, and (3) Record of amendments to the acceptable solutions, verification methods, and handbooks.

Until the 1990s, the Building Code of Australia contained prescriptive-based regulation for fire safety. In October 1996 the Australian Building Codes Board (ABCB) launched the new code which contains about 20% of clauses with performance-based fire safety provisions to undertake fire design to meet fire safety objectives. These guidelines acknowledge that designers may have to meet goals outside of the main performance requirement of life safety to include property protection and continuity of operations.

Brazil did not have specific requirements for fire safety provisions until the 1940s, when the Brazilian Association of Technical Regulations (ABNT) was commissioned. Still, special attention to fire protection, through codal provisions, was only given after two major fires that occurred in the 1970s, in the city of São Paulo, which left more than 350 people dead and over 600 people injured. As a result, few fire-related provisions were incorporated in response to the rise in number of residential fires in Brazil. It was not until the 1990s that serious progress was made in the area of fire safety through the establishment of the Brazilian Committee for Fire Protection. It is interesting to note that the enforcement of fire safety measures is still not addressed in the Brazilian model building code and is conducted by each state through the *Corpo de Bombeiros da Polícia Militar*, or *State's Fire Brigade of the Military Police*. It should be noted that fire protection provisions in Brazilian building codes are prescriptive-based (Tavares, 2009).

Japan, since 1950, had operated under a highly prescriptive building code system: the Building Standards Law (1989). However, inadequacy in available fire safety provisions have led to limited flexibility in architectural design. As a result, *The Total Fire Safety Design System of Buildings*, with performance-based fire safety provisions, was developed and adopted in the mid-1980s. This system comprises five subsystems: total fire safety, prevention of fire outbreak and spread, smoke control and evacuation, fire resistance, and fire safety for dwellings; each of which has four components: fundamental requirements, technical standards for engineering evaluation, prediction method of relevant fire phenomena, and concepts of testing methods.

In India, the National Building Code (Bureau of Indian Standards, 2016) and associated building standards (referred to as bylaws) lay down specifications and provisions for regulating fire safety in buildings. The formulation of NBC and associated standards follow three main aspects: (1) fire prevention, (2) life safety, and (3) fire protection. The NBC (Part 4) provides comprehensive recommendations on minimum requirements for fire protection. These recommendations relate to the demarcation of fire zones, restrictions on construction of buildings in each fire zone, classification of buildings based on occupancy, needed fire resistance ratings for structural and nonstructural components in different types of building (occupancies), and other restrictions and requirements necessary to minimize danger to life from fire, smoke, and fumes. Much of the fire resistance provisions in the NBC are adopted from original British codes and standards and are perceptive in nature.

2.5.7 Limitations of Codes and Standards

The current provisions for fire safety in most building codes and standards are mostly prescriptive-based, hence these provisions do not yield realistic assessment of fire performance of structures. As an example, current fire resistance ratings for structural members listed in U.S. codes and standards in the form of *prescriptive-based* tables and charts are derived from standard fire tests following ASTM E119 specifications. This standard was developed 100 years ago, and as such prescriptive-based provisions have little relevance to current (or modern) construction and do not yield economic or optimum fire safety design, especially in the case of special buildings and structures. Further, the prescriptive-based provisions are not keeping pace with latest advancements, whether in terms of structural design, architectural layout, or new construction materials.

Another major drawback of current prescriptive provisions in fire codes and standards is their inability to allow designers to develop unique and innovative engineering solutions based on intended level of fire performance needed to mitigate effects of fire hazard. The provisions in codes continue to promote a passive design philosophy where designers are required to chase a predetermined safety level that is only applicable for idealized laboratory conditions and rarely reflects performance under actual fire, loading, and restraint conditions.

Performance-based provisions, on the other hand, can overcome many of the limitations associated with prescriptive-based guidelines by establishing concise fire safety goals and objectives that capture realistic fire phenomenon and behavior as encountered in structures. Further, performance-based provisions allow designers to develop alternative but innovative and unique solutions to mitigate fire hazard and also facilitate the application of advanced analysis and modern technologies to arrive at

Code type	Advantages	Disadvantages
Prescriptive-based codes	• Straightforward evaluation of compliance with established requirements • No requirements for high level of engineering expertise	• Requirements specified to meet fictional level of safety • Complex terminology and structure of code • Do not promote cost-effectiveness • Do not provide scope for innovation • Provisions/solutions are too general and do not take specific conditions of fire, loading, restraint, etc. into account
Objective-based codes	• Include overall objectives or goals for achieving fire safety, and provide some guidance on solutions to achieve required level of safety • Permit use of alternative solutions	• Still a new concept and needs improvements with respect to defining objectives • Provide only limited approaches for developing alternative designs (solutions) • Not fully adopted in most countries
Performance-based codes	• Specify clear fire safety goals and leave means of achieving these goals to the designer • Allow alternative solutions for achieving innovation and cost-effective designs • Facilitate use of advanced analysis and computer models • Facilitate harmonization across various regions and countries and thus promote trade	• Difficult to establish validity of solution and define quantitatively • Need high level of training and expertise to undertake analysis • Difficult to evaluate compliance with established requirements • Need computer models and large amount of input parameters (fire, loading, etc.)

Table 2.5 Advantages and Disadvantages of Prescriptive-, Objective-, and Performance-Based Codes (Provisions) for Fire Safety

cost-effective solutions. Performance-based provisions, due to their engineering rationale and scientific sounding, can better set the groundwork for international standardization and trade, and can respond better to stakeholder needs.

Despite many advantages with performance-based framework for fire safety, tools for undertaking performance-based fire design continue to lag, and thus it is not easy for designers and regulators to undertake performance-based fire safety design in practice. Due to its engineering nature and reliance on complex computer models, performance-based analysis requires a large number of input parameters, often selected through simplified assumptions. Performance-based analysis also requires a high level of training and expertise on the part of designers to perform the analysis and to interpret results of fire engineering simulations. Perhaps one of the major challenges of performance-based provisions is the lack of an approved framework to establish the equivalency (i.e., fire safety) of their proposed solutions to that required by prescriptive codes.

In between prescriptive-based and performance-based fire safety provisions, a middle ground is the use of *objective-based* provisions, where fire safety can be evaluated through the use of some level of calculations methods, but not entirely performance-based. In objective-based provisions the objectives and purpose of building code requirements are clearly defined and understandable[4] such that designers can easily identify *intent* (propose) of a code requirement and hence can better design an alternative solution to satisfy fire safety objectives. Further, objective-based provisions not only describe details on design and construction aspects to achieve the codal requirements, through inclusion of *Functional Statements*, but also provide *Acceptable Solutions*, defined as solutions meeting the qualitative performance of established acceptable solutions, to achieve such objectives and functional statements. In some cases, alternative solutions are also described and provided to aid designers and engineers. Overall, objective-based provisions are evolving in some countries but still not fully adopted in most countries in the world. Table 2.5 lists the main advantages and disadvantages of prescriptive-, objective-, and performance-based codes.

[4] Objectives and purpose in prescriptive- and performance-based codes are not fully clear, nor highlighted.

2.6 Evaluating Fire Resistance

When fire grows to full size, the role of structural members becomes critical to withstand the adverse effects of fire on the structure. Therefore, structural fire safety is a key consideration in building design, since structural stability and integrity is the last line of defense in the event of fire. Building codes account for this through specifying passive fire protection requirements for structural members in the form of fire resistance ratings (of 1/3, 3/4, 1, 1 1/2, 2, 3, 4 hours), and a designer must ensure that a given structural member meets the fire rating requirement. Fire resistance of a structural member can be evaluated by either applying perspective-based approaches or using rational calculation methods.

2.6.1 Prescriptive Approaches

Perspective-based approaches evaluate fire resistance of a member based on *tabulated data, simplified rules,* or *calculation methods* derived from standards fire tests. These prescriptive approaches take into consideration only a limited number of parameters where the analysis is carried out at *sectional* or *member level* (and not at system level). This approach has been widely adopted due to the fact that for most types of buildings, the level of needed fire performance and safety is known from past experience. Also, due to the complexity in the application of advanced methods (and novel fire calculation techniques), prescriptive approaches provide simple and often conservative solutions to achieve the required level of fire resistance. Prescriptive-based approaches, due to their simplistic nature, do not account for actual design variables present in a structure (i.e., fire, loading, restraint, etc.) and realistic failure modes, which can often lead to unrealistic fire performance assessment, especially when fire growth characteristics are not similar to that under a standard fire.

2.6.2 Rational Methods

A rational methodology refers to the application of *engineering-based calculation approaches* for evaluating fire resistance (performance) of structures, taking into account realistic conditions, or close to that, present during a fire scenario. The rational methods can be applied at *sectional, member,* and/or *system* (whole building) level, with complexity increasing from sectional to member to system level. Under rational methods, all critical factors (such as specific material and geometric properties of a member, boundary conditions, fire intensity, load level, restraint conditions, etc.) can be accounted for. Although a number of rational methods have been developed for fire engineering over the last four decades, because these methods are constantly evolving, application of many of these rational methods is still a relatively new concept , and the methods are in need of validation and benchmarking.

The discussion in this chapter clearly shows that fire resistance evaluation, whether carried out through prescriptive-based approaches or rational methodology, is a complex procedure. The latter chapters in this book are aimed at providing needed information to designers, academics, and practitioners for applying prescriptive and/or rational engineering approaches for fire resistance analysis of structures at the sectional, member, or system level.

2.7 Emerging Trends and Research Needs

In the last few decades there have been significant advances on developing further understanding of fire safety requirements in buildings. There is still, however, a number of specific areas where further research is needed to advance the state of the art within this area.

2.7.1 Effect of Situational Awareness in Evacuation under Fire Conditions

High-rise buildings can be subjected to natural or manmade hazards during their service life which may necessitate emergency evacuation. Efficient evacuation of occupants in a high-rise building is a growing concern especially in the event of an emergency situation such as fire in a building given that the spread of fire and smoke may result in inaccessible exit routes, reduced visibility, and congestion, which in turn can impede evacuation. The geometric characteristics of the egress system—namely, the size, number, and arrangement of exit paths (stairs)—determine the efficiency of the evacuation process. Prescriptive requirements for geometric characteristics of emergency stairs are usually stipulated in the building codes such as the International Building Code (IBC, 2015) and are prescriptive in nature. Such provisions only establish the minimum requirements for the design of egress system and often do not account for the behavior of humans during emergency situations and any distribution from the occupants in trying to put off the fire, and so on. Therefore, for efficient evacuation process, it becomes crucial to design a performance-based robust egress system that ensures safe and fast evacuation in the time of emergencies as encountered during a fire incident (Kuligowski and Bukowski, 2004).

In a recent study, Naser and Kodur (2018) showed that incorporating cognitive abilities in a critical structure, to enhance resiliency, can improve the efficiency of evacuation. A framework of incorporating sensors into structural members to provide situational awareness during emergency evacuation was explored in this work. Situational awareness allows occupants and first responders to comprehend the severity of the disaster in identifying the location of the people stuck within a building and the nature of fire growth. Such structural awareness measures can also point to vulnerable areas (or routes) in a building, can provide a future projection of the status of the building, and can guide evacuation process and firefighting. While the current building and egress system design typically follows the prescriptive approaches laid out in building codes, it may be highly beneficial to strategically design the egress system within a building to strategically take into account story height, building layout, and number and arrangement of exit paths. For this, further research is needed on the development of an intelligent, robust, and low energy consumption sensor framework, advancement of algorithms that can identify fire breakout and track occupants' progress of evacuation and first responders' location, and communication tools to ensure continuous communication between evacuees and authorities.

2.8 References

ACI/TMS Standard. *Code Requirements for Determining Fire Resistance of Concrete and Masonry Construction* (2014). Farmington Hills, MI: American Concrete Institute, 2014.

AISC. *Steel Construction Manual.* Chicago, IL: American Institute of Steel Construction, 2017.

American Society of Civil Engineers (ASCE). *Standard Calculation Methods for Structural Fire Protection.* Vol. 29. Reston, VA: American Society of Civil Engineers, 2007.

American Society of Civil Engineers (ASCE). *Minimum Design Loads for Buildings and Other Structures.* Vol. 7. Reston, VA: American Society of Civil Engineers, 2016.

American Wood Council. National Design Specification® (NDS®) for Wood Construction. *Design Values for Wood Construction: A Supplement to the 2015 Edition, National Design Specification for Wood Construction.* Leesburg, VA: National Forest Products Association, 2015.

ASTM. *Standard Test Methods for Fire Tests of Building Construction and Materials, ASTM E119.* West Conshohocken, PA: American Society for Testing and Materials, 2016.

Building Standard Law. Ministry of Construction. *Kentikubutsu no Sogu Koka Sekkeihov* [*Total Fire Safety Design System of Buildings*] Tokyo: The Building Center of Japan, 1989.

Bureau of Indian Standards, National Building Code of India, 2016.

Drysdale, Dougal. *An Introduction to Fire Dynamics.* Hoboken, NJ: Wiley, 2011.

Fitzgerald, Robert W., and Brian J. Meacham. *Fire Performance Analysis for Buildings.* Hoboken, NJ: Wiley, 2017.

Fletcher, Ian A., Audun Borg, Neil Hitchen, and Stephen Welch. "Performance of Concrete in Fire: A Review of the State of the Art, with a Case Study of the Windsor Tower fir," 2006.

Gosselin, Guy C. *Structural Fire Protection: Predictive Methods.* Institute for Research in Construction, National Research Council of Canada, Ottawa, Canada, 1987.

Guntrip, Laura. "Fire Safety in Care Homes: A Big Problem under Harsh Scrutiny." *Nursing and Residential Care* 19, no. 8 (2017): 468–71.

IBC, ICC. "International Building Code." International Code Council, Inc. (formerly BOCA, ICBO and SBCCI) 4051 (2015): 60478–5795.

ISO (2012) ISO 834-1:1999. *Fire Resistance Tests–Elements of Building Construction—Part 1: General Requirements.* Geneva: International Organization for Standardization.

Khaliq, Wasim, Venkatesh Kodur, and Nikhil Raut. "Comparative Fire Performance of High Strength Concrete Columns with Different Types of Fiber Reinforcement." *Applications of Structural Fire Engineering* (2015): 493–500.

Khorasani-Zavareh, Davoud, and Mohammadreza Shokouhi. "Collapse of the Plasco Building due to Fires and Its Lessons Learnt." *Safety Promotion and Injury Prevention* 5, no. 3 (2018): 120–24.

Kuligowski E, Bukowski RW. Design of occupant egress systems for tall buildings. CIB World Building Congress, 2004.

Naser, M. Z., and V. K. R. Kodur. "Cognitive Infrastructure: A Modern Concept for Resilient Performance under Extreme Events." *Automation in Construction* 90 (2018): 253–264.

National Fire Protection Association (NFPA). *NFPA 5000: Building Construction and Safety Code.* Quincy, MA: National Fire Protection Association, 2018.

NBCC. *National Building Code of Canada.* Ottawa: Canada, National Research Council of Canada, 2015.

Precast/Prestressed Concrete Institute. *PCI MNL-124, Design for Fire Resistance of Precast/Prestressed Concrete.* 3rd ed. (MNL-124-11). Chicago, IL: Precast/Prestressed Concrete Institute.

SFPE. *SFPE Handbook of Fire Protection Engineering, 5th edition.* Quincy, MA: Springer, National Fire Protection Association, 2016.

SFPE. *SFPE Engineering Standard on Calculating Fire Exposures to Structures.* Gaithersburg, MD: Society of Fire Protection Engineers, 2016.

Tavares, Rodrigo Machado. "An Analysis of the Fire Safety Codes in Brazil: Is the Performance-Based Approach the Best Practice?" *Fire Safety Journal* 44, no. 5 (2009): 749–55.

UL 263, Standard for Fire Tests of Building Construction and Materials. Northbrook, IL: Underwriters Laboratories, 2011.

CHAPTER 3

Fire Growth and Fire Severity

3.1 General

This chapter presents an overview of fire growth and fire spread in buildings. The basics of heat transfer mechanisms associated with fire spread in buildings are reviewed. The various types of fuels present in different building occupancies, together with the combustion process, that lead to full-size fire in a building are explained. In addition, various fire development models ranging from simple empirical relations to advanced fire models that can be adopted to estimate fire intensity and temperature progression at different stages of fire are presented. Finally, the concept of equivalent fire severity to compare a real fire exposure to a standard fire scenario is discussed.

3.2 Heat Transfer Mechanisms

Heat transfer is the exchange of thermal energy between physical objects (or systems) within an environment. During fire exposure, the extent of heat transfer to structural members is governed by the mechanisms of heat transfer, thermal properties of consistent materials, and boundary conditions between various mediums and objects. The rate of heat transfer is dependent on the temperature of the emitting and the receiving objects (or systems), as well as on the properties of the transferring medium (i.e., air). In materials which are good conductors of heat, such as steel, heat is transferred by interactions involving free electrons (Wickström, 2016). In poor heat conductors such as concrete, heat is transferred by vibration of the molecular lattice.

The three fundamental modes of heat transfer are conduction (diffusion), convection, and radiation. The rate and extent of heat transfer is dependent not only on the properties/characteristics of the medium under which heat is transferred but also on the conditions present in the objects/mediums. The two conditions commonly present are either a *steady* state or a *transient* state condition.

3.2.1 Conduction

Conduction is the primary mechanism for heat transfer in solid materials and is the main influencing factor in the ignition of solid surfaces. Solids are often classified based on their ability to transfer heat. A solid that is a good conductor allows temperature to transfer easily within its microstructure. Materials that conduct ions, protons or electrons, are capable of transferring heat. Examples of good conductors are metals such as steel and copper. On the other hand, examples of poor conductors (or insulators) are concrete, plastic, and wood.

The rate at which thermal energy is conducted as heat within a body (e.g., steel beam) or between two bodies (e.g., steel beam compositely attached to concrete slab) is a function of the temperature difference (temperature gradient) between the cooler and hotter surfaces of the beam and the properties of the conductive medium (e.g., air) through which the heat is transferred. The thermal properties needed for heat transfer calculations in solid objects include density, ρ (kg/m^3), specific heat, C_p (J/kg.K), and thermal conductivity, k (W/m.K). These properties are temperature-dependent and can vary significantly at elevated temperatures. Table 3.1 shows the thermal properties of some commonly used construction materials.

The thermal resistance offered by a material for heat transfer is expressed in terms of thermal diffusivity, $\alpha = k/\rho C_p$ (m^2/s), and thermal inertia,[1] $I = \sqrt{k\rho c_p}$ (W^{2}s/m^{4}K^2). Materials with low thermal diffusivity conduct more heat than materials with high thermal diffusivity. Depending on the variation in heat with respect to time, the rate of conduction can be either in a steady state or in a transient state.

The rate of conduction (heat transfer) within solids can vary depending on the environmental (boundary) conditions, namely, steady state and transient state. In a steady-state situation, temperature

[1]It should be noted that materials with low thermal inertia ignite at a fast rate because their surface temperature increases rapidly.

Material	Density, ρ (kg/m^3)	Specific Heat, C_p (J/kg.K)	Thermal Conductivity, k (W/m.K)	Thermal Diffusivity, $\alpha = k/\rho C_p$ (m^2/s)	Thermal Inertia, $I = \sqrt{k\rho c_p}$ (W^{2}s/m^{4}K^2)
Concrete	1900–2300	800–1000	0.2	7.6×10^{-7}	6.15×10^2
Steel	7800	450	50	1.4×10^{-5}	1.32×10^4
Wood	500	1000	0.13	2.6×10^{-7}	2.55×10^2
Brick	1750	1000	0.5	3.2×10^{-7}	9.35×10^2
Aluminum	2700	900	218	6.4×10^{-5}	2.3×10^4
Copper	8900	386	150–190	1.1×10^{-4}	2.45×10^4
Fiber-based insulating board	229	2090	0.041	8.6×10^{-8}	1.40×10^2
Gypsum plaster	800	1440	0.48	4.1×10^{-7}	7.44×10^2

Table 3.1 Thermal Properties of Common Materials Used in Construction Applications (Hurley et al., 2015)

difference driving the conduction (heat transfer) is constant, so after attaining an equilibration state, the spatial distribution of temperatures (temperature field) in the conducting object does not change any further. Thus, the rate of heat transfer by conduction is directly proportional to the temperature gradient between any given two points in the object. This proportionality is known as thermal conductivity k. Thus, heat transfer ($\dot{q}''$) through conduction in one-dimensional space is given by

$$\dot{q}'' = k\frac{dT}{dx} \tag{3.1}$$

where $\dot{q}''$ is the heat flow per unit area (W/m^2), k is the thermal conductivity (W/m.K.), T is the temperature (°C or K), and x is the distance between two points in heat flow direction (m).

Sectional temperature within a body (object), in general, varies with time, as well as point of location within a body. In rectangular coordinates, this variation in location is expressed as $T(x, y, z, t)$, where (x, y, z) indicates variation in the x-, y-, and z-directions, and t indicates variation with time. Thus, under transient heat flow condition, one-dimensional heat transfer by conduction (with no internal heat being released) can be expressed as

$$\frac{\delta^2 T}{\delta^2 x} = \frac{1}{\alpha}\frac{\delta T}{\delta t} \tag{3.2}$$

where t is the time (s) and α is the thermal diffusivity (m^2/s).

The above heat transfer equations can also be extended to two- and three-dimensional space. These equations can be solved using simple formulas (i.e., utilizing lumped heat capacity equations), design charts (graphical aid from standard fire tests), and/or numerical analysis (i.e., one-, two-, or three-dimensional finite difference or finite element analysis).

3.2.2 Convection

Convection is the second mechanism of heat transfer and is the most important heat transfer mechanism in fluids and gases. Convection allows heat to transfer through the movement of fluids, either gases or liquids. Heat transfer through convection is a dominant mechanism in building fires as flame spread and vertical transport of smoke (and hot gases) to the ceiling, or outside of a window from a room fire, occurs through this mechanism. Convective heat transfer is usually taken to be directly proportional to the temperature difference between a medium and an object (or within a fluid) as

$$q'' = h\Delta t \tag{3.3}$$

where q'' (W/m^2) is the heat flow per unit area, ΔT is the temperature difference between the surface of object (°C or K), and h is the convective heat transfer coefficient (W/m^{2}K) that depends on geometry of the surface, nature of heat flow, fire intensity, and thickness of boundary condition. A typical value of h used to represent convective heat transfer from fire zone to structural members in buildings ranges between 20 and 25 W/m^{2}K in standard fire conditions and between 25 and 50 W/m^{2}K under hydrocarbon fire (high intensity) exposure (Eurocode 1, 2002). It is worth noting that recent experimental tests have showed that convective heat transfer in real fires can significantly vary between 15 and 75 W/m^{2}K depending on the intensity of flames (Veloo and Quintiere, 2013).

3.2.3 Radiation

Radiation is the transfer of thermal energy by electromagnetic waves traveling through a vacuum or transparent medium. Radiative heat transfer is the most dominant mechanism in building fires. Radiation also contributes to rapid fire growth and fire spread by transferring thermal energy between flames to fuel surfaces, hot smoke to building objects, and flames from one building to an adjacent building.

The extent of radiative heat transfer is dependent on the characteristics of the two objects or surfaces (emitter or receiver) linked in the heat transfer. The radiant heat flux, q'' (W/m^2), at a point on a receiving surface is given by

$$q'' = \phi \varepsilon_e \sigma T_e^4 \tag{3.4}$$

where ϕ is a configuration or view factor that accounts for how much of the emitter is seen by the receiving surface, ε_e is an emissivity of the emitting surface, σ is the Stefan-Boltzmann constant (5.67×10^{-8} W/m^{2}K^4), and T_e is the absolute temperature of the emitting surface (K).

The heat transfer occurring through radiation between two surfaces is illustrated in Fig. 3.1. In the case of two parallel surfaces (emitter and receiver), this view factor (configuration factor), ϕ, can be expressed as

$$\phi = \left(\frac{1}{90}\right)\left[\frac{X}{\sqrt{1+x^2}}\tan^{-1}\left(\frac{y}{1+x^2}\right) + \frac{y}{\sqrt{1+y^2}}\tan^{-1}\left(\frac{x}{1+y^2}\right)\right] \tag{3.5}$$

where $x = H/2r$, $y = W/2r$, and H and W are the height and width of the rectangular source, as shown in Fig. 3.1. Other expressions and design aids for the configuration factor can be found in Drysdale (2011).

If the distance r is large relative to the size of the emitting surface, then the view factor (Eq. 3.5) can be simplified to

$$\phi = \frac{A_1}{\pi r^2} \tag{3.6}$$

where A_1 is the area of emitting surface ($A_1 = HW$). Drysdale (2011) provides various view factors for many situations and values of emissivity.

Thus, in this case, the resulting heat flow q'' (W/m^2) from the emitting surface to the receiving surface (given in Eq. 3.4) gets simplified to

$$q'' = \phi \varepsilon \sigma (T_e^4 - T_r^4) \tag{3.7}$$

where T_r is the absolute temperature of the receiving surface (K) and ε is the resultant emissivity of the two surfaces, taking into consideration the emissivity of receiving and emitting surfaces. The emissivity can be

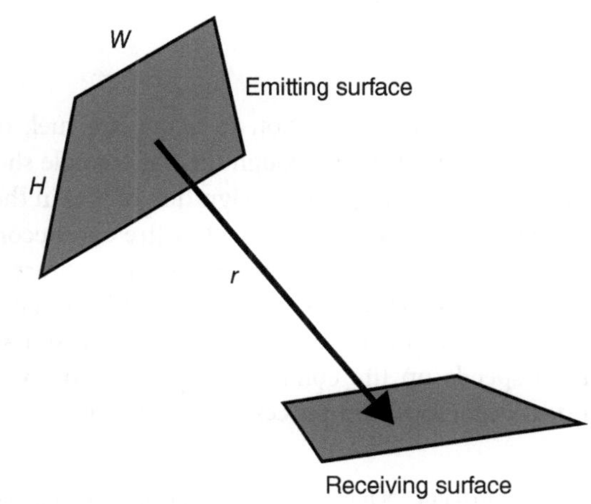

FIGURE 3.1 View factor, emitting and receiving surfaces.

Material	Emissivity (ε)
Aluminum (oxidized)	0.25
Asbestos board	0.96
Basalt	0.72
Brick, red rough	0.93
Clay	0.91
Concrete	0.85
Copper-nickel alloy, polished	0.059
Glass, smooth	0.92–0.94
Gypsum	0.85
Masonry, plastered	0.93
Stainless steel	0.40
Steel, galvanized—Old	0.88
Steel, galvanized—New	0.23
Wood Oak, planned	0.885

Table 3.2 Typical Values of Emissivity of Common Construction Materials (Drysdale, 2011; SFPE, 2015)

measured using simple devices such as Leslie's cube or thermal radiation detectors (i.e., a thermopile or a bolometer). The resultant emissivity between two objects can also be calculated using Eq. 3.8:

$$\varepsilon = \frac{1}{\left(\frac{1}{\varepsilon_e} + \frac{1}{\varepsilon_r} - 1\right)} \tag{3.8}$$

where ε_r is the emissivity of the receiving surface and ε_e is the emissivity of the emitting surface.

The value of emissivity ranges from 0 to 1, where 1 is assigned to a "black body" radiator. In a fire incident occurring in buildings, the emissivity of flames[2] (radiating objects) vary between 0.7 and 1, while the emissivity of typical structural members made of steel or concrete vary between 0.7 and 0.85, respectively (CEN, 2004; CEN, 2005). Table 3.2 shows typical values of emissivity for common construction materials.

Although emissivity of a given object can be assumed to be of a constant value, for simplicity the emissivity is also a function of the composition of the radiating object and can change during different stages of fire growth. For instance, surface emissivity of zinc-coated steel elements in the fire equals 0.32, which is less than half of the value of emissivity for a steel element without surfacing ($\varepsilon_{steel} = 0.7$). When galvanized steel (zinc-coated steel) reaches 450°C, zinc material melts and steel becomes the emitting surface, which significantly changes the emissivity value at that stage of fire (Jirku and Wald, 2015).

3.3 Fire and Heat

3.3.1 General

The components necessary for combustion to occur are fuel, oxygen, and a source of ignition. This combustion process can be illustrated through the fire triangle shown in Fig. 3.2. Fire in a room develops due to the burning process resulting from the ignition of fuel in the presence of oxygen. Once combustion has begun, with ample fuel and oxygen present, a fire can become *self-sustaining*. As the fuel burns, it creates more heat, and this increased heat raises the temperature of other objects (possible fuels) to their ignition temperature. To support further combustion, additional oxygen is drawn into the fire zone. The oxygen, in turn, increases the burning rate and more fuel becomes involved in the combustion process.

While oxidation speeds up the combustion process in an object, another process simultaneously occurs. The chemical decomposition process occurs in the object under heating conditions. The rise in

[2]Emissivity can also vary due to nature of fuel type. For example, emissivity of flames due to burning of alcohol, petrol, kerosene and benzene are 0.066, 0.36, 0.37 and 0.7, respectively. The emissivity of flames due to burning of wood can be estimated by, where fth is the flame thickness (Beyreis et al., 1971).

Figure 3.2 Illustration of combustion process through "Fire triangle."

temperature causes the substance (object) to degrade (or thermally decompose), which results in the emittance of vapors and gases. This interaction continues until either all the fuel has been consumed or all the oxygen has been used up or the heat has been dissipated so that the temperature of the fuel drops below its ignition temperature, causing the fire to *self-extinguish*.

3.3.2 Fuel and Ventilation

Fuel refers to any material that has carbon in its composition. Most of the contents in buildings are composed of derivatives of organic materials that primarily have carbon in their composition. Organic materials are by-products of plants (such as cotton and wood) and petrochemicals (such as plastics/polymer composites). Therefore, most of the materials that form part of the building structure, lining materials, and the contents of the building (e.g., furniture) constitute as fuel. All of these materials have hydrocarbon molecules that are mainly made of carbon and bonding atoms such as oxygen and nitrogen. Depending on its composition, each material has a certain level of heat energy stored in it which gets released during the combustion process.

Calorific Value

Once ignition (combustion) occurs, the heat release rate from fuel depends on the nature of the burning objects, in terms of material composition, and the amount of air (oxygen) available. The amount of heat released during the complete combustion of a unit mass of fuel with unlimited ventilation is referred to as its *calorific value* (or heat of combustion). Most solid, liquid, and gaseous fuels have a calorific value between 15 and 50 MJ/kg. Typical calorific values (ΔH_c in MJ/kg) for materials and fuels present in buildings are given in Eurocode 1 (2002) and reproduced in Table 3.3.

The calorific heat release values listed in the table are for dry materials. If materials contain a significant level of moisture (as in the case of wood or gypsum), the moisture reduces the calorific value and needs to be accounted for. Hence, the *effective calorific* value, $\Delta H_{c,n}$ (MJ/kg), can be computed as

$$\Delta H_{c,n} = H_c(1 - 0.001m_c) - 0.025m_c \tag{3.9}$$

Material	Calorific Value (MJ/kg)
Alcohol	27–33
Cellulose	15–18
Coal	28–34
Gasoline	43–44
Grain	16–18
Foam rubber	34–40
Kerosene	43.2
Petroleum	43.7
PVC (polyvinyl chloride)	16–17
Wood	16–20
Wool	21–26

Table 3.3 Calorific Values of Common Materials (Eurocode 1, 2002)

where m_c is the moisture content (as a percentage by total weight), calculated from

$$m_c = \frac{100m_d}{100 + m_d} \tag{3.10}$$

where m_d is the moisture content as a percentage of the dry weight. The effect of moisture on calorific value can be illustrated by considering wood. Wood contains different levels of moisture depending on the species type and the condition of the wood. The typical heat of combustion of oven-dry wood ranges between 12 and 21 MJ/kg (Eurocode 1, 2002). Considering a sample of wood with a moisture content of $m_d = 10\%$ and $H_c = 19$ MJ/kg, and using Eqs. 3.9 and 3.10, the effective calorific value of this sample of wood is 18.6 MJ/kg.

Knowing the calorific value of fuel and mass of fuel, the maximum possible energy, E (MJ), that can be released by burning of that fuel can be computed as

$$E = M\Delta H_c \text{ for dry fuel} \tag{3.11}$$

or

$$E = M\Delta H_{c,n}, \text{ for fuel containing moisture} \tag{3.12}$$

where M is the mass of the fuel (kg).

Fire Load

Most of the contents present in buildings or other structures contain hydrocarbons, and these have thermal energy stored in them. For fire load calculations, fire load in a compartment is expressed in terms of fire load energy density (FLED), e_f(MJ/m^2), which can be computed as

$$e_f = E/A_f \tag{3.13}$$

where A_f(m^2) is the floor area of the compartment.

Since lining materials that form compartment boundaries in a room (walls, floors, and ceiling) can be a significant source of fuel, the fuel from various lining materials is to be taken into account in fire load calculations. Thus, total fire load, expressed in terms of energy density per total room bounding surfaces, e_t(MJ/m^2), is given by

$$e_t = E/A_t \tag{3.14}$$

where A_t (m^2) is the total area of the bounding surfaces of the room (i.e., floor, ceiling, and walls, including window openings[3]). Comparing e_f and e_t, it can be seen that $e_f > e_t$ by the ratio A_f/A_t. This difference between e_f and e_t can be large enough to make it important to distinguish which fuel load density is being used in any given situation.

FLED is a sensitive property that mainly depends on the occupancy type. In order to standardize FLED for fire design in buildings, a large number of surveys have been conducted in buildings to arrive at standardized fuel load energy density for various occupying types. For instance, the New Zealand Building Code (MBIE, 2007) assumes design values of 400, 800, and 1200 MJ per square meter of floor area for residential, office, and retail occupancies, respectively. On the other hand, Eurocode 1 (2002) classifies structures into nine classes, with fire load ranging from 122 to 1824 MJ/m^2 of floor area. Eurocode 1 (2002) recommends that the design fire load be taken at the 80th percentile. In addition to Eurocode recommendations, a number of surveys have reported average fire fuel loads in different occupancies. Table 3.4 lists fire load densities for different occupancies, compiled from a number of sources.

Design fire loads should be determined in a similar way to loads for structural design under external loading events such as wind, snow, or earthquakes. The design fire load should include both *fixed* and *moveable fire* loads and should also reflect the possible maximum fire load, with less than 10% probability being exceeded in 50 years. In order to arrive at a safe design, it is recommended that selected fire design load should be in the 80th to 95th percentile values of surveyed fire loads. At this range, and with a coefficient

[3]There are six surfaces in a rectangular (or square) room.

Source	Occupancy Type	Average (MJ/m²)	80% Fractile
Eurocode 1 (2002)	Dwelling	780	948
	Hospital (room)	230	280
	Hotel (room)	310	377
	Library	1500	1824
	Office	420	511
	Classroom of a school	285	347
	Shopping center	600	730
	Theater (cinema)	300	365
	Transport (public space)	100	122
CIB (1986)	Office	420	680
	Dwelling	640	750
	Hotel	345	420
	School	285	360
Culver (1976)	General/clerical offices	598	898
	Conference rooms	425	714
	File, storage rooms	1125	1968
New Zealand Building Code (MBIE, 2007)	Residential	400	486
	Office	800	972
	Retails	1200	1458
National Fire Protection Association (2011)	Government buildings	641	779
	Private buildings	720	874

Table 3.4 Fire Loads in Various Occupancies

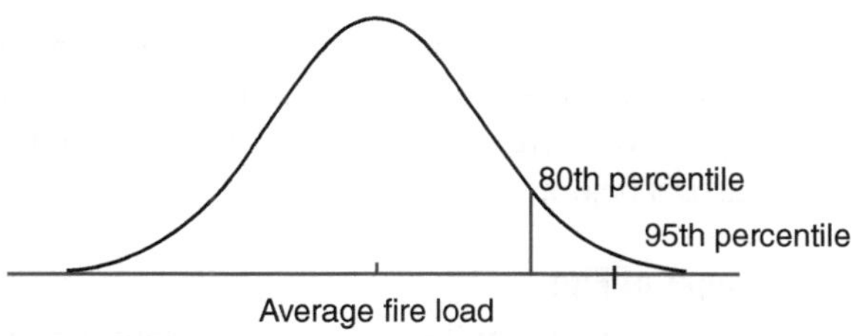

Figure 3.3 Representation of fire design load.

of variation of 50% to 80% of the average value, the 90th percentile value will be 1.65 to 2.0 times the average value listed in Table 3.4 (Buchanan and Abu, 2017). Figure 3.3 shows the probability density distribution functions of fire loads.

Probability of Fire Ignition

The random occurrence of accidental (rare) events such as fires can be modeled using *Poisson distribution*, where the occurrence of events (fires) is statistically independent. The ignition resulting in a fire can be modeled with a mean rate of occurrence, λ_{Ign} (Phan et al., 2010). This rate of ignition is a function of a number of parameters including floor area, age of the building, and level of maintenance, among others. The mean rate of fire occurrence is summarized in Table 3.5 for several common building occupancies (CIB, 1986; SFPE/SEI, 2003). Following ignition, the likelihood of a fully developed fire with the potential to cause significant structural damage depends on the presence (together with the timely activation) of active fire protection systems (i.e., suppression systems) as well as the quick arrival and response of firefighters. The probability of ignition developing into flashover in common occupancies ranges from 0.01% to 10%.

3.4 Fire Development in a Compartment

The process of fire development can be presented in terms of fire growth in a *room* or *compartment*, since ignition starts in one compartment and then spreads into various compartments in a building. As discussed

Occupancy	Mean Rate × 10^{-6}/m²/yr	Probability of Occurrence of Fully Developed Fire
Office	1–2	10^{-3}–10^{-2}
Dwelling	2–5	10^{-1}
Hotel/School	0.5	10^{-3}–10^{-2}
Commercial lot	1	10^{-2}

Table 3.5 Probability of Fire Occurrences (CIB, 1986; SFPE/SEI, 2003).

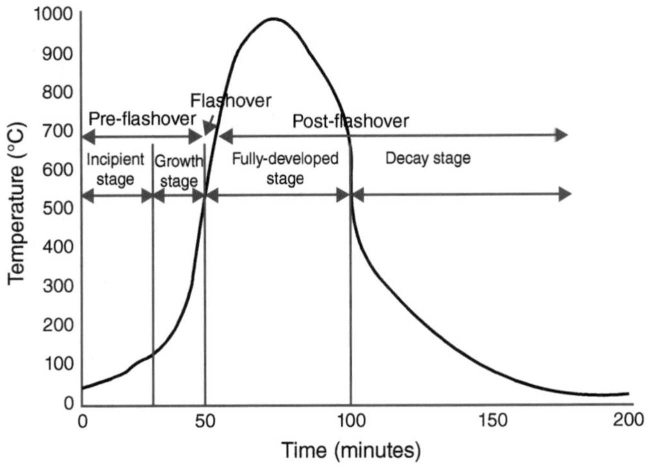

Figure 3.4 Time-temperature curve, for the full fire development process inside a typical room, with no suppression.

in Chap. 2, fire development in a compartment can be grouped under four distinct stages: ignition (also known as incipient), pre-flashover (growth period), post-flashover (fully developed fire), and cooling (decay). The overall process of fire development in a typical compartment is fully explained by Drysdale (2011) and is summarized below.

3.4.1 Different Stages of Fire

Regardless of fuel type, room size, ignition cause, and extent of ventilation present, the overall fire growth phenomenon in a compartment is quite similar and occurs in the four stages grouped under *incipient, growth, fully developed,* and *decay* periods. Despite grouping fire development into four "stages," the actual fire growth and decay process is continuous, with each stage smoothly transitioning from one to the next. In addition, while this holds true in most building fires, it is not easy to clearly demonstrate the transition point from one stage to another in actual fire incidents. Also, only in controlled environments, such as in laboratory fire tests, various fire stages can be clearly distinguished. The complete process of fire development inside of a typical room can be represented through a time-temperature curve as shown in Fig. 3.4.

The *incipient stage* represents initial ignition of fuel in a room wherein heating of fuel occurs through early combustion and has not yet significantly impacted the environment inside the compartment (through the release of heat and toxic gases). Once combustion occurs, the growth of fire is largely dependent on the characteristics and configuration of the fuel involved, the geometry of the room, and the available ventilation. During this initial phase, radiant heat from burning objects heats up adjacent fuel items and thus the process of pyrolysis starts. In the meantime, a plume of hot gases and flames released from combustion rise from the burning zone and mixes with the cooler air within the room.

During the incipient stage, fire has unique characteristics of being small, weak, and local due to low thermal energy. Hence, temperatures in the room are low and combustion can often be controlled or extinguished through portable fire extinguishers without the need for protective clothing or breathing apparatus. However, if the fire continues to grow, additional fuel (objects) in the room will start to ignite, which increases the buildup (volume) of layers of hot gases under the room ceiling. The burning of this additional fuel will release a larger amount of heat (thermal) energy, leading to an overall raise in temperatures. At this stage, fire has moved beyond its incipient stage and with adequate oxygen intake will continue to grow fairly quickly. This marks the transition to the next stage, the growth stage. At this point, fire presents an immediate danger to safety of occupants.

In the *growth stage*, the rate of thermal energy released by the burning objects will continue to increase. This increase in heat energy surges the volume and temperature of the hot gas layers (collected near celling) as well as compartment pressure. The increase of pressure in the hot (upper) layers causes them to push down within the compartment and out through any openings or voids such as in ceiling. Since the pressure of the cool layer underneath hot layers is low, this results in inward movement of air from outside of the compartment. At the intersection line where hot and cold layers meet, the pressure is neutral, and this is commonly referred to as the *neutral plane*.

Also in this stage, a fire can continue to grow through flame spread or by ignition of other fuel/materials within the compartment. As flames of the plume reach the ceiling, flames can bend and begin to extend horizontally. Pyrolysis[4] products as well as flammable by-products of incomplete combustion in the hot gas layer will ignite and continue this horizontal extension across the ceiling. This phenomenon is known as *rollover* and is an indicator of impending flashover (Peacock et al., 1999).

Flashover refers to the transition state at which point all fuel present in the room starts burning, and this point represents a sudden transition from a growing (local) fire to a fully developed fire. Flashover in a typical building compartment occurs when the temperature in the compartment reaches 500 to 600°C, which is equivalent to a heat flux[5] of 15 to 20 kW/m^2 (Babrauskas, 1979; Purkiss and Long-Yuan, 2013). In the flashover stage, conditions in the compartment become life-threating due to high levels of temperature and the presence of toxic gases which cannot be easily extinguished by occupants.[6] During flashover, hot gases can push out (break) the openings in the compartment such as a door leading to another room at a substantial velocity (larger than that at the growing stage), which marks transition of fire into the fully developed stage.

In the growth stage of a fire, the rate of burning (heat release) is controlled by the nature of the fuel and the extent of ventilation (oxygen), and temperature and heat flux in a room are extremely high due to burning of all exposed surfaces in the room, since the fire is transitioning to a fully developed stage. Once the *fully developed stage* is reached, the rate of burning in the room is controlled by the available ventilation. Fire temperature in the compartment at this stage is in the range of 600 to 1000°C. At this stage, the fire can spread through multiple compartments (rooms) or floors. In high-rise buildings, occupants present elsewhere (away from burning compartment) may not fully realize or sense the fire until fire transitions into a fully developed fire. Thus, to ensure life safety, fire detection and alarming systems must alert occupants at the early stages of the fire (incipient or growth stage) to evacuate. Due to high room temperatures developing, structural members start to experience temperature-induced degradation in capacity.

As the fuel available within the compartment is consumed or the oxygen runs out, the combustion process starts to decline and the overall temperature in the room starts to decay slowly. At this point, fire transitions from a fully developed fire into a decaying fire. In the *decay stage*, the level of combustion is controlled by the extent of fuel available, referred to as the *fuel-controlled fire*, or extent of oxygen available, referred to as the *ventilation-controlled fire*. If either fuel or ventilation runs out, combustion subsides and the temperature in the room decays due to transfer of heat to the surrounding environment.

3.5 Pre-Flashover Fire Calculations

In the *pre-flashover period*, combustion is local and restricted to a small area of the compartment; therefore, only localized burning occurs. The overall rise in temperature within the bounded fire compartment is low, and hence there may not be clear signs of fire, especially from outside the compartment. In some cases, the burning of items near a window can cause glass to break, which increases air supply and accelerates combustion. In some incidents, human intervention (such as accidently opening a door) could also accelerate the burning process and the occurrence of flashover. Hence, activation of active fire protection systems such as alarms and sprinklers during the pre-flashover period is critical to prevent fire growth, limit fire spread, and ensure quick evacuation of occupants.

---

[4]Thermal decomposition of materials at elevated temperatures.

[5]At flashover sufficient heat energy is being transferred to each square foot of floor to raise the temperature of half a liter of water between 26 and 40.5°C every minute.

[6]Still, alert occupants can escape to safety if smoke or flames are not blocking egress path and escape routes.

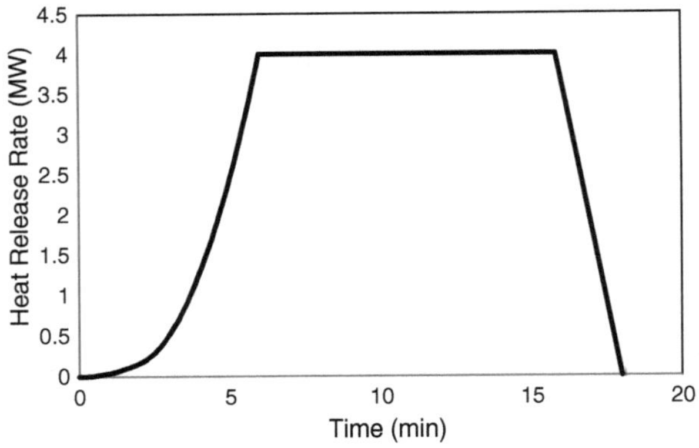

Figure 3.5 Typical t-squared fire for a single burning object.

Therefore, in the design of active protection systems, heat release rate, rate of fire growth, and temperature rise should be properly accounted for in order to arrive at the proper activation time for detection and controlling of a fire. For determining the severity of pre-flashover fire, simple hand calculations or advanced computer models (i.e., zone models) can be applied to evaluate fire growth characteristics. One such simplified method to quantify heat release rate of pre-flashover fires is the t-squared approach.

3.5.1 T-Squared Fires

The fire load present in a compartment is expressed in terms of energy density—that is, fire load per unit floor area. Each object (or fuel item) in the compartment has a certain amount of thermal energy (E), and upon ignition, this energy gets released over time, t. The heat release follows a function in which the heat release rate is directly proportional to the square of the burning time. This trend is referred to as a *t-squared fire* and can be represented through a parabolic curve, as plotted in Fig. 3.5. Once peak heat release rate (Q_p) is attained, the object continues to burn in a steady state with a constant peak heat release rate until its thermal energy dissipates. This t-squared fire growth function can be applied to evaluate fire growth characteristics in a room at the pre-flashover stage. Thus, t-squared design fires are often utilized as input to room fire models to evaluate fire development in a compartment. In Fig. 3.5, a representative heat release rate (Q) as a function of time (t) of a fuel item is plotted in which the area bound by the Q-t curve is the total thermal energy (E) present in the object.

The heat release rate at time (t) in the parabolic portion of the curve can be expressed as

$$Q = (t/k)^2 = (\alpha t)^2 \tag{3.15}$$

where k is a growth constant (s/MW$^{1/2}$) and α is a fire intensity coefficient (MW/s^2). The growth constant k (or fire intensity coefficient, α) accounts for the rate of fire growth, which is different for different fuel objects, depending on combustion characteristics. In fire protection engineering, this growth constant k is taken as 600, 300, 150, and 75 s/MW$^{1/2}$ for slow, medium, fast, and ultrafast burning fires, respectively. Correspondingly, α equals 2.93×10^{-6}, 1.17×10^{-6}, 4.66×10^{-5}, and 1.874×10^{-4} for slow, medium, fast, and ultrafast fires, respectively.

Slow fires represent burning objects with a slow heat release rate as in the case of densely packed (thick) wood logs. Medium fires correspond to fires with a medium heat release rate, such as those arising from the burning of thinly packed solid wood panels or individual furniture items with small amounts of plastics. Fast fires, as in the case of burning of some upholstered furniture, comprise high stacked wood pallets and represent rapid burning. Finally, ultrafast fires correspond to very rapid burning of highly combustible items (such as densely stacked plastics or liquid fuels, e.g., gasoline). Figure 3.6 shows t-squared fire curves for slow, medium, fast, and ultrafast fires.

Figure 3.6 shows the resulting heat release rates for furniture in an office fire with slow, medium, fast, and ultrafast fire growth rates for a peak heat release rate of 2 MW. All four fuel items have the same amount of thermal energy. But depending on burning characteristics, the heat release rate and the extent of a steady state of burning are different in each case.

T-squared fires can be used to describe the heat release rate for burning of a single or multiple fuel objects. In a compartment, fire can spread from the first burning object to a second object through flame contact or by radiant heat transfer. The time to ignition of a second object depends on the intensity of

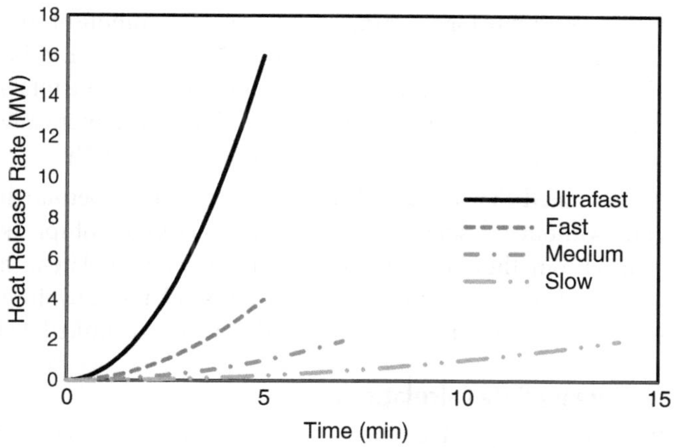

Figure 3.6 Heat release rate for a fire in office furniture with slow, medium, fast, and ultrafast fire growth rates.

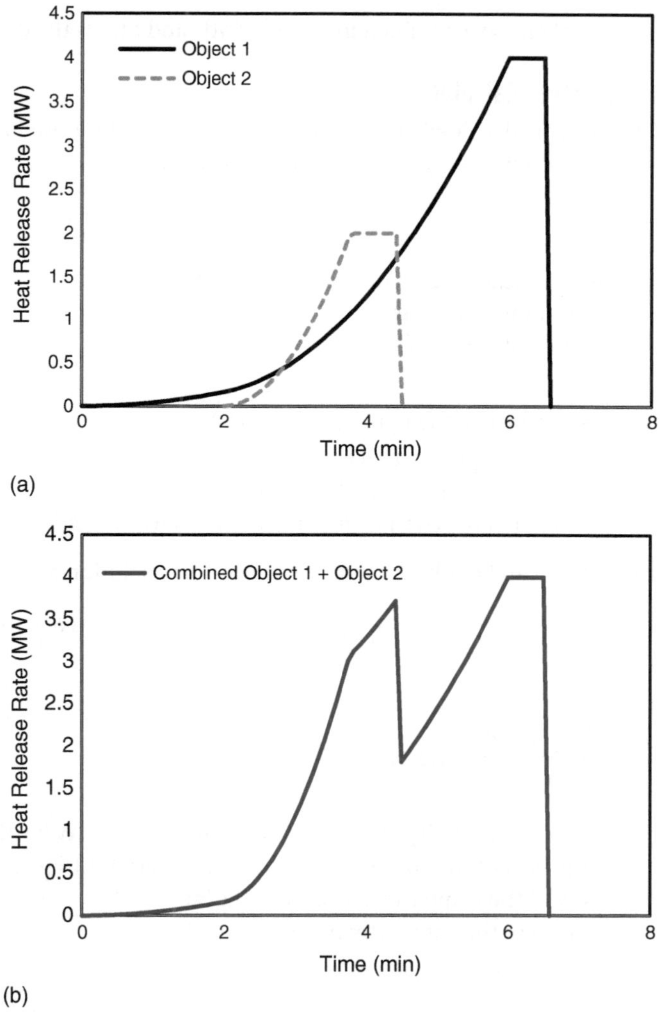

Figure 3.7 T-squared heat release rate (fire) separately and combined for two objects. (a) Burning of individual objects and (b) combined burning of two objects.

radiation from the flame and distance between the two objects. Knowing the time to ignition of the second object, and individual Q_p-t curve of two objects, the heat release rate of two Q_p curves can be summed (at each point of time) to obtain the total heat release rate resulting from burning of these two objects. The combined Q_p-t curve, representing the burning of a number of objects, is the total thermal energy that is possible from burning two objects as shown in Fig. 3.7b. For instance, in Fig. 3.7a, the first fuel object burns with a medium growth rate for 4.5 minutes, followed by 1 minute of steady burning at peak heat release rate of 2.5 MW, while the second object ignites after 2 minutes of lag, burning at a fast growth rate for 1.5 minutes followed by steady burning at 2.0 MW for 1 minute. The total heat release can be evaluated as the sum of heat release from two objects and can be obtained numerically by adding each point on two Q_p-t curves. The resulting combined heat release is shown in Fig. 3.7b.

It should be noted that t-squared design fires assume open-air burning conditions with unlimited ventilation and without any suppression through intervention of sprinklers or firefighters. However, when objects burn in a room, they burn differently, since different fuel items have different ignition temperatures and burning rates. The burning rate in fuel objects is enhanced by different heat transfer mechanisms, i.e., conduction, convection, and fuel radiation, and is generally limited by the available ventilation.

3.5.2 Heat Release Rate Calculations

Knowing the maximum possible energy, E (MJ), that can be released during combustion over specific period (usually taken in seconds), the heat release rate can be calculated. This heat release when, averaged over the burning duration (t), represents the average heat release rate, Q(MW), and is given as

$$Q = E/t \tag{3.16}$$

where E is the total energy contained in the fuel (MJ) and t (s) is the duration of the burning.

3.5.3 Temperature Calculations

Beyler (1991) developed a simple relation to estimate temperature rise in a compartment without ventilation by applying a nonsteady energy balance differential equation:

$$mc_p \frac{dT}{dt} = \dot{Q} - h_k A_T \Delta T_g \tag{3.17}$$

where $\dot{Q}$ = Energy (heat) release rate of the fire (kW)
m = Mass of the gas in the compartment (kg)
c_p = Specific heat of gas (kJ/kg·K)
$\Delta T_g = T_g - T_0$
T_g = Temperature of the upper gas layer (K)
T_0 = Ambient temperature (K)
h_k = Effective heat transfer coefficient (kW/m²K)
A_T = Total area of the compartment enclosing surfaces (m²)

The above equation can be solved using the closed-form solution as

$$\Delta T_g = 2\frac{K_2}{K_1}\left(K_1\sqrt{t} - 1 + e^{-K_1\sqrt{t}}\right) \tag{3.18}$$

where

$$K_1 = \frac{2(0.4\sqrt{kpc})A_T}{mc_p} \quad \text{and} \quad K_2 = \frac{\dot{Q}}{mc_p}$$

where k = Thermal conductivity of the compartment surface (kW/m·K)
c = Specific heat of the compartment surface material (kJ/kg·K)
ρ = Density of the compartment surface (kg/m³)
δ = Thickness of the compartment surface (m)
t = Exposure time (s)

It should be noted that a number of other methods are available in the literature for estimating pre-flashover compartment fire temperatures (McCaffrey et al., 1981; Foote et al., 1986; Deal and Beylert, 1990; Peatross and Beyler, 1994). These methods require complex set of calculations and are to be carried out in a number of steps.

3.6 Flashover Fire Calculations

Flashover occurs when local burning spreads to all available fuel within the compartment. When a fire grows without any intervention, the temperature in the hot upper gas layer will increase, and at a critical heat flux level of 15 to 20 kW/m^2 (or over 600°C temperature), all exposed combustible items in the room will begin to burn. At this stage, there is a rapid increase in the heat release rate and correspondingly room temperature. This transition from a localized fire to the burning of all combustible items in a room represents flashover. Keep in mind that flashover, by definition, can only occur in an enclosed space (i.e., room/compartment), since hot gases do not get trapped in an open environment.

Certain conditions must be met for the flashover to occur. For example, there must be sufficient fuel and ventilation for fire to develop into a significant size, the ceiling must be able to trap hot gases, and the geometry of the room must allow radiant heat flux from the hot layers to reach critical ignition levels at the level of the fuel items (floor). From an analysis on a large number of experimental fires, Walton and Thomas (1995) observed that flashover can only occur if heat from fire reaches a critical value of heat release Q_{fo} (MW). Although this value is very approximate and depends on the size, shape, lining materials, and location of a fire within the room, Q_{fo} can be approximated as

$$Q_{fo} = 0.0078A_t + 0.378A_v\sqrt{H_v} \tag{3.19}$$

where A_t is the total internal surface area of the room (m^2), A_v is the area of the window opening (m^2), and H_v is the height of the window opening (m).

For a t-squared burning, time to attain flashover, t_{fo} (s), can roughly be estimated as

$$t_{fo} = k\sqrt{Q_{fo}} \tag{3.20}$$

This relation conservatively assumes that all windows break at the time of flashover, at about 600°C.

3.7 Post-Flashover Fire Calculations

Beyond flashover, the behavior of fire and associated burning characteristics change dramatically in a compartment, since the very high temperature and radiant heat fluxes cause all exposed and combustible surfaces to burn (provided there is sufficient oxygen). In this period, the rate of temperature rise in the compartment is high, as the rate of heat release from burning reaches its peak. Attaining maximum temperatures of over 1000°C is quite possible in the burning compartment, and this leads to air flow, smoke, and gases to become very turbulent. The temperature rise continues until the rate of generation of volatiles from the fuel subsides as the rate of fuel consumption decreases.

Structural members in a compartment experience a high temperature rise during the post-flashover stage. In fact, it is during the post-flashover stage that structural members are exposed to the most severe effects of fire and where loss of structural capacity and integrity or even collapse is most likely to occur. After peak temperature is attained, fire enters the decay phase, during which the temperature within the compartment starts to decrease as the rate of combustion slows down. It should be noted that owing to thermal inertia, cross-sectional temperatures in structural members will continue to rise for a short period of time into the decay period; in other words, there will be a time lag before the sectional temperature in structural members starts to decay. In the post-flashover stage, the behavior of fire depends on the availability of fuel and oxygen and, as such, fires can be grouped under fuel-controlled or ventilation-controlled fires. Most fires are fuel controlled in the initial stages of fire and during final stages (decay phase) of fire.

3.7.1 Fuel-Controlled Fire

Fires in large, well-ventilated rooms containing fuel items with a limited area of combustible surfaces tend to be fuel controlled, and the rate of burning in this case is similar to that occurring for fuel items burning in open air, with the additional radiation from the hot upper layer, ceiling surfaces, or hot walls. The average heat release rate in a fuel-controlled fire (Q_{fuel}) can be calculated provided the total fuel load and duration of burning are known. For typical residential fires with a free burning duration (t_b) of 20 minutes, and acting as fuel-controlled fires, Law (1973) proposed a crude equation for evaluating heat release rate Q_{fuel} (MW):

$$Q_{fuel} = E / 1200 \tag{3.21}$$

where E is the total thermal energy from fire load (MJ).

When the burning rate of a fuel item is controlled by available surface area (wall and floor lining), Drysdale (2011) estimated the heat release rate Q_{fuel} (MW) to be

$$Q_{fuel} = q_i'' A_{fuel} \frac{\Delta H_c}{L_v} \tag{3.22}$$

where q_i'' is the incident radiation reaching the surface of the fuel (MW/m²) in a post-flashover fire and is generally assumed to be about 70 kW/m². A_{fuel} is the exposed fuel surface area (m²), ΔH_c is the heat of combustion (MJ/kg), and L_v is the heat of gasification (MJ/kg)—that is, the energy required to pyrolize a unit mass of fuel, ranging from 1.7 to 5.9 MJ/kg for wood and 1.2 to 3.7 MJ/kg for plastics.

For the burning of thin and thick slabs of wood under a fuel-controlled fire, Babrauskas (1979) proposed the heat release rate, Q_{fuel} (MW), to be evaluated as

$$Q_{fuel} = v_p \rho A_{fuel} \Delta H_c \tag{3.23}$$

where ρ is the density of the fuel (kg/m³) and v_p (m/s) is the surface regression rate of 0.5 to 0.6 mm/min and can be approximated as 8.5 to 10.0×10^{-6} (for thick slabs) or $2.2 \times 10^{-6} D^{-0.6}$ (for thin slabs), where D is the thickness of the slab of wood (m).

Knowing the total amount of thermal energy possible, E (MJ), and assuming constant (average) heat release rate, Q_{fuel} (MW), the duration of burning, t_b (s), is

$$t_p = \frac{E}{Q_{fuel}} \tag{3.24}$$

In both ventilation and fuel-controlled fires, all of the combustible materials in the room may not fully (and immediately) burn. So, an efficiency factor, ranging from 0.5 to 0.9 (of fuel fraction), is often applied to reduce the available fuel for burning.

The most important input parameter needed for evaluating structural response under fire conditions is the variation in room temperature (fire) with time. Due to variations in room geometry, ventilation conditions, and fuel characteristics, measured temperatures in post-flashover fires in room fire tests have shown a wide scatter. Figure 3.8 shows time-temperature curves measured in room fire tests, together with the ISO 834 standard time-temperature curve adopted for fire-resistance testing. It can be seen that fire growth curves in real rooms have a distinct decay phase to represent limited fuel or ventilation, or actions of firefighting, while this decay phase is not present in the ISO 834 standard fire curve.

Based on temperature measurements in room fire tests, Law (1983) proposed the following empirical relation for calculating maximum temperature, T_{max} (°C), attained in the compartment for a fire scenario:

$$T_{max} = 6000(1 - e^{-0.1\Omega})/\sqrt{\Omega} \tag{3.25}$$

$$\text{where } \Omega = \frac{A_t - A_v}{A_v \sqrt{H_v}} \tag{3.26}$$

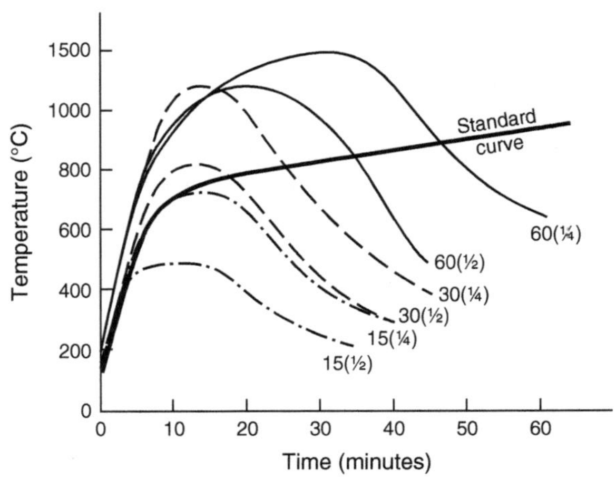

Figure 3.8 Measured time-temperature curves in typical room fires. (*Butcher et al., 1966.*)

If the compartment has a low fuel load, this maximum temperature, T_{max} (°C), may not be reached and reduces to T (°C), which is given as

$$T = T_{max}(1 - e^{-0.05\psi}) \tag{3.27}$$

$$\text{where } \psi = \frac{L}{\sqrt{A_v(A_t - A_v)}} \tag{3.28}$$

where L is the wood-equivalent fire load (kg).

The effect of various fuel load levels on fire growth characteristics was investigated through experiments by Magnusson and Thelandersson (1970). Results from these tests indicate that the behavior of fire in a room depends highly on ventilation and available fuel load levels. Data from these tests was used to derive the famous *Swedish time-temperature curves* by applying heat balance principles. These curves, reproduced in Fig. 3.9, show variation in compartment temperature with time for varying load and ventilation factors.

Each group of Swedish fire curves is for a specific fuel load level (taken as MJ/m² of the total surface area) and vary based on a specific ventilation factor (F_v). In each group, the rate of burning is the same, since it is controlled by the opening size, but at later stages, an increasing fuel load leads to longer and hotter fires before the decay period begins. The curves shown in Fig. 3.9 can be rearranged to show time-temperature variation for a constant fuel load, but with varying ventilation factor (openings). For instance, Fig. 3.10 shows that well-ventilated fires (higher F_v) burn faster and attain higher temperature over a shorter duration than poorly ventilated fires (lower F_v), where the peak temperatures are low but the fire burns over a longer duration.

The decay rate in a fuel-controlled fire depends mainly on the shape and material type of the fuel, the size of ventilation openings, and the thermal properties of lining materials. Hence, the rate of temperature decay in a post-flashover fire is not easy to predict. In general, when the surface area of burning reduces, all fires become fuel controlled. In the case of liquid fuels, the burning phase will end suddenly due to complete combustion occurring in liquid fuels. However, for solid fuels like wood, the decay phase is longer due to incomplete combustion initially.

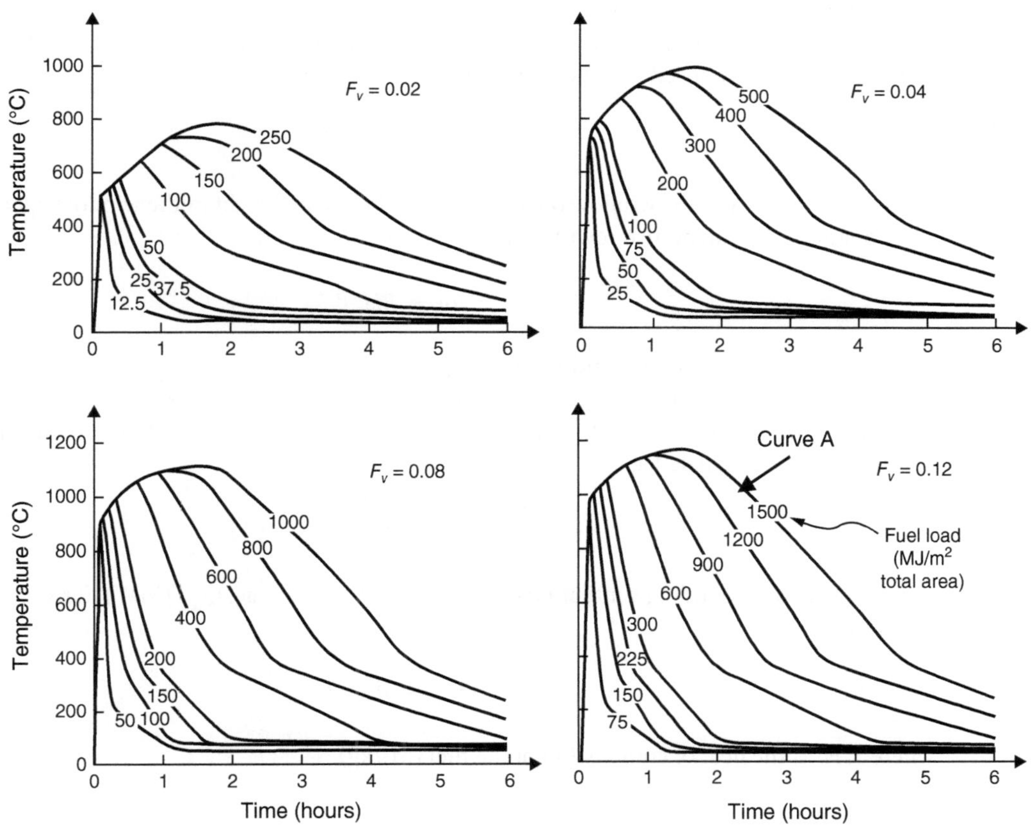

FIGURE 3.9 Time-temperature curves for different ventilation factors and fuel load. (*Magnusson and Thelandersson, 1970.*)

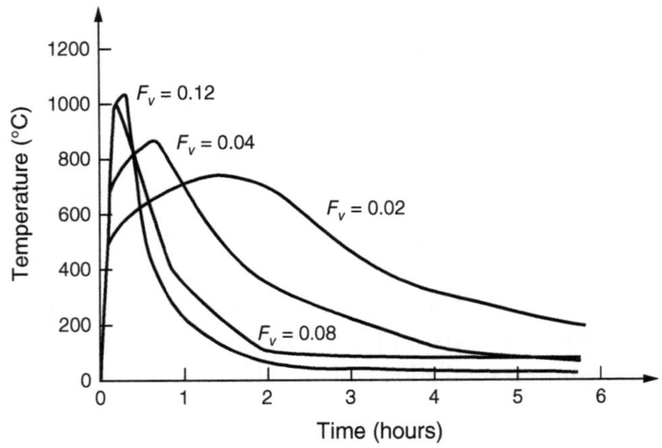

FIGURE 3.10 Time-temperature curves for varying ventilation and constant fuel load (200 MJ/m² total surface area).

3.7.2 Ventilation-Controlled Fire

In many fire scenarios, available ventilation in the compartment can be limited during the post-flashover stage, and hence ventilation becomes the controlling factor of fire growth. These fires are referred to as *ventilation-controlled fires*, and the rate of burning in such fires depends on the size and shape of ventilation openings. Due to the limited amount of air in the room, flames from the fire tend to extend outside the openings, looking for oxygen, to burn vigorously. The rate of burning in a ventilation-controlled fire is also limited by the volume of cold air entering (and correspondingly to that of the hot gases leaving the room). Ventilation effects can also be seen in large openings (such as windows in the roof) that allow rapid heat loss from the fire compartment. Further, linings on walls and floors of the compartment made of materials with low thermal conductivity and high heat capacity (e.g., concrete) insulate the compartment, and this leads to retaining higher temperature levels in the decay period. On the other hand, linings made of materials with high thermal conductivity and low specific heat (the same as in composites/metals) can store less heat, leading to a rapid rate of decay.

To characterize ventilation-controlled fires in typical room conditions, with mostly wood-based fuels, Kawagoe (1958)[7] proposed the rate of burning, $\dot{m}$ (kg/s), to be

$$\dot{m} = 0.092 A_v \sqrt{H_v} \tag{3.29}$$

where A_v is the area of the window opening (m²) and H_v is the height of the window opening (m).

Equation 3.29 can be also rewritten as

$$\dot{m} = BH_v^{1.5} \tag{3.30}$$

where B is the breadth of the window opening.

This equation shows that the burning rate is largely dependent on the size of the window opening.

If the total available mass of fuel, $[M_f(\text{kg})]$, is known, then the duration of the burning period (t_b in seconds) can be calculated as

$$t_b = \frac{M_f}{\dot{m}} \tag{3.31}$$

And the corresponding ventilation-controlled heat release rate, Q_{vent} (MW), for steady-state burning is given as

$$Q_{vent} = \dot{m} \Delta H_c \tag{3.32}$$

where ΔH_c is the heat of combustion of the fuel (MJ/kg).

[7]It should be noted that Kawagoe's equation may not be appropriate for fuels other than wood.

Knowing total thermal energy content of fuel available for combustion, E (MJ), the duration of burning period (t_b, in seconds) in ventilation-controlled fires is given by

$$t_b = \frac{E}{Q_{vent}} \tag{3.33}$$

Due to the complex nature of fire growth characteristics and various assumptions used in the development of these principal relations, it should be noted that the burning rate calculations in the above equations may not always lead to precise values. For instance, calculations of heat release rates can be inaccurate due to unknown portion of pyrolysis products that burn outside of the room (e.g., flames extending through broken windows) rather than inside the room. Other conditions that may lead to an inaccurate estimation of fire growth characteristics relate to the fact that fuel loads (items) may not be uniformly distributed in the room, and some portions of certain fuel types may not be fully combustible. Despite these approximations, Kawagoe's equation still forms the basis of most post-flashover ventilation-controlled fire calculations.

Based on a large number of small-scale compartment fires with wood cribs, Law (1983) proposed a more refined equation to quantify burning rate. This equation is a general form to Kawagoe's equation in which the burning rate is expressed as

$$\dot{m} = 0.18A_v\sqrt{\frac{H_vW}{D}}(1 - e^{-0.036\Omega}) \tag{3.34}$$

$$\Omega = \frac{A_t - A_v}{A_v\sqrt{H_v}} \tag{3.35}$$

where W is the compartment width (m), D is the compartment depth (m), and A_t is the total area of the internal surfaces of the compartment (m^2).

Law's equation, given by Eq. 3.34, applies to compartments with windows located on one wall, since it does not distinguish between the windows' size and location (or windows located on more than one wall). This equation is given in Eurocode 1 (2002) to calculate the burning rate and to assess flame height from compartment windows.

Law's expression, when used for square compartments (with $\frac{A_v\sqrt{H_v}}{A_t} = 0.05$ and $\Omega = 20$), tends to predict a similar burning rate to that obtained from Kawagoe's equation. However, if Law's equation is used to evaluate the burning rate in shallow (rectangular) rooms with small windows (or openings), then Law's equation predicts a higher burning rate than that predicted by Kawagoe's equation.

The amount of ventilation in a fire compartment, a controlling factor in ventilation-controlled fires, can be expressed through a ventilation factor [F_v (m$^{1/2}$)] as

$$F_v = \frac{A_v\sqrt{gH_v}}{A_t} \tag{3.36}$$

where A_v is the area of the window opening (m^2) H_v is the height of the window opening, g is the velocity of gas (m/s), assumed to be 1 (m/s), and A_t is the total area of the internal surfaces of the compartment (m^2).

In cases where there is more than one opening (window) in the compartment, the above equations can still be used if A_v is adjusted to account for the total area of all window openings (m^2). In this scenario, H_v also needs to be adjusted to account for weighted average height (m) of all windows and doors (m). Thus,

$$H_v = (A_1H_1 + A_2H_2 + \ldots)/A_v \tag{3.37}$$

$$A_v = A_1 + A_2 + \ldots = B_1H_1 + B_2H_2 + \ldots \tag{3.38}$$

$$A_t = 2(l_1l_2 + l_1H_r + l_1H_r) \tag{3.39}$$

The terms B_i (m) and H_i (m) are the breadth and height of the windows, l_1 and l_2 are the floor plan dimensions, and H_r (m) is the room height.

In the case that vents (openings) in the ceiling are present in a compartment, such openings increase ventilation to fire and thus need to be accounted for to better represent the room ventilation level. The following expression accounts for the effect of vents in the ceiling in the ventilation factor:

$$\left(A_v\sqrt{H_v}\right)_{fict} = A_v\sqrt{H_v} + 2.3A_h\sqrt{h} \tag{3.40}$$

where A_h (m^2) is the horizontal opening area in ceiling and h (m) is the vertical distance from window opening mid-height to compartment ceiling. This equation is applicable only for values of $\frac{A_h\sqrt{h}}{A_v\sqrt{H_v}}$ in the range 0.3 to 1.5.

3.8 Time-Temperature Relations

The severity of fire exposure has a significant influence on the performance of structural members under fire conditions. This severity of fire can be expressed through time-temperature relations that describe fire development starting from the initial ignition stage to the end of the cooling stage. These time-temperature relations form the most important input parameter for fire resistance analysis of structures and can be grouped under two different categories, standard and design, depending on the approximation used in their derivation.

3.8.1 Standard Fire

Standard fire, representing a standardized fire exposure, is the most widely used fire scenario for comparative fire performance evaluation of building materials and structural members. This fire scenario is derived based on the assumption that fire growth occurs in a typical compartment mostly due to the burning of cellulosic products (wood) and does not include any decay phase. This fire curve is most widely used for regulating materials, products, and assemblies in buildings.

In order to enforce fire safety regulations, each country requires structural members and assemblies to satisfy a certain level of fire rating based on the type and occupancy of buildings. Fire test standards specify standard time-temperature curves to be used for establishing fire resistance ratings for structural members. The building codes further reference applicable fire test standards with specifications for undertaking such fire tests. The most widely used fire standard in the United States is ASTM E119, and this standard is similar to other recognized fire test standards such as UL 263 and NFPA 251 (United States), ISO 834 (Europe as well as International), CAN/ULC-SI01 (Canada), DIN 4102 (Germany), BS 476 Part 20 (United Kingdom), and AS 1530 Part 4 (Australia). These standards specify a standard fire curve through a number of time-temperature discrete points (or continues functions)[8] for establishing fire ratings of structural members of assemblies. The fire curves adopted in some of the fire test standards are compared in Fig. 3.11.

For evaluating fire resistance of structural members when the fire exposure on the member is from outside a burning compartment, Eurocode proposes a separate fire curve referred to as *external fire*.

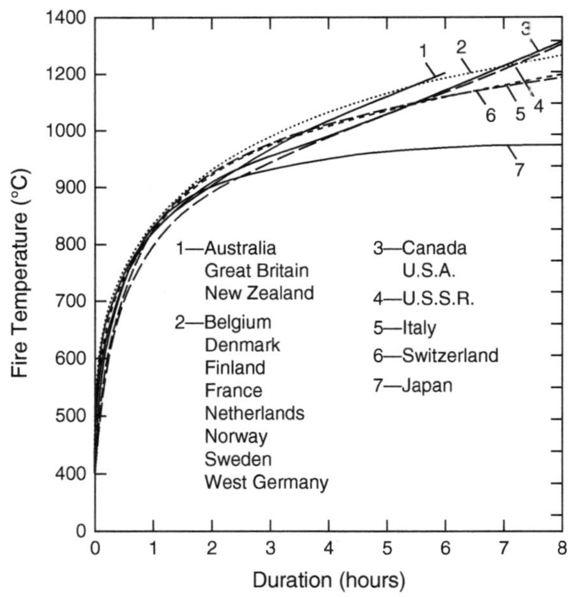

FIGURE 3.11 Fire curves adopted in fire test standards around the world. (*DiNenno, 2008; printed with permission from SFPE.*)

[8]A complete discussion on standard fire curves is provided in Chap. 5.

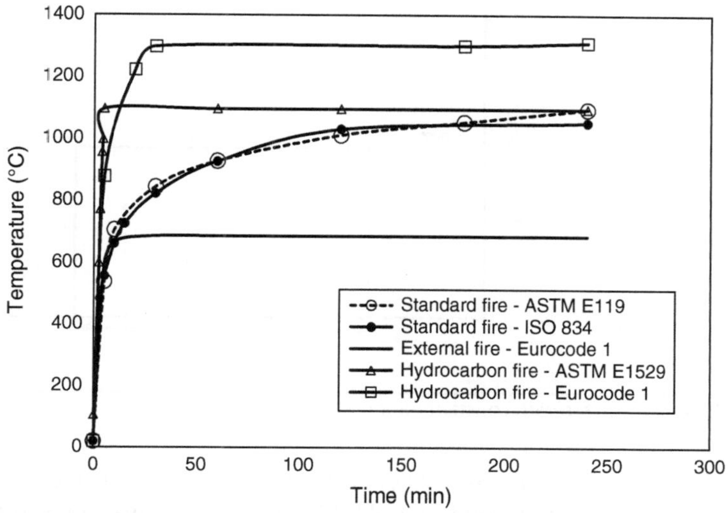

Figure 3.12 Comparison of standard, hydrocarbon, and external fire curves in different test standards (ASTM E119, 2016; Eurocode 1, 2002; ISO 834, 2012).

Temperatures in this fire scenario are lower than a conventional standard fire due to the fact that no radiative feedback occurs as (1) the fire is burning outside of a compartment and (2) a layer of hot gases does not accumulate but rather spreads away from the burning compartment.

In the case of a fire resulting from petrochemicals and plastics, fire growth is much faster and consequently is much more rapid than that represented by a cellulosic standard fire curve. This became much more evident in the late 1980s as a result of the number of fires in constructions that were housing petroleum products (e.g., offshore platforms for oil production). Hence, a more appropriate temperature-time curve defined by a standardized *hydrocarbon* fire curve was adopted. Some of the widely used hydrocarbon fire curves include ASTM E1529, UL 1709, as well as Eurocode hydrocarbon fire. Figure 3.12 shows a comparison between hydrocarbon, external, and standard fires. A few observations can be noted through a comparison of these curves. For instance, a hydrocarbon fire curve is the most severe, followed by a cellulosic (standard) fire curve and then an external fire curve. In all of these fire curves, fire temperature increases throughout the duration of the fire exposure and does not decay, despite the fact that most fires decay after a certain length of time due to a lack of either fuel or ventilation or as a result of firefighting actions.

3.8.2 Design Fires

There is a growing realization that the above-described standard fires do not represent actual fire exposure scenarios in buildings and other structures. One of the primary drawbacks in standard fire curves is the absence of the decay (cooling) phase as present in a real fire. As mentioned, a decay phase is always present in any fire scenario simply due to burnout of fuel, limited ventilation, or firefighting activities. To evaluate the realistic performance of structures under fire conditions, arriving at a *realistic* fire exposure scenario that accounts for fuel load, compartment features, and ventilation level is necessary. Simplified relations for calculating temperature-time curves based on fuel load and ventilation characteristics are available in fire design standards. Depending on the complexity, these relations are grouped under hand calculations, published curves, and parametric fire equations.

Empirical Relations

Constant Peak Temperature One simple approach that can be used in fire resistance evaluations is to assume that the fire temperature is constant throughout the burning period (as shown in Fig. 3.13). In this crude approach, the maximum temperature and duration of the burning period are estimated using equations described previously, such as *t-squared* fires, in Sec. 3.5.1. This approach does not distinguish incipient, growth, or decay stages of fire development in a compartment or temperatures other than the peak temperature reached during the burning stage.

Swedish Fire Curves The second approach to represent a realistic fire scenario is through the use of Swedish fire curves discussed in Sec. 3.7.1. In this approach, and for a given fuel load and ventilation factor, temperature-time points in a compartment (or room) can be directly deduced from similar figures to that

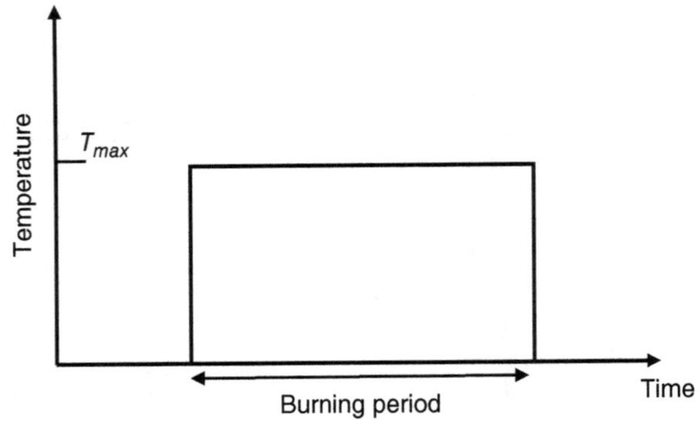

FIGURE 3.13 Design fire with constant temperature.

in Fig. 3.9. In case temperature at 2 hours of fire exposure is needed for burning in a room with a fuel load of 1500 MJ/m^2 and ventilation factor of 0.12, room temperature can be read from Fig. 3.9 as 1080°C. The full temperature-time curve for this room would be that identified as *curve A*.

Parametric Temperature-Time Equation

Eurocode 1 (2002), for a given fuel load, ventilation openings, and type of lining materials in the compartment, gives temperature-time relation to represent a compartment fire. This relation is referred to as a *parametric fire,* and the equation is derived to give a good fit to the burning period of fire as in Swedish fire curves. The resulting time-temperature equation is given as

$$T = 1325(1 - 0.324e^{-0.2t^*} - 0.204e^{-1.7t^*} - 0.472e^{-19t^*}) \tag{3.41}$$

$$\text{with } t^* = \Gamma t \tag{3.42}$$

where t^* is the fictitious factor, t is the actual time (hours), and Γ is given by

$$\Gamma = \frac{(F_v/F_{ref})^2}{(b_v/b_{ref})^2} \tag{3.43}$$

where b is the square root of thermal inertia $= \sqrt{k\rho C_p}$ (Ws$^{1/2}$/m^{2}K) under a given parametric fire and b_{ref} is the square root of the reference thermal inertia $= \sqrt{k\rho C_p}$ under standard fire, F_v is the ventilation factor ($\sqrt{m}$) given by $F_v = \dfrac{A_v\sqrt{H_v}}{A_t}$ under a given parametric fire, and F_{ref} is the reference value of the ventilation factor under standard fire.

Eurocode 1 gives values of F_{ref} and b_{ref} as 0.04 and 1160, respectively. These reference values are calculated based on exposure to the standard fire ISO 834 in a compartment. Applying the Eurocode 1 values, for F_{rehf} and b_{ref}, Eq. 3.43 simplifies to

$$\Gamma_{EC} = \frac{(F_v/0.04)^2}{(b_v/1160)^2} \tag{3.44}$$

Recent research has shown that the realistic value of $b_{ref} = 1160$ can underestimate the fire temperature predicted in Eq. 3.41. Feasey and Buchanan (2000) proposed a revised value of $b_{ref} = 1900$ to better estimate fire temperature in Eq. 3.41. Therefore, Eq. 3.44 can be rewritten as

$$\Gamma_{Design} = \frac{(F_v/0.04)^2}{(b_v/1900)^2} \tag{3.45}$$

Eurocode 1 also specifies relations for calculating duration of burning, t_d (hours), in a compartment, and this is given as

$$t_d = 0.00013\frac{e_t}{F_v} \tag{3.46}$$

$$t_d = 0.00013\frac{E}{A_v\sqrt{H_v}} \tag{3.47}$$

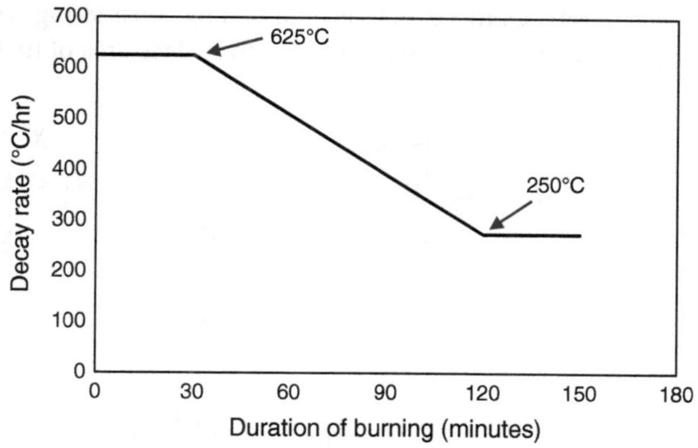

Figure 3.14 Temperature decay rate as per Eurocode parametric fires (Eurocode 1, 2002). *(Reproduced from CEN.)*

where e_t is the fuel load (MJ/m^2 total surface area) and E is the total energy of the fuel (MJ) in the room.

Eurocode 1 also gives recommendations to evaluate temperature in the cooling phase of fire. In this recommended approach, a rate of decay to peak fire temperature is applied until the temperature in the burning compartment reaches ambient temperature levels. This decay rate $(dT/dt)_{ref}$ is 625°C/hr for fires with a burning period less than half an hour, and 250°C/hr for fires with a burning period greater than 2 hours as shown in Fig. 3.14. When burning period is between 30 and 120 minutes, an interpolation between the two decay rates can be used as shown in Fig. 3.14.

Feasey and Buchanan (2002) argued that the Eurocode formulation for the cooling phase of parametric fire may be unsatisfactory for large or small openings. Also, they pointed out that parametric equations can predict higher peak temperatures for insulated compartments and/or compartments with large ventilation openings. As a result, Feasey and Buchanan proposed an alternate equation to calculate decay phase of parametric fire as

$$\frac{dT}{dt} = \left(\frac{dT}{dt}\right)_{ref} \frac{\left(\sqrt{F_v}/0.04\right)}{\left(\sqrt{b}/1900\right)} \tag{3.48}$$

Matsuyama et al. Method

Matsuyama et al. (1998) method is derived as a simple equation that can easily predict the temperature inside the compartment, the rate of heat loss due to ventilation, and the rate of heat loss due to compartment boundary. This equation is used in the Japanese standard for evaluating temperatures in a fuel-controlled and/or ventilation-controlled fire. The fire temperature (T) in the compartment and fire duration (t_D) are estimated as

$$T = 1280\left(\frac{Q}{\sqrt{A_T\sqrt{k\rho c}}\sqrt{A_o\sqrt{H_o}}}\right)^{2/3} t^{1/6} + T_\infty \tag{3.49}$$

and

$$t_D = \frac{1}{60}\frac{F_L A_r}{Q} \tag{3.50}$$

where Q = Heat release rate by combustion (MW)
A_T = Internal surface area of compartment enclosure (m^2)
$\sqrt{k\rho c}$ = Thermal inertia of compartment enclosure (kWs$^{1/2}$/m$^2\cdot$K)
A_o = Area of window opening (m^2)
H_o = Height of window opening (m)
T_∞ = Initial and ambient temperatures (°C)
F_L = Fire load density (MJ/m^2)
A_r = Floor area of the room (m^2)
t_D = Fire duration (min)

The heat release rate, Q, in Eq. 3.49 is obtained by considering an additional parameter, *burning type index, X*, which is a fraction of ventilation factor to surface area of fuel. Q and X are given by

$$Q = A_{fuel}\begin{cases} 1.6X & X \le 0.081 \\ 0.13 & 0.081 \le X \le 0.1 \\ 2.5Xe^{-11X} + 0.048 & 0.1 \le X \end{cases} \quad (3.51)$$

$$X = \frac{A_o\sqrt{H_o}}{A_{fuel}} \quad (3.52)$$

where

$$A_{fuel} = 0.26F_L^{1/3}A_{room} \quad (3.53)$$

Computer Models

Over the last four decades, a number of computer models have been developed for calculating the heat release rate and fire development (including fire growth, temperature rise, etc.) in the post-flashover stage of room fires. CFAST (Peacock, 2012), OZone (Cadorin et al., 2001), and COMPF2 (Babrauskas, 1979), among others, are the widely used programs in fire engineering applications. All these computer models are based on heat balance principles in which heat produced from combustion is assumed to be equal to the summation of heat absorbed by the compartment, heat losses to surrounding objects, and heat dissipated through openings, and is expressed as

$$\dot{q}_C = \dot{q}_W + \dot{q}_R + \dot{q}_L \quad (3.54)$$

where $\dot{q}_C$ is the heat produced by combustion of the fuel, $\dot{q}_W$ is the heat conducted into the surrounding structure, $\dot{q}_R$ is the heat radiated through the opening, and $\dot{q}_L$ is the heat carried out of the opening by convection of hot gases and smoke as illustrated in Fig. 3.15.

The early version of computer models assumed the burning process in a compartment to be a single well-mixed reactor (zone), and these were referred to as *single-zone models*. In single-zone models, fire undergoes thermal energy exchange with compartment surroundings and the energy released during the combustion process is obtained by applying the first law of thermodynamics to the system (i.e., fire and compartment). By definition, single-zone models do not account for specific fire ignition location within a room and assume that combustion (burning of fuel) to generate fire temperatures are uniform throughout the depth of the compartment. Due to these assumptions, single-zone models assume heat flow to be identical on all compartment boundaries (including the ceiling). *COMPF2* is an example of single-zone computer program, developed by Babrauskas (1979), and is used for calculating compartment temperatures in post-flashover room fires. This program evaluates the heat release rate, taking into account fire

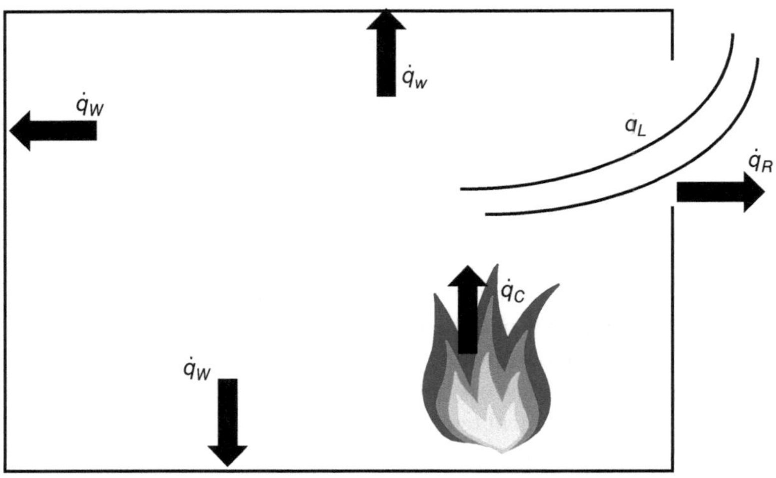

Figure 3.15 Heat balance for a post-flashover room fire.

characteristics (i.e., compartment ventilation and fuel load), as well as key characteristics of fuel types (i.e., porosity of wood). This program can also calculate the rate of temperature reduction in the decay period for varying types and shapes of fuel.

Two-zone models can better represent the actual burning characteristics in a compartment and overcome some of the drawbacks in one-zone models. In two-zone models, the compartment environment is assumed to consist of two zones, an unburning zone and a burning zone. Since these zones are two distinct thermodynamic systems, thermal energy and fuel mass interact at the interface of these two zones as well as their common surroundings (walls, ceiling). The burning rate is numerically computed by solving equations resulting from the laws of thermodynamics applied to the two zones. Two-zone models[9] are often not applied due to convergence issues arising from excessive turbulence resulting from interaction of fire and compartment boundaries (such as windows).

OZone, developed by Cadorin et al. (2001), is an example of a computer program that utilizes a two-zone model to evaluate the evolution of gas (room) temperatures in a compartment during a fire event. This time-temperature plot generated from OZone can be used as input to calculate temperature rise in structural members present in the compartment. This program can also evaluate fire resistance of structural steel members such as beams and columns following provisions of Eurocode 3. Although recent versions of OZone utilize both one- and two-zone models, the main limitations of this program are that it does not include a pyrolysis model and that it can only analyze individual (single) compartments in which room geometry is restrained to four walls and number of openings to three. In lieu of OZone, other two-zone computer models such as FASTLite (Portier et al., 1996) and FIERAsmoke (Yager, 1999) can be used for evaluating gas temperature in a room.

Alternatives to one-zone and two-zone models, *multizone models*, in which the compartment is divided into a number of zones, can be used for fire growth calculations. Each zone in a multizone model is assumed to form an independent thermal system. Thus, heat transfer analysis in multizone models is taken one step further by considering energy and mass balances over several zones and therefore predicts fire characteristics that are closer to the actual fire growth phenomenon.

Another approach to evaluate fire growth and spread under various burning and boundary conditions (including fuel type, compartment configuration, etc.) is through *computational fluid dynamics* (CFD). CFD is a branch of fluid mechanics that uses advanced numerical analysis to simulate the interaction of flames, fuel, and gases with surfaces defined by boundary conditions. CFD models do not divide fire compartments into zones, but rather into a set of grids (i.e., mesh similar to that used in discretizing a structure for finite element analysis) and then solve heat balance equations at the level of each grid. CFD models, due to their complexity and large number of input parameters, require a high level of expertise for analysis and interpretation of simulation results, i.e., temperature fields, gas velocity, flame and smoke spread, and so on. Commercial software programs such as ANSYS and ABAQUS, as well as Dynamics Simulator developed by the National Institute of Science and Technology, USA (McGrattan et al., 2004), can be applied to undertake CFD analysis for evaluating fire growth calculations.

3.9 Fire Severity

The main criterion to be applied in designing structures for fire safety is to ensure that the fire resistance of a structural member is greater than the severity of the fire to which the member is exposed to:

$$\text{Fire resistance} \geq \text{Fire severity} \tag{3.55}$$

Domain	Units	Fire Resistance	Fire Severity
Time	Minutes or hours	Time to reach failure	≥ Fire duration as calculated or specified by code
Temperature	°C	Temperature of the member to cause failure	≥ Maximum temperature reached in the member during fire
Strength	kN or kN.m	Reduced load capacity of the structural member at elevated temperature	≥ Applied load on the member during fire

Table 3.6 Methods for Comparing Fire Resistance with Fire Severity

[9] Two-zone models are computationally expensive.

Fire resistance of a structural member is its ability to resist collapse or fire spread during exposure to a fire of a specified severity. Fire severity, on the other hand, is a measure of the destructive impact of fire on the structural member, including resulting restrained forces arising from high temperature effects that could cause failure of the member. There are three main verification methods for comparing fire severity against fire resistance to ensure Eq. 3.55 is satisfied. This verification may be in the time domain, temperature domain, and strength domain, as shown in Table 3.6.

3.9.1 Need for Fire Severity

The type of fire exposure has a significant influence on fire resistance of a given structure. However, from a practical standpoint, an infinite number of fire exposures are possible in a given compartment and therefore it becomes almost impossible to evaluate fire resistance of a member, specifically through fire tests, under all possible fire scenarios. This fact has motivated researchers to develop the concept for *fire severity equivalence* in order to relate adverse effects experienced by a structural member during any possible (real) fire exposure to that experienced by the member under one standard fire exposure under which the member was tested to evaluate fire resistance. There are four main domains to establish fire severity equivalence: time, temperature, strength, and energy (Kodur et al., 2010). A review of these domains and their applicability to different practical situations follows.

3.9.2 Methods for Deriving Equivalent Fire Severity

The Conseil Internationale du Bâtiment pour la Recherche l'Étude et la Documentation (International Council for Building Research Studies and Documentation, or CIB, 1986) proposed three possible levels of analysis, with increasing levels of complexity—H_1, H_2, and H_3—that can be applied to evaluate fire performance of structural members. According to this proposal, the analysis can be carried out by representing the fire scenario and structural member (or system) at different levels of idealization as shown in Fig. 3.16.

The first and the simplest level, H_1, relates the fire to the same temperature-time relation as generated in a standard fire test. As will be discussed in Chap. 5, a standard fire test allows fire resistance assessment of either simple structural members or subassemblies. Theoretically, full structures can also be tested in a standard fire setup if suitable test facilities are available. At this simple level, calculation methods may also be applied to evaluate fire resistance at the member, subassembly, or system level under standard fire

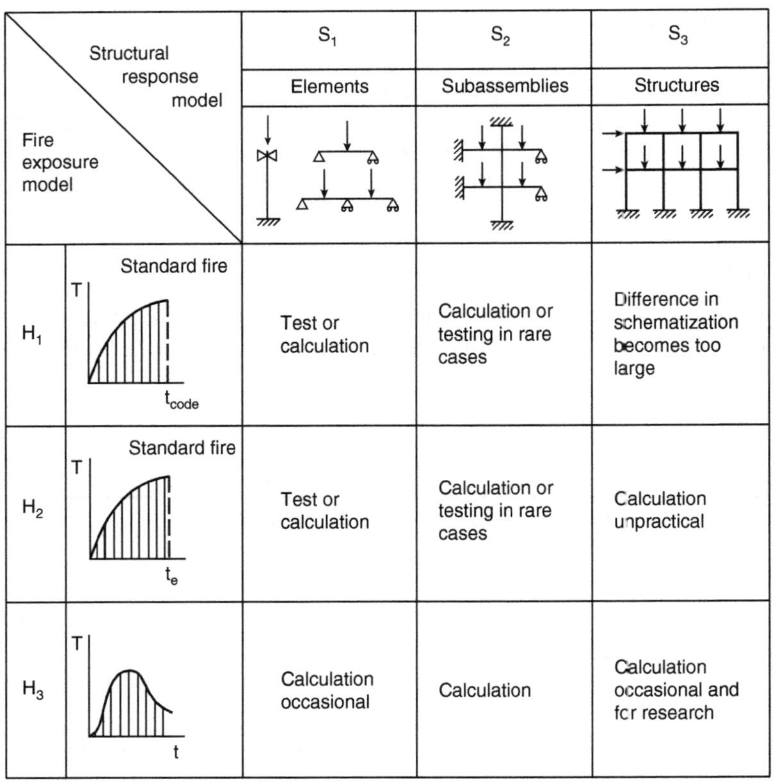

Figure 3.16 Representation of fires and structure for fire resistance analysis at various levels (CIB, 1986). (*Courtesy of Elsevier.*)

exposure. H_1 is the most commonly used fire exposure for fire resistance assessment of structural members under perspective-based code environment.

The second level of analysis, H_2, is applied when the exposure to a given fire can be equated to an equivalent duration of exposure on the standard fire, called equivalent time, or t_e. *Equivalent time* is the exposure time to the standard fire that is equivalent to a complete burnout of a real fire in the same room. This is an advancement over the first level of analysis and is currently being used to extend fire resistance developed under standard fire exposure to members under different fire exposure.

The third level, Fire exposure H_3, represents a realistic fire (i.e., complete burnout) that occurs in a room without any fire suppression. At this level, designers can apply standard or real fires to represent a temperature-time curve.

At all fire exposure levels, fire resistance of a structure can be assessed considering a single member, a subassembly, or a whole structure. While evaluation under levels H_1 and H_2 will be sufficient for most structures, H_3 assessment should be carried out for innovative or complex structures (or for research purposes).

3.9.3 Domains for Expressing Fire Severity Equivalence

The concept of fire severity equivalence was originally developed by Ingberg (1928), and later researchers have proposed additional approaches and empirical formulas for evaluating equivalent fire severity in a structural member. These approaches are based on failure considering four domains: time, temperature, strength, and energy.

Time Domain

The simplest approach to establish (or verify) that a structural member survives a given fire with a known fire exposure severity is in the *time domain* and can be expressed as

$$t_{fail} \geq t_s \tag{3.56}$$

where t_{fail} is the required rating of the structural member (as listed in the published listing of ratings or hand/numerical calculation methods) and t_s is the fire duration or fire severity, usually a time of standard fire exposure either specified by code/standard or calculated for a real fire using an equivalent time of standard fire exposure. For example, IBC requires a beam in a building to have 2-hour fire rating ($t_{fail} = 2$ hours), and the duration of a fire being considered is to be less than 2 hours.

Temperature Domain

The second approach to verify that a structural member survives a given fire (where its severity is known) is through the *temperature domain.* This verification of fire severity can be expressed as

$$T_{fail} \geq T_{max} \tag{3.57}$$

where T_{fail} is the temperature that causes structural or thermal failure of a structural member or barrier and T_{max} is the maximum temperature reached in the same member during exposure to a given fire. This maximum temperature can be measured during fire testing or calculated through thermal analysis. It should be noted that T_{max} depends on the function of the fire-exposed member under consideration. While failure temperature in the case of a structural member is the critical temperature at which the stability (load-carrying capacity) is lost, failure temperature for barriers, such as walls and floors, is governed by the unexposed side temperature to meet insulation failure criteria so as to limit fire spread to other areas. Verification in the temperature domain is mainly suitable for steel structural members and non-load-bearing members (e.g., walls).

Strength Domain

Strength domain, unlike temperature domain, is more applicable for evaluating fire severity equivalence in a load-bearing structural member (i.e., beam, column, etc.) made of any construction material. Fire severity under strength domain can be evaluated through comparing applied loading present on the member during fire with the reduced load capacity of a structural member at any given fire exposure time such that

$$R_f \geq U_f \tag{3.58}$$

where R_f is the minimum load capacity (i.e., moment capacity) reached during a fire or the required load capacity at a certain time specified by codes and U_f is the applied load on the member during fire event. The

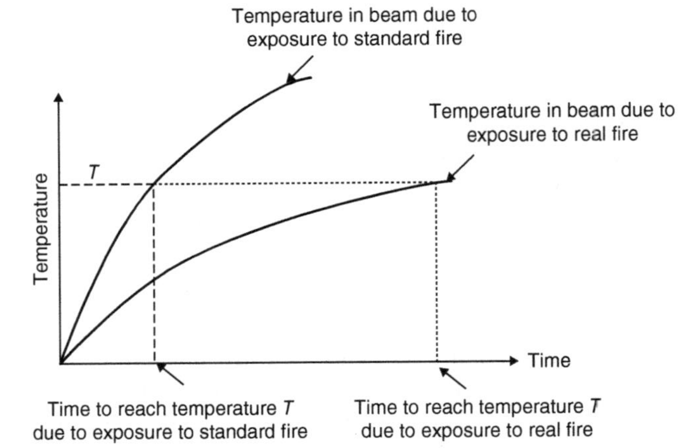

Figure 3.17 Equivalence of fire severity based on peak temperature attained in a structural member.

sectional capacity in a structural member can be evaluated through hand calculations or thermostructural analyses, while the applied loading effects can be calculated using load combinations specified in national codes and standards for fire conditions (ASCE, 2016).

Energy Domain

Equivalent fire severity can also be established through the *energy domain*, wherein equivalency between the effect of a given fire exposure and that of a standard fire exposure is established through comparing representative thermal energy applied on the structural member, as

$$E_{sf} \geq E_{df} \tag{3.59}$$

where E_{sf} is the energy required to induce failure during a fire and E_{df} is the thermal energy required to induce adverse effects on a structural member. The verification in the energy domain considers the level of deformations (deflections)[10] undergone in the member due to fire exposure and is more representative of fire performance of a member than just considering the rise in sectional temperatures alone. The verification in the energy domain can be established once magnitude of deformation under a given fire exposure exceeds deformation limit (i.e., span/30) specified in building codes and fire test standards, and at this point failure is said to occur in the member (Kodur et al., 2010).

3.9.4 Estimation of Equivalent Fire Severity

As discussed earlier, equivalent fire severity is a concept to relate the severity of a real fire exposure on a structural member to that of standard fire exposure on the structural member (Law, 1997). In simple terms, equivalent fire severity is basically the time of standard fire exposure that produces the same effect on a structural member (or assembly) as that of real fire exposure (see Fig. 3.17). Estimation of such equivalent fire severity is useful in the fire design process to extend published fire-resistance ratings on structural members, derived from standard tests, into designs where real fire exposures are considered. There are four main methods to relate a real fire exposure to a standard fire exposure: equal area concept, maximum temperature concept, minimum load capacity concept, and time-equivalent formulas.

Equal Area Concept

The *equal area concept*, developed by Ingberg (1928), establishes time equivalency between a design (real) fire to that of a standard fire simply by equating the area under the time-temperature curve of these fires. In other words, if the area under the time-temperature curve of the design (real) fire equals that of the area in the time-temperature curve of a standard fire, then both fires are considered to have similar equivalency.

The equal area concept is a simple, easy-to-use method, and in fact this concept is often used to correct the fire ratings from a standard fire-resistance test when the time-temperature curve in a furnace does not exactly follow that specified in standard fire exposure testing standards,[11] as can be seen in Fig. 3.18. Still,

[10]The deformation a member undergoes under fire conditions can be evaluated through simplified approaches (i.e., moment-curvature) or advanced analysis (i.e., finite element method).

[11]A detailed example will be covered in Chap. 5.

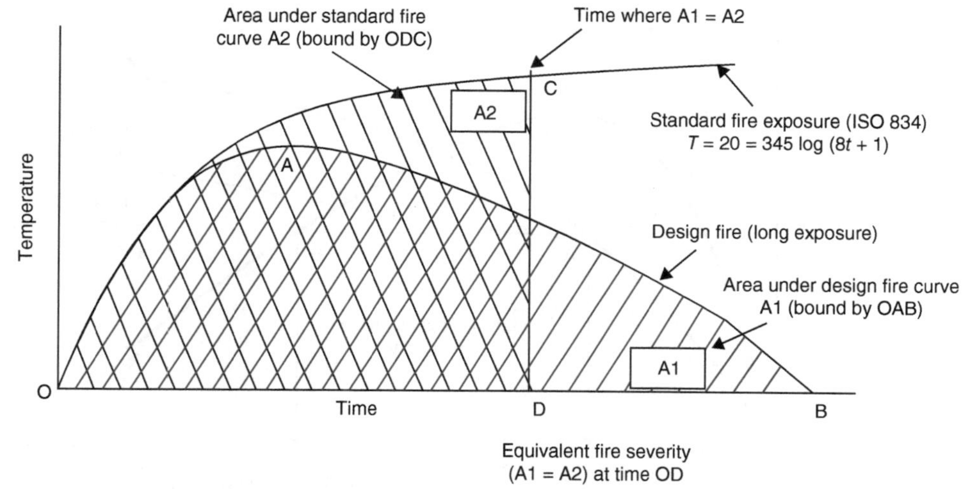

FIGURE 3.18 Equal area concept for deriving equivalent fire severity.

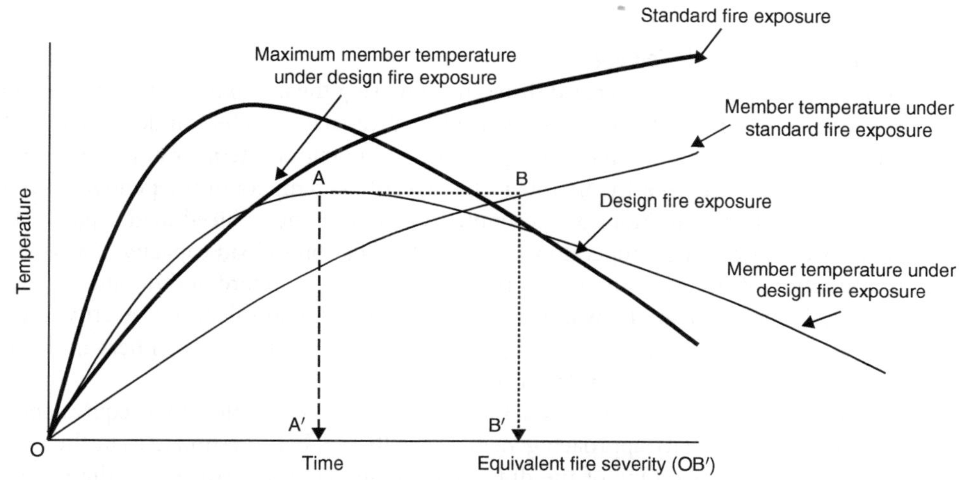

FIGURE 3.19 Maximum temperature concept for deriving equivalent fire severity.

the concept of equal area suffers from the fact that it does not have a rational engineering basis and may give an incorrect assessment of equivalency between fires with significantly varying peak temperatures and exposure durations. For example, comparing an area bound by a time-temperature curve of a standard fire to the corresponding area bound by a realistic fire underestimates variation in the heat transfer in the case of *short hot fire* and overestimates the heat transfer as in the case of *long cold fire*.

Maximum Temperature Concept

A more refined approach for establishing time equivalency, *maximum temperature concept*, was proposed by Law (1973) and then Pettersson (1973). In this concept, time equivalency is derived by equating peak temperature attained in a structural member resulting from exposure to a design fire to the peak temperature reached in the member under standard fire exposure as shown in Fig. 3.19. This concept was mostly established for protected steel structural members to estimate time equivalency from exposure to a standard fire test with that exposed to any design fire scenarios. Although initially derived for protected steel elements, this method can be extended to derive time equivalency for reinforced concrete members. In this case, time equivalency can be established through comparing peak temperature attained in steel reinforcement under a given fire exposure (point A) to the peak temperature reached under exposure to the standard fire (point B).

Unlike the equal area concept, the maximum temperature approach has some basis in engineering. However, there are a few drawbacks to this approach as well. In instances where maximum steel temperatures in the structural member used for computing the time equivalency are much higher or lower than those which would cause failure in a particular member, applying maximum temperature of this concept may lead to unconservative time equivalency.

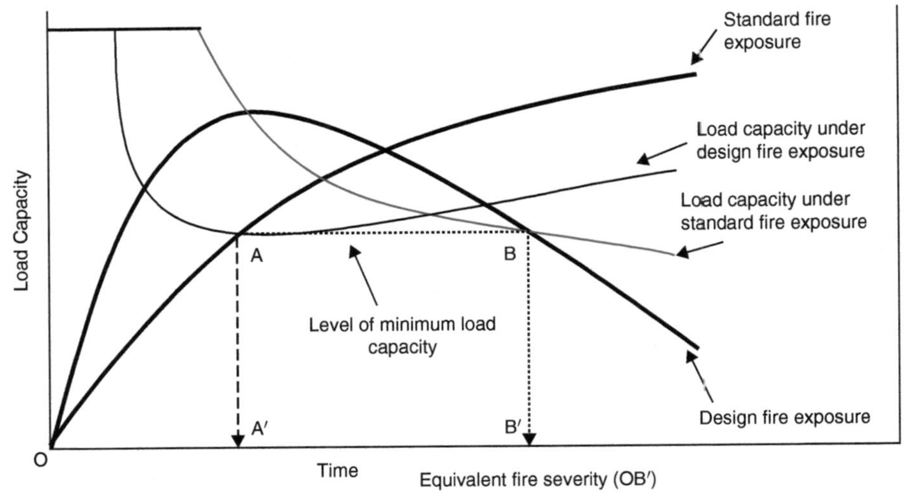

Figure 3.20 Minimum load capacity concept for deriving equivalent fire severity.

Minimum Load Capacity Concept

Minimum load concept is derived through comparing the *minimum load capacity* attained under a design fire scenario to the corresponding reduced load capacity reached under a standard fire exposure. This approach is more rational than the equal area and maximum temperature concepts since various critical factors such as fire severity, member characteristics, temperature-dependent material properties, load level, and support conditions are accounted for in deriving reduced load capacity. Figure 3.20 illustrates this concept by tracing temperature-induced degradation in load capacity of a typical structural member as a function of fire exposure under two fire scenarios, standard and design. As the load capacity of the structural member degrades with fire exposure time, the equivalent fire severity is defined as that point in time at which the minimum load capacity attained in structural members under design fire (point A) equals that attained under standard fire exposure (point B).

Since this concept involves equating reduced load capacities, time equivalency obtained using the minimum load capacity approach generally results in a better estimate than the equal area or maximum temperature concepts. One of the drawbacks of applying this concept is the need for detailed strength analysis to calculate reduced capacity, which may require the number of input parameters into analysis, and this can also require significant computational effort. Further, the minimum load capacity concept does not take into consideration the deformation limit state in the member which may be governing failure limit state under some conditions.

Maximum Deformation Concept

In this concept, time equivalency of a structural member subjected to a design fire is established by comparing the *maximum deformation* (deflection) reached in a structural member under a design fire to the same level of deformation reached under a standard fire exposure. The time equivalency is established by applying the deformation failure limit state. This limit state often governs failure of flexural members (such as beams) due to large deflections occurring under fire conditions. Recently, Kodur et al. (2010) derived time equivalency by equating thermal energy required to attain failure through deflection limit states in reinforced concrete beams exposed to standard and design fires. Figure 3.21 illustrates this concept in which time equivalency is established when the level of maximum deformation reached in a beam exposed to a design fire equals that attained under a standard fire exposure.

Establishing time equivalency through the maximum deformation concept can lead to realistic predictions of equivalent fire severity, as this approach incorporates the adverse effects of fire-induced thermal energy on structural members. However, evaluating deformations (or deflections) under fire conditions is a tedious process and require the use of advanced analysis (such as finite element method).

Time-Equivalent Formulas

Though the above described concepts are useful in establishing time equivalency between real fire and standard fire exposure, the effort involved in calculating needed parameters to establish equivalency (specifically deflections or moment capacity) is quite high. To overcome some of the drawbacks associated with these

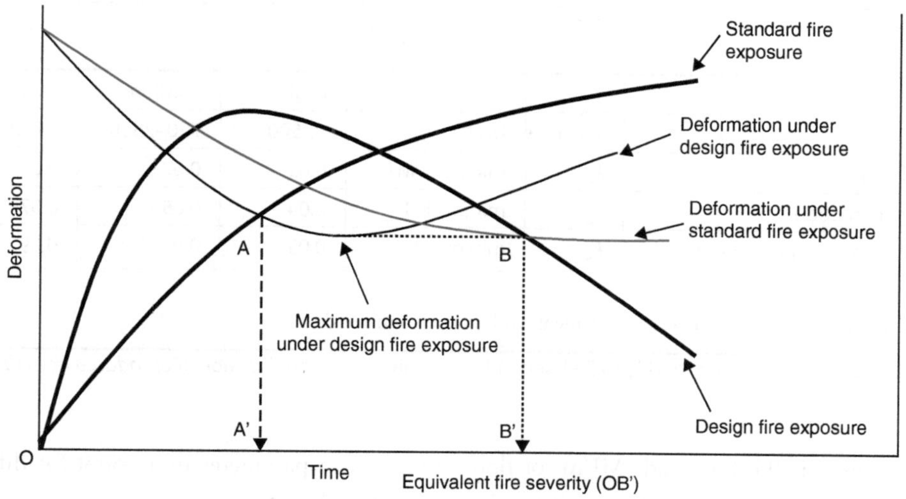

FIGURE 3.21 Maximum deformation concept for deriving equivalent fire severity.

approaches, simplified empirical relations have been proposed to establish equivalency between standard and design fire exposures (such as the CIB, 1986; Law, 1983, Eurocode 1, 2002; and Kodur et al., 2010).

CIB Formula A simplified relation was proposed by Pettersson (1973) to evaluate time equivalency between a standard and a design fire[12] using maximum temperature concept. This relation takes into consideration the main parameters: ventilation and fuel load in the compartment in evaluating equivalent fire severity. In this formula, equivalent time, t_e (min), of a real fire exposure to that of a standard fire exposure (ISO 834) is given by

$$t_e = k_c \, w \, e_f \tag{3.60}$$

where e_f is the fuel load (MJ/m^2 of floor area), k_c is a parameter to account for different linings of the compartment (listed in Table 3.7), and w is the ventilation factor (m$^{-0.25}$) given by

$$w = \frac{A_f}{\sqrt{A_v A_t \sqrt{H_v}}}$$

Law Formula Law (1973) proposed a formula similar to that proposed by Pettersson (1973), and its basis is the concept of maximum temperature. Law derived this formula using data from fire tests on structural members with varying sizes of compartments carried out by Thomas and Heselden (1972). This formula is a function of the fire load, internal area of the compartment, and ventilation area. The time equivalency is given as

$$t_e = \frac{e_f A_f}{\Delta H_c \sqrt{A_v (A_t - A_v)}} \tag{3.61}$$

It should be noted that both Law and CIB formulas can only be valid for compartments with vertical openings in their walls. These formulas cannot be applied to rooms with horizontal openings (vents) in the roof. Both equations generally predict similar equivalent time of exposure to that of a standard fire exposure, with some exceptions wherein Law's formula tends to predict slightly higher time-equivalency values.

Eurocode Formula The above-discussed time-equivalency formulas were revised and then incorporated into Eurocode 1. The modified equation for equivalent time of exposure from a design (real) fire to that of a standard fire (specifically ISO 834), t_e (min), is given as

$$t_e = k_b \, w \, e_f \tag{3.62}$$

[12] This formula was adopted in CIB (1986).

$b = \sqrt{b\rho c_p}$						
			High	Medium	Low	
	Term	Unit	>2500	720–2500	<720	General*
CIB	k_c	min.m$^{2.25}$/MJ	0.05	0.07	0.09	0.1
Eurocode	k_b	min.m^2/MJ	0.04	0.055	0.07	0.07
Large compartment	k_b	min.m^2/MJ	0.05	0.07	0.09	0.09

* The general case is used for compartments with unknown materials.

Table 3.7 Values of k_c and k_b to Be Used in Time-Equivalent Formulas Table (*Reproduced from CEN*)

where e_f is the fuel load (MJ/m^2 of floor area), k_b is a parameter to account for different linings of the compartment, and w is the ventilation factor (m$^{-0.25}$) given by

$$w = \left(\frac{6.0}{H_r}\right)^{0.3}\left[0.62 + \frac{90(0.4 - \alpha_v)^4}{1 + b_v a_h}\right] > 0.5 \tag{3.63}$$

where H_r is the fuel compartment height (m), and

$$0.025 \leq \alpha_v = \frac{A_v}{A_f} \leq 0.25 \tag{3.64}$$

$$\alpha_h = \frac{A_h}{A_f} \leq 0.20 \tag{3.65}$$

$$b_v = 12.5(1 + 10\alpha_v - \alpha_v^{\,2}) \tag{3.66}$$

where A_f is the floor area of the compartment (m^2), A_v is the area of vertical openings in walls (m^2), and A_h is the area of horizontal openings in roofs (m^2).

The values of k_c and k_b depend on the thermal properties of construction materials that form the linings on the compartment. These two factors have slightly different values and units because of the different ventilation factors in their respective formulas. Table 3.7 lists values of k_c and k_b to be used in the different time-equivalent formulas.

While both CIB and Eurocode formulas may give similar results for compartments with similar geometry (mainly for small rooms with tall windows), in case of large compartments with high ceilings and low window heights, the Eurocode formula yields lower time-equivalency predictions than the CIB formula.

Energy-Based Time-Equivalency Method Kodur et al. (2010) proposed a simplified formula to evaluate time equivalency of structural members. This formula is based on the principle of *equivalent energy*, and as such it establishes energy equivalency between standard and design fire exposures. The *energy-based concept* is derived based on the assumption that two fires will have the same fire severity if they transfer the same amount of thermal energy to a structural member. The amount of thermal energy transferred to a structural member exposed to fire is related to the heat flux on the fire-exposed boundaries of the structural member through convection and radiation.

The equation to evaluate time equivalency between a standard and a design fire ($t_{e(design)}/t_{e(standard)}$) is given as

$$t_{\mathrm{e(design)}} = (1.6 - 0.0004 \times \mathrm{T}_{\max})\, t_{\mathrm{e(standard)}} \tag{3.67}$$

where $t_{e(design)}$ is the time equivalent computed from maximum deflection method, $t_{e(standard)}$ is the time equivalent computed from equivalent energy method, and T_{max} is the maximum temperature of design fire exposure.

Unlike equal area and maximum temperature concepts which were validated for steel members, the energy-based approach was validated against reinforced concrete beams only.

Limitations It should be stressed that the time-equivalent concept and associated formulas are crude approximations developed to help in the fire design process. These methods were mainly derived through

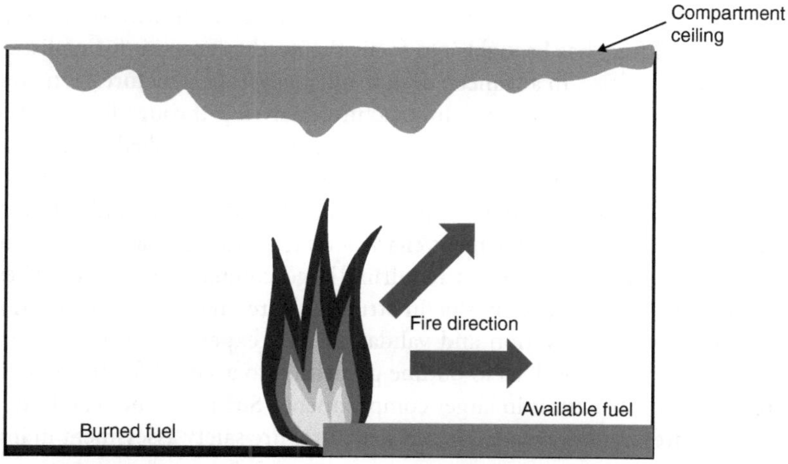

Figure 3.22 Traveling fires.

observations from fire tests carried out in small rooms, or in the case of the maximum temperature concept, on fire tests on protected steel members. Hence, time-equivalent concepts and formulas may not be applicable for highly varying real fire scenarios as compared to standard fire scenarios or to large-sized rooms or to compartments with large plastics in furnishings. Further, the above-discussed formulas do not take into consideration the effect of member continuity, composite construction (as in insulating functionality and capacitive protection), and so on. As such, these methods may not be suitable in a number of practical scenarios encountered in actual buildings.

3.10 Emerging Trends and Research Needs

In the last few decades, there have been significant developments with respect to advancing the understanding of fire dynamics, fire growth, and fire severity. However, there are still a number of specific areas relating to fire characterization where further research and development efforts are needed to advance the state of practice in the structural fire engineering field. This section highlights some of the emerging trends and research needs relating to traveling fires and fire growth models.

3.10.1 Traveling Fires

Currently, much of the advanced analysis for fire response simulation of structures is based on the unrealistic assumption of uniform fire distribution within a compartment (or enclosure). In reality, when a fire breaks out, the combustion process and burning conditions will not be uniform throughout the compartment, but rather highly turbulent and dynamic depending on fuel type, surface characteristics, and available ventilation. As a matter of fact, a closer observation of fire growth and development in larger compartments clearly show that fire tends to travel as a function of compartment geometry, fuel distribution, and ventilation (Rein et al., 2007) (see Fig. 3.22). Because of the dynamic and turbulent nature of fires, flames spread in both directions within an enclosure. This mode of fire spread results in varying temperature distribution across structural members that are part of the compartment, not only with time but also with spatial location. Such varying fire growth characteristics, specifically the traveling nature of fire, if not properly accounted for, can lead to inaccurate prediction of fire performance of structures.

There have been limited efforts to incorporate the traveling fire phenomenon into modeling the fire performance of structures (Rein et al., 2007; Rackauskaite et al., 2017). The early work on traveling fires in buildings dates back to the 1990s when Clifton (1996) proposed a model for accounting behavior and movement of fire in large enclosures. In this model, a large compartment is divided into a number of segments. Once fire in a segment reaches the burning stage (fully developed) as measured through fire temperatures attained, the fire is said to move to the neighboring segment. This model is a crude representation of the complex process of fire development and movement in large enclosures. Recently a few other studies proposed new CFD-based fire models to account for the traveling nature of fires in the structural fire engineering arena (Rein et al., 2007; Stern-Gottfried and Rein, 2012).

Recent investigation into high-profile fire incidents, such as those in the World Trade Center buildings and Plasco building, has brought into focus the need to account for realistic fire scenarios, including the traveling nature of fire, in advanced design and analysis of structures. One way to incorporate realistic fire exposure, including traveling fires, in structural analysis is through the use of computational fluid dynamics (CFD) based models. Although CFD analysis is routinely applied to study combustion characterizations and fire growth modeling, its use in advanced analysis of structures is not fully explored, as this type of analysis requires various input parameters (i.e., flame speed, combustion rate, air levels, etc.) that are dependent on specific compartment characteristics, fuel type, and other associated parameters, as well as on a complex set of calculations requiring large computation power. These factors are hindering the adoption of traveling fire scenarios in structural fire engineering applications. Further research efforts, through theoretical derivation and validation with experimental measurements, are needed to develop needed parameters, as well as to outline guidelines to account for traveling fire scenarios in undertaking fire resistance of structures in larger compartments. Such guidance can lead to adopting the phenomenon of traveling fires in performance-based structural fire safety assessment in important buildings.

3.10.2 Fire Models for Realistic Fire Representation

Much of the relations for fire growth characteristics (including time-temperature relations) presented in this chapter are based on research undertaken on fires in traditional compartments, representative of the 1960s building occupancies. These compartments tended to be of much smaller sizes, with few openings of smaller dimensions as compared to modern-day buildings, and contained mostly wood-based furnishings and lining materials that constituted most of the fuel. Over the last few decades, compartment features in modern buildings have significantly changed. The configuration and layout of modern buildings, as well as the contents, are vastly different. As the buildings are being designed to cater to functional (open) architectural concepts, compartments in recent years tend to have larger openings and hence and as a result can house higher fuel loads with large amounts of plastic-based furnishings and lining materials.

The fire growth characteristics, from ignition to the decay stage, in a structure is highly dependent on fuel type, amount of ventilation, and compartment size. Other key features that can influence fire growth, including floor plan topology, presence of active fire protection systems (e.g., sprinklers), and environmental conditions (smoking/cooking habits etc.), among others, are still not considered in evaluating fire growth. Thus, there is a need to develop improved, updated, and representative fire models that better represent realistic fire growth phenomenon, encompassing pre-flashover, growth, and post-flashover burning and decay stages, occurring in modern buildings.

Such models should also take into account features of modern structures such as various fuel types, compartment characteristics, inflammable lining/cladding materials, active suppression systems, and human interventions (and occurrence of other factors such as window breakage). Fire growth models need to address the relation between localized burning and the mechanics of fire development and travel within an enclosure as well as between parallel stories. While traditional fire growth models predict homogeneous temperature conditions and uniform burning throughout the fire compartment, advanced fire models can provide an in-depth insight into the fire growth phenomenon such as ignition probability, fire size, burning duration, metrics to measure fire severity, and so on. These improved fire growth models should be validated for different occupancy types using the findings of experiments and numerical simulations.

3.11 References

American Society of Civil Engineers (ASCE). "Minimum Design Loads and Associated Criteria for Buildings and Other Structures, ASCE/SEI 7-16." 2016.

ASTM E119. *Standard Test Methods for Fire Tests of Building Construction and Materials, ASTM E119.* West Conshohocken, PA: American Society for Testing and Materials, 2016.

Babrauskas, Vyto. "COMPF2–A Program for Calculating Post-Flashover Fire Temperatures Final Report." NBS TN 991 (1979).

Beyler, C. L. "Analysis of Compartment Fires with Overhead Forced Ventilation." *Fire Safety Science* 3, no. 10 (1991): 291.

Beyreis, J. R., H. W. Monsen, and A. F. Abbasi. "Properties of Wood Crib Flames." *Fire Technology* 7, no. 2 (1971): 145–155.

Buchanan, Andrew H., and Anthony Kwabena Abu. *Structural Design for Fire Safety.* Hoboken, NJ: Wiley, 2017.

Butcher, Edward Gordon, T. B. Chitty, and L. A. Ashton. *The Temperature Attained by Steel in Building Fires.* HM Stationery Office, 1966.

Cadorin, J. F., D. Pintea, and J. M. Franssen. "The Design Fire Tool Ozone V2. 0—Theoretical Description and Validation on Experimental Fire Tests." *Rapport interne SPEC/2001_01 University of Liege* (2001).

CEN. *Eurocode 2: Design of Concrete Structures. Part 1-2: General Rules—Structural Fire Design. EN 1992-1-2.* Brussels: European Committee for Standardization, Brussels: 2004.

CEN. *Eurocode 3: Design of Steel Structures. Part 1-2: General Rules—Structural Fire Design. EN 1993-1-2.* Brussels: European Committee for Standardization, Brussels: 2005.

CIB. Design Guide—Structural Fire Safety, CIB-W14. *Fire Safety Journal* 10, no. 2 (1986): 75–138.

Clifton, George Charles. *Fire Models for Large Firecells.* HERA Report R4-83., Manukau City, New Zealand: HERA, 1996.

Culver, C. Survey results for fire loads and live loads in office buildings. Vol. 85. US Department of Commerce, National Bureau of Standards, 1976.

Deal, Scot, and Craig Beylert. "Correlating Preflashover Room Fire Temperatures." *Journal of Fire Protection Engineering* 2, no. 2 (1990): 33–48.

DiNenno, Philip J. *SFPE Handbook of Fire Protection Engineering.* Quincy, MA: SFPE, 2008.

Drysdale, Dougal. *An Introduction to Fire Dynamics.* Hoboken, NJ: Wiley, 2011.

Eurocode 1 (EN 1991-1-2) (English). *Eurocode 1: Actions on Structures. Part 1-2: General Actions —Actions on Structures Exposed to Fire* [Authority: The European Union Per Regulation 305/2011, Directive 98/34/EC, Directive 2004/18/EC], 2002.

Feasey, Roger, and Andrew Buchanan. "Post-flashover Fires for Structural Design." *Fire Safety Journal* 37, no. 1 (2002): 83–105.

Foote, K. L., and P. J. Pagni. "Temperature Variations for Forced-Ventilated Compartment Fires." In *Fire Safety Science: Proceedings of the First International Symposium*, vol. 1, p. 139. Boca Raton, FL: CRC Press, 1986.

Hurley, Morgan J., Daniel T. Gottuk, John R. Hall Jr, Kazunori Harada, Erica D. Kuligowski, Milosh Puchovsky, John M. Watts Jr, and Christopher J. Wieczorek, eds. *SFPE Handbook of Fire Protection Engineering.* Berlin/Brandenburg: Springer, 2015.

Ingberg, Simon H. "Tests of the Severity of Building Fires." *NFPA Quarterly* 22, no. 1 (1928): 43–61.

ISO 834-1:1999. *Fire Resistance Tests—Elements of Building Construction—Part 1: General Requirements.* Geneva: International Organization for Standardization, 2012.

Jirku, Jiri, and Frantisek Wald. "Influence of Zinc Coating to a Temperature of Steel Members in Fire." *Journal of Structural Fire Engineering* 6, no. 2 (2015): 141–146.

Kawagoe, K. *Fire Behavior in Rooms.* Building Research Institute, Ministry of Construction (Japan). No. 27. Report, 1958.

Kodur, V. K. R., P. Pakala, and M. B. Dwaikat. "Energy Based Time Equivalent Approach for Evaluating Fire Resistance of Reinforced Concrete Beams." *Fire Safety Journal* 45, no. 4 (2010): 211–220.

Law, Margaret. *Prediction of Fire Resistance. Paper No. 2 of Symposium No. 5: Fire Resistance Requirements for Buildings–A New Approach.* London: Department of the Environment and Fire Offices Committee Joint Fire Research Organisation, Her Majesty's Stationary Office, 1973.

Law, Margaret. "Basis for the Design of Fire protection of Building Structures." *Structural Engineer* 61 (1983): 25–33.

Law, Margaret. "A Review of Formulae for T-equivalent." *Fire Safety Science* 5 (1997): 985–996.

Magnusson, Sven Erik, and Sven Thelandersson. "Temperature-Time Curves of Complete Process of Fire Development." *Bulletin of Division of Structural Mechanics and Concrete Construction* 16 (1970).

Matsuyama, Ken, Takashi Fujita, Hideki Kaneko, Yoshifumi Ohmiya, Takeyoshi Tanaka, and Takao Wakamatsu. "A Simple Predictive Method for Room Fire Behavior." *Fire Science and Technology* 18, no. 1 (1998): 23–32.

MBIE. *New Zealand Building Code Handbook,* 3rd ed. Wellington, New Zealand: Department of Building and Housing, 2007.

McCaffrey, B. J., J. G. Quintiere, and M. F. Harkleroad. "Estimating Room Temperatures and the Likelihood of Flashover Using Fire Test Data Correlations." *Fire Technology* 17, no. 2 (1981): 98–119.

McGrattan, Kevin, Simo Hostikka, Jason Floyd, Howard Baum, Ronald Rehm, William Mell, and Randall McDermott. "Fire Dynamics Simulator (version 5), Technical Reference guide." *NIST Special Publication* 1018, no. 5 (2004).

National Fire Protection Association. NFPA 557 Standard for Determination of Fire Loads for Use in Structural Fire Protection Design. NFPA, 2011.

Peacock, Richard D. "CFAST." (2012).

Peacock, Richard D., Paul A. Reneke, Richard W. Bukowski, and Vytenis Babrauskas. "Defining Flashover for Fire Hazard Calculations." *Fire Safety Journal* 32, no. 4 (1999): 331–345.

Peatross, Michelle J., and Craig L. Beyler. "Thermal Environment Prediction in Steel-bounded Preflashover Compartment Fires." *Fire Safety Science* 4 (1994): 205–216.

Pettersson, Ove. *The Connection Between a Real Fire Exposure and the Heating Conditions According to Standard Fire Resistance Tests: With Special Application to Steel Structures.* Lund, Sweden: Lund Institute of Technology, Division of Structural Mechanics and Concrete Construction, 1973.

Phan, Long T., Therese P. McAllister, John L. Gross, and Morgan J. Hurley. "Best Practice Guidelines for Structural Fire Resistance Design of Concrete and Steel Buildings." NIST technical note 1681 (2010).

Portier, Rebecca W., Richard D. Peacock, and Paul A. Reneke. "FASTLite: Engineering Tools for Estimating Fire Growth and Smoke Transport (NIST SP 899)." *Special Publication (NIST SP)-899* (1996).

Purkiss, John A., and Long-Yuan Li. *Fire Safety Engineering Design of Structures.* Boca Raton, FL: CRC Press, 2013.

Rackauskaite, Egle, Panagiotis Kotsovinos, Ann Jeffers, and Guillermo Rein. "Structural Analysis of Multi-storey Steel Frames Exposed to Travelling Fires and Traditional Design Fires." *Engineering Structures* 150 (2017): 271–287.

Rein, Guillermo, Xun Zhang, Paul Williams, Ben Hume, Alex Heise, Allan Jowsey, Barbara Lane, and Jose L. Torero. "Multi-story Eire Analysis for High-rise Buildings." *11th Interflam, London* (September 2007): 605–616.

SFPE/SEI. *Designing Structures for Fire.* Published for the Society of Fire Protection Engineers. Quincy, MA: DesTech Publications, Inc., 2003.

Stern-Gottfried, Jamie, and Guillermo Rein. "Travelling Fires for Structural Design–Part I: Literature Review." *Fire Safety Journal* 54 (2012): 74–85.

Thomas, P. H., and A. J. M. Heselden. Fully-Developed Fires in Single Compartments. Fire Research Station, Building Research Establishment, Borehamwood, Hertfordshire, UK 1972, pp. 410–423.

Veloo, Peter S., and James G. Quintiere. "Convective Heat Transfer Coefficient in Compartment Fires." *Journal of Fire Sciences* 31, no. 5 (2013): 410–423.

Walton and Thomas (1995), "Estimating Temperatures in Compartment Fires" (Chapters 3–6). In *SFPE Handbook of Fire Protection Engineering.* Gaithersburg, MD: Society of Fire Protection Engineers.

Wickström, Ulf. *Temperature Calculation in Fire Safety Engineering.* Berlin/Brandenburg: Springer, 2016.

Yager, B. *FIERA Smoke Movement Model Technical Report.* Ottawa: National Research Council Canada, 1999.

3.12 Nomenclature

$\dot{q}''$	Heat flow per unit area (W/m²)
$\dot{q}_C$	Heat produced by combustion of the fuel
$\dot{q}_L$	Heat carried out of the opening by convection of hot gases and smoke
$\dot{q}_R$	Heat radiated through the opening
$\dot{q}_W$	Heat conducted into the surrounding structure
A_{fuel}	Exposed fuel surface area (m²)
F_{ref}	Reference value of the ventilation factor
L_v	Heat of gasification (MJ/kg)
T_e	Absolute temperature of the emitting surface (K)
$\dot{m}$	Rate of burning of wood fuel (kg/s)
q_i''	Incident of radiation reaching the surface of the fuel (MW/m²) in a post-flashover fire
t^*	Fictitious duration
t_d	Duration of parametric fire
ΔH_c	Calorific values (MJ/kg)
$\Delta H_{c,n}$	Effective calorific value
A_t	Total area of the bounding surfaces of the room (i.e., floor, ceiling, and walls, including window opening)
A_v	Area of the window opening (m²)
B	Breadth of the window opening
C_p	Specific heat (J/kg.K)
D	Compartment depth (m)
E	Maximum possible energy (MJ)
e_f	Fire load energy density (FLED) per floor area (MJ/m²)
e_t	Total fire load
F_v	Ventilation factor ($m^{1/2}$)
g	Velocity of gas (m/s), assumed to be 1 (m/s)
h	Convective heat transfer coefficient (W/m²K)
H_v	Height of the window opening (m)
k	Growth constant ($s/MW^{1/2}$)
k	Thermal conductivity (W/m.K)
k_c	Parameter to account for different linings of the compartment
$k\rho C_p$	Thermal inertia (W^2s/m^4K^2)
L	Wood-equivalent fire load (kg)
m_c	Moisture content
m_d	Moisture content as a percentage of the dry weight
M_f	Total mass of fuel available (kg)
Q	Average heat release rate (MW)
Q_{fo}	Critical values of heat release (MW)
Q_{fuel}	Heat release rate (MW)
Q_{vent}	Ventilation-controlled heat release rate (MW)
R_f	Minimum load capacity reached during a fire or the load capacity at a certain time specified by codes
$s_{lim,1}$	Limiting thickness (mm)
t	Duration of the burning

t_b	Duration of the burning period (s)
t_{code}	Fire resistance or required fire severity
t_e	Equivalent time of exposure to an ISO 834 fire test (min)
t_{fail}	Time to failure of a building member, usually a fire-resistance rating
T_{fail}	Temperature which would cause failure of building members (thermal or structural failure)
t_{fo}	Time to flashover (s)
T_{max}	Maximum temperature (°C)
T_{max}	Maximum temperature reached in building members during a fire or the temperature at a certain time specified by codes
T_r	Absolute temperature of the receiving surface (K)
t_s	Fire duration or fire severity
U_f	Applied load at the time of a fire
v_p	Surface regression rate (m/s)
W	Compartment width (m)
w	Ventilation factor ($m^{-0.25}$)
$\alpha = k/\rho C_p$	Thermal diffusivity (m^2/s)
ρ	Density (kg/m^3)
σ	Stefan-Boltzmann constant (5.67×10^{-8} W/m^2K^4)
Ω	Opening factor ($m^{-1/2}$)
ΔT	Temperature difference between surface of solid and fluid (°C or K)
α	Fire intensity coefficient (MW/s^2)
ϕ	Configuration or view factor

3.13 Numerical Examples

3.1

Calculate the heat release rate (Q_p) for slow, medium, fast, and ultrafast burning fires given that the growth constant, k, is taken as 600, 300, 150 and 75 s/MW$^{1/2}$, respectively. Assume burning has a duration of 5 minutes.

Solution

5 minutes $= 5 \times 60 = 300$ seconds

Slow fire: $Q_p = (t/k)^2 = (300/600)^2 = 0.25$ MW

Medium fire: $Q_p = (t/k)^2 = (300/300)^2 = 1.0$ MW

Fast fire: $Q_p = (t/k)^2 = (300/150)^2 = 4.0$ MW

Ultrafast fire: $Q_p = (t/k)^2 = (300/75)^2 = 16.0$ MW

3.2

Calculate the radiant heat flux (q'') from burning in a compartment with a configuration view factor (ϕ) of 1.0, flames temperature of 700°C, and an emissivity (ε) of 0.8.

Solution

Emitter temperature: $T = 700°C = 973$ K
Configuration view factor: $\phi = 1.0$
Emissivity: $\varepsilon = 0.8$
Stefan-Boltzmann constant: $\sigma = 5.67 \times 10^{-8}$ W/m^2K^4

Radiant heat flux: $q'' = \phi \varepsilon_e \sigma T_e^4 = 1.0 \times 0.8 \times 5.67 \times 10^{-8} \times \dfrac{(973)^4 - (293)^4}{1000} = 40.3\ kW/m^2$

3.3

Calculate the required heat release rate (Q_{fo}) to cause flashover in a compartment of 5 m long × 4 m wide × 3 m high. Assume that this compartment has one window of 1.5 m high × 3 m wide and apply Thomas's flashover criterion.

Solution

Length of room: $l_1 = 5$ m, Width of room: $l_2 = 4.0$ m, Height of room: $H_r = 3$ m, Height of window: $H_v = 1.5$ m, Width of window $B = 3.0$ m

Area of internal surfaces: $A_t = 2(l_1 l_2 + l_1 H_r + l_2 H_r)$

$$A_t = 2(5 \times 4 + 5 \times 3 + 4 \times 3) = 94 \text{ m}^2$$

Area of windows: $A_v = BH_v = 3 \times 1.5 = 4.5 \text{ m}^2$

Thus, heat release rate to cause flashover: $Q_{fo} = 0.0078 A_t + 0.378 A_v \sqrt{H_v}$

$$Q_{fo} = (0.0078 \times 94) + 0.378 \times 4.5\sqrt{1.5} = 2.8 \text{ MW}$$

3.4

Calculate the ventilation heat release rate (Q_{vent}) for a post-flashover fire in a room 6 m × 4 m in floor area, and 3 m high, with one window 2 m high × 3 m wide. Assume ventilation-controlled conditions and fuel made of wood with heat of combustion (ΔH_c) of 17 MJ/kg. Also, calculate the duration of burning (t_b) if the available fuel load energy density is 900 MJ/m² floor area. Compare your evaluated ventilation heat release rate (Q_{vent}) with prediction from Law's equation.

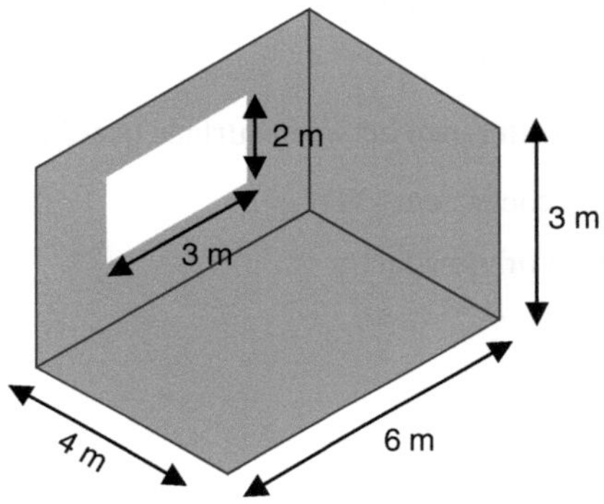

Solution

Window area: $A_v = 2 \times 3 = 6 \text{ m}^2$

Rate of burning: $\dot{m} = 0.092 A_v \sqrt{H_v} = 0.092 \times 6\sqrt{2} = 0.781$ kg/s

Heat of combustion: $\Delta H_c = 17$ MJ/kg

Heat release rate: $Q_{vent} = \dot{m}\Delta H_c = 0.781 \times 17 = 13.3$ MW

Fuel load energy density: $e_f = 900$ MJ/m²

Floor area: $A_f = 6 \times 4 = 24 \text{ m}^2$

Total thermal energy: $E = e_f A_f = 900 \times 24 = 21{,}600$ MJ

$$\text{Duration of burning: } t_b = \frac{E}{Q_{vent}} = \frac{21600}{13.3} = 1624 \text{ s} \approx 27.6 \text{ min}$$

Using Law's equation

Room width: $W = 6.0$ m, Room depth: $D = 4.0$ m

Window area: $A_v = 2 \times 3 = 6 \text{ m}^2$

$$\text{Opening factor: } \Omega = \frac{A_t - A_v}{A_v\sqrt{H_v}} = \frac{108 - 6}{6\sqrt{2}} = 12 \text{ m}^{-1/2}$$

Rate of burning: $\dot{m} = 0.18A_v\sqrt{\frac{H_vW}{D}}(1 - e^{-0.036\Omega})$

$$\dot{m} = 0.18 \times 6 \times \sqrt{\frac{2 \times 6}{4}}(1 - e^{-0.036\times12}) = 0.65 \text{ kg/s}$$

Heat release rate can be evaluated as: $Q_{vent} = \dot{m}\Delta H_c = 0.65 \times 17 = 11.15 \text{ MW}$

Duration of burning can be evaluated as: $t_b = \frac{E}{Q_{vent}} = \frac{21600}{11.05} = 1954.75 \text{ s} \approx 32.6 \text{ min}$

3.5

Using Babrauskas's expressions, calculate the fuel-controlled heat release rate in a compartment with 20 m^2 of wood burning under two conditions: thick slabs of wood burning and thin wood slabs (of 40 mm thick) burning. Assume wood density to be 350 kg/m^3 and having heat of combustion of 17 MJ.

Solution

For thick slabs of wood, Regression rate: $v_p = 9.0 \times 10^{-6}$ m/s, Density: $\rho = 350$ kg/m^3, Area of fuel: $A_{fuel} = 20$ m^2, Heat of combustion: $\Delta H_c = 17$ MJ/kg
Heat release rate can be evaluated as: $Q_{fuel} = v_p \rho A_{fuel} \Delta H_c$

$$Q_{fuel} = 9 \times 10^{-6} \times 350 \times 20 \times 17 = 1.7 \text{ MW}$$

For thin wood slab of 40 mm thickness, Thickness of slab: $D = 0.04$ m
Regression rate: $v_p = 2.2 \times 10^{-6} D^{-0.6}$

$$v_p = 2.2 \times 10^{-6} \times 0.04^{-0.6} = 1.5 \times 10^{-5} \text{ m/s}$$

Heat release rate: $Q_{fuel} = v_p \rho A_{fuel} \Delta H_c$

$$Q_{fuel} = 1.5 \times 10^{-5} \times 350 \times 20 \times 17 = 1.8 \text{ MW}$$

3.6

Calculate the maximum temperature that can be reached in a room 6 m × 4 m of 3 m high, with one window 2 m high × 3 m wide using Law's equation. Assume burning wood has a heat of combustion (ΔH_c) of 17 MJ/kg and a thermal energy of 15,000 MJ. Compare your work to that using Swedish curves.

Solution

Room width: $W = 6.0$ m, Room depth: $D = 4.0$ m, Area of windows: $A_v = 6.0$ m^2, Area of internal surfaces: $A_t = 108$ m^2

Opening factor: $\Omega = \frac{A_t - A_v}{A_v\sqrt{H_v}} = \frac{108 - 6}{6\sqrt{2}} = 12 \text{ m}^{-1/2}$

Rate of burning: $\dot{m} = 0.18A_v\sqrt{\frac{H_vW}{D}}(1 - e^{-0.036\Omega})$

$$\dot{m} = 0.18 \times 6 \times \sqrt{\frac{2 \times 6}{4}}(1 - e^{-0.036\times12}) = 0.657 \text{ kg/s}$$

Heat release rate can be evaluated as: $Q_{vent} = \dot{m}\Delta H_c = 0.657 \times 17 = 11.2 \text{ MW}$

Duration of burning can be evaluated as: $t_b = \frac{E}{Q_{vent}} = \frac{15000}{11.2} = 1339.3 \text{ s} \approx 22.3 \text{ min}$

Maximum temperature:

$$T_{max} = 6000(1 - e^{-0.1\Omega}) / \sqrt{\Omega} = 6000\frac{1 - e^{-0.1\times12}}{\sqrt{12}} = 1210°\text{C}$$

Check reduction factor for fuel load:

Fuel load (wood equivalent): $L = E/\Delta H_c = 15000/17 = 882.4$ kg

Temperature parameter: $\psi = \frac{L}{\sqrt{A_v(A_t - A_v)}} = \frac{882.4}{\sqrt{6(108 - 6)}} = 35.7$

Reduced maximum temperature: $T = T_{max}(1 - e^{-0.05\psi}) = 1210(1 - e^{-0.05\times 35.7}) = 1004.1°C$

Swedish Curves

Fuel load (based on total bounded area in compartment): $e_t = \dfrac{E}{A_t} = \dfrac{15{,}000}{108} = 138.9\ \text{MJ/m}^2$

Ventilation factor: $F_v = \dfrac{A_v\sqrt{H_v}}{A_t} = \dfrac{6\sqrt{2}}{108} = 0.078\ \text{m}^{-1/2}$

From Swedish curves (see Fig. 3.9), temperature values can be roughly interpolated for $F_v = 0.8$ (≈ 0.78) and $e_t = 138.9$, giving a temperature of about 830°C after 22.3 minutes of fire exposure.

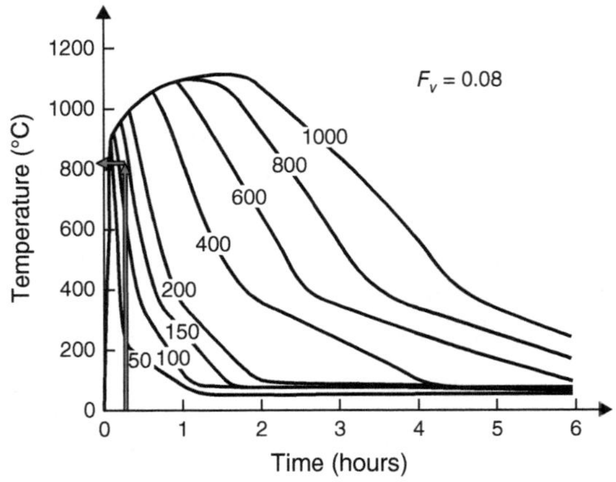

3.7

Using CIB, Law, and Eurocode formulas, compare the fire severity arising from burning of fuel of 1000 MJ/m² in a compartment with 3.0 m high and of floor area = 20 m² (4.0 m × 5.0 m). Assume that there is one window that is 3.0 m wide and 2.0 m high. The room is constructed from concrete. Assume net calorific value of wood: $\Delta H_c = 17$ MJ/kg.

Solution

CIB Formula

Length of room: $L_1 = 6.0$ m, Width of room: $L_2 = 4.0$ m, Floor area: $A_f = L_1 L_2 = 5 \times 4 = 20$ m², Height of room: $H_r = 3.0$ m

Fuel load (floor area): $e_f = 1000$ MJ/m²

Area of internal surfaces:

$$A_t = 2(l_1 l_2 + l_1 H_r + l_2 H_r) = 2(5 \times 4 + 5 \times 3 + 4 \times 3) = 94\ \text{m}^2$$

Thermal inertia of concrete: $b = \sqrt{k\rho C_p} = 1391 \dfrac{Ws^{1/2}}{m^2 K}$

Conversion factor (from Table 3.7): $k_c = 0.07$ min.m$^{2.25}$/MJ, Window height: $H_v = 2.0$ m, Window width: $B = 2.0$ m

Window area: $A_v = H_v B = 2 \times 2 = 4$ m²

Ventilation factor: $F_v = \dfrac{A_v\sqrt{H_v}}{A_t} = \dfrac{4\sqrt{2}}{94} = 0.06\ \text{m}^{-1/2}$

$$w = \frac{A_f}{\sqrt{A_v A_t \sqrt{H_v}}} = \frac{20}{\sqrt{4(94)\sqrt{2}}} = 0.867$$

Equivalent fire severity: $t_e = k_c w e_f = 0.05 \times 0.867 \times 1000 = 43.35$ min

Law Formula

Net calorific value of wood: $\Delta H_c = 17$ MJ/kg

Equivalent fire severity: $t_e = \dfrac{e_f A_f}{\Delta H_c \sqrt{A_v (A_t - A_v)}} = \dfrac{1000 \times 20}{17\sqrt{4(94 - 4)}} = 62$ min

Eurocode Formula

Thermal conductivity: $k = 1.6$ W/m-K, Density: $\rho = 2200$ kg/m^3, Specific heat: $c_p = 880$ J/kg-K

Thermal inertia of concrete: $b = \sqrt{k\rho C_p} = 1760 \dfrac{\text{Ws}^{1/2}}{\text{m}^2\text{K}}$

Conversion factor (from Table 3.7): $k_c = 0.055$ min.m$^{2.25}$/MJ, Window height: $H_v = 2.0$ m, Window width: $B = 3.0$ m

Window area: $A_v = H_v B = 2 \times 2 = 4$ m^2

$$0.025 \le \alpha_v = \frac{A_v}{A_f} = \frac{4}{94} = 0.43 \le 0.25$$

$$b_v = 12.5(1 + 10\alpha_v - \alpha_v^{\,2}) = 12.5(1 + 10 \times 0.43 - 0.43^2) = 63.9$$

$$\text{Ventilation factor: } w = \left(\frac{6}{H_r}\right)^{0.3}\left[0.62 + \frac{90(0.4 - \alpha_v)^4}{1 + b_v a_h}\right] > 0.5$$

$$w = \left(\frac{6}{3}\right)^{0.3}\left[0.62 + \frac{90(0.4 - 0.43)^4}{1 + 63.9 \times 0}\right] = 0.76\ \text{m}^{0.3} > 0.5$$

Equivalent fire severity: $t_e = k_b\, w\, e_f = 0.055 \times 0.76 \times 1000 = 42$ min

3.8

Two buildings, located side by side, are to be designed to prevent the occurrence of fire in one of them due to a fire in the other building. The building code stipulates that ASTM E-119 fire is to be followed and the maximum duration of fire that can occur in the building is 30 minutes. Find the minimum distance between two buildings so that ignition does not occur in the second building due to fire in the first building. Assume the following in the analysis:

Emissivity factor = 0.9
Stefan-Boltzmann constant = 5.67×10^{-8} W/m^{2}K^4 and
Maximum radiant heat flux = 12 kW/m^2
Opening (window) area = 3 × 3 m

Solution

$t_{max} = 30$ minutes
$\varepsilon = 0.9$
$\sigma = 5.67 \times 10^{-8}$ w/m^{2}k^4
$q'' = 12$ w/m^2
Opening area, $A_1 = 3 \times 3 = 9$ m^2
ASTM E119

$$T(°\text{C}) = 750(1 - e^{-3.79\sqrt{0.169}}) + 170.4\sqrt{0.169} + 20$$

$$T = 20 + 698.77 + 120.5 = 839.27°\text{C}$$

By radiation (for heat transfer in adjacent building):

$q'' = \sigma\varepsilon\phi T^4$

$12 \times 10^3 = \phi \times 9 \times 5.67 \times 10^{-8} \times ((839.27 + 273)^4 - (20 + 273)^4)$

$12{,}000 = \phi \times 9 \times 5.67 \times 10^{-8} \times 1.52 \times 10^{12}$

$\phi = 0.154$

$\phi = A_1/\pi r^2 = 9/\pi r^2$

$r = 2.9$ m

CHAPTER 4

Properties of Materials at Elevated Temperatures

4.1 Overview[1]

This chapter outlines the characteristics of common materials used in construction applications. The key properties that influence fire resistance performance, together with the test methods used to deduce these properties, are discussed. In addition, the trends for the variation in thermal, mechanical, and other material-specific properties with temperature of commonly used construction materials are presented.

4.2 Characteristics of Building Materials

4.2.1 Classification

Based on composition, construction materials can be classified as either homogeneous or heterogeneous. *Homogeneous* materials refer to those with uniform composition and properties throughout their volume and are rarely found in nature. On the other hand, *heterogeneous* materials have different composition and varying properties throughout the volume. Most construction materials are heterogeneous, yet their heterogeneity is often glossed over when practical problems are dealt with.

For example, concrete is made of a number of components (e.g., aggregate, sand, cement, etc.), thus the heterogeneity of concrete is easily noticeable. Other heterogeneities related to the microstructure of materials, that is, their grain and pore structures, are rarely detectable by the naked eye, such as those present in steel or other metals. The microstructure depends greatly on the way the materials are formed. In general, materials formed by solidification from a melt show the highest degree of homogeneity. The result of the solidification is normally a *polycrystalline* material, comprising polyhedral grains of crystals, which are equiaxial and randomly oriented. Severe cold working in metals may produce an elongated grain structure and crystals with preferred orientations.

Noncrystalline solids are called amorphous materials, and examples include gels and glasses. Gels are formed by the coagulation of a colloidal solution. Glasses (vitreous materials) are solids with a liquidlike, grainless submicroscopic structure with a low crystalline order. Upon heating, they will go through a series of phases of decreasing viscosity.

Synthetic polymers (plastics) are made of long macromolecules created by polymerization from smaller repeating units (monomers). In the case of *thermoplastic* materials, the mobility of the molecular chains increases upon heating. Such materials soften, much like glass. In some other types of plastics, called *thermosetting* materials, polymerization also produces cross-bonds between the molecular chains. These cross-bonds prevent the loosening of the molecular structure and the transition of the material into a liquidlike state.

Some building materials (e.g., gypsum, brick) are formed from a wet, plastic mass or from compacted powders by firing. The resulting product is a polycrystalline solid with a well-developed pore structure. Two important building materials, concrete and gypsum, are formed by mixing finely ground powders (and aggregates) with water, which solidifies by hydration. The cement paste in a concrete has a highly complex microstructure, interspersed with very fine, elaborate pores.

Most building materials can be treated as isotropic materials. Isotropic materials are assumed to possess the same properties in all directions. Exceptions are some of the advanced composite materials, such as

[1] Most of the information presented in this chapter is taken from Kodur and Harmathy, Properties of Building Materials, Chapter 9 in *SFPE Handbook of Fire Protection Engineering*, 5th ed., Springer, New York, 2016. Please refer to the *SFPE Handbook* for full details on properties of construction materials at elevated temperatures.

fiber-reinforced polymers (FRP), and wood, which possess varying properties in different directions and are classified as anisotropic materials.

Among the material properties, those that are unambiguously defined by the current composition and phase are referred to as structure-insensitive. Some others depend on the microstructure of the solid or on its previous history. These materials are called structure-sensitive.

4.2.2 Porosity and Moisture Sorption

The performance of a material at elevated temperatures is dependent on the chemical composition and molecular structure of the material. One of the factors that can influence properties of materials at elevated temperatures is the presence of moisture (water). The composition and extent of water present in a material is commonly described by two terms, porosity and moisture sorption.

In general, most construction materials are in solid form but contain some pores (e.g., cured concrete and/or masonry). What is commonly referred to as a solid material is actually all the material within its visible boundaries. Clearly, if the solid is porous, the so-called solid material consists of at least two phases: (1) a solid-phase matrix and (2) a gaseous phase (namely, air) in the pores within the solid matrix. In addition, there is a third phase (of liquid or liquidlike nature) present within the solid material. This liquid phase is made of moisture either absorbed from the atmosphere to the pore surfaces or held in the pores by capillary condensation. The third phase is always present if the pore structure is continuous. It should be noted that pores of some foamed plastics are not readily accessible to atmospheric moisture (e.g., discontinuous pores).

The pore structure of materials is characterized by two properties: *porosity, P* ($m^3 \cdot m^{-3}$), the volume fraction of pores within the visible boundaries of the solid, and *specific surface, S* ($m^2 \cdot m^{-3}$), the surface area of the pores per unit volume of the material. For a solid with continuous pore structure, the porosity is a measure of the maximum amount of water the solid can hold when saturated. The specific surface and, to a lesser degree, porosity together determine the moisture content the solid holds in equilibrium with given atmospheric conditions.

At equilibrium conditions, the relationship between water content and equilibrium humidity of a porous material can be displayed graphically by a curve, the so-called *sorption isotherm*. For each humidity value, a sorption isotherm indicates the corresponding water content value at a given, constant temperature. A sorption isotherm usually has two branches: (1) an *adsorption branch*, obtained by monotonically increasing the relative humidity of the atmosphere from 0% to 100% through very small equilibrium steps; and (2) a *desorption branch*, obtained by monotonically lowering the relative humidity from 100% to 0%. Derived experimentally, the sorption isotherms offer some insight into the nature of the material's pore structure (Harmathy, 1967, 1970).

For heterogeneous materials consisting of solids of different sorption characteristics (e.g., concrete, consisting of cement paste and aggregates), the sorption isotherms can be estimated using the simple mixture rule (with $m = 1$):

$$\chi^m = \sum_i v_i \pi_i^m \tag{4.1}$$

where χ = Material property for the composite
χ_I = Material property for the composite's ith constituent
v_i ($m^3 \cdot m^{-3}$) = Volume fraction of the ith constituent
m (dimensionless) = Constant that has a value between −1 and +1

Most building materials are porous in microstructure and therefore capable of holding moisture. For instance, concrete (or more accurately, the cement paste in the concrete) and wood, because of their large specific surfaces, can hold water in amounts substantial enough to be taken into consideration in fire performance assessments. The amount of moisture held within a material depends on the atmospheric conditions. Because the presence of moisture may have a significant and often unpredictable effect on the properties of materials at temperatures below 100°C, it is imperative to conduct all property tests on specimens brought into a moistureless "reference condition" by some drying technique prior to the test. The reference condition is normally interpreted as that attained by heating the test specimen in an oven at 105°C until its weight shows no change. A few building materials, however, among them all gypsum products, may undergo irreversible physicochemical changes when held at that temperature for an extended period. To bring them to a reference condition, specimens of these materials should be heated in a vacuum oven at some lower temperature level (e.g., at 40°C in the case of gypsum products).

4.2.3 Survey of Construction Materials

The number of building materials has dramatically increased during the past few decades. In the last three decades, *high-performing materials*, such as FRP, high-strength concrete (HSC), high-strength steel, and engineered wood, have been developed to achieve cost-effectiveness in civil engineering construction. Although many of these high-performing materials possess superior properties at ambient temperatures, the same cannot be said of their performance at elevated temperatures. In these materials additional complexities such as temperature-induced spalling in HSC, temperature-induced buckling in high-strength steel, and delamination in engineered wood can arise which may severely impact fire performance of a structural member made of these newly developed materials.

According to Harmathy (1993), there are two main types of construction materials, combustible (burnable) and noncombustible (nonburnable). The mechanical and thermal properties of these materials are of primary interest to a designer concerned with the structural performance of a building during a fire. From the point of view of their performance in fire, building materials can be divided into the following groups:

1. *Group L (load-bearing) materials.* Materials capable of carrying high stresses (structural loads), usually in tension or compression. With these materials, the mechanical properties related to behavior in tension and/or compression are of principal interest.
2. *Group L/I (load-bearing/insulating) materials.* Materials capable of carrying moderate stresses and, in fire, providing thermal protection (insulation) to Group L materials. With Group L/I materials, the mechanical properties (related mainly to behavior in compression) and the thermal properties are of equal interest.
3. *Group I (insulating) materials.* Materials not designed to carry a load. Their role in fires is to resist (limit) the transmission of heat through building elements and/or to provide insulation to Group L or Group L/I materials. With Group I materials, only the thermal properties are of interest.
4. *Group L/I/F (load-bearing/insulating/fuel) materials.* Group L/I materials have good thermal insulation properties but may become fuel in fire (e.g., wood, FRP).
5. *Group I/F (insulating/fuel) materials.* Group I materials have no load-bearing capacity but good insulating properties. However, they can become fuel in fire.

By necessity, only a few commonly used materials will be discussed in this chapter, namely, structural steel, light-gauge steel, and reinforcing/prestressing steel in Group L; concrete (including fiber-reinforced concrete), brick, and masonry in Group L/I; wood and FRP in Group L/I/F (or Group I/F and L/F); and gypsum and fire insulation materials in Group I.

4.3 Material Properties at Elevated Temperatures

The behavior (and performance) of a structural member exposed to fire is dependent, in part, on the thermal, mechanical, and deformational properties of its constituent materials. The calculation methods and numerical modeling techniques for predicting the process of temperature-dependent deterioration of building components (under fire conditions) have been rapidly evolving in recent years. The key input information into these calculation methods and numerical models are high-temperature properties of materials. Data generation on elevated temperature properties of materials have not kept pace. For most practical fire-related design applications, information on the properties of building materials in the temperature range of interest, from 20 to 800°C, is required. In this temperature range, most construction materials are not stable. Upon heating, they undergo physicochemical changes ("reactions" in a generalized sense), accompanied by transformations in their microstructure and changes in their properties. For example, the behavior of concrete at 500°C is completely different from that at room temperature.

The thermophysical and mechanical properties of most construction materials change substantially within the temperature range associated with building fires. At elevated temperatures, many building materials undergo physicochemical changes. Most of the properties associated with construction materials are temperature dependent and sensitive to testing method parameters such as heating rate, strain rate, temperature gradient, and so on. Harmathy (1983) stated that the lack of adequate knowledge of the behavior of construction materials at elevated temperatures is the most disturbing trend in fire safety engineering. There has been a tendency to use *notional* (also called "typical," "proprietary," "empirical," etc.) values for material properties in numerical computations—in other words, values that ensure agreement between experimental and analytical results. Harmathy warned that this practice might lead to a proliferation of theories that lack general validity. In cases where such an approach is followed, numerical predictions can be unconservative (e.g., underestimate fire resistance), leading to inadequate selection/design/detailing of structural components/systems.

Clearly, the generic information available on the properties of building materials at room temperature is seldom applicable in scenarios where design for fire safety is critical. It is imperative, therefore, that the fire safety practitioner knows how to extend, based on a priori considerations, the utility of the scanty data that can be gathered from the technical literature. Also, knowledge of unique material-specific characteristics at elevated temperatures, such as spalling in concrete, local buckling in steel, or charring in wood, is critical to determine the fire performance of a structural member. In order to provide necessary information for fire designers and researchers on the behavior of construction materials under fire conditions, the following sections discuss various temperature-dependent properties of main materials used in civil infrastructure.

4.3.1 Thermal Properties

The material properties that influence temperature rise and distribution in a structural member are its thermal expansion, mass loss, density and porosity, specific heat, thermal conductivity, and thermal diffusivity. These properties depend on the composition and characteristics of the constituent materials.

Thermal Expansion

The thermal expansion characterizes the expansion (or shrinkage) of a material caused by heating (or cooling) and is defined as the expansion (shrinkage) of a unit length of a material when undergoing a rise (drop) of one degree in temperature. The expansion is considered to be positive when the material elongates and is considered negative (shrinkage) when it shortens. This expansion or shrinkage in a material due to temperature changes can be traced through a dilatometric curve. This dilatometric curve is a record of the fractional change of a linear dimension of a solid at a steadily increasing or decreasing temperature. With mathematical symbolism, the dilatometric curve is a plot of

$$\frac{\Delta l}{l_o} \text{ as a function of } T \tag{4.2}$$

where $\Delta \ell = \ell - \ell_0$ and ℓ_0 (m) and ℓ (m) are the changed and initial dimensions of the solid, respectively; the latter usually is taken at room temperature. $\Delta \ell$ reflects not only the linear expansion or shrinkage of the material but also the dimensional effects brought on by possible physicochemical changes (i.e., reactions).

The heating of the solid, as a rule, usually takes place at a predetermined rate of 5°C.min^{-1}. Because the physicochemical changes proceed at a finite rate and some of them are irreversible, a dilatometric curve obtained from the heating of material rarely coincides with that obtained during cooling. A steady rise or decline in the slope of the dilatometric curve can be interpreted due to sluggish reactions. On the other hand, discontinuities in the slope indicate very fast reactions. For example, heating the material at a rate higher than 5°C.min^{-1} usually causes the reactions to shift to higher temperatures and to develop faster.

Thermal expansion of a material is usually defined by a coefficient of linear thermal expansion, α (m.m^{-1}.K^{-1}),

$$\alpha = \frac{1}{l}\frac{dl}{dT} \tag{4.3}$$

Since $\ell = \ell_0$, the coefficient of linear thermal expansion is, for all intents, the tangent to the dilatometric curve. Thus, for solids that are isotropic in a macroscopic sense, the coefficient of volumetric expansion is approximately equal to 3α.

The thermal expansion can be measured with a dilatometric apparatus, and this apparatus is capable of producing curves that show the expansion of the materials over a range from 20 to 1000°C. Usually a small sample size (of 76.2 mm long and about 13 by 13 mm in cross section) subjected to a small spring load is needed to measure thermal expansion and generate dilatometric curve (Maxwell, 1904). Thermal expansion can also be measured using a thermomechanical analyzer (TMA). This equipment monitors thermal expansion properties of a specimen against time or temperature under a specified programmed atmosphere following ASTM E831 or ISO 11359 test procedure (Kodur and Shakya, 2013).

Mass Loss

The mass loss is often used to express the loss of mass at elevated temperatures. The mass loss with temperature is usually measured by means of the thermogravimetric analyzer (TGA) technique (ASTM E1868, 2015). A specimen of known mass is heated at a constant temperature, while its mass is continuously measured as a function of time. At the end of a predetermined time interval or temperature, the mass loss of the specimen is calculated as a percent of the original (initial) mass. The loss of mass can be tracked and recorded through a thermogravimetric curve, and this curve represents the fractional variation in the mass of a solid at steadily increasing or decreasing temperature. The curve can be mathematically symbolized by

$$\frac{M}{M_o} \text{ as a function of } T \tag{4.4}$$

where M and M_0 (kg) are the changed (lost) and original masses of the solid, respectively, and T is the temperature in the material. The original mass is the mass measured at room temperature. A heating rate of 5°C.min^{-1} is to be used for mass loss measurements.

A thermogravimetric curve reflects reactions accompanied by a loss or gain of mass, but, naturally, it does not reflect changes in the materials' microstructure or crystalline order; $M/M_0 = 1$ is the thermogravimetric curve for a chemically inert material. An increase in the rate of heating usually causes those features of the thermogravimetric curve that are related to chemical reactions to shift to higher temperatures and to develop faster.

Figure 4.1 shows the variation in mass loss for concrete in the temperature range from 20 to 1000°C. The thermogravimetric curves shown in this figure were obtained by a Dupont 951 thermogravimetric analyzer (Dupont, 1977), using specimens of 10 to 30 mg in mass, placed in a nitrogen atmosphere (Harmathy, 1983). The rate of temperature rise was 5°C.min^{-1}.

Density and Porosity

The density, ρ (kg.m^{-3}), is the mass of a unit volume of the material, composed of the solid itself and the air-filled pores, in an oven-dry condition. Assuming that the material is isotropic with respect to its dilatometric behavior, its density at any temperature can be calculated from the thermogravimetric and dilatometric curves as follows:

$$\rho = \rho_o \frac{\left(\dfrac{M}{M_o}\right)_T}{\left(\left[1+\dfrac{\Delta l}{l_o}\right]^3\right)_T} \tag{4.5}$$

where ρ_o (kg.m^{-3}) is the density of the solid at the reference temperature (usually room temperature) and the T subscript indicates values pertaining to temperature T in the thermogravimetric and dilatometric curves.

The density of composite solids (such as concrete) at room temperature can also be calculated by means of the mixture rule in its simplest form (Eq. 4.1 with $m = 1$):

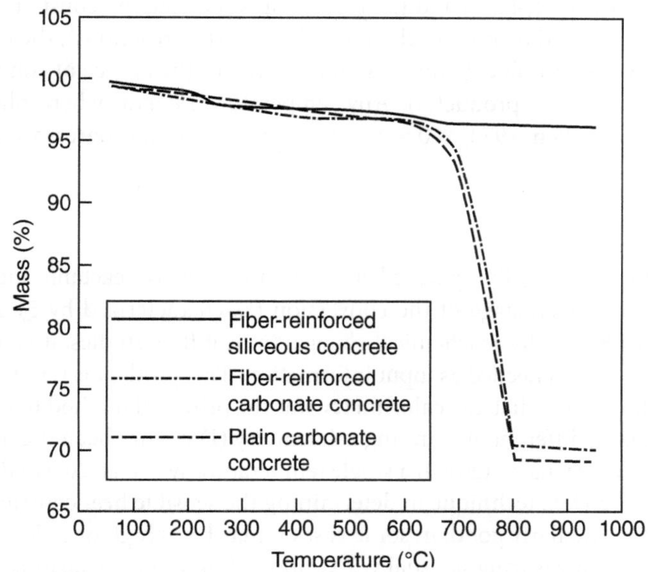

FIGURE 4.1 Mass loss of various concrete types as a function of temperature.

$$\rho = \sum_i v_i \rho_i \tag{4.6}$$

where ν and ρ are volume and density, and i subscript relates to information on the ith component in the mixture (material). Please note that at elevated temperatures, the expansion of the components is subject to constraints, and therefore the mixture rule can yield only a crude approximation.

If, as usual, the composition is given in mass fractions rather than in volume fractions, the volume fractions can be obtained as

$$v_i = \frac{w_i / \rho_i}{\sum_i w_i / \rho_i} \tag{4.7}$$

where w_i is the mass fraction of the ith component (kg.kg^{-1}).

True density, ρ_t (kg.m^{-3}), is the density of the solid in a poreless condition. Such a condition is nonexistent for many building materials and therefore may be a theoretical value derived on crystallographic considerations, or determined by some standard technique, for example, ASTM C135 (ASTM Test Method C135±86, 2007). The porosity (P) of a material is the fraction of its bulk volume occupied by voids. The relationship between the porosity and density is

$$P = \frac{\rho_t - \rho}{\rho_t} \tag{4.8}$$

The overall porosity of a composite material consisting of porous components is evaluated as

$$P = \sum_i v_i P_i \tag{4.9}$$

where, again, the i subscript relates to the ith component of the material.

The pore size of the voids in the composite material can be determined by mercury intrusion porosimetry method (Rootare, 1970). This method involves evacuating the gas from the pores and then forcing mercury into a sample by gradually increasing the pressure on the mercury.

Specific Heat

The specific heat of a material describes the amount of heat required to raise a unit mass of the material by a unit temperature. A calorimetric curve describes the variation with temperature of the apparent specific heat of a material at constant pressure, c_p (J·kg^{-1}·K^{-1}). The apparent specific heat is defined as

$$c_p = \frac{\delta h}{\delta T_p} \tag{4.10}$$

where h is enthalpy (J·kg^{-1}) and the p subscripts indicate the constancy of pressure. If the heating of the solid is accompanied by physicochemical changes (i.e., reactions), the enthalpy becomes a function of the reaction progress variable, ξ (dimensionless), that is, the degree of conversion at a particular temperature from reactant(s) into product(s). Any temperature interval where physicochemical change takes place (Harmathy and Allen, 1973) is $0 \leq \xi \leq 1$, and the reaction progress variable is

$$c_p = \bar{c}_p + \Delta h_p \frac{d\xi}{dT} \tag{4.11}$$

where c_p (J·kg^{-1}·K^{-1}) is the specific heat for that mixture of reactants and (solid) products that the material consists of at a given stage of the conversion (as characterized by ξ) and Δh_p (J·kg^{-1}) is the latent heat associated with the physicochemical change. In heat flow studies, it is usually the ρc_p product (J·m^{-3}·K^{-1}) rather than c_p that is needed as input information. This product is referred to as volumetric specific heat.

Until the 1980s, adiabatic calorimetry was the principal method to study the relationship of c_p versus T. Since the 1980s, differential scanning calorimetry (DSC) has been the most commonly used technique for mapping the c_p versus T curve in a single temperature sweep at a desired rate of heating. Unfortunately, the accuracy of the DSC technique in determining the sensible heat contribution to the apparent specific heat may not be particularly good; in some cases it may be as high or as low as ±20%. The rate of temperature rise adopted in measuring is usually 5°C·min^{-1}, but at higher heating rates, the peaks in the DSC curve tends to shift to higher temperatures and become sharper. For specific heat measurements temperatures above 600°C, a high-temperature differential thermal analyzer (DTA) can also be used. Harmathy (1983), with the aid of a Dupont 910 differential scanning calorimeter, developed calorimetric curves for a number of materials by placing the samples, 10 to 30 mg in mass, in a nitrogen atmosphere.

Figure 4.2 shows variations in apparent specific heat at temperature intervals due to physicochemical instability. The apparent specific heat consists of sensible heat and latent heat contributions. The latter contribution will result in extremities in the calorimetric curve. These extremities will be a maximum if the reaction is endothermic and a minimum if it is exothermic.

In recent years, the hot disk method is utilized to measure simultaneously three heat transfer properties (thermal conductivity, thermal diffusivity, and specific heat) in a single test, unlike other existing test apparatus which requires separate equipment for measuring thermal conductivity and specific heat. The hot disk thermal analyzer utilizes the transient plane source (TPS) technique and uses a double-spiral-shaped sensor element to measure the specific heat of materials (Kodur and Khaliq, 2011). Thermal properties can be measured as per ISO 22007-2 procedure (ISO 22007-2, 2015).

Materials that undergo exothermic reactions may yield negative values in the calorimetric curve. A negative value for c_p indicates that, at the applied (and enforced) rate of heating, the rate of evolution of reaction heat exceeds the rate of absorption of sensible heat by the material. In natural processes, the apparent specific heat can never be negative, because the heat evolving from the reaction is either scattered to the surroundings or, if absorbed by the material, causes a very fast temperature rise. If the heat of reaction is not very high, obtaining nonnegative values for c_p can be achieved by suitably raising the scanning rate. For this reason, some materials undergoing exothermic reactions must be tested at rates of heating higher than 5°C·min^{-1}, often as high as 50°C·min^{-1}.

It should be noted that if experimental (measured) information is not available, the c_p versus T relationship can be calculated from data on heat capacity and heat of formation for all the components of the material (including reactants and products), tabulated in a number of handbooks (Eitel, 1952; Perry, 1953). Examples of calculations for deriving specific heat are presented in Harmathy (1993), where information was developed for the apparent specific heat versus temperature relation for a cement paste and four kinds of concrete. In addition, Eurocodes 2 and 3 provide constitutive relationship of specific heat for various types of concrete, reinforcing steel, structural steel, and wood. These relationships are further discussed in Chaps. 7, 8, and 10 of this book.

Thermal Conductivity

The temperature rise in a member, as a result of heat flow, is a function of the thermal conductivity of the material. Heat transmission solely by conduction can occur only in poreless, nontransparent solids. In porous solids (most building materials), the mechanism of heat transmission is a combination of conduction, radiation, and convection. (If pore size is less than about 5 mm, the contribution of pores to convective heat transmission is negligible.) The thermal conductivity of porous materials is a convenient empirical factor that makes it possible to describe the heat transmission process with the aid of the Fourier law. This empirical factor will depend not only on the conductivity of the solid matrix but also on the

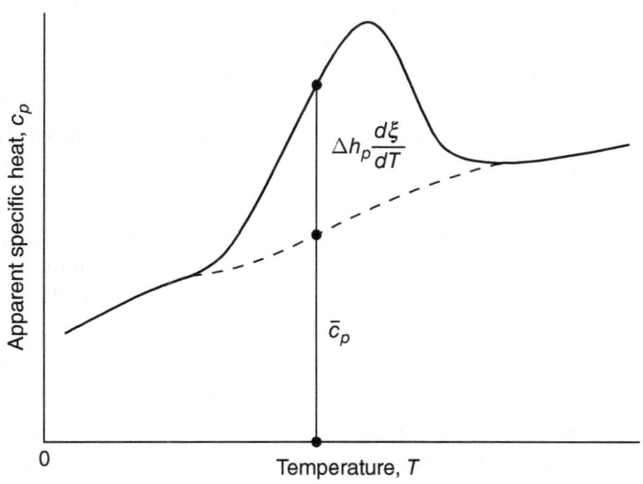

Figure 4.2 The apparent specific heat.

porosity of the solid and the size and shape of the pores. At elevated temperatures, because of the increasing importance of radiant heat transmission through the pores, conductivity becomes sensitive to temperature gradient.

Because measured values of the thermal conductivity depend, to some extent, on the temperature gradient employed in the test, large discrepancies may be found in thermal conductivity data reported by various laboratories. A thermal conductivity value yielded by a particular technique is applicable only to heat flow patterns similar to that characteristic of the technique employed. Among various methods, the hot wire method (ASTM C1113, 2013) and the guarded heat flow meter method (ASTM E1530, 2016) are commonly used for measuring thermal conductivity of various materials. The hot wire method is used to measure the thermal conductivity of "refractories" such as insulating bricks, concrete, and powder or fibrous materials. Because it is basically a transient radial flow technique, isotropic specimens are required. In the guarded heat flow meter method, a small sample of the material to be tested is held under a compressive load between two polished metal surfaces, each controlled at a different temperature. The lower surface is part of a calibrated heat flux transducer. As heat flows from the upper surface through the sample to the lower surface, an axial temperature gradient is established in the stack. By measuring the temperature difference across the sample along with the output from the heat flux transducer, thermal conductivity of the sample can be determined with known thickness. These two test methods for measuring thermal conductivity can also be applied for measuring conductivity at higher temperatures up to 850°C and above, but measuring at higher temperatures becomes cumbersome and costly. The thermal conductivity can be measured following ISO 22007-2 procedure and using the hot disk thermal analyzer described earlier.

Experimental data indicate that porosity is not a greatly complicating factor as long as it is not more than about 0.1. With most insulating materials (such as some types of gypsum), however, the porosity may be 0.8 or higher. Conduction through the solid matrix may be an insignificant part of the overall heat transmission process; therefore, using the Fourier law of heat conduction in analyzing heat transmission may lead to deceptive conclusions. If the solid is not oven-dry, a temperature gradient will induce migration of moisture, mainly by an evaporation condensation mechanism (Harmathy, 1969). The migration of moisture is usually, but not necessarily, in the direction of heat flow and manifests itself as an increase in the apparent thermal conductivity of the solid. Furthermore, even oven-dry solids may undergo decomposition (mainly dehydration) reactions at elevated temperatures. The sensible heat carried by the gaseous decomposition products as they move in the pores adds to the complexity of the heat flow process. At present there is no way of satisfactorily accounting for the effect of simultaneous mass transfer on heat flow processes occurring under elevated temperature conditions.

The thermal conductivity of layered, multiphase solid mixtures depends on whether the phases lie in the direction of, or normal to, heat flow and is determined using the simple mixture rule (DeVries, 1952). At higher temperatures, because of radiative heat transfer through the pores, the contribution of the pores to the thermal conductivity of the solid must not be disregarded (Kingery, 1960).

The thermal conductivity of solids is a structure-sensitive property. For crystalline solids, the thermal conductivity is relatively high at room temperature and gradually decreases as the temperature rises. On the other hand, for predominantly amorphous solids, the conductivity is low at room temperature and increases slightly with the rise of temperature. The conductivity of porous crystalline materials may also increase at very high temperatures (>600°C) because of the radiant conductivity of the pores.

The thermal conductivity of materials such as concrete or brick can be measured, in the temperature range between 20 and 800°C, using a non-steady-state hot wire method or hot disc method as per ISO 22007-2 procedure. The thermal conductivity values at discrete temperature levels can be plotted to obtain a curve. Both the steady-state and variable-state techniques of measuring thermal conductivity require the stabilization of a pattern of temperature distribution (and thereby a certain microstructural pattern) in the test sample prior to the test. The test results can be viewed as points on a continuous thermal conductivity versus temperature curve obtained by an imaginary scanning technique performed at an extremely slow scanning rate. Because each point pertains to a more or less stabilized microstructural pattern, there is no way of knowing how the thermal conductivity would vary in the course of a physicochemical process developing at a finite rate and varying microstructure.

On account of the nonreversible microstructural changes brought about by heating, the thermal conductivity of building materials (and perhaps most other materials) is usually different in the heating and cooling cycles. Open and solid circles are used in the figures to identify thermal conductivity values obtained by stepwise increasing and stepwise decreasing the temperature of the sample, respectively. Also, often the thermal conductivity of a material is taken as invariant with respect to the direction of heat flow.

Thermal Diffusivity

The thermal diffusivity of a material is defined as the ratio of thermal conductivity to the volumetric specific heat of the material. It measures the rate of heat transfer from an exposed surface of a material to the inside. The larger the diffusivity, the faster the temperature rise at a certain depth in the material. Similar to thermal conductivity and specific heat, thermal diffusivity varies with temperature rise in the material. Thermal diffusivity, ϑ, can be calculated using the relation

$$\vartheta = \frac{k}{\rho c_p} \tag{4.12}$$

where k = Thermal conductivity
ρ = Density
c_p = Specific heat of the material

Thermal Inertia

Thermal inertia is defined as the responsiveness (sensitivity) of a material to variations in temperature (i.e., the degree of rapidness or slowness with which the temperature of a body approaches that of its surroundings). Thermal inertia of a material is dependent on its absorptivity, its specific heat, its thermal conductivity, and its dimensions, among other factors. For simplicity, the thermal inertia of a material is expressed as the square root of the product of the material's bulk thermal conductivity (k) and volumetric heat capacity (ρc_p), where the latter is the product of density (ρ) and specific heat capacity (c_p). Thermal inertia is a useful parameter to gauge the thermal resistance of a material to temperature variation.

$$I = \sqrt{k \rho c_p} \tag{4.13}$$

4.3.2 Mechanical Properties

The mechanical properties that govern the fire performance of a structural member are stress-strain relations, modulus of elasticity, ductility, and strength of the constituent materials, and these properties vary as a function of temperature.

There are three steady-state test methods to determine high-temperature material properties: stressed, unstressed, and residual (Fu-Ping et al., 2004; Phan, 1996). Figure 4.3 illustrates the main difference in these three test methods to evaluate mechanical properties. In the case of the stressed test method, specimens are preloaded to a certain load, usually 30% to 60% of the room-temperature ultimate compressive strength of concrete. Specimens are then exposed to the desired temperature and further loaded to failure.

In the unstressed test method, the test specimen is heated to the target temperature without any preload. Once equilibrium conditions are met and the target temperature is reached throughout, the specimen is loaded in increments till failure. In the case of the residual test method, a test specimen is heated to a target temperature till reaching steady-state, and then allowed to cool down to room temperature. Specimen is then loaded to failure at ambient temperature to obtain the residual strength.

Stress-Strain Relations

The mechanical properties of solid materials are derived from conventional tensile or compressive strength tests. The strength properties are usually expressed in terms of stress-strain relations, which form input data into mathematical (or numerical) models for evaluating fire resistance of structural members. Figure 4.4 shows the variation in stress, σ (Pa), with increasing strain (deformation), ε ($m \cdot m^{-1}$), for a metallic material, while it is strained (deformed) in a tensile test at a more or less constant loading rate (i.e., constant crosshead speed), usually of the order of 1 $mm \cdot min^{-1}$. Generally, because of a decrease in the strength and increase in ductility of the material, the slope of the stress-strain curve decreases with increasing temperature.

Tests for evaluating high-temperature strength property are carried out mainly either in a transient or a steady state. In transient-state tests, the specimen is subjected to a constant load and then exposed to uniform increasing temperature. In this transient-state test, the heating rate has a great influence on the strain rate and thus different heating rates produce different strain rates in the material. In steady-state tests, the specimen is heated to a specific target temperature, and after that, a load is applied to evaluate. Steady-state tests are generally faster and easier to conduct than transient-state tests. Stress and strain values are recorded continuously under constant temperature. Further, the test specimen can be subjected

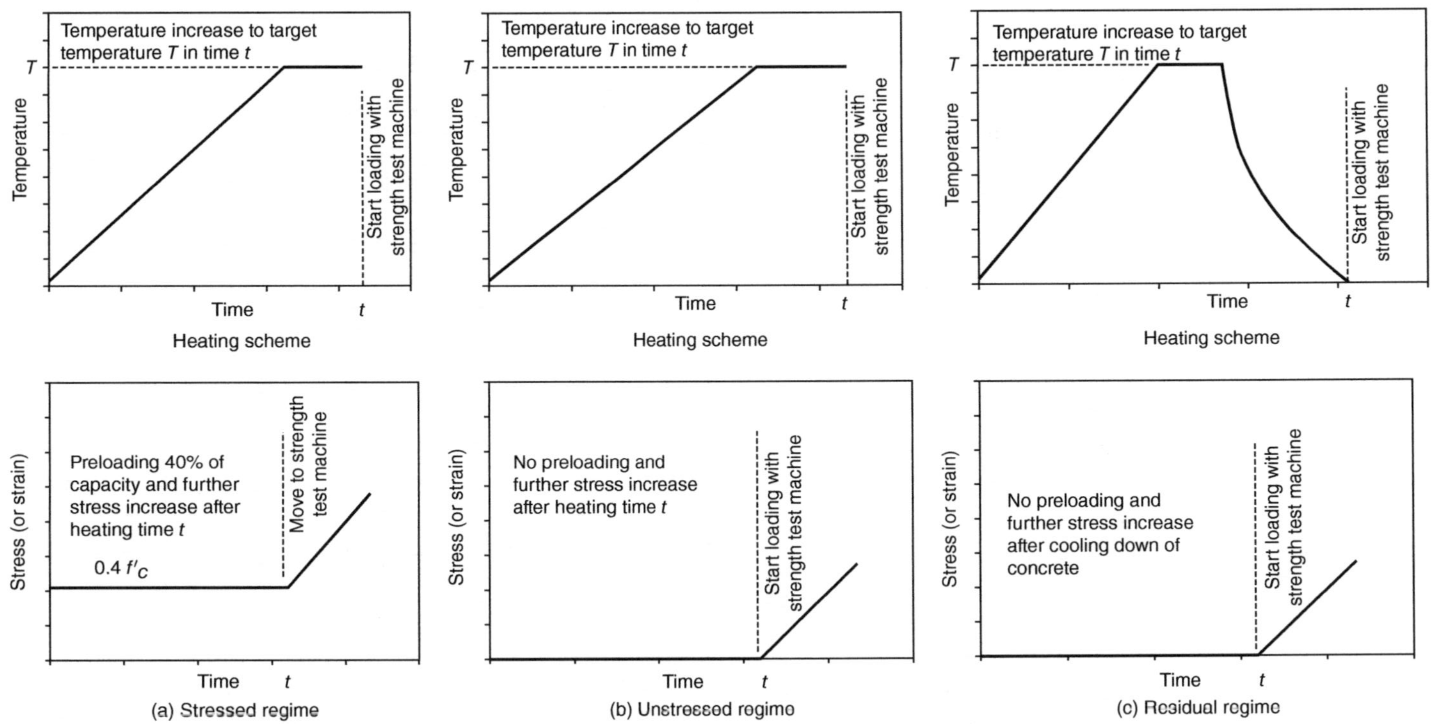

Figure 4.3 Various testing regimes for evaluating mechanical properties at elevated temperatures.

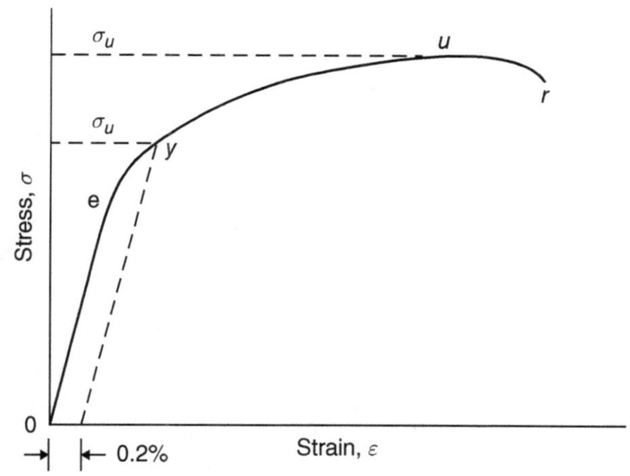

FIGURE 4.4 Stress-strain curve (strain rate is roughly constant).

to failure either through the load-controlled (loading rate is constant) or through the deformation-controlled technique (strain rate is constant) (Anderberg, 1978). Despite the fact that the strain rate has a significant effect on test results, a large amount of test data represented in the literature does not provide any information on strain rates for the tests that were carried out. Therefore, test standards are still concerned with defining limits for strain rates in tests (Anderberg, 1978; Cooke, 1988).

There is limited guidance on test procedures for undertaking mechanical properties at elevated temperatures. Often room temperature test methods are extended to measure elevated temperature properties. Recently, RILEM International Union of Laboratories and Experts in Construction Materials, Systems and Structures (RILEM is from the name in French) has provided test parameters and procedures for mechanical property tests of concrete within temperature ranges of 20 to 750°C. The strength at high temperatures can be determined either in the hot state (at maximum attained temperature) or in the residual state (after cooling down to ambient temperature). Various mechanical properties such as modulus of elasticity, yield strength, ultimate strength, and ultimate strain can be derived from the stress-strain tests carried out in any of the three regimes shown in Fig. 4.3.

Modulus of Elasticity, Yield Strength, Ultimate Strength

The modulus of elasticity is a measure of the ability of the material to resist deformation and is defined as the ratio of the deforming stress to the strain in the material. The modulus of elasticity of a material decreases with increasing temperature and is usually obtained from the slope of a stress-strain curve.

The tensile or compressive strength of the material is expressed by means of its yield strength or ultimate strength. Often the strength at an elevated temperature is expressed as a percentage of the compressive (or tensile) strength at room temperature. Figure 4.5 shows the variation in strength with temperature (ratio of strength at elevated temperature to that at room temperature) for concrete, steel, wood, and FRP. For all four materials, the strength decreases with increasing temperature; however, the rate of strength loss is different due to the different composition and characteristics of these materials.

For materials such as concrete, compressive strength is of main interest, because concrete has very limited tensile strength at room and at higher temperatures. However, for materials such as steel and wood, both compressive and tensile strengths are of equal interest in fire resistance evaluation.

Section 0-*e* of the curve in Fig. 4.4 represents the elastic deformation of the material, which is instantaneous and reversible. The modulus of elasticity, *E* (Pa), is the slope of that section. Between points *e* and *u* the deformation is plastic, nonrecoverable, and quasi-instantaneous. The plastic behavior of the material is characterized by the yield strength at 0.2% offset, σ_y (Pa), and the ultimate strength, σ_u (Pa). After some localized necking (i.e., reduction of cross-sectional area in a tensile test), the test specimen ruptures at point *r*. The modulus of elasticity is said to be a structure-insensitive property.

For metals of similar metallurgical characteristics, the stress-strain curve can be reproduced at room temperature at a reasonable tolerance, and the shape of the curve does not depend significantly on the crosshead speed. However, at sufficiently high temperatures (above 300°C), the material undergoes plastic deformation, even at constant stress, and the *e*-*r* section of the stress-strain curve will depend markedly on the crosshead speed.

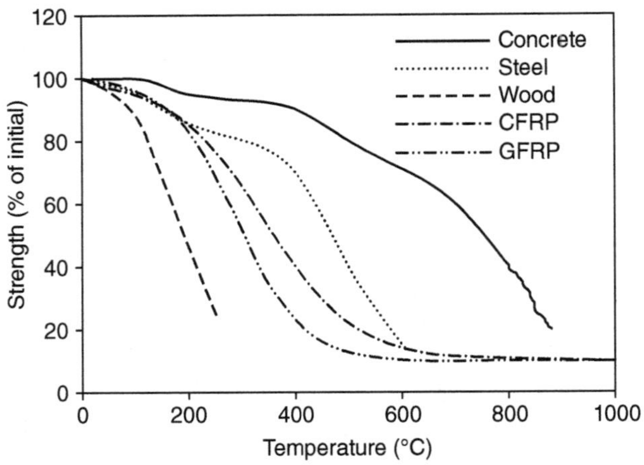

FIGURE 4.5 Variation in strength with temperature for different materials.

The modulus is often measured as the tangent modulus to the initial linear elastic portion of the stress-strain curve. At elevated temperatures, it is difficult to accurately measure strains due to the complexities with high-temperature strain gauges. This is often overcome by utilizing the dynamic method (ASTM E1876, 2015), which can compute the elastic modulus of materials at elevated temperatures. Metal (steel) specimens possess specific mechanical resonant frequencies that are determined by the elastic modulus, mass, and geometric property of test specimens. Therefore, the dynamic elastic properties of steel can be evaluated if the geometry, mass, and mechanical resonant frequencies of a steel specimen are measured. Dynamic elastic modulus is determined using the resonant frequency in either the flexural or the longitudinal mode of vibration. The accuracy and feasibility of this dynamic method have been validated by Wang et al. (2012).

4.3.3 Deformational Properties

Deformation properties are highly dependent on the chemical composition, the type of aggregate, and the chemical and physical reactions that may occur in the material during heating (Schneider, 1985). Deformation properties include creep strain and transient strain, and the temperature has significant influence on these properties.

Creep

Creep, often referred to as creep strain, is defined as the time-dependent plastic deformation of the material and is denoted by ε_t (m·m^{-1}). At normal stresses and ambient temperatures, the deformation due to creep is not significant. The adverse effect of creep can be apparent at higher stress levels and at elevated temperatures. At these conditions, the rate of deformation caused by creep effects can be substantial (Kodur and Dwaikat, 2010). Hence, the main factors that influence creep are temperature, stress level, and duration of temperature exposure (Hurley et al., 2015).

In a creep test, the variation in ε_t is recorded against time, t (h), at constant stress (more accurately, at constant load level) and at constant (elevated) temperature T (K). A typical strain-time curve is shown in Fig. 4.6*a*. The total strain, ε (m·m^{-1}), is evaluated as

$$\varepsilon = \frac{\sigma}{E} + \varepsilon_t \tag{4.14}$$

Figure 4.6*a* shows that there are three dominant stages in the creep stress-strain response of a material. In this figure, the initial (*0-e*) section of the strain-time curve represents the instantaneous elastic (and reversible) part of the curve; the rest is creep, which is essentially nonrecoverable. The creep in a material can be grouped under three stages. The creep increases at a fast rate at first (primary creep, section e-s_1 in Fig. 4.6*a*), then proceeds for a long time at an approximately constant rate (secondary creep, section s_1-s_2), and finally accelerates at a rapid rate until rupture occurs (tertiary creep, section s_2-r). The curve becomes steeper (creep strain amplifies) if the test is conducted either at a higher load (stress) level or at a higher temperature.

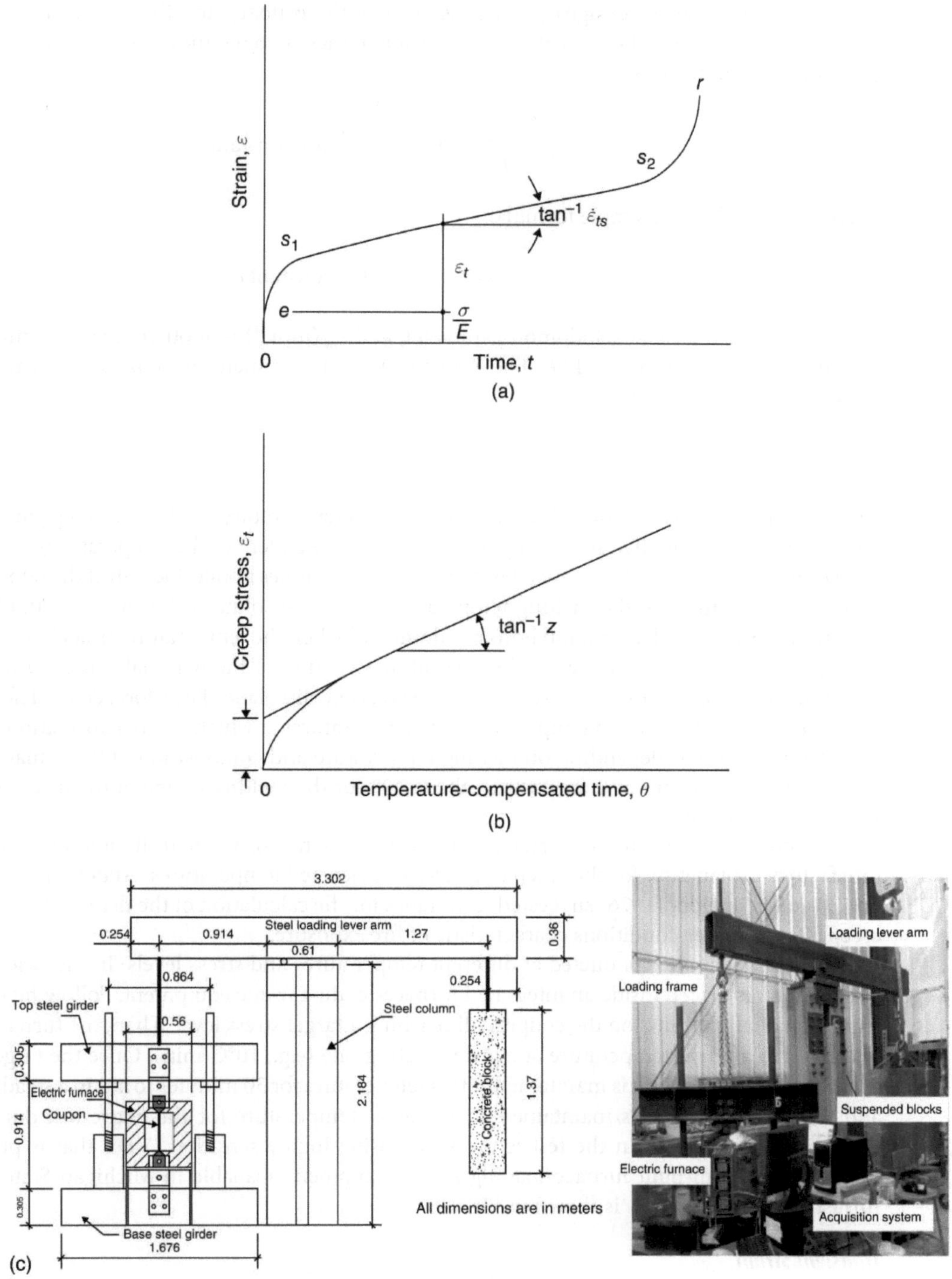

FIGURE 4.6 (a) Creep strain vs. time curve (T = constant; $\sigma \approx$ constant); (b) creep strain vs. temperature-compensated time curve ($\sigma \approx$ constant); (c) custom-built creep test equipment for high-temperature creep tests.

Dorn's (1955) concept is particularly suitable for dealing with deformation processes developing at varying temperatures. Dorn eliminated the temperature as a separate variable by the introduction of a new variable: the "temperature-compensated time," θ (h), defined as

$$\theta = \int_0^t e^{\Delta H_c / RT} dt \tag{4.15}$$

where ΔH_c (J·kmol^{-1}) is the activation energy of creep, and R (J·kmol^{1}·K^{-1}) is the gas constant.

From a practical (or design) point of view, only the primary and the secondary creep strain are of importance. It has been shown that the creep strain in these two regimes can be described by the following equation (Harmathy, 1967):

$$\varepsilon_t = \frac{\varepsilon_{to}}{ln2} cosh^{-1}\left(2^{Z\theta/\varepsilon_{to}}\right), \text{ for constant } \sigma \tag{4.16}$$

or approximated by the simple formula:

$$\varepsilon_t \approx \varepsilon_{to} + Z\theta, \text{ for constant } \sigma \tag{4.17}$$

where Z (h^{-1}) is the Zener-Hollomon parameter, and ε_{t0} (mm^{-1}) is another creep parameter, the meaning of which is explained in Fig. 4.6*b*. The Zener-Hollomon parameter is defined as (Zener and Hollomon, 1944):

$$Z = \varepsilon_{ts} e^{\Delta \dot{H}_C/RT} \tag{4.18}$$

where $\dot{\varepsilon}_{ts}$ ($mm^{-1}\cdot h^{-1}$) is the rate of secondary creep at a temperature T. The two creep parameters, Z and ε_{t0}, are functions of the applied stress only (i.e., they are independent of the temperature).

For most materials, 50% of stress level under creep becomes noticeable only if the temperature is higher than about one-third of the melting temperature (on the absolute scale). As an example, in the case of structural steel, creep deformation becomes dominant when the temperature in steel exceeds about 500°C. At high temperatures, vacancies in the crystalline structure of the material can diffuse into dislocation. This causes the dislocation to move faster to an adjacent slip plane. Therefore, creep deformations in steel accelerate with an increase in temperature. The temperatures at which creep deformations accelerate differ in different materials depending on melting temperature and composition of these materials. In general, creep can be significant at temperatures above 40% of the melting temperature of the material (0.4 T_m), where T_m is in kelvins.

The creep of concrete or mortar is due to the presence of water in its microstructure. There is no satisfactory explanation for the creep of concrete at elevated temperatures. Anderberg and Thelandersson (1976) and Schneider (1976) suggested techniques for the calculation of the deformation of concrete (with creep effects) under conditions characteristic of fire exposure.

Creep tests can be conducted at different temperatures and stress levels. In the case of steel material, steel coupon is placed inside an integrated furnace-loading frame equipment. Following this, a predefined load is gradually applied on the coupon till reaching a target stress level. Then, the furnace is turned on to be heated to a target temperature at a given heating rate (e.g., 10°C/min). Once the target temperature is attained, the steel coupon is maintained at this temperature for 30 minutes to attain a steady-state condition and then the specimen is maintained at the target temperature for the entire test duration. The creep displacement generated in the test is measured using highly sensitive LVDT that is placed outside the furnace. A custom-built furnace-loading frame equipment available in Michigan State University Civil Infrastructure Laboratory is shown in Fig. 4.6*c*.

Transient Strain

Transient strain in cementitious materials (such as cement) occurs during the first-time heating of concrete and is independent of time. It is essentially caused by thermal incompatibilities between the aggregate and the cement paste (Kodur, 2014). Transient strain of concrete, similar to that of high-temperature creep, is a complex phenomenon and is influenced by factors such as temperature, moisture content, loading, and mix proportions.

A review of literature indicates that there is very limited information on creep and transient strain of concrete at elevated temperatures (Kodur et al., 2014). However, based on the limited available data in literature, Anderberg and Thelandersson (1976) developed constitutive models for creep and transient strain in concrete at elevated temperatures. According to Anderberg and Thelandersson, the creep strain and transient strain of concrete at elevated temperatures are given by the following two equations, respectively:

$$\varepsilon_{cr} = \beta_1 \frac{\sigma}{f_{c,T}} \sqrt{t}\, e^{d(T-293)} \tag{4.19}$$

$$\varepsilon_{tr} = k_2 \frac{\sigma}{f_{c,20}} \varepsilon_{th} \tag{4.20}$$

where ε_{cr} = Creep strain
ε_{tr} = Transient strain
$\beta_1 = 6.28 \times 10^{-6}\ s^{-0.5}$
$d = 2.658 \times 10^{-3}\ K^{-1}$
T = Concrete temperature (°K) at time t (s)
$f_{c,T}$ = Concrete strength at temperature T
σ = Stress in the concrete at the current temperature
k_2 = A constant ranges between 1.8 and 2.35
ε_{th} = Thermal strain
$f_{c,20}$ = Concrete strength at room temperature

A recent study reviewed a number of transient strain (and creep) models and showed how high-temperature creep strains increase significantly at temperatures above 500°C and stress and loading levels above 50%. Temperature-induced creep strain has significant influence on the resulting axial displacements in concrete, specifically at latter stages of fire exposure.

4.3.4 Material-Specific (Special) Properties

Material-specific properties refer to specific properties that occur in certain construction materials at both room and elevated temperature. However, these properties can be predominant at elevated temperatures. The material-specific properties include spalling in concrete, local buckling in steel, charring in wood, and delamination in wood or FRP.

Critical Temperature

In building materials, such as steel and FRP, the determination of failure of a structural member exposed to fire is often linked to the *critical temperature* of the material the structural member is made of. An example of this is failure of steel structural members (columns or beams), which is linked to the critical temperature in steel (usually around 550°C). This temperature corresponds to a 50% loss of yield strength in steel.

The critical temperature is defined as the limiting temperature in the material at which the structural member loses much of its design load-carrying capacity and can no longer support the applied load. When this critical temperature is reached in the material, the safety factor for the structural member against failure under strength limit state becomes less than 1.

North American standards (ASTM E119, 2016) assume a critical or failure temperature of 538°C for structural steel. It is a typical failure temperature for columns under full design load (representing dead plus live load). This temperature is taken as the failure temperature in the calculation of fire resistance of steel members. However, in actuality, the strength failure in a structural member may occur at a higher or lower steel temperature depending on the load intensity.

This concept of critical temperature is also used for reinforced and prestressed steel in concrete structural members for evaluating fire resistance. The fire resistance ratings in concrete members are generally obtained through the provision of minimum member dimensions and minimum thickness of concrete cover. The minimum concrete cover thickness requirements are intended to ensure that the temperature in the reinforcement does not reach its critical temperature for the required duration. For reinforcing steel, the critical temperature is taken as 593°C, whereas for prestressing steel the critical temperature is 426°C and these temperatures represent the limit for 50% of strength loss (Lie, 1992).

The critical temperature concept is also used for FRP reinforcement, which is used as internal reinforcement or as external reinforcement for strengthening applications. Since FRP composites comprise two key elements, the fibers and thermosetting polymer matrix, the properties of such materials depend on the type and composition of FRP. The critical temperature in FRP is also defined as the temperature at which FRP loses 50% of its initial strength. This temperature is in the range of 250 to 325°C depending on the types of fiber used in the FRP (Yu and Kodur, 2014).

Fire-Induced Spalling

Fire-induced spalling is defined as the breakage of chunks (or pieces) of concrete from the surface of concrete structural member exposed to high and rapidly rising temperatures, such as those experienced during a fire event. Spalling can occur in the initial stages of fire exposure and this often occurs as violent *mini explosions*. Spalling can also occur when large cracks develop, resulting from fire exposure, wherein pieces fall off the surface. The consequences of spalling may be limited as long as the extent of the damage is small, but extensive spalling may lead to early loss of stability and integrity of the member. Unlike other properties, spalling is hard to measure during fire incidents or fire tests but is often quantified through volumetric loss in the member following fire tests (or actual fire incidents).

Although spalling might occur in all concretes, high-strength concrete (HSC) is more susceptible than conventional normal-strength concrete (NSC), because of its low permeability and low water-cement ratio. Hence, spalling is one of the major concerns in new types of concrete such as high-strength concrete HSC and UHPC (strength in the range of 70 to 120 MPa), and should be properly accounted for in evaluating fire performance. Based on review of a number of fire tests on HSC specimens, it has been reported that spalling can often be of explosive nature (Danielsen, 1997). Spalling in typical NSC and HSC columns is compared in Fig. 4.7 using the data obtained from full-scale fire tests on loaded columns (Kodur and Sultan, 1998a). It can be seen that the HSC column experiences significant fire-induced spalling, while the extent of spalling in NSC column is almost nil.

Spalling is believed to be caused by the buildup of pore pressure during heating. The extremely high water vapor pressure, generated during exposure to fire, cannot escape due to the high density (and low permeability) of HSC, and this pressure buildup often reaches the saturation vapor pressure. At 300°C, the pressure reaches approximately 8 MPa; such internal pressures are often too high to be resisted by the HSC mix having a tensile strength of approximately 5 MPa. The drained conditions at the heated surface, and the low permeability of concrete, lead to strong pressure gradients close to the surface in the form of the so-called "moisture clog" (Anderberg, 1997). When the vapor pressure exceeds the tensile strength of concrete, chunks of concrete fall off from the structural member. Additional discussion on spalling phenomenon, together with different types of spalling and influence on fire resistance of structural members, is provided in Chap. 8.

Charring

Charring is the process of formation of a layer of char at the fire-exposed surface of a wood member during exposure to fire. The charring process also occurs in other members, such as FRP and some types of plastics, when subjected to heating. When exposed to heat, wood undergoes thermal degradation (pyrolysis), the conversion of wood to char and gas, resulting in a reduction in the density of the wood. The temperature at which charring occurs in wood is between 280 and 300°C in most species of wood (ASCE, 1992). The charred layer does not have any strength left. Therefore, fire resistance of a wood structural member depends on the extent of cross-section loss due to charring; that is, the remaining strength in the uncharred portion of wood. Charring is influenced by a number of parameters, the most important ones being density, moisture content, and contraction of wood. The influence of the moisture content and density of the wood on the charring rate is illustrated in Fig. 4.8 for Douglas fir (wood species) exposed to

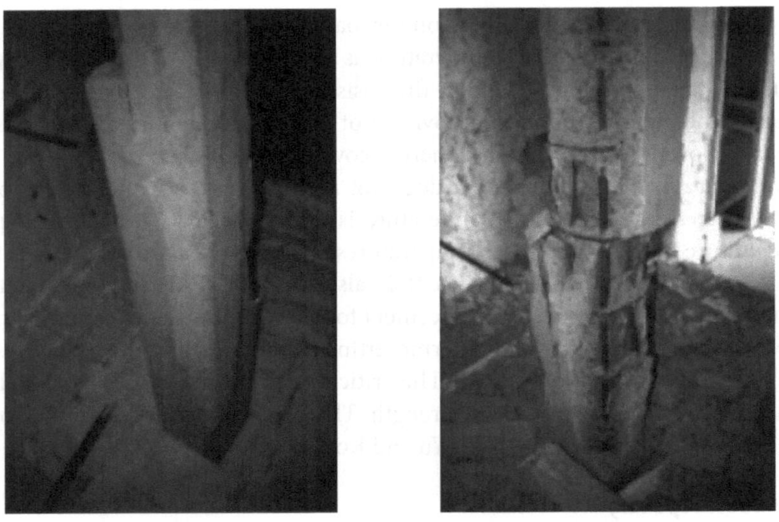

(a) (b)

Figure 4.7 Spalling in NSC and HSC columns after exposure to fire (Kodur and Sultan, 1998a): (a) normal-strength concrete column and (b) high-strength concrete column.

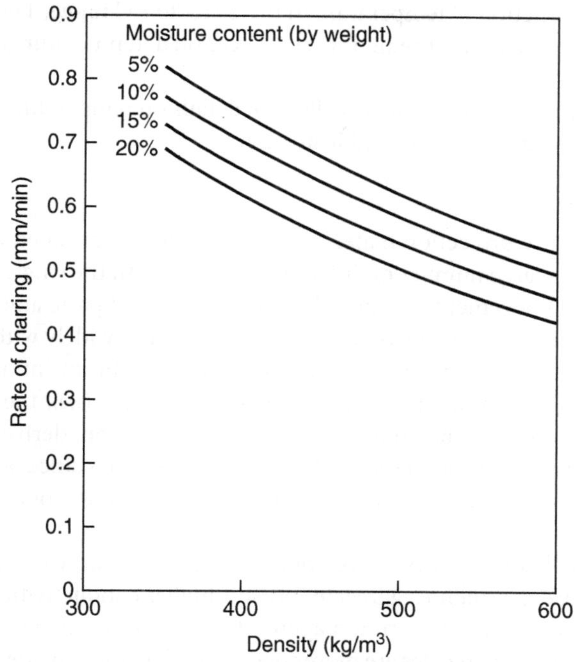

Figure 4.8 Rate of charring in Douglas fir as a function of its density (dry condition) for various moisture contents when exposed to ASTM standard fire.

a standard fire. It can be seen that the charring rate decreases with increasing density of the wood and also with increasing moisture content.

The charring rate, a critical parameter in determining the fire resistance of a wood structural member, is defined as the rate at which wood is converted to char (in millimeters per minute). The charring rate is often measured through cone calorimeter or Ohio State University (OSU) heat release rate apparatus. When a wood product burns at a constant rate of heat release per unit area, the boundary between the pyrolysed material and the intact wood, that is, the pyrolysis front, proceeds to the wood in depth direction. Since all pyrolysing wood can be considered to char, the charring rate, β, corresponds to the propagation rate of the pyrolysis front. Charring rate decreases with increasing density. The average rate of charring transverse to the grain is approximately 0.65 mm/min. The charring rate parallel to the grain of wood is approximately twice the rate when it is transverse to the grain. Further discussion on charring, including charring rates for different species of wood, is provided in Chap. 10.

4.4 Properties of Building Materials

Information on the properties of building materials at elevated temperatures is scattered throughout the literature. There are a few publications, however, that may be particularly valuable for fire safety design professionals. A book by Harmathy (1961); the ASCE manual on structural fire protection (ASCE, 1992) and Eurocodes 2 (CEN, 2004a), 3 (CEN, 2005a), 4 (CEN, 2005b), and 5 (CEN, 2004b); and Chap. 9 of the SFPE handbook (Kodur and Harmathy, 2016) present a wealth of information on concrete, steel, wood, brick, gypsum, and various plastics. The thermal properties of 31 building materials are surveyed in an NRCC report by Harmathy (1983) and review article by Kodur et al. (2010). The mechanical and thermal properties of concrete are discussed in an (ACI) guide (ACI, 1983), in reports by Bennetts (1981), and Schneider (1985), and in another review paper by Kodur et al. (2010). Those of steel are surveyed in the ACI guide, in Bennetts's report, and in reports by Anderberg (1983) and Kodur (2014). Information on the thermal conductivity of more than 50 rocks (potential concrete aggregates) is presented in a paper by Birch and Clark (1940). Information on thermal properties of insulation materials can be found in a report by NIST and that on spray-applied fire-resistive material (SFRM) can be found in the work of Shakya and Kodur (2013) and Arablouei and Kodur (2016). The relations for thermal and mechanical properties, at elevated temperatures, for some building materials are listed in the ASCE structural fire protection manual as well as various chapters of this book, namely Chaps. 7, 8, 9, 10, and 11. In most cases these properties are expressed as a function of temperature in the 20 to 800°C range. These values can be used as input data in mathematical models for predicting cross-sectional temperatures and fire performance of structural members.

The following sections provide collective details on temperature-dependency of concrete, steel, wood, masonry, composites, gypsum, and insulation.

4.4.1 Concrete

Concrete, from fire protection materials points of view, is grouped under Group L/I materials. The word *concrete* covers a large number of different materials, with the single common feature that they are formed by the hydration of cement. Because the hydrated cement paste amounts to only 24 to 43 volume% of the constituents present, the properties of concrete may vary widely with the aggregates used.

The aggregate type and moisture content have significant influence on the specific heat of concrete. The usual range of variation in the volume-specific heat (i.e., the product ρc_p) for normal-weight and lightweight concretes is shown in Fig. 4.9. This information, derived by combining thermodynamic data with thermogravimetric observations (Harmathy, 1993), has since been confirmed by differential scanning calorimetry. Experimental data are also available on few concretes and some of their constituents (Harmathy, 1970).

The thermal conductivity (k) of concrete depends mainly on the nature of its aggregates. Concretes made with dense, crystalline aggregates possess higher conductivities than those made with amorphous or porous aggregates. Among common aggregates, quartz has the highest conductivity; therefore, concretes made with siliceous aggregates are more conductive than those made with carbonate (limestone) aggregates.

Derived from theoretical considerations, the solid curves in Fig. 4.10 describe the variation in thermal conductivity with temperature for four concretes. In deriving these curves, two concretes (see curves 1 and 2) were visualized to represent limiting cases among normal-weight concretes, and the other two (see curves 3 and 4), limiting cases among lightweight concretes. The data points in Fig. 4.10 represent experimental data. They reveal that the upper limiting case is probably never reached with aggregates in common use and that the thermal conductivity of lightweight concretes may be somewhat higher than predicted on theoretical considerations.

Further experimental information on the thermal conductivity of some normal-weight and many lightweight concretes is available in the literature (Harmathy, 1970).

Data generated in tests have been used to develop temperature-dependent relations for thermal properties of concrete, and these relations are given in the ASCE manual (ASCE, 1992) and Eurocode 2 (CEN, 2004a) and other references (Kodur, 2014). These properties are generally expressed as a function of temperature in the Eurocode, and as a function of aggregate type and temperature in the ASCE manual. These high-temperature thermal property relations for different concrete types are presented in Chap. 8.

Previous studies have clearly shown that the thermal conductivity and heat capacity of concrete are significantly influenced by the moisture content and concrete porosity (Harmathy, 1970; Kodur and Khaliq, 2011). However, such interdependencies are not taken into consideration in most thermal property

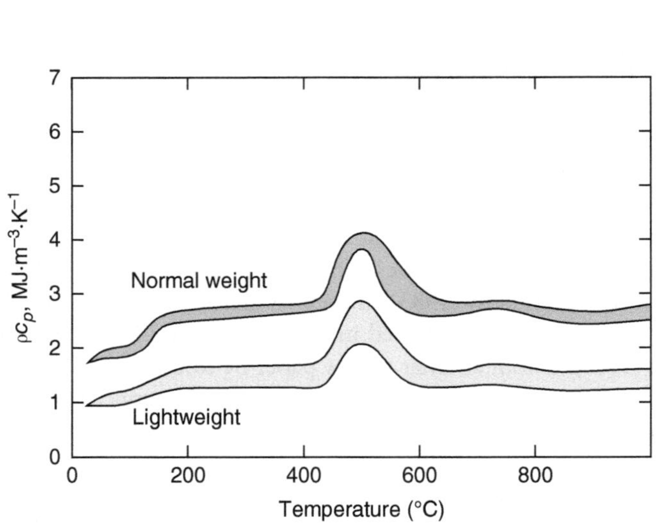

FIGURE 4.9 Usual ranges of variation for the volume-specific heat of normal-weight and lightweight concretes (Harmathy, 1970).

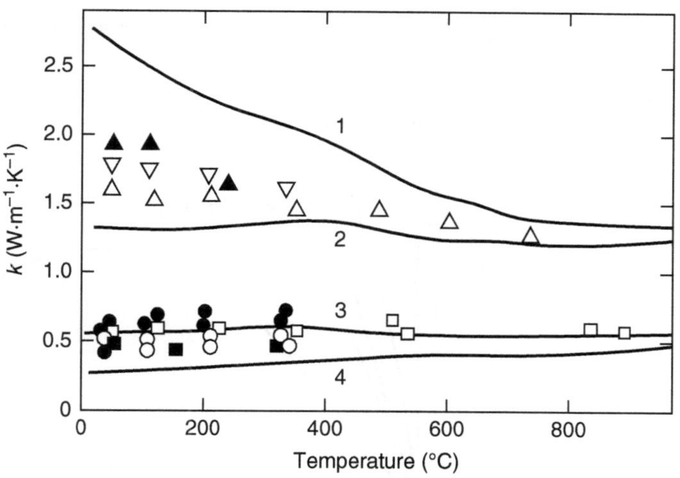

Figure 4.10 Thermal conductivity of four "limiting" concretes and some experimental thermal conductivity data. Symbols: ▼, ▽—various gravel concretes; ●—expanded slag concretes; ■, □—expanded shale concretes; ○—pumice concrete. (Harmathy, 1970; Harmathy and Allen, 1973.)

relations. Therefore, the current high temperature material models may not fully represent accurate thermal properties of new types of concrete, which possess moisture content and also lower porosity, as compared to conventional concretes.

Until the 1980s, the compressive strength of concrete was around 20 to 50 MPa, which is referred to as normal-strength concrete (NSC). In recent years, concrete with a compressive strength in the range 50 to 120 MPa is being widely used in practice and is referred to as high-strength concrete (HSC). Depending on the density, concretes are usually subdivided into two major groups: (1) normal-weight concretes with densities in the 2150- to 2450-kg·m^{-3} range and (2) lightweight concretes with densities between 1350 and 1850 kg·m^{-3}. Fire safety practitioners subdivide the normal-weight concretes into silicate (siliceous) and carbonate aggregate concretes, according to the composition of the principal aggregate. Also, a small amount of discontinuous fibers (steel or polypropylene fibers) is often added to the concrete mix to achieve superior properties; this concrete is referred to as fiber-reinforced concrete (FRC). In this section, the different properties of concrete are discussed under three groups: NSC, FRC, and HSC.

Normal-Strength Concrete

A good amount of information is available in the literature on the mechanical properties of various types of normal-strength concrete. This information is summarized by reports from Bennetts (1981) and Schneider (1985), the ACI guide (ACI, 1989), the ASCE fire protection manual (ASCE, 1992), Harmathy's book (Harmathy, 1993), and Kodur and Harmathy (2016). Figure 4.11 shows the stress-strain curves for a lightweight concrete with expanded shale aggregate at room temperature (24°C) and a few elevated temperature levels (Harmathy and Berndt, 1966). The shape of the curves may depend on the time of holding the test specimen at the target temperature level before the compression test.

The modulus of elasticity (E) of conventional concretes at room temperature falls within a very wide range, 5.0×10^3 to 35.0×10^3 MPa, dependent mainly on the water-cement ratio in the mixture, the age of concrete, the method of conditioning, and the amount and nature of the aggregates. Cruz (1966) found that the modulus of elasticity decreases rapidly with the rise of temperature, and the fractional decline does not depend significantly on the type of aggregate (in Fig. 4.12, E_0 is the modulus of elasticity at room temperature). From other surveys (Bennetts, 1981), it appears, however, that the modulus of elasticity of normal-weight concretes decreases faster with the rise of temperature than that of lightweight concretes.

The compressive strength, σ_u, of NSC may also vary within a wide range. Compressive strength is influenced by the same factors as the modulus of elasticity. For conventionally produced normal-weight and lightweight concretes, the strength at room temperature is usually between 20 and 50 MPa.

Information on the variation in the compressive strength with temperature is presented in Fig. 4.13 (for a silicate aggregate concrete), Fig. 4.14 (for a carbonate aggregate concrete), and Fig. 4.15 (for two lightweight aggregate concretes, one made with the addition of natural sand) (Abrams, 1971). $(\sigma_u)_0$ in the figures stands for the compressive strength of concrete at room temperature. In some experiments, the

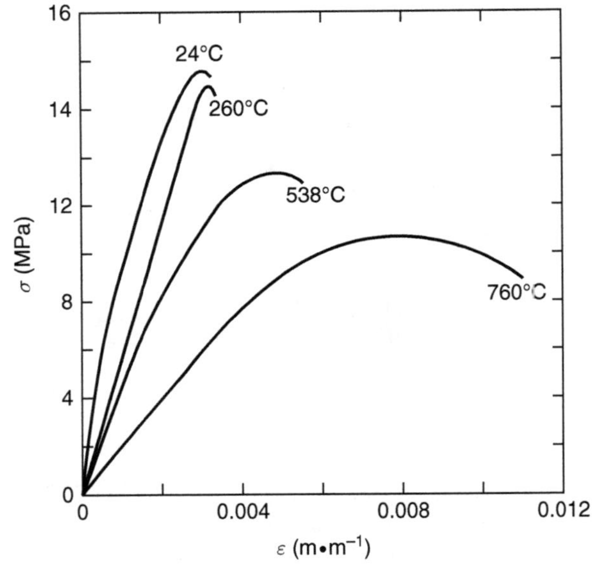

Figure 4.11 Stress-strain curves for a lightweight masonry concrete at room and elevated temperatures (Harmathy and Berndt, 1966).

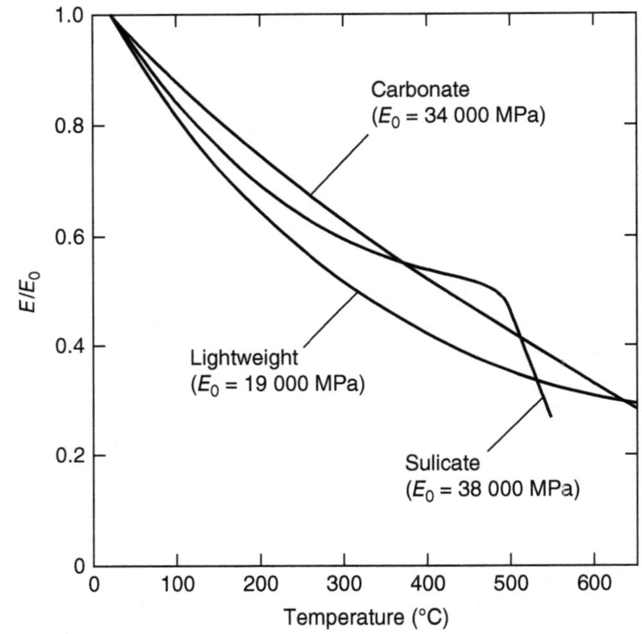

Figure 4.12 The effect of temperature on the modulus of elasticity of concretes with various aggregates (Makelainen and Miller, 1983).

specimens were heated to the test temperature without load (see curves labeled "unstressed"). In other tests they were heated under a load of about 40% of the ultimate strength (see curves labeled "stressed"). Again, in additional tests they were heated to the target temperature without load, then cooled to room temperature and stored at 75% relative humidity for six days, and finally tested at room temperature (see curves labeled "unstressed residual").

Based on the data from mechanical property tests, researchers have proposed different strength-temperature relating for mechanical properties of concrete. The most widely used relations are presented in Eurocode 2 (CEN, 2004a), the ASCE manual of practice (1992), and Kodur et al. (2004). The ASCE model is valid for NSC, the Eurocode model is valid for both NSC and HSC, and the Kodur et al. model is valid for HSC. These relations give the rate of degradation of concrete as a function of temperature only, and without any consideration to variations in other significant parameters such as rate of loading, heating, and material

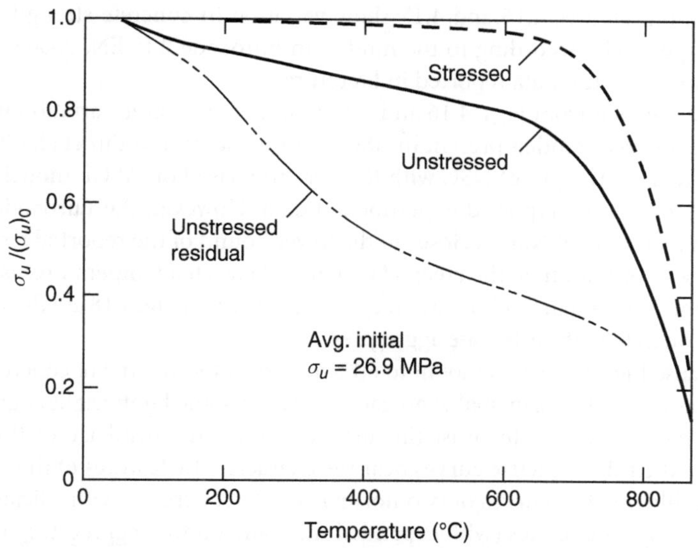

FIGURE 4.13 The effect of temperature on the compressive strength of a normal-weight concrete with silicate aggregate (Abrams, 1971).

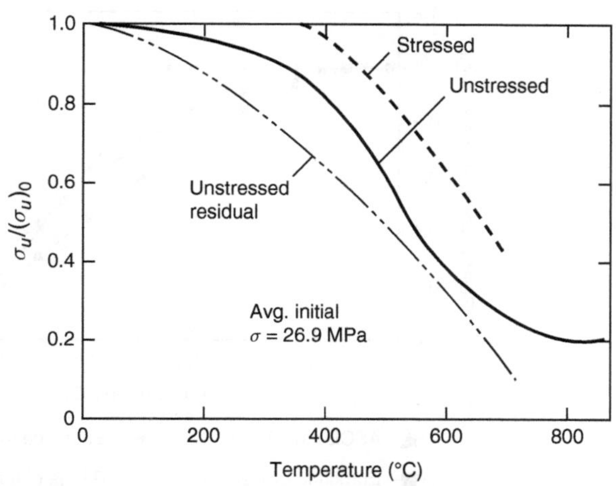

FIGURE 4.14 The effect of temperature on the compressive strength of normal-weight concrete with carbonate aggregate (Abrams, 1971).

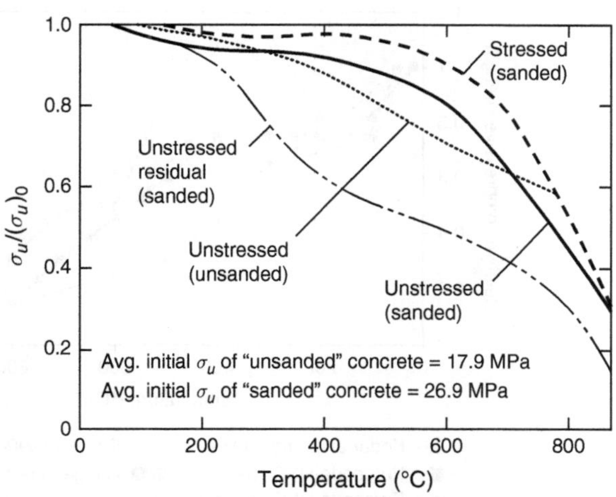

FIGURE 4.15 The effect of temperature on the compressive strength of two lightweight concretes (one with natural sand) (Abrams, 1971).

composition. Figures 4.16 and 4.17 show variation in concrete strength with temperature for NSC and HSC, respectively, according to the models in Eurocode 2 (CEN, 2004a), ASCE (1992), and Kodur et al. (2004) and from test data reported in literature.

It can be seen from Figs. 4.16 and 4.17 that there is a large variation in the reported test data and also from the different models present in ASCE, Eurocode, and Kodur et al. (2004). For example, the reduction in compressive strength of NSC with temperature based on ASCE model (1992) seems to be closer to the upper bound of the reported experimental data. However, the Eurocode 2 (CEN, 2004a), curve for the strength reduction of NSC is closer to the lower bound of the reported test results.

Some information on the creep of concrete at elevated temperatures is available from the work of Cruz (1966) and MareÂchal (1972). The creep curves shown in Fig. 4.18 are those recorded by Cruz for a normal-weight concrete with carbonate aggregates.

Because the aggregates amount to 60% to 75% of the volume of concrete, the dilatometric curve usually resembles that of the principal aggregate. However, some lightweight aggregates, for example, pearlite and vermiculite, are unable to resist the almost continuous shrinkage of the cement paste on heating, and therefore their dilatometric curves bear the characteristic features of the curve for the paste.

The dilatometric curves of two normal-weight concretes (with silicate and carbonate aggregates) and two lightweight concretes (with expanded shale and pumice aggregates) are shown in Fig. 4.19 (Harmathy and Allen, 1973). These curves were obtained in the course of a comprehensive study performed on 16 concretes.

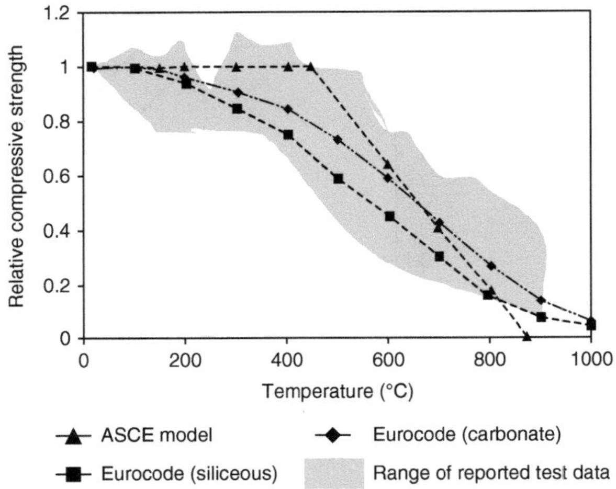

FIGURE 4.16 Variation in relative compressive strength of normal-strength concrete as a function of temperature.

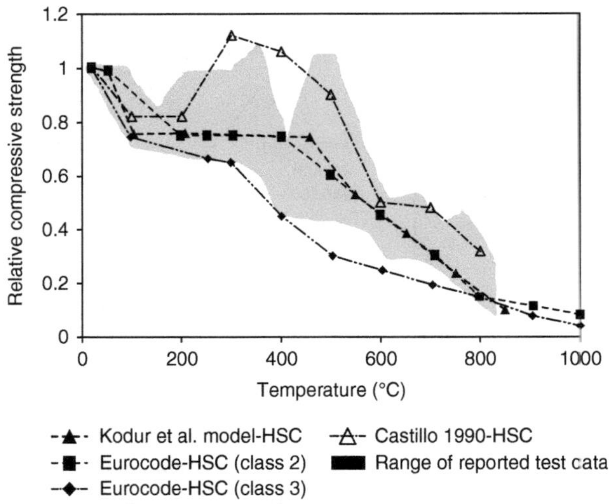

FIGURE 4.17 Variation in relative compressive strength of high-strength concrete as a function of temperature.

The results of dilatometric and thermogravimetric tests were combined to calculate the volumetric heat capacity (ρc_p) versus temperature relation for these four concretes, as shown in Fig. 4.20. The partial decomposition of the aggregate is responsible for a substantial drop (above 700°C) in the density of concretes made with carbonate aggregate.

In reinforced concrete structures, the bond between rebars and concrete (at elevated temperatures) plays a major role in determining the fire resistance of structural members. Diederichs and Schneider (1981) investigated the variation in bond strength between deformed and plain rebars and concrete as a function of temperature. They found that the bond strength reduction follows the same pattern as compressive strength for deformed and rusted plain bars. However, higher reduction in bond strength was observed for plain reinforcing bars. They also found that the bond strength at elevated temperature increases

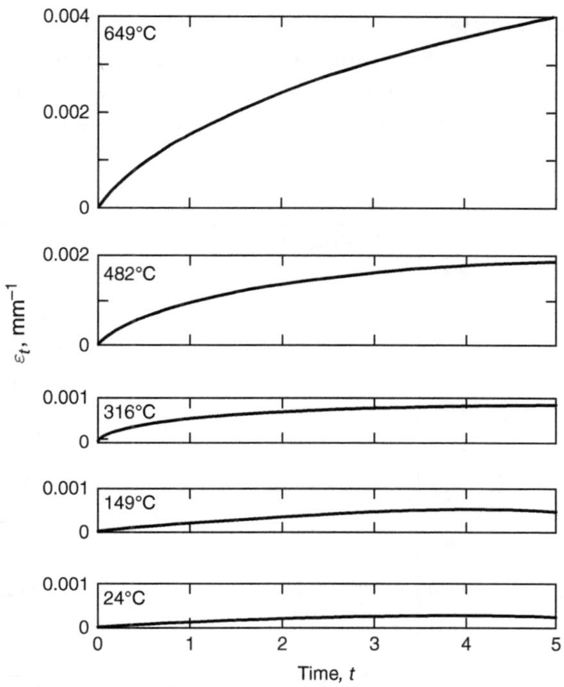

FIGURE 4.18 Creep of a carbonate aggregate concrete at various temperature levels (applied stress: 12.4 MPa; compressive strength of the material at room temperature: 27.6 MPa).

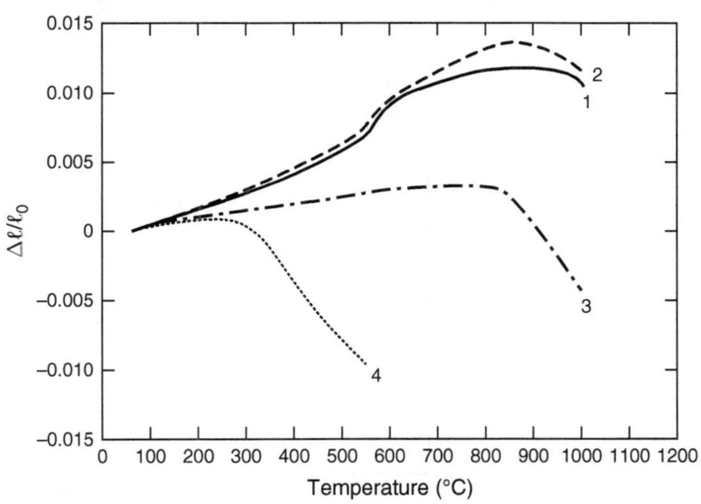

FIGURE 4.19 Dilatometric curves for two normal-weight and two lightweight concretes (Harmathy and Allen, 1973): (1) normal-weight concrete with silicate aggregate, (2) normal-weight concrete with carbonate aggregate, (3) lightweight concrete with expanded shale aggregate, and (4) lightweight concrete with pumice aggregate.

with decreasing coefficient of thermal expansion of concrete, which is significantly influenced by the type of aggregate. Diederichs and Schneider also inferred that the water-cement ratio and the bar diameter have a minor effect on the bond (strength) between steel rebars and concrete. Figure 4.21 illustrates the variation in bond strength as a function of temperature for reinforced and prestressed concrete.

More recently, Kodur and Agrawal (2017) studied temperature-induced bond deterioration at the rebar-concrete interface arising from the degradation in the mechanical properties of two materials, coupled with thermal cracking resulting from differential thermal expansion in the bonded region. They noted that bond strength decreases with increase in temperature (beyond 400°C) and the rate of bond strength reduction is higher than the compressive strength of concrete. Kodur and Agrawal also inferred that heating rate, saturation period, and stress level during heating influence bond strength at elevated temperatures. In addition, reduction in bond strength directly correlates with concrete cover thickness—in other words, the lower the thickness cover, the higher the reduction in bond strength.

Fiber-Reinforced Concrete

Steel and polypropylene discontinuous fibers are the two most common fibers added to the concrete mix to improve structural properties of concrete. Studies have shown that polypropylene fibers in a concrete mix are quite effective in minimizing spalling in concrete under fire conditions (Bilodeau et al., 1998).

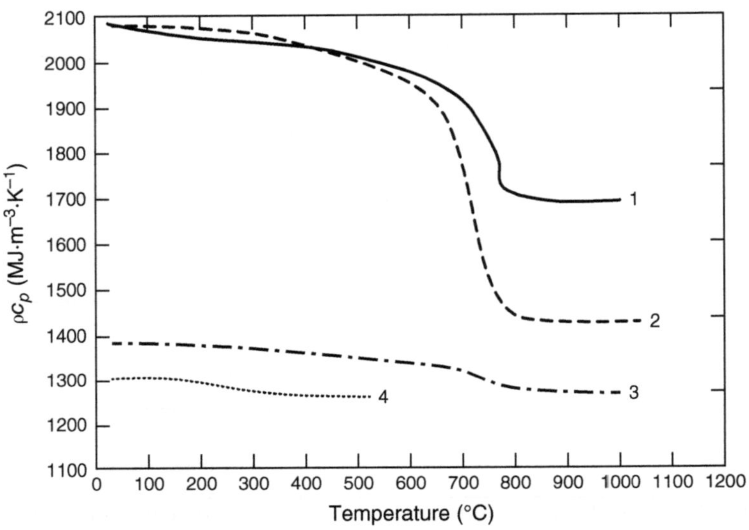

FIGURE 4.20 Volumetric heat capacity of two normal-weight and two lightweight concretes: (1) normal-weight concrete with silicate aggregate, (2) normal-weight concrete with carbonate aggregate, (3) lightweight concrete with expanded shale aggregate, and (4) lightweight concrete with pumice aggregate.

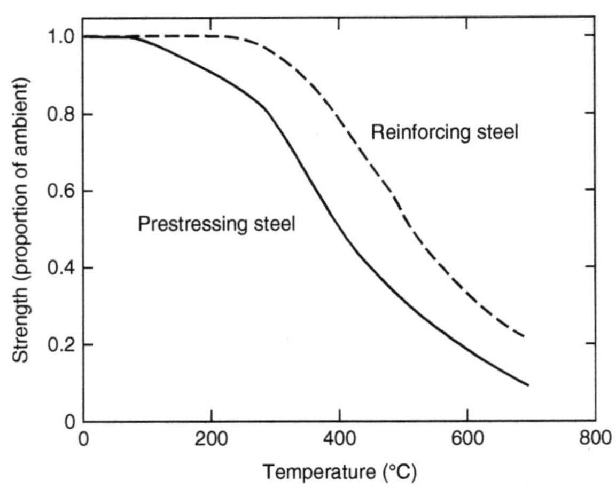

FIGURE 4.21 Variation in bond strength as a function of temperature for reinforced and prestressed concrete (Diederichs and Schneider, 1981).

The polypropylene fibers melt at a relatively low temperature of about 170°C and create channels for the steam pressure in concrete to escape. This prevents the mini explosions that cause spalling in concrete members. Based on various studies, the amount of polypropylene fibers needed to minimize spalling is about 0.1% to 0.25% (by volume). The polypropylene fibers were found to be most effective in HSC members made with normal-weight aggregate. The addition of polypropylene fibers also improves certain mechanical properties, such as tensile strength, ductility, and ultimate strain, at room temperature.

Overall, Lie and Kodur (1996) have carried out detailed experimental studies and developed dilatometric and thermogravimetric curves for various types of steel fiber-reinforced concrete (SFRC). Based on these studies, they have also developed relations for expressing thermal and mechanical property variations in SFRC over a temperature range of 0 to 1000°C (Jumppanen et al., 1986). There is limited information on thermal and mechanical properties of concrete with polypropylene fibers. The thermal properties of SFRC, at elevated temperatures, are similar to those of plain concrete.

Mechanical properties of SFRC at elevated temperatures are more beneficial to achieve higher fire resistance than that of plain concrete. There is some information available on SFRC's properties at elevated temperatures. The effect of temperature on the compressive strength for two types of SFRC slightly is shown in Fig. 4.22. The strength of both types of SFRC exceeds the initial strength up to about 400°C. This is in contrast to the strength of plain concrete, which decreases slightly with temperatures up to 400°C. Above 400°C, the strength of SFRC decreases at an accelerated rate.

The effect of temperature on the tensile strength of steel fiber-reinforced concrete (carbonate aggregate) is compared to that of plain concrete in Fig. 4.23 (Purkiss, 1984). The tensile strength of SFRC decreases at a lower rate than that of plain concrete throughout the temperature range, with the strength being significantly higher than that of plain concrete up to about 350°C. The increased tensile strength delays the propagation

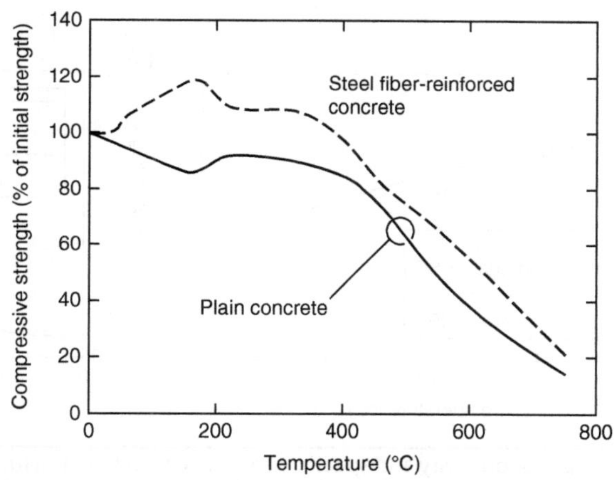

Figure 4.22 Effect of temperature on compressive strength of steel fiber-reinforced concrete.

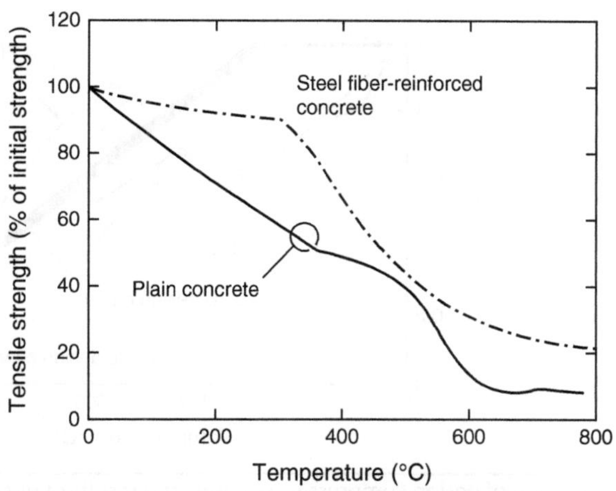

Figure 4.23 Effect of temperature on tensile strength of steel fiber-reinforced concrete.

of cracks in SFRC structural members and is highly beneficial when the member is subjected to flexural stresses. Further, under fire conditions, higher tensile strength of SFRC at elevated temperatures helps to withstand tensile stresses generated from pore pressure and can prevent fire-induced spalling in concrete.

The type of aggregate has a significant influence on the tensile strength of steel fiber-reinforced concrete. The decrease in tensile strength for carbonate aggregate SFRC is higher than that for siliceous aggregate SFRC.

High-Strength Concrete

Kodur and Sultan (1998b) have presented detailed experimental data on thermal properties of HSC (for both plain and SFRC types). The type of aggregate has significant influence on the thermal properties of HSC at elevated temperatures. Figure 4.24 shows the thermal conductivity and specific heat of HSC, made of siliceous and carbonate aggregates, as a function of temperature. Based on the test data, Kodur and Sultan (2003) proposed relations for expressing thermal conductivity, specific heat, thermal expansion, and mass loss of HSC as a function of temperature. The variation in thermal expansion with concrete temperature for siliceous and carbonate aggregate HSC is similar to that of NSC, with the aggregate having a strong influence. Overall, the thermal properties of HSC, at elevated temperatures, are similar to those of NSC.

The strength of concrete has significant influence on the mechanical properties of concrete. The mechanical and deformation properties of HSC vary differently with temperature than those of NSC. This variation is more pronounced for mechanical properties, which are affected by these factors: compressive strength, moisture content, density, heating rate, percentage of silica fume, and porosity (core structure) (Kodur and McGrath, 2006). The available information on the mechanical properties of HSC at elevated temperatures is presented in a review report by Phan (1996) and Kodur (2014).

Figure 4.25 shows the comparison of compressive strength of NSC and HSC types, together with CEB and European design curves for NSC. The loss in compressive strength with temperature is higher

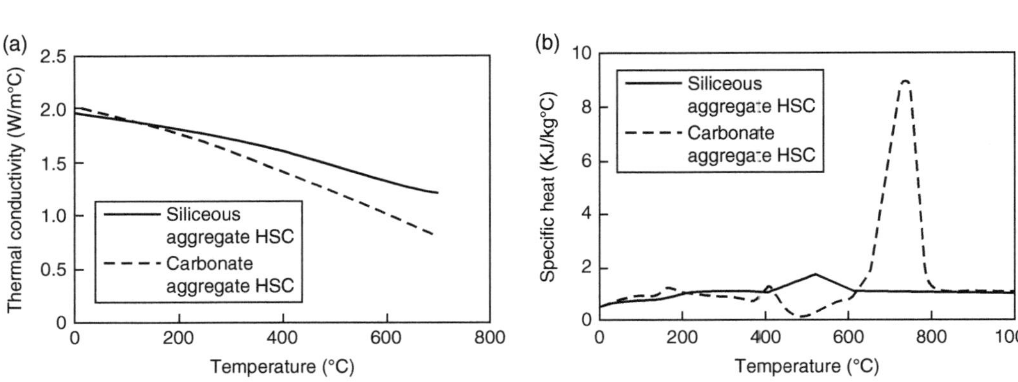

FIGURE 4.24 Thermal conductivity and specific heat capacity of HSC as a function of temperature: (a) thermal conductivity of high-strength concrete and (b) specific heat of high-strength concrete.

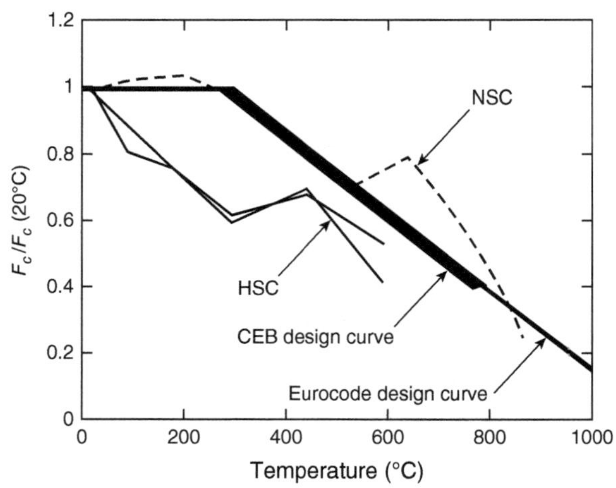

FIGURE 4.25 Comparison of design compressive strength and results of unstressed tests of lightweight aggregate concrete (Phan, 1996).

for HSC than that for NSC up to about 450°C. The difference between compressive strength versus temperature relationships of normal-weight and lightweight aggregate concrete is not significant. However, HSC mixture with silica fume has higher compressive strength loss with increasing temperature than HSC mixture without silica fume. Based on a series of high-temperature material property tests, Cheng et al. (2004) have proposed a set of stress-strain relationships for HSC as a function of temperature. The variation, with temperature, in modulus of elasticity and tensile strength of HSC is similar to that of NSC.

HSC, due to low porosity, is more susceptible to fire-induced spalling than NSC, and such explosive spalling may occur when HSC is exposed to rapidly rising temperature experienced under severe fire conditions. Hence, one of the major concerns for the use of HSC is regarding its behavior in fire, particularly the occurrence of spalling at elevated temperatures. One of the key properties that influences the extent of spalling in a concrete member is the porosity or permeability of concrete for predicting spalling and porosity varies with temperature. Figure 4.26 shows the variation in porosity with temperature for NSC and HSC. The data in this figure are taken from the measurements of porosity after exposure to different temperatures (Noumowe et al., 1996).

The spalling in HSC can be minimized by creating pores through which water vapor can be relieved before vapor pressure, and corresponding tensile stress, reaches critical value. This is usually achieved through adding polypropylene fibers to the HSC (Bilodeau et al., 1998), and in some cases steel fibers can also be effective in mitigating fire-induced spalling. Also, Kodur and McGrath (2006) have shown that spalling in HSC column can be minimized to a significant extent by providing bent ties as lateral confinement. Figure 4.27 illustrates conventional and improved tie configuration for minimizing spalling in HSC columns.

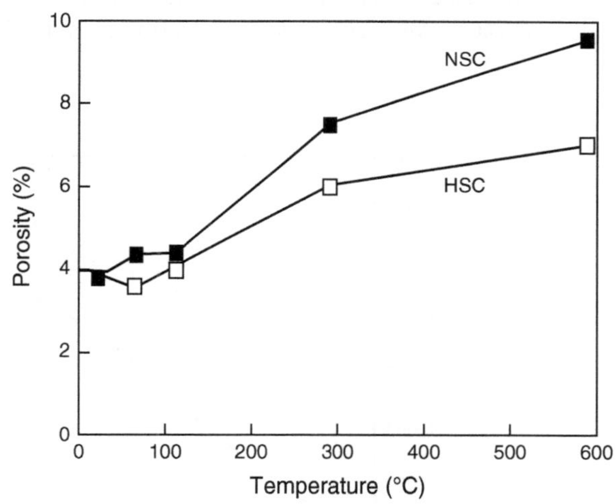

Figure 4.26 Porosity of HSC and NSC as a function of temperature.

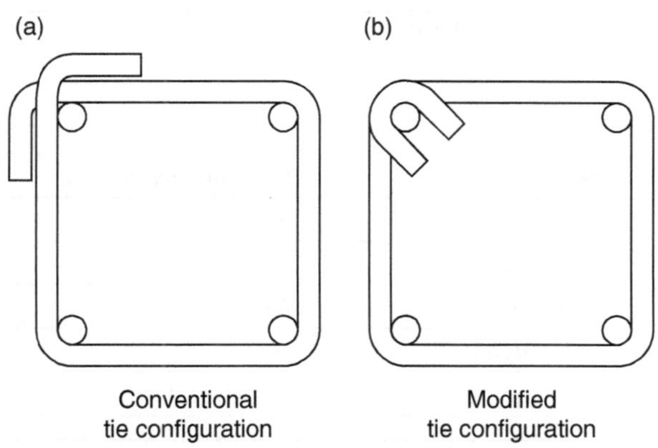

Figure 4.27 Modified tie configuration for achieving higher fire resistance in concrete structures.

4.4.2 Steel

Steel is a Group L material and extensively used in construction industry in various forms; structural shapes (columns, beams), as reinforcement (reinforcing or prestressing steel) in concrete structures; and as light gauge steel studs/joists (wall and floor systems). The steels most often used in building construction are either hot-rolled or cold-drawn. The steel used for structural shapes and reinforcing bars used in concrete are hot-rolled, low-carbon, ferrite-pearlite steels. They have a randomly oriented grain structure, and their strength depends mainly on their carbon content. The prestressing steel, wires and strands, used in prestressing concrete are usually made from cold-drawn, high-carbon, pearlitic steels with an elongated grain structure, oriented in the direction of the cold work. Light-gauge steel, made from cold-formed steel, finds wide applications for lightweight framing in buildings, such as walls and floors.

The density (ρ) of steel is about 7850 kg·m^{-3}. Its coefficient of thermal expansion (β) is a structure-insensitive property and for an average carbon steel, β is 11.4×10^{-6} m·m^{-1}·K^{-1} at room temperature. The dilatometric curve shown in Fig. 4.28 is applicable to most of the common steels. The curve reveals substantial contraction of the material at about 700°C, which is associated with the transformation (phase change of steel) of the ferrite-pearlite structure into austenite.

Being a structure-sensitive property, the thermal conductivity of steel is not easy to define. For carbon steels it usually varies within the range of 46 to 65 W·m^{-1}·K^{-1}.

Equations for various properties of steel, as functions of temperature, are available in the ASCE structural fire protection manual (ASCE, 1992) and in Eurocode 3 (CEN, 2005a). In the ASCE manual, the same set of relationships is applicable for thermal properties of both structural and reinforcing steels. However, separate relationships for stress-strain and elasticity are given for the two steels with slightly conservative values for structural steel. Also, Poh (2001) proposed a general stress-strain equation that expresses stress explicitly in terms of strain in a single continuous curve.

The critical temperature of steel, defined as the temperature at which yield strength reduces to 50% of ambient value, is often used as a benchmark for determining the failure of structural members exposed to fire. The critical temperature for various types of steels is given in Table 4.1.

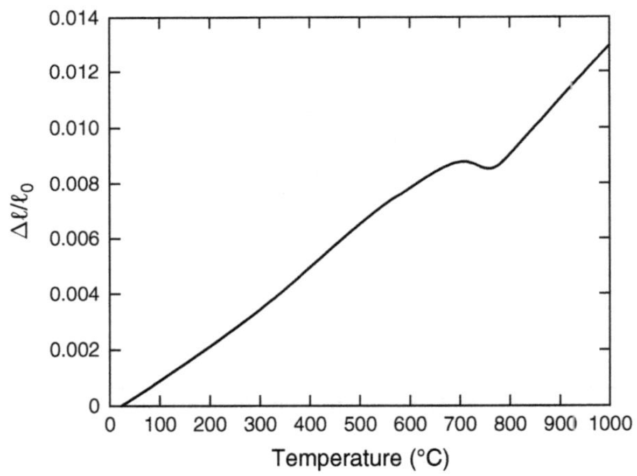

FIGURE 4.28 Dilatometric curve for steel.

Steel	Standard/Reference	Critical Temperature
Structural steel	ASTM	538°C
Reinforcing steel	ASTM	593°C
Prestressing steel	ASTM	426°C
Light-gauge steel	EC 3	350°C
	Gerlich et al.	400°C

TABLE 4.1 Critical Temperature for Various Types of Steel

Information on the mechanical properties of two typical steels [structural steel (ASTM A36) and a prestressing wire (ASTM A421)] is presented in Figs. 4.29 through 4.31 and in Table 4.2 (Harmathy and Stanzak, 1970). Figures 4.29 and 4.30 are stress-strain curves at room temperature (24°C and 21°C, respectively) and at various elevated temperatures. Figure 4.31 shows the effect of temperature on the yield and ultimate strengths of the two steels.

Creep at elevated temperature can be significant in the case of steel structures due to significant degradation in mechanical properties. Table 4.2 presents information on the effect of stress on the two creep parameters, Z and ε_{t0}. Because creep is a very structure-sensitive property, the creep parameters may show a substantial spread, even for steels with similar characteristics at room temperature. The application of the creep parameters in evaluating failure time in fire-exposed structural failure is discussed by Hamilton and Crosser (1962).

The modulus of elasticity (E) for most steels is about 210 GPa at room temperature. Figure 4.32 shows the variation in modulus of elasticity with temperature for structural steels (ECCS, 1983) and steel reinforcing bars (Anderberg, 1978). (In Fig. 4.32, $E0$ is the modulus of elasticity at room temperature, while E_T is the modulus at any given temperature.)

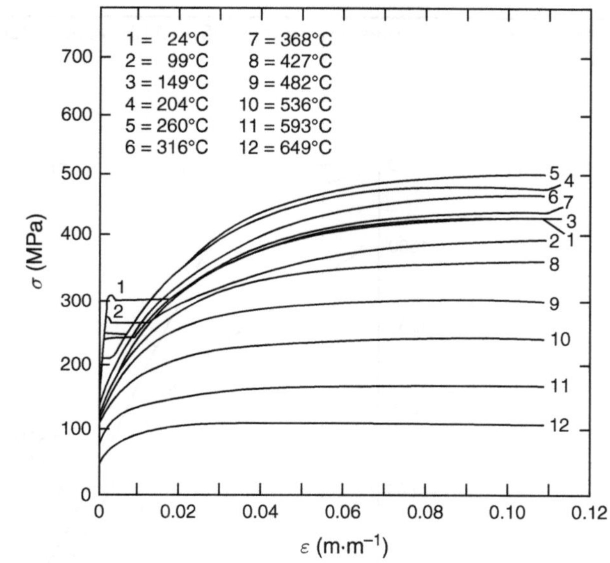

Figure 4.29 Stress-strain curves for a structural steel (ASTM A36) at room temperature and elevated temperatures (Harmathy and Stanzak, 1970).

Steel	$\Delta H_c/R$ (k)	$\varepsilon_{t0}(\sigma)$ (m·m⁻¹)	$Z(\sigma)$ (h⁻¹)
ASTM A36	38,890	$3.258 \times 10^{-17}\sigma^{1.75}$	$2.365 \times 10^{-20}\sigma^{4.7}$
			if $\sigma \leq 103.4 \times 106$
			$1.23 \times 10^{16} \exp(4.35 \times 10^{-8}\sigma)$
			if $103.4 \times 10^6 \leq \sigma \leq 310 \times 10^6$
ASTM A421	30,560	$8.845 \times 10^{-9}\sigma^{0.67}$	$1.952 \times 10^{-10}\sigma^{3}$
			if $\sigma \leq 172.4 \times 10^6$
			$8.21 \times 10^{13} \exp(1.45 \times 10^{-8}\sigma)$
			if $172.4 \times 10^6 \leq \sigma \leq 690 \times 10^6$

σ is stress measured in Pa.

Table 4.2 Creep Parameters for a Structural Steel and Prestressing Steel

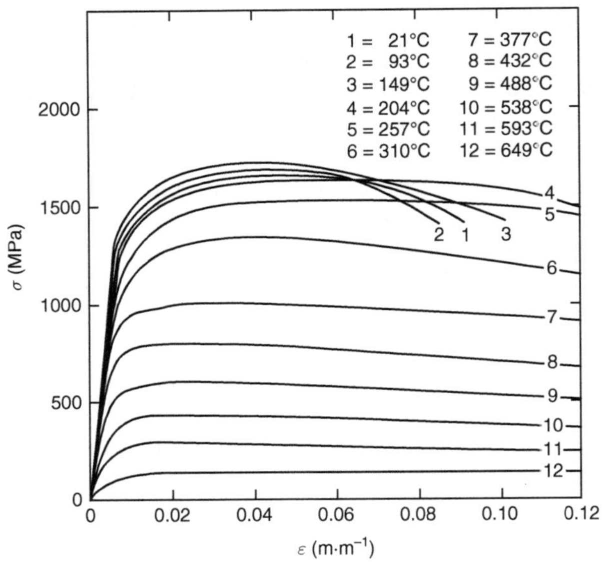

Figure 4.30 Stress-strain curves for prestressing steel (ASTM A421) at room temperature and elevated temperatures (Harmathy and Stanzak, 1970).

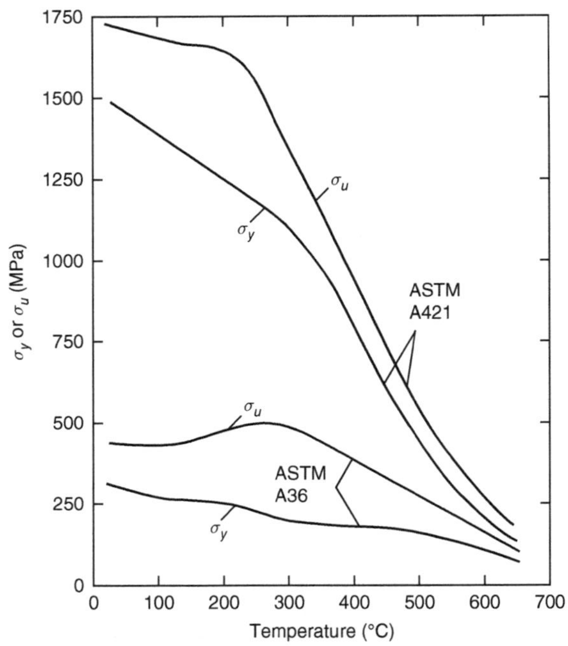

Figure 4.31 Ultimate and yield strength of structural steel (ASTM A36) and prestressing steel (ASTM A421) at elevated temperatures (Harmathy and Stanzak, 1970).

The above-discussed high-temperature properties are generally applicable to conventional carbon (mild) steel whose chemical composition consist of iron, carbon, manganese, sulfur, and phosphorous. In recent years, a number of new types of steels are available and made by adding alloys such as nickel, titanium, boron, and chromium. These alloys influence durability characteristics as well as the mechanical properties of steel. For example, molybdenum, chromium, and niobium can increase the fire resistance property of steel, while chrome and nickel can enhance corrosion resistance (Li et al., 2003). Current design rules on fire resistance of steel structures (EC3, 2005; BS: 5950, 2003) are mainly based on experimental data on mild (carbon) steel and do not account for specific property variations in new types of alloy steels.

Recent research by Wang et al. (2012, 2014) clearly shows that high-strength steel (Q460 and Q690) exhibits a slower loss of strength and modulus throughout a 20 to 800°C temperature range as compared to

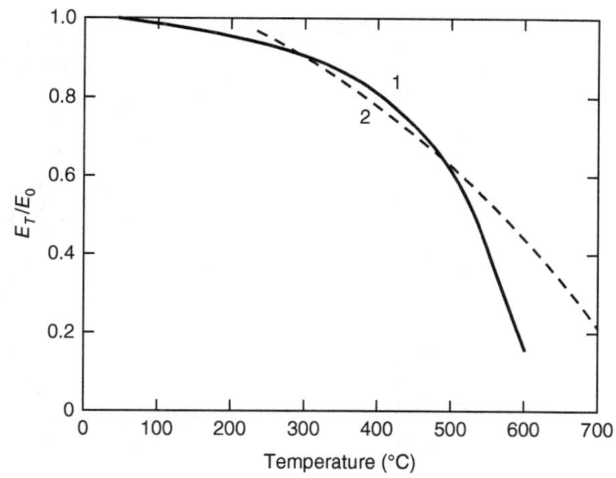

FIGURE 4.32 The effect of temperature on the modulus of elasticity of (1) structural steel and (2) reinforcing steel.

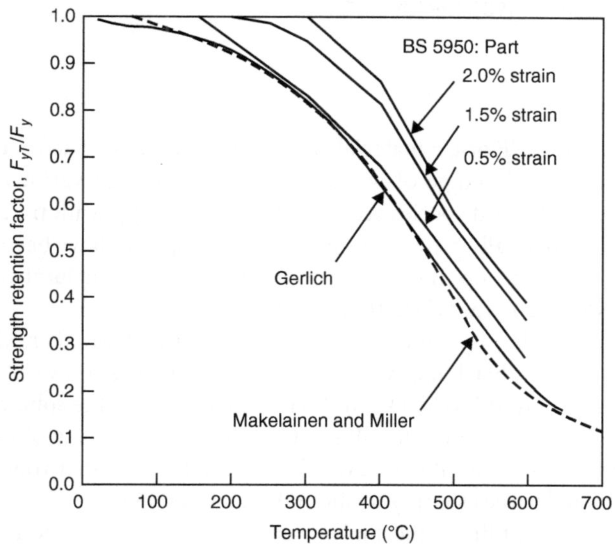

FIGURE 4.33 Reduction of the yield strength of cold-formed light-gauge steel at elevated temperatures (Gerlich, 1995; Makelainen and Miller, 1983; Gerlich et al., 1996).

mild steel. This is mainly due to the presence of chromium and niobium, which improves the fire resistance properties of steel.

Tests by Kodur et al. (2010) have also shown that the type of heat treatment has significant influence on strength properties of steel—for example, annealing and normalizing produces normal-strength steel, whereas quenching and tempering produces high-strength steel. High-strength steel, produced using quenching and tempering process and used in bolts (A490 bolts), possesses slightly lower thermal conductivity than that of conventional mild steel.

The temperature-dependent properties of cold-formed light-gauge steel are slightly different from those of hot-rolled structural steel. Gerlich (1995) and Makelainen and Miller (1983), based on steady-state and transient tests on cold-formed steel tension coupons (cut from studs) and galvanized sheets, proposed relationships for yield strength and modulus of elasticity. Figure 4.33 shows the variation in yield strength corresponding to 0.5%, 1.5%, and 2% strains of light-gauge steel at elevated temperatures. The BS 5950 curves represent a conservative 95% confidence limit (i.e., a 5% chance that strength would fall below the curve), whereas the other two curves are representative of mean test data. Figure 4.34 shows the variation in modulus of elasticity of light-gauge steel at elevated temperatures. The modulus, E_T, represents the tangent modulus at low stress levels (or initial tangent modulus), because steel stress-strain relations become increasingly nonlinear at elevated temperatures.

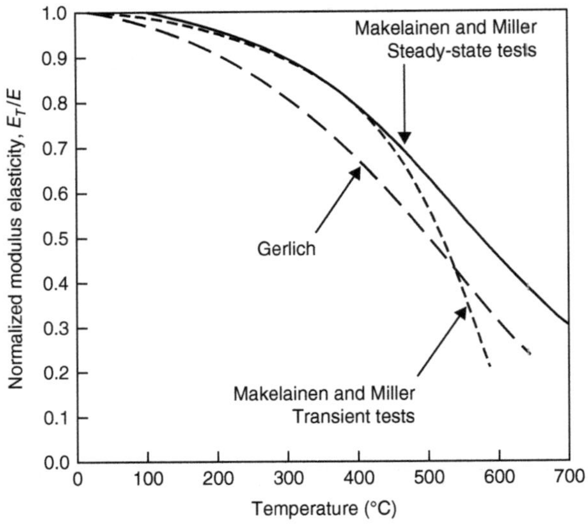

Figure 4.34 Modulus of elasticity of cold-formed light-gauge steel at elevated temperatures (Gerlich, 1995; Makelainen and Miller, 1983; Gerlich et al., 1996).

4.4.3 Wood

Wood is a Group L/I/F or I/F material and is widely used as structural members in residential and low-rise constructions. A wide range of wood species are grown in various countries (and regions), but only a limited number of wood species are used in structural applications. In the United States, about 180 wood species are commercially grown, but only about 25 species have been assigned working stresses in timber structures. The two groups most extensively used as structural lumber in the United States are the Douglas firs and the southern pines (Schaffer, 2007).

The oven-dry density (ρ) of commercially important woods range from 300 kg·m^{-3} (white cedar) to 700 kg·m^{-3} (hickory, black locust). The density of Douglas firs varies from 430 to 480 kg·m^{-3} and that of southern pines from 510 to 580 kg·m^{-3}. The true density of the solid material that forms the walls of wood cells (ρ_t) is about 1500 kg·m^{-3} for all kinds of wood. The density of wood decreases with temperature; the density ratio (ratio of density at elevated temperature to that at room temperature) drops to about 0.9 at 200°C and then declines sharply to about 0.2 at about 350°C.

The coefficient of linear thermal expansion (β) of typical wood ranges from 3.2 × 10^{-6} to 4.6 × 10^{-6} m·m^{-1}·K^{-1} along the grain and from 21.6 × 10^{-6} to 39.4 × 10^{-6} m·m^{-1}·K^{-1} across the grain (Wangaard, 1958). Wood shrinks at temperatures above 100°C, because of the reduction in moisture content. Lie (1992) reported that the amount of shrinkage can be estimated as 8% in the radial direction, 12% in the tangential direction, and an average of 0.1% to 0.2% in the longitudinal direction. The dilatometric and thermogravimetric curves of a pine with a 400 kg·m^{-3} oven-dry density are shown in Fig. 4.35 (Harmathy, 1983).

The thermal conductivity (k) across the grain of this southern pines was measured to be 0.86 to 1.07 W·m^{-1}·K^{-1} between room temperature and 140°C (Anderberg and Thelandersson, 1976). The thermal conductivity increases initially up to a temperature range of 150 to 200°C, then decreases linearly up to 350°C, and finally increases again beyond 350°C.

Figure 4.36 shows the apparent specific heat for the same pine, as a function of temperature. The accuracy of the curve [developed by differential scanning calorimeter (DSC)] is somewhat questionable. However, it provides useful information on the nature of decomposition reactions that take place between 150 and 370°C.

Wood is an orthotropic material, so the strength and stiffness properties in longitudinal and transverse directions are influenced by grain orientation. The mechanical properties of wood are influenced by moisture content, rate of charring, and grain orientation. The modulus of elasticity (E) at room temperature of air-dry, clear wood along the grain varies from 5.5 × 10^3 to 15.0 × 10^3 MPa, and its crushing strength (σ_u) varies from 13 to 70 MPa. These properties are related and roughly proportional to the density, regardless of the wood species (Gross, 1975).

The mechanical properties of wood deteriorate significantly with a rise in temperature. Figure 4.37 shows the variation in the modulus of elasticity and compressive strength of oven-dry, clear wood with temperature (Gerhards, 2007; Schaffer, 2007). (E_0 and $[\sigma_u]_0$ in the figure are modulus of elasticity and compressive

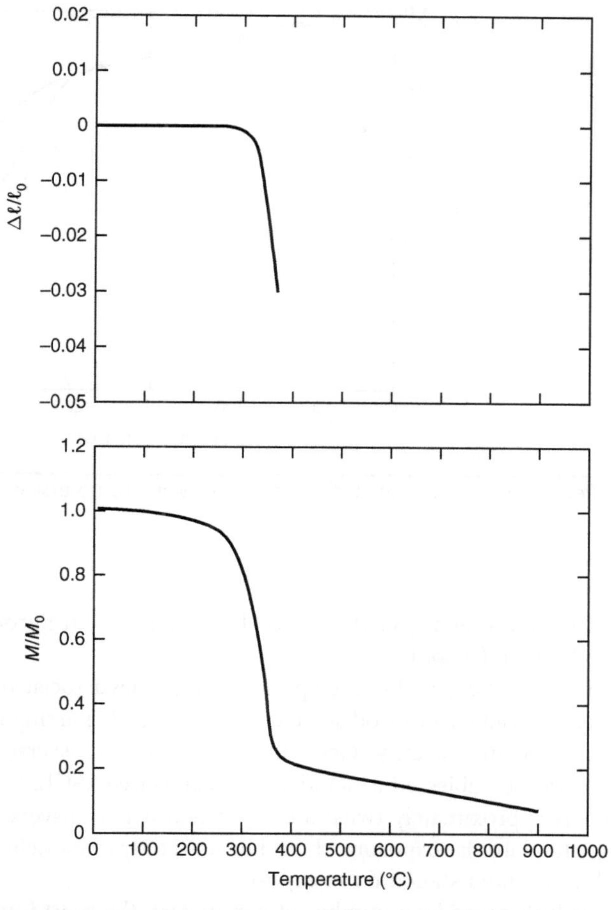

Figure 4.35 Dilatometric and thermogravimetric curves for a pine of 400 kg·m^{-3} density.

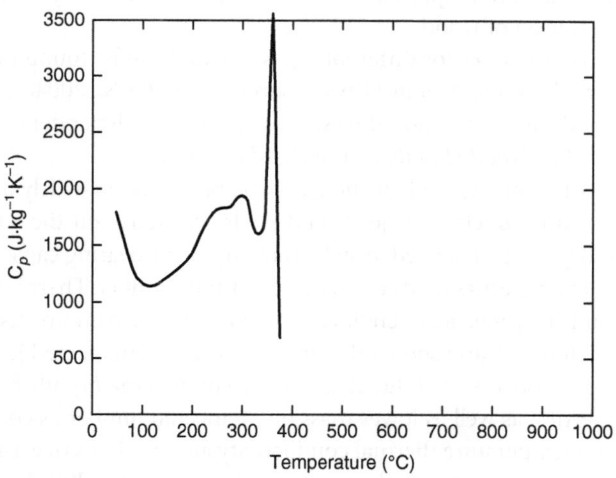

Figure 4.36 Apparent specific heat for a pine of 400 kg·m^{-3} density (Harmathy, 1983).

strength at room temperature, respectively.) The modulus of elasticity decreases slowly with temperature up to about 200°C, when it reaches about 80%, and then the decline is more rapid. The compressive strength also drops linearly to about 80% at about 200°C, and then the drop is more rapid—to about 20% around 280°C.

The tensile strength of wood exhibits behavior similar to that of compressive strength, but the decline in tensile strength with temperature is less rapid. The moisture content plays a significant role in determining the extent of strength and stiffness reduction, with increased moisture content leading to higher reduction. There is limited information on stress-strain relations for wood. The temperature-dependent relations for

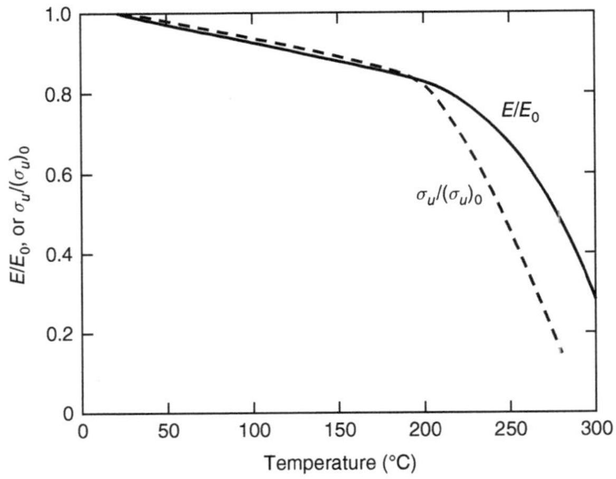

Figure 4.37 Effect of temperature on modulus of elasticity and compressive strength of wood (Gerhards, 1981; Schaffer, 1977).

reduced stiffness and design strength of wood can be found in Eurocode 5 (CEN, 2004a), (Part 1.2) and is given in this book in Chap. 10.

Charring is one of the main high-temperature properties associated with wood and should be considered in predicting fire resistance of wood structures. The rate of charring is influenced by the radiant heat flux or, alternatively, the fire severity. Generally, a constant transverse-to-grain char rate of 0.6 mm/min can be used for woods subjected to standard fire exposure (ASCE, 1992). The charring rate parallel to the grain of wood is approximately twice the rate when it is transverse to the grain. These charring rates, derived under standard fire exposure, should be used only in modeling the performance of wood sections' furnace conditions under standard fire exposure.

Charring is influenced by a number of parameters, the most important ones being density, moisture content, and contraction of wood. It is reasonable to modify the charring rate of 0.6 mm/min to approximately 0.4 mm/min for moist dense wood or to 0.8 mm/min for dry and light wood. The fire retardant chemicals often applied to reduce flame spread in wood may only delays the charring process slightly until ignition of wood.

Specific charring rates for different types of wood can be found in "Structural Fire Protection" (ASCE, 1992) and Bénichou and Sultan (1999). Eurocode 5 (CEN, 2004a), gives general guidance for charring depth in a wood member exposed to standard fire. The dependency of charring rate on the radiant heat flux is discussed in *Wood Handbook* (Gerhards, 1981).

In recent years new types of *engineered wood* products are widely used in residential construction. These engineered wood products (e.g., joists and studs) capitalize on the strength of wood and the efficiency of the sectional shape (e.g., I-shaped joists) to enhance load-bearing capacity at ambient conditions, while at the same time reducing the mass and cost of the structural member. There is very limited data on high-temperature thermomechanical properties of engineered lumber and also on fire resistance of engineered joists and studs. Limited research has clearly shown the fire resistance of engineered joists to be significantly lower than that of conventional wood joists (Kodur et al., 2017). This was mainly attributed to poor thermal, mechanical. and charring properties, as well as lower mass of engineered lumber as compared to traditional wood products. Typically, room temperature thermal conductivity and modulus of elasticity of engineered lumber are higher than other types of wood due to the presence of compressed plies. A comparison of charring rates indicates that engineered lumber has a higher rate of charring rate than conventional wood (Kodur et al., 2017).

4.4.4 Masonry and Brick

Bricks belong in the L/I group of materials and is extensively used in walls of low-rise residential and industrial buildings construction. Also, bricks together with building stone such as granite, limestone, and concrete block are grouped under *masonry* construction materials. Traditionally, the term brick referred to a unit composed of clay and laid in mortar. A brick can be composed of clay-bearing soil, sand, and lime, or concrete materials. Masonry and brick units are joined through mortar. Masonry is a durable construction material and its quality depends on type of mortar used to join individual masonry/brick units, workmanship, and the pattern in which the units are assembled. Three types of mortars are permitted for use in civil construction: Type N, Type S, and Type M mortars, and these types vary according to their bond and compressive strength. Grout is often used to fill brick and hollow units or spaces between wythes of solid unit masonry. Grout increases the compressive, shear, and flexural strength of the masonry (or brick) walls and also bonds steel reinforcement and masonry together.

The thermal properties of bricks and masonry, including thermal conductivity and specific heat, are often assumed to be similar to that of concrete. The density (ρ) of bricks ranges from 1660 to 2270 kg·m^{-3}, depending on the raw materials used in the manufacture, and on the molding and firing technique. The true density of the material (ρ_t) is somewhere between 2600 and 2800 kg·m^{-3}, slightly higher than that of normal-density concrete. Eurocode 6 (2005c) recommends using a density of 900 to 2400 kg·m^{-3} depending on the type of raw material used in making the brick. For example, bricks made of calcium silicate and lightweight aggregate concrete have a density that varies between 1600 to 2000 and 600 to1000, kg·m^{-3}, respectively. The mortar material can have a density between 1500 to 1750 kg·m^{-3}.

As shown in Fig. 4.38, the specific heat of brick slightly increases with a rise in temperature and is sensibly independent of the density of the brick (Malhotra, 1982). On the other hand, thermal conductivity of masonry increases with bricks of higher density (see Fig. 4.39).

At room temperature, the coefficient of thermal expansion (α) for clay bricks is about 5.5×10^{-6} m·m^{-1}·K^{-1}, which is very close to concrete. The dilatometric and thermogravimetric curves for a clay brick of 2180 kg·m^{-3} density are shown in Fig. 4.40 (Harmathy, 1983). The variation with temperature in the specific heat and the thermal properties of clay brick is shown in Figs. 4.40 to 4.42.

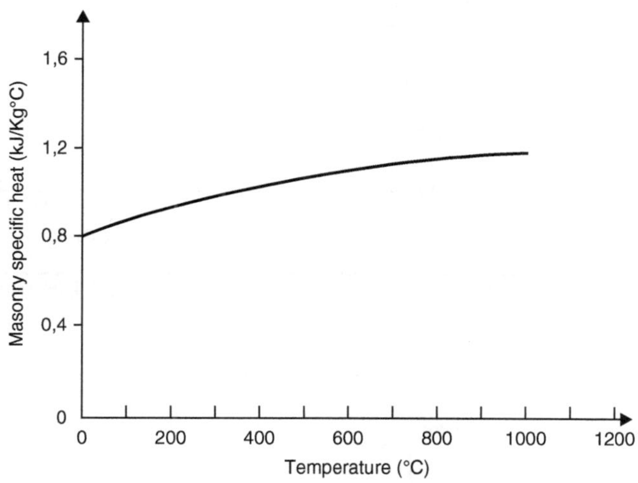

FIGURE 4.38 Variation in the specific heat of masonry with temperature.

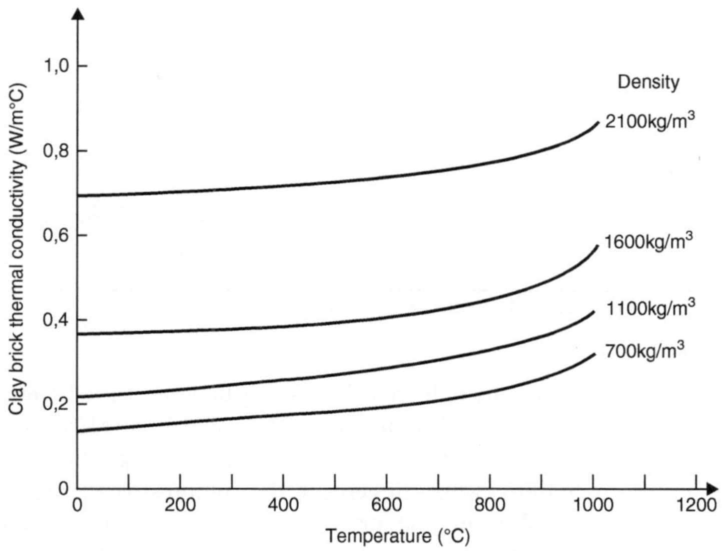

FIGURE 4.39 Variation in the thermal conductivity of masonry with temperature.

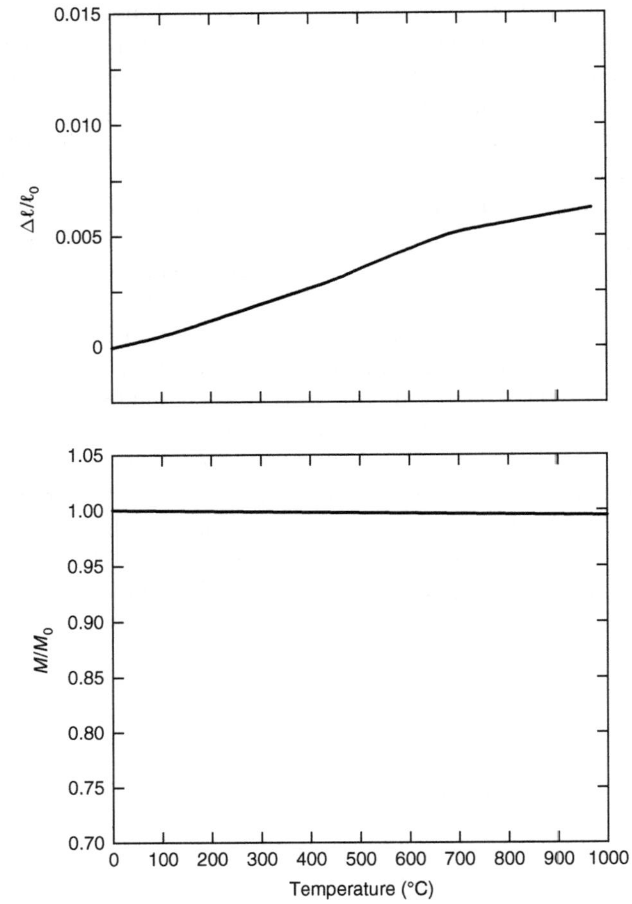

FIGURE 4.40 Dilatometric and thermogravimetric curves for a clay brick (Harmathy, 1983).

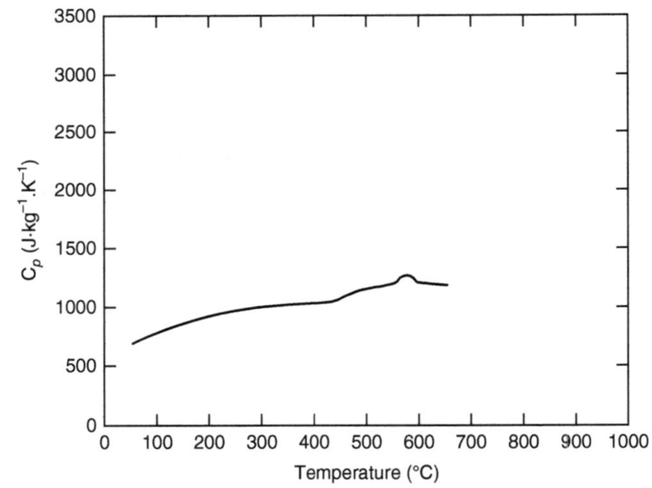

FIGURE 4.41 Apparent specific heat of a clay brick (Harmathy, 1983).

The modulus of elasticity of brick (E) is usually between 10×10^3 and 20×10^3 MPa. Its compressive strength (σ_u) varies in a very wide range, from 9 to 110 MPa; 50 MPa may be regarded as the average value (McBurney and Lovewell, 1933). This value is an order of magnitude greater than the stresses allowed in the design of grouted brickwork. In general, the compressive strength of bricks decreases slowly and reaches to 85% of initial strength at 400°C; this trend is similar to that of plain concrete. According to Eurocode 6 (2005c) masonry units generally become less brittle at elevated temperatures, with a prolonged strain corresponding to the maximum peak stress. In fact, bricks made of autoclaved aerated concrete

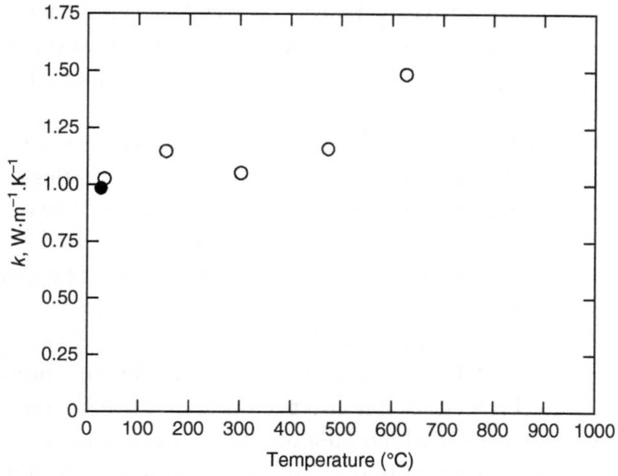

Figure 4.42 Thermal conductivity of a clay brick. Symbols: O—heating cycle, ●—after cooling. (Harmathy, 1983.)

(AAC) and/or calcium silicate can achieve slightly higher compressive strength at elevated temperature up to 500 to 700°C. Overall, limited research on properties of masonry and brick at elevated temperature is due to the fact that masonry is rarely considered for important load-bearing roles in buildings (Harmathy, 1983).

4.4.5 FRP Composites

In recent years, there has been a growing interest in the use of fiber-reinforced polymer (*FRP*) composites in civil engineering construction (especially for retrofitting and strengthening applications). This is mainly due to advantages, such as high strength and durability (resistance to corrosion), that FRP offers over traditional construction materials, as well as ease of installation in repair and strengthening of structures. FRP composites are composed of two key elements: the fiber (glass, carbon, or aramid) and a thermosetting or thermoplastic polymer matrix such as epoxy, vinyl ester, phenolic, or polyester resin. The commonly used fiber types in FRP for critical infrastructure are carbon fiber-reinforced plastic (CFRP), glass fiber-reinforced plastic (GFRP), and aramid fiber-reinforced plastic (AFRP) composites. FRPs are combustibles and are similar to wood in that they burn when exposed to fire and can be classified as an L/I/F type material.

A common feature of FRP is that fiber and binding polymer matrix form a macroscopically orthotropic layer of material with much improved mechanical properties along the direction of fibers, as compared to that in transverse direction. Due to these features, FRPs exhibit unique "orthotropic" properties when compared to concrete or steel. The strength and stiffness of an FRP are governed by the properties of fibers and matrix, the fiber cross-sectional area, and the method of manufacturing; these properties vary in different directions (Rasheed, 2014).

Fibers are often selected to have high stiffness, high ultimate strength, low variation in strength between individual fibers, and uniform diameter. On the other hand, the matrix is the binding material for a group of fibers and generally has poor thermal and mechanical properties. Matrix materials are chosen such that they are thermally compatible with the fibers in order to reduce magnitude of stresses resulting from differential thermal expansion. Matrix materials also have much lower density than the fibers, and this is to get reduced overall weight of the composite (FRP). In general, matrix materials can be grouped into two broad categories: thermoplastics and thermosetting resins. Both thermoplastics and thermosets are characterized by low thermal conductivity property.

FRP can be used as internal reinforcement (as an alternative to traditional reinforcing bars) and/or as external reinforcement in forms, such as wrapping and sheeting or near surface-mounted reinforcement (bars or strips) for the rehabilitation and strengthening of concrete members (Yu and Kodur, 2014). One of the main impediments to using FRPs in buildings is its poor fire resistance properties (Kodur et al., 2005).

There are some major differences associated with FRP as a material. The properties of FRP depend on the type and composition of fiber and matrix, and the availability of various types of these fibers and matrix combinations makes it difficult to establish "generic" properties at elevated temperatures. The material properties are controlled by the fibers in the longitudinal direction and by the matrix in the transverse direction. In addition to thermal and mechanical properties, factors such as burning, charring, evolution of smoke, and toxicity in fire also play a significant role in determining the overall fire performance of FRP-incorporated structures. A summary of typical mechanical properties for various types of FRPs, in comparison to other commonly used construction materials, at room temperature, is presented in Table 4.3.

Exposure to high temperature has detrimental effects on the behavior of FRP composites and produces severe degradation in its mechanical and deformational properties, i.e., reduction of strength and stiffness, and increase in deformability, thermal expansion, and creep. The degradation can be quite rapid at above 100°C as the glass transition temperature of the matrix is reached.

The glass transition temperature is often considered as the upper limiting of temperature for thermal stability of FRP composites. This glass transition temperature varies with the type of resin used and is in the range of 70 to 90°C. From a strength consideration, GFRP strength and stiffness properties reduce to 25% at temperatures of 250°C (Gates, 1993). The critical temperature concept used for steel in evaluating fire resistance of steel structures has also been utilized for FRP-reinforced structures. For FRP, the critical temperature, temperature at which 50% of tensile strength is lost, is much lower than that for steel and depends on the composition of fibers and matrix. Wang and Kodur have developed critical temperature limits for glass and carbon FRP reinforcing rebars and, accordingly, the critical temperature for GFRP is 250°C which is lower than that for CFRP given as 325°C (Wang et al., 2007).

Information on properties of FRP has been reviewed by Yu (2013). The thermal properties of FRP are dependent on direction, fiber type, fiber orientation, fiber volume fraction, and laminate configuration. Table 4.4 shows thermal properties for different types of FRP at room temperature. In the longitudinal direction, the thermal expansion of FRPs is lower than that of steel. However, in the transverse direction, it is much higher than that of steel. At room temperatures, FRPs, with the exception of carbon fibers, in general have low thermal conductivity.

Information on the thermal properties of FRP at elevated temperatures, especially in the range of 300 to 500°C, is limited, which is likely because such information is proprietary to the composite materials'

Material	Modulus of Elasticity E_1 (MPa)*	Modulus of Elasticity E_2 (MPa)**	Tensile Strength σ_{t1} (MPa)	Comp. Strength σ_{c1} (MPa)	Shear Modulus G (MPa)	Shear Strength S (MPa)	Poisson's Ratio v	Tensile Strength σ_2 (MPa)	Comp. Strength σ_{c2} (MPa)
GFRP (glass/epoxy)	55,000	18,000	1050	1050	9000	42	0.25	28	140
GFRP (glass/epoxy) unidirectional	42,000	12,000	700	—	5000	72	0.30	30	—
CFRP (carbon/epoxy) unidirectional	180,000	10,000	1500	—	7000	68	0.28	40	—
CFRP (graphite/epoxy)	207,000	5200	1050	700	2600	70	0.25	40	120
Boron/epoxy	207,000	21,000	1400	2800	7000	126	0.30	84	280
ARP (aramid/epoxy) unidirectional	76,000	8000	1400	—	3000	34	0.34	12	—
Mild steel	200,000	—	550	240	—	380	—	—	—
Concrete (normal strength)	31,000	—	~4	40	—	~7	0.15–0.20	—	—
Wood (Douglas fir)	9800	—	69	—	—	—	—	—	—

*E_1 = modulus of elasticity in longitudinal direction, **E_2 = modulus of elasticity in transverse direction.

TABLE 4.3 Properties of Various FRP Composites and Other Materials at Room Temperature manufacturers. Also, there is limited information on the evolution of smoke and toxins in FRP composites exposed to fire.

Thermal expansion of FRP reinforcement at elevated temperatures varies in longitudinal and transverse directions, and further the coefficient of thermal expansion is highly dependent on the type of fiber, resin, and volume fraction of fiber. The longitudinal coefficient of thermal expansion is dominated by the properties of the fiber, while the transverse coefficient is dominated by the properties of the resin. Figure 4.43*a*,*b* shows longitudinal and transverse coefficients of thermal expansion for typical GFRP and

Material	Coefficient of Thermal Expansion (Unidirectional) (α: 10^{-6} m·m$^{-1\times}$°C)		Thermal Conductivity k (W·m^{-1}° C^{-1})	
	Longitudinal α_L	Transverse α_T	Longitudinal k_L	Transverse k_T
Glass/epoxy (S-glass)	6.3	19.8	3.46	0.35
Glass/epoxy (E-glass: 63% fiber)	7.13	—	—	—
Carbon/epoxy (high modulus)	−0.9	27	48.4–60.6	0.865
Carbon/epoxy (ultra-high modulus)	−1.44	30.6	121.1–129.8	1.04
Boron/epoxy	4.5	14.4	1.73	1.04
Aramid/epoxy (Kevlar 49)	−3.6	54	1.73	0.73
Concrete	6.16		1.36–1.90	
Steel	10.8–18		15.6–46.7	
Epoxy	—	54–90	—	0.346

Table 4.4 Thermal Properties of Various FRPs and Other Materials at Room Temperature

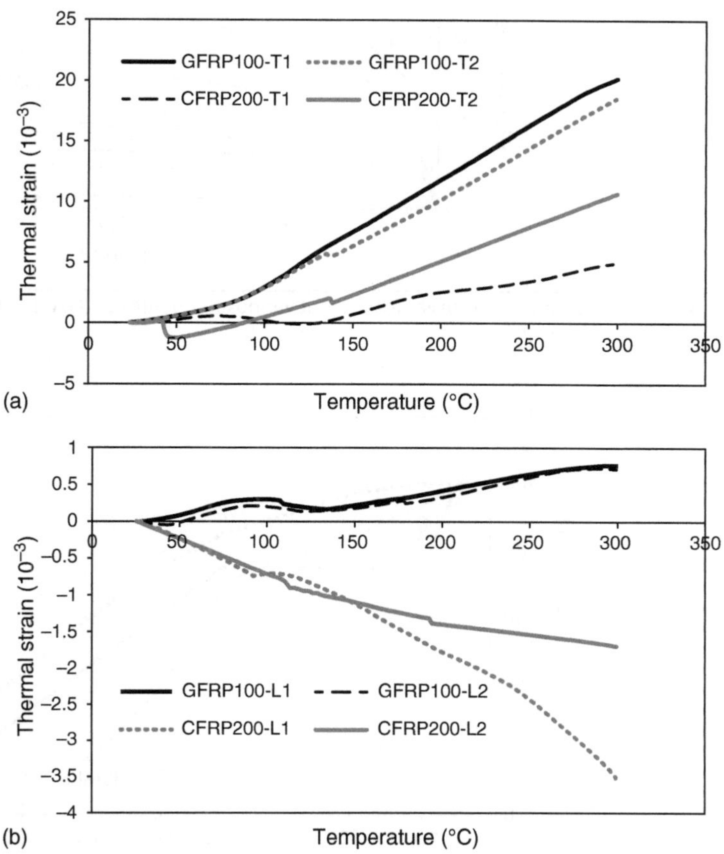

Figure 4.43 Variation in thermal strain in GFRP and CFRP in (a) longitudinal and (b) transverse directions as a function of temperature.

CFRP bars. Note that usually there is a change in expansion rate at around glass transition temperature (T_g), indicating FRP reinforcement experiences different coefficients of thermal expansion before and after phase change (T_g). In transverse direction, the dimension of GFRP and CFRP rebars increase with temperature, and GFRP undergoes higher thermal expansion than that of CFRP. However, in longitudinal direction, GFRP rebar slightly expands with temperature, but CFRP rebar contracts with increase in temperature. The coefficients of thermal expansion in transverse direction for GFRP and CFRP rebars can be taken to be 64.5 and 7.79×10^{-6}/C, respectively, while the corresponding coefficients of thermal expansion in longitudinal direction are 2.48 and -7.6×10^{-6}/C, respectively (Yu and Kodur, 2014).

The stress-strain relations, from the studies conducted by Gates (1991), for a CFRP composite (IM7/5260) are shown in Fig. 4.44 for various temperatures. It can be seen that the tensile strength of IM7/5260 composite reduces to approximately 50% at about 125°C and to about 75% at a temperature of 200°C. The strain level for a given stress is also higher with the increase in temperature. Recently, research has reported high temperature strength and stiffness properties of glass and carbon FRP rebars; full details of the tests are given in Wang et al. (2007), as well as in Yu and Kodur (2014).

The variation in strength with temperature (ratio of strength at elevated temperature to that at room temperature) for FRP along with that of other traditional construction materials is shown in Fig. 4.45. The rate of strength loss is much higher for FRP than for concrete and steel, resulting in a 50% strength loss at about 200°C.

The bond between FRPs and concrete (or between FRP layers or lap splices in multiple layup applications) is critical for transferring loads in a structural member from concrete to FRP reinforcement.

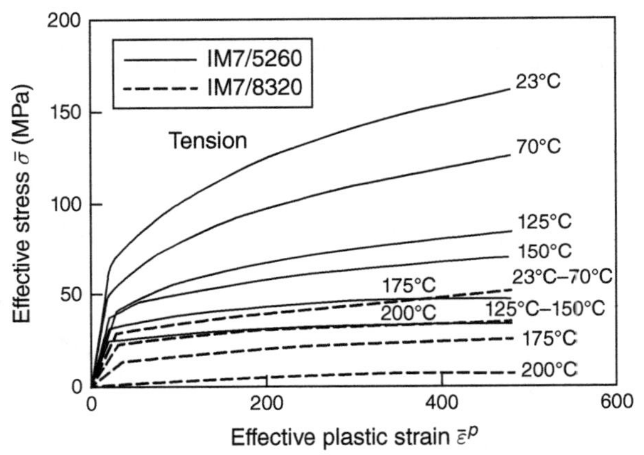

FIGURE 4.44 Tensile stress-strain curves for CFRP at various temperatures (Gates, 1991).

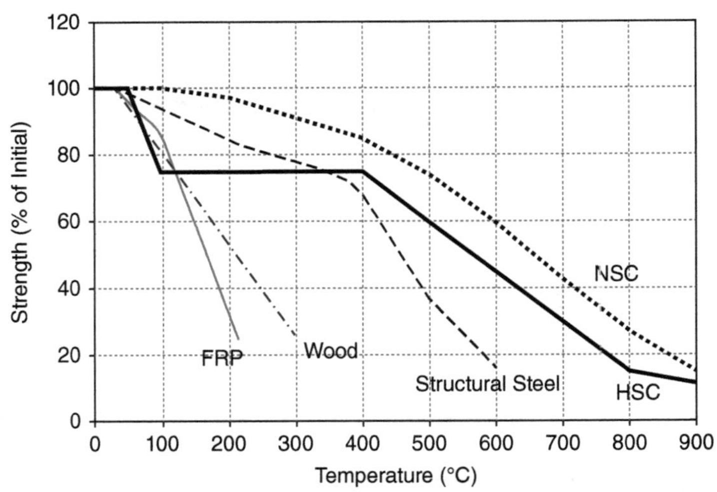

FIGURE 4.45 Variation in degradation in strength properties in various construction materials.

This load transfer occurs through the polymer resin matrix and thus relies heavily on the mechanical properties of the polymer. When the matrix material deteriorates (softens) at temperatures above the specific polymer's glass transition temperature, T_g, a significant loss of bond properties occurs, resulting in a loss of interaction between FRP and concrete. The glass transition temperature of commonly used polymer matrix materials is typically in the range of 70 to 90°C.

Data on temperature-dependent bond properties of FRP plates and bars, used in external strengthening or as internal reinforcement, has been reported by Yu and Kodur (2014). These studies have indicated a significant decrease in bond strength, to about 10% of room temperature bond strength (i.e., temperatures close to or above T_g; 100 to 150°C). This reduction in bond strength is attributed to changes in the properties of the polymer matrix at the interface of FRP bars or strips.

4.4.6 Gypsum

Gypsum (calcium sulfate dihydrate: $CaSO_4 \cdot 2H_2O$) is grouped as a Group I material from fire protection terminology. Gypsum products are used extensively in building construction in the form of boards, including wallboard, formboard, and sheathing, or in powder form by mixing gypsum with other ingredients such as vermiculite to form a fire insulation material. Gypsum board is produced by mixing water with plaster of paris (calcium sulfate hemihydrate: $CaSO_4 \cdot ^1\!/_2H_2O$) or with Keene's cement (calcium sulfate anhydrite: $CaSO_4$). The interlocking crystals of $CaSO_4 \cdot 2H_2O$ are responsible for the hardening of the material. The core of a gypsum board is fabricated with plaster of paris, into which weight- and set-controlling additives are mixed. Furthermore, plaster of paris, with the addition of aggregates (such as sand, pearlite, vermiculite, or wood fiber), is used in wall plaster as base coat, and Keene's cement (neat or mixed with lime putty) is used as finishing coat.

There can be significant variation in fire performance of different types of gypsum board based on the type and the formulation of the core, which varies from one manufacturer to another. A gypsum board with inherent fire resistance property achieved through the gypsum in the core is referred to as regular gypsum. When the core of the gypsum board is modified with special core additives or with enhanced additional properties, to improve the inherent fire resistance of regular gypsum board, this kind of gypsum board is grouped under fire-rated gypsum boards. Gypsum board, based on composition and performance, is grouped into various types, such as regular gypsum board, type C or type X gypsum board, Super type C, and type XXX gypsum board. Type C (or type X) gypsum board has higher fire resistance than regular gypsum board, and Super type C and type XXX have much higher fire resistance than type C (or type X) system.

ASTM C1396 (2017) describes two types of gypsum board, regular and type X, each processing different degrees of fire resistance properties. Type X gypsum board, with higher fire resistance properties, is typically used to achieve fire-resistance rated structural members (walls, floors, columns, etc.). Type XXX gypsum board meets and/or exceeds the requirements of type X drywall specified by ASTM C 1396. Super type C is appropriate for commercial applications requiring extended fire resistance performance that exceeds the type X classification. This product was developed with the addition of mineral additives integrated into the product's core to enhance the gypsum core's performance at elevated temperature.

Gypsum is an ideal fire protection material. The water inside the gypsum plays a critical role in defining its thermal properties and exhibiting its resistance to fire conditions. On heating, gypsum ($CaSO_4 \cdot 2H_2O$) will lose the two water (H_2O) molecules at temperatures between 125 and 200°C. The heat of complete dehydration is 0.61×10^6 J/kg of gypsum. Due to substantial absorption of energy in the dehydration process, a gypsum layer applied to the surface of a building element (walls, floors, etc.) is capable of markedly delaying the heat penetration into the underlying structural member.

The thermal properties of the gypsum board vary depending on the composition of the core. The variation with temperature in the volumetric specific heat (ρc_p) of pure gypsum has been illustrated by Harmathy (1961), based on information reported in the literature (Ljunggren, 1960; West and Sutton, 1954). The thermal conductivity of gypsum products is difficult to assess, owing to large variations in their porosities and the nature of the aggregates. A typical thermal conductivity for gypsum plaster board of about 700 $kg{\cdot}m^{-3}$ density is 0.25 $W{\cdot}m^{-1}{\cdot}K^{-1}$. Figures 4.46 and 4.47 illustrate the typical variation in thermal conductivity and the specific heat, respectively, of the gypsum board core with temperature. The plots reflect the expressions proposed by Sultan (1996), based on tests conducted on type X gypsum board specimens. The specific heat measurements were carried out at a heating rate of 2°C/min. The dehydration of gypsum results in two peaks that appear in the specific heat curve at temperatures around 100°C and 650°C. The peak values are slightly variant to those reported earlier by Harmathy (1968); this may be due to the differences in gypsum composition, used in different test specimens (from different proprietary boards).

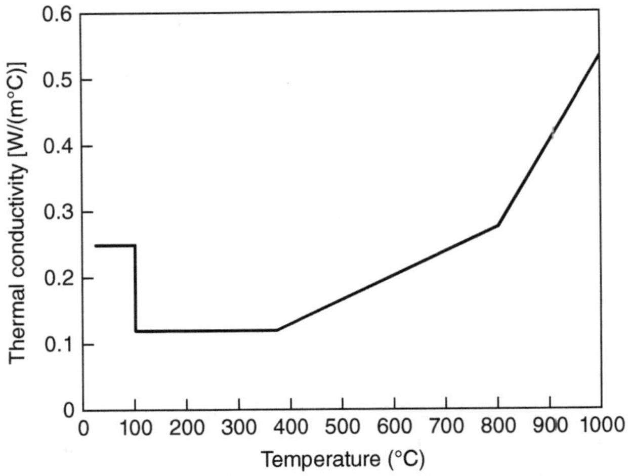

FIGURE 4.46 Thermal conductivity of type X gypsum board core as a function of temperature (Sultan, 1996).

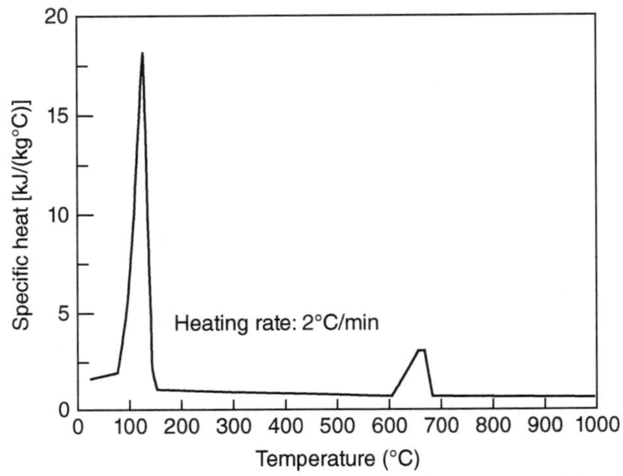

FIGURE 4.47 Specific heat of type X gypsum board core as a function of temperature (Sultan, 1996).

The coefficient of thermal expansion (β) of gypsum products varies between 11.0×10^{-6} and 17×10^{-6} $m{\cdot}m^{-1}{\cdot}K^{-1}$ at room temperature, depending on the nature and amount of aggregates used. The dilatometric and thermogravimetric curves of a fire-resistant (type X or type C) gypsum board of 678 $kg{\cdot}m^{-3}$ density are shown in Fig. 4.48.

There is not much information in the mechanical properties of the gypsum board at elevated temperatures, because these properties are difficult to obtain experimentally and also because the mechanical strength of gypsum board at room temperature (as well as elevated temperature) is very low and is often neglected. The Gypsum Association (2019) lists typical mechanical properties, at room temperature, for some North American gypsum board products. The attachment details (screw spacing, orientation of gypsum board joints, stud joist spacing, etc.) of gypsum board to structural members can have a noticeable effect on the fire performance of the gypsum board, and therefore due consideration should be given to detailing aspects such as screw/nail types, their spacing, and the depth of penetration.

4.4.7 Insulation

Insulation is grouped under Group I material and is commonly used as a fire protection material both for load-bearing structural members such as steel columns and beams, and for lightweight framing assemblies such as floors and walls (constructed of steel or wood). Provisions of such fire insulation helps delay the temperature rise of structural members, thereby enhancing fire resistance of the member. There are a number of fire insulation materials available in the market. Common insulation materials used for fire

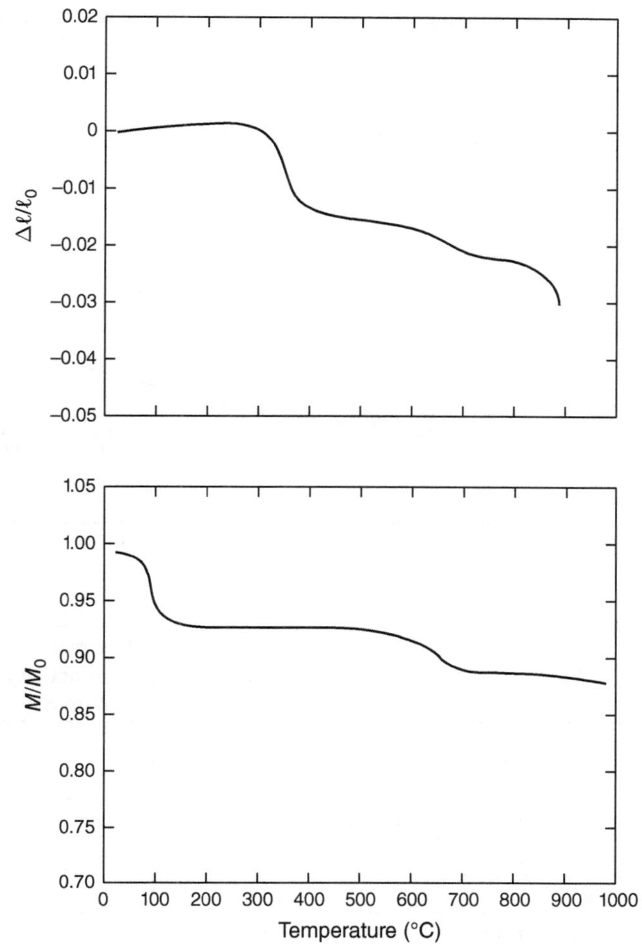

FIGURE 4.48 Dilatometric and thermogravimetric curves of type C (or type X) gypsum board of 678 kg·m^{-3} density.

protection include intumescent paints, spray mineral fibers, gypsum boards, SFRM, and compressed fiber board. For achieving fire resistance in lightweight construction assemblies (i.e., walls and floors), mineral wool and glass fiber are the two most widely used insulation materials.

The thermal properties of fire insulation play an important role in determining the fire resistance of a member. There is limited information available on the temperature-dependent thermal properties of various types of insulation. Figure 4.50a shows the variation in thermal conductivity with temperature for glass fiber and rock fiber insulation types obtained from Bénichou and Sultan (1999). It can be seen that thermal conductivity of these insulation types is very small at ambient conditions and then increases with a rise in temperature. The differences in thermal conductivity values at higher temperatures are mainly due to variations in the chemical composition of fiber.

Full-scale fire resistance tests on walls and floors have shown that the mineral fiber insulation performs better than glass fiber insulation. This is mainly because glass fiber melts in the temperature range of 700 to 800°C and cannot withstand direct fire exposure. The melting point for mineral fiber insulation is higher. The density of glass fiber is about 10 kg/m^3 and is much lower than that of rock fiber, which is about 33 kg/m^3.

The mineral wool insulation, when installed tightly between the studs in walls or joists in floors, significantly enhances fire resistance of non-load-bearing steel stud walls because it acts as an additional fire barrier after the fire-exposed gypsum board falls off from the walls or floors (Sultan, 1995). On the other hand, cavity insulation slows down the flow of heat through the wall assembly and can cause an accelerated temperature rise in the fire-exposed gypsum board.

The above-listed thermal properties for fire insulation are at room temperature, and they can vary significantly with temperature and also with insulation composition, which can vary for different trade names (from different commercial manufactures) among the same type of insulation (e.g., SFRM).

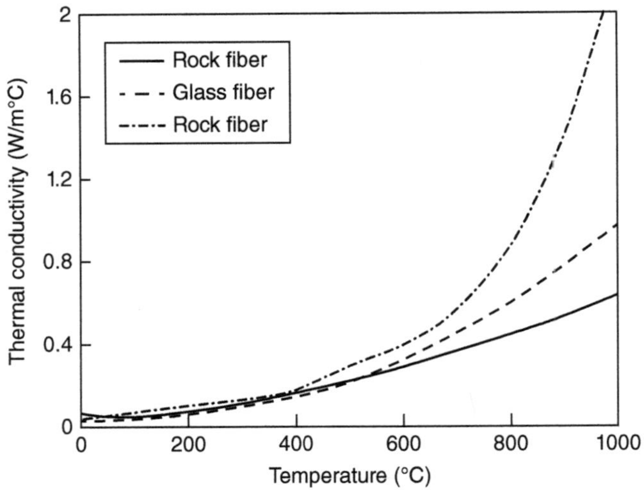

Figure 4.49 Thermal conductivity of insulation as a function of temperature (Bénichou and Sultan, 1999).

Material	Density ρ (kg/m³)	Thermal Conductivity k (W/m·K)	Specific Heat c (J/kg·K)	Equilibrium Moisture Content %
Sprayed mineral fiber	300	0.12	1200	1
Perlite or vermiculite plaster	350	0.12	1200	15
High-density perlite or vermiculite plaster	550	0.12	1200	15
Fiber silicate or fiber calcium silicate	600	0.15	1200	3
Gypsum plaster	800	0.2	1700	20
Mineral wool, fiber silicate	150	0.2	1200	2

Table 4.5 Properties of Some Commonly Used Fire Insulation Materials

However, in current practice fire resistance of insulated structural (steel) members is evaluated by considering only room temperature thermal properties of fire insulation (Kodur et al., 2011). This is mainly due to the lack of reliable data on the effect of temperature on thermal properties of fire insulation.

Another common form of fire insulation applied on load-bearing steel structural members to achieve required fire resistance is SFRM. SFRM, through its superior thermal protection, delays the temperature rise in steel. SFRM, available under different trade names, offers several advantages such as cost-effectiveness, ease of application, and light weight over other types of fire insulation, and therefore is widely used as fire insulation (or fireproofing) to steel structures. SFRM is mainly composed of base materials such as gypsum, cementitious and mineral fiber, and other additives such as vermiculite.

The thermal properties of some of the commonly used fire insulation martials are given in Table 4.5. It should be noted that these values are average property values and can vary depending on the manufacturer (or proprietary) and on the proportions of different constituent materials. Also, the moisture content of the insulation material, at any given time, has an effect on the thermal properties.

Figure 4.49*a* shows the variation in thermal conductivity with temperature for three types of commercially available SFRM (A, B, and C) generated in a recent research study (Kodur and Shakya, 2013). The thermal conductivity of three SFRM types at room temperature is in the range of 0.07 and 0.2 W/m·K. This variation is due to the variation in their densities and also to the composition of ingredients in each type. The trends in the figure further indicate that temperature has a significant effect on the thermal conductivity of SFRM. This variation in thermal conductivity at higher temperatures is primarily governed by changes in moisture content and density of different SFRM types.

Insulation materials such as SFRM experience shrinkage at higher temperatures, as opposed to expansion phenomenon in materials such as steel, concrete, and wood. The variation in thermal strain for three types of SFRM is plotted as a function of temperature in Fig. 4.50. This variation in thermal strain

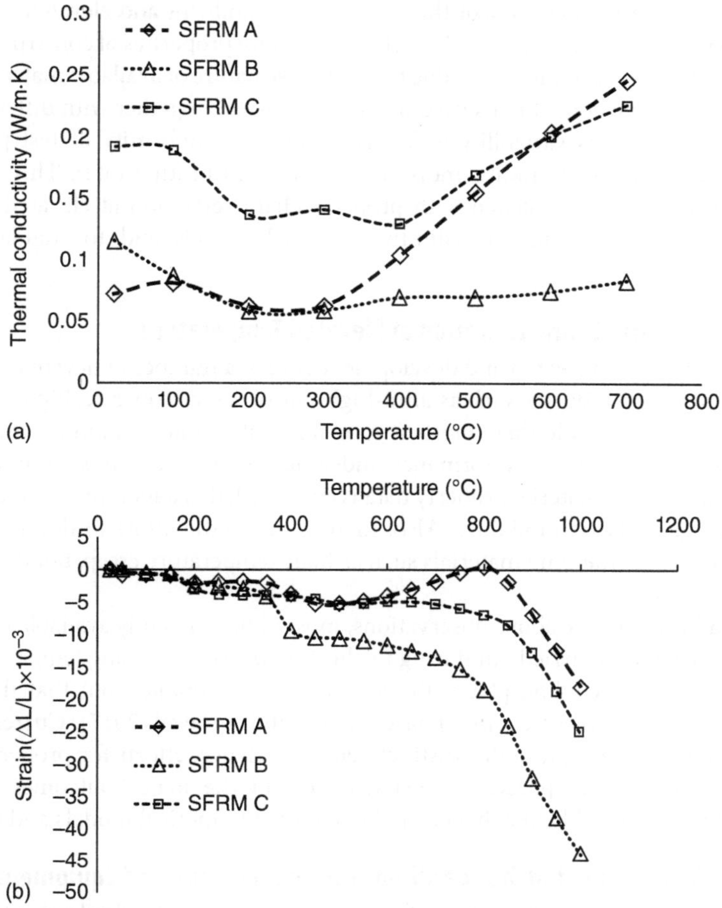

Figure 4.50 Effect of temperature on (a) thermal conductivity and (b) thermal contraction of different SFRMs.

Insulation Type	Density (Kg/m³)		Decrease in density (%)
	Room temp. (20°C)	700°C	
SFRM A	298	241.3	19.0
SFRM B	423.2	349.8	17.3
SFRM C	451.8	381.2	15.6

Table 4.6 Density of SFRM at Room Temperature and After Exposure to 700°C

with temperature is also attributed to changes in moisture content. However, the loss of moisture (content) only accounts for the shrinkage phenomenon that occurs in the 100 to 400°C range. The intermediate expansion resulting in an increase in thermal strains in the 400 to 800°C range is dictated by the expansion of intumescent material, such as vermulite, which is added to SFRM to counteract shrinkage, and the percentage of vermiculite in SFRM has major influence on the level of contraction.

The change in density for three types of SFRM at ambient conditions and after exposure to 700°C is presented in Table 4.6 (Kodur and Shakya, 2013). There is a decrease in density in all three types of SFRM at 700°C, which is predominantly due to the loss of moisture. This decrease in density in SFRM is comparable to that in gypsum, and attributed to dehydration reactions, which take place at elevated temperature (Park et al., 2010).

4.5 Emerging Trends and Research Needs

This chapter presented a number of studies that carried out temperature-dependent thermal and mechanical property characterization of various traditional construction materials. While there is an abundance of information on the behavior of these materials at ambient and elevated temperatures, it is worth noting that there also is large variability in high-temperature properties of construction materials, such as concrete and wood. This variability arises due to the diverse nature of available materials as well as their composition (such as different constituents used in concrete mixes), together with different test setups and equipment, among others. This variability also arises from nonuniformity of test parameters and environmental conditions (such as humidity, moisture content, and heating rate). This section highlights some of the research directions associated with property characterization at elevated temperature, and the need to develop high-temperature constitutive material models and to standardize testing procedures and equipment.

4.5.1 Property Characterization at Elevated Temperature

Due to significant research and development efforts, a number of new materials are finding applications in the construction industry such as ultra-high-performance concrete, high-strength steel, engineered wood, and composites. While there exists a good amount of information on the behavior of such materials at room temperature, their performance under elevated temperatures has not been investigated yet. Even in cases where some material property data is evaluated, this data is mostly focused on mechanical properties (strength and elastic modulus). Also, there is very little effort to develop expressions for properties of traditional and emerging materials such as high-temperature creep, bond, pore pressure, or charring rates (Kodur, 2014).

In addition to the above observations, most of the currently available constitutive material models are derived from tests carried out during the heating phase (i.e., rising temperature of specimen) with little or no regard to the cooling phase. Recent studies have pointed out that the construction materials have distinct behavior in the cooling phase (Kodur and Agrawal, 2017). Currently adopted material property expressions fail to capture this distinct behavior. Hence, efforts for property characterization under both heating and cooling phases are needed. One of the major outcomes of high-temperature property characterization will help enhance predictability of numerical models and rational methods/techniques.

4.5.2 Development of Standardized Testing Procedure and Equipment

To ensure proper characterization of material properties, a standard test is the key. Such procedures are put in place to ensure consistent recoding of measured data points and to allow reproducibility and verification of measured properties among various researchers. In most countries/regions there exist standards (such as ASTM or ISO) for measuring mechanical properties at ambient temperatures. As established in this chapter, there is lack of test methods, equipment, and test procedures for characterizing materials at elevated temperatures.

The lack of standardized material characterization procedure (and equipment) at elevated temperatures has led material scientists and researchers to develop their own test methods. As a result, there is significant variations in testing setups, specimen size, heating rates, and data collection techniques, and this translated into development of number of constitutive material models. These variations in high-temperature constitutive models present a challenge to a designer wishing to evaluate (or model) fire resistance of civil structures. Since the use of different constitutive models can result in different fire resistance predictions, there is a clear need to develop standardized testing procedures in order to unify current state of fire resistance evaluation.

4.6 References

Abrams, Melvin S. *ACI SP 25*. Detroit, MI: American Concrete Institute, 1971.

American Concrete Institute (ACI). *Guide for Determining the Fire Endurance of Concrete Elements, ACI-216-89*. Detroit, MI: American Concrete Institute, 1989.

Anderberg, Yngve. "Mechanical Properties of Reinforcing Steel at Elevated Temperatures," *Tekniska Meddelanden*, 36, Lund, Sweden 1978.

Anderberg, Yngve. "Properties of Materials at High Temperatures Steel." *LUTVDG/TVBB-3008-SE* 3008 (1983).

Anderberg, Yngve, and Sven Thelandersson. "Stress and Deformation Characteristics of Concrete at High Temperatures. 2. Experimental Investigation and Material Behaviour Model." *Bulletin of Division of Structural Mechanics and Concrete Construction, Bulletin 54* (1976).

Anderberg, Yngve. "Spalling Phenomena of HPC and OC." In *NIST Workshop on Fire Performance of High Strength Concrete in Gaithersburg*. Gaithersburg, MD: National Institute of Standards and Technology, pp. 69–73. 1997.

Arablouei, Amir, and Venkatesh Kodur. "Modeling Delamination of Fire Insulation from Steel Structures Subjected to Blast Loading." *Engineering Structures* 116 (2016): 56–69.

ASCE. "Structural Fire Protection: Manual of Practice," No. 78, American Society of Civil Engineers, New York, 1992.

ASTM C1396, C1396/C1396M–17. *Standard Specification for Gypsum Board.* West Conshohocken, PA: American Society for Testing and Materials, 2017.

ASTM C1113/C1113M–09. *Standard Test Method for Thermal Conductivity of Refractories by Hot Wire (Platinum Resistance Thermometer Technique).* West Conshohocken, PA: American Society for Testing and Materials, 2013.

ASTM E119-16a. *Standard Test Methods for Fire Tests of Building Construction and Materials.* West Conshohocken, PA: American Society for Testing and Materials, 2016.

ASTM E1530-11. *Standard Test Method for Evaluating the Resistance to Thermal Transmission of Materials by the Guarded Heat Flow Meter Technique.* West Conshohocken, PA: American Society for Testing and Materials, 2016.

ASTM E1868-10. *Standard Test Methods for Loss-On-Drying by Thermogravimetry.* West Conshohocken, PA: American Society for Testing and Materials, 2015.

ASTM E1876-15. *Standard Test Method for Dynamic Young's Modulus, Shear Modulus, and Poisson's Ratio by Impulse Excitation of Vibration.* West Conshohocken, PA: American Society for Testing and Materials, 2015.

ASTM Test Method C135±86. *2007 Annual Book of ASTM Standards,* 15.01. Philadelphia, PA: American Society for Testing and Materials, 2007.

Bénichou, Noureddine, and M. A. Sultan. *Fire Resistance of Lightweight Wood-Framed Assemblies: State-of-the-Art Report.* Ottawa: Institute for Research in Construction, 1999.

Bennetts, I. D. *Report No. MRL/PS23/81/001.* Clayton, Australia: BHP Melbourne Research Laboratories, 1981.

Bilodeau, A., V. M. Malhotra, and G. C. Hoff. "Hydrocarbon Fire Resistance of High Strength Normal Weight and Light Weight Concrete Incorporating Polypropylene Fibres." In *International Symposium on High Performance and Reactive Powder Concretes,* 271–296, Sherbrooke, QC, 1998.

Birch, Albert Francis, and Harry Clark. "The Thermal Conductivity of Rocks and Its Dependence upon Temperature and Composition." *American Journal of Science* 238, no. 8 (1940): 529–558.

British Standards Institution. BS 5950. "Structural Use of Steelwork in Building," Part 8. In *Code of Practice for Fire Resistant Design.* London: British Standards Institution, 2003.

Buchanan, Andrew H., and Anthony Kwabena Abu. *Structural Design for Fire Safety.* Hoboken, NJ: Wiley, 2017.

Castillo, C. and A. J. Durrani, "Effect of Transient High-Temperature on High-Strength Concrete," *ACI Materials Journal 87,* no. 1 (1990): 47–53.

CEN. *Eurocode 2: Design of Concrete Structures—Part 1-2: General Rules—Structural Fire Design. EN 1992-1-2.* Brussels: European Committee for Standardization, 2004a.

CEN. *Eurocode 5: Design of Timber Structures—Part 1-2: General—Structural Fire Design. EN 1995-1-2.* Brussels: European Committee for Standardization, 2004b.

CEN. *Eurocode 3: Design of Steel Structures—Part 1-2: General Rules—Structural Fire Design. EN 1993-1-2.* Brussels: European Committee for Standardization, 2005a.

CEN. *Eurocode 4: Design of Composite Steel and Concrete Structures—Part 1-2: General Rules—Structural Fire Design. EN 1994-1-2.* Brussels: European Committee for Standardization, 2005b.

CEN. *Eurocode 6: Design of Masonry Structures—Part 1-2: General Rules—Structural Fire Design, EN 1996-1-2.* European Committee for Standardization, Brussels, 2005c.

Cheng, Fu-Ping, V. K. R. Kodur, and Tien-Chih Wang. "Stress-strain Curves for High Strength Concrete at Elevated Temperatures." *Journal of Materials in Civil Engineering* 16, no. 1 (2004): 84–90.

Cooke, Gordon ME. "An Introduction to the Mechanical Properties of Structural Steel at Elevated Temperatures." *Fire Safety Journal* 13, no. 1 (1988): 45–54.

Cruz, C. R. "Elastic Properties of Concrete at High Temperatures." *Journal of the PCA Research and Development Laboratories* 8 (1966): 37–45.

Cruz, C. R. Technical report relating to Creep in Concrete at Elevated Temperatures. *Journal, PCA Research and Development Laboratories* 8 (1966): 37.

Danielsen, Ulf, T. A. Hammer, H. Justnes, and S. Smeplass. "Marine Concrete Structures Exposed to Hydrocarbon Fires." *Report, SINTEF-The Norwegian Fire Research Institute* (1997): 56–76.

DeVries, D. A. *Problems Relating to Thermal Conductivity.* Bulletin de l'Institut International du Froid, Annexe-1, Louvain, Belgique (1952): 115.

Diederichs, U., and Ulrich Schneider. "Bond Strength at High Temperatures." *Magazine of Concrete Research* 33, no. 115 (1981): 75–84.

Dorn, John Emil. "Some Fundamental Experiments on High Temperature Creep." *Journal of the Mechanics and Physics of Solids* 3, no. 2 (1955): 85–116.

Dupont. *Thermogravimetric Analyzer (TGA).* Wilmington, DE: Dupont Instruments, 1977.

ECCS. *European Recommendations for the Fire Safety of Steel Structures. European Convention for Construction Steelwork, Tech. Comm. 3.* New York: Elsevier, 1983.

Eitel, Wilhelm. *Thermochemical Methods in Silicate Investigation.* New Brunswick, NJ: Rutgers University Press, 1952.

Gates, Thomas S. "Effects of Elevated Temperature on the Viscoplastic Modeling of Graphite/Polymeric Composites." In *High Temperature and Environmental Effects on Polymeric Composites.* ASTM International, 1993.

Gerhards, Charles C. "Effect of Moisture Content and Temperature on the Mechanical Properties of Wood: An Analysis of Immediate Effects." *Wood and Fiber Science* 14, no. 1 (2007): 4–36.

Gerlich, J. T. "Design of Loadbearing Light Steel Frame Walls for Fire Resistance." ME paper, Christchurch, New Zealand: University of Canterbury, 1995.

Gypsum Association. Gypsum Board: Typical Mechanical and Physical Properties, *GA-235-2019.* Washington, DC: Gypsum Association, 2019.

Gross, H. "High-Temperature Creep of Concrete." *Nuclear Engineering and Design* 32, no. 1 (1975): 129–147.

Hamilton, R. L., and O. K. Crosser. "Thermal Conductivity of Heterogeneous Two-Component Systems." *Industrial & Engineering Chemistry Fundamentals* 1, no. 3 (1962): 187–191.

Harmathy, T. Z. *ASTM STP301*. Philadelphia, PA: American Society for Testing and Materials, 1961.

Harmathy, T. Z. *ASTM STP422*. Philadelphia, PA: American Society for Testing and Materials, 1967.

Harmathy, T. Z. "Simultaneous Moisture and Heat Transfer in Porous Systems with Particular Reference to Drying." *Industrial & Engineering Chemistry Fundamentals* 8, no. 1 (1969): 92–103.

Harmathy, T. Z. "Thermal Properties of Concrete at Elevated Temperatures." *Journal of Materials*, 1970.

Harmathy, T. Z. "Determining the Temperature History of Concrete Constructions Following Fire Exposure." *Journal Proceeding* 65, no. 11(1968): 959–964.

Harmathy, T. Z. *DBR Paper No. 1080, NRCC 20956*. Ottawa: National Research Council of Canada, 1983.

Harmathy, T. Z. *Fire Safety Design and Concrete*. Harlow, UK: Longman Scientific and Technical, 1993.

Harmathy, T. Z., and L. W. Allen. "Thermal Properties of Selected Masonry Unit Concretes." *Journal Proceedings*, 70, no. 2 (1973): 132–142.

Harmathy, T. Z., and John Earl Berndt. "Hydrated Portland Cement and Lightweight Concrete at Elevated Temperatures." *Journal Proceedings*, 63, no. 1(1966): 93–112.

Harmathy, T. Z., and W. W. Stanzak. *ASTM STP464*. Philadelphia, PA: American Society for Testing and Materials, 1970.

Hurley, Morgan J., Daniel T. Gottuk, John R. Hall Jr, Kazunori Harada, Erica D. Kuligowski, Milosh Puchovsky, John M. Watts Jr, and Christopher J. Wieczorek, eds. *SFPE Handbook of Fire Protection Engineering*. Springer, 2015.

ISO/DIS 22007-2, Plastics – Determination of thermal conductivity and thermal diffusivity – Part 2: Transient plane source method. 2015.

Jumppanen, U-M., Ulrich Diederichs, and Konrad Hinrichsmeyer. *Material Properties of F-Concrete at High Temperatures*, VTT Research Report, No. 452, Technical Research Center of Finland, Espo, 1986.

Kingery, W. D. "Introduction to Ceramics." New York: John Wiley & Sons, 1960, pp. 483–486.

Kodur, V. R. "Properties of Concrete at Elevated Temperatures." *ISRN Civil Engineering*, Hindawi Publishing Corporation, pp. 1-28, 2014.

Kodur, V. K. R. "Properties of Concrete at Elevated Temperatures." *ISRN Civil Engineering* 2014, no. 468510, (2014): 15 pages.

Kodur, V. K. R., and A. M. Shakya. "Effect of temperature on thermal properties of spray applied fire resistive materials." *Fire Safety Journal* 61 (2013): 314–323.

Kodur, V. K. R., and Ankit Agrawal. "Effect of Temperature Induced Bond Degradation on Fire Response of Reinforced Concrete Beams." *Engineering Structures* 142 (2017): 98–109.

Kodur, V. K. R., and M. M. S. Dwaikat. "Effect of High Temperature Creep on the Fire Response of Restrained Steel Beams." *Materials and Structures* 43, no. 10 (2010): 1327–1341.

Kodur, V. K. R., M. M. S. Dwaikat, and Rustin Fike. "High-Temperature Properties of Steel for Fire Resistance Modeling of Structures." *Journal of Materials in Civil Engineering* 22, no. 5 (2010): 423–434.

Kodur, V. K. R., and T. Z. Harmathy. "Properties of Building Materials." In *SFPE Handbook of Fire Protection Engineering*. New York: Springer, 2016, pp. 277–324.

Kodur, V. K. R., and Wasim Khaliq. "Effect of Temperature on Thermal Properties of Different Types of High-strength Concrete." *Journal of Materials in Civil Engineering* 23, no. 6 (2011): 793–801.

Kodur, V. K. R., R. Kumar, and D. Baingo. *Fire Resistance of FRP Reinforced Concrete Slabs*. Ottawa, Canada: Institute for Research in Construction, 1998.

Kodur, V. K. R., and R. McGrath. "Effect of Silica Fume and Lateral Confinement on Fire Endurance of High Strength Concrete Columns." *Canadian Journal of Civil Engineering* 33, no. 1 (2006): 93–102.

Kodur, V. K. R., and M. A. Sultan. "Structural Behaviour of High Strength Concrete Columns Exposed to Fire." (1998a): 217–232.

Kodur, V. K. R., and M. A. Sultan. "Thermal Properties of High Strength Concrete at Elevated Temperatures." *ACI Special Publications* 179 (1998b): 467–480.

Kodur, V. K. R., and M. A. Sultan. "Effect of Temperature on Thermal Properties of High-Strength Concrete." *Journal of Materials in Civil Engineering* 15, no. 2 (2003): 101–107.

Kodur, V. K. R., T. C. Wang, and F. P. Cheng. "Predicting the Fire Resistance Behaviour of High Strength Concrete Columns." *Cement and Concrete Composites* 26, no. 2 (2004): 141–153.

Kodur, V. K. R., Luke A. Bisby, and Mark F. Green. "Experimental Evaluation of the Fire Behaviour of Insulated Fibre-Reinforced-Polymer-Strengthened Reinforced Concrete Columns." *Fire Safety Journal* 41, no. 7 (2006): 547–557.

Kodur, V. R., Bisby, L. A; Green, M. F. "Preliminary Guidance for the Design of FRP-Strengthened Concrete Members Exposed to Fire." *Journal of Fire Protection Engineering* 17, no. 1 (2007): 5-26.

Kodur V. K. R, M. M. S, Dwaikat, and M. B. Dwaikat. "High-Temperature Properties of Concrete For Fire Resistance Modeling of Structures," *ACI Materials Journal* 105, no. 5 (2008): 517–527.

Kodur, V., Stein, J. Fike, Rustin, and Tabbador. M. "Comparative Fire Performance of Traditional Lumber and Engineered Wood Joists." *Journal of Structural Fire Engineering* 8, no. 1 (2017): 2-13.

Li, Guo-Qiang, Shou-Chao Jiang, Ying-Zhi Yin, Kai Chen, and Ming-Fei Li. "Experimental Studies on the Properties of Constructional Steel at Elevated Temperatures." *Journal of Structural Engineering* 129, no. 12 (2003): 1717–1721.

Lie, T. T., ed. "Structural Fire Protection." New York, NY: American Society of Civil Engineers, 1992.

Lie, T. T., and V. K. R. Kodur. "Thermal and Mechanical Properties of Steel-Fibre-Reinforced Concrete at Elevated Temperatures." *Canadian Journal of Civil Engineering* 23, no. 2 (1996): 511–517.

Makelainen, P., and K. Miller. *Mechanical Properties of Cold-Formed Galvanized Sheet Steel Z32 at Elevated Temperatures*. Helsinki: Helsinki University of Technology, 1983.

Malhotra, H. L. "Report on the Work of Technical Committee 44PHT 'Properties of Materials at High Temperatures'." *Matériaux et Construction* 15, no. 2(1982): 161–170.

Ljunggren, Pontus. "Determination of Mineralogical Transformations of Gypsum by Differential Thermal Analysis." *Journal of the American Ceramic Society* 43, no. 4 (1960): 227.

MareÂchal, J.C. *ACI SP 34*. Detroit, MI: American Concrete Institute, 1972.

Maxwell, James Clerk, and Joseph John Thompson. *A Treatise on Electricity and Magnetism*. Vol. 2. Clarendon, 1904.
McBurney, J. W., and C. E. Lovewell. "Strength, Water Absorption and Weather Resistance of Building Bricks Produced in the United States." *Proceeding of ASTM*, 33, no. II, (1933): 636–650.
Noumowe, A. N. "Thermal Stresses and Water Vapour Pressure of High Performance Concrete at High Temperature." In *4th Int. Symp. on Utilization of High-strength/High-performance Concrete*, Paris, vol. 2 (1996): 561–570.
Park, Seul×Hyun, Samuel L. Manzello, Dale P. Bentz, and Tensei Mizukami. "Determining Thermal Properties of Gypsum Board at Elevated Temperatures." *Fire and Materials* 34, no. 5 (2010): 237–250.
Perry, John H. *Chemical Engineers' Handbook*. New York: McGraw-Hill, 1953: p. 533.
Phan, L. T. *Fire Performance of High-Strength Concrete: A Report of the State-of-the-Art*. Gaithersburg, MD: National Institute of Standards and Technology, 1996.
Poh, K. W. "Stress-Strain-Temperature Relationship For Structural Steel." *Journal of Materials in Civil Engineering* 13, no. 5 (2001): 371–379.
Purkiss, J. A. "Steel Fibre Reinforced Concrete at Elevated Temperatures." *International Journal of Cement Composites and Lightweight Concrete* 6, no. 3 (1984): 179–184.
Rasheed, Hayder A. *Strengthening Design Of Reinforced Concrete with FRP*. Boca Raton, FL: CRC Press, 2014.
Rootare, H. M. "A Review of Mercury Porosimetry." In *Advanced Experimental Techniques in Powder Metallurgy*, pp. 225–252. Boston, MA: Springer, 1970.
Schaffer, Erwin. "State of Structural Timber Fire Endurance." *Wood and Fiber Science* 9, no. 2 (2007): 145–170.
Schneider, U. "Behaviour of Concrete under Thermal Steady State and Non-Steady State Conditions." *Fire and Materials* 1, no. 3 (1976): 103–115.
Schneider, Ulrich, ed. *Properties of Materials at High Temperatures - Concrete*. Kassel, Germany: Kassel University, 1985.
Shakya, A. M., and V. K. R. Kodur. "Effect of Temperature on the Mechanical Properties of Low Relaxation Seven-Wire Prestressing Strand." *Construction and Building Materials* 124 (2016): 74–84.
Sultan, M. A. "Effect of Insulation in the Wall Cavity on the Fire Resistance Rating of Full-Scale Asymmetrical (1 2) Gypsum Board Protected Wall Assemblies." *Internal Report, National Research Council of Canada*, Ottawa, ON, Canada (1995): 545–550.
Sultan, Mohamed A. "A Model for Predicting Heat Transfer through Noninsulated Unloaded Steel-Stud Gypsum Board Wall Assemblies Exposed to Fire." *Fire Technology* 32, no. 3 (1996): 239–259.
Wang W., Kodur V. K. R., Yang X., and Li G., "Experimental Study on Local Buckling of Axially Compressed Steel Stub Columns at Elevated Temperatures." *Thin-Walled Structures* 82 (2014): 3-45.
Wang, Wei-yong, Bing Liu, and Venkatesh Kodur. "Effect of Temperature on Strength and Elastic Modulus pf High-Strength Steel." *Journal of Materials in Civil Engineering* 25, no. 2 (2012): 174–182.
Wang, Y. C. P. M. H. Wong, and V. Kodur. "An Experimental Study of the Mechanical Properties of Fibre Reinforced Polymer (FRP) and Steel Reinforcing Bars at Elevated Temperatures." *Composite Structures* 80, no. 1 (2007): 131–140.
Wangaard, "Section 29." In *Engineering Materials Handbook*, C.L. Manrtell (ed.). New York, NY: McGraw-Hill, 1958.
West, Richard R., and Willard J. Sutton. "Thermography of gypsum." *Journal of the American Ceramic Society* 37, no. 5 (1954): 221–224.
Yu, Baolin. *Fire Response of Reinforced Concrete Beams Strengthened with Near-Surface Mounted FRP Reinforcement*. East Lansing, MI: Michigan State University, 2013.
Yu B. and Kodur, V. K. R., "Effect of Temperature on Strength and Stiffness Properties of Near-Surface Mounted FRP Reinforcement", *Journal of Composites Part B: Engineering* 58 (2014): 510-517.
Yu, Baolin, and Venkatesh Kodur. "Effect of Temperature on Strength and Stiffness Properties of Near-Surface Mounted FRP Reinforcement." *Composites Part B: Engineering* 58 (2014): 510–517.
Zener, Clarence, and John Herbert Hollomon. "Effect of Strain Rate upon Plastic Flow of Steel." *Journal of Applied Physics* 15, no. 1 (1944): 22–32.

4.7 Nomenclature

a	Material constant, dimensionless
b	Constant, characteristic of pore geometry, dimensionless
c	Specific heat ($J{\cdot}kg^{-1}{\cdot}K^{-1}$)
$\bar{c}$	Specific heat for a mixture of reactants and solid products ($J{\cdot}kg^{-1}{\cdot}K^{-1}$)
E	Modulus of elasticity (Pa)
h	Enthalpy ($J{\cdot}kg^{-1}$)
Δh	Latent heat associated with a "reaction" ($J{\cdot}kg^{-1}$)
ΔH_c	Activation energy for creep ($J{\cdot}kmol^{-1}$)
k	Thermal conductivity ($W{\cdot}m^{-1}{\cdot}K^{-1}$)
L_v	Heat of gasification of wood
ℓ	Dimension (m)
$\Delta\ell = \ell - \ell_0$	

m	Exponent, dimensionless
M	Mass (kg)
n	Material constant, dimensionless
P	Porosity ($m^3 \cdot m^{-3}$)
q_n	Net heat flux to char front
R	Gas constant (8315 $J \cdot kmol^{-1} \cdot K^{-1}$)
S	Specific surface area ($m^2.m^{-3}$)
t	Time (h)
T	Temperature (K or °C)
v	Volume fraction ($m^{-3} \cdot m^3$)
w	Mass fraction ($kg \cdot kg^{-1}$)
Z	Zener-Hollomon parameter (h^{-1})
ϑ	Thermal diffusivity
α	Coefficient of linear thermal expansion ($m \cdot m^{-1}$)
γ	Expression defined by Eq. 4.3, dimensionless
β_0	Charring rate (mm/min)
δ	Characteristic pore size (m)
ε	Emissivity of pores, dimensionless
ε	Strain (deformation) ($m \cdot m^{-1}$)
ε_{t0}	Creep parameter ($m \cdot m^{-1}$)
$\dot{\varepsilon}_{ts}$	Rate of secondary creep ($m.m^{-1} \cdot h^{-1}$)
θ	Temperature-compensated time (h)
ξ	Reaction progress variable, dimensionless
χ	Material property (any)
ρ	Density ($kg \cdot m^{-3}$)
σ	Stress; strength (Pa)
σ	Stefan-Boltzmann constant (5.67×10^{-8} $W \cdot m^{-2} \cdot K^{-4}$)
g	Glass transient (temperature)
a	Of air
I	Of the *i*th constituent
p	At constant pressure
s	Of the solid matrix
t	True
t	Time-dependent (creep)
T	At temperature T
u	Ultimate
y	Yield
0	Original value, at reference temperature

CHAPTER 5

Fire Resistance Evaluation through Testing

5.1 Overview

This chapter provides an overview of concepts and procedures for undertaking fire resistance tests on structural members, components, and assemblies. The steps involved in undertaking standardized fire resistance tests on structural members are outlined. The relevant test methods, procedures, and conditions for undertaking such tests are summarized. The performance criteria that are to be applied for evaluating failure of structural members under fire conditions are discussed. In addition, rules of thumb and extension principles for extending measured fire resistance data to other similar structural members are discussed.

5.2 Standardized Test Methods

The most common approach for evaluating fire resistance of building elements, such as beams, columns, walls, and floors, is through standardized fire resistance tests. The main purpose of these tests is to obtain fire resistance ratings of structural members for use in building construction. Data from such tests can be utilized to establish comparative fire resistance ratings of different structural members (as well as building components) and also for the development of new products or construction methods.

5.2.1 Overview of Fire Tests

The basic provisions for standard fire tests were developed in the early 20th century. While there have been minor revisions in some of the provisions, the test methods remain essentially unchanged to date. Generally, the process of conducting standard fire tests includes the following:

1. A sponsor reaches a contract with a testing laboratory to undertake (individual or a series of) fire resistance tests on a structural member/assembly.
2. The testing laboratory witnesses the construction (preparation) of the test specimen to ensure that the specimen is designed and built as per the specifications in applicable design codes and standards.
3. The structural member is then tested in a certified testing facility and under standard fire conditions as per the governing fire test standard. During the test, data and observations are collected.
4. If the test is successful, based on compiled observations and test data, the testing laboratory assigns a fire rating for the member and issues a proprietary report.
5. The listed assembly can then form the basis for obtaining approvals from the local building authority for meeting fire rating requirements in buildings and is deemed as satisfying the required fire ratings if an assembly similar to the tested assembly is used in buildings.

Fire tests are conducted in specially designed fire-testing furnaces with specific features and dimensions. During the test, the member or assembly is to be subjected to a standard fire exposure, i.e., a time-temperature curve, in which the temperature is increasing for the entire duration of the test. The performance of tested members or assemblies is measured in terms of fire resistance, which is basically the ability of a member to withstand exposure to standard fire conditions. Fire resistance is usually expressed as the duration (time) during which a building element (e.g., structural member, component, or assembly) meets the specified criteria of performance. The tests are to be conducted as per the testing procedures laid out in fire test standards adopted in each country.

Based on the time taken to reach failure, a fire (resistance) rating is assigned for the tested member by rounding down to the nearest 30-minute interval up to 2 hours, after which hourly intervals are used.[1] The three failure criteria that are to be satisfied in standard fire resistance tests include insulation (or barrier) criterion to limit the temperature rise and fire propagation, stability (or strength) criterion to prevent collapse, and integrity criterion to limit flame (fire) spread.

5.2.2 Evolution of Fire Testing Standards

The idea of undertaking tests for evaluating fire performance of structural members was conceived in the United States in the last decade of the 19th century. The first recorded fire tests in North America were conducted in 1890 in Denver, Colorado, with the aim to evaluate the fire performance of floor systems intended for use in the Denver Equitable Building. In 1894, William H. Merrill established the *Underwriters Laboratories Inc.* in Chicago, Illinois, to comparatively rate structural members based on fire performance, so as to enhance fire safety in buildings and help eliminate the adverse effects of fires. A few years later, a test method for undertaking fire tests was developed for the first time at a meeting held during the International Fire Prevention Congress in 1903. This test method emphasized the ability of building components and structural members to endure a certain *duration* of standard fire exposure.

As a result of the Baltimore fire in 1904, the *American Society for Testing and Materials*, or ASTM, established in 1898, organized Committee P.[2] This committee provided the formal structure that contributed to the development of fire test methods and prepared a proposal for fire endurance testing of floors. During the period from 1906 through 1907, the U.S. Geological Survey used the gas-fired fire testing station at Columbia University to undertake a number of fire tests on building components made of different materials. These tests were reported by R. L. Humphrey in *Geological Survey Bulletin* 370 in 1909. In 1908, the ASTM accepted the results of these tests. A year later, the ASTM also adopted a fire test method for partitions used in buildings.[3] Work from these fire tests, together with other tests, for the protection of structural columns eventually led to the drafting of a testing protocol in 1918 that was the forerunner of the ASTM E119 test standard. After numerous minor revisions, the fire testing methods and procedures have evolved into the comprehensive ASTM E119 standard.

Then, in 1946, delegates from 25 countries met in London and decided to create a new organization named the *International Organization for Standardization (ISO)*. This organization was officially commissioned on February 23, 1947, with the objective "to facilitate the international coordination and unification of industrial standards." Currently, the ISO is a network of the national standards institutes of 157 countries; therefore, the ISO is able to act as a bridging organization in which a consensus can be reached on solutions that meet both the requirements of business and the broader needs of society. The ISO has a specific standard for undertaking fire resistance tests on different building components (i.e., ISO 834), and this standard has been adopted in a number of countries, for example, the United Kingdom. Many other countries have their own fire test standards. A brief listing of fire testing standards adopted in Europe, China, Russia, and Canada, among other countries, is given in Table 5.1. For further information on fire resistance testing, background, and history, readers are encouraged to refer to the paper from Babrauskas and Williamson (1978).

5.2.3 Current Standards for Fire Resistance Tests

Fire resistance tests are carried out according to specifications laid out in national or international fire test standards. The most widely used international standard for the determination of fire resistance is the ISO 834 standard. In the United States, three different test standards, ASTM E119, NFPA 251, and UL 263, are used for evaluating fire resistance of structural members. Many European and Asian countries have adopted ISO 834; whole other countries such as Japan, Australia, and New Zealand have their own standards. Standard fire tests are conducted on single structural members such as a beam, column, or floor/ceiling assembly in a furnace environment where the member (or assembly) is subjected to a standard fire exposure. The fire exposure is specified in terms of temperature-time curve in applicable national

[1] In Canada an additional cutoff point is added at 45 minutes for the first hour.

[2] The predecessor of current committee (dealing with fire tests) E5.

[3] Parallel efforts were carried out in 1914 by the National Fire Protection Association (NFPA). The NFPA Committee on Fire Resistive Construction brought the fire endurance test method forward for action, and a revised form of the method was adopted in 1918 as the NFPA 251 standard.

Standard	Country	Title
ASTM E119	United States	Standard Test Methods for Fire Tests of Building Construction and Materials
NFPA 251		Standard Methods of Tests of Fire Resistance of Building Construction and Materials
UL 269		Standard for Fire Tests of Building Construction and Materials
ISO 834	International	Fire Resistance Tests—Elements of Building Construction
BS476	United Kingdom	Fire Tests on Building Materials and Structures Method for Determination of the Fire Resistance of Elements of Construction (General Principles)
EN 1363	Europe	Fire Resistance Tests: General Requirements
CAN/ULC-S101-07	Canada	Standard Methods of Fire Endurance Tests of Building Construction and Materials
DIN 4102	Germany	Fire Behaviour of Building Materials and Building Components
GB 8624	China	Fire Tests to Building Materials and Products
IS 3809	India	Fire Resistance Tests of Structures
GOST R 1182	Russia	Fire Tests for Construction Materials and Products
NBR 10636	Brazil	Fire Resistance Tests of Building Elements
DS/INSTA 412	Denmark	Fire Tests—Building Products
AS 1530 Part 4	Australia	Methods for Fire Tests on Building Materials, Components and Structures Fire Resistance Tests of Elements of Construction

Table 5.1 Fire Testing Standards for Structural Members in Different Countries

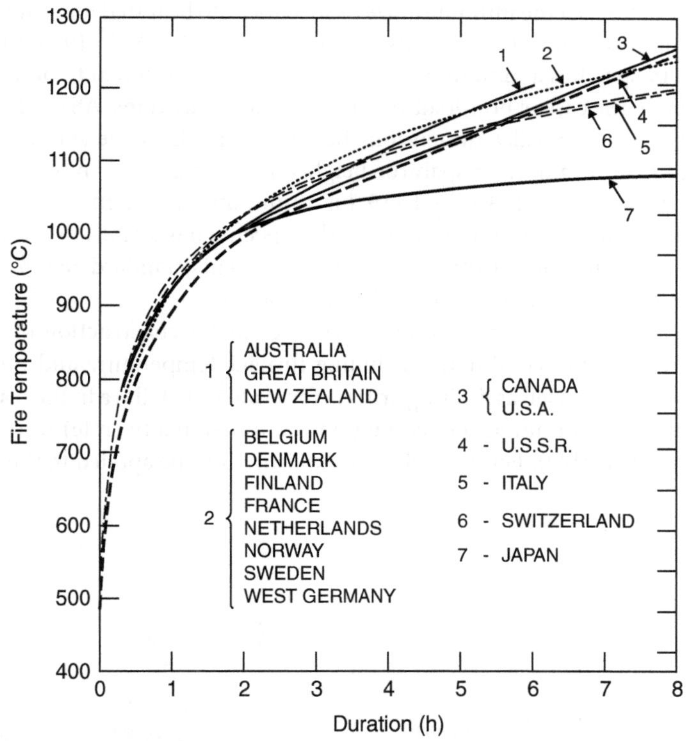

Figure 5.1 Fire curves adopted by different standards around the world. *(From* SFPE Handbook *and printed with permission of SFPE.)*

international fire test standards. The specified fire exposures in most standards are quite similar but have minor variations, as shown in Fig. 5.1.

For the purpose of fire testing, the test specimens must be constructed such that it is representative of the structural member used in a typical building. The test specimen must also be fabricated to ensure the normality of its quality and condition at the time of testing. The minimum dimensions of the member, dictated by the size of the test furnace as specified in the test standard, varies with the type of the structural member (e.g., walls, columns, beams, etc.).

5.3 Procedure for Undertaking Standard Fire Resistance Tests

The current prescriptive-based specifications in most countries require fire resistance evaluations to be based on the results of *full-scale* fire (resistance) tests. Although full-scale tests are expensive, for many years they have been considered essential to establish fire (resistance) ratings on structural members. This is due to the fact that small-scale fire tests may not be able to capture the effects during fire of potential problems caused by connections, shrinkage, deflections, gaps between panels of lining materials, and so forth on structural performance. A brief description of fire tests, test procedures, specimen conditions, instrumentation, and required documentation on fire test is provided in this section.

5.3.1 Fire Test Furnace

Fire resistance tests are carried out by exposing a test specimen to a standard fire under controlled conditions in a fire test furnace. A typical fire test furnace is composed of a heating chamber enclosed with in a large steel box lined with fire bricks or a ceramic fiber blanket. The furnace is equipped with a set of burners fueled by gas or by fuel oil, an exhaust chimney, and several thermocouples for measuring gas (fire) temperature. A typical layout of a fire furnace is shown in Fig. 5.2.

The fire test furnaces are built specifically to test one specific type of assembly (e.g., floors) to test different types of test specimens (e.g., beams, columns, etc.). Some test furnaces can be tipped from a horizontal to vertical orientation to test either a wall or a floor. Special test furnaces are available in some commercial laboratories for testing members or subassemblies under nonstandard test conditions. Figure 5.3 shows some of the fire test furnaces for testing different structural members.

Most testing standards do not specify construction details of the fire furnace but provide recommendations of the minimum size of specimens to be tested for evaluating fire resistance. For example, the minimum size of wall specimens specified both in ASTM E119 and ISO 834 standards is 3.0 × 3.0 m. ASTM E119 specifies a minimum size of roof and slab specimen to be 16 m^2, with neither dimension less than 3.7 m. In the case of non-load-bearing walls and partitions, ASTM E119 specifies a minimum exposed area of 9 m^2, with smaller dimensions being 2.7 m. The same standard also requires a beam and/or a column to have a minimum span (or height) of 3.7 m and 2.7 m, respectively. ISO 834 specifies similar requirements to that of ASTM E119 for minimum specimen sizes to be tested under standard fire conditions. While the minimum roof and slab specimen is 3 m × 4 m, the minimum specimen size for non-load-bearing walls and partitions is 3 m × 3 m. This standard requires beams and columns to have a minimum span (or height) of 4 m and 3 m, respectively.

While fire test standards do not specify details for construction of the fire furnace, these standards specify fire exposure conditions, including furnace temperature and furnace pressure to be maintained during a fire test, as well as limiting criteria for evaluating failure in the tested specimen. Fire test standards specify exposure conditions depending on the functionality (role) of the specimen (structural member, component, assembly). For example, fire exposure is to be applied to the underside of a floor assembly, to

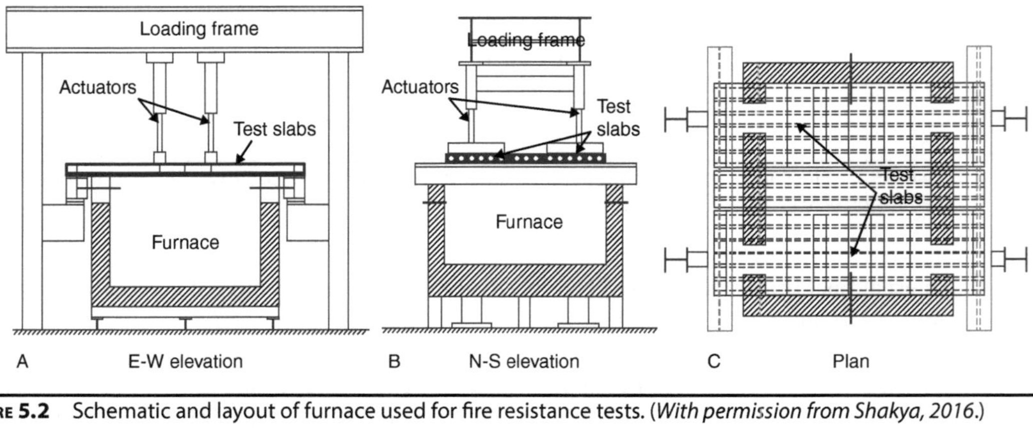

FIGURE 5.2 Schematic and layout of furnace used for fire resistance tests. (*With permission from Shakya, 2016.*)

one side of a wall assembly, to three sides of a beam, and to all four sides of a column. For this reason, specific furnace types are utilized to undertake fire tests on each type of structural member (column, slab, beam, or wall). As mentioned above, fire test furnaces are designed to test specific structural members and only a few test furnaces are configured to test different types of structural members (beam, column, and/or slab).

Furnaces are to be lined with high-temperature refractory bricks or fiberglass insulation. These refractory bricks are of a low-density material (approximately 775 kg/m^3) and have a maximum operating temperature of approximately 1425°C. The refractory bricks, due to their high thermal inertia, relative to the fire exposure period, can absorb significant amounts of heat during the initial period of the fire test (first 15 minutes), producing a dominantly convective heat environment within the test furnace. Once the brick temperature matches with the furnace air temperature, the furnace environment within the furnace transitions to a highly radiative environment. On the other hand, furnaces lined with ceramic fiber insulating material have a faster heating rate due to good insulating properties (i.e., less heat loss/absorption) of ceramic fibers (Harada et al., 1998).

Until the 1980s, furnaces were mostly heated with liquid fuels, such as heavy oil or kerosene. In recent years gaseous fuels such as natural gas or propane are mostly used as the fuel. Cooke (1994) noted that the oil-fired furnace produces a more thermally severe furnace environment as compared to the natural gas-fired environment. Cooke (1994) also noted that heat absorption is lower for gaseous fuels and higher for liquid fuels.

The fuel pumped into the chamber is ignited using a number of burners. The number of burners depends on the size (volume) of the furnace chamber. The "burners" in the test furnace should be strategically located so as to generate uniform temperature throughout the furnace.[4] Two basic types of burners are currently used in fire resistance test furnaces: premixed burners and diffusion burners. Premixed burners carefully control the amount of fuel and combustion air (oxygen) injected into the burner and into the test furnace, resulting in a very uniform flame shape and heating conditions. On the other hand, the air-gas mixture can be adjusted to suit a range of furnace conditions, providing operational flexibility not available with diffusion burners. Premixed burners also produce high gas velocities (turbulence) inside the furnace, which is desired to produce an environment similar to that of fully developed compartment fires.

(a) Wall furnace at National Research Council, Canada (Reproduced with the permission of the National Research Council of Canada, copyright holder.)

(b) Floor furnace at National Research Council, Canada (Reproduced with the permission of the National Research Council of Canada, copyright holder.)

(c) Tongji University, China

Figure 5.3 Typical test furnace for undertaking fire resistance tests.

[4] This is due to the fact that once ignited, burners typically produce a large flame plume, which, depending on the relative location of the test specimen to the burners, may result in undesirable localized heating effects.

5.3.2 Placement and Boundary Conditions of Test Specimens

Usually a test specimen is constructed and instrumented in a frame outside the furnace, where the specimen can be cured (as in concrete specimens, or steel specimens with applied insulation). This also facilitates collecting pretest measurements such as relative humidity of concrete, moisture content of wood, and so on. Following curing, the specimen is then placed in the frame of the furnace for fire testing. Some testing standards also require structural members to be tested with similar conditions to that in actual buildings; for example, beams are to be tested with some level of support restraints. However, most test furnaces are not designed to provide restraints, and hence most of the horizontal elements are tested with simply supported conditions. Once the specimen is properly installed, the fire test starts with the ignition of burners and then the controlling of fuel and oxygen levels to produce a standard time-temperature curve as that prescribed in a fire testing standard.

The test furnaces are equipped with devices for measuring temperature and pressure within the furnace. These devices are also used to monitor performance of furnace components such as fans, burners, and exhaust system. One such instrumentation is a *thermocouple*, which is an electrical device consisting of two dissimilar conductors forming electrical junctions at differing temperatures. A thermocouple produces a temperature-dependent voltage as a result of the thermoelectric effect, and this voltage can be interpreted to measure temperatures. Thermocouples can measure temperature whether inside of furnace or at the surface of the test member or within a tested structural member.

The structural response of a fire-tested member is grouped in terms of deflection of deformations monitored though *linear variable differential transformers* (LVDTs). Further, digital photos or videos have additional value, especially in recording lateral or torsional deflections. Subsequent image analysis can provide quantitative deflection data as well.

5.3.3 Fire Exposure

Traditionally, most building fires have been considered to occur with the bulk of the combustible material in a compartment to be of cellulosic nature. Therefore, the standard furnace temperature-time curve was established with this fuel in consideration. The time-temperature relation is specified in the ASTM E119 and ISO 834 standards for fire exposures as

For ASTM E119 fire:

$$T(°C) = 750(1 - e^{-0.49\sqrt{t}} + 22\sqrt{t}) + 20 \tag{5.1}$$

For ISO 834 fire:

$$T = 345 \log (8t + 1) + 20 \tag{5.2}$$

where T is the furnace temperature (°C) and t is the time (minutes).

It can be seen from Table 5.2 that ASTM E119 fire exposure gives slightly higher temperatures than that of ISO 834 during the first 30 minutes. While temperature-time curves in most codes and standards are

Time (minutes)	ASTM E119/ NFPA No. 251 (°C)	ISO 834 (°C)
0	20	20
5	538	576
10	704	678
30	843	842
60	927	945
90	978	1006
120	1010	1026
180	1052	1110
240	1093	1153
480	1260	1257

Table 5.2 Discrete Time-Temperature Values for ASTM E119 and ISO 834 Standard Fire Curves

Time (minutes)	External Fire (°C)	RWS (°C)	Hydrocarbon (°C)
0	20	20	20
5	600	1140	930
10	660	1200	1140
30	680	1300	1297
60	680	1350	1300
90	680	1300	1300
120	680	1200	1300
180	680	1200	1300

Table 5.3 Discrete Points for Typical Standard Fire Curves for Testing Building Elements

similar, there are very minor variations in the temperature levels, specifically between ASTM E119 and ISO 834. Table 5.2 lists temperatures at various times to be maintained in a furnace during a fire test, after which, temperatures achieved in the ISO 834 become greater than that of ASTM E119 fire curve.

The time-temperature curves in the table represent a scenario for burning inside of a compartment and as such are only applicable for structural members located in a compartment. In the case of fire occurring outside of the compartment/separating walls, an external fire scenario is to be used to evaluate the fire performance of exterior members. Eurocode 1 provides an example of this fire scenario in order to simulate flames (plumes) generated either from the inside of the respective fire compartment or from a compartment situated below or adjacent to the respective external wall. Eurocode 1 refers to this as an *external fire* and gives a relation to represent the time-temperature curve as

$$T(^\circ\text{C}) = 660(1 - 0.686e^{-0.32t} - 0.313e^{-3..8t}) + 20 \tag{5.3}$$

If the combustibles are composed of hydrocarbon products, then a severe burning condition will persist. The fire scenario, as well as the procedure for testing structural members under such hydrocarbon fires, is specified in ASTM E1529 (as a *hydrocarbon,* or HCM, fire curve) or as a RijksWaterStaat, or RWS, fire curve. Table 5.3 lists discrete temperature-time points on external and hydrocarbon fire curves.

The ASTM E1529 hydrocarbon fire curve was first introduced in the early 1990s when the petroleum industry needed to develop a new fire exposure scenario as a result of multiple failures of fire-proofed structural steel members under petroleum spill fires. This fire curve is designed to apply a sudden and intense thermal shock by instantaneously applying a thermal energy equivalent to 158 kW/m^2 to the tested element. The relation to represent the hydrocarbon curve is given as

$$T(^\circ\text{C}) = 1080(1 - 0.325e^{-0.167t} - 0.675e^{-2.5t}) + 20 \tag{5.4}$$

The *RWS* was established by the Ministry of Transport in the Netherlands in 1979 for the evaluation of protecting materials in tunnels. Fire exposure in tunnels can be quite severe due to geometry and also to possible involvement of chemical fuels. Therefore, an RWS fire curve is very severe and describes a scenario of a fire accident in a road tunnel where a 50 m^3 gasoline-carrying tanker explodes. This fire curve is represented by discrete time and temperature points rather than temperature-time relations. Figure 5.4 and Table 5.3 compare these fire curves.

Pressure inside of a furnace is another parameter that can affect fire test conditions and thus the performance of the test specimen. Testing standards specify the specific pressure to be maintained in a furnace during a fire test. Most fire tests conducted in the United States are carried out with negative pressure conditions.[5] This is due to the fact that the ASTM E119 test does not specify any specific pressure requirements. However, ISO 834 specifies a positive pressure of 10 Pa in the testing of horizontal members (beams, floors, and so on). In addition, ISO 834 also specifies a mean value of 8 Pa per meter height in assessing the furnace pressure conditions (with a maximum pressure not exceeding 20 Pa on the top of the test specimen). Other standards such as BS 476 specify other pressure requirements in the furnace during

[5] Positive pressure can force flames through any cracks.

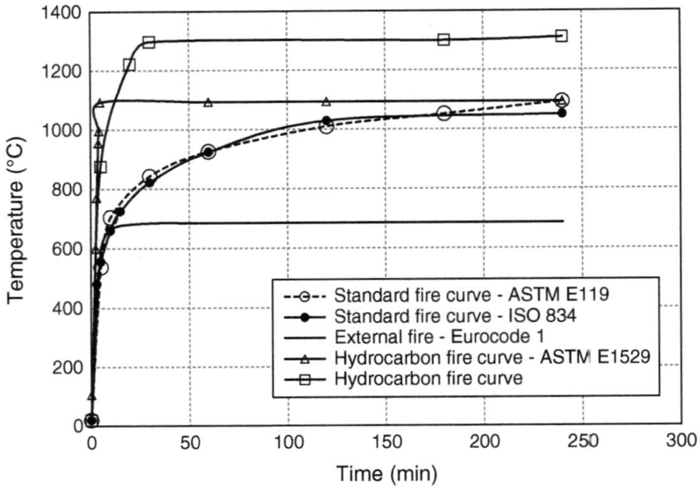

FIGURE 5.4 Time-temperature curves for various fire exposure scenarios.

a fire test. For example, this standard requires maintaining a pressure gradient of 8.5 Pa per meter across the height of the vertical test specimen.

5.3.4 Loading

The fire resistance of a structural member is significantly influenced by the loading present on the structural member. Test standards, such as ASTM E119, require that structural members be loaded during a fire resistance test. The applied load level shall be equivalent to the maximum load allowed by nationally recognized structural design criteria, unless limited design criteria are specified and then a corresponding reduced load is applied (ASTM E119, 2016). It is generally accepted that the load level present during the occurrence of fire can be lower[6] than that at ambient conditions. Therefore, loadings are often given in terms of dead load (self-weight of the specimen) as well as a portion of the expected live load (30% to 50% of a typical live load), since the presence of high wind or earthquake, among other types of loading, during a fire is not common.

5.3.5 Failure Criterion

The fire resistance of a structural member, as discussed above, is the duration at which point the member ceases to perform its intended function. This functionality varies from one member type to another, and the failure criterion depends on the type of structural member. Therefore, in a fire test, fire exposure on the test specimen with applied loading is to be continued until failure occurs, or until the specimen has withstood the test conditions for a time equal to that specified in the condition of acceptance. Fire test standards specify three main failure criteria for defining failure of a structural member in a standard fire test:

- *Insulation* (barrier) criterion to limit fire spread through temperature rise
- *Stability* (strength) criterion to prevent collapse of structural members through loss of load-bearing capacity or stability
- *Integrity* criterion to limit fire spread through passage of flames or hot gases.

For certain structural members, codes specify required fire resistance ratings separately to meet each of these criteria. For instance, a load-bearing wall may have a specified fire resistance rating (FRR) of 60/60/60, meaning 60 minutes of fire resistance is required under each of the three criteria. However, for a non-load-bearing wall, the specified fire resistance might only be –/60/60, indicating that this wall needs to meet only integrity and insulation limit states and not stability (which means zero), since no load has to be carried by this wall.

It should be noted that some standards require successful completion of the *hose-stream test* as an additional integrity. For example, ASTM E119 states that when required, the fire exposure is followed by the application of a specified standard fire hose stream applied in accordance with the practice specified in

[6] Since fire is a rare event, not all load types (e.g., wind, earthquake, etc.) may necessarily be present at the time of a fire event.

Member/Criteria	Stability	Integrity	Insulation
Beam	X	—	—
Column	X	—	—
Load-bearing wall	X	X	X
Non-load-bearing wall	—	X	X
Floor/ceiling	X	X	X
Window	—	X	X
Doors	—	X	X

TABLE 5.4 Endpoint Criteria for Determining Fire Rating of Tested Structural Members

ASTM E2226 (2015). When the conditions of acceptance require a hose stream test, a test specimen is subjected to the impact, erosion, and cooling effects of a water hose stream (for walls), immediately after exposure to fire. The water stream is to be applied through a 64-mm hose discharging through a *National Standard Playpipe* and 6 m away from the fire-tested specimen. The water pressure applied varies with the fire rating of the member. For example, 207 kPa of water pressure is to be applied to a specimen that completed a fire rating of 1 to 4 hours. For specimens with fire resistance that exceeds 4 hours, the water pressure increases to 310 kPa.

Table 5.4 lists failure criteria to be satisfied by different structural members in a fire test. Beams and columns, for example, are required only to demonstrate the stability criterion for the fire resistance period. However, load-bearing walls and floors have to meet all three limiting criteria of strength, integrity, and insulation. Non-bearing walls when used for a fire separation function (compartmentation), together with windows and floors, only have to meet insulation and integrity criteria. In addition to the above criteria, in the case of walls, doors, and windows, additional criteria (e.g., the hose stream test as discussed above) are to be applied. The pressure of the water is often calibrated depending on the required fire rating, and water is generally applied to the fire-exposed side only.

The insulation/integrity criterion is meant to test the ability of a barrier (wall or floor) to contain fire spread from a fire compartment (i.e., room) to another compartment through temperature rise on the unexposed face of the barrier member. For this criterion to be satisfied, a peak temperature rise at any one point on the unexposed side of the barrier needs to be less than 180°C, or the average temperature rise, measured at nine strategically placed points on the unexposed surface, should be less than 140°C. The integrity criterion is aimed at preventing fire spread through the passage of flames or hot gases across the barrier member (i.e., wall or floor) to the next compartment.

On the other hand, the stability criterion is to ensure that a structural member maintain its load-bearing function and stability throughout the duration of the fire test. For this performance criteria to be satisfied, the testing must continue to carry the applied load on the member. For certain members, some codes also specify deflection (or deformation) limits to be satisfied during a fire test. According to ASTM E119, a specimen attains failure when either the deflection or limiting rate of deflection limit is exceeded. The limit for deflection in a member is $L^2/400d$, where L = span length of the beam (mm) and d = effective depth of the beam (mm). The limit for the rate of deflection between two successive increments is $\frac{L^2}{9000d}$. BS 476 specifies failure of a member when the maximum deflection of a horizontal member (beam) exceeds $L/20$ at any fire exposure time, or when the rate of deflection exceeds a $\frac{L^2}{9000d}$ limit. In vertically loaded members (columns), vertical contraction (deformation) is limited to $h/100$ (mm) or rate of contraction is limited to $3h/1000$ (mm/min), where h is the initial height of the member (mm).

5.3.6 Reporting of Test Results

There is some guidance in test standards for interpreting and reporting the data generated in standard fire tests. In the case of ASTM E119, for example, test results are expressed in the time duration of fire resistance to the nearest integral minute. The report should also include observations of details of the behavior of the structural member/assembly during the test and after the furnace fire is extinguished, including information on temperature-time curve, deformation, spalling, cracking, burning of the test specimen or its component parts, continuance of flaming, and any production of smoke.

It is worth noting that the ASTM E119 standard states that the furnace temperature must follow the profile of the standard fire with a maximum discrepancy of the area under the time-temperature curve within 10% of the standard (E119) time-temperature curve for the first 1 hour, within 7.5% for the next hour (i.e., up to 2 hours and within 5% after 2 hours). Thus, the most important condition of a successful fire test is to ensure measured temperature-time points in the furnace follow closely that of specified temperature-time points in the standard fire curve. In case the measured temperature-time points (or curve) are higher (or lower) than the standard fire curve, the measured fire resistance period needs to be adjusted through a correction factor, *C*.

The correction factor is to be calculated taking into account the difference between measured and specified temperature-time curves in a fire test and is given as

$$C = 2I(A - A_{st})/3(A_{st} + L_g) \tag{5.5}$$

where C = Correction in the same units as I

I = Indicated fire-resistance period

A = Area under the curve of indicated average furnace temperature for the first three-fourths of the indicated period

A_{st} = Area under the standard furnace curve for the same part of the indicated period

L_g = Lag correction in the same units as A and A_s (30°C·hr, or 1800°C·min)

Example 5.1 illustrates the use of this correction.

5.4 Measured Results in Fire Tests

Standard fire tests are carried out to derive fire resistance ratings to facilitate the use of fire-rated structural members in buildings (and other constructions). The main data collected in a standard fire test are fire temperatures, time to failure of the member, cross-sectional temperature, deformations (or deflections), and observations during the fire exposure. The cross-sectional temperatures are collected only at limited locations in a member, including the unexposed side of a barrier member, while deformations, if collected, are measured at a specific location such as at the midspan deflection of beams. Data collected from tests is used to interpret and assign a fire rating to the tested member. The test results are reported as the time, to the nearest minute, of the resistance of the assembly to fire exposure, as defined in that fire testing standard. The final fire rating indicates the limiting time for the failure criterion/criteria rounded down to the nearest appropriate classification (rating) of 30, 45, 60, 90, 120, 180, or 240 minutes. The standard test is only intended to register the performance up to the desired rating period of standard fire exposure and not intended to evaluate the actual performance of the (failed) structural member beyond the rating period.

Fire ratings derived from standard fire tests for structural members are to be certified by a recognized laboratory or agency for acceptance in practice. For instance, testing agencies such as UL maintain a list of approved fire resistance ratings of various assemblies. Industry organizations (such as the Gypsum Association) also maintain listings of approvals for their products. Fire resistance ratings derived from tests can be *proprietary* or *generic*.

5.4.1 Proprietary Ratings

In the case of fire resistance tests commissioned by manufacturers of the proprietary materials, the ratings of members made of proprietary products become *proprietary*. Such proprietary ratings can only be applied to specific products and cannot be applied to products (or material types) other than that of the tested assembly/member. In general, listings only apply to a duplicate of prototype design used in the test and any changes can nullify the rating. A typical proprietary rating for fire-rated gypsum board walls (with type-X gypsum protection) is shown in Fig. 5.5.

Individual listing agencies, such as UL, often provide a unique designation, sectional details, construction specifications, conditions of the carried-out fire test, and so on for fire-rated members. For example, beams are grouped under classifications N and O, and columns are grouped under X, Y, and Z classifications. Each classification designation is then followed by a membrane protection number that

Typical 1-Hour Fire-Rated Wall

GA FILE NO. WP 1009	PROPRIETARY*	1 HOUR FIRE	55 to 59 STC SOUND

GYPSUM WALLBOARD, STEEL STUDS

One layer ⅝" proprietary type X paper faced or glass mat faced gypsum panel product (sound board) applied parallel to each side of 3⅝" steel studs 24" o.c. with 1" Type S drywall screws 8" o.c. at vertical Joints and 12" o.c. at floor and ceiling runners and intermediate studs.

Joints staggered 24" on each side and on opposite sides. Sound tested with studs space 16" o.c. and with 3½" glass fiber friction fit in stud space. **(NLB)**

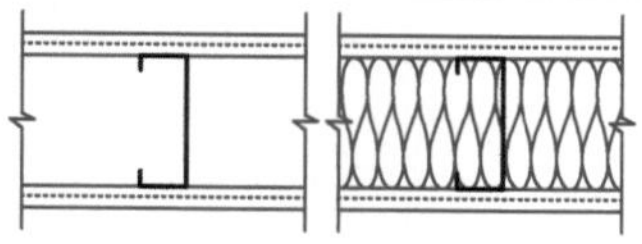

PROPRIETARY GYPSUM BOARD

Temple-Inland - ⅝" Type X ComfortGuard Sound Deadening Gypsum Board

Thickness: 4⅞ *
Approx. Weight: 5 psf
Fire Test: UL R6937, 06NK17962, 2-17-07; 09NK05706, 4-13-09; U465
Sound Test: OL 11-0606. 6-15-11

Typical 2-Hour Fire-Rated Wall

WALLS AND INTERIOR PARTITIONS, NONCOMBUSTIBLE

GA FILE NO. WP 1451	PROPRIETARY*	2 HOUR FIRE	60 to 64 STC SOUND

GYPSUM WALLBOARD, STEEL STUDS

Base layer ⅝" proprietary type X gypsum wallboard applied parallel or at right angles to each side of 2½" proprietary steel studs 24" o.c. with 1" Type S drywall screws 24" o.c. **Face** layer ⅝" type X gypsum wallboard applied parallel or at right angles to each side with 1⅝" Type S drywall screws 16" o.c. Face layer horizontal joints offset 12" from base layer horizontal joints. When face layer is applied at right angles to framing locate first screw 1¼" from board edge and locate second screw 2" from board edge.

Sound tested with horizontal resilient furring channels 24" o.c. and glass fiber insulation in stud cavity. **(NLB)**

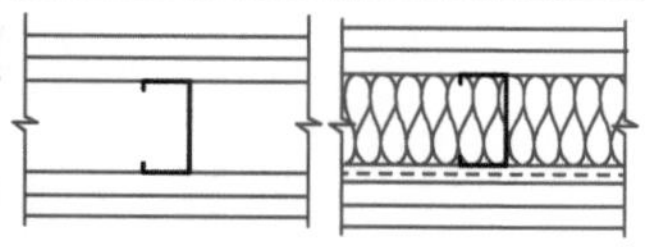

PROPRIETARY GYPSUM BOARD

National Gypsum Company - ⅝" Gold Bond® Brand FIRE-SHIELD® Gypsum Board

Thickness: 5* (Fire) 5½" (Sound)
Approx. Weight: 9 psf
Fire Test: UL R3501, 07NK12306, 8-24-07. UL Design V450
Sound Test: RAL TL05-081, 4-14-05

FIGURE 5.5 Proprietary rating for fire-rated gypsum walls.

varies from 000 to 899. The type of applied insulation (i.e., SFRM) follows this designation (600 to 899). Unprotected members do not have insulation, and their membrane protection number varies from 900 to 999. For example, the listing shown in Fig. 5.6 is for a steel column (Group X) protected with SFRM insulation (No. 701). Hence, the designation of this member is X701. This listing is for a fire-rated steel column with a fire rating of 1 to 4 hours.

5.4.2 Generic Fire Data

In *generic* ratings, the fire resistance of members infers the applicability of ratings to members made of a full range of generic materials (such as conventional concrete), with no reference to a specific manufacturer. Such generic ratings are listed as addendums in building codes and standards such as IBC, NBC, ACI/TMS 216.1, and so on. Samples of these listing are presented in Tables 5.5 and 5.6, which show listings of minimum cover for steel columns encased in normal-weight precast covers as well as ratings for concrete slabs published by the ACI/TMS 216.1 (ACI/TMS Standard, 2014).

5.5 Limitations of Standard Fire Resistance Tests

Although standard fire tests provide useful data to derive fire resistance ratings, the tests themselves have a number of drawbacks. For example, fire test data cannot shed light on the performance of a structural member under realistic fire loading or on restraint conditions as present during an actual fire incident in a building. Furthermore, fire resistance ratings determined through a standard fire can be overly conservative in some cases, leading to irrational and expensive fire resistance solutions. Another major

Design No. X701

October 26, 2017

Ratings – 1, 2, 3 and 4 Hr.

*** Indicates such products shall bear the UL or cUL Certification Mark for jurisdictions employing the UL of cUL Certification (such as Canada), respectively.**

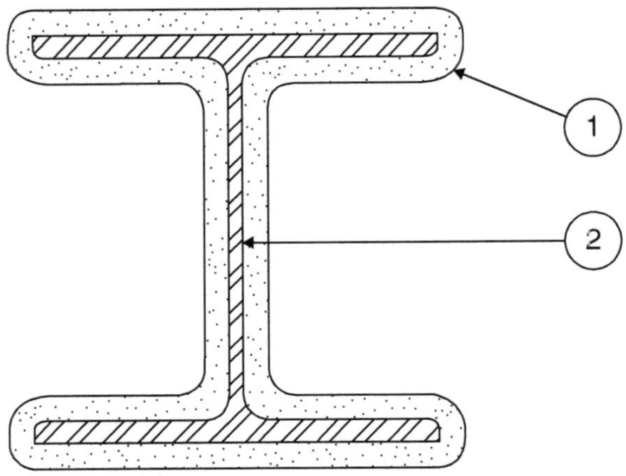

1. Spray-Applied Fire Resistive Materials* – See table below for appropriate thickness. Min avg and min ind density of 15/14 pcf respectively. Min avg and min ind density of 19/18 pcf respectively for Types 7GP and 7HD. Min avg and min ind density of 40/36 pcf for Type AV650. Min avg and min ind density of 50/45 pcf for Type AV800. Min avg and min ind density of 22/19 pcf respectively for Types Z-106 and Z106/HY. Min avg and min ind density of 40/36 pcf for Type Z-146. For method of density determination, refer to Design Information Section, preceding these designs.

Rating Hr	Min Thkns In.
4	2–1/2
3	1–11/16
2	1–1/8
1–1/2	7/8
1	11/16

The thicknesses contained in the table below are applicable when the Spray-Applied Fire Resistive Materials applied to columns flange tips are reduced to one-half that shown in the table below:

Rating Hr	Min Thkns In.
4	2–7/8
3	2–1/16
2	1–3/8
1–1/2	1–1/16
1	13/16

ARABIAN VERMICULITE INDUSTRIES – Types AV650 and AV800 evaluated for Exterior Exposure.

Figure 5.6 UL listing for an SFRM insulated steel column.

Structural Shape	Fire Resistance Rating (hr)			
	1	2	3	4
W14×233	1.5	1.5	2.5	3.5
W12×190	1.5	1.5	2.5	3.5
W10×112	1.5	2	3	4
W8×21	1.5	3	4	4.5

Table 5.5 Listing for Minimum Cover for Steel Columns Encased in Normal-Weight Precast Covers (*Authorized Reprint from ACI, 216.1, 2014*)

Structural Shape	Minimum Equivalent Thickness for Fire-Resistance Rating (mm)			
	1	2	3	4
Siliceous	89	127	157	178
Carbonate	81	117	145	168
Lightweight	64	91	112	130

Table 5.6 Listing for Concrete Wall/Slab Thickness (*Authorized Reprint from ACI, 216.1, 2014*)

limitation of standard fire resistance tests is the quantity and quality of data collected. Most often, the data collected in standard fire tests are only until the endpoint criterion is attained (i.e., the desired fire rating period) without any consideration to the actual response of the member beyond the required fire rating period.

Furthermore, undertaking fire tests is an expensive proposition, both in terms of the specimen preparation and the cost of the actual test. The rating derived from a specific fire test is applicable to the member tested and not to other members and/or exposure conditions. As discussed, fire test furnaces are restricted in size; thus, it is generally impossible to test actual-size members as used in practice. On the same note, only isolated structural members, with idealized end support conditions, can be tested in a fire testing facility (i.e., furnace). Due to furnace size and instrumentation limitations, all possible load redistribution mechanisms during fire exposure cannot be taken into consideration in evaluating fire performance. For example, most specimens are fire-tested with idealized end conditions—specifically, beams and slabs with simply supported ends and columns with some indeterminate degree of fixity arising from the loading platens in the test rig. These end conditions may not represent actual restraints (boundary) conditions in real structures.

Perhaps the most notable drawback with regard to fire testing is the reproducibility of fire test results carried out whether in the same fire furnace or in a different fire furnace. This is due to the fact that replicate tests are rarely undertaken, since (1) relevant fire test standards do not mention the need for reproducibility of tests, and (2) undertaking fire tests are costly and time-consuming. In one notable study, Dotreppe et al. (1995) reported the results on a pair of identical reinforced concrete columns tested in the same furnace, with one column attaining a fire resistance of 84 minutes and the other similar column yielding at 138 minutes. It is not fully clear why the anomaly occurred, but these results do indicate the possible dangers of relying heavily on a single fire resistance test usually used to establish fire ratings in codes. In a different study, Witteveen and Twilt (1981) tested concrete-filled hollow steel sections at different furnaces and reported differences in the behavior due to the loading arrangements and furnace characteristics. To minimize differences arising from variations in differences to heating applied on the test specimen (due to the furnace environment), BS EN 1363-1 (British Standards Institute, 2012) specifies that the furnace linings shall consist of insulation materials with a density less than 1000 kg/m^3 with a minimum thickness of 50 mm composed of at least 70% of the internally exposed surface.

5.6 Extension Principles

In the early years (until the 1970s), testing was the only method to evaluate the fire resistance of structural members and assemblies. In a fire test, resistance of a building element to withstand the effects of fire is evaluated for a specific size of the specimen, properties of constituent materials, and loading conditions. For any other condition or added feature, a new test is required, which can be costly and time-consuming. To overcome this drawback, a number of rules of thumb based on logical intuition and established principles have been proposed over the years. Many of these rules have been deduced based on results and observations from fire tests on building members, and therefore they are applicable for typical members used in building construction. These rules have been verified against a limited data set, collected in fire tests on structural members, with some logical reasoning. They are mainly intended to deduce relative fire performance of a structural member to enable the preselection of members (with relevant insulation configuration) for undertaking standard fire resistance tests.

Many of the proposed rules of thumb are specific to a material type used in construction or the geometric dimensions of structural member. Thus, these rules can be grouped under two categories depending on the dominant parameter influencing fire performance: *material properties* and *geometric dimensions* of the member.[7] These extension rules are briefly discussed with rationale applied to arrive at such rules.

5.6.1 Extension Principles Based on Material Properties

For Steel Members

Guideline 1 Fire resistance (R) of a protected or unprotected steel element (or member) increases with an increase in the ratio of the sectional capacity (i.e., yield strength of steel) of the member to the applied loading (present during fire) if failure is not governed by the thermal criterion (limiting temperature).

---

[7] It should be noted that in all cases it is assumed that the variations do not introduce higher stresses in load-bearing elements. Additional information for the basis of several of the extension rules discussed here can be found in Harmathy (1965), Stanzak and Lie (1973), Gustaferro and Selvaggio (1967), and Lie (1972).

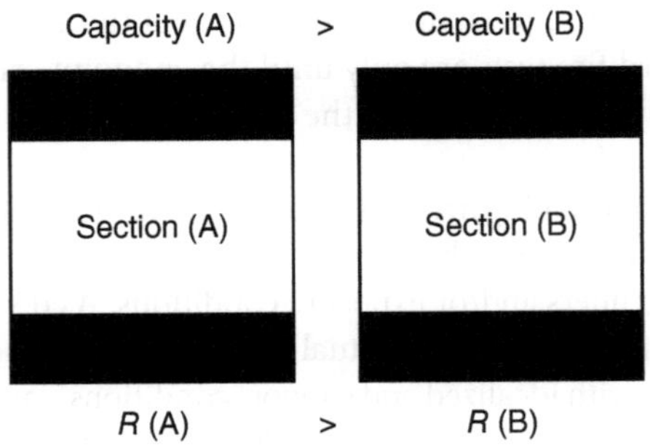

Guideline 2 Fire resistance of a steel member increases with an increase in weight *(W)* to the heated perimeter (*D*[8]) of a steel section. The heated perimeter is the portion of the perimeter of the cross section that could be exposed to fire. This is mainly attributed to the fact that with increased weight (volume) of the steel section, more heat is required to increase the temperature of the steel member to reach failure (or critical) temperature. Conversely, if the sectional perimeter through which the heat is applied to the steel member decreases, it will take less time to reach the failure temperature.

The applicability and accuracy in the prediction of the above rule are dependent on the shape of the steel section. For example, for wide flange shapes and angles, the area through which heat is supplied to the steel is less than that corresponding to the developed heated perimeter, because the fire-exposed area is reduced between the flanges or legs. Therefore, the validity of the rule is limited to sections with similar shapes.

$$\left(\frac{W}{D}\right)_A > \left(\frac{W}{D}\right)_B$$

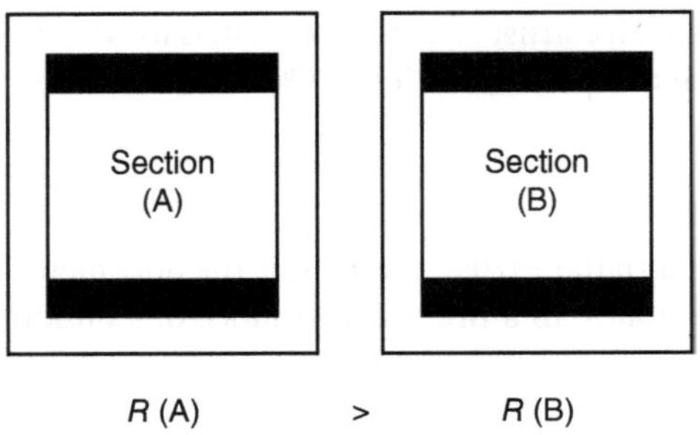

For Concrete Members

Guideline 1 Fire resistance of a concrete member increases with the compressive strength of the concrete, if no fire-induced spalling of concrete occurs in the member.

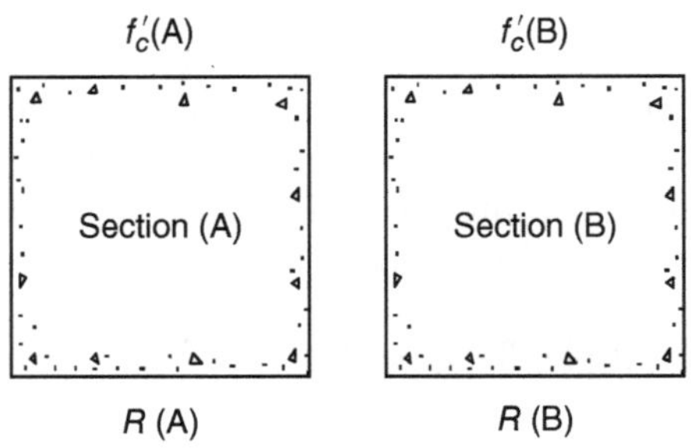

Guideline 2 Structural members made of carbonate aggregate concrete have a higher fire resistance as compared to members made of siliceous aggregate concrete.[9] The higher fire resistance in carbonate

[8] Further discussion of heated perimeter is provided in Chaps. 6 and 7.

[9] Carbonate aggregates are derived from limestone, dolomite, and limerock, while siliceous aggregates are predominantly derived from silica, granite, and sandstone.

aggregate concrete is due to the fact that calcination of carbonate aggregates requires high amounts of thermal energy and therefore some amount of heat from fire is utilized for this purpose. Hence, concrete made of carbonate aggregates retains about 75% of its compressive strength at about 500°C, while the strength of concrete containing siliceous aggregate decreases to 55% at that same temperature. Further, thermal conductivity of siliceous aggregate concrete (1.6 W/m.K) is slightly higher than that of carbonate aggregate (1.3 W/m.K), and therefore carbonate aggregate concrete experiences a slower rise in temperature than that of siliceous aggregate concrete (Kodur, 2014).

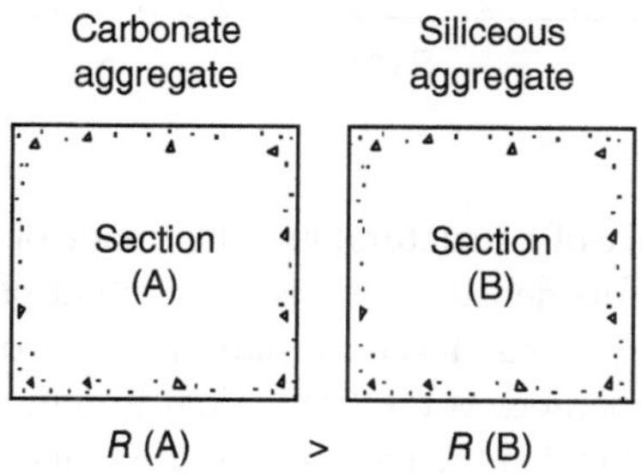

Guideline 3 The use of carbonate aggregate in place of siliceous aggregate in concrete is beneficial in realizing higher thermal fire resistance in fire separating (barrier) members.

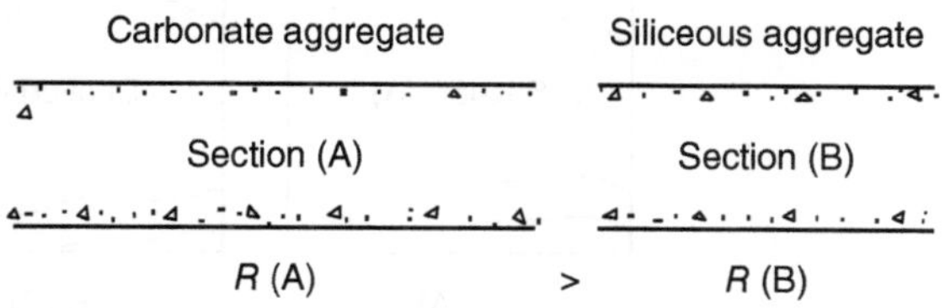

Guideline 4 Fire resistance of a member made of lightweight concrete is higher than an identical member made of normal-weight concrete. This is attributed to the fact that thermal properties of lightweight concrete are more favorable than those of normal-weight concrete from the point of view of heat transmission. Therefore, the rate of temperature rise in a member made of lightweight concrete is slower than that made of normal-weight concrete.

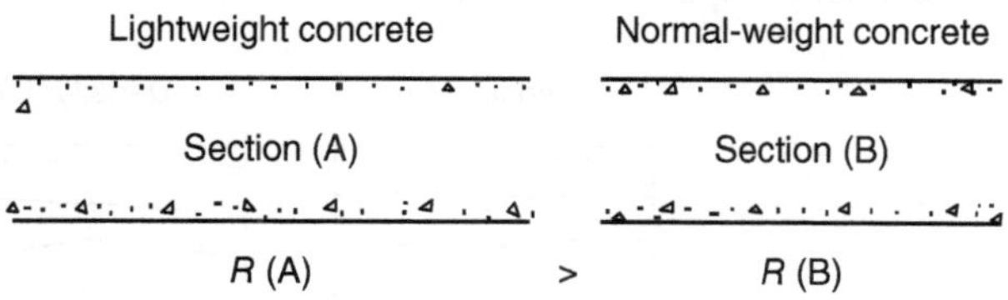

Guideline 5 Fire resistance of multilayer reinforced or prestressed concrete members, such as slabs and floors, increases if, in the bottom layer, carbonate aggregate concrete is placed instead of siliceous aggregate concrete. This is due to the fact that thermal properties of carbonate aggregate concrete are more favorable than those of siliceous aggregate concrete from the point of view of heat transmission. Therefore, it takes more time to reach limiting temperature in reinforcing or prestressing steel if the reinforcements are protected by layers of carbonate concrete than if they are protected by layers of siliceous concrete.

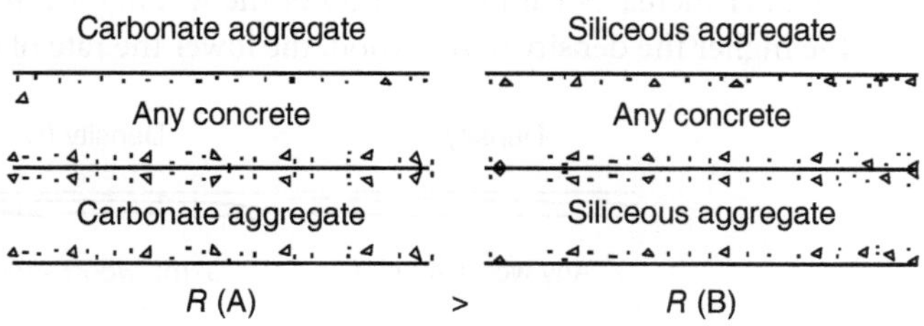

Guideline 6 Fire resistance of multilayer reinforced concrete slabs or floors increases if, in the bottom layer, lightweight concrete is placed instead of normal-weight concrete. This is due to similar analogy as in

Guideline 4 where thermal properties of lightweight concrete are more favorable than those of normal-weight concrete from the point of view of heat transmission across concrete layers.

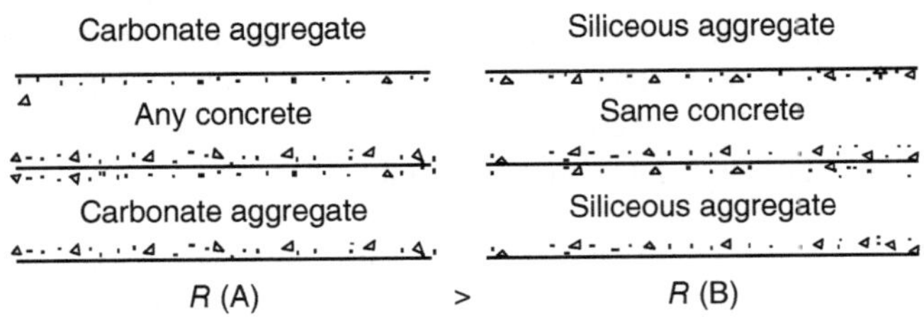

For Timber Members

Guideline 1 Fire resistance of a structural member made of timber increases with compressive (and tensile) strength of the wood, if its density or moisture content of wood does not decrease. Under fire exposure, timber members undergo char layer formation that grows in thickness with time. Consequently, fire resistance of a timber member is the time taken to reduce the original (uncharred) section to a critical (reduced) section size at which the member can no longer support the applied load.

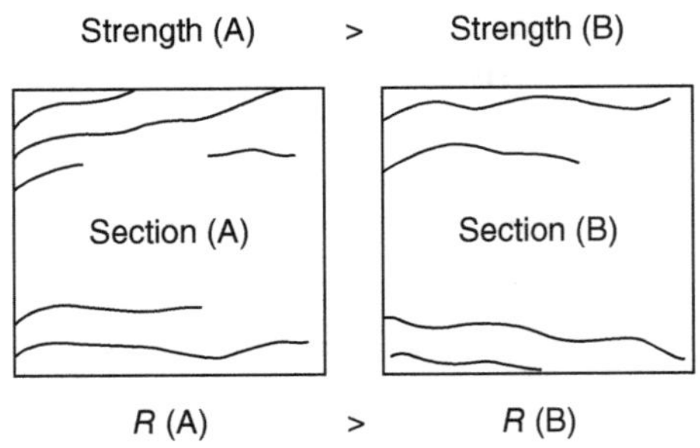

Guideline 2 Fire resistance of a structural member made of timber increases with an increase in wood density, for the same species of wood. The higher the density of the wood, the lower the rate of charring of wood under fire exposure, and thus it takes more time to reduce the original section of the element to a critical size at which the member can no longer support the applied load. Because different species may differ in the properties that determine the fire resistance of the element, such as strength, this rule is only valid for when members of species of wood are compared.

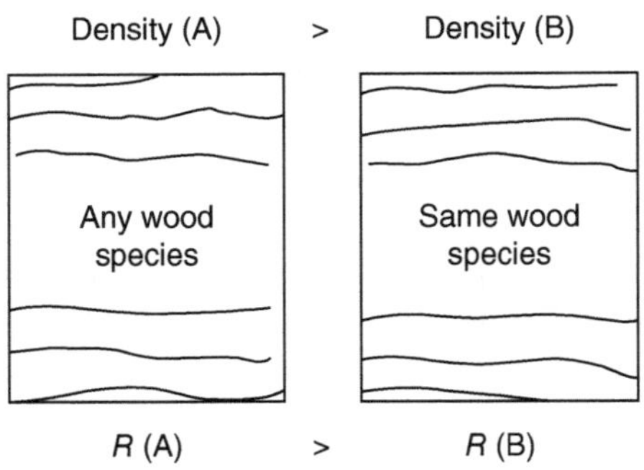

Guideline 3 In the case of compartmentation (barrier) members, such as floors and slabs, the thermal fire resistance of a member increases with the density of the wood, for the same species of wood. This is due to the fact that the higher the density of the wood, the lower the rate of burning through the wall or floor.

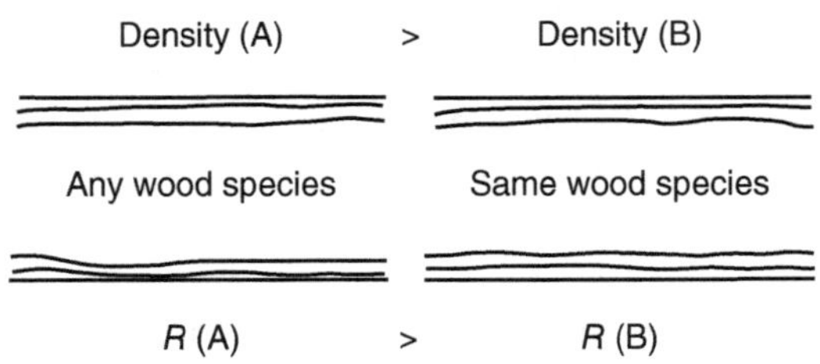

5.6.2 Extension Principles Based on Member Dimensions

Guideline 1 Fire resistance of a reinforced concrete member increases with an increase in cover thickness to the steel reinforcement. Similarly, fire resistance of an insulated steel member increases with an increase in insulation thickness.

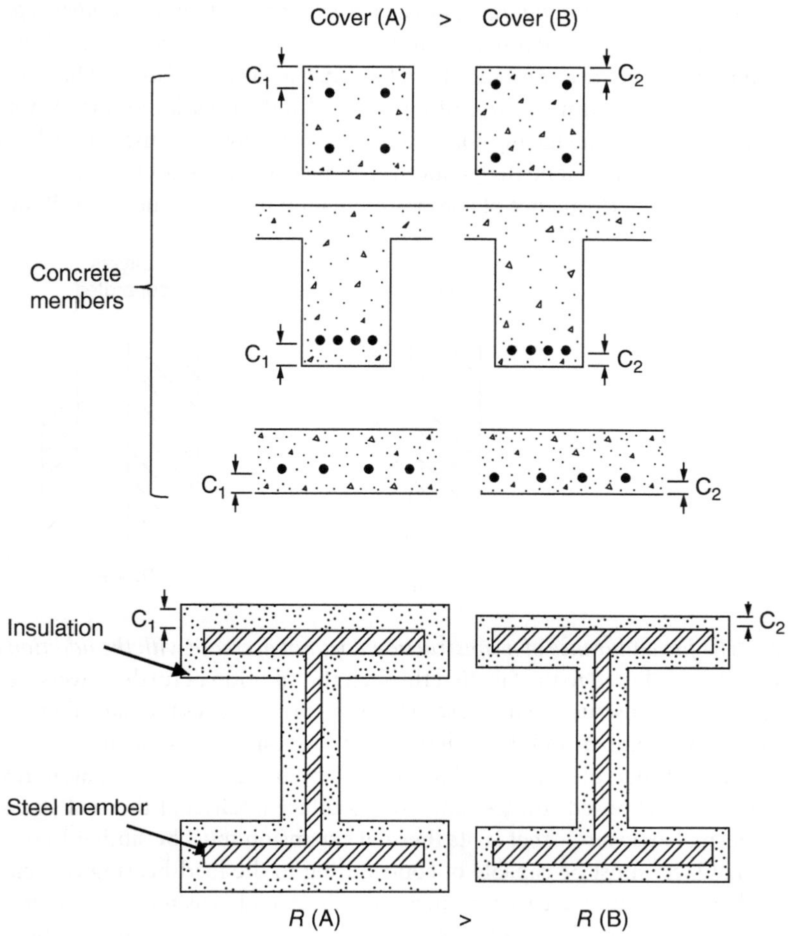

Guideline 2 Fire resistance of a reinforced concrete member increases with an increase in the percentage of steel reinforcement (A_s).

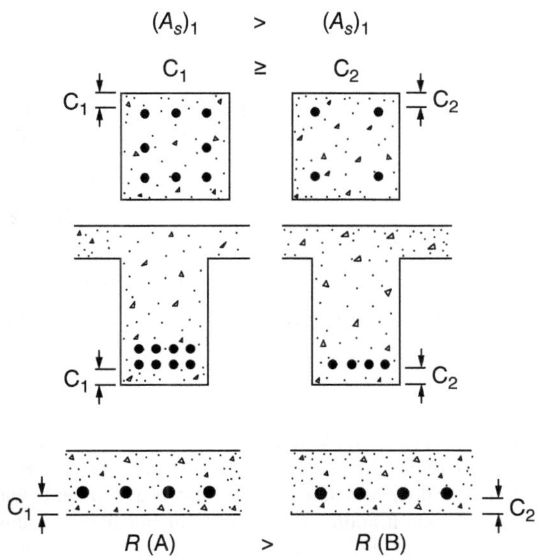

5.6.3 Harmathy's 10 Rules for Fire Resistance[10]

Harmathy (1965) has proposed 12 simple rules for extending the fire resistance of a tested specimen to other similar assemblies. The presented text for each rule is reproduced in italics, as expressed by Harmathy in the original source.

Rule 1 *The "thermal" fire endurance of a construction consisting of a number of parallel layers is greater than the sum of the "thermal" fire endurances characteristic of the individual layers when exposed separately to fire.* This rule implies that the fire resistance (*R*) of a construction consisting of layers made from a noncombustible material equals the sum of fire resistance of all individual layers made from the same materials. As an example, a concrete slab having combined layers, e.g., doubling the original thickness, would lead to having twice the fire resistance if no fire-induced spalling in concrete occurs. This rule is very helpful in gauging the comparative fire performance of compartmentation members such as walls and floors.

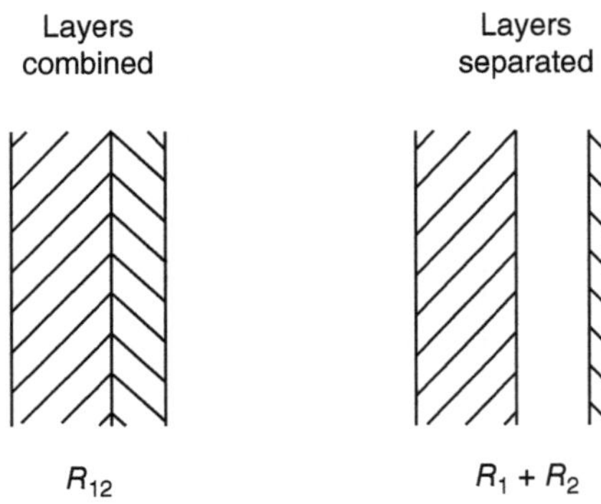

Rule 2 *The fire endurance of a construction does not decrease with the addition of further (additional) layers.* This can be attributed to the fact that through the addition of further layers, resistance to heat flow and heat capacity of the construction increase, which, in turn, reduces the rate of temperature rise on the unexposed surface of the member, and this enhances the fire resistance of the member. However, a close examination will reveal that this rule may be subject to certain limitations. An apparent restriction is that the new layer, applied on the exposed surface, must not produce additional thermal stresses on the member resulting from thermal exposure—that is, its thermal expansion must be similar to those of the adjacent layer. Each additional layer must be capable of contributing to the load-bearing capacity of the construction at least as much as is required to carry the increased dead load (arising from that layer). If this requirement is not fulfilled, the allowable live load must be reduced by an amount equal to the weight of the additional layer.

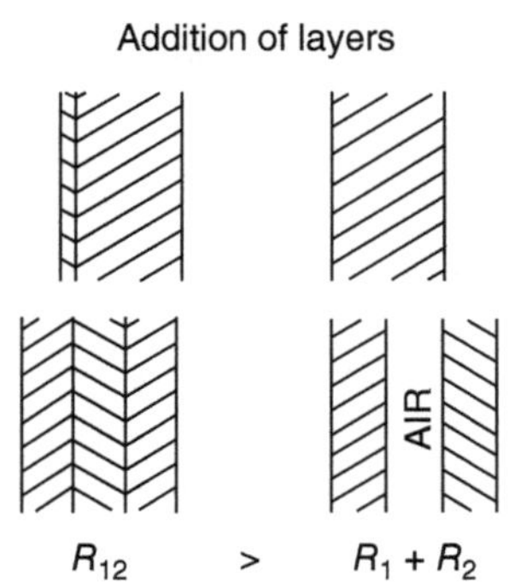

Rule 3 *The fire endurance of constructions containing continuous air gaps or cavities is greater than the fire endurance of similar constructions of the same weight, but containing no air gaps or cavities.* For example,

[10]These rules are represented, in italics, as stated by Harmathy (1965), and further rationale for these rules are discussed as well. In the context of these rules, "fire endurance" refers to fire resistance as described. Fire resistance of a structural member (wall, beam, column, slab, or floor) is defined as the ability of a member to withstand the effects of fire so as to maintain load-bearing capacity, integrity function, and/or to act as an insulation barrier against spread of fire during the entire duration of fire. Similarly, "thermal fire endurance" is defined as the duration (in time) required for a structural member to reach the thermal limit state, i.e., critical temperature. Please refer to Harmathy (1965) for more information and details.

the addition of empty or filled cavities can enhance the fire resistance of a structure. Filling cavities with concrete can improve fire resistance due to the good thermal capacity property of concrete.

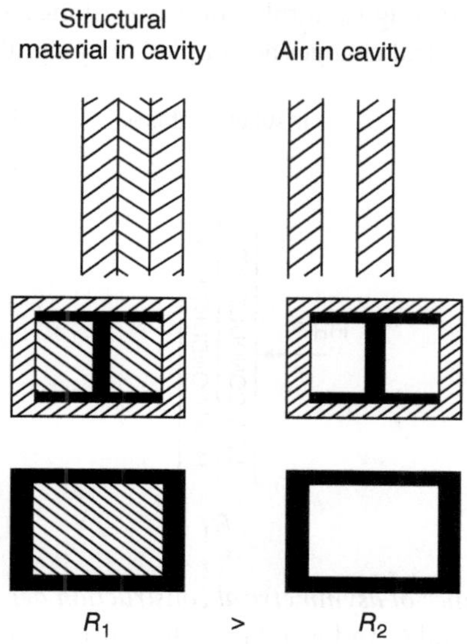

Rule 4 *The farther an air gap or cavity is located from the exposed surface, the more beneficial its effect on the fire endurance.* Since air is a very poor insulator, the addition of a cavity on the fire-exposed surface of a member will delay the temperature rise and thus loss of capacity in the member. The addition of such a gap (or void), especially when located in a region close to fire source, can reduce the rate of temperature rise across the layers.

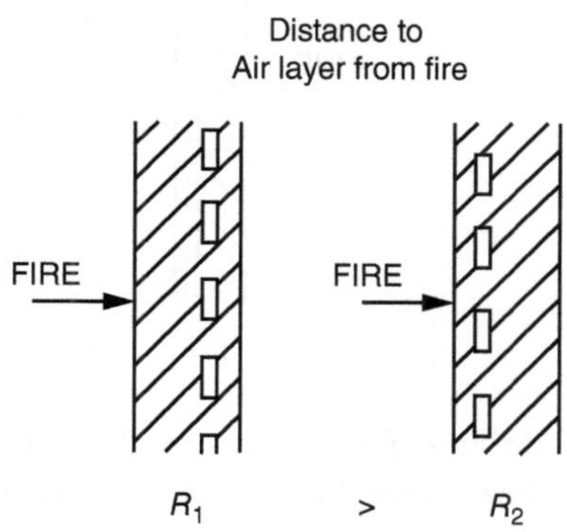

Rule 5 *The fire endurance of a construction cannot be increased by increasing the thickness of a completely enclosed air layer.* If the thickness of the air layer is larger than 12 mm (1/2 in.), the heat transfer through the air layer depends only on the temperature of the bounding surfaces, and is practically independent of the distance between the two insulation surfaces. It should be noted that this rule is not applicable if the air layer is not completely enclosed—in other words, if there is a possibility of fresh air entering the gap (e.g., from top or bottom) at an appreciable rate.

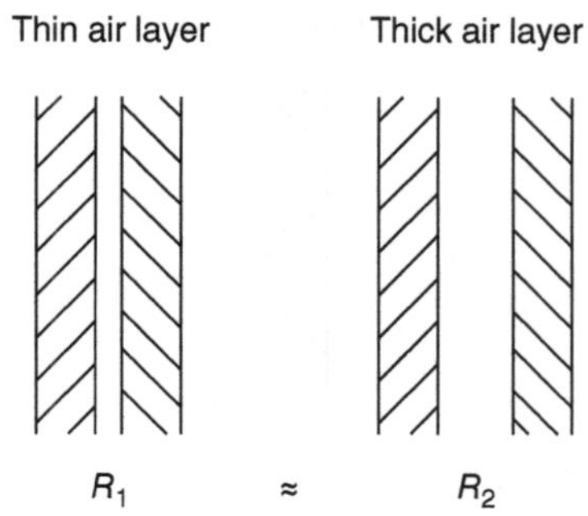

Rule 6 *Layers of materials of low thermal conductivity are better utilized on that side of the construction on which fire is more likely to happen.* As an example, in the case of a floor, adding a layer of insulation from underneath the floor, rather than from the top, can improve fire resistance of the floor. The validity of this rule has been demonstrated through a number of fire resistance tests. The rule may not be applicable to materials undergoing physicochemical changes accompanied by significant heat absorption or heat evolution.

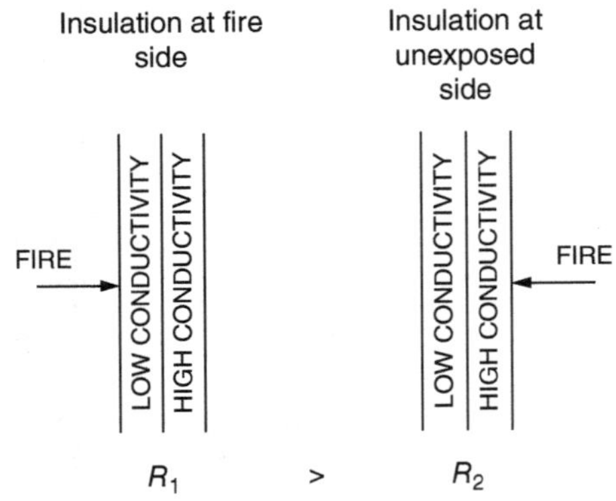

Rule 7 *The fire endurance of asymmetrical construction depends on the direction of heat flow.* This rule is a companion to Rules 4 and 6, and is to take into consideration the location of air gaps or cavities, as well as the sequence of different cross-sectional layers of a structural member. As in the case of Rule 6, this rule can be helpful in placing insulation layers strategically (e.g., underneath a floor, than on top of a floor) to achieve better fire resistance.

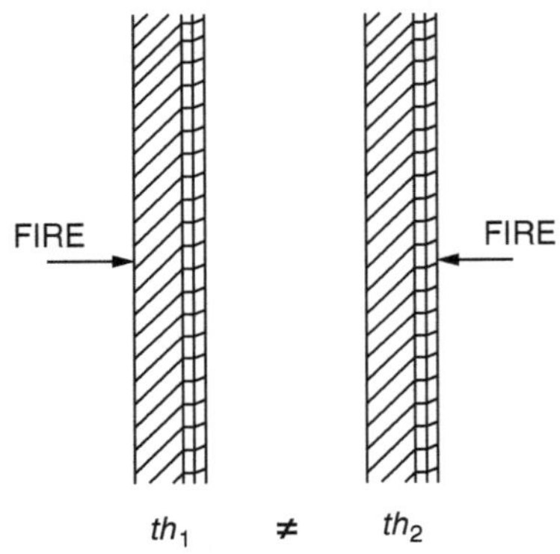

Rule 8 *The presence of moisture, if it does not result in explosive spalling, increases the fire endurance of a structural member.* The flow of heat through the cross section is greatly hindered by the absorption of heat associated with moisture desorption. Test results have shown that in some cementitious-based materials, fire resistance may increase as much as 8% for each 1% (by volume) of moisture in the construction. As the latent heat required for the desorption of moisture is roughly proportional to the amount of moisture present in a unit volume of the material, there is no direct relation between the relative humidity (in the pores) of the material and the increase in fire resistance. As pointed out by Shorter and Harmathy (1965), materials of low permeability (dense concretes) are liable to undergo explosive spalling if the moisture content is higher than a critical value. The permeability of "well-cured" concrete members is several orders of magnitude lower than that of members during early ages; thus, with aging, concretes become less vulnerable to spalling.

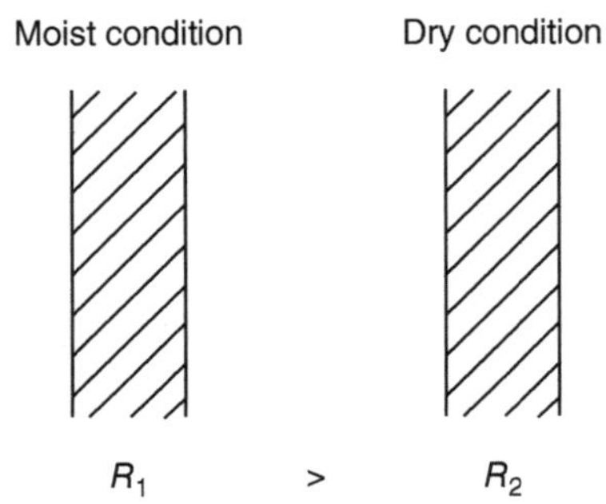

Rule 9 *Load-supporting elements, such as beams, girders, and joists, yield higher fire endurances when subjected to fire endurance tests as parts of floor, roof, or ceiling assemblies than they would when tested separately.* When load-bearing members are tested separately, the imposed load is constant and equal to the design load throughout the test. Under such configurations they cannot redistribute forces (as compared to when tested as part of a multielement assembly), thus isolated members cannot yield higher fire endurance than they do when tested as parts of a floor, roof, or ceiling assembly.

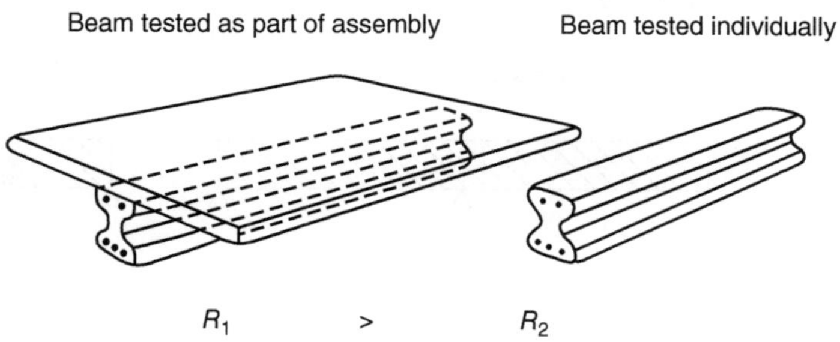

Rule 10 *Reduction of the length of a column or the span of a beam or floor is beneficial for the structural fire resistance of these members.* Under fire exposure, the capacity of a load-bearing member reduces gradually until it can no longer support the load on it. The capacity of a member increases if its span or length is reduced.[11] Therefore, for a given load, the time to failure or the fire resistance increases if the length or span of the member is reduced.

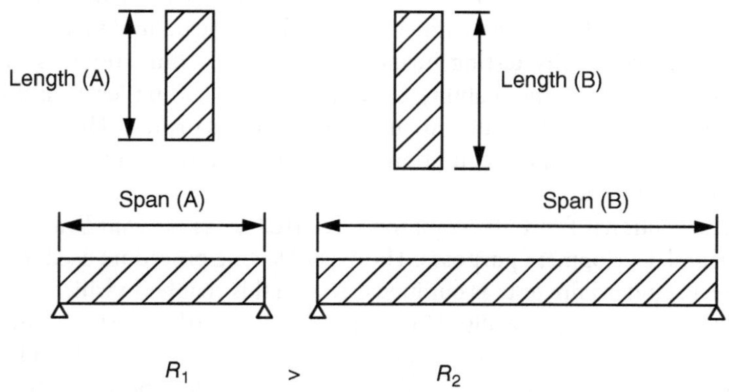

Rule 11 *The structural fire resistance of a member increases with reduction of the load to which it is subjected.* During exposure to fire, the capacity of a member reduces gradually until it can no longer support the applied load. The lower the load, the lower the capacity needed to support the actions of the external load. Therefore, the time to failure or the fire resistance increases if the applied load on the member is reduced.

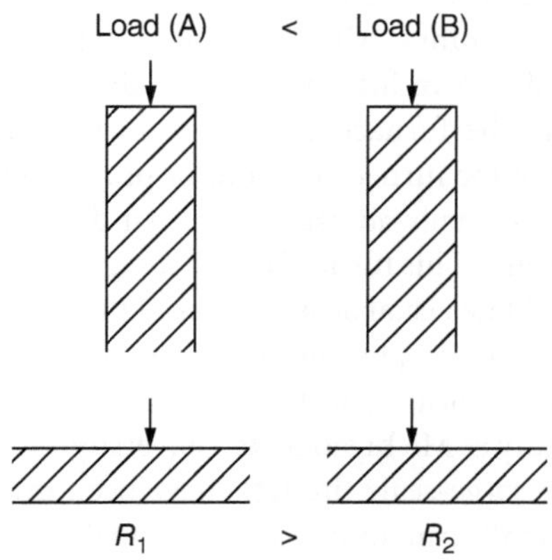

[11] In some instances, the capacity of a member reduces if its span or length is reduced due to instability arising from shear force, as well as instability arising from fire-induced local buckling.

Rule 12 *The structural fire resistance of continuous floor slabs or beams is greater than that of simply supported floors or beams.* Analogues to Rule 9, when a continuous member is heated from below due to fire exposure, a negative temperature-induced moment is created which reduces the positive moment in the span. Owing to the counteracting moments in the span, stresses in the lower reinforcement (i.e., at midspan support location) in continuous concrete members or in the lower part of steel members will be reduced, and this will lead to an increase in the critical temperature level (rebars, or steel) at which failure will occur, resulting in higher fire resistance.

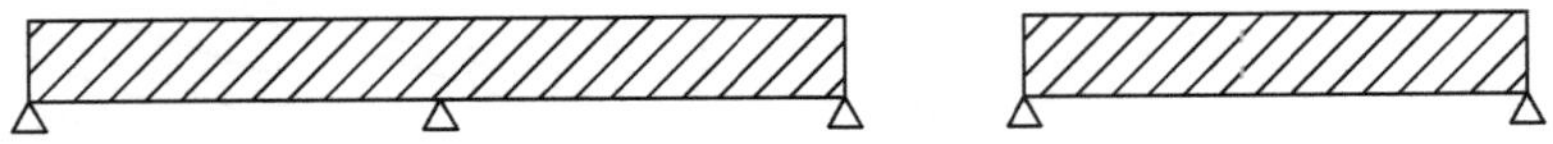

5.7 Resistance Evaluation under Nonstandard Fire Conditions

Much of the discussion presented in preceding sections covered procedures to carry out fire resistance tests under standard fire conditions. In some cases, fire resistance tests can be conducted under nonstandard fire scenarios (often loosely referred to as real or design fire conditions) in order to evaluate realistic response of structural members (or components). The outcome of such fire tests is not to derive fire ratings of the tested structural member or assembly (frame) but rather to observe and evaluate member behavior under fire exposure conditions as occurring in buildings. In many cases, and due to the limitations of the fire testing facility, full-scale member sizes (or larger structural systems such as portal frames) cannot be tested in a standard fire testing furnace facility. Other limitations associated with standard fire testing facilities, for instance, the inability to apply realistic loading levels and simulate representative restraint conditions, can hinder the ability to carry out a standard fire test. Thus, to evaluate the response of structural members (or system) under conditions encountered close to that in practice, an alternative testing regime needs to be followed.

Unconventional fire tests, commonly referred to as *nonstandard fire evaluations*, are undertaken for research and development purposes. The overall testing procedure used in nonstandard fire tests is similar to that under standard fire tests as outlined earlier in this chapter in which a specimen is prepped, installed, and then tested in a fire testing facility. However, in nonstandard fire tests, the member is tested under nonstandard fire exposure wherein conditions, such as loading (level), restraint, and so on, are different from that in standard fire tests. Further, additional data (such as cross-sectional temperatures, strains, and deflections) is collected during nonstandard fire tests for the duration of the test until failure occurs in the member, rather than only a limited set of data collected in standard fire tests until the required rating period is reached.

The civil infrastructure laboratory at Michigan State University houses a medium-scale fire testing furnace to carry out nonstandard fire resistance tests. This fire test furnace has been specially designed to produce varying conditions of fire scenarios (temperature-time curves) and structural loading, similar to that which occur during a standard or nonstandard (real) fire. The furnace is made of a steel framework supported by four steel columns, with the furnace chamber inside the framework. This chamber is 2.44 m wide, 3.05 m long, and 1.78 m high and can produce a maximum heat energy of 2.5 MW using six gas burners located within the furnace. Inside the furnace, six type-K chromel-alumel thermocouples are mounted on the walls of the furnace to monitor progression of temperature. During the course of the fire test, the gas supply is manually adjusted such that the furnace temperature follows a predetermined fire curve (standard or design). This furnace is also designed with two small view ports, provided on either side of furnace walls, to facilitate visual observations of fire-exposed members. This fire test furnace is designed to test horizontal members (i.e., beams and slab), as well as vertical members (columns). Vertical loading can be applied on columns, beams, or slabs using a one-, two-, or five-point loading setup system. Figure 5.7 shows the fire test facility at Michigan State University.

In a few instances, large-scale fire tests have been carried out on full-scale buildings to evaluate the performance of the overall structural system. The main differences in such tests arise from the geometry of the specimen, fire exposure, load level, restraint conditions, and so on. The early tests to evaluate global fire response of structural systems was carried out by the American Iron and Steel Institute (AISI) and the National Building Standards (NBS) in the early 1980s. These tests were carried out on a two-story, four-bay (4.9 m × 6.1 m) composite steel-framed building. Then in the 1990s a series of tests referred to as Cardington

Figure 5.7 Structural fire test facility at MSU's Civil Infrastructure Laboratory.

fire tests were carried out on a steel building. The success of these tests led to the undertaking of a new set of tests but on a concrete building in 2001 (Bisby et al., 2013).

The Cardington tests were inspired following a major fire incident in a steel-framed office building (Broadgate building in London) in which the steel structure withstood severe fire without much structural fire damage. Following this incident, British Steel's Swinden Technology Centre (British Steel) teamed up with the European Coal and Steel Community (ECSC), TNO (The Netherlands), and CTICM (France) to undertake a series of fire tests in a full-scale steel-framed building similar to the Broadgate building. These tests were carried out on a modern, eight-story composite steel-framed structure built within the Building Research Establishment (BRE) large-scale test facility at Cardington. The tested steel-framed building was a composite steel and concrete structure designed to meet the BS 5950 design code. The structure was also checked for compliance with the provisions of the EC3 ENV 1993-1-1, which was the prestandard for Eurocode 3 at the time. The building was designed as a typical commercial office building with three 21 m × 45 m bays by five bays in plan, and had a total height of 33 m. The structure also incorporated a central elevator shaft and two stair wells at either end of the building. Figure 5.8 shows the floor plan of Cardington test building along with the configuration of various fire tests.

In total, seven fire tests on this structure were reported (British Steel, 1999). The first test investigated the fire response of a 9-m-long composite steel beam in which the end connections were kept in the heating zone. This test noted the development of runaway deflection at midspan sections as well as the occurrence of a tensile failure in the beam-to-column end-plate connections due to contraction of the steel in the cooling phase of the fire. The second test was used to study the behavior of a single-story frame with three beams and four columns. The main observation of the second test was of notable local buckling of steel columns at around 670°C. The third and fourth tests investigated the behavior of a membrane's actions and load transfer within the first floor of the Cardington building.

In the fifth and sixth tests, a large compartment, located between the second and third floors of the building, was uniformly fire-loaded with wooden cribs at 40–45 kg/m^2. This test examined the fire response of uninsulated steel beams (columns and connections in the compartment were fire-protected). Results from the fifth test indicated the occurrence of local buckling near the beam-to-beam connections. The seventh and final test examined tensile membrane action and documented the robustness and behavior of well-designed steel connections during fire.

Similar tests to the Cardington fire tests were also carried out in Europe, the United States, and Australia (Proe and Thomas, 2010), and new tests are planned to be undertaken at the current large-scale extension to the National Fire Research Laboratory (NFRL) at National Institute of Standards and Technology (NIST) in the United States.

The new facility at NIST provides an additional 1965 m^2 of floor space to the existing laboratory. This fire testing facility can generate a maximum heat release rate of 20 MW and accommodate 9-m-high (2 story × 2 bay × 3 bay) structural systems or components. Gravity loading can be applied using hydraulic actuators, fixed loads, or a combination of the two. Fire scenarios involving gas, liquid, wood cribs, or

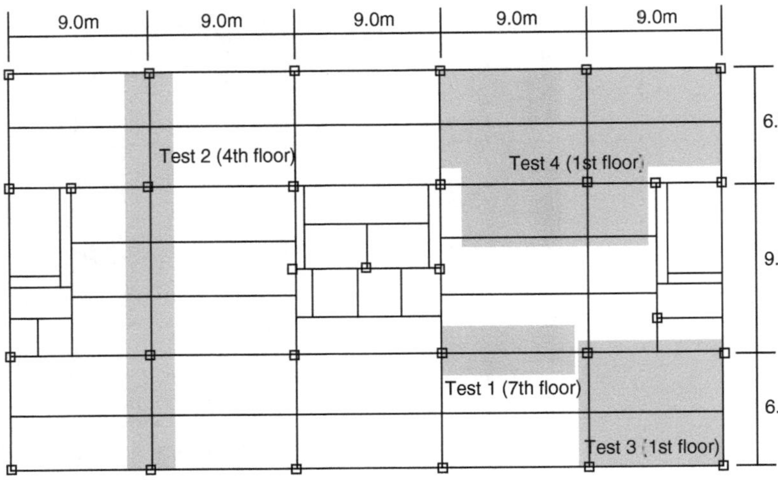

Figure 5.8 Typical floor layout of the Cardington building. (*Adapted from Bisby et al., 2013.*)

Figure 5.9 Fire test facility at National Institute of Standards and Technology (NIST). (*Courtesy of NIST.*)

actual building contents can be applied to simulate various building fire conditions. The test area consists of a 486 m^2 strong floor with multiple anchor points and a 9-m-high strong wall. The strong wall can act to stabilize a test specimen to prevent uncontrolled failure, to provide lateral restraint, or to laterally load a structure. A schematic of this fire test facility at NIST is shown in Fig. 5.9.

5.8 Fire Resistance of Nonstructural Members

Integrity and insulation properties of a compartment can be breached during a fire event if *nonstructural members*, such as doors, windows, walls, and cladding (that form the compartment), do not exhibit sufficient thermal resistance. These barrier elements are of different sizes and shapes, and have specific installation detailing than those of structural members such as beams and columns. As a result, the fire resistance of these barriers elements is to be evaluated under representative conditions present in buildings. In the case of nonstructural members, failure can occur through either insulation (i.e., reaching a specific temperature on the unexposed side), or integrity (e.g., formation of holes and large cracks). To address variations and special characteristics of each of these building components, specific testing procedures are to be adopted, and these are detailed in fire test standards. Table 5.7 lists applicable standards for fire testing of such assemblies.

Doors in high-rise buildings are to have minimum fire ratings, as doors play an important role in minimizing fire spread in a building.[12] The test method for evaluating fire resistance of a door is specified

[12] Fire spread in the floor (from one compartment to another) is often facilitated through open doors (frequently left open by evacuees).

Component	Standard No.	Fire Testing Standard
Doors	UL 10B	Fire tests of door assemblies (neutral or negative pressure)
	UL 10C	Positive fire pressure tests of door assemblies
	NFPA 252	Standard methods of fire tests of door assemblies
	ISO 3008-1	Fire resistance tests—door and shutter assemblies—part 1: general requirements
Windows	UL 9	Standard for fire tests of window assemblies
	NFPA 257	Standard on fire test for window and glass block assemblies
	ASTM E1966-15	Standard test method for fire-resistive joint systems
	ISO 3009	Fire-resistance tests—elements of building construction— glazed elements
Cladding	NFPA 285	Standard fire test method for evaluation of fire propagation characteristics of exterior non-load-bearing wall assemblies containing combustible components
	ASTM E2707-15	Standard test method for determining fire penetration of exterior wall assemblies using a direct flame impingement exposure
	BS 8414	Fire performance of external cladding systems
Ducts	ASTM E2336-16	Standard test methods for fire resistive grease duct enclosure systems
	ISO 6944-1	Fire containment—elements of building construction—part 1: ventilation ducts
Joints	ASTM E1966-15	Standard test method for fire-resistive joint systems
Non-load-bearing walls	ASTM E2707-15	Standard test method for determining fire penetration of exterior wall assemblies using a direct flame impingement exposure
	NFPA 285	Standard fire test method for evaluation of fire propagation characteristics of exterior non-load-bearing wall assemblies containing combustible components
	ISO 834-8	Fire-resistance tests—elements of building construction—part 8: specific requirements for non-load-bearing vertical separating elements

Table 5.7 Applicable Standards for Fire Testing of Non-Load-Bearing Components

in standards such as NFPA 252, ISO 3008-1, and UL 10B and 10C. Fire doors have solid cores and can be designed with inherent fire resistance. However, main weaknesses in a door from fire penetration point of view are handles, hinges, and door edges. Fire-rated doors are tested in similar fashion to non-load-bearing wall assemblies (i.e., under positive/negative pressure). It should be noted that glazed doors are only required to meet integrity requirements. To meet such requirements, a standard 100 × 100 × 19 mm cotton pad should not ignite due to radiating heat arising from flames, spreading through cracks in a fire-tested door. In addition, any glass on door panels must be fire-resistant glass or wired glass.

In the evaluation of the fire performance of a door, the door is mounted on a wall assembly and tested under a specified fire exposure. The door must remain in place (and closed; see Fig. 5.10). A door assembly is exposed to a standard fire, and the fire is controlled to achieve a specified temperature and pressure throughout a specified time period. During the fire test, no flaming on unexposed side can occur for the first 30 minutes, but "light flaming" near the edges is permitted after the first 30 minutes (IBC, 2015). The integrity of the door assembly is evaluated using a cotton wool pad test when the average unexposed surface temperature of a door assembly is less than 361°C above ambient and openings are created by the fire exposure. Other temperature limits may be imposed by building codes (e.g., average limiting temperature measured at 3 points <230°C on the unexposed side in the first 30 minutes). The required fire resistance for a door varies and it is 20, 30, 45 minutes, 1, 1½, and 3 hours. Soon after the completion of the fire endurance test of the door assembly, it is to be subjected to a hose stream test to ensure the integrity of the door.

The Steel Door Institute (SDI) has other requirements specific to fire doors, as listed in Table 5.8. Under fire exposure, typical glass softens/melts at about 600 to 800°C and is assumed to crack/break when exposed to thermal shock at much lower temperatures. Thus, SDI also specifies the use of 6.5-mm wire glass and ceramic glass for window glazing in fire-rated doors. It should be noted that double glazing tends to remain in place much longer than single layers of glass. Fire-resistant barrier glazing is assembled with special glass, such as wired or ceramic glass.

Similar to fire doors, windows need to meet fire resistance requirements to minimize fire spread from one compartment to another. In the United States, the specific standard test methods for evaluating fire performance of windows are laid out in ASTM E1966, NFPA 257, and UL 9 standards. The tested window

Figure 5.10 Fire test on a door (immediately post-fire exposure). (*Courtesy of Achim Hering.*)

Wall Rating (hr)	Door Rating (hr)	Description and Use
4	3	Openings in walls that separate buildings or divide a single building into designated fire areas
2	1.5	Openings used in enclosures of vertical communication or egress paths through the building
2	1.5	Openings in a wall with a potential for severe fire exposure from the exterior of the building
1	¾	Openings in a wall with a potential for light fire exposure from the exterior of the building
1	1	Doors dividing occupancies in a building

Table 5.8 Requirements for Fire Doors as per Steel Door Institute Specifications (*Adapted from Steel Door Institute*)

Opening Class	Rating (hr)	Typical Use	Rated Glass Window Area (cm^2)
A	3	In walls separating buildings of fire areas	645
B	1–1.5	Opening in 1- to 2-hr rated enclosures	645
C	⅓–¾	Opening in room-corridor partitions	8360 (with maximum dimension of 137 cm)
D	1.5	Exterior walls (severe exposure)	
E	¾	Exterior walls (light exposure)	

Table 5.9 Listing of Opening Classes and Fire Rating Required in Window/Door Application (*Adapted from IBC*)

needs to have a cross-sectional area of 9.2 m^2 and to be mounted in a wall test furnace. NFPA 80 provides a listing of different opening classes and fire ratings required for windows/doors (see Table 5.9). The evaluation criterion followed in the case of windows is for the tested window to remain in place with no flame spreading to unexposed surface and the maximum opening cannot exceed the width of the frame. A hose stream test often follows fire exposure in which 70% of glass needs to be in place (and no more than 5% lost from any glazed area) when water pressure is applied to the window.

Air ducts are another building component that need to satisfy fire resistance requirements in buildings, especially if the air duct is passing through a barrier (wall). Air ducts not only can compromise the fire resistance of the barrier but also can be a potential path for the spread of fire throughout the building. Hence, most building codes require ducts to be provided with specific fire ratings. The fire resistance of ducts can be evaluated following ASTM E2336 (ASTM International, 2016) provisions. In this standard, an air duct is tested in a combustion chamber, with a tubular L-shaped assembly, capable of delivering the similar temperature level to that expected in a compartment (building). In typical building applications, the fire resistance of ducts can be improved through insulating material such as ceramic fiber and gypsum boards. Fire resistance of air ducts can also be improved by placing a fire damper inside the duct. Such dampers are designed to automatically close in case of fire and can also control smoke movement.

5.9 Emerging Trends and Research Needs

In the last few decades there have been limited advances with respect to testing procedures, methods, instrumentation, and data collection devices for undertaking fire tests on structural members/components. However, these advances have not transformed into testing practices, either as part of standard fire tests or as part of testing undertaken for research. Some of the areas where further advancement is needed with respect to fire resistance tests are discussed in the following sections.

5.9.1 Hybrid Fire Testing

Hybrid testing refers to the coupling of physical testing of an isolated structural member while numerically simulating subassemblages (of surrounding structural members) that interact with the member in order to simulate the response of a full structural system. Such hybrid testing has been successfully applied for tracing the seismic response of structural systems (Reinhorn et al., 2005). In a typical seismic-based hybrid test, the overall structure is grouped into experimental and computational substructures. The interface forces between the tested (real) substructure (beam or column) and the computational (simulated) substructure are imposed by actuators (or a shake table). The measured displacements, generated in experiments, are fed back to the computational model. The test continues until the structural member fails, at which point the behavior of the whole structure is assumed to be captured through numerical simulation.

Only recently the feasibility of hybrid testing has been explored to a limited extent to evaluate the performance of structures under fire conditions (Mostafaei, 2013). In this study a single structural member (column) was tested under fire conditions, and numerical simulation was carried out to trace the response of the remaining part of the whole structure (building) that the fire-tested column is part of without the need to physically test the entire structure. From a practical standpoint, the hybrid testing concept can offer significant benefits, since testing the entire structure is almost impossible due to the cost, time, and complexities involved with such testing under fire conditions. Hybrid fire testing can shed insights into fire performance at the system level, specifically by taking into consideration interaction between fire-exposed and non-fire-exposed members that are part of the structural system. In other words, hybrid testing provides a higher level of comprehensiveness (such as 3D stress maps) and realistic response evaluation compared to traditional full-scale fire resistance testing on isolated members.

While hybrid testing presents a novel approach for evaluating the fire performance of large (or full-scale) structural systems, the application of this technique in the structural fire engineering arena is still relatively new and not fully developed. There is a need to advance the concept and procedure for hybrid testing to better fit into fire resistance evaluations. Specific test methods and numerical approaches for coupling real-time physical testing with numerical modeling, by properly accounting for various structural interactions, is needed. Appropriate simulation techniques (including solvers and algorithms), in conjunction with guidance on proper substructuring of the whole system, discretization, and so on, are also needed to facilitate numerically efficient and computationally feasible analysis techniques (Sauca, 2017).

5.9.2 Modernizing Testing Procedures and Standardization

The current approach of testing structural elements in fire test furnaces has numerous drawbacks, as discussed in Sec. 5.5. The most serious drawback in the standardized test method is that critical factors governing the fire response of a structural member, namely the cooling phase of fire exposure, structural interactions arising from adjacent members, and restraint conditions as present during a fire in structures, are not accounted for in fire resistance evaluations. Other major drawbacks include minimum requirements for data collection during the fire test, which in turn limits any understanding of fire performance of tested structural members, or considerations to all possible limit states in evaluating failure.

The genesis of standard fire test methods dates back to the early 20th century, and they have remained mostly unchanged throughout the past 100 or so years. While standard fire tests generally provide a conservative estimation of fire resistance, it is aimed at evaluating the comparative fire performance of a structural member rather than evaluating the actual fire performance under realistic fire exposure scenarios. Originally these standard fire tests were carried out mostly for defining fire ratings of structural members and assemblies. However, in the last two decades, a large number of fire tests have been carried out by researchers for validating numerical models and calculation methods. So, today there is a significant change in the purpose of undertaking such tests. Standard test methods and procedures must be modernized to introduce realistic conditions, as present in a structure, in evaluating the actual fire performance of structural members.

There is scope for exploring fire resistance tests at scaled levels (carried out at small-sized specimens) to allow extending data measured in scaled fire tests to full-sized structural members. For example, techniques that can transform the structural response of fire-tested isolated simply supported beams to predict the response of beams connected to other structural members are desired. Further, additional data collection requirements should be introduced, as part of standardized test methods and procedures, via redundant and different instrumentations types in order to better capture the performance of fire-tested specimens, especially with regard to capturing complex failure modes. Results from recent studies show the possibility of failure of the structural member in the cooling rather than the heating stage (Kodur and Agrawal, 2016), clearly indicating the need to examine the performance of fire-exposed structural members over the entire range of fire exposure (including the cooling phase), loading, and restraint scenarios.

5.9.3 Instrumentation

Despite significant advances in recent years in sensor technology for monitoring response of structures under ambient conditions, there is very little data, except temperature, that can be accurately measured in structures under fire conditions. In fact, the major drawback with standard fire testing continues to be limited instrumentation that is mounted in the furnace or on the tested member to capture critical response parameters during the entire duration of fire exposure. When a structural member is subjected to fire, the member continually experiences changes in its deflection, restraint forces, strains and stresses, temperatures, cracking, and delamination patterns throughout the fire exposure duration. As such, installing proper instrumentation that can maintain its full functionality at elevated temperatures for monitoring fire performance of fire-tested structural members and assemblies is of critical importance. Other than thermocouples, conventional instrumentation currently used in fire tests includes strain gauges and displacement transducers (to a limited extent), and these devices cannot capture the full range of response parameters, including strains, lateral deflection, restraint forces, and so on.

In the last decade or so, some gauges have been developed to measure strains in structural members at elevated temperatures. However, these high-temperature strain gauges cannot capture measurements under rapid heating rates (as that encountered in actual fires) or when sectional temperature exceeds 400°C (Kodur et al., 2012). Thus, there is an urgent need for developing instrumentation and sensors that can monitor thermal and structural changes occurring in structural members under high-temperature exposure. More specifically, high-temperature strain gauges and displacement transducers that can adequately perform at temperatures in the range of 1000°C with high reliability and durability are needed. Also, modern techniques such as laser and digital image correlation for capturing crack formation, extent (degree) of spalling, and strain mapping to better capture response of structural systems tested under fire conditions are needed.

5.10 References

ACI/TMS Standard. *Code Requirements for Determining Fire Resistance of Concrete and Masonry Construction Assemblies (ACI 216.1-14/TMS-0216-14)*. Farmington Hills, MI: American Concrete Institute, 2014.

ASTM E119. *Standard Test Method for Fire Tests of Building Construction and Materials, American Society for Testing and Materials*. West Conshohocken, PA: ASTM, 2016.

ASTM E2226-15. *Standard Practice for Application of Hose Stream*. West Conshohocken, PA: ASTM International, 2015.

ASTM International. E2336-16 Standard Test Methods for Fire Resistive Grease Duct Enclosure Systems. West Conshohocken, PA: ASTM International, 2016. doi: https://doi.org/10.1520/E2336-16

Babrauskas, Vytenis, and Robert Brady Williamson. "The Historical Basis of Fire Resistance Testing—Part II." *Fire Technology* 14, no. 4 (1978): 304–316.

Bisby, Luke, John Gales, and Cristián Maluk. "A Contemporary Review of Large-Scale Non-standard Structural Fire Testing." *Fire Science Reviews* 2, no. 1 (2013): 1.

British Standards Institute. *BS EN 1363-1: 2012 Fire Resistance Tests. General Requirements*. London: BSI, 2012.

British Steel. *The Behaviour of Multi-Storey Steel Framed Buildings in Fire*. Rotherham, UK: British Steel, 1999, p. 82

Cooke, Gordon. "Can Harmonisation of Fire Resistance Furnaces Be Achieved by Plate Thermometer Control?" *Fire Safety Science* 4 (1994): 1195–1207.

Dotreppe, Jean-Claude, Jean-Marc Franssen, and Yves Vanderzeipen. "A Straightforward Calculation Method for the Fire Resistance of Reinforced Concrete Columns." First European Symposium on Fire Safety Science. Zürich, Switzerland: ETH Zürich, 1995.

Gustaferro, A. H., and S. L. Selvaggio. "Fire Endurance of Simply-Supported Prestressed Concrete Slabs." Portland Cement Association Research department Bulletin, No. 212, 1967.
Harada, Kazunori, Hideaki Masuda, Masashi Yoshida, Takeshi Motegi, and Masateru Sudo. "Heat Transfer in Fire Resistance Furnaces—Experimental Investigation of the Variation of Heat Impact due to Furnace Construction and Control Thermocouples." *Bulletin of Japan Association for Fire Science and Engineering* 48, no. 2 (1998): 53–61.
Harmathy, T. Z. "Ten Rules of Fire Endurance Rating." *Fire Technology* 1, no. 2 (1965): 93–102.
IBC, ICC. "International Building Code." International Code Council, Inc.(formerly BOCA, ICBO and SBCCI), 2015.
Kodur, Venkatesh. "Properties of Concrete at Elevated Temperatures." *ISRN Civil Engineering*, article ID 468510, 2014.
Kodur, Venkatesh K., and Ankit Agrawal. "Critical Factors Governing the Residual Response of Reinforced Concrete Beams Exposed to Fire." *Fire Technology* 52, no. 4 (2016): 967–993.
Kodur, V. K. R., Maria Garlock, and Nestor Iwankiw. "Structures in Fire: State-of-the-Art, Research and Training Needs." *Fire Technology* 48, no. 4 (2012): 825–839.
Lie, T. T. *Fire and Buildings*. London: Applied Science Publishers Ltd, 1972.
Mostafaei, Hossein. "Hybrid Fire Testing for Assessing Performance of Structures in Fire—Application." *Fire Safety Journal* 56 (2013): 30–38.
Proe, D., and I. Thomas. "Planning for a Large-Scale Fire Test on a Composite Steel-Frame Floor System." In *Proceedings of the 6th International Conference on Structures in Fire*, pp. 390–397. East Lansing, MI: Michigan State University, 2010.
Reinhorn, Andrei M., Mettupalayam V. Sivaselvan, Z. Liang, Xiaoyun Shao, M. Pitman, and S. Weinreber. "Large Scale Real Time Dynamic Hybrid Testing Technique—Shake Tables Substructure Testing." *Proceedings of the International Conference on Advances in Experimental Structural Engineering* 1 (2005): 457–464.
Sauca, Ana. "Development and Implementation of a Methodology for Hybrid Fire Testing Applied to Concrete Structures with Elastic Boundary Conditions." PhD dissertation, Université de Liège, Liège, Belgique, 2017.
Shakya, Anuj Man. *Flexural and Shear Response of Precast Prestressed Concrete Hollowcore Slabs Under Fire Conditions*. Michigan State University, 2016.
Shorter, G. W., and T. Z. Harmathy. "Moisture Clog Spalling." *Proceedings of the Institution of Civil Engineers* 20 (1965): 75–90.
Stanzak, W. W., and Tiam Tjoan Lie. "Fire Resistance of Unprotected Steel Columns." *Journal of the Structural Division* 99, no. 5 (1973): 837–852.
Witteveen, Jelle, and Leen Twilt. "A Critical View on the Results of Standard Fire Resistance Tests on Steel Columns." *Fire Safety Journal* 4, no. 4 (1981): 259–270.

5.11 Nomenclature

T	Temperature of the emitting surface (°C)
t	Duration of fire
A_s	Area of steel (mm²)
L	Length (span) of a member (mm)
d	Effective depth of a member (mm)
R	Fire resistance
th	Cover thickness (mm)
I	Indicated fire-resistance period
A	Area under the curve of indicated average furnace temperature for the first three-fourths of the indicated period
A_{st}	Area under the standard furnace curve for the same part of the indicated period
L_g	Lag correction in the same units as A and A_s [30°C.h (1800°C·min)]

5.12 Numerical Examples

5.1

ABC architectural company wants to use a new wall assembly in a mega housing project and the proposed load-bearing walls require a 60-minute fire rating. To determine the fire rating, ABC carries out a fire test in a private fire test laboratory as per ASTM E119 fire standard specifications. The assembly fails in 72 minutes. However, during the fire test, due to poor control, the measured furnace temperatures turn out to be higher/lower than the specified temperatures in the standard as follows:

Time (min)	5	10	20	30	40	50	60	70
Temperature higher/lower than standard fire (%)	−10	−5	0	8	12	16	6	4

Did the fire test on the wall assembly meet the E119 criterion to be successful? If not, can any remedial measures be applied to make use of the results from the test? Answer the question through required engineering calculations.

Solution

As mentioned in Sec. 5.3.6, the ASTM E119 standard states that the furnace temperature must follow the profile of the standard fire with a maximum discrepancy of the area under the time-temperature curve within 10% of the standard (ASTM E119) time-temperature curve for the first hour, within 7.5% for next hour—that is, up to 2 hours and within 5% after 2 hours. Figure E5.1 plots the standard ASTM E119 as well as the applied fire curve.

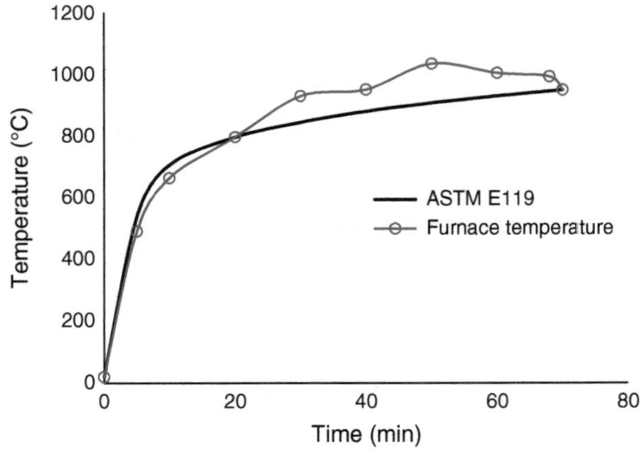

FIGURE E5.1 Comparison of the furnace temperature and the standard fire.

The area under the curve can be evaluated by summing the area of the number of trapezoids (the curve is considered a straight line within the interval). The area of each segment is given as

$$A = \frac{T_n + T_{n-1}}{2} \times (t_n - t_{n-1})$$

Following is the equation used for interpolation:

$$Y = \frac{Y_2 - Y_1}{X_2 - X_1} \times (X - X_1) + Y_1$$

Using these equations, the area under the curve can be computed as tabulated in Table E5.1 and shown in Fig. E5.2.

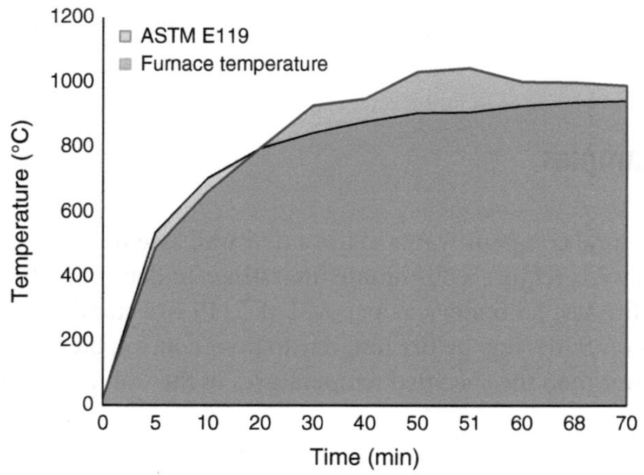

FIGURE E5.2 Area under furnace temperature and standard fire curves.

	E119		Furnace					
Time (min)	Temperature (°C)	Total Areas under the Curve (°C min)*	Furnace Temperature Discrepancy % from E119	Temperature (°C)	Discrete Areas under the Curve	Total Area under the Curve (°C min)	Variation	Remark
0	20	—	0	20	—	—	—	—
5	538	1290	-9	489.58	1273.95	—	—	—
10	704	4300	-6	661.76	2878.35	—	—	—
20	795	11,650	0	795	7283.8	—	—-	—
30	843	19,650	10	927.3	8611.5	—	—	—
40	878	28,060	8	948.24	9377.7	—	—	—
50	905	36,780	14	1031.7	9899.7	—	—	—
51	907.2	37,668	15	1043.28	1037.49	40,362.49	—	—
60	927	45,740	8	1001.16	9199.98	49,562.47	-8.36	<10% *OK*
68	938	53,052	6.5	998.97	8000.52	57,562.99	-8.50	< 7.5% *Not OK*
70	942	54,910	5	989.1	1988.07			

*ASTM E119 provides values for area under the standard fire curve for various exposure times.

Table E5.1 Evaluation of the Area under the Time-Temperature Curves

The tested assembly failed at 68 minutes; hence, the duration of the fire test is calculated as 68/60 = 1.13 hour.

The measured temperature yielded the curve area to be 8.36% (<10%) greater than the standard fire curve area for duration up to 1 hour and 8.50% (<7.5%) greater than the standard fire curve area up to 68 minutes, or 1.13 hour. Based on the ASTM E119 standard, the discrepancies are acceptable within and beyond 1 hour.

The correction factor for the indicated fire resistance can be calculated as follows:

$$\text{Correction in the same units as } I\ (C) = 2\frac{I(A - A_s)}{3(A_s + L)}, \text{ according to ASTM E119 standard,}$$

where I = Fire-resistance period (68 min)

A = Area under the curve of indicated average furnace temperature for the first three-fourths (3/4) of the indicated period = at ¾ × 68 min = at 50 min = 39325°C·min

A_s = Area under the standard furnace curve for the same part of the indicated period (at 50 min) = 37705 °C min

L = Lag correction in the same units as A and A_s [30°C·h (1800°C·min)] = 1800°C·min

$$C = 2 \times \frac{68(37668 - 40362)}{3(40362 + 1800)} = 3 \text{ min}$$

Therefore, the corrected fire resistance is 71 mm (68 + 3).

5.2

Using Appendix A of ACI/TMS 216.1 (ACI/TMS Standard, 2014), select the appropriate concrete cover thickness for the following steel columns protected with a layer of concrete masonry of 1600 kg/m^3 density.

W10×54 (1-hour rating)
W12×40 (3-hour rating)
W14×82 (2-hour rating)

Solution

W10×54 (1-hour rating), equivalent thickness = 25.4 mm
W12×40 (3-hour rating), equivalent thickness = 77.2 mm
W14×82 (2-hour rating), equivalent thickness = 47 mm

5.3

Using Table 721.2.2.1 in the International Building Code (2015), select the minimum slab thickness to satisfy 2- and 4-hour fire ratings for a slab made of siliceous and carbonate aggregates.

Solution

Slab made of siliceous aggregates: 127 mm (for 2-hour rating) and 178 mm (for 4-hour rating)
Slab made of carbonate aggregates: 117 mm (for 2-hour rating) and 168 mm (for 4-hour rating)

5.4

Using Appendix B in Eurocode 6, select the minimum wall thickness to satisfy 60- and 180-minute fire ratings for clay masonry separating load-bearing (single-leaf) walls.

Solution

Minimum wall thickness required:

90 mm (for 2-hour rating) and 170 mm (for 4-hour rating)

CHAPTER 6

Fire Resistance Evaluation through Calculation Methods

6.1 Overview

This chapter presents the application of calculation methods for evaluating the fire resistance of structural members. The different approaches that can be applied for fire resistance evaluation of structural members and systems are discussed. The steps involved in undertaking fire resistance calculations, at various levels of analysis, are outlined. The geometric and material properties to be input into computer programs for evaluating the thermal and structural response of members are discussed, and the main response parameters that can be generated from such an analysis are presented.

Figure 6.1 lays out different approaches that can be applied for evaluating the fire resistance of structures. These approaches are broadly grouped under two categories: *prescriptive-based approaches* and *rational engineering approaches*. In each category there are different methods for deriving simply the fire resistance of a member or detailed response of a structure under fire conditions. The procedures in undertaking fire resistance evaluation through each of these methods, together with drawbacks and limitations, are discussed in this chapter.

6.2 Prescriptive Approaches

Traditionally, a fire resistance evaluation of a structural member was carried out through physical testing by subjecting the member to a standard fire exposure for a specified duration in a furnace by adopting a procedure similar to that described in Chap. 5. Based on the performance of the tested member under fire exposure, the member is assigned a fire rating (e.g., 1 hour, 2 hours, etc.) for the duration of fire exposure sustained in the fire test.[1] Such fire ratings, derived from standard fire tests, are listed in documents or directories maintained by testing organizations, code authorities, or materials manufacturers. These listings, or fire resistance ratings, can be grouped under three main categories: *generic ratings*, which apply to traditional (general) material types; *proprietary ratings*, which are linked to a specific (proprietary) material, assembly, or manufacturer; and more recently, approved *correlation methods*[2] derived using fire test data by applying extension principles as discussed in Chap. 5.

6.2.1 Listings

Most countries allow fire resistance of commonly used structural members or assemblies to be deduced by direct reference to the ratings generated through standard fire resistance tests carried out on similar structural members or assemblies. Based on test data from standard fire tests, fire ratings for similar types of structural or nonstructural members and assemblies can also be derived via *extension rules*, and in some cases via *expert evaluation* or *opinion*. Such fire resistance listings are maintained as directories by certified (or recognized) testing agencies, trade organizations, and/or industrial manufacturers. In North America, independent testing organizations such as Underwriters Laboratories (Underwriters Laboratories, 2012) and Southwest Research Institute (Southwest Research Institute, 2012) maintain a registry of approved assemblies with assigned fire resistance ratings. These ratings for structural members are derived based on

[1] Members or assemblies that are unable to meet the test criteria are not given a fire rating.

[2] In some unique scenarios, expert opinions can also be sought by testing and/or approving authorities to check or interpolate results (fire ratings) obtained through calculation methods.

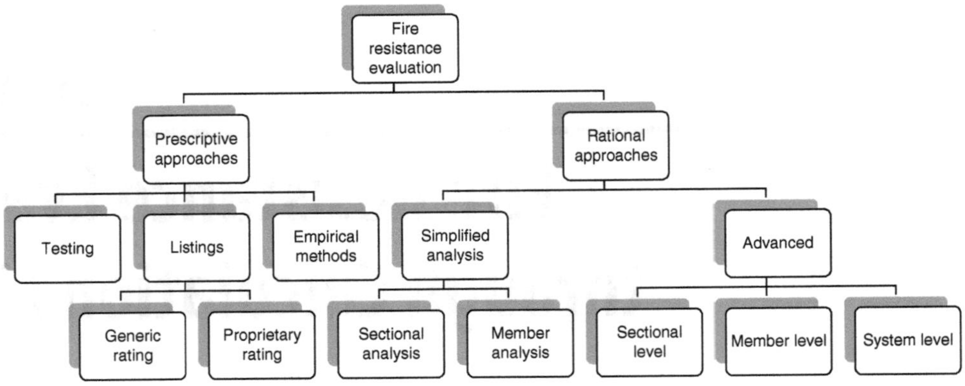

Figure 6.1 Approaches for fire resistance evaluation.

Steel Column Size	Concrete/Masonry Density (kg/m³)	Minimum Equivalent Thickness, *teq*, in mm for Fire-Resistance Rating of Concrete-Masonry-Protected Steel Column			
		1 hour	2 hours	3 hours	4 hours
W14×82	1280	20.1	40.4	58.7	75.7
W12×72	1600	23.1	47.8	68.6	87.1
W10×33	1922	34.8	62.0	84.6	104.6
W8×18	1762	36.1	63.5	85.6	105.2

Table 6.1 Fire Resistance of Concrete Masonry-Protected Steel Columns (*Authorized Reprint from ACI 216/1, 2014*)

results of fire tests carried out in accordance with fire testing standards such as UL 263, ASTM E119, NFPA 251, or ISO 834. Fire *listings* generally fall into two categories: generic ratings or proprietary ratings.

Generic fire resistance ratings are assigned to structural members and assemblies in general, without reference to a particular manufacturer, and are applicable to members made of generic material types (e.g., normal-weight concrete). Such generic ratings are listed in national codes and standards (e.g., IBC, various Eurocodes, ACI 216.1) and have a wide range of applicability. This type of rating is also published by trade organizations and industries (such as the Gypsum Association), which maintain listings of approved fire-rated products manufactured or used by their associated firms. Generic ratings are widely used in practice due to ease of fire resistance evaluation.

For example, codes and standards list generic ratings for fire protection of structural steel members encased with concrete [see one such example in Table 6.1 taken from ACI/TMS Standard (2014)]. These tables provide the minimum equivalent thickness of concrete masonry required for achieving certain fire ratings in concrete masonry-protected steel columns of different sizes. This is applicable for general concrete masonry but does not provide specific property requirements of concrete except for density. In a similar manner, generic listings for achieving certain fire ratings in different steel or wood members protected with gypsum boards are also available in codes and standards. In these listings, the thickness and number of layers of a particular type of gypsum board (Type X or Type C) to achieve certain fire ratings in a structural member can be obtained. It is worth noting that generic ratings are often conservative and are only applicable to standard fire exposure; they do not take into account a number of influencing parameters, such as load level, restraint conditions, and so on.

Proprietary listings refer to fire resistance ratings derived through fire tests on structural members, made of proprietary products (or assemblies) or material types (of a specific composition). As such, proprietary ratings are applicable only to that member with a configuration of a tested specific assembly having similar material characteristics, construction procedure, and erection/assembly methods. Proprietary ratings cannot be extended to other members made of a similar material type, but with a different chemical composition, due to differences in materials and/or installation methods that can impact the fire resistance of the member. Proprietary ratings may be less conservative than generic ratings because they relate to specific (defined) products/material types.

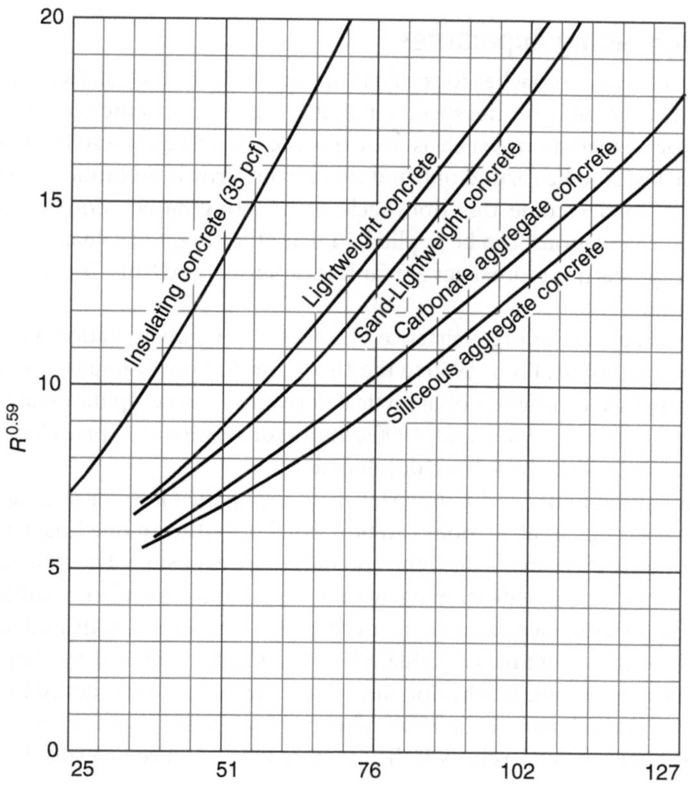

FIGURE 6.2 Values of $R^{0.59}$ to be used in Eq. 6.1. (*With permission from ASCE.*)

6.2.2 Correlation or Empirical Methods

In lieu of undertaking standard fire tests or relating to listings, fire resistance of different structural members can be obtained via *correlation methods.* These correlation methods are developed by applying the extension rules discussed in Chap. 5 or through curve fitting techniques to results obtained from standard fire resistance tests. These correlation equations mostly assume failure to occur when a critical temperature is reached in a member. For example, once the temperature in steel reaches 550°C, a steel structural member is assumed to lose much of its strength and stiffness and is not capable of sustaining applied loading on the member. The duration to reach that limiting temperature in a steel member is denoted as the resistance of this member.

Another example of the correlation-based approach is the one present in ACI 216.1 (ACI/TMS Standard, 2014). In this correlation, the fire resistance rating of a concrete wall (or slab) made of individual wythes can be evaluated as

$$R = (R_1^{0.59} + R_2^{0.59} + \cdots + R_n^{0.59})^{1.7} \tag{6.1}$$

where R = fire resistance rating of assembly in minutes and $R_1, R_2, \ldots, R_n$ = fire resistance rating of individual wythe in minutes. Values of $R_n^{0.59}$ for individual wythes of concrete for use in the formula can be obtained from charts given in ACI 216.1 and are shown in Fig. 6.2.

Similar correlation methods applicable to concrete, steel, composite, and wood structural members are present in Eurocodes, ACI 216.1, the ASCE manual, and other fire codes and standards and are discussed in respective chapters.

Correlation methods, due to their simplicity, are easy to apply and require very little effort for evaluating the fire resistance of common structural members (made of traditional materials). However, the applicability of these methods is limited to a certain range of parameters; in addition, these methods can provide only fire ratings of structural members and not detailed response of the member in the entire range of fire exposure. These correlation methods do not take into consideration the effect of realistic load level, restraint conditions, cooling phase of fire, and so forth on the fire response of the member.

6.3 Rational Engineering Approaches

The prescriptive approaches for evaluating the fire resistance of a structural member through fire tests or listings are applicable for the specific member type, the specific material the member is composed of, and the specific test conditions such as fire exposure, load level, and so on. For any other member or condition, a new fire test is to be carried out, and this can be a costly and time-consuming exercise for evaluating fire resistance. To overcome this drawback, engineering-based *rational approaches,* which utilize a set of numerical calculations, can be applied, in lieu of testing, to evaluate fire resistance. Once developed, such rational engineering approaches are far less expensive and time-consuming to apply than undertaking new fire resistance tests.

For evaluating the fire resistance of structural members, rational approaches vary in complexity and range of applicability. From 1960 to the 1980s, most of the rational approaches were developed through the use of numerical techniques or analytical derivations, as computational tools were not readily available for use. Starting in the 1980s and 1990s, numerous rational approaches that utilize finite element-based numerical techniques have been developed.

Rational methods can be undertaken at a *simplified* level or at an *advanced* level of analysis. At both levels, fire resistance calculations can be carried out either at one final (fire exposure) time increment or in a series of time increments until the end of the fire exposure. The first step of analysis is the calculation of sectional temperatures developing within a structural member resulting from fire exposure. With the knowledge of cross-sectional temperatures, the loss in strength and stiffness properties in constituent materials of the structure is evaluated. By taking the reduced mechanical properties into account, the degradation in structural behavior, in terms of deflection and reduced load-carrying capacity, and so forth, of the fire-exposed member is then evaluated. Once these response parameters are known, different failure limit states are applied to check if the member fails or survives a specified fire exposure.

6.3.1 Simplified Approaches

In the last three decades, numerous *analytical and numerical approaches* have been developed for determining temperature rise and its distribution in a member, as well as the associated loss of capacity in a fire-exposed structural member. A set of simplified equations or the use of simple formulae or graphs is applied to determine the temperature distribution and loss of capacity due to fire exposure. Normally, ordinary algebraic equations or empirical equations are sufficient for making calculations to evaluate the rise in sectional temperature or the level of temperature-induced capacity degradation or deformation in a structural member. Through applying these simple equations, a designer can check if a structural member fails or resists fire exposure at any given time-step—for example, after two hours of fire exposure in a beam. Fire resistance evaluation through these simplified approaches requires only moderate effort, but this type of analysis yields rather a conservative estimation. This type of calculation could be applied for evaluating fire resistance of a member based on sectional analysis, typically at critical sections of a structural member. Figure 6.3 shows a flowchart illustrating various steps involved in fire resistance evaluation through simplified approaches.

The simplest form of this type of approach is through the use of design aids/charts. Temperature distribution within a structural member with specific geometrical features can be obtained from published figures or charts given in ACI 216.1 (ACI/TMS Standard, 2014), Eurocode 2, or other national codes/standards and design conditions. Figure 6.4 illustrates two such temperature profile plots, one for a concrete section of a beam and the other for any segment in a concrete section at various depths of the member. The temperature rise in steel reinforcement and/or in concrete at a given depth in a concrete section of a member can be read out from these charts at any given fire exposure time. With the knowledge of the temperature in concrete layers and rebar, the reduced strength in these materials can be deduced, which can then be used to calculate the reduced flexural or shear or axial capacity of the member at that fire exposure time. For example, the temperature rise in a corner steel rebar located at 25 mm from the edge of a concrete beam with dimensions similar to those shown in Fig. 6.4*a* can reach 800°C after 90 minutes of exposure to a standard fire. Once the rebar and concrete temperature and corresponding reduced strengths are known, the moment capacity after 90 minutes of fire exposure can be evaluated by applying the relevant capacity equation. A detailed example covering the application of this approach is given in Chap. 8.

In lieu of charts and graphs, empirical equations are available to estimate temperature rise in fire-exposed structural members. An example of one such correlation equation is the *best-fit method* for evaluating temperature in a steel section when subjected to fire exposure. This equation was developed

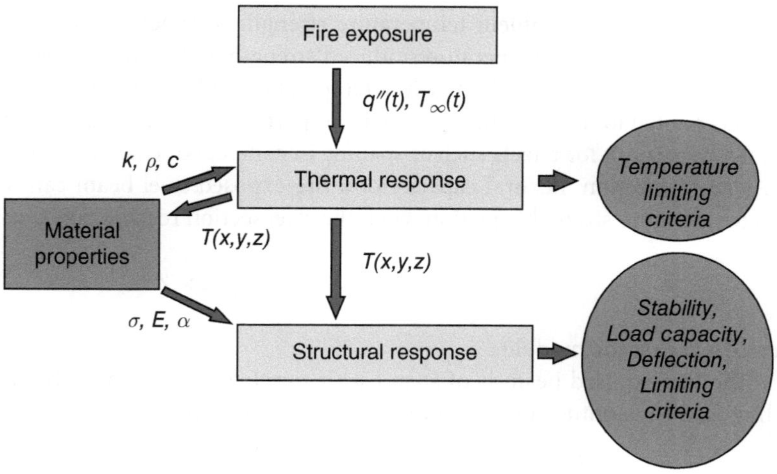

Figure 6.3 Flowchart illustrating fire resistance evaluation through simplified approaches.

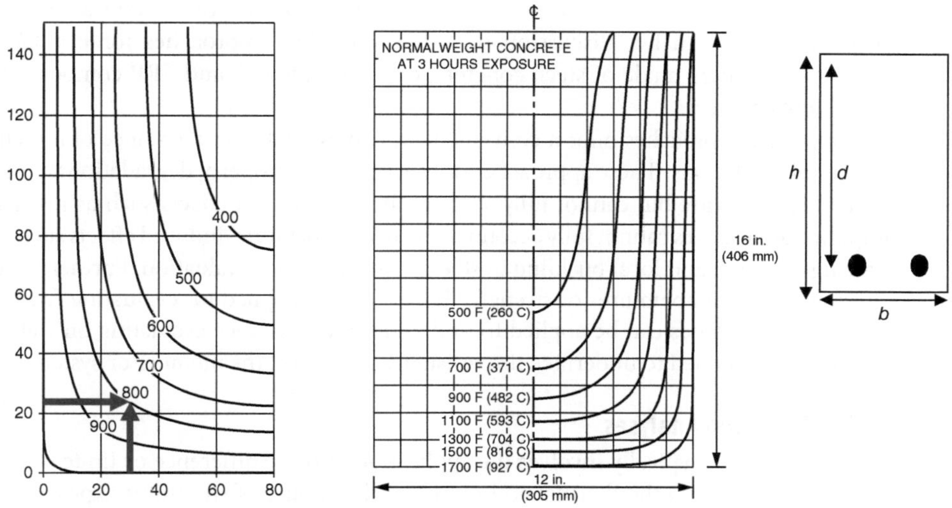

(a) Temperature profiles (°C) after 90 minutes of exposure to ISO 834 in a beam with height of 300 mm and width of 160 mm (Eurocode 2) (Reproduced from CEN)

(b) Temperature distribution in a normal-weight concrete rectangular unit with height of 406 mm and width of 305 mm at 3 hours of fire exposure to ASTM E119 (Authorized reprint from ACI, 216/1. (2014))

Figure 6.4 Temperature profiles in typical fire-exposed concrete beams.

through correlation of temperature data in fire tests. The time to reach a limiting temperature in a fire-exposed steel section is expressed in terms of key parameters influencing temperature rise. For the case of an unprotected steel structural member exposed to fire, time to failure, t, is expressed in terms of surface area and volume of steel section and is given as

$$t = \frac{0.54(T_{\text{lim}} - 50)}{\left(F/V\right)^{0.6}} \tag{6.2}$$

where F is the surface area of unit length (m^2), V is the volume of steel in unit length (m^3), and t is the time to reach a limit temperature. Full details on the applicability of this method, together with limitations, are presented in Sec. 7.6.2.

Fire resistance analysis in the strength domain can also be carried out through simplified approaches. In this case, room temperature capacity equations governing structural behavior (i.e., flexural, axial, torsional, etc.) are extended to elevated temperature, with due consideration to temperature-induced loss in strength and modulus properties of constituent materials. Room temperature capacity equations are modified by replacing ambient temperature strength parameters (i.e., $f_{y,20°C}$, $f'_{c,20°C}$, etc.) by a reduction factor (k) to account for temperature-induced strength and modulus degradation such that $f_{T°C} = k_{fyT°C} \times f_{y,20°C}$ and $E_{T°C} = k_{E,T°C} \times E_{20°C}$. Similarly, a temperature-induced reduction in the cross section of a section can be accounted for by reducing geometric properties of a structural member—i.e., plastic modulus, Z, or area, A—to account for effects such as spalling in concrete or charring in wood.

As an illustration, flexural capacity of a fire-exposed steel beam can be evaluated by extending the room temperature strength equation when the steel section reaches 500°C as:

$$M_f = k_{500°C} \times f_{y,20°C} \times Z \tag{6.3}$$

where Z is the plastic modulus.

If the fire-exposed beam is of span l with simple supports and subjected to a uniformly distributed loading w, the magnitude of the bending moment during fire exposure is

$$M_{fire} = w \times l^2/8 \tag{6.4}$$

Once the bending moment present during a fire, M_{fire}, exceeds the reduced moment capacity due to fire effects, M_f, failure is said to occur in the beam. The time to reach this point marks the fire resistance of this particular beam. Further discussion on other simplified approaches for evaluating fire resistance of structural members made of steel, concrete, composite, wood, and FRP composites will be presented in subsequent chapters.

Simplified approaches offer a convenient way to evaluate fire resistance of a member, but they have a number of limitations. These approaches only account for property degradation in materials resulting from elevated temperature but cannot fully account for fire-induced effects such as restraint forces. Also, these simplified approaches cannot fully account for other temperature-induced effects such as spalling of concrete, charring of wood, and local buckling in steel in fire resistance evaluation. Furthermore, only limited failure criterion (either temperature- or strength- based) can be applied for evaluating fire resistance. In addition, simplified approaches can be applied to undertake fire resistance evaluating only at a cross-sectional level and cannot be applied to undertake fire resistance analysis at the member or system level.

6.3.2 Advanced Analyses

Advanced analysis typically involves the application of finite difference or finite element- based numerical techniques for tracing the thermal and mechanical response of structures exposed to fire. Such numerical models can account for a number of influencing factors in evaluating response over the entire range of fire exposure, and thus, the fire resistance of structures can be predicted with more precision. The analysis can be carried out at the sectional level of a member or at the member/assembly level in a structure, or at the system level by incorporating the whole structure.[3] Further, it is possible to account for geometric and material nonlinearities, temperature-induced property degradation in materials, and other high-temperature effects, such as creep deformation, restraint forces, and spalling effects. Further, for all possible failure modes inducing temperature strength (i.e., axial, flexure, or shear), deflection limit states can be applied to evaluate the failure of a member. In addition, advanced analysis can accommodate different fire scenarios' (with full consideration due to the cooling phase) support conditions, loading conditions, and other varying factors.

Advanced analyses are carried out through the use of sophisticated computer programs which utilize finite element or finite difference formulations for heat transfer and structural analysis. While the finite difference technique predicts the thermal or structural behavior of a structure member at discrete (spatial) points by solving the approximation of derivatives between successive points within the domain, the finite element method discretizes the domain of a structure member into small elements and solves for temperatures and/or deformation at the nodes of these elements. Overall, numerical models based on advanced analysis can simulate the response of fire-exposed structures as close as possible to the scenario that exists in the real structure. Some of the commonly used finite element software readily available for fire resistance analysis include ABAQUS (2017), ANSYS (2018), and SAFIR (Franssen, 2011).

Typically, analysis is carried out at various time increments, by dividing the total fire exposure time into a number of time-steps. The geometry of the fire-exposed structure is to be defined, and then the structure is to be discretized *(meshed)* into a number of *elements*. Each element is composed of a number

[3]Please refer to Fig. 3.16, which further breaks down analysis levels.

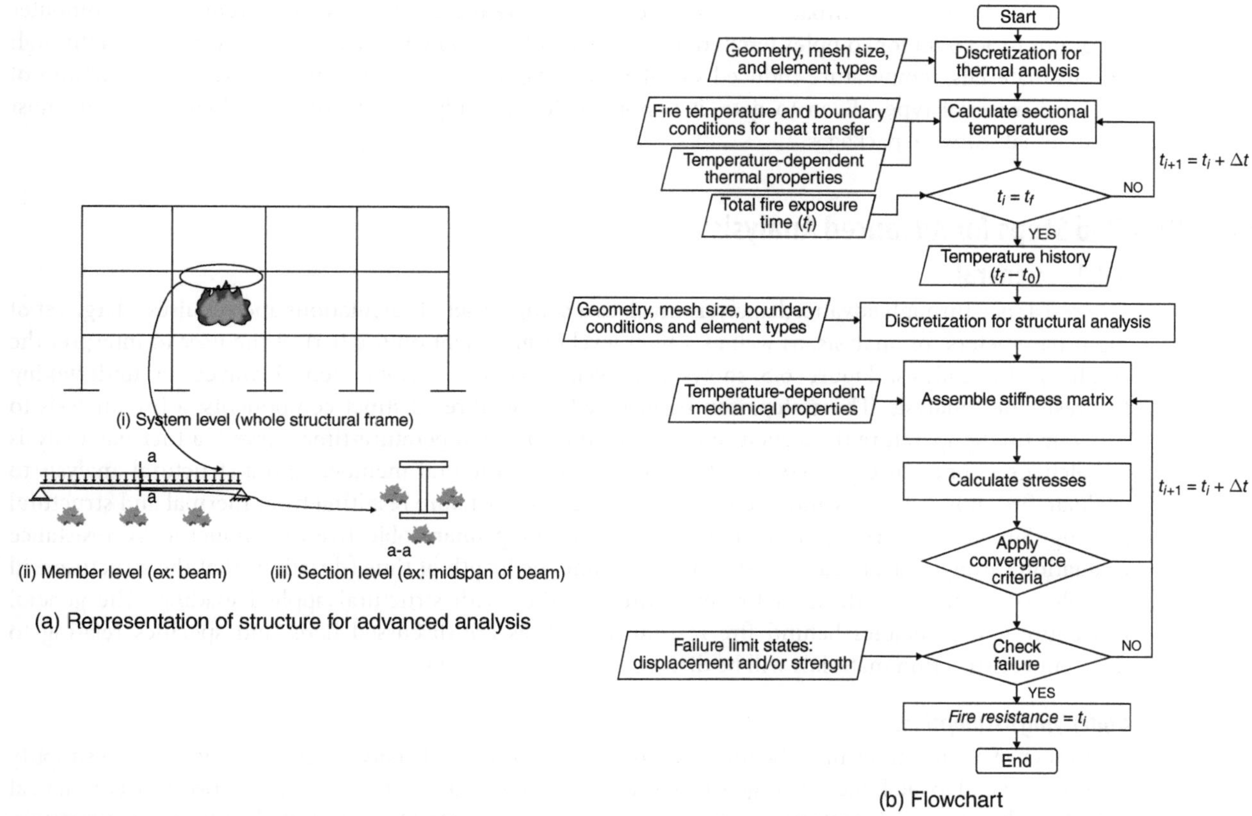

Figure 6.5 Flowchart illustrating calculation steps for fire resistance evaluation through advanced analysis. (*With permission from Aziz, 2015.*)

of nodes that forge the element boundary. The number of elements utilized to mesh a given finite element model is often governed by the complexity in the geometry of the analyzed structure and the level of accuracy needed in predictions.

Depending on the geometry of the structure and sophistication adopted in the analysis, the selection of element type as well as the number of elements can vary. Available element types vary from one computer program to another. Fire resistance in advanced analysis can be undertaken at the cross-sectional, member, or system level (refer to Fig. 3.16). This is reflected in the discretization adopted, as well as boundary conditions defined for the analysis. In the case of fire resistance analysis at the system level, all critical factors affecting response of the fire-exposed structure at the global level, such as composite action between slab and beam, tensile membrane action, fire-induced restrain forces, and redistribution of forces/stresses and member interactions, can be taken into account. These factors can enhance the fire resistance of a member by providing alternate load paths to transfer loads from fire-weakened members to other, cooler parts of the structure. One needs to keep in mind that accounting for such factors, which requires specific input parameters as well as element types, assurance of numerical stability, and so on to capture their effects on the member and restraint boundary conditions, can complicate the analysis. Recent works carried out by Jiang et al. (2015) and Nguyen et al. (2016) provide some insights into complexities and strategies to be adopted in undertaking advanced analysis for fire resistance of structures.

The advanced analysis is carried out in two stages: thermal and structural calculations, as shown in Fig. 6.5. The thermal and structural analyses can be coupled or uncoupled and are to be carried out at various time increments (time-steps). Temperature-dependent material properties are to be input into the analysis. The thermal analysis generates cross-sectional temperature distribution within the section or member (or system) at each time increment. These nodal temperatures are then applied as thermal-body load on the structure to be analyzed. Then the second stage, structural analysis, is carried out wherein a set of equilibrium and compatibility equations are solved so as to generate deformations and stresses resulting at various nodal points of the structure. These stresses and deformations are compared against failure limit states such as flexural axial deflection to establish the fire resistance of the fire-exposed member. Once generated, if sectional temperatures, stresses, deformation, or capacity exceed any specified limit state, failure is said to have been attained in the structure. Further details on various factors to be considered in advanced analysis are provided in Sec. 6.4.

One of the major drawbacks of advanced analysis is the limitations in currently used computer programs to account for complex temperature-dependent phenomena such as moisture migration through porous materials, temperature-induced shrinkage effects, spalling, or charring that occur in structures of different material types. Even to date, it is not possible to fully capture some of these effects in most commercial software packages.

6.4 Detailed Steps for Advanced Analysis

6.4.1 General

Fire resistance through advanced analysis involves a complex set of calculations and requires a large set of input parameters for analysis, as well as a high level of judgment on the part of the user to interpret the results from analysis. Figure 6.5 shows a flowchart of the various steps involved in undertaking fire resistance analysis. These steps can be grouped under three distinct components: a fire analysis to generate fire temperature the structure is subjected to (i.e., temperature-time curve[4]), a thermal analysis to calculate temperature progression (history) within a structural member, and a structural analysis to evaluate fire-induced forces and stresses developed in the structure resulting from thermal and structural loading. Similar to carrying out a standard fire test, the primary objective to conduct a fire resistance evaluation through advanced analysis is to determine the length of time (duration) that the structure will be able to sustain the effects of fire exposure, together with structural applied loading. The general procedure and principles behind fire resistance analysis are discussed here, and specifics relating to different construction materials will be discussed in Chaps. 7 to 11.

Underlying Assumptions

A number of assumptions must be made in undertaking advanced analysis; these are required to simplify the problem and to facilitate convergence in recalling a numerical solution. The assumptions cover practical aspects with regard to discretization of member geometry, uncertainties associated with high-temperature material properties and constitutive laws, varying loading/support conditions, fire exposure scenario, and so on. A fundamental assumption, as in the case of structural analysis, is that plane sections before bending remain plane after bending. Other assumptions, directly related to structural fire engineering, include fire-induced vapor movement (and moisture movement) within a concrete or wood member to not significantly affect the response of a structure under fire conditions.

6.4.2 Analysis Procedure

Fire resistance analysis is carried out by incrementing fire exposure time in time-steps. These time-steps can be equal increments (e.g., 30 seconds) or unequal increments (e.g., the first few time-steps are 5 minutes each, and then closer to the end of fire exposure, smaller time-steps of 1 minute can be adopted) to facilitate smooth convergence. The first step in the analysis is to evaluate fire temperature surrounding a structure (within a compartment). This can be calculated following temperature-time relations discussed in Chap. 3 or by assuming a standard fire exposure such as ASTM E119 or ISO 834.

For fire resistance analysis, the structure is to be discretized into a set of elements bound by nodes. There are two stages of analysis: thermal and structural.[5] Thermal analysis traces the development of cross-sectional temperature (and distribution) in a structure as a function of fire exposure time. Once completed, these nodal temperatures are then applied to the structure as thermal loading. At each time-step of analysis, temperatures and stresses as well as deflection of the structural member are generated. The resulting stresses can be further integrated to generate reduced sectional capacity of a member as a given time-step.

These response parameters—cross-sectional temperatures, capacity, and deflections—are checked against allowable failure limits to see if the members have experienced failure. For example, in the case of beams, the reduced moment capacity at a given time-step is compared against the bending moment arising from structural loading to see if the member has failed. Similarly, in the case of columns (compression members), the axial and buckling capacity are to be compared against loading effects. If the failure is attained, at a time-step, the total duration of that time-step is taken as the fire resistance of the member. If the failure is not reached, the time is incremented and the whole analysis is repeated. It should be noted

[4]This component is often carried out in a separate analysis using methods described in Chap. 3.

[5]In a typical fire resistance analysis, the fire temperature could be assumed to follow that of a standard fire, and thus the need for evaluating a temperature-time history of fire may not be necessary.

that any additional failure limiting states, including deflection, shear force, and local buckling, can also be checked against applicable limit states to determine failure of the structure.

For undertaking advanced analysis, due consideration should be given to a number of details that influence the outcome of fire resistance analysis. These details relate to element types, material constitutive models, restraint conditions, convergence criteria, and failure limit states. Due to different features in available computer software, some of these input parameters (such as element types, material damage models, etc.) might vary. A full discussion of these details, together with some recommendations, is provided in the next section.

6.4.3 Heat Transfer Calculations

To evaluate the response of a structure under fire exposure, temperature rise within a structural member is evaluated first through heat transfer analysis. This step is important since the ability of a *barrier* to maintain its insulation property as well as a *structural member* to continue its load-carrying capability depends on temperatures generated within its cross section. In the case of a non-load-bearing member, output from heat transfer analysis can be directly utilized to assess the average temperature rise on the unexposed side and determine the failure state of the member. In the case of a load-bearing structural member, temperature (and thermal profile) obtained from heat transfer analysis is an essential input to the structural model for calculating member capacity, deformations, and stresses.

Heat transfer calculations require knowledge of the geometry of the element, temperature-dependent thermal properties of constituent materials, convective and radiation heat transfer coefficients at the exposed and unexposed boundaries of the structure, and thermal properties of the surrounding air. It should be noted that materials with high thermal conductivity may be assumed to have uniform temperature across their cross section—that is, in steel flanges or webs. However, in materials with low thermal conductivity such as concrete and wood, it is important to know the thermal gradients developed within concrete depth/width, since such thermal gradients can influence temperature rise of reinforcing steel, vapor/moisture migration, and development of thermal stresses and creep deformations.

Heat transfer from the fire source to surfaces of the structural member occurs through convection and radiation mechanisms. On the other hand, heat transfer within the solid portion of the member occurs mainly through conduction. In the case of structural members/assemblies with voids, heat transfer can also occur via a combination of convection, radiation, and conduction mechanisms. Fire exposure along the length of the structural member is generally assumed to be uniform in fire resistance analysis. This assumption simplifies the complex heat transfer phenomenon and reduces a three-dimensional heat transfer problem into a two-dimensional problem. Therefore, the governing partial differential heat transfer equation within the cross section of a structural member can be written as

$$\rho c_{(T)} \frac{dT}{dt} = \nabla \cdot (k_{(T)} \nabla T) \tag{6.5}$$

where ρ = density, c = specific heat, ∇ = the spatial gradient operator, k = conductivity matrix, T = temperature, and t = time.

Since the main mechanism of heat transfer at the fire-structural member interface is through convection and radiation, the heat flux at this interface due to convection and radiation can be expressed through the following equation:

$$q_b = (h_{con} + h_{rad})(T - T_f) \tag{6.6}$$

where h_{con} = the convective heat transfer coefficient and h_{rad} = the radiative heat transfer coefficient, which is defined as

$$h_{rad} = 4\sigma\varepsilon(T^2 + T_f^2)(T + T_f) \tag{6.7}$$

where T_f = fire temperature, σ = Stefan-Boltzman constant = 5.67×10^{-8} (W/m^2.K^4), and ε = emissivity factor, which is related to the "visibility" of the exposed surface of the structural member to the fire.[6]

[6]Emissivity is defined as the ratio of the energy radiated from a material's surface to that radiated from a blackbody (a perfect emitter) at the same temperature and wavelength and under the same viewing conditions. It is a dimensionless number between 0 (for a perfect reflector) and 1 (for a perfect emitter). Emissivity is influenced by fuel type, flame color, distance from radiant body, etc. Further details on emissivity, from a fire dynamics point of view, can be found in Drysdale (2011).

In numerical heat transfer analysis, fire is assumed to be originating from a point source (i.e., perfect black body) radiating heat equally in all spatial directions. Thus, to account for heat loss to the surrounding environment, as fire travels from its source to the structural member, it is assumed that only 70% of the radiant heat reaches the structural member surface. Therefore, an emissivity factor of 0.7, as specified in Eurocode 1 (2002), can be assumed in fire resistance analysis. The Eurocodes list material emissivity factors as 0.7 for concrete, 0.8 for steel, and 0.8 for timber, respectively. For radiative heat transfer from the fire (flames), a conservative emissivity value of 1.0 is used for design purposes. For standard fire exposure, Eurocode 1 (2002) recommends a value of 25 $W/m^2 \cdot K$ as the convective heat flux coefficient at the exposed surface. This value changes to 35 $W/m^2 \cdot K$ for a parametric fire and 50 $W/m^2 \cdot K$ for a hydrocarbon fire exposure. The convective heat flux coefficient at the unexposed face of a member is taken as 4 $W/m^2 \cdot K$ when only convective losses are considered or 9 $W/m^2 \cdot K$ when radiative losses are included.

According to Fourier's law, the governing heat transfer within a structural member is through conduction and can be expressed as

$$k\left(\frac{\partial T}{\partial y}n_y + \frac{\partial T}{\partial z}n_z\right) = -q_b \tag{6.8}$$

where n_y and n_z are components of the vector normal to the boundary in the plane of the cross section. The right-hand side of Eq. 6.8 depends on the type of imposed boundary condition. For example, in the case of a beam, where the structural member is exposed to fire from three sides, two types of boundary equations are to be considered in thermal analysis:

- On fire-exposed boundaries (surface), the heat flux is governed by

$$q_b = -h_f(T - T_f) \tag{6.9}$$

- On unexposed boundaries, the heat flux governed by

$$q_b = -h_0(T - T_0) \tag{6.10}$$

where h_f and h_0 are the heat transfer coefficients on the fire side and the cold side, respectively, and T_f and T_0 are temperatures at the fire-exposed side and the cold side, respectively.

Thus, the nodal temperatures of any element are related by the shape function matrix (N) to arrive at the temperature in the element:

$$T = \{N\}^T \cdot T_e \tag{6.11}$$

Then, the above heat transfer equation, subjected to appropriate boundary conditions, can be discretized as (Cook et al., 1974):

$$C_e^t \dot{T}_e + K_e^t T_e = Q_e \tag{6.12}$$

where C_e^t = specific heat matrix, K_e^t = thermal "stiffness" matrix and is the sum of conductivity and convection matrices, T_e = nodal temperature, and Q_e = applied nodal thermal load and is composed of the convective and radiative heat fluxes.

The above equations are suitable for analysis under a standard fire scenario, as well as in a realistic fire scenario wherein a decay phase is assumed to start once the burning reaches a maximum temperature. Examples of design fires are Swedish fire curves and the parametric fire curves specified in the Eurocode 1 (2002), discussed in detail in Chap. 3.

A deficiency in most heat transfer analysis software packages is their inability to account for mass transfer such as the transport of water or water vapor through materials within their internal pore structure (i.e., concrete, wood). Moisture movement has an influence on the fire performance of materials such as gypsum plaster, concrete, and wood, as described by Fredlund (1979). Although most software programs do not directly model shrinkage of the material or ablation of the material from fire-exposed surfaces, such effects can usually be simulated by modifying temperature-dependent thermal conductivity and specific heat property to reflect shrinkage and ablation.

6.4.4 Strength Calculations

The nodal temperatures generated from thermal analysis are to be supplied as an input to the strength (structural) analysis, to assess the structural response of a structure. To perform strength analysis, most finite element packages utilize virtual work principles. According to these principles, any virtual change of internal strain energy must be balanced by a change in the external work due to the applied loading.

$$\delta U = \delta V \tag{6.13}$$

where U = strain energy and V = external work.

The variation in strain energy (δU) can be evaluated as

$$\delta U = \int_v \{\delta\varepsilon\}\sigma dv \tag{6.14}$$

For structural members subjected to fire conditions, the strain vector ($\delta\varepsilon$) is the sum of thermal (ε_{th}), mechanical (ε_m), and creep (ε_{cr}) strains, and σ is the stress.

$$\varepsilon = \varepsilon_m + \varepsilon_{th} + \varepsilon_{cr} \tag{6.15}$$

Variation in external work (δV) due to the applied nodal forces (F_e^n) can be computed by assuming a variation in nodal displacement $\{\delta u\}$ as

$$\delta V = \{\delta u\}^{\mathrm{T}}\{F_e^n\} \tag{6.16}$$

The nodal displacements of the finite elements (u_e) are related to the nodal displacement field through shape functions matrix (N) as follows:

$$u_e = \{N\}^T \cdot u \tag{6.17}$$

Thus, the virtual work equation (Eq. 6.13) can be rewritten in matrix form as

$$K_e u_e - F_e^{th} = F_e^n \tag{6.18}$$

where K_e = element stiffness matrix and F_e^{th} = element thermal load vector.

In solving the above set of equations, displacements and stress levels at nodal points can be generated. These response parameters can then be checked against all possible failure limit states to evaluate failure of a fire-exposed structure being analyzed.

6.4.5 Failure Limit States

The response parameters generated from thermal and structural analysis can be utilized to check the stability and integrity of a structural member under fire exposure. For evaluating realistic failure of a structural member, all possible limiting criteria are to be applied. The different failure limit states include thermal (temperature), strength, and deformation and need to be checked against applicable limiting values to evaluate failure of a structural member at each time-step. Failure is said to occur once any of these limiting states is exceeded, as can be seen in Fig. 6.5.

For satisfying strength limit state, reduced sectional capacity in a structure (member) at any given fire exposure time is to be checked against the actions acting on the structure. Sectional capacity (i.e., moment, shear, axial capacity) at each time-step is evaluated utilizing internal bending, shear, and axial stresses generated from the above strength calculations in advanced analysis as outlined in the preceding section. These stresses, generated at distinct nodes across the length of the structural member, have to be integrated across the depth of the section. Most commercial packages do not directly output sectional capacity. Such calculations are to be performed by writing a separate script within the computer model or via a separate subroutine or spreadsheet calculations. Once the stresses across the depth of the section are obtained, these stresses can be used to calculate associated sectional moment, shear, or axial capacity. To maintain equilibrium, the calculated moment, shear, or axial capacities (effect of actions) are then compared against the bending moment and shear force due to applied loading acting on the member during fire exposure. If the actions exceed that of the corresponding sectional capacity, failure is said to have occurred in the structural member. Alternatively, if the internal stresses in the member exceed that of the reduced strength and elastic modulus properties, failure is said to occur.

In certain types of members, deflection limit state can be the governing criterion and thus this state is also to be applied to evaluate failure of such fire-exposed horizontal or vertical members. Further, under severe fire conditions, deflections can increase at a rapid pace just prior to failure of the member, due to temperature-induced softening in materials and also high-temperature creep effects. Such a rapid rise in the rate of deflection indicates a threat to the stability of a structure and imminent failure. Therefore, the rate of deflection can be applied as a limiting state for evaluating failure of the member. When the structural member attains a limiting *deflection* or limiting *rate of deflection*, as that given in fire codes or standards such as ISO 834 or ASTM E119, the structural member is said to attain failure.

Similarly, the local buckling limit state can also be checked at each time-step by comparing member slenderness to that of temperature-dependent slenderness limits (Naser and Kodur, 2016). For instance, slenderness of plates (e.g., flanges and web in a W-shaped steel section) in columns or beams can be checked against temperature-dependent slenderness limits at each time-step of analysis. Once the sectional slenderness exceeds the compactness (or noncompactness) slenderness limit, local buckling is said to occur and sectional capacity (i.e., flexural and/or shear) is adjusted to account for the loss arising from local buckling. The user can define other limit states, such as extent of spalling in concrete members or level of charring in wood members, to define failure.

6.4.6 Critical Parameters to Be Considered in Advanced Analysis

Since the advanced analysis approach aims at tracing a detailed and accurate response of structures under fire conditions, the analysis involves a complex set of calculations with due consideration to a large set of parameters such as selection of computer models, discretization, and material models (Garlock and Glassman, 2014). This section presents additional details on the critical factors in undertaking advanced fire resistance analysis.

Selection of Computer Model

Finite element method (FEM) based computer programs, utilized for advanced fire resistance analysis, can be general-purpose software packages (i.e., can solve multiphysics problems) or special-purpose software packages specifically coded for fire resistance analysis. Some of the widely used general-purpose general computer programs are ANSYS, ADINA, DIANA, and ABAQUS. In lieu of commercial programs, specially developed computer software for fire resistance applications, such as SAFIR and VULCAN, can also be applied to undertake advanced analysis. The special-purpose software is primarily developed by researchers at universities, specifically for fire resistance analysis, and may not be as user-friendly as those of commercially available package. Overall, the analysis procedure using aforementioned computer models is very similar and includes two stages of analysis: thermal and structural.

A side note to consider, whether general-purpose or special-purpose software is used, is that these software packages need to be properly validated or verified for structural fire engineering problems. A key advantage of general-purpose finite element packages is that they have vast libraries of elements for discretization, built-in solution techniques, built-in elevated temperature properties for various material types, and preprocessing and post-processing capabilities that can aid in inputting data for analysis or interpreting results from analysis. Perhaps most importantly, these software packages have well-articulated documentation and customer support that can be consulted for resolving any issues with the analysis.

Since special-purpose software is developed by smaller companies or university researchers, it may not be user-friendly or updated and improved continuously. Further, many of the special-purpose programs are specially developed for certain material types and for analysis under certain fire scenarios (i.e., standard fires). These programs may not have sufficient capabilities to handle discretization of complex structural shapes or simulation of large-scale structures (system level), nor being validated at that scale. This contrasts with general-purpose software, which often provides an easy-to-use user interface and a wide range of solver types.

Discretization of Structure

The first step for undertaking fire resistance analysis through finite element software is to define the geometry of the analyzed structure and then discretize the structure into smaller segments or elements. Separate element types are to be selected for thermal and strength analysis. In both analyses, temperatures and stresses are calculated at the nodes of discretizing elements. Better approximations of temperature and stresses at any location in a structure can be obtained by optimizing meshes, as very fine meshes may result in increased computational time with little improvement in accuracy.

Computer Program	Thermal Analysis			Structural Analysis		
	Concrete/Wood	Steel	Others	Concrete/Wood	Steel	Others
ANSYS	SOLID70	SHELL131	SURF153 (radiation/ convection), LINK8 (steel reinforcement)	SOLID65, SOLID185	SHELL181	SOLID45 (steel supports, composites), SPRING40 (bond-slip effects)
ABAQUS	DC3D8, DC2D4	DS4	T3D2 (steel reinforcement)	C3D8	S4	CONN3D2 (bond slip effects), Spring1 (restraint conditions)
DIANA	HX8HT		BQ4HT (convection)	CHX60, HX24L	Q20SF, CT30S	Q24IF (Bond-slip)
ADINA	21-node brick	Shell conduction element	2D conduction	27-node brick	MITC9	2-node linear, 3D truss element (for steel reinforcement)
SAFIR	SOLID	SHELL	BEAM, BLOCK (for supports)	SOLID	SHELL	BEAM, BLOCK (for supports)

Table 6.2 Element Types for Discretization of Structures for Fire Resistance Analysis through Commercial Finite Element Programs

Depending on the type of software selected for analysis, the available element types for discretization can vary. The typical elements available in the ANSYS elemental library for thermal analysis in structural fire engineering problems are SHELL131, SOLID70, LINK33, and SURF152. SHELL131 is a layered shell element having in-plane and through-thickness thermal conduction capability and can be used to simulate thin structures such as steel members. On the other hand, SOLID70 is a cubic thermal element with conduction capability and can be used to discretize solidlike structural members (i.e., made of concrete, wood, etc.). LINK33 is a uniaxial element with the ability to conduct heat between its nodes and is used to simulate reinforcement (reinforcing or prestressing steel bars/strands embedded in a concrete beam, slab, or column). SURF152 is a surface thermal element capable of simulating heat conduction, convection, and radiation. For heat transfer modeling, SURF152 is overlaid on top of SHELL131 or SOLID70 elements to simulate convection and radiation mechanisms of heat transfer from the fire source to a structural member. Table 6.2 lists different element types available in some of the commercial computer software for structural fire engineering applications. Figure 6.6 shows element types commonly available in commercial programs.

For simulating structural response, the structure is to be discretized with structural elements available within the software elemental library. For example, for a composite beam-slab assembly, the different components—steel beam, concrete slab, steel reinforcement, and shear studs—are to be properly discretized. In the ANSYS software package, different structural components can be discretized using suitable elements, namely SHELL181, SOLID65, LINK8, BEAM188, COMBIN39, COMBIN14, CONTA173, and TARGE170.

The SHELL181 element is formulated to capture local buckling of flanges and web, and as such it is often used to discretize thin steel plates (i.e., in a beam or column). SOLID65 is adopted to discretize concrete slab, since this element is capable of accounting for cracking and crushing of concrete. LINK8 is used to simulate reinforcing steel embedded in concrete slab. COMBIN14 is a spring element used to model the bond-slip behavior between steel reinforcement and surrounding concrete. In order to accurately simulate shear stud, BEAM188 can be used and assumed to be welded to the top flange of steel beams. To simulate the bond-slip action at the interface between shear studs and steel beam, COMBIN39 can be used. Other elements can also be used to simulate bond behavior such as nonlinear surface-to-surface contact elements (CONTA174 and TARGE170). This contact pair can be fully bonded to simulate the full composite action between the concrete slab and steel beam, or it can be partially bonded to account for the slip that occurs between the concrete slab and steel beam in case of partial composite action. Further details on discretization of structures into elements in ANSYS for fire resistance analysis can be found elsewhere (Kodur and Naser, 2015).

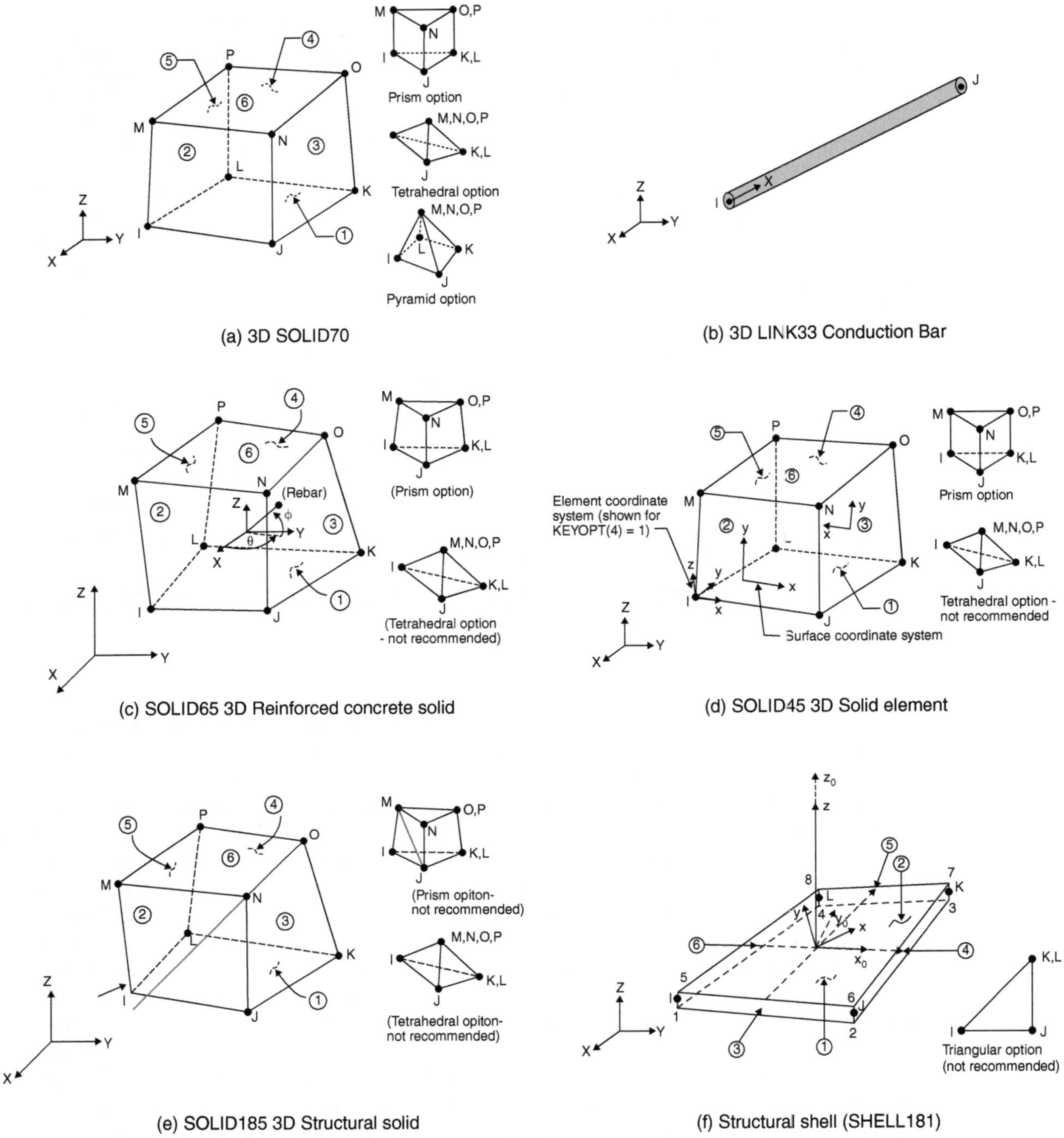

FIGURE 6.6 Various elements that can be adopted for discretization of a structure for fire resistance analysis. (*From ANSYS software.*)

In the case of ABAQUS, the thermal model for structural steel, concrete and reinforcement can be discretized using DS4 (shell element), DC3D8 (brick element), and DC1D2 (link element), respectively. For the structural analysis, a structure can be discretized using S4 shell elements (for steel sections), continuum brick elements (C3D8), and two-noded link elements (T3D2) for concrete and reinforcing/prestressing steel, respectively. C3D8 element has eight nodes with three degrees of freedom, namely, three translations in the x, y, and z directions. This element can be used for 3D modeling of solids with or without reinforcement, and it is capable of accounting for cracking of concrete in tension, crushing of concrete in compression, creep, and large strains. T3D2 elements are used to model one-dimensional reinforcing/prestressing bars or rods that are assumed to deform by axial stretching only. They are pin-jointed at their nodes; only translational displacements and the initial position vector at each node are used in the discretization. Similar elements in other software packages such as DIANA and SAFIR are listed in Table 6.2.

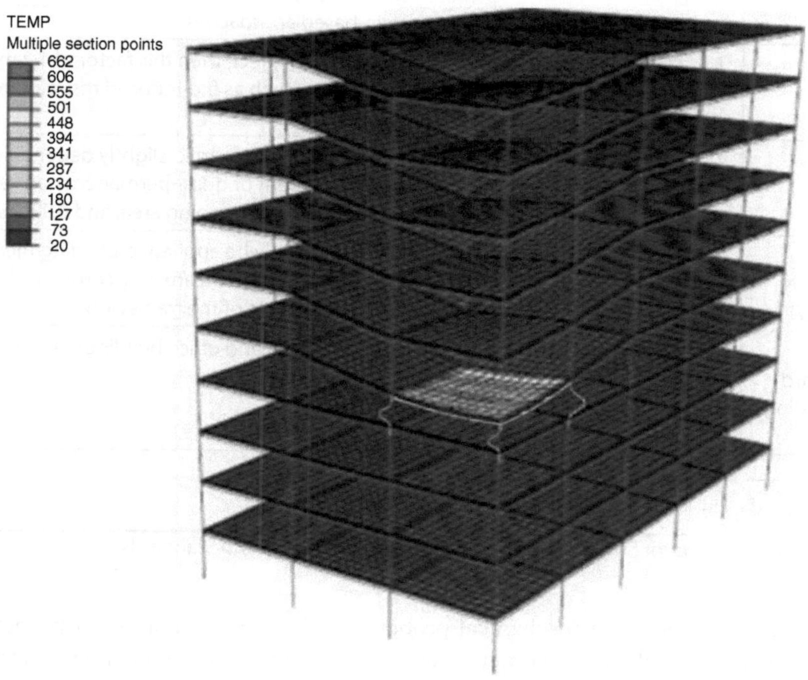

Figure 6.7 Three-dimensional model of a composite frame for fire resistance analysis. (*With permission from Elsevier.*)

In order to carry out a "system-level" analysis, varying structural configurations and materials that are part of entire structure need to be accounted for. This may lead to the use of various element types with differing geometrical and material properties throughout the structure. In whole-structure analysis, it must be recognized that the performance of each structural member depends on its interactions with the surrounding structure. For example, loss of strength and stiffness of one highly stressed member leads to redistribution of loads/stresses to other neighboring (less stressed) members. Thus, numerical analysis of the whole structure needs to allow interaction between fire-weakened members and the adjacent (cooler) members. Figure 6.7 shows a three-dimensional model of a composite frame for fire resistance analysis (Agarwal and Varma, 2014).

Analysis at the system level can take into account all possible critical factors affecting response of the overall structure system, not just at a critical section or in a specific member, as in cross-sectional or member-level analysis. Factors such as composite action between slab and beam, tensile membrane action, fire-induced restraint forces, redistribution of forces/stresses, and member interactions, which can influence response of a structural system, can be given full consideration at the system level. These factors enhance fire resistance of a structural member (or system) by providing alternate load paths to transfer loads from fire-weakened members to other, cooler parts of the structure. One needs to keep in mind that accounting for such factors can be a hectic task, since a large set of additional input parameters as well as specific element types must be considered, together with resolving convergence issues, to capture the full effect on the behavior of structures under fire conditions. Recent work by Nguyen et al. (2016) provides some insights into complexities associated with this level of analysis.

Material Properties at Elevated Temperatures

In advanced analysis, due consideration should be given to variation in thermal, mechanical, and deformation properties at elevated temperature, as the behavior of structural members exposed to fire is highly dependent on these properties of materials. Relations for variation in high-temperature properties of different construction materials are specified in codes and standards (e.g., ASCE manual and Eurocodes), as well as listed in Chaps. 7 to 11. Degradation of material properties at elevated temperature can be approximated at various levels. At the first level, in the simplest form, some of the thermal properties can be input as constant values throughout the 20 to 800°C temperature range; however, strength and modulus degradation with temperature has to be provided as input to the model to give full consideration to

Code	Load Combination	Remarks
ASTM E119 (2016)	$U = D + L$	Mostly used for evaluating fire resistance of members through perspective-based approaches
ASCE 7	$U = (1.2 \text{ or } 0.9)\, D + A_k + 0.5\, L + 0.2S$	If D has a stabilizing effect, then this factor of 1.2 may be reduced to 0.9. A_k = load or load effect (such as fire-induced restraint) resulting from extraordinary event A (fire)
Eurocode 1	$U = 1.0\, D + P\, L$	Load combination can be varied slightly depending on whether the fire load is taken to be of frequent or quasi-permanent nature. $P = 0.5$ for residential and office buildings, 0.7 for shopping area, and 0.9 for storage areas
British Code (British Standards Institution, 1987)	$U = D + L + 0.33W$	Wind loads should only be applied to building members where the height to eaves is greater than 8 m and are only considered when checking the design of the primary elements of the framework
Australia/New Zealand Standard (Australian Building codes, 1996)	$U = 1.0\, D + M\, L$	$M = 0.4$ for residential and office buildings and 0.6 for occupancies with semipermanent live loads

U = factored loads specified in common fire codes, D = dead load, and L = live load.

Table 6.3 Load Combinations for Structural Fire Design in Different Codes and Standards

temperature effects on mechanical properties. The implementation of properties at this level would be sufficient for evaluating moment capacity in a fire-exposed member, such as the case described in Sec. 6.3.1, specifically Eq. 6.3.

At the second level, actual relations for temperature-dependent thermal properties and stress-strain properties need to be provided as input for advanced analysis. Finally, at the most complex level, additional properties such as high-temperature creep, spalling, debonding, and so on that influence fire resistance of structures under certain conditions need to be explicitly considered in the finite element models to simulate realistic behavior of a structure. As an example, to capture the realistic behavior of a steel structure in compression and tension, temperature-dependent multilinear stress-strain relations with von Mises plasticity yielding criterion and isotropic hardening plasticity model are to be supplied as input data to fire resistance analysis. Similarly, for capturing the full plastic behavior of concrete, Eurocode 2 temperature-dependent stress-strain relations combined with the Williams and Warnke (1975) constitutive material model are to be supplied as input data. Such constitutive models can take into account the spread of plasticity of concrete in both compression and tension regimes at elevated temperature exposure. The compressive plastic behavior of concrete can be defined using isotropic multilinear compressive stress-strain curves that vary with temperature.

Structural Parameters

Structural parameters can significantly influence the response of a structure under fire exposure, and these factors have to be duly accounted for in advanced analysis. The parameters that have significant influence include load intensity, material phenomena (spalling, charring, debonding, and local buckling), structural interaction (composite action, stress redistribution, tensile membrane action, etc.), and restraint conditions at supports.

The level of loading present during a fire incident constitutes only part of the ultimate factored loads that a structural member is designed for at ambient conditions. This is due to the fact that fire is an accidental loading event and during fire all other extreme loading events (that are taken into consideration at ambient conditions) are unlikely to occur simultaneously. This rationale has been accepted in many codes and standards, and few codes provide guidance on load combinations to be adopted by structural fire design. However, there is still a wide-ranging variance in the level of loading specified in these codes for fire design. Table 6.3 lists load combinations specified in ASCE 7, Eurocode, and the Australia/New Zealand Standard under fire conditions.

Other structural parameters such as level of composite action developed at the interface of concrete slab and a steel beam or the bond between rebars (steel or FRP) and surrounding concrete can be accounted for in advanced analysis. Similarly, the bond-slip action between FRP reinforcement and concrete in FRP-strengthened reinforced concrete can also be included. For example, the horizontal shear force-slip that develops in a concrete slab between shear studs and concrete can be simulated using force-slip relations based on nonlinear constitutive relation proposed by Ollgaard et al. (1971). The relation is derived for ambient conditions and can also be extended to elevated temperature through modifying the factor Q_u.

$$V_h = Q_u[1 - e^{-4.75S}] \tag{6.19}$$

where V_h is the shear force, Q_u is the strength of the studs calculated as $Q_u = 0.43A_s\sqrt{E_c f_c'} \leq 0.7A_s f_u$; A_s is the cross-sectional area of the shear studs, E_c is the elastic modulus of the concrete taken as $4600\sqrt{f_c'}$ (in MPa), f_c' is compressive strength of the concrete, and f_u is ultimate strength of the studs. S is slip length, and the maximum slip length can be obtained from the CEB-FIP (1993) model code.[7]

In advanced analysis, specific finite element types can be used to model the bond-slip between the longitudinal steel bars and concrete, or between the FRP reinforcement and epoxy material. The bond-slip behavior between the steel reinforcement and concrete, and FRP rebars and concrete can be accounted for based upon the first segment (i.e., stiffness) of the CEB-FIP (1993) model defined as

$$\tau = \tau_u\left(\frac{s}{s_u}\right)^{0.4} \tag{6.20}$$

where τ is the bond stress at a given slip (s), τ_u is the ultimate bond stress, and s_u is the ultimate slip, corresponding to τ_u. The values of τ_u and s_u depend on the reinforcement bar type and surrounding material. The values of τ_u for the steel reinforcement embedded in concrete are often taken as $\sqrt{f_c'}$ (CEB-FIP, 1993). The values τ_u and s_u for FRP bars embedded in epoxy resin can be taken based on actual material property test data.[8]

This bond stress-slip relations input as material models *spring* (or *combine*) elements available in commercial software such as ASYS and ABAQUS to derive longitudinal stiffness *(k)* for evaluating the bond-slip action by applying following equation proposed by Nie et al. (2004):

$$k = \frac{\pi}{s_u} p d_r N_r \tau_u \left(\frac{L_1 + L_2}{2}\right) \tag{6.21}$$

where p is the horizontal distance between the bars, d_r is the diameter of reinforcements, N_r is the number of bars, and L_1 and L_2 are the lengths of two adjacent bar elements (LINK8).

The debonding effect under elevated temperature exposure, as that in the case of strengthened and protected concrete structures, can be modeled using the spring or cohesive elements. In order to accurately model the bond-slip behavior, the ultimate shear stress (τ_{max}) and its corresponding slip (displacement) (s_u) must be determined first. The evaluation of these parameters can be carried out as per the equation developed by Lu et al. (2005):

$$\tau = \begin{cases} \tau_{max}\sqrt{\dfrac{s}{s_0}} & \text{when } s \leq s_0 \\ \tau_{max} e^{-\alpha\left(\frac{s}{s_0}-1\right)} & \text{when } s > s_0 \end{cases} \tag{6.22}$$

where s_0 is the local slip at τ_u, β_w is the width ratio factor, f_t is the tensile strength of concrete, G_f is the interfacial fracture energy, and S_f is the local slip when bond stress reduces to zero. These factors can be calculated using the following expressions:

$$s_0 = 0.0195\beta_w^2 f_t, \ \alpha = \frac{1/G_f}{\tau_{max}s_0} - \frac{2}{3}, \ G_f = 0.308\beta_w^2\sqrt{f_t}, \ \beta_w = \sqrt{\left(2.25 - \frac{w_f}{S_f}\right) / \left(1.25 + \frac{w_f}{S_f}\right)},$$

$$\tau_{max} = 1.5\beta_w f_t$$

Another approach to account for temperature-induced interfacial bond-slip action in concrete or steel structures is through the implementation of cohesive elements in advanced analysis. Xu and Needleman

[7]Can be set to 1.27 mm for typical applications.
[8]Anwarul (2009) showed that τ_u and s_u for FRP bars are 15.24 MPa and 0.11 mm, respectively.

(1994) derived an exponential cohesive zone material model. This model has two segments, an increasing segment up to the ultimate shear stress (τ_{max}) and corresponding slip (s_u), and then descends with a softening response up to the slip at failure (s_f), which is assumed to be equal to four times (s_u). The interface elements were placed along the interface of the FRP/concrete, in the longitudinal direction of the member's x-axis or at the interface of fire insulation and steel surface in a structural member. This model is implemented in a variety of generic software such as ANSYS. Recent work of Arablouei and Kodur (2016) explains the applicability of cohesive zone models to elevated temperatures.

The restraint conditions (i.e., boundary conditions) can be imposed through specifying constraints to the element (or nodes) of a discretized structure. The degree of freedom can be restrained in the x, y, and/or z direction depending on the actual boundary condition in the member being analyzed. In many instances, especially in the modeling of a single structural member, a member can have one or more planes of symmetry. If a member (i.e., beam or column) is symmetrical along the longitudinal and transverse directions, then only a quarter model of the structure is to be analyzed. This can significantly reduce the total number of elements, which in turn will lead to less computational effort. For making use of symmetry, a structural member (or system) needs to meet symmetry conditions in all criteria, namely geometry, boundary conditions, material layout, and loading (including fire exposure).

Numerical Stability and Convergence to Solution

Fire resistance analysis involves a highly nonlinear set of calculations that can lead to convergence issues resulting from numerical instability due to diminishing stiffness, approaching zero or a negative value, arising from singularity or high levels of plasticity phenomenon at elevated temperature. This is due to the occurrence of large deformations (arising from temperature-induced strength and stiffness degradation or creep effects, or resulting from the loss of cross section from spalling or charring phenomenon). Further, in some cases, the ultimate plastic strains can be exceeded due to stress concentrations, and numerical technique may not converge to a solution due to large loading steps. Therefore, in advanced analysis, numerical problems will occur in converging to a solution, specifically closer to the failure of a member. In order to avoid solution instability, a numerical tolerance limit ranging between 0.05 and 0.2 can be applied (ANSYS, 2018). More specifically, in thermal analysis, the thermal instability can be avoided as long as variation in successive equilibrium iterations between time-steps is less than 0.1°C (Naser and Kodur, 2018). These convergence issues vary from one problem to the other and also vary with the number of parameters accounted for in fire resistance analysis. The users have to develop a feeling for each set of analysis based on physical behavior of specific fire-exposed structural member and need to carry out a sensitivity analysis to determine appropriate convergence criteria that can lead to numerical solution.

Output (Response) Parameters

Advanced analysis generates large amounts of response parameters relating to thermal and structural response of a structure under fire exposure. The common response output parameters from computer software such as ANSYS or ABAQUS are nodal temperature, nodal deformations, stresses and strains, and deflections, among others, at each time-step of the analysis. The output parameters can also include temperature distributions (such as thermal gradient, temperature contours), crack growth (in concrete and masonry members), and local buckling contours (in steel members). Using these output parameters, strength-related response parameters including sectional capacity (i.e., moment, shear, axial), stability parameters, and so on can be derived to identify different failure modes under fire exposure. Please refer to Sec. 6.4.5 to revisit the discussion of how to derive some of these parameters utilizing output results from advanced analysis.

6.4.7 Case Study

The applicability of the above-discussed advanced analysis procedure in tracing the response of a typical structural member under the combined effects of fire exposure and structural loading is through analysis of a steel beam using ANSYS 14.0 computer software. Critical parameters such as geometric and material nonlinearities, temperature-dependent material properties, and various failure limit states are accounted for in tracing the response. Steps involved in the analysis are briefly discussed below, and full details on the application of advanced analysis for fire resistance evaluation can be found in Kodur and Naser (2014), as well as in Kodur and Alogla (2017).

Geometry and Discretization

For the analysis, a steel beam (shown in Fig. 6.8*a*) made of W16×31 section with a flange width of 140.6 mm and overall depth of 404 mm is selected from the AISC design manual. The flange and web thicknesses of

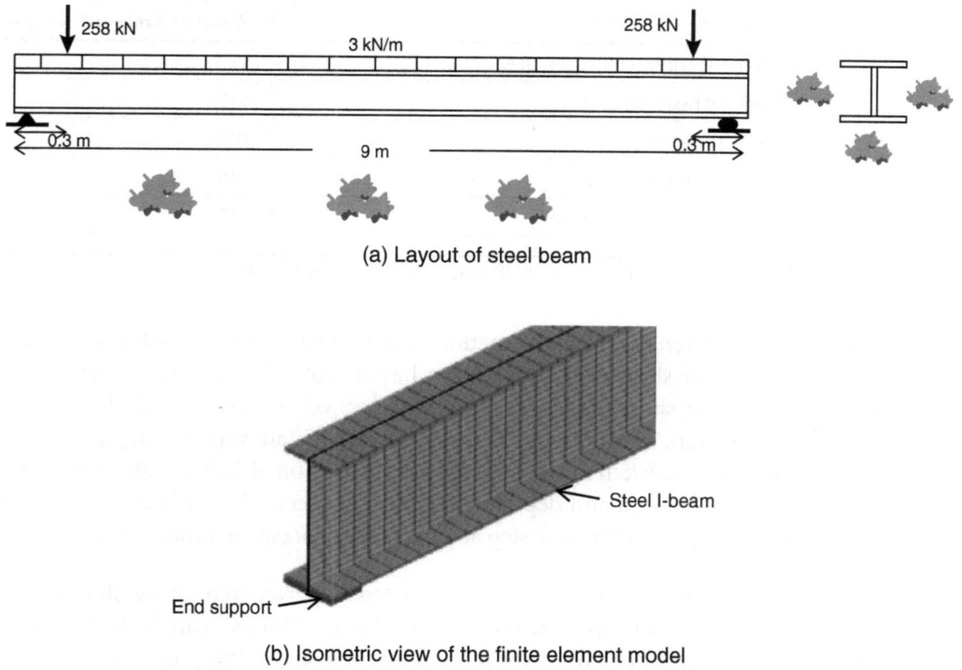

Figure 6.8 Geometry and discretization for the steel beam selected for analysis.

this section are 11.2 mm and 7 mm, respectively. The beam is made of Grade 345 (MPa) steel and is assumed to have continuous lateral support provided along its 9-m span. The beam is subjected to uniformly distributed loading (3 kN/m) together with two concentrated loads of 258 kN applied close to the supports which represent about 50% of flexural capacity at room temperature (see Fig. 6.8*a*). This beam is exposed to a standard fire, similar to that of the ASTM E119 standard fire. The analysis was carried out in ANSYS at incrementing time-steps until the failure occurred in the beam.

The selected beam is discretized to generate a three-dimensional finite element model with geometry of a typical hot-rolled built-in steel I-section. For this, different thermal and structural element types available in ANSYS 14.0 are utilized. SOLID70 and SURF152 elements are used as thermal elements to simulate heat transfer under fire conditions. SOLID70 is a thermal element with conduction capability, while SURF152 is a four-noded thermal element capable of simulating heat conduction, convection, and radiation. Typically, SURF152 is overlaid on top of SOLID70 elements to simulate convection and radiation effects from the fire exposure zone to the structural member. SOLID185, which has eight nodes, is used for modeling three-dimensional solid structures. This cubic element has three degrees of freedom at each node—translations in the nodal *x*, *y*, and *z* directions—and it can account for large deflections and large strain effects.

Upon completing thermal analysis, SOLID70 thermal elements can be transformed into SOLID 185 structural elements for undertaking strength (structural) analysis. Such transformations are necessary to account for temperature-induced degradation in properties in constituent materials. The finite element model is meshed using equal side quadrilateral-type mesh to ensure uniformity. The discretized beam has 35,000 elements. An isometric view of the discretized beam with elements is shown in Fig. 6.8*b*.

High-Temperature Material Properties and Constitutive Laws The various room temperature properties of steel used in the beam are listed in Table 6.4. Eurocode 3 recommended thermal and mechanical property relations to be utilized to derive high-temperature property relations for structural steel, and this is provided as input to ANSYS. These property relations are listed in Table 6.4 and also discussed in Chap. 7. The nonlinearity in steel material is accounted for through a bilinear elastoplastic constitutive material model based on the von Mises plasticity yield criterion.

Failure Limit State In the analysis, different limiting criteria—namely, thermal, strength, and deflection limit states—are considered for evaluating failure of the beam at each incrementing time-step. ANSYS outputs internal bending and shear stresses, at individual nodes, and these stresses are integrated across

[9]The latest edition of the ASTM E119 adopts new limits for deflection ($L_c^2/400d$), where L_c is the clear span of the beam.

Property	Value at Ambient Temperature
Thermal conductivity (W/m.K)	51
Specific heat (J/kg.K)	450
Density (kg/m³)	7900
Yield strength (MPa)	345
Modulus of elasticity (GPa)	210

Table 6.4 Properties of Steel (Beam) at Room Temperature

the depth of the section (using supplementary external subroutines) to derive sectional moment and shear capacity. Flexural (or shear) failure is said to have occurred at a section at any given time-step once the bending moment (or shear force) resulting from applied loading exceeds the moment (or shear) capacity at that critical section. In addition, the deflection limit state was also applied to evaluate failure at each time-step. Accordingly, when the beam attained a deflection of $L/20$ or rate of deflection reaches $L^2/9000d^9$, where L and d are the span and depth of the beam, respectively, the beam is said to have attained failure. The duration to the preceding time-step at which failure occurred is taken as the fire resistance of the beam.

Results from Advanced Analysis The various output parameters generated from ANSYS analysis—temperatures, stresses, midspan deflection, and failure modes—can be used to draw inferences on the response of the beam under fire exposure. Figure 6.9 shows the progression of temperatures at the top flange, bottom flange, and web in the steel beam as a function of fire exposure time. It can be seen that the steel temperature in the web and bottom flange increases at a higher rate with fire exposure time compared to that in the top flange. Temperature in the web reaches 500°C at 10 minutes into fire exposure and further reaches 760°C at 20 minutes, and this rapid increase in sectional temperatures is mainly due to the high thermal conductivity of steel, as well as the lower slenderness (mass) of steel sections.

Figure 6.9*b*, *c* shows the degradation of moment and shear capacity of the analyzed steel beam with fire exposure time, along with bending moment and shear force resulting from applied loading present during the fire exposure. These plotted moment and shear capacities are at critical sections: location of midspan section for moment and location of point loading for shear. As expected, moment and shear capacity in steel beams starts to degrade after 9 minutes into fire exposure. This is due to the steel temperature in the lower flange and web exceeding 400°C, as can be seen in Fig. 6.9*a*. As the temperature rise continues, further degradation of moment and shear capacity takes place until the failure of the beam. This beam fails once shear capacity falls below applied shear loading at about 12 minutes (and failure through flexure occurs at 17 minutes). This failure is accompanied with temperature-induced local buckling which occurs at 11.6 minutes, as can be seen in the ANSYS output sectional profile in Fig. 6.10*d*. Figure 6.9*e* shows the predicted midspan deflection in this beam as a function of fire exposure time. The midspan deflections are very low until about 13 minutes; then deflections increase at a rapid pace prior to failure of this beam as per the BS 476 deflection limit. As discussed in Chap. 5, British Standard (BS-476) recommends a deflection limit of ($L/20$), which was applied to evaluate failure.

Advanced analysis also generates other response parameters such as instability effects that can be used to evaluate critical temperature-induced phenomena that a structure undergoes prior to failure. Figure 6.9*e* shows the onset of local buckling in the web around 11.6 minutes into fire exposure. Such temperature-induced local buckling in the web produces further reduction in shear capacity, and this can accelerate failure of a beam under fire conditions. Effects such as these in structural members under fire exposure are complex phenomena and often counterintuitive. Recent experiments and advanced analysis clearly show that effects such as temperature-induced local buckling can significantly reduce sectional capacity under certain sectional and loading geometry, and thus should be given due consideration in evaluating fire response of structures (Glassman et al., 2016; Reis et al., 2016).

6.5 Emerging Trends and Research Needs

In the last three decades, there have been rapid advances in the availability of computational tools and software for undertaking structural analysis at ambient conditions. However, the progress of computational tools for fire resistance analysis has been slow, and there are still limitations on the capability of such tools for structural fire engineering analysis. Some of the areas where further advancement is needed, with respect to fire resistance analysis through calculation methods, are discussed in the following section.

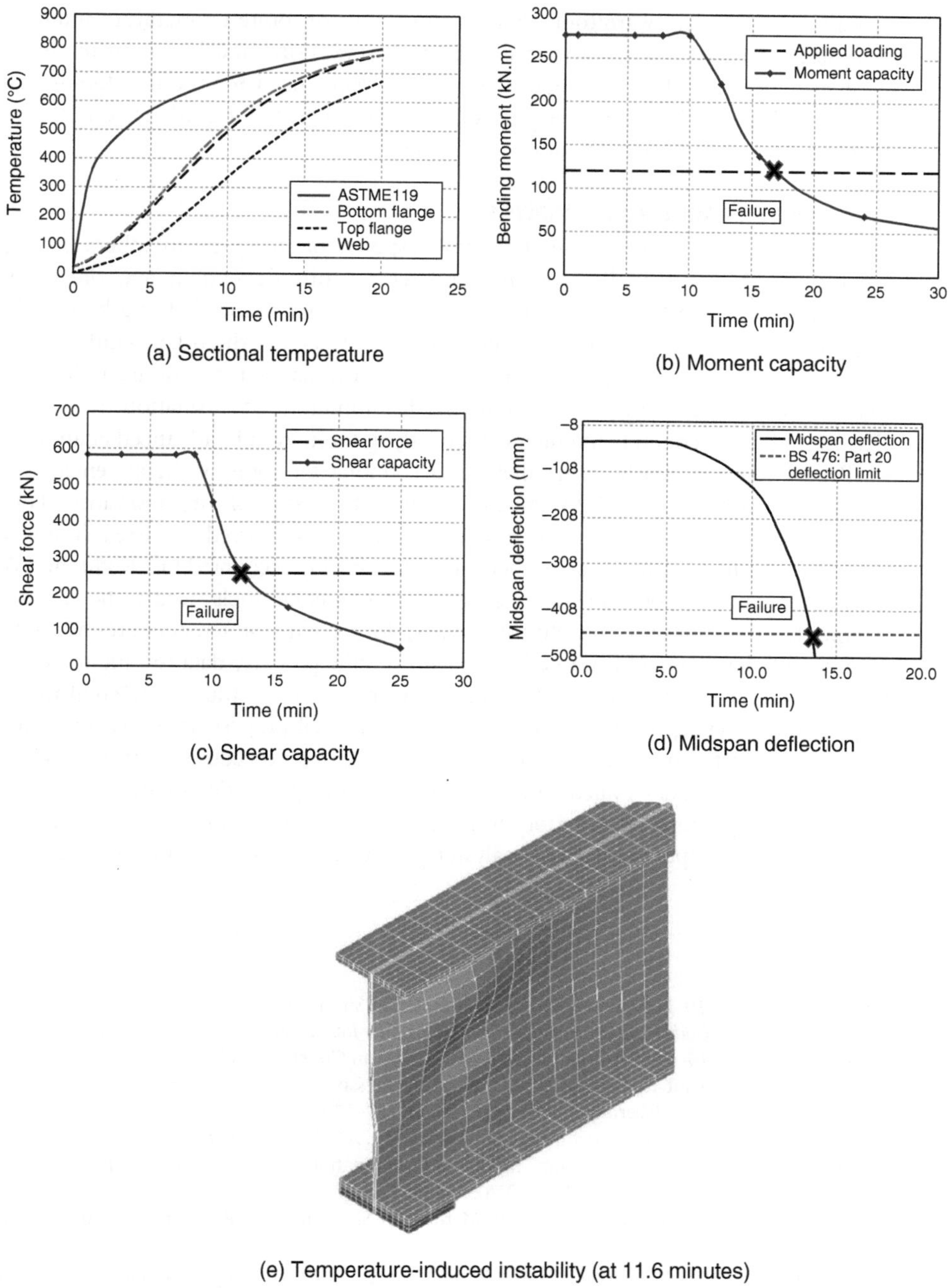

Figure 6.9 Comparison of predicted and measured temperature and deflections as a function of fire exposure time. (*With permission from ASCE.*)

6.5.1 Computer Software for Advanced Analysis

The advanced analysis procedure is to be applied for tracing the realistic response of a structure under fire conditions. Most currently available computer software for structural fire engineering analysis still cannot fully account for specific temperature-induced phenomena such as spalling in concrete, charring in wood, and creep and local buckling in steel structures. Not accounting for such temperature-dependent phenomena can lead to inaccurate response predictions. Moreover, output results from most special-purpose programs, developed for structural fire engineering applications, still need to be post-processed, via supplementary tools to derive critical response parameters to evaluate failure of the analyzed structure.

Other major drawbacks with currently available software tools are the poor validation of these tools, lack of guidance on modeling temperature-induced effects, and no benchmark solutions to check the output trends (Ferreira et al., 2018). The lack of well-validated computer packages and guidance for use is hindering wider acceptance of computational approaches for fire resistance analysis in design practice. To overcome this limitation, researchers should work toward developing newer software or extending currently available software tools that specifically incorporate temperature-induced effects in advanced analysis. Also, such tools have to be validated against experimental data at various levels of analysis (section, member, and system). In addition, there is an urgent need for the development of user manuals and benchmarking solutions for easier use of software progress for guidance on interpretation of results by designers and practitioners (Kodur et al., 2012).

6.5.2 Input Parameters for Analysis

Predicting accurate thermal and structural response of structural members through advanced analysis depends on the level of discretization and temperature-dependent material properties of constituent materials adopted in the analysis. Currently, temperature-induced strength and stiffness degradation in properties of materials is fully accounted for in analysis. On the other hand, specific critical parameters such as moisture migration, fire-induced spalling in concrete, temperature-induced local buckling in steel sections, high-temperature creep effects, delamination of insulation in steel, and debonding of FRP-strengthened concrete members are mostly ignored even in advanced analysis.

Recent research clearly shows that not accounting for some of the aforementioned parameters in fire resistance analysis can lead to unconservative estimations of fire resistance. For example, neglecting high-temperature creep effects in fire resistance analysis of reinforced concrete columns was shown to give 16% higher fire resistance than the case where creep is explicitly accounted for (Kodur and Alogla, 2017). Thus, in order to enhance response predictions, advanced numerical tools need to be supplied with realistic input parameters (e.g., material constitutive models) to account for temperature-induced phenomena. Further research and development is needed to develop specific property data relating to delamination of insulation, pore pressure progression in concrete, high-temperature creep strain in different materials, and so on. Such property data can be generated through small-scale material-property tests, in conjunction with observations in large-scale fire tests. There is a critical need for developing test procedures, instrumentation, and actual properties for undertaking tests at elevated temperatures. Once such properties are measured and benchmarked against actual fire tests, property relations for variation in these properties with temperature can be developed, and such relations will form input to advanced analysis for evaluating realistic response of structures under fire conditions.

6.6 References

ABAQUS. Version 6.10. Providence, RI: Dassault Systèmes Simulia Corporation, 2017.

ACI/TMS Standard. *Code Requirements for Determining Fire Resistance of Concrete and Masonry Construction Assemblies (ACI 216.1-14/TMS-0216-14)*. Farmington Hills, MI: American Concrete Institute, 2014.

Agarwal, Anil, and Amit H. Varma. "Fire-Induced Progressive Collapse of Steel Building Structures: The Role of Interior Gravity Columns." *Engineering Structures* 58 (2014): 129–140.

ANSYS. ANSYS Metaphysics. Version 19. Canonsburg, PA: ANSYS Inc., 2018.

Arablouei, Amir, and Venkatesh Kodur. "Modeling Delamination of Fire Insulation from Steel Structures Subjected to Blast Loading." *Engineering Structures* 116 (2016): 56–69.

ASCE/SEI/SFPE 29-05. Standard Calculation Method for Structural Fire Protection. Reston, VA: American Society of Civil Engineers, 2005.

ASTM E119. *Standard Test Method for Fire Tests of Building Construction and Materials*. West Conshohocken, PA: American Society for Testing and Materials, 2016.

Aziz, Esam Mohammed. *Response of Fire Exposed Steel Bridge Girders*. Michigan State University, 2015.

Board, Australian Building Codes. *Building Code of Australia*. CCH Australia, 1996.

British Standards Institution. BSI S 476-3:1987. *Fire Tests on Building Materials and Structures—Part 20: Method for Determination of the Fire Resistance of Elements of Construction (General Principles)*. UK: BSI, 1987.

Comite Euro-International du Beton (CEB-FIP). *CEB-FIP Model Code 1990. Bulletin D'Information No. 213/214 (Concrete Structures)*. Lausanne, Switzerland: CEB-FIP, 1993.

Cook, Robert D., David S. Malkus, and Michael E. Plesha. *Concepts and Applications of Finite Element Analysis*. Vol. 4. New York: Wiley, 1974.

Drysdale, Dougal. *An Introduction to Fire Dynamics*. Hoboken, NJ: Wiley, 2011.

Eurocode 1. *EN 1991-1-2: Actions on Structures. Part 1–2: General Actions—Actions on Structures Exposed to Fire*. Brussels, Belgium: European Committee for Standardization, 2002.

Garlock, Maria E. Moreyra, and Jonathan D. Glassman. "Elevated Temperature Evaluation of an Existing Steel Web Shear Buckling Analytical Model." *Journal of Constructional Steel Research* 101 (2014): 395–406.

Glassman, Jonathan D., Maria E. Moreyra Garlock, Esam M. Aziz, and Venkatesh K. Kodur. "Modeling Parameters for Predicting the Postbuckling Shear Strength of Steel Plate Girders." *Journal of Constructional Steel Research* 121 (2016): 136–143.

Ferreira, Joao Daniel Romeiro, Thomas Gernay, and Jean-Marc Franssen. "Discussion on a Systematic Approach to Validation of Software for Structures in Fire." Paper presented at the 10th International Conference on Structures in Fire, Ulster University, Belfast, June 2018.

Fredlund, Bertil. *Structural Design of Fire-exposed Rectangular Laminated Wood Beams with Respect to Lateral Buckling*. Lund, Sweden: Lund Institute of Technology, Department of Structural Mechanics, 1979.

Islam, Anwarul. "Effective Methods of Using CFRP Bars in Shear Strengthening of Concrete Girders." *Engineering Structures* 31, no. 3 (2009): 709–714.

Jiang, Jian, Lingzhu Chen, Shouchao Jiang, Guo-Qiang Li, and Asif Usmani. "Fire Safety Assessment of Super Tall Buildings: A Case Study on Shanghai Tower." *Case Studies in Fire Safety* 4 (2015): 28–38.

Kodur, V. K. R., and S. M. Alogla. "Effect of High-Temperature Transient Creep on Response of Reinforced Concrete Columns in Fire." *Materials and Structures* 50, no. 1 (2017): 27.

Kodur, V. K. R., and M. Z. Naser. "Effect of Shear on Fire Response of Steel Beams." *Journal of Constructional Steel Research* 97 (2014): 48–58.

Kodur, V. K. R., and M. Z. Naser. "Effect of Local Instability on Capacity of Steel Beams Exposed to Fire." *Journal of Constructional Steel Research* 111 (2015): 31–42.

Kodur, V. K. R., Maria Garlock, and Nestor Iwankiw. "Structures in Fire: State-of-the-Art, Research and Training Needs." *Fire Technology* 48, no. 4 (2012): 825–839.

Lu, X., J. Teng, L. Ye, and J. Jiang, J. "Bond-Slip Models for FRP Sheets/Plates Bonded to Concrete." *Engineering Structures* 27, no. 6 (2005) 920–937.

Naser, M. Z., and V. K. R. Kodur. "Factors Governing Onset of Local Instabilities in Fire Exposed Steel Beams." *Thin-Walled Structures* 98 (2016): 48–57.

Naser, M. Z., and V. K. R. Kodur. "Response of Fire Exposed Composite Girders under Dominant Flexural and Shear Loading." *Journal of Structural Fire Engineering* 9, no. 2 (2018): 108–125.

Nguyen, Ha, Ann E. Jeffers, and Venkatesh Kodur. "Computational Simulation of Steel Moment Frame to Resist Progressive Collapse in Fire." *Journal of Structural Fire Engineering* 7, no. 4 (2016): 286–305.

Nie, J., J. Fan, and C. Cai. "Stiffness and Deflection of Steel-Concrete Composite Beams under Negative Bending." *Journal of Structural Engineering* 130, no. 11 (2004). doi: 10.1061/(ASCE)0733-9445(2004)130:11(1842).

Ollgaard, J. G., R. G. Slutter, and J. W. Fisher. "Shear Strength of Stud Connectors in Lightweight and Normal Weight Concrete." *AISC Engineering Journal* 8 (1971): 55–64.

Reis, André, Nuno Lopes, Esther Real, and Paulo Vila Real. "Numerical Modelling of Steel Plate Girders at Normal and Elevated Temperatures." *Fire Safety Journal* 86 (2016): 1–15.

Standards New Zealand. *Concrete Structures Standard Part 1—The Design of Concrete Structures.* NZS 3101. Wellington: Standards New Zealand, 2006.

Southwest Research Institute. *Directory of Listed Products for the Construction Industry.* San Antonio, TX: Fire Technology Department, Southwest Research Institute, 2012.

Underwriters Laboratories. *Fire Resistance Directory.* Northbrook, IL: Underwriters Laboratories, 2012.

Williams, B., V. K. R. Kodur, M. F. Green, and L. Bisby. "Fire Endurance of Fiber-Reinforced Polymer Strengthened Concrete T-Beams." *ACI-Structural Journal* 105, no. 1 (2008): 60–67.

Willam, K.J., E. D. Warnke. Constitutive model for the triaxial behavior of concrete. In: Proceedings, international association for bridge and structural engineering, 19, ISMES. 1975. p. 174.

Xu, X-P., and A. Needleman, "Numerical Simulations of Fast Crack Growth in Brittle Solids." *Journal of the Mechanics and Physics of Solids* 42 (1994): 1397–1434.

6.7 Nomenclature

L Depth of the structural member

d Span and depth of the structural member

ρ Density

c Specific heat

∇ Spatial gradient operator

k Conductivity matrix

T Temperature

t Time

h_{con} Convective heat transfer coefficient; equals 25 W/m$^2\cdot$K for external, ASTM E119, and ISO 834 fires as recommended in Eurocode (Eurocode 1, 2002)

h_{rad} Radiative heat transfer coefficients

T_f Fire temperature

σ Stefan-Boltzman constant $= 5.67\times10^{-8}$ (W/m$^2\cdot$K^4)

ε Emissivity factor, which is related to the "visibility" of the exposed surface of structural member to the fire

C_e^t Specific heat matrix

K_e^t Thermal "stiffness" matrix; sum of conductivity and convection matrices

T_e Nodal temperature

Q_e Applied nodal thermal load; composed of the convective and radiative heat fluxes

U Strain energy = External work

K_e Element stiffness matrix

F_e^{th} Element thermal load vector

6.8 Numerical Examples

6.1

A non-load-bearing concrete wall of 140 mm thickness is fabricated with two layers: a base layer of 60-mm siliceous aggregate concrete and a second layer of 80-mm carbonate aggregate concrete (as shown in the figure below). Evaluate the fire resistance of this wall by applying a prescriptive-based approach as per (a) ACI 216.1, (b) Eurocode 2, and (c) New Zealand (NZS 3101-1) codes.

Solution

(a) As per ACI 216.1 provisions, fire resistance of walls/floors can be evaluated by summing *R* values for different layers.

Using $R^{0.59}$ values in Fig. 6.2 (from ACI 216.1), fire resistance can be evaluated as

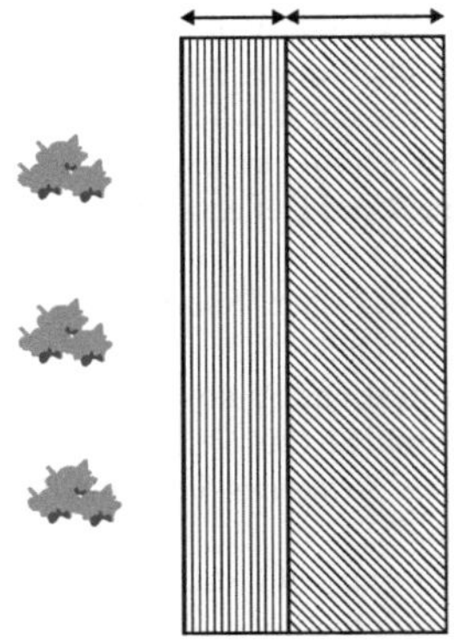

$$R_1^{0.59} = 8 \text{ (for 60 mm of carbonate aggregate concrete layer)}$$

$$R_2^{0.59} = 11 \text{ (for 80 mm of siliceous aggregate concrete layer)}$$

Thus, fire resistance (R) of a multilayered wall is given as: $R = (R_1^{0.59} + R_2^{0.59} + \cdots + R_n^{0.59})^{1.7}$

$$R = (8 + 11)^{1.7} = 149.2 \text{ minutes}$$

(b) As per Eurocode 2 provisions:

Fire resistance of a non-load-bearing wall is only required to meet the thermal insulation criterion I and integrity criterion E. The minimum wall thickness should not be less than that given in Table E6.1, which gives fire resistance for siliceous aggregate (no specific table for carbonate aggregate). The given wall can be assumed to be made of one layer of 140-mm thickness. Since siliceous aggregate concrete has a lower fire resistance, the fire rating obtained from this table can be conservative.

Standard Fire Resistance	Minimum Wall Thickness (mm)
1	2
EI 30	60
EI 60	80
EI 90	100
EI 120	120
EI 180	150
EI 240	175

Table E6.1 Minimum Wall Thickness of Non-Load-Bearing Walls

Using Table E6.1:
The tabulated thickness is for 120 mm (no specific table for 140 mm thickness). So, fire resistance of 140-mm-thick wall ≈ fire resistance of a 120-mm-thick wall = 120 minutes.

Alternatively, interpolation can be done between 120- and 150-mm thickness to achieve fire resistance. Using interpolation, fire resistance for a 140-mm-thick wall is 160 minutes.

(c) As per New Zealand (NZS 3101-1) code provisions:

Fire resistance of a wall to meet insulation criteria can be obtained from Table E6.2 (given in NZS 3101-1).

Fire Resistance Rating (minutes)	Effective Thickness (mm)
30	60
60	75
90	95
120	110
180	140
240	165

Table E6.2 Minimum Effective Thickness for Insulation of Non-Load-Bearing Walls

Using Table E6.2:
Unlike Eurocode 2, the NZS 3101-1 does not specify the type of concrete. Hence, the wall is also assumed to be made of one layer of 140-mm thickness.
Fire resistance of wall = 180 minutes

6.2

A simply supported steel beam of W18×40 section, made of A992 steel and spanning 10 m, is exposed to design a fire scenario as shown in Fig. E6.2. The geometrical and material properties are as given in the figure. The beam is insulated with 25-mm-thick insulation and subjected to a bending moment, M_f, of 230 kN·m and shear force, V_f, of 200 kN. Evaluate if the beam can withstand fire exposure when the temperature in the beam reaches 600°C after 60 minutes of fire exposure.

As per AISC manual (2017), W18×40 has the following properties:

$$Z = 1490 \text{ mm}^3,\ d = 454.6 \text{ mm},\ t_w = 8 \text{ mm},\ f_y = 345 \text{ MPa}$$

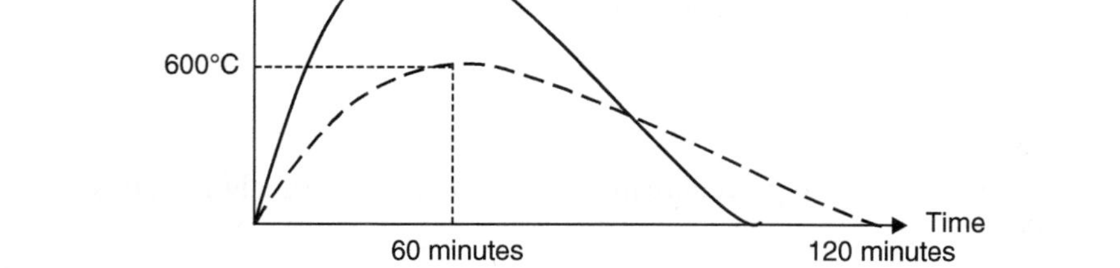

Figure E6.2 Steel beam and temperature-time history of fire-exposed steel beam.

Solution

It is given as per Fig. E6.2 that the average temperature rise in steel section is 600°C, at 60 minutes.

(1) *Reduced moment capacity at 600°C:*

$$M_{600°C} = \Phi_f \times f_{y,600°C} \times Z \qquad (\Phi_f = 1, \text{ for fire deign})$$

$$f_{y,600°C} = k_f \times f_y = 0.47 \times 345 = 162 \text{ MPa } (k_f = 0.47, \text{ from Eurocode 3 as shown in Chap. 7})$$

$$M_{600°C} = 162 \times 1490 = 241.4 \text{ kN.m} \quad > M_f = 230 \text{ kN.m (beam fails in flexure mode)}$$

(2) *Shear capacity at 600°C:*

$$V_{600°C} = \Phi_f \times 0.6 f_{y,600°c}\, d\, t_w C_v \qquad (\Phi_f = 1, \text{ for fire deign})$$

$$V_{600°C} = 0.6 \times 162 \times 454.6 \times 8 \times 1.0 = 353.5 \text{ kN } > V_f = 200 \text{ kN (beam does not fail in flexure mode)}$$

where $C_v = 1.0$.
Therefore, the beam fails before reaching 600°C, i.e., prior to 60 minutes.

6.3

Evaluate through advanced analysis fire resistance of a composite girder made of W24×62 section (standard hot-rolled section) in the AISC manual. The girder has a flange width, flange thickness, web depth, and web thickness of 179, 15, 610, and 11 mm, respectively, as shown in Fig. E6.3. The steel section is fabricated using A572 Grade 345 steel. The concrete slab, cast along the full length of the steel girder, has a depth and width of 140 and 815 mm, respectively, and is made of concrete with a compressive strength of 45 MPa. Assume that the beam is loaded with a gravity load composed of 50% of its room temperature capacity at midspan and is subjected to ASTM E119 fire exposure.

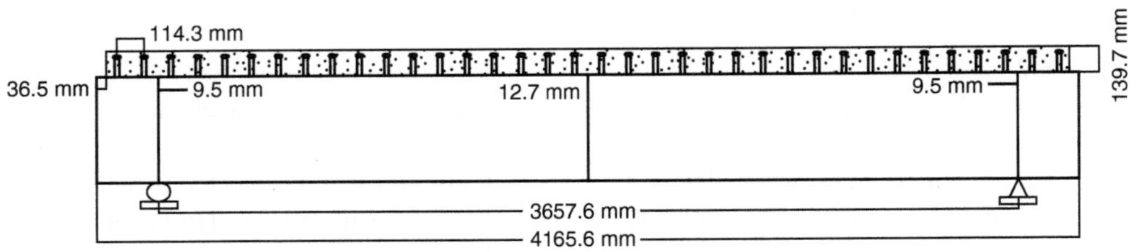

(a) Composite girder

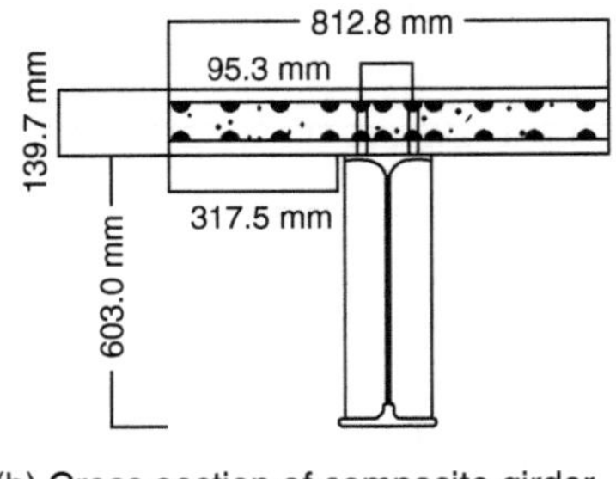

(b) Cross section of composite girder

Figure E6.3 Schematic of composite girder.

Solution

Following the outlined procedure in Sec. 6.4, fire resistance analysis on this girder can be carried out using ANSYS.

Step 1. Heat transfer analysis of the composite steel-concrete girder is carried out using two types of elements available in ANSYS, namely SOLID70 and SURF152. For the discretization of the girder, slab, and stiffeners, SOLID70 elements are used.
Step 2. In the structural analysis, the bottom flange, web, top flange, and stiffeners of the steel girder were modeled with SHELL181 elements, and the concrete slab was modeled with SOLID185 elements.
Step 3. High-temperature property relations of concrete and structural steel are input into the analysis as per Eurocode 2 and Eurocode 3, respectively.
Step 4. Heat transfer analysis gives temperature rise across the depth of the beam, as shown in the temperature profile plot.

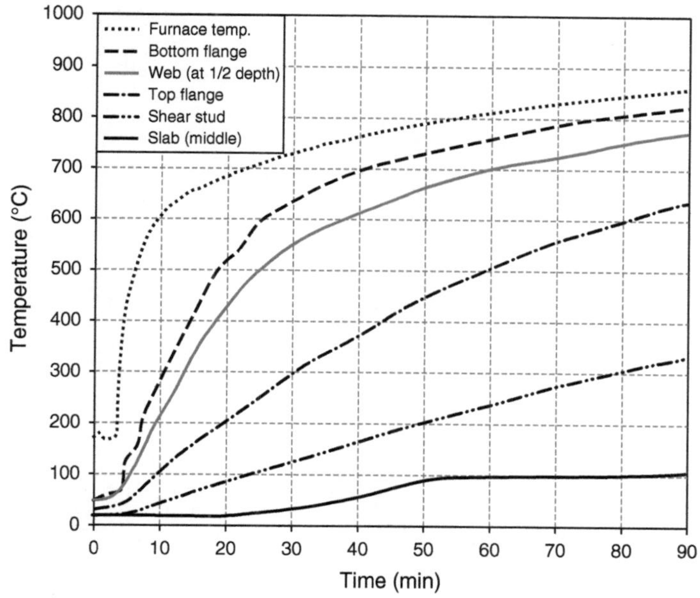

Step 5. Structural analysis gives the midspan deflection time history as a function of fire exposure. Also, reduction in moment or shear capacity with time can be evaluated by integrating stresses (output from ANSYS) across the depth of section.

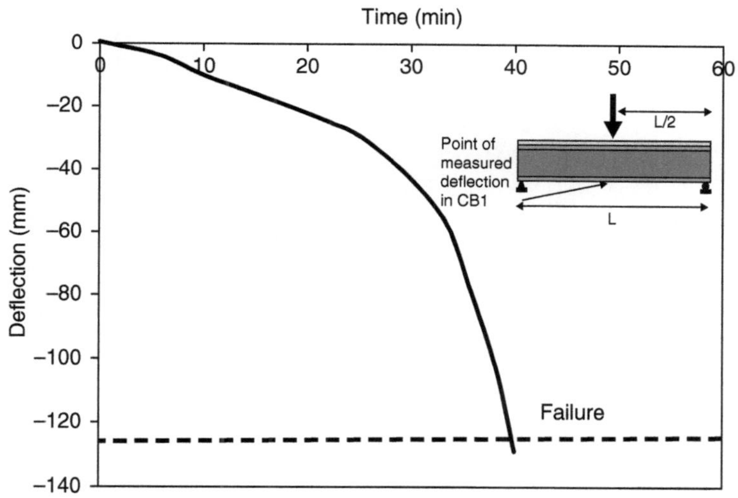

Step 6. Failure time (fire resistance) of the girder is 40 minutes, as resulting deflection at 40 minutes exceeds limiting criteria.

Refer to the following paper for full details on undertaking this analysis:

Aziz, E., V. Kodur, J. Glassman, and M. Garlock. "Behavior of Steel Bridge Girders under Fire Conditions." *Journal of Constructional Steel Research* 106 (2015): 11–22.

CHAPTER 7

Steel Structures

7.1 Overview

This chapter presents an overview of the behavior of steel structures under fire conditions. Temperature-dependent thermal and mechanical properties of steel and fire insulation needed for evaluating fire performance of steel structures are presented. The different strategies that can be implemented for achieving required fire resistance through various protection methods are discussed. Finally, detailed steps for evaluating fire resistance of steel structures through tabulated, simplified, and advanced calculation methods are described.

7.2 Properties at Elevated Temperature

7.2.1 General

Steel, as a material, possesses excellent mechanical properties in terms of strength, stiffness, ductility, and so on and thus is widely used in the construction industry. Steel is classified under different types and grades based on its chemical composition, thermophysical properties, and fabrication process. The most commonly used steel types in structural applications are carbon steel, high-strength low-alloy (HSLA) steels, heat-treated carbon steels, stainless steel, and cold-formed steels (CFS). Carbon steel is mainly used in building applications due to high strength and high ductility properties. HSLA steels are mainly used in bridge construction, since they can provide better performance in term of strength, weldability, and corrosion/weather resistance than carbon steel. Stainless steel is used in special applications where improved performance is required (e.g., under fire conditions or high corrosion environments). CFS, on the other hand, is popular nowadays in building construction due to its ease of fabrication and low energy consumption.

Steel, when exposed to elevated temperatures, as encountered during a fire incident, undergoes a number of physiochemical processes that essentially change its composition, microstructure, and thermal and mechanical properties. Therefore, the variation in these properties of steel with temperature is to be given due consideration for evaluating fire resistance of steel structures. For more details on basic phenomenon relating to steel properties at ambient conditions and elevated temperature, the reader is encouraged to review Chap. 4.

7.2.2 Property Relations for Steel

A rise in temperature in steel causes changes to its crystalline structure, which also affect thermal and mechanical properties of steel: thermal conductivity, specific heat, stress-strain response, yield strength, and modulus of elasticity. These temperature-induced microstructural changes produce degradation in thermal and mechanical properties of structural steel. The variations in these properties of steel, in the range of 20 to 800°C, must be accounted for in fire resistance calculations.

Based on material property tests at elevated temperatures, temperature-dependent specific heat and thermal conductivity of steel property relations for carbon steel are deduced and given in fire codes and standards, as well as in other publications (Lie, 1992; CEN, 2005). A critical review of these property relations was undertaken by Kodur et al. (2010), who reported a noticeable variation between the different sets of data available in the literature and attributed this difference to variations in test methods, heating conditions, and data collection techniques.

The thermal conductivity of carbon steel is around 46 W/m·°C at room temperature and decreases with temperature in an almost linear fashion. On the other hand, the specific heat of steel, which is at room temperature 450 J/kg·°C, increases with a rise in temperature due to individual atoms in steel moving farther apart, thus achieving a higher energy state. Further, at around 750°C a spike in specific heat occurs as a result of a phase change in steel in which the atoms transition from a face-centered to a body-centered cubic structure. Similar to other metals, the thermal strain of steel linearly increases with a rise in temperature. ASCE (1992) and CEN (2005) provide empirical relations for thermal properties of steel; these are shown in Table 7.1 for carbon steel and stainless steel. Generally, the thermal properties of different types of structural steel are similar to those of carbon steel.

The variation in yield strength and elastic modulus of carbon steel with temperature, as specified in different codes and standards, is shown in Fig. 7.1. The degradation in mechanical properties can be attributed to the increased thermal energy provided from heat, which activates the nucleus of the iron atoms in steel, causing the atoms to move farther apart due to the rising temperature. This leads to decreased bond strength, which in turn reduces the yield strength and elastic modulus. The significant variation reported in different studies, similar to the one observed in thermal properties, can be attributed to variation in test setup, testing regimes (i.e., steady-state, transient, etc.), and test procedure that leads to different test measurements, thereby leading to differences in relations of mechanical properties in different codes and standards. As can be seen in Fig. 7.1, most published data show that steel loses about 50% of its strength at about 550°C. This temperature is referred to as the *critical temperature* of steel.

Of the various property relations presented in Fig. 7.1, three distinct models widely used in structural fire engineering practice are CEN (2005), ASCE (1992) manual, and Poh's (2001) model. A comparison of relations in Fig. 7.1 shows that the Eurocode 3 model predicts less reduction in the yield strength of steel with temperature compared to the ASCE model. However, the Eurocode model predicts a higher reduction in the elastic modulus of steel with temperature as compared to the ASCE model. Also, the Eurocode model assumes no reduction in yield strength of steel up to 400°C, while the ASCE model assumes a loss of 30% at 400°C. Figure 7.1 also shows that the Eurocode 3 relation distinguishes between two limits: the proportionality limit and the yield limit. The proportionality limit in the Eurocode is the end of the linear portion of the stress-strain curve, after which point the stress-strain response remains elastic but becomes nonlinear. The yield limit in the Eurocode is the point after which the stress-strain variation becomes both nonlinear and inelastic. This trend in the stress-strain response is not captured in the ASCE material model, wherein a sharp transition is present between linear-elastic and inelastic zones. The proportionality limit present in the Eurocode stress-strain curves at elevated temperatures is to capture viscoelastic behavior that is partly due to creep effect. Therefore, CEN (2005) stress-strain curves partly account for high-temperature creep effects at elevated temperatures.

For advanced fire resistance analysis of steel structures, temperature-dependent stress-strain relations are to be provided as input data into computational models. As Eurocode 3 temperature stress-strain relations were developed based on transient tests under slow heating rates on steel coupons conducted by Twilt and Witteveen (1974) and Anderberg (1988), these relations account for partial creep effects. Therefore, when used in fire resistance analysis, the Eurocode constitutive relations generally produce slightly more flexible responses as compared to ASCE[1] or Poh constitutive models. On the other hand, Poh's (2001) generalized temperature-stress-strain relations for structural steel are based on a larger set of test data that accounts for specific features, such as the yield plateau and the effect of strain hardening. The Poh's temperature-dependent stress-strain relations have been validated against specified relations in Eurocode 3 and other standards, and have been shown to give more realistic predictions of fire resistance when used in conjunction with a high-temperature explicit creep model. Tables 7.2 and 7.3 present the ASCE (1992), CEN (2005), and Poh (2001) relations in a tabulated form for structural, stainless steel, and cold-form steel, respectively. Table 7.4 lists temperature-dependent reduction factors for higher strength steels (with application in bolts and welds) as per Eurocode 3 and AISC provisions.

Typical stress-strain response of carbon steel is illustrated in Fig. 7.2. It can be seen from this figure that there is a well-defined yield point in the stress-strain curve of steel up to 300°C but it ceases to exist beyond this temperature. This is due to the onset of softening and degradation of yield strength and elastic modulus properties of steel. Due to this distinct behavior, effective yield strength (end of the elastic phase) of steel at elevated temperatures is often taken at 1% to 2% of proof strength (see Fig. 7.3) rather than the conventional ambient value of 0.2% yield stress (Kirby and Preston, 1988). For example, ASCE manual and Poh stress-strain relations consider a sharp yield point as an end of the linear-elastic phase, while the Eurocode standard considers "proportionality limit" as an end of the linear phase. The end of elasticity in the Eurocode is the yield point and is defined as the offset stress (also known as proof stress) corresponding to 0.2% strain.

Creep is defined as the time-dependent plastic strain under constant stress and temperature. At room temperature and under service load levels, creep deformations of steel are insignificant; however, at elevated

[1]Less information is known about test conditions and data that were used to derive ASCE temperature-stress-strain curves of steel. Also, there is limited test data on property relations for HSLA and other types of steels. More discussion on these issues can be found in Kodur and Aziz (2015).

		ASCE Manual (ASCE, 1992)	Eurocode 3 (CEN, 2005)
Structural Steel	Thermal Conductivity (W/m·°K)	For $T \leq 900°C$ $k_s = -0.022T + 48$ For $T > 900°C$ $k_s = 28.2$	For $T < 800°C$ $k_s = 54 - 3.33x10^{-2}T$ For $T \geq 800°C$ $k_s = 27.3$
	Specific Heat (J/kg·K)	For $T \leq 650°C$ $c_s = \frac{(0.004 \times T + 3.3) \times 10^6}{\rho_s}$ For $650°C < T \leq 725°C$ $c_s = \frac{(0.068 \times T + 38.3) \times 10^6}{\rho_s}$ For $725°C < T \leq 800°C$ $c_s = \frac{(-0.086 \times T + 73.35) \times 10^6}{\rho_s}$ For $T > 800°C$ $c_s = \frac{4.55 \times 10^6}{\rho_s}$	For $T < 600°C$: $c_s = 425 + 7.73 \times 10^{-1}T -$ $1.69 \times 10^{-3}T^2 + 2.22 \times 10^{-6}T^3$ For $600°C \leq T < 735°C$ $c_s = 666 + \frac{13002}{738 - T}$ For $735°C \leq T < 900°C$ $c_s = 545 + \frac{17820}{T - 731}$ For $900°C \leq T \leq 1200°C$ $c_s = 650$
	Coefficient of Thermal Expansion (per °C)	For $T < 1000°C$ $\frac{\Delta \ell}{\ell} = (0.004T + 12) \times 10^{-6}$ For $T \geq 1000°C$ $\frac{\Delta \ell}{\ell} = 16 \times 10^{-6}$	For $T < 750°C$ $\frac{\Delta \ell}{\ell} = 1.2 \times 10^{-5}T + 0.4 \times 10^{-8}T^2$ -2.416×10^{-4} For $750°C \leq T \leq 860°C$ $\frac{\Delta \ell}{\ell} = 1.1 \times 10^{-2}$ For $860°C < T \leq 1200°C$ $\frac{\Delta \ell}{\ell} = 2 \times 10^{-5}T - 6.2 \times 10^{-3}$
Stainless Steel		Eurocode 3 (CEN, 2005)	
	Thermal Conductivity (W/m·K)	$k = 14.6 + 1.27 \times 10^{-2}T$	
	Specific Heat (J/kg·K)	$c = 450 + 0.28 \times T - 2.91 \times 10^{-4}T^2 + 1.34 \times 10^{-7}T^3$	
	Coefficient of Thermal Expansion (per °C)	$\frac{\Delta \ell}{\ell} = (16 + 4.79 \times 10^{-3}T - 1.243 \times 10^{-6}T^2) \times (T - 20)10^{-6}$	

Table 7.1 Relations for Thermal Properties of Steel as a Function of Temperature

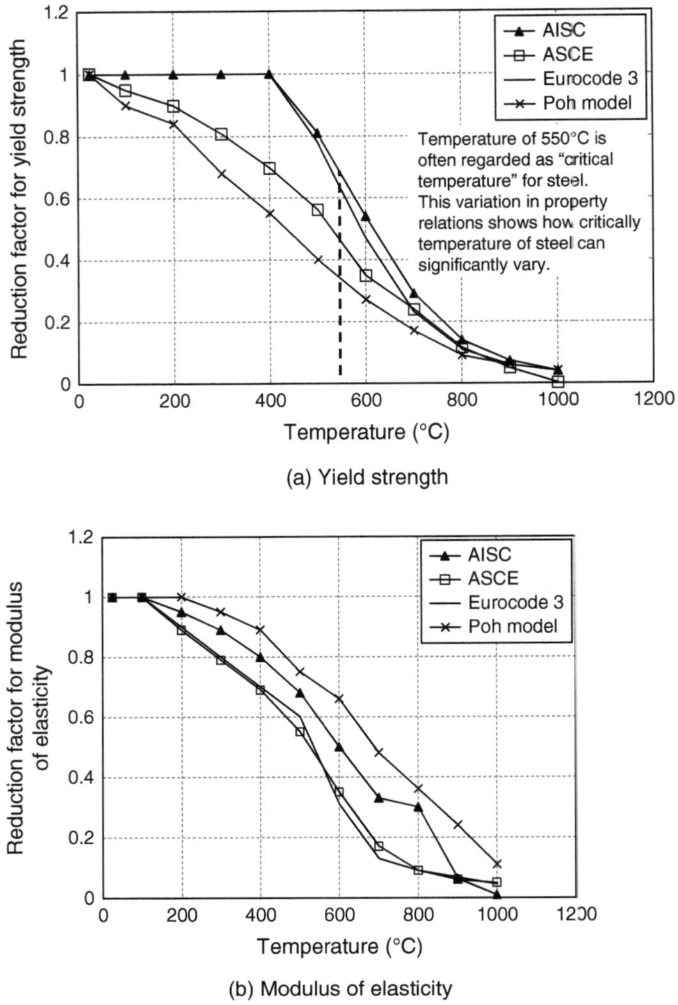

FIGURE 7.1 Degradation in yield strength and modulus of elasticity of steel at elevated temperature.

temperatures, creep deformations accelerate and may affect the global response of structures. Generally, creep deformations in steel become noticeable at temperatures above 400°C (Kirby and Preston, 1988). At present, most fire resistance analyses are carried out using Harmathy's high-temperature creep model that is mainly based on Dorn's creep theory (Dorn, 1955; Harmathy, 1967), as well as that of Fields and Fields (1989). It is worth noting that Eurocode 3 states that *the effects of transient thermal creep need not be given explicit consideration* [Clause 4.3.3(4) of CEN (2005)]. However, the *ASCE Manual of Practice* (ASCE, 1992) states that high-temperature creep should be accounted for in fire resistance analysis through one of the two options. The first option is to use *effective* temperature-stress-strain curves derived from transient-state tests at relevant heating and strain rates, while the second option is to use specific high-temperature creep models developed for structural steel. The widely used creep models for structural steel are listed in Table 7.5. Further discussion on transient creep effects can be found in Chap. 4 and in the recent work of Kodur and Aziz (2015), as well as in Morovat et al. (2014).

7.2.3 Property Relations for Fire Insulation

When exposed to fire conditions, steel structures undergo a rapid rise in steel temperature due to the low specific heat, high thermal conductivity, and lower mass of steel. Thus, steel members are to be protected with fire insulation such as gypsum-based products or cementitious-based spray-on insulation to achieve required fire resistance. As insulation materials have low strength or stiffness properties, only thermal properties are of interest from a fire resistance consideration. As in the case of other materials, properties of insulation also change with temperature. However, there is limited data on temperature-dependent properties of insulation materials. Often, room-temperature thermal properties, as provided in manufacturers' listings, are used for evaluating fire resistance of steel structures (see Table 7.6). This practice can lead to unconservative fire resistance values (Kodur and Shakya, 2013; Kodur and Arablouei, 2015).

	ASCE Manual (ASCE, 1992)	Eurocode 3 (CEN, 2005)	Poh (2001)
Temperature-Stress-Strain Relations	$\sigma_s = \begin{cases} \varepsilon \le \varepsilon_p \\ \quad E_{s,T} \times \varepsilon \\ \varepsilon > \varepsilon_p \\ \quad (c_1\varepsilon + c_2)F_{y,T} - \frac{c_3 \times F_{y,T}^2}{E_{s,T}} \end{cases}$ where $\varepsilon_p = \frac{c_2F_{y,T} - c_3F_{y,T}^2}{E_{s,T} - c_1F_{y,T}}$	$\sigma_s = \begin{cases} \varepsilon \le \frac{F_{p,T}}{E_{s,T}} \\ \quad E_{s,T} \times \varepsilon \\ \frac{F_{p,T}}{E_{s,T}} < \varepsilon < c_1 \\ \quad F_{p,T} - c + \frac{b}{a}\sqrt{a^2 - (c_1 - \varepsilon)^2} \\ c_1 < \varepsilon < c_3 \text{ and } T < 400^\circ\text{C} \\ \quad \frac{(F_{u,T} - F_{y,T})\varepsilon + c_3F_{y,T} - c_1F_{u,T}}{(c_3 - c_1)} \\ c_3 < \varepsilon \le c_4 \text{ and } T < 400^\circ\text{C} \\ \quad F_{u,T} \\ c_4 < \varepsilon < c_5 \text{ and } T < 400^\circ\text{C} \\ \quad F_{u,T} \times \frac{(c_5 - \varepsilon)}{(c_5 - c_4)} \\ c_2 < \varepsilon \le c_4 \text{ and } T \ge 400^\circ\text{C} \\ \quad F_{y,T} \\ c_4 < \varepsilon < c_5 \text{ and } T \ge 400^\circ\text{C} \\ \quad F_{y,T} \times \left[1 - \frac{(\varepsilon - c_2)}{(c_3 - c_2)}\right] \\ \varepsilon \ge c_5 \\ \quad 0.0 \end{cases}$ where $a^2 = (c_1 - F_{p,T}/E_{s,T})(c_1 - F_{p,T}/E_{s,T} + c/E_{s,T})$ $b^2 = c(c_1 - F_{p,T}/E_{s,T})E_{s,T} + c^2$ $c = \frac{(F_{y,T} - F_{p,T})^2}{(c_1 - F_{p,T}/E_{s,T})E_{s,T} - 2(F_{y,T} - F_{p,T})}$	$\sigma_s = \frac{\varepsilon}{a\vert\varepsilon\vert}\left\{ a - \vert a\vert + \left[\frac{(\beta_1 - \beta_2)b}{\left(1 + \left\vert\frac{(\beta_1 - \beta_2)b}{\beta_3}\right\vert^{\beta_5}\right)^{\frac{1}{\beta_5}}} + \beta_2 b\right] \bullet \left[1 + \frac{\vert b\vert - \vert(b - 0.001)\vert}{0.001}\right] \right\}$ where $a = E_{s,T}\vert\varepsilon\vert + F_{y,T}, \quad b = \vert\varepsilon\vert - \beta_4 - \frac{F_{y,T}}{E_{s,T}}$ $\frac{\beta_1}{E_s} = \begin{cases} T \le 500^\circ\text{C} \\ \quad \frac{c_6T + c_7}{(1 + \vert c_8T + c_9\vert^5)^{0.2}} + c_{10}T + c_{11} \\ T > 500^\circ\text{C} \\ \quad \frac{E_{s,T}}{E_s} \end{cases}$ $\frac{\beta_2}{E_s} = \frac{c_{12}T + c_{13}}{(1 + \vert c_{14}T + c_{15}\vert^5)^{0.2}} + c_{16}T + c_{17}$ $\frac{\beta_3}{F_y} = \begin{cases} T \le 500^\circ\text{C} \\ \quad \frac{c_{24}T + c_{25}}{(1 + \vert c_{26}T + c_{27}\vert^5)^{0.2}} + c_{28}T + c_{29} \\ T > 500^\circ\text{C} \\ \quad \frac{F_{y,T}}{F_y} \end{cases}$ $\frac{\beta_4}{(F_y/E_s)} = \frac{c_{30}T + c_{31}}{(1 + \vert c_{32}T + c_{33}\vert^5)^{0.2}} + c_{34}$ $\beta_5 = \frac{c_{35}T + c_{36}}{(1 + \vert c_{37}T + c_{38}\vert^5)^{0.2}} + c_{39}T + c_{40}$
Elastic Modulus	$\frac{E_{s,T}}{E_s} = \begin{cases} 1.0 + \frac{T}{c_9 \ln\left(\frac{T}{c_{10}}\right)}, & T \le 600^\circ\text{C} \\ \frac{c_{11} - c_{12}T}{T - c_{13}}, & T > 600^\circ\text{C} \end{cases}$	$\frac{E_{s,T}}{E_s} = \begin{cases} 1.0, & T < 100^\circ\text{C} \\ c_6T + c_7, & 100^\circ\text{C} \le T < 500^\circ\text{C} \\ c_8T + c_9, & 500^\circ\text{C} \le T < 600^\circ\text{C} \\ c_{10}T + c_{11}, & 600^\circ\text{C} \le T < 700^\circ\text{C} \\ c_{12}T + c_{13}, & 700^\circ\text{C} \le T < 800^\circ\text{C} \\ c_{14}T + c_{15}, & 800^\circ\text{C} \le T < 1200^\circ\text{C} \\ 0.0, & 1200^\circ\text{C} \le T \end{cases}$	$\frac{E_{s,T}}{E_s} = \frac{c_1T + c_2}{(1 + \vert c_3T + c_4\vert^5)^{0.2}} + c_5$
Yield Strength	$\frac{F_{y,T}}{F_y} = \begin{cases} 1.0 + \frac{T}{c_4 \ln\left(\frac{T}{c_5}\right)}, & T \le 600^\circ\text{C} \\ \frac{c_6 - c_7T}{T - c_8}, & 600^\circ\text{C} < T \end{cases}$	$\frac{F_{p,T}}{F_y} = \begin{cases} 1.0, & T < 100^\circ\text{C} \\ c_{16}T + c_{17}, & 100^\circ\text{C} \le T < 400^\circ\text{C} \\ c_{18}T + c_{19}, & 400^\circ\text{C} \le T < 500^\circ\text{C} \\ c_{20}T + c_{21}, & 500^\circ\text{C} \le T < 600^\circ\text{C} \\ c_{22}T + c_{23}, & 600^\circ\text{C} \le T < 700^\circ\text{C} \\ c_{24}T + c_{25}, & 700^\circ\text{C} \le T < 800^\circ\text{C} \\ c_{26}T + c_{27}, & 800^\circ\text{C} \le T < 1200^\circ\text{C} \\ 0.0, & 1200^\circ\text{C} \le T \end{cases}$ $\frac{F_{y,T}}{F_y} = \begin{cases} 1.0, & T < 400^\circ\text{C} \\ c_{28}T + c_{29}, & 400^\circ\text{C} \le T < 500^\circ\text{C} \\ c_{30}T + c_{31}, & 500^\circ\text{C} \le T < 600^\circ\text{C} \\ c_{32}T + c_{33}, & 600^\circ\text{C} \le T < 700^\circ\text{C} \\ c_{34}T + c_{35}, & 700^\circ\text{C} \le T < 800^\circ\text{C} \\ c_{36}T + c_{37}, & 800^\circ\text{C} \le T < 900^\circ\text{C} \\ c_{38}T + c_{39}, & 900^\circ\text{C} \le T < 1200^\circ\text{C} \\ 0.0, & 1200^\circ\text{C} \le T \end{cases}$ $\frac{F_{u,T}}{F_y} = \begin{cases} c_{40}, & T < 300^\circ\text{C} \\ c_{41}T + c_{42}, & 300^\circ\text{C} \le T < 400^\circ\text{C} \\ 1.0, & 400^\circ\text{C} \le T \end{cases}$	$\frac{F_{y,T}}{F_y} = \frac{c_{18}T + c_{19}}{(1 + \vert c_{20}T + c_{21}\vert^5)^{0.2}} + c_{22}T + c_{23}$

Table 7.2 Relations for Mechanical Properties of Structural Steel as a Function of Temperature as per Eurocode 3

	ASCE Manual (ASCE, 1992)	Eurocode 3 (CEN, 2005)	Poh (2001)
Coefficients to Be Used in Above Equations	$c_1 = 12.5$; $c_2 = 0.975$; $c_3 = 12.5$; $c_4 = 900$; $c_5 = 1{,}750$; $c_6 = 340$; $c_7 = 0.34$; $c_8 = 240$; $c_9 = 2{,}000$; $c_{10} = 1{,}100$; $c_{11} = 690$; $c_{12} = 0.69$; and $c_{13} = 53.5$	$c_1 = 0.02$; $c_2 = 0.02$; $c_3 = 0.04$; $c_4 = 0.15$; $c_5 = 0.20$; $c_6 = -1 \times 10^{-3}$; $c_7 = 1.1$; $c_8 = -2.9 \times 10^{-3}$; $c_9 = 2.05$; $c_{10} = -1.8 \times 10^{-3}$; $c_{11} = 1.39$; $c_{12} = -4 \times 10^{-4}$; $c_{13} = 0.41$; $c_{14} = -2.25 \times 10^{-4}$; $c_{15} = 0.27$; $c_{16} = -1.933 \times 10^{-3}$; $c_{17} = 1.193$; $c_{18} = -0.6 \times 10^{-3}$; $c_{19} = 0.66$; $c_{20} = -1.8 \times 10^{-3}$; $c_{21} = 1.26$; $c_{22} = -1.05 \times 10^{-3}$; $c_{23} = 0.81$; $c_{24} = -2.5 \times 10^{-4}$; $c_{25} = 0.25$; $c_{26} = -1.25 \times 10^{-4}$; $c_{27} = 0.15$; $c_{28} = -2.2 \times 10^{-3}$; $c_{29} = 1.88$; $c_{30} = -3.1 \times 10^{-3}$; $c_{31} = 2.33$; $c_{32} = -2.4 \times 10^{-3}$; $c_{33} = 1.91$; $c_{34} = -1.2 \times 10^{-3}$; $c_{35} = 1.07$; $c_{36} = -5 \times 10^{-4}$; $c_{37} = 0.51$; $c_{38} = -2 \times 10^{-4}$; $c_{39} = 0.24$; $c_{40} = 1.25$; $c_{41} = -2.5 \times 10^{-3}$; and $c_{42} = 2$	$c_1 = -1.4 \times 10^{-3}$; $c_2 = 1.001$; $c_3 = 2.74 \times 10^{-3}$; $c_4 = -1.959$; $c_5 = 0.493$; $c_6 = 0.01334$; $c_7 = 7.6705$; $c_8 = 8 \times 10^{-3}$; $c_9 = -4.6$; $c_{10} = 1.6 \times 10^{-4}$; $c_{11} = -1.6975$; $c_{12} = -6.4 \times 10^{-5}$; $c_{13} = 0.03616$; $c_{14} = 0.0222$; $c_{15} = -12.556$; $c_{16} = 4 \times 10^{-6}$; $c_{17} = -3.8 \times 10^{-4}$; $c_{18} = -1.25 \times 10^{-3}$; $c_{19} = 0.50625$; $c_{20} = 3.175 \times 10^{-3}$; $c_{21} = -1.2857$; $c_{22} = -2 \times 10^{-4}$; $c_{23} = 0.63425$; $c_{24} = 7.4 \times 10^{-3}$; $c_{25} = -0.925$; $c_{26} = 0.0133$; $c_{27} = -1.667$; $c_{28} = -2.4 \times 10^{-3}$; $c_{29} = 0.046$; $c_{30} = -0.05$; $c_{31} = 10$; $c_{32} = 0.01$; $c_{33} = -2$; $c_{34} = 5$; $c_{35} = -3 \times 10^{-3}$; $c_{36} = 1.125$; $c_{37} = 0.0133$; $c_{38} = -5$; $c_{39} = 1 \times 10^{-3}$; and $c_{40} = 0.275$

Table 7.2 Relations for Mechanical Properties of Structural Steel as a Function of Temperature as per Eurocode 3 (*Continued*)

	Stainless Steel*						Cold-Form Steel
	Grade 1.4301		Grade 1.4401 / 1.4404		Grade 1.4571		
Temperature (°C)	$k_{E,T}$	$k_{fy0.2,T}$	$k_{E,T}$	$k_{fy0.2,T}$	$k_{E,T}$	$k_{fy0.2,T}$	$k_{fy0.2,T}$
20	1	1	1	1	1	1	1
100	0.96	0.82	0.96	0.88	0.96	0.89	1
200	0.92	0.68	0.92	0.76	0.92	0.83	0.89
300	0.88	0.64	0.88	0.71	0.88	0.77	0.78
400	0.84	0.60	0.84	0.66	0.84	0.72	0.65
500	0.80	0.54	0.80	0.63	0.80	0.69	0.53
600	0.76	0.49	0.76	0.61	0.76	0.66	0.3
700	0.71	0.40	0.71	0.51	0.71	0.59	0.13
800	0.63	0.27	0.63	0.40	0.63	0.50	0.07
900	0.45	0.14	0.45	0.19	0.45	0.28	0.05
1000	0.20	0.06	0.20	0.10	0.20	0.15	0.03

*Grade 1.4301 (widely known as 304) is suitable for rural, urban, and light industrial sites, while grade 1.4401/1.4404 (widely known as 316/316L) is for more highly alloyed grades and will perform well in marine and industrial sites. Grade 1.4571 (commonly known as 316Ti) has inherent improvements to mechanical strength, especially at elevated temperatures above about 600°C.

Table 7.3 Reduction Factors for Yield Strength and Modulus of Elasticity of Stainless Steel and Cold-Form Steel as Per Eurocode 3 (*Reproduced from CEN*)

More information regarding temperature-dependent properties of insulation materials can be found in Chap. 4 and NIST, and specific property data can be found in published literature. Temperature-dependent properties of spray-applied fire resistive material (SFRM)-based insulation such as CAFCO-300, Carboline Type-5MD, and Tyfo WR-AFP were generated by Kodur and Shakya (2013), and these are listed in Table 7.7.

Insulation materials such as concrete and masonry can also be used to protect structural members made of steel. Relations for high-temperature thermal properties of concrete and masonry are discussed in Chap. 8. For various other insulation materials, thermal properties can be found in design guides and some standards.

Temperature (°C)	Eurocode 3		Temperature (°C)	AISC
	Bolts	Welds		Bolts (Group A and Group B)*
20	1	1	20	1
100	0.968	1	93	0.97
150	0.952	1	150	0.95
200	0.935	1	200	0.93
300	0.9030	1	320	0.88
400	0.775	0.876	430	0.71
500	0.550	0.627	480	0.59
600	0.220	0.378	540	0.42
700	0.100	0.130	650	0.16
800	0.067	0.074	760	0.08
900	0.033	0.018	870	0.04
1000	0	0	980	0.01

*Group A = A325 MPa and Group B = A490 MPa.

Table 7.4 Reduction Factors for Yield Strength of High-Strength Steel Used in Bolts and Welds as per Eurocode 3 and AISC

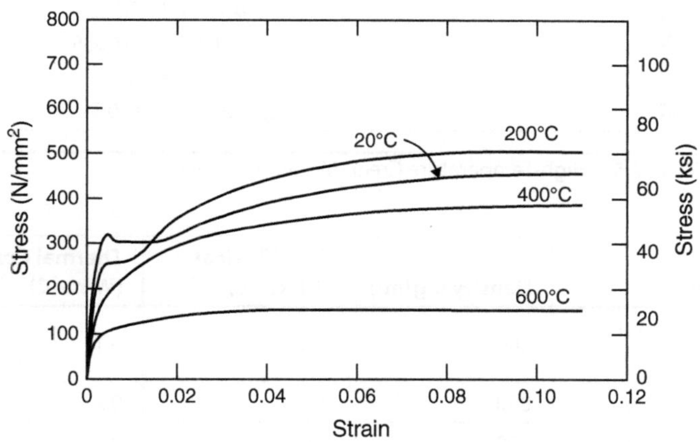

Figure 7.2 Stress-strain response of carbon steel (of Grade 240 MPa) at elevated temperature.

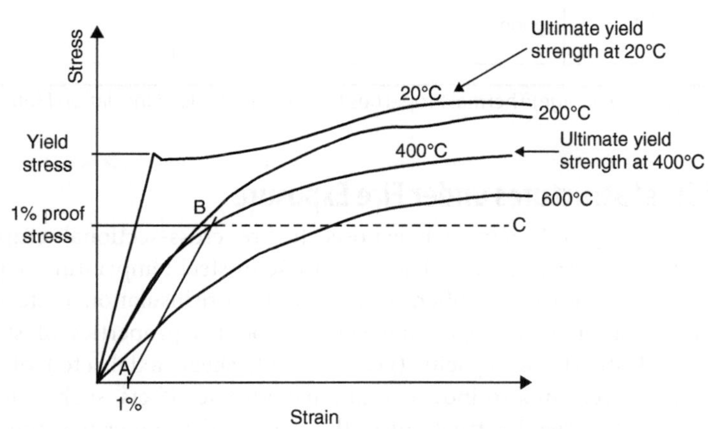

Figure 7.3 Yield stress definition.

<table>
<tr><th></th><th>Creep Model</th></tr>
<tr><td>Dorn-Harmathy Modified Creep Model</td><td>$$\frac{\partial \varepsilon_{cr}}{\partial t} = Z \times e^{\left(\frac{-Q}{RT_R}\right)} \times \begin{cases} \coth^2\left[\dfrac{3 \times t \times Z \times e^{\left(\frac{-Q}{RT_R}\right)}}{\varepsilon_{cr,0}}\right], & \text{for time hardening} \\ \coth^2\left[\dfrac{\varepsilon^*_{cr}}{\varepsilon_{cr,0}}\right], & \text{for strain hardening} \end{cases}$$
where
$$Z = \begin{cases} 0.026 \times \sigma_s^{4.7}, & \sigma_s \leq 15000 \text{ psi} \\ 1.23 \times 10^{16} \times e^{0.0003\sigma_s}, & 15000 \text{ psi} < \sigma_s \leq 45000 \text{ psi} \end{cases}$$
and σ_s is the stress in steel; $\varepsilon_{cr,0} = 1.7 \times 10^{-10}\, \sigma^{1.75}$; $Q/R = 70{,}000°$ Rankine; $T_R =$ Degrees Fahrenheit $+ 459.67$ (Rankine); $t =$ time (hr) and $\varepsilon^*_{cr} = \sum\left|\frac{\partial \varepsilon_{cr}}{\partial t}\right|\Delta T$</td></tr>
<tr><td>Fields and Fields Model</td><td>$$\varepsilon_c = at^b \sigma^c$$
where t is time and σ is stress. The parameters a, b, and c are temperature-dependent material properties derived by Fields and Fields (1989) as
$$a = 10^{-(a_0 + a_1 T)}$$
where $a_0 = 6.1$, $a_1 = 0.00573$ for $T \leq 500°C$
and $a_0 = 13.25$, $a_1 = -0.00851$ for $T \geq 500°C$
and
$$b = 10^{-(b_0 + b_1 T)}$$
$$b_0 = -1.1,\ b_1 = 0.0035$$
$$c = c_0 + c_1 T$$
$$c_0 = 2.1,\ c_1 = 0.0064$$</td></tr>
</table>

Table 7.5 High-Temperature Creep Models for Steel

Material	Density (kg/m³)	Specific Heat (kJ/kg·°C)	Thermal Conductivity (W/m·°C)	Moisture Content (% wt)
Sprayed Mineral Fiber	300	1.05	0.1	1.0
Gypsum Plaster	800	1.7	0.2	20.0
Mineral Fiber Sheets	500	1.5	0.25	2.0
Aerated Concrete	600	1.2	0.3	2.5
Lightweight Concrete	600	1.2	0.8	2.5
Normal-Weight Concrete	2200	1.2	1.7	1.5

Table 7.6 Typical (Average) Thermal Properties for Commonly Used Insulation Materials (Malhotra, 1984)

7.3 Behavior of Steel Structures under Fire Exposure

When a steel structural member is exposed to fire, cross-sectional temperatures in the steel member increase with fire exposure time. The rate of rise in steel temperature depends on fire severity, surface area (perimeter) of the steel member, and thickness of fire insulation on steel. The rise in steel temperatures leads to degradation in strength and elastic modulus properties of steel, which in turn results in degradation of structural capacity (i.e., moment, shear, axial, etc.) of the steel member. A rise in temperature in steel can also induce additional adverse effects, such as local buckling or creep strains, which may lead to large deformations in the structural member that can lead to the abrupt failure of a fire-exposed steel member.

Thermal Conductivity: K_t (W/m · K)		
CAFCO-300	$k_t = 0.0778 - 0.000054\,T$	$20°C \le T \le 300°C$
	$k_t = -0.08 + 0.000469\,T$	$300°C \le T \le 700°C$
Carboline Type-5MD	$k_t = 0.121 - 0.000319\,T$	$20°C \le T \le 200°C$
	$k_t = 0.0468 + 0.000050\,T$	$200°C \le T \le 700°C$
Tyfo WR-AFP	$k_t = 0.207 - 0.000318\,T$	$20°C \le T \le 200°C$
	$k_t = 0.147 - 0.000035\,T$	$200°C \le T \le 400°C$
	$k_t = 0.0054 + 0.000321\,T$	$400°C \le T \le 700°C$
Specific Heat: c_p (J/kg · K)		
CAFCO-300	$c_p = 3236 + 4.16\,T$	$20°C \le T \le 200°C$
	$c_p = 6815 - 12.60\,T$	$200°C \le T \le 400°C$
	$c_p = 1645 - 0.36\,T$	$400°C \le T \le 700°C$
Carboline Type-5MD	$c_p = 1627 + 22.3\,T$	$20°C \le T \le 100°C$
	$c_p = 4446 - 5.05\,T$	$100°C \le T \le 400°C$
	$c_p = -1336 + 9.37\,T$	$400°C \le T \le 700°C$
Tyfo WR-AFP	$c_p = 643 + 1.93\,T$	$20°C \le T \le 200°C$
	$c_p = 1241 - 0.924\,T$	$200°C \le T \le 400°C$
	$c_p = 195 + 1.71\,T$	$400°C \le T \le 600°C$
	$c_p = 1826 - 1.08\,T$	$600°C \le T \le 700°C$
Thermal Strain: ε_t $(\Delta L/L) \times 10^{-3}$		
CAFCO-300	$\varepsilon_t = (-0.228 - 0.00616\,T)\,10^{-3}$	$20°C \le T \le 300°C$
	$\varepsilon_t = (4.45 - 0.0201\,T)\,10^{-3}$	$300°C \le T \le 500°C$
	$\varepsilon_t = (-15.50 + 0.0196\,T)\,10^{-3}$	$500°C \le T \le 800°C$
	$\varepsilon_t = (77.1 - 0.0943\,T)\,10^{-3}$	$800°C \le T \le 1000°C$
Carboline Type-5MD	$\varepsilon_t = (0.525 - 0.0112\,T)\,10^{-3}$	$20°C \le T \le 200°C$
	$\varepsilon_t = (-0.745 - 0.00689\,T)\,10^{-3}$	$200°C \le T \le 300°C$
	$\varepsilon_t = (17.40 - 0.065\,T)\,10^{-3}$	$300°C \le T \le 400°C$
	$\varepsilon_t = (-4.43 - 0.0124\,T)\,10^{-3}$	$400°C \le T \le 700°C$
	$\varepsilon_t = (65.2 - 0.108\,T)\,10^{-3}$	$700°C \le T \le 1000°C$
Tyfo WR-AFP	$\varepsilon_t = (0.211 - 0.00873\,T)\,10^{-3}$	$20°C \le T \le 150°C$
	$\varepsilon_t = (1.17 - 0.0183\,T)\,10^{-3}$	$150°C \le T \le 300°C$
	$\varepsilon_t = (-1.57 - 0.00688\,T)\,10^{-3}$	$300°C \le T \le 500°C$
	$\varepsilon_t = (-5.68 + 0.00096\,T)\,10^{-3}$	$500°C \le T \le 700°C$
	$\varepsilon_t = (43.0 - 0.0646\,T)\,10^{-3}$	$700°C \le T \le 1000°C$
Mass Loss: m_t (%)		
CAFCO-300	$m_t = 0.464 - 0.0112\,T$	$20°C \le T \le 125°C$
	$m_t = 23.1 - 0.185\,T$	$125°C \le T \le 200°C$
	$m_t = -12.1 - 0.00466\,T$	$200°C \le T \le 300°C$
	$m_t = -0.10 - 0.0433\,T$	$300°C \le T \le 400°C$
	$m_t = -12.90 - 0.0117\,T$	$400°C \le T \le 600°C$
	$m_t = -7.49 - 0.0193\,T$	$600°C \le T \le 775°C$
Carboline Type-5MD	$m_t = 0.21 - 0.00698\,T$	$20°C \le T \le 100°C$
	$m_t = 25.8 - 0.247\,T$	$100°C \le T \le 175°C$
	$m_t = -15.7 - 0.00434\,T$	$175°C \le T \le 275°C$
	$m_t = -6.96 - 0.0346\,T$	$275°C \le T \le 375°C$
	$m_t = -17.6 - 0.00753\,T$	$375°C \le T \le 775°C$
Tyfo WR-AFP	$m_t = 0.434 - 0.0136\,T$	$20°C \le T \le 100°C$
	$m_t = 8.77 - 0.108\,T$	$100°C \le T \le 200°C$
	$m_t = -8.98 - 0.0115\,T$	$200°C \le T \le 600°C$
	$m_t = 9.53 - 0.0409\,T$	$600°C \le T \le 775°C$

Table 7.7 Temperature-Dependent Thermal Property Relations of Different Types of SFRM Fire Insulation

The behavior of a typical steel structural member under fire exposure is illustrated through tracing the response of an uninsulated steel beam as depicted in Fig. 7.4*a*. This beam is subjected to uniformly distributed loading. When exposed to a standard fire (i.e., ISO 834), the beam experiences temperature rise with increasing fire exposure time as shown in Fig. 7.4*b*. The rapid rise in temperature in a steel beam, due to the high thermal conductivity and low specific heat of steel, leads to degradation in strength and elastic modulus properties, resulting in a gradual reduction in moment capacity of this beam, as shown in Fig. 7.4*c*. When the degrading moment or shear capacity falls below that from the applied moment (or shear force due to loading), failure of the beam can occur. In the case of this specific beam, the degrading moment capacity at the critical midspan section falls below that of the applied bending moment at about 13 minutes of fire exposure, leading to the failure of the beam.

The illustration clearly shows the adverse effects of temperature on the behavior of steel beams under fire conditions and how such effects can lead to early failure. This vulnerability of steel structures to fire can be mitigated for the most extent through the provision of fire insulation on structural members. For example, if the above steel beam was insulated with 25-mm-thick spray-applied vermiculite fire insulation (SFRM) and exposed to the same loading and fire conditions, the average temperature rise in the beam would be limited to 600°C at about 120 minutes (see Fig. 7.4*d*). In the case of this specific insulated beam, the degrading moment capacity at the critical midspan section falls below that of the applied bending moment at about 135 minutes of fire exposure, leading to the failure of the beam (see Fig. 7.4*e*). Therefore, the beam will meet a *fire rating* of two hours as per the prescriptive approach.

Although failure in most flexural members is governed by flexural mode, failure can also occur under other modes, such as large deflections, local buckling, and shear-moment or axial force-moment interaction, and these modes must be given due consideration in evaluating failure under elevated temperature exposure. For example, a number of recent studies have pointed out that fire-induced forces can cause significant compression forces in the early stages of fire but then turn into tensile forces toward later stages (Pakala et al., 2012). Since these forces are generated from elevated temperature effects, such forces on a member can significantly alter failure mode from a ductile mode (i.e., flexural) at room temperature into a brittle failure (compressive buckling) under fire conditions.

The above-described scenario, illustrated for steel beams, is equally applicable for other structural members such as columns, trusses, and girders, and due consideration needs to be given to the different load-carrying mechanisms—for instance, compression (buckling effects) in columns. For realistic fire resistance assessment, relevant failure modes for specific structural member types need to be checked. Also, failure of a structural member under fire conditions is a function of various factors including geometrical features (e.g., slenderness), load level, fire scenario, and restraint conditions, among others, and how these factors can influence the development of temperature-induced forces and stresses within a fire-exposed member.

7.4 Strategies for Achieving Required Fire Resistance

7.4.1 General

Steel structures experience a rapid rise in temperature when exposed to fire due to the high thermal conductivity and low specific heat of steel. This rise in temperature leads to the loss of strength and elastic modulus properties of steel, leading to degradation in the load-bearing capacity of the member. As a result, steel structural members can experience failure as early as 20 minutes into a fully grown fire. Since building codes specify 1 to 4 hours of fire resistance ratings to structural members, appropriate strategies are to be developed to enhance failure time from 20 minutes to the required level of fire resistance.

7.4.2 Strategies for Enhancing Fire Resistance

The common strategy for enhancing fire resistance of a steel structural member is to limit temperature rise to below the critical temperature of steel. This can be achieved through two strategies: external *insulation* or a *capacitive* mechanism. Insulating a steel member involves placing external insulation in the form of a box around the steel section or in a contour layout surrounding the steel (Fig. 7.5). The box layout is adopted when prefabricated boards are used, while the contour protection strategy is utilized when fire insulation material is sprayed or wrapped on the section (see Fig. 7.5*b*). The common insulation materials used for fire protection of steel structures are gypsum vermiculite SFRM products, cementitious masonry materials, and intumescent paints. These materials exhibit low thermal conductivity and high specific heat and hence act as a thermal barrier that slows down temperature rise in steel to a great extent.

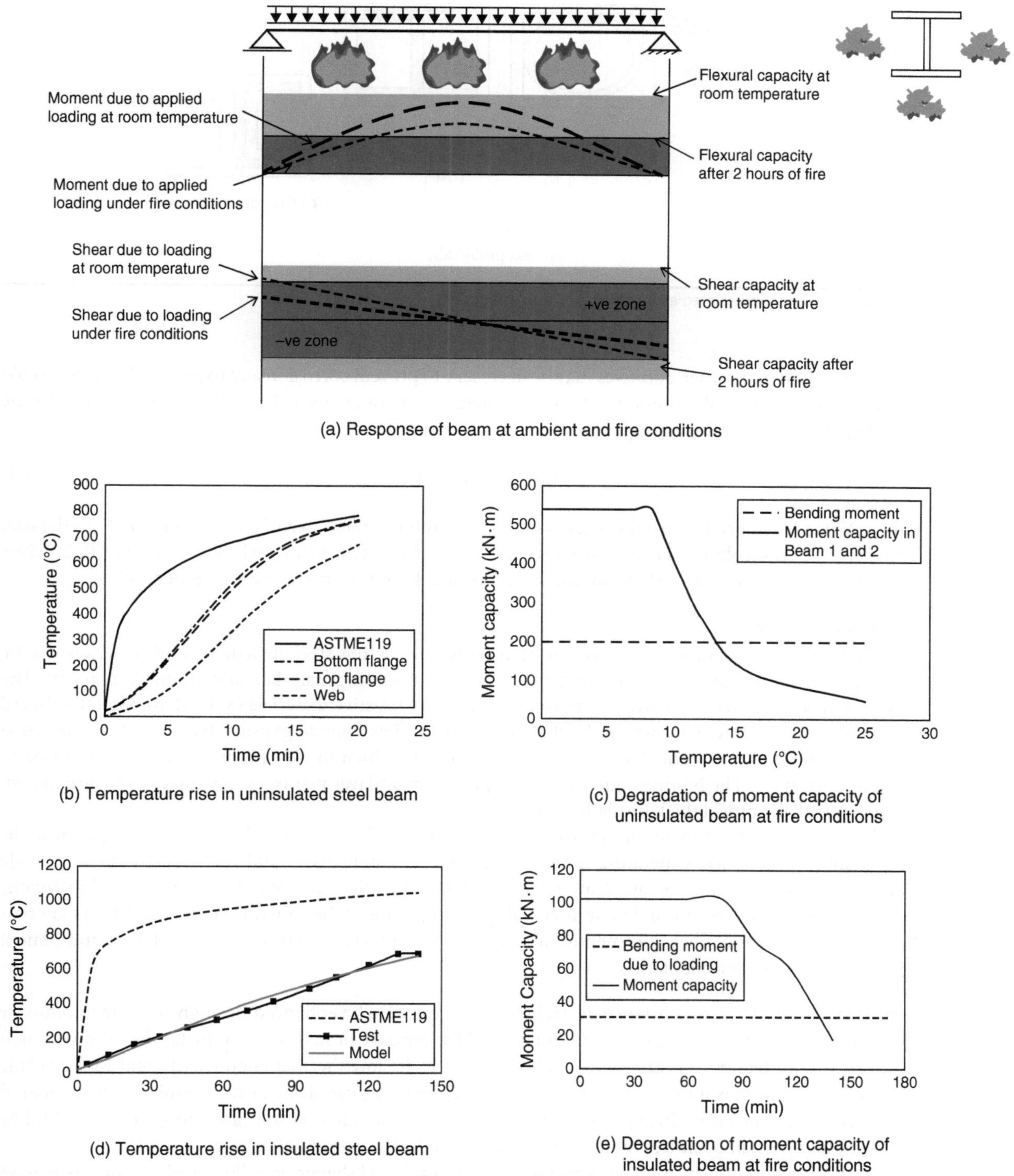

FIGURE 7.4 Behavior of a typical steel beam under fire conditions.

Alternatively, a capacitive protection strategy can be adopted for enhancing fire resistance of steel structures in the case of certain steel shapes such as hollow structural steel (HSS) sections. In this strategy, concrete can be filled within HSS section columns. Under fire conditions, concrete absorbs heat from the steel section and helps keep the steel cooler for a longer duration, as well as helps to carry a portion of the load. Alternatively, circulating water in an HSS column during a fire incident will keep the steel member cool under fire exposure.

When concrete filling is adopted, such concrete filling can be designed to solely act as a thermal (capacitive) insulator or to carry part of a load in conjunction with steel.[2] If water filling is used, a plumbing system is required to avoid excessive pressures when the water is heated and to ensure a proper water

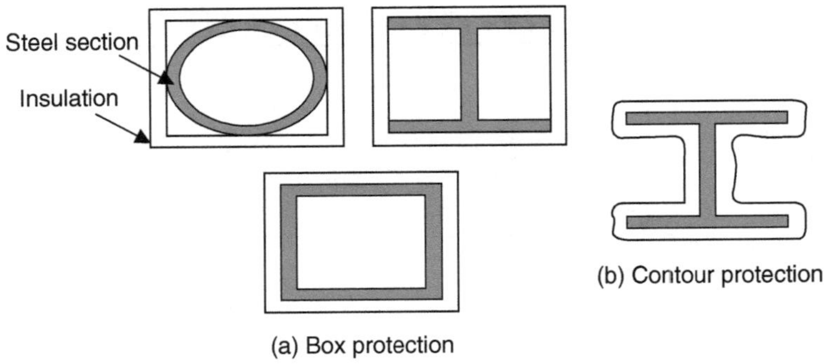

Figure 7.5 Protection strategies for steel structural members.

circulation system. Further, additives may be necessary to prevent corrosion and to prevent freezing in cold climates (Kodur and MacKinnon, 2000). The quantity of water needed to fill an HSS section can be quantified using the following equation:

$$V_w = 3.92Aq \times 10^{-7} \tag{7.1}$$

where V_w is the required external storage water (m^3), A is the surface area of column (m^2), and q is the heat transferred to the column during a fire test per unit surface area (kJ/m^2) which equals 15,070, 225,260, 580,960, 785,460, and 1,914,640 kJ/m^2 for ¾, 1, 2, 3, and 4 hours of fire exposure, respectively.

Insulation Protection

A variety of fire insulation materials can be used for protecting steel structures, and materials can be grouped under three categories: insulting materials, heat-absorbing materials, and intumescent paints. The commonly used insulation materials are spray-applied fire resistive materials (SFRM), mineral fiberboard products, and ceramic wool wraps. Heat-absorbing materials include gypsum-based or concrete-based products, and these materials have high moisture content, which in turn absorbs heat when exposed to high temperatures. The intumescent coatings, applied as multilayer paints, expand upon exposure to high temperatures, forming a thick insulating layer.

For achieving optimum levels of fire performance, fire insulation materials should be noncombustible, non-smoke or toxic gas generating, possess low thermal conductivity and high specific heat, provide uniformity and consistency in application, be sufficiently durable with good stickability to steel members, be lightweight (low density), and be aesthetically pleasing. Some of the widely used fire insulation materials in practice are mineral fiber or cementitious spray-applied materials, gypsum wallboard, Portland cement concrete, and intumescent paints.

Concrete Encasement Fire resistance of steel structures can also be enhanced through *concrete or masonry encasement* (shown in Fig. 7.6). In fact, until the 1970s, concrete or masonry was by far the most common form of fire protection for structural steelwork. Due to its low thermal conductivity and high specific heat, concrete (or masonry), when encased on a steel member, can delay heat transmission to structural members. Fire protection through concrete or masonry encasement is still often adopted in the Middle East and many Asian countries due to wider availability and low production costs.

In this method of insulation, concrete is casted around steel shapes or hollow steel sections to achieve desired fire resistance rating. The required thickness of concrete encasement can be arrived at through tabulated data or thumb rule methods specified in design manuals (ACI, 2014; CEN, 2005). In lieu of ordinary concrete (or masonry) encasement, autoclaved aerated concrete (AAC)[3] can also be adopted. ACC is produced through pressurized heat treatment of raw materials, which results in a finished product that is up to five times the volume of the raw materials and has high air content up to 80%. Steel structural members with ACC encasement can achieve up to a 4-hour fire rating as per UL directory.

Encased concrete is typically assumed to provide only thermal protection to steel and not add any load-carrying capacity. However, in recent years, a number of empirical correlations have been developed to

[2]Full details on this aspect of concrete filling option are provided in Chap. 9.

[3]ACC is a specialty concrete made of sand, cement, lime, gypsum, and water with an expansion agent.

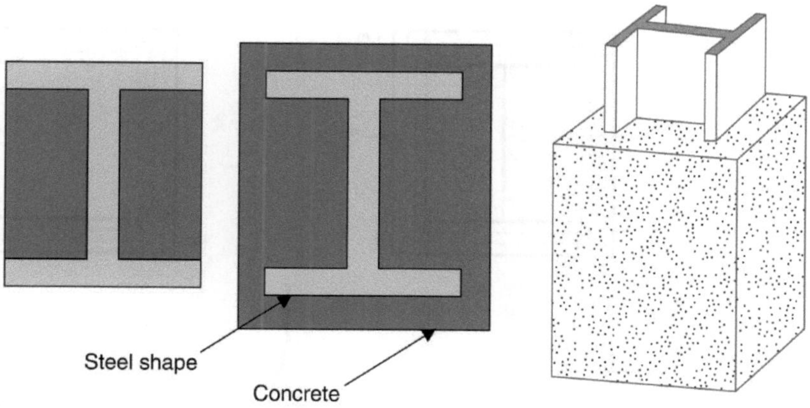

Figure 7.6 Concrete encasement of steel.

implicitly account for the contribution of concrete to load-carrying capacity, arising from composite action between concrete and steel (Fellouh et al., 2017). An in-depth discussion of these is given in Chap. 9.

Concrete encasement on steel structures has other benefits as well: it can enhance durability, impact resistance, corrosion resistance, and abrasion resistance. However, concrete or masonry encasement adds significantly to the weight of the structure (due to high density of concrete/masonry). Other disadvantages include longer curing times and poor aesthetics. Therefore, in recent years an alternative encasing for steel structures includes such insulating materials as lightweight boards and sprays.

Membrane Protection (Board and Mat Systems) Another feasible option to enhance fire resistance of steel structures is through the provision of external layers of gypsum board or mat systems on steel shapes. Two types of *membrane protection*—mineral fiberboards and mat products—can be utilized. Mineral fiberboards are usually semirigid, while the mat products are fully flexible. The main ingredients in these insulation systems are gypsum, fiber-reinforced calcium silicate, vermiculite-sodium silicate, and mineral fiber materials. Gypsum board, available in a range of thicknesses and sizes, is composed of a noncombustible core (usually of not less than 65% gypsum) and paper-laminated surfaces to form flat sheets.

Gypsum contains a large amount of water (up to 25%), and as a result, when a gypsum board is exposed to fire, much of the heat is utilized for water crystallization, thus limiting temperature rise in the board and also in the protected steel member. In other words, when a gypsum board is subjected to elevated temperatures, crystallization of water gradually increases and this process consumes large amounts of energy, thus delaying heat transmission through dehydrating the board. Therefore, gypsum board effectively acts as a fire barrier until most of its water content is driven out. The thermal conductivity of gypsum boards is in the range of 0.03 W/m·K to 0.05 W/m·K.

Gypsum boards come in various types, sizes, and brands. The gypsum boards that are usually used for fire protection are classified as fire-rated boards and are grouped under *Type X* and *Type C* in North America. Type X and Type C have a specially formulated gypsum core in order to provide higher fire resistance than regular gypsum board. Type C gypsum products exhibit superior fire performance characteristics, as compared to Type X products, that can exceed ASTM fire rating requirements. Fire ratings of up to 4 hours can be achieved through the use of Type X and Type C when applied in multiple layers.

One of the best features of membrane protection is the size and speed of installation on steel members. Such board products can be used in wall or floor assemblies to provide an envelope around steel trusses. Board systems are easy to install and can be finished with decorative paint. Board systems can be glued or screwed together or to a metal (or wood) framing around the structural member (especially columns). Thus, the characteristics of nails, screws, or glues, as well as detailing and spacing arrangement of fasteners, are most critical in achieving required fire resistance. Detailed specifications for the attachment of board products to steel members are provided by board manufactures and as part of fire rating listings (Stanzak and Lie, 1973; PABCO, 1984; AISI, 1991). Figure 7.7 illustrates commonly used attachments of gypsum wallboard, with and without steel covers, to steel columns.

One advantage of using gypsum boards or mats is that they are not wet systems during installation, and thus result in less impact on construction time lines and other trades. These systems are of relatively low cost and provide smooth surface finishes when compared to other fire protection options.

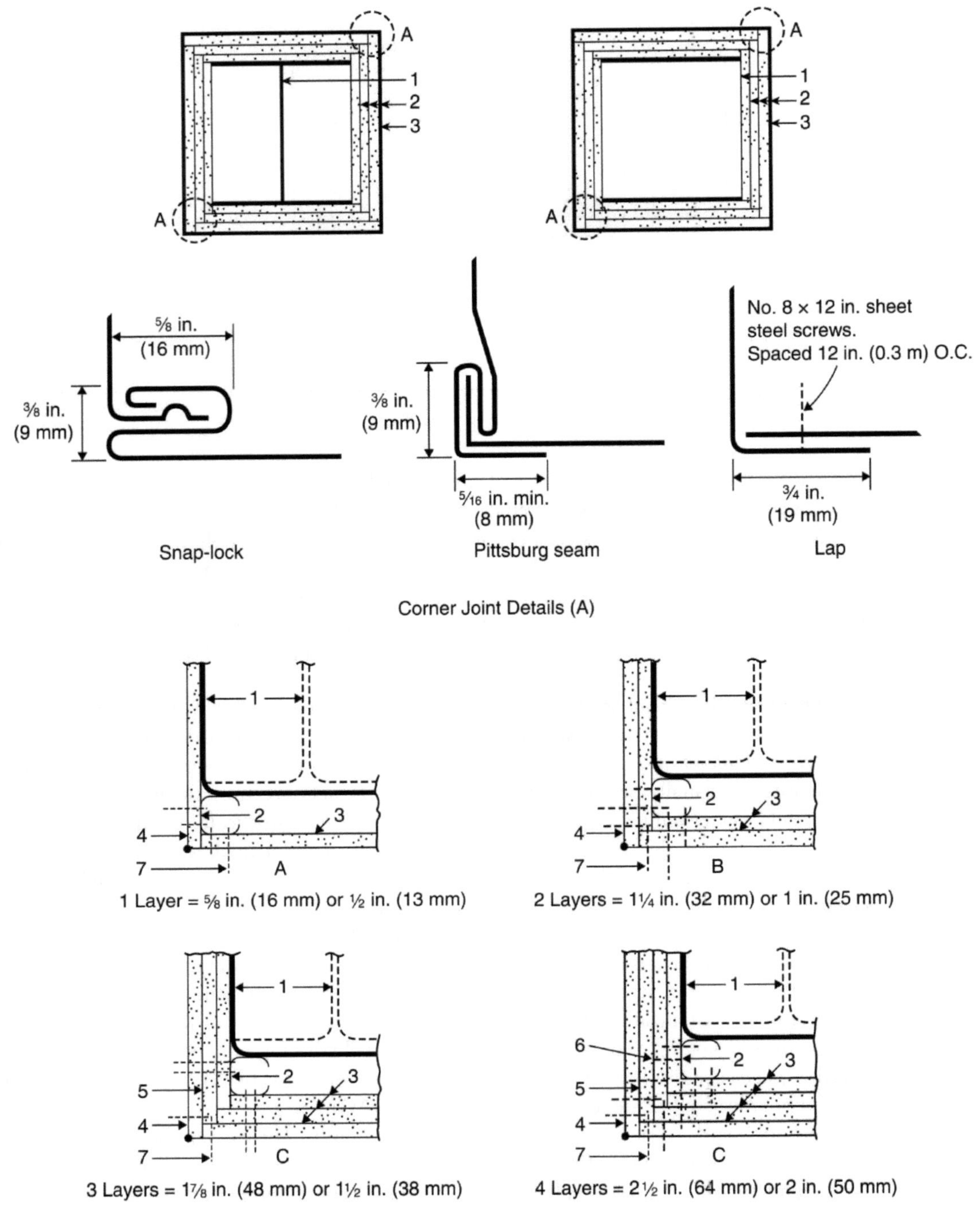

Figure 7.7 Attachment mechanisms of gypsum wallboards to steel columns. (*Milke, 2016.*)

Spray-Applied Fire Resistive Materials Currently the preferred method of fire insulation on steel structural members is *spray-applied fire resistive materials* (SFRMs). SFRM is composed of cement-based mineral fibers such as glass or cellulosic fibrous reinforcing substances such as cementitious plasters, magnesium oxychloride cements, or cementitious-based products. SFRMs can be mineral based or cementitious based. Mineral-based SFRM is a mixture of mineral fibers, binders (usually, Portland cement based), air, and water. This mixture is to be applied with spray-on equipment that feeds the dry mixture of mineral fibers and various binding agents to a spray nozzle, where it gets mixed with water as it is sprayed on to steel surface to be protected (see Fig. 7.8*a*). In the final cured form, the mineral fiber SFRM is usually lightweight and essentially noncombustible, possesses poor heat conduction, and can provide up to 4 hours of fire rating.

(a) SFRM application process on a steel column

(b) SFRM application process on a steel beam

Figure 7.8 Application of SFRM to steel members. (*AISI, 1974*.)

Cementitious-based SFRMs are composed of a binder material (i.e., calcined gypsum and Portland cement) mixed with aggregates (vermiculite and/or perlite) and other additives and foaming agents such as magnesium oxychloride, magnesium oxysulfate, calcium aluminate, calcium phosphate, or ammonium sulfate. Cementitious SFRMs are classified based on their density and can be grouped as low density (about 240 kg/m^3), medium density (320 to 430 kg/m^3), and high density (640 to 1280 kg/m^3) products. For unique applications, such as columns in offshore structures, higher density SFRM products ranging between 1600 and 2400 kg/m^3 are preferred.

SFRM products are very versatile in applicability and can be applied with ease on columns, beams, and trusses (see Fig. 7.8), as well as on complex regions such as connections or steel brackets with limited free space and low surface area/perimeter. Prior to application of SFRM, steel surfaces should be carefully prepared since it is hard to adequately control quality of installation (i.e., uniformity of thickness) in the field. One of the main concerns of using spray-on materials is their durability (due to their softness and lightness) and bond (stickability) to steel members.[4] The required thickness of SFRM needed to achieve 1 to 4 hours of fire ratings may be found in manufacturers' product sheets or design guides or listings. It should be noted that a few of the earliest SFRM products in the United States contained asbestos, which is no longer allowed due to its carcinogenic nature that can lead to health issues.

Intumescent Coating Applying *intumescent (paint) coating* on steel surface is another effective strategy to achieve required fire resistance in steel structural members. These coatings, applied as thin layers on steel structural members, have the appearance of a thick film or paint (see Fig. 7.9). The intumescent paints are composed of two ingredients: a resin binder and a mixture of chemicals that decomposes and releases a gas when heated. Under fire exposure, intumescent paint undergoes an endothermic carbonation reaction that results in the production of an inert gas. This reaction is triggered at temperature levels closer to the melting of resin. The release of inert gas causes the resin to melt so as to develop a thick layer of char on the steel surface, which insulates the steel member from fire exposure. This expanded char layer possesses low thermal conductivity, thus limiting the temperature rise in the steel. Intumescent paints may expand as much as 15 to 30 times of their initial thickness during fire exposure. Depending on the thickness level, the paints can provide fire ratings in the range of half an hour to 4 hours in steel structures.

[4]The ICBO (1997) provides test methods for testing stickability (adhesion and cohesion) of spray-on fire protection material.

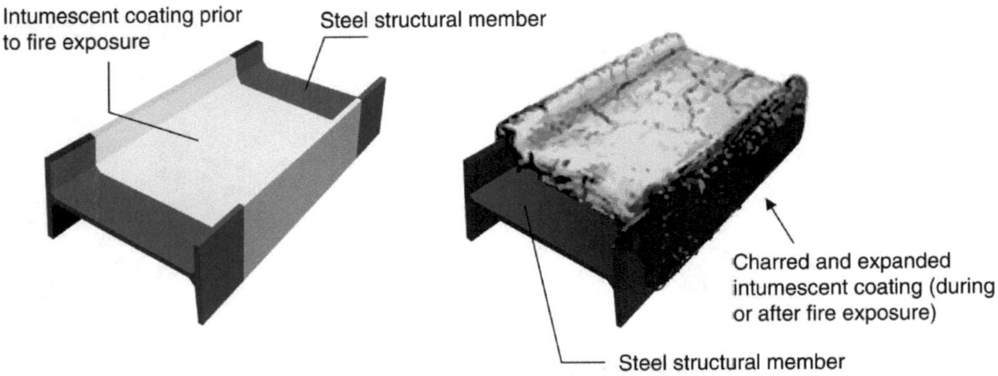

Figure 7.9 Intumescent paint (coating) on steel member (before and after fire exposure)

Two distinct types of intumescent coatings are available: water or solvent based (referred to as thin film coatings) and epoxy based (referred to as mastics or thick film coatings). Water/solvent-based coatings are thin (usually in 0.5- to 5-mm thickness range) and mostly intended for controlled environment inside buildings, although some systems are available for external exposures. Epoxy-based coatings are thicker than water-based coatings (up to 45 mm thickness) and are mostly used in outdoor application and petrochemical installations where the exterior environment, as well as fire severity, can be harsh. The intumescent coating is applied in multiple layers of coatings having different properties and functionality. The base coat (i.e., primer) is formulated to provide a strong bond (adhesion) to the steel member, while the top coat (i.e., sealing) is to have a durable and aesthetically appealing finished surface. Due to the need to apply intumescent coatings in different layers, the application can be lengthy and laborious. From research point of view, intumescent coatings are proprietary products and their thermal properties have been studied only to a limited extent. Lawson and Oshatogbe (1999) reported measurements of the thermal conductivity of 13 different intumescent chars that averaged 0.096 ($\pm$0.033) W/m·°C.

Newer types of coatings for fire protection of steel structures include ablative, subliming, and aerogels, which have been developed in the last three decades. Ablative coatings are those that gradually sublime during fire exposure due to absorbing heat energy that changes the virgin solid coating into a gas composite. This action prevents heat transmission to the steel surface the ablative coating is applied to. These coatings are similar to intumescent paints but require a complex application procedure which results in relatively high cost. In the case of subliming compounds, these coatings have an active ingredient that absorbs heat as it changes from solid to gas phase. Aerogels, on the other hand, are typically made of silica, and have lattice structure and large porosity. Aerogels make very good insulators, as they do not thermally degrade or generate toxic fumes.

The main advantages of coatings are their lower density and smaller thickness (in the range of a portion of millimeter) and thus do not add any dead weight to the protected structure. Further, intumescent coatings can be quickly applied, are quite durable, are easily repaired, and are good at covering complex structural shapes. They are relatively thin and often aesthetically acceptable and can be given any colored finish. However, the main disadvantages of these coatings can be summarized by the fact that they have a high relative cost when compared to spray-on systems (SFRM). Similar to application of SFRM, appropriate surface preparation is needed prior to installation and requires regular inspections. Also, during inspection process, it can be difficult to verify the existence of intumescent coating (or assess level of applied thickness) at later dates, post-installation.

Exterior Steel (Flame Shields) In certain cases where required fire resistance is low (30 to 60 minutes), steel sheets, also referred to as *flame shields*, wrapped around a structural member can provide some level of fire resistance. Flame shields are intended to reduce the incident radiant heat flux on steel members by preventing direct flame impingement. The effectiveness of flame shields to protect exposed spandrel beams was first examined by Seigel (1970). An example of this method can be seen in a deep steel beam in a 54-story United States Steel Corporation building in New York City, which is shown in Fig. 7.10. In this building, a 14-gauge sheet steel (in terms of lightweight steel cladding) was used as the flame shield.

Flame shields can be placed on standoffs to create an air gap which enhances fire resistance.[5] When used in the form of wrapping sheets (e.g., UL Designs X101, X203, X526) and wire ties (e.g., UL Designs

[5]Revisit Harmathy's rules in Chap. 5.

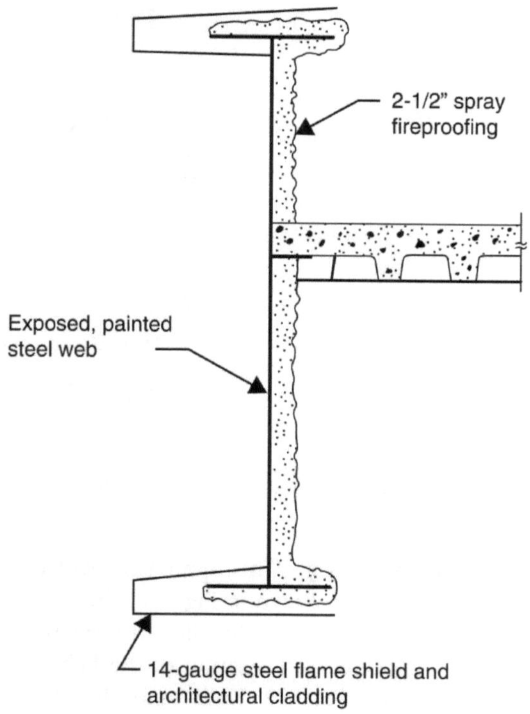

Figure 7.10 Flame shield for fire protection of spandrel girder on the United States Steel Corporation building in New York City.

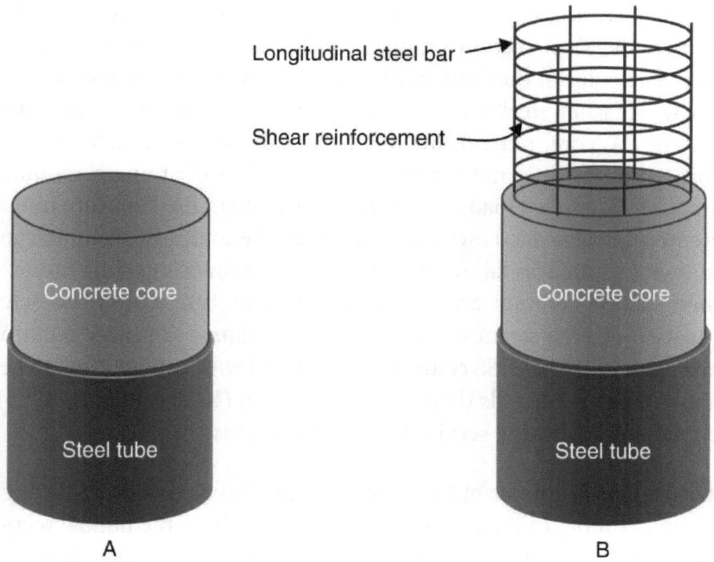

Figure 7.11 Typical illustration of a concrete-filled HSS (CFHSS) column.

Y714, XR614, X204, X205), steel shields provide protective (through radiation based and/or reflection based) shielding that can enhance fire resistance of steel structures. Polished metal sheets have also been used to protect spandrel girders by reflecting the radiated heat away from the spandrel in local exposures such as that encountered in mills and industrial plants.

Capacitive Protection

An alternative to external insulation, *capacitive protection* strategy, can be utilized to enhance fire resistance of certain steel structural shapes. This method works on the principle of utilizing an infill protective material with high heat capacity, to act as heat sink so as to absorb heat from the heated steel member. Two common examples of capacitive protection are concrete filling and water filling in hollow structural section steel members (Lie and Kodur, 1996).

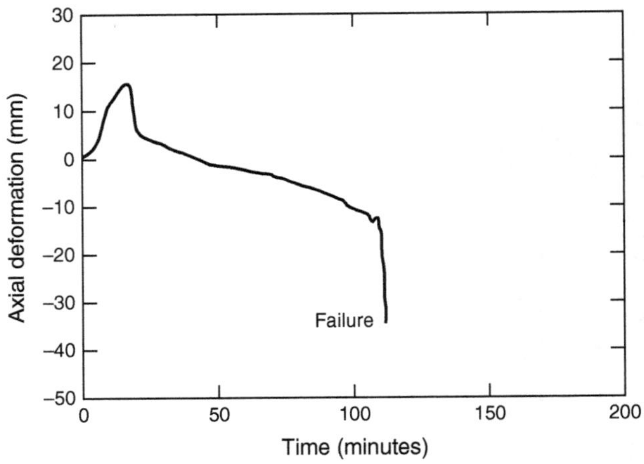

FIGURE 7.12 Behavior of concrete-filled HSS columns (with various concrete types) exposed to fire.

Concrete Filling This approach involves filling interior of a steel HSS (tubular or pipe sections) with concrete as shown in Fig. 7.11. In this arrangement, heat generated in steel resulting from fire exposure gets transferred to the filled material (concrete or liquid).

A typical illustration of the behavior of concrete-filled hollow structural steel (CFHSS) column is shown in Fig 7.12 (Kodur and MacKinnon, 2000). At room temperature, the external load applied on the CFHSS column is resisted by both concrete and steel.[6] When this column is exposed to fire, steel gets heated, but at a slower rate since much of the heat in steel gets dissipated to interior concrete due to the high specific heat of concrete. Initially, steel carries most of the structural load during the early stages of fire, since the steel section (HSS) expands more rapidly than the concrete core. However, with a rise in steel temperature, the steel section gradually yields and the contribution of steel to load-carrying capacity decreases. The column then enters the contraction phase at some point between 20 and 30 minutes after exposure to fire when steel temperature reaches around 500°C. Temperatures within concrete at this point, especially in the core, remain low due to low thermal conductivity and high specific heat of concrete. Therefore, concrete does not experience a large loss of strength modulus and as such can carry higher percentage of the applied load. With further increase in fire exposure time and thus sectional temperatures, concrete strength also decreases and, ultimately, the composite column can no longer support the load. The column then experiences failure through compression or buckling mode.

When CFHSS columns are exposed to fire, moisture within concrete evaporates and turns to steam. Accumulation of such steam within the core of column can cause bursting of HSS sections. Therefore, to minimize such bursting, HSS columns, when filled with concrete, are to be provided with two vent holes of 12.7 mm at regular intervals (heights) of every 3 m (Kodur and Lie, 1995). Attention should be made not to block these holes during service life of the columns.

Water Filling Fire resistance of HSS columns can also be enhanced through filling the hollow section with liquid (water) filling. In the case of liquid (water) filling, the hollow section acts both as the load-bearing structure and as the water container. The liquid used for protection is an aqueous solution (i.e., water-based) that has a number of additives added primarily for overcoming freezing, corrosion, and biological growth. The temperature rise within the HSS section during exposure to fire gets transferred to water, and this keeps the steel section cool for a prolonged duration. During a fire incident, fresh water is continuously circulated through the HSS section and thus the steel section is kept cool for longer duration. A schematic arrangement of components of a water-filled HSS columns, composed of pipes to connect the columns, a water storage tank, and a number of associated valves, is shown in Fig. 7.13. The cooling effect is facilitated through the absorption of heat by water from steel, and then removal of heat through vaporization of heated water and with circulation of fresh water to the HSS. If sufficient heat is delivered to the liquid, vaporization can be expected, which enhances the efficiency of the heat removal process. Since heat is constantly transferred from the steel to the capacitive material, the temperature in the HSS remains low (below 200°C) for a considerable length of time.

[6]Chapter 9 will revisit the load-carrying mechanisms in CFHSS columns in more details.

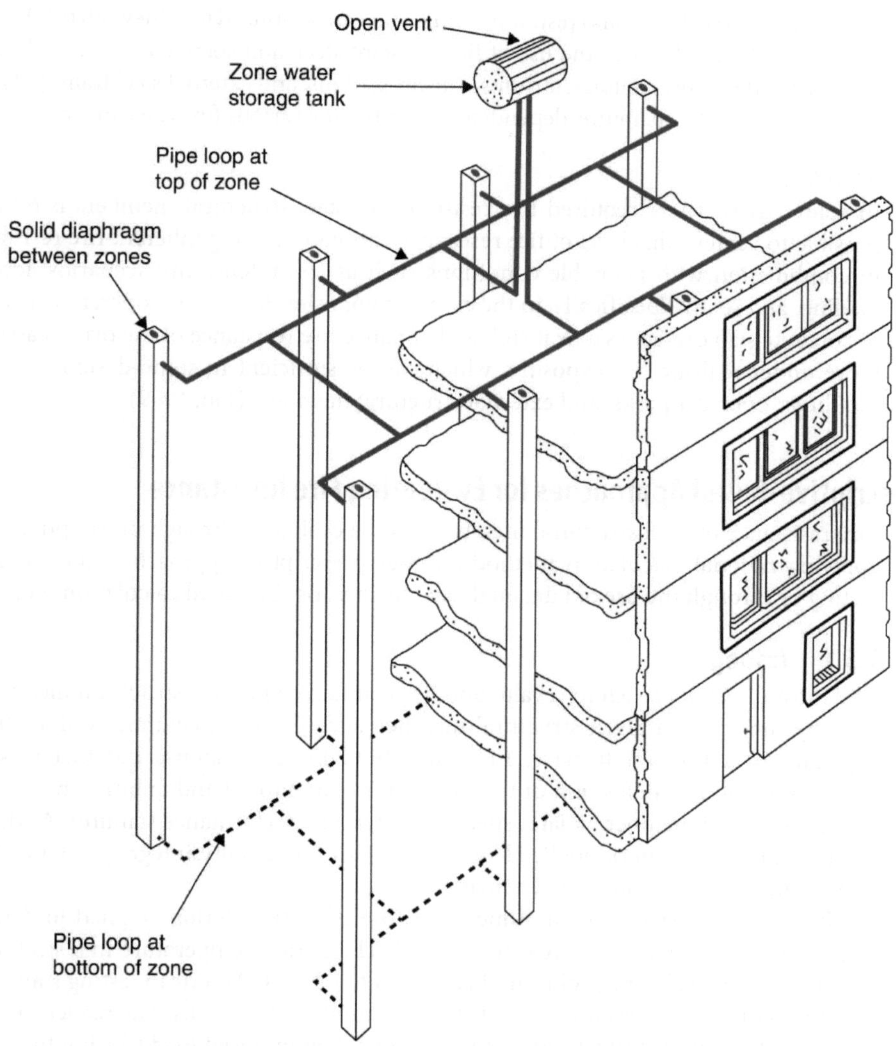

FIGURE 7.13 Schematic arrangement of a liquid-filled-column fire protection system.

There are two different ways to have water filling in the steel sections: permanently filling water in structural members or circulating the water through the member only when a fire breaks out. In the latter system, a fire detection system is required, and once fire is detected, water is circulated through the HSS section. In unreplenished systems, the attainable fire resistance time is limited and depends on the total water content (including any reservoir tank) and on the shape of the heated structure. In systems where the water is constantly being circulated, the fire resistance can be much higher. Water cooling by natural flow is mainly used for vertical or inclined structural members in order to ensure the circulation of water. Due to the complex arrangement needed for this type of fire protection strategy, water filling has been employed only in very limited applications. One example of this application is the 64-story United States Steel Corporation building in Pittsburgh (Bond, 1975).

Fire-Resistant Steel

In the last two decades, there have been efforts to develop *fire-resistant steel*[7] so as to achieve fire resistance without any external fire insulation. In this type of steel, the composition of steel is modified with the addition of additives such as molybdenum chromium, molybdenum, and other alloying elements to improve the yield point at high temperatures. As a result, the yield strength of steel starts at much higher temperatures—close to 600°C. Therefore, fire-resistant steels, as compared to carbon steels, retain higher yield strength at elevated temperatures. Thus, in cases where required fire resistance is in the order of 60 to 90 minutes, the inherent resistance of fire-resistant steel, together with the presence of favorable conditions such as less severe fire

[7]Fire-resistant Grade SM490A steel was developed in Japan and has a yield strength of 325 MPa and a tensile strength of 490 to 610 MPa.

exposure, low load levels, and restraint conditions, can be utilized to achieve 60 to 90 minutes of fire resistance. Japan is in the forefront in the use of fire-resistant steel, and some common applications are in multistory open car parks, sport facilities, atriums, railway stations, and external steel frames (Muratov et al., 2007). Sha (2013) has given temperature-dependent properties for various fire-resistant and heat-resistant[8] steels.

Unprotected Steel

In some cases, where required fire resistance in steel structural members is 60 minutes or lower, it is possible to achieve this level of fire resistance through utilizing inherent fire resistance properties of steel in combination with favorable conditions such as less intense fire scenarios, lower levels of structural loading, and so on. Specifically, in the case of composite steel beam-concrete slab assemblies, the presence of concrete slab can act as a heat sink and enhance fire resistance of the overall assembly to 60 minutes or more under realistic fire exposure, which can be sufficient in some design situations, such as low-rise buildings, open car parks, and external structural members (Liu, 1999).

7.5 Prescriptive-Based Approaches for Evaluating Fire Resistance

Fire resistance of steel structural members can be evaluated through prescriptive-based approaches or by applying rational calculation methods. Under prescriptive approaches, the evaluation can be through testing or through the use of tabulated data, listings, or simplified calculation methods.

7.5.1 Testing

The most direct approach for evaluating fire resistance of a steel structural member is to measure failure time by subjecting a given structural member, such as beam, column, wall, or floor, to a standard fire exposure in a fire test furnace. In such a test, limited response parameters such as cross-sectional temperature and/or deflection of the member are monitored and failure time is evaluated as the time at which the tested member violates specified criteria of performance (failure). As discussed in Chap. 5, the three performance criteria applied to evaluate failure are stability, integrity, and insulation (i.e., temperature rise on the unexposed surface of a member).

In the case of steel structural members, the stability criterion, applied to determine failure, is often linked to *critical temperature* reached in steel. The critical temperature in a steel member varies with the member type (e.g., beams, columns, bar joists, etc.). North American testing standards (e.g., ASTM E119) assume a critical temperature of 538°C for structural steel columns. The reader should note that the failure of a steel structure is influenced by the level of actual imposed load (during fire exposure), member end restraint (axial and rotational), and other factors, as discussed previously. Eurocode 3 provides critical temperature of steel members in a tabulated form, and a sample is listed in Table 7.8. This critical temperature in Eurocode 3 accounts for the level of applied loading.

In recent years, steel stud walls and steel joist floors, fabricated from steel sections (i.e., I-joists or studs), are widely used, sometimes in conjunction with wood construction. These assemblies mainly consist of uniformly spaced load-bearing steel studs and insulated with gypsum wall board or other protective membrane on either side of wall or floor. Fire rating listings for these assemblies are contained in codes such as the National Building Code of Canada (NBCC). Table 7.9 illustrates one such load-bearing steel wall assembly, *S10a*, in which C-shaped steel studs spaced at 400 mm or 600 mm o.c. (off center), filled with 89-mm thick absorptive material, and protected with 15.9-mm type X gypsum board on both sides can achieve a 1-hour fire rating under a standard fire exposure. Similarly, floor assembly, *F40a*, in which a subfloor of 15.5 OSB is laid out on top of C-shaped steel studs spaced at 400 mm or o.c. filled without absorptive material and protected with two 15.9 mm type X gypsum board layers on ceiling side can promote a 1-hour fire rating under a standard fire exposure.

Listings and Tabulated Data

As discussed in Chap. 5, data generated in standard fire tests is utilized to derive fire resistance ratings of structural members and assemblies. The derived ratings are listed in the form of a directory for use in practice, commonly referred to as listings. Each listing represents a certain fire rating assigned to a specific structural member and lists a required insulation thickness, insulation type, structural configuration, and installation specifications to achieve that level of fire rating. Fire resistance derived through listings and tabulated data do not provide realistic fire performance of a steel member, but provide a comparative assessment of fire resistance of the member under standard fire exposure. A typical listing for structural

[8]Heat-resistant steel has a much higher resistance to temperature-induced creep effects under fire conditions.

Member	Critical Temperature (°C)	
	ASTM	Eurocode 3 (Varies with Load Level)
Columns	538	711°C—for load level of 22% 496°C—for load level of 80%
Beams	593	
Open web joists	593	

Table 7.8 Critical Temperature Limit for Evaluating Failure in Steel Structural Members Exposed to Fire (*Reproduced from CEN*)

(a) Walls

Type of Wall	Wall Number	Description	Fire-Resistance Rating		Typical Sound Transmission Class
			Load-bearing	Non-Load-bearing	
• *Load-bearing Steel* Studs • 0.84 mm to 1.52 mm Thickness	S10	• 41 mm × 92 mm load-bearing steel studs spaced 400 mm or 600 mm o.c. • with or without cross-bracing on one side • with or without absorptive material • 2 layers gypsum board on each side	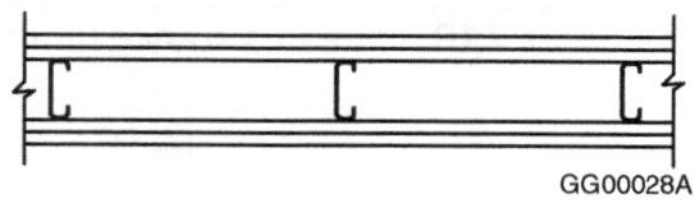		
	S10a	S10 with • 89 mm thick absorptive material • 15.9 mm Type × gypsum board	1 h	–	38
	S10b	S10 with • 89 mm thick absorptive material • 12.7 mm Type × gypsum board	45 min [1 h]	–	38
	S10c	S10 with • 89 mm thick absorptive material • 12.7 mm regular gypsum board	–	–	36

(b) Floors

Type of Assembly	Assembly Number	Description	Fire-Resistance Rating	Typical Sound Transmission Class (STC)	Typical Impact Insulation Class (IIC)
• Cold-Formed-Steel Floor Joists	F40	• subfloor of 15.5 mm plywood, OSB or waferboard, or 17 mm tongue and groove lumber • on steel joists spaced not more than 600 mm o.c. • with or without absorptive material in cavity • 2 layers of gypsum board on ceiling side	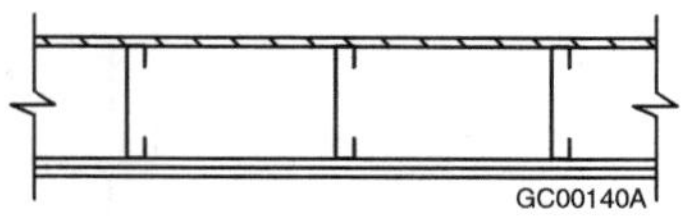		
	F40a	F40 with • steel joists spaced 400 mm o.c. • no absorptive material in cavity • 15.9 mm Type × gypsum board	1 h	34	31
	F40b	F40 with • steel joists spaced 600 mm o.c. • no absorptive material in cavity • 15.9 mm Type × gypsum board	45 min	35	31
	F40c	F40 with • steel joists spaced 400 mm o.c. • absorptive material in cavity • 15.9 mm Type × gypsum board	45 min	36	34

Table 7.9 Sample for Fire Rating for Lightweight Steel Walls and Floors (*Reproduced with Permission of the National Research Council of Canada, Copyright Holder*)

Member	Steel Section	W (Weight)	D* (Heated Perimeter)	W/D	Required Dry Film Thickness (in.) for Fire Rating Duration 60 Minutes	90 Minutes	120 Minutes
I section column (fully exposed)	W6x12	12	26.67	0.45	0.129	0.247	N/A
	W10x39	39	50.00	0.78	0.072	0.134	0.198
	W10x49	49	58.33	0.84	0.067	0.126	0.185
I section beam (with top flange protected concrete)	W6x12	12	22.64	0.53	0.093	0.118	0.192
	W10x39	39	41.94	0.93	0.076	0.099	0.161
	W10x49	49	48.51	1.01	0.072	0.095	0.154

*D (heated perimeter) is to be calculated as per the discussion in Sec. 7.5.3.1.

Table 7.10 Fire Resistance Rating for Structural Steel Columns and Beams from UL Fire Directory

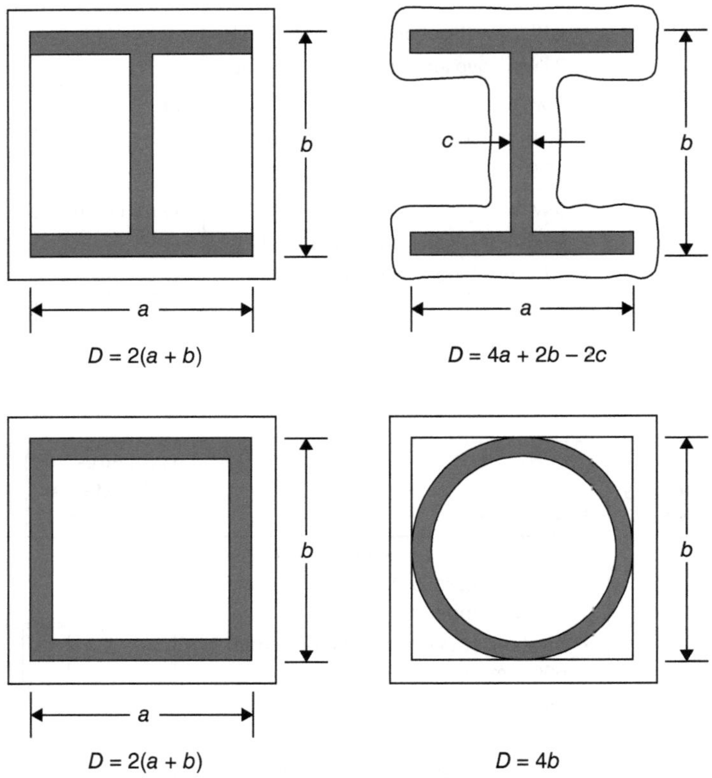

Figure 7.14 Calculation of heated perimeter in fire-protected steel columns with four-side exposure.

steel columns and beams from UL fire directory is shown in Table 7.10 wherein the required dry film thickness (DFT) of intumescent coating taken for various fire ratings is tabulated.

7.5.2 Correlation Methods

Fire resistance of steel structural members derived from listings or tabulated data has numerous limitations in terms of practical applications, given that tabulated data can only be applied for members of similar geometric shapes and material configuration, as well as for the same type of insulation (often proprietary in nature). As such, the use of tabulated data for deriving fire ratings is limited to a narrow range of structural shapes and members. As discussed in Chap. 6, starting in the 1960s and 1970s, data developed from standard fire tests was extended to other structural members through *correlation equations*. Such correlation equations were originally published in trade association manuals, building codes, or directories maintained by testing agencies such as Underwriters Laboratories (UL). In the case of steel structures, these empirical relations mainly correlate fire resistance of a structural steel member to its weight (*W*) and

Structural Shape (metric)	Structural Shape (imperial)	W/D	Required Thickness to Achieve Certain Fire Resistance Rating		
			1 hour	2 hours	3 hours
W100×19	W4×13	**0.79**	25.40	38.10	57.15
W150×22	W6×15	**0.63**	25.40	38.10	57.15
W250×73	W10×49	**1.23**	12.70	28.58	47.63

Table 7.11 Thickness Requirements to Achieve Fire Rating in Steel Columns

heated parameter (*D*), among other influencing parameters. Figure 7.14 illustrates the calculation of heated perimeter (*D*) for different steel shapes insulated in box or contour profile. The weight (*W*) of a steel section per unit weight can be obtained from design manuals, such as the American Institute of Steel Construction (AISC). The heated perimeter depends on the dimensions of steel section, the profile pattern of the protection system, and the number of fire-exposed sides.

The correlation to estimate fire resistance of unprotected or protected steel columns can be arrived at using a ratio of *W*/*D* for wide-flange sections or *A*/*P* for hollow steel sections. The *W*/*D*[9] is the ratio of weight per linear foot (or meter) to the heated perimeter of the steel at the protection interface (or the perimeter of the section if steel is not unprotected).[10] A similar ratio to *W*/*D* is used in a hollow steel section and is referred to as *A*/*P*,[11] which is the ratio of cross-sectional area divided by the heated perimeter.

Steel Columns

A steel section with larger *W*/*D* ratio (or larger *A*/*P* ratio) experiences slower temperature rise in the steel and, hence, requires longer duration to reach critical (failure) temperature limit state. This has been verified through fire tests and observations in real fire incidents. Therefore, fire resistance of a steel member is a function of *W*/*D* or (*A*/*P*) ratios of the member and can be used for extending the fire resistance of a tested steel member to other members.

Membrane Protection Fire resistance of a steel structural member protected with membrane insulation is a function of not only insulation thickness but also the *W*/*D* ratio of the member. Therefore, steel members with higher *W*/*D* ratio will have greater fire resistance for the same level of insulation thickness. Thus, steel sections with a certain *W*/*D* ratio that have a certain fire rating can be substituted by shapes with greater *W*/*D* ratios while maintaining the same thickness of protection used in larger section and listed in fire listings. However, such substitution may lead to uneconomical design, as shapes with larger *W*/*D* ratios require less fire protection material than shapes with smaller *W*/*D* ratios (for the same level of fire resistance). Thus, in order to derive optimum thickness for fire insulation, correlation relations can be employed through interpolating data derived from standard fire resistance tests. Table 7.11 presents one such table showing required (optimum) member thickness to achieve 1-hour, 2-hour, and 3-hour fire resistance ratings in steel columns of varying *W*/*D*. It can be seen from this table that W250×73 section—which has a *W*/*D* ratio of 1.23 that is much larger than that of W150×22 (0.63)—only requires an insulation thickness of 12.7 mm (as opposed to 25.4 mm in W150×22) to achieve a 1-hour fire rating.

In lieu of the above type of tables, correlation equations are also available for evaluating fire resistance of steel columns (ASCE, 1992). One such equation for evaluating fire resistance of steel columns protected with Type X gypsum-based insulation is as follows:

$$R = 1.60\left[\frac{h(W'/D)}{2}\right]^{0.75} \qquad \text{(7.2a—SI units)}$$

$$R = 2.17\left[\frac{h(W'/D)}{2}\right]^{0.75} \qquad \text{(7.2b—FPS units)}$$

[9]For calculation in metric units, the *W*/*D* ratio is often substituted by the M/D ratio, where M is the mass of steel in kg/m and *D* is the heated perimeter in m. $M/D = W/D/0.017$.

[10]It should be noted that W/D-based correlation methods are only applicable for shapes with $W/D \leq 0.215$ ($W/D \leq 3.65$) in International System of Units (French: Système international d'unités, SI) and traditional units, respectively.

[11]$A/P = W/D \times 144 / 490$.

where R is the fire resistance in hours, h is the total thickness of gypsum wallboard in mm (in.), D is the heated perimeter of structural steel column in mm (in.), W' is the total weight of steel column and gypsum wallboard protection in kg/m (lbs/lin ft), $W' = W + 0.0008\,hD$ (SI units) and $W' = W + 50hD/144$ (FPS units), and W is the average weight of steel columns in kg/m (lbs/lin ft).

Contour Protection As mentioned in the "Insulation Protection" section earlier in the chapter, spray-applied fire resistive material (SFRM) can be applied with ease on different shapes, including W-shape pipe and tubular sections as well as other sections, to enhance fire resistance (see Fig. 7.14). The fire resistance of a steel column, protected with SFRM, is a function of thickness of SFRM, its thermal properties, and W/D ratio of steel section. Stanzak and Lie (1973), based on elementary heat transfer principles, proposed an equation to estimate thickness of spray-applied protection material required to achieve a specific fire resistance in steel columns protected with SFRM-type insulation:

$$R = \left(C_1 \frac{W}{D} + C_2\right)h \qquad \text{(7.3a—SI units)}$$

$$R = \left(C_3 \frac{W}{D} + C_4\right)h \qquad \text{(7.3b—FPS units)}$$

where R is the fire resistance in hours, h is the thickness of spray-applied fire resistive material in mm (in), D is the heated perimeter of structural steel column in mm (in.) as calculated in Fig. 7.14, W is the average weight of steel column in kg/m (lbs/lin ft), and C_1 (or C_3) and C_2 (or C_4) are the material-dependent constants for wide-flange columns (SFRM). The constants C_1 and C_2 (or C_3 and C_4) are to account for thermal properties of SFRM which vary with the type of SFRM being applied and is generally provided by SFRM manufacturers or in design standards.

The values of constants in Eq. 7.3 are established based on data from at least two fire resistance tests carried out on two different column sizes (smallest and largest sizes in that shape) exposed to standard fire in accordance with *Appendix P* of the Standard Building Code (1988). In deriving values for constants, for the smallest column the first fire test is intended to develop a minimum fire rating and the second fire test on a smaller column is to quantify the maximum thickness needed of fire protection material. For a larger column, the first test is conducted to achieve a minimum intended thickness of fire protection material and the second test to develop a maximum desired fire resistance rating for a steel column. The constants C_1, C_2, C_3, and C_4 can also be provided by the manufacturer.

The above correlation equations are generally applicable in evaluating fire resistance for most hot-rolled steel shapes under standard fire exposure. However, the following limitations apply for the use of the above empirical equations in evaluating fire resistance. Equations 7.1 and 7.2 are applicable to a given column when its W/D ratio lies between the W/D ratio of the smallest tested column and the largest tested column, as well as when the calculated thickness of insulation lies between maximum and minimum thickness provided in the tested columns. Further, these equations are only applicable for wide-flange columns and steel shapes with open cross section (channels, angles, and structural tees).

Concrete or Masonry Protection Encasement of steel structural members with concrete or masonry significantly enhances fire resistance of steel members due to their good fire resistance properties. Empirical equations for evaluating fire resistance of wide-flange (W-shaped) steel beams or columns with concrete or masonry encasement are available in ASCE 29 (2005). These equations were originally derived by Lie and Harmathy (1974). Depending on the thickness of encased masonry or concrete, steel structural members can achieve 1 to 4 hours of fire rating.

In deriving these expressions, failure was said to occur in steel columns when steel attains a critical (failure) temperature of 538°C as per ASTM E119 limiting criterion. Fire resistance of steel columns protected with concrete (including precast concrete) or masonry, as specified in ASCE 29 (2005), is given as

$$R = R_0(1 + 0.03m) \qquad \text{(7.4)}$$

where R_0 is calculated as

$$R_0 = 0.17\left(\frac{W}{D}\right)^{0.7} + 0.28\frac{h^{1.6}}{k_c^{0.2}}\left[1 + 26\left(\frac{H}{\rho_c c_c h(L+h)}\right)^{0.8}\right] \qquad \text{(7.4a—SI units)}$$

$$R_0 = 1.22\left(\frac{W}{D}\right)^{0.7} + 0.0027\frac{h^{1.6}}{k_c^{0.2}}\left[1 + 31{,}000\left(\frac{H}{\rho_c c_c h(L+h)}\right)^{0.8}\right] \quad \text{(7.4b—FPS units)}$$

where R = Fire resistance (in hours) at equilibrium moisture conditions
R_0 = Fire resistance (in hours) at zero moisture content
m = Equilibrium moisture content of concrete, by volume (%)
D = Heated perimeter of the steel column, m (in.)
h = Thickness of concrete or equivalent thickness (for masonry) in m (in.)
k_c = Ambient temperature thermal conductivity of concrete in W/m·K (Btu/hr·ft·°F)
H = Ambient temperature thermal capacity of steel column = 0.46 W·kJ/m·K (0.11 W Btu/ ft·°F)
ρ_c = Concrete or masonry density in kg/m³ (lbs/cu.ft)
c_c = Ambient temperature specific heat of concrete or masonry in kJ/kg·K (Btu/ft·°F)
L = Interior dimension of one side of square concrete or masonry box protection in m (in.)

Refer to Figure 7.15 on how to calculate some of the sectional dimensions.

The above equations are developed based on the assumption that concrete/masonry provides only thermal insulation and does not provide any contribution to load-carrying capacity.

In recent years hollow concrete or masonry units are gaining popularity to reduce dead weight of concrete/masonry (see Fig. 7.15). The use of such units can accelerate pace of construction as these units are often prefabricated and do not require formwork or in situ casting.

Notes

1. If the box protection is not square, or if the concrete cover thickness is not constant, h and L shall be taken as the average value, i.e., $h = ½(h_1 + h_2)$ and $L = ½(L_1 + L_2)$.

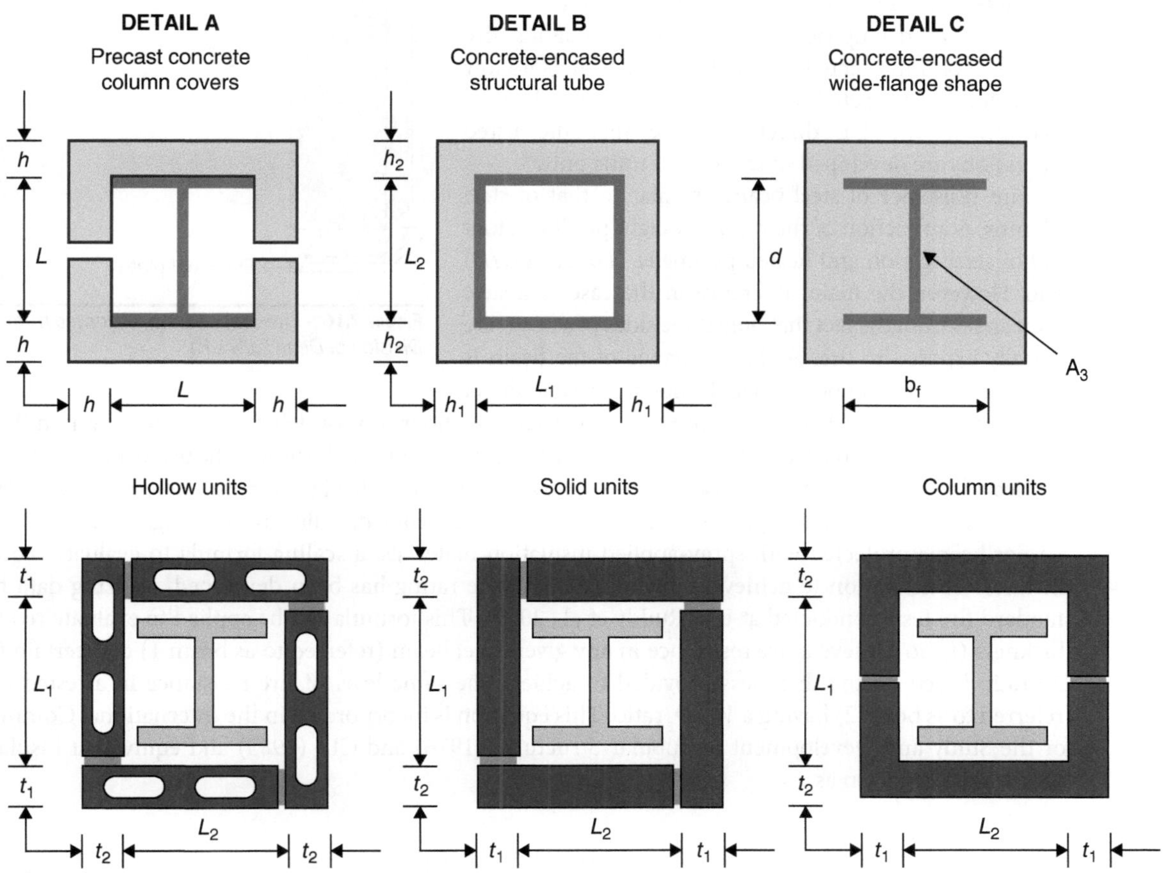

Figure 7.15 Fire protection through concrete-encased or hollow concrete (or masonry units).

2. If the steel column is completely encased in concrete, with all reentrant spaces filled, the thermal capacity of the concrete within the reentrant space may be added to the thermal capacity of the steel column, increasing the value of H to

$$H = 0.46W + \frac{\rho_c c_c}{1{,}000{,}000}(b_f d - A_s) \qquad \text{(7.5a—SI units)}$$

$$H = 0.11W + \frac{\rho_c c_c}{144}(b_f d - A_s) \qquad \text{(7.5b—FPS units)}$$

where b_f is the flange width of the steel column in mm (in.), d is the depth of the steel column in mm (in.), and A_s is the cross-sectional area of the steel column in mm² (in²).

i. $L = ½(L_1 + L_2)$

ii. h in equation is based on the equivalent thickness of the masonry unit:

a. If solid, h = lesser of (t_1, t_2).

b. If hollow, h = {lesser of (t_1, t_2)} × {percent solid of the unit} (where percent is expressed as a decimal).

Another option available for achieving required fire resistance is through the use of the *Fire-Trol column* (Fire-Trol, 2006). The Fire-Trol column, shown in Fig. 7.16, is a prefabricated, fireproofed unit consisting of a load-bearing steel column (A), encased in a special proprietary insulating material (B), which is permanently protected by an outer non-load-bearing steel shell (C). Fire-Trol columns are rated (prescriptive requirements) and labeled by Underwriters Laboratories, Inc., for fire retardant classifications of 2, 3, and 4 hours. Columns are completely shop-fabricated and shipped to the job site ready to erect.

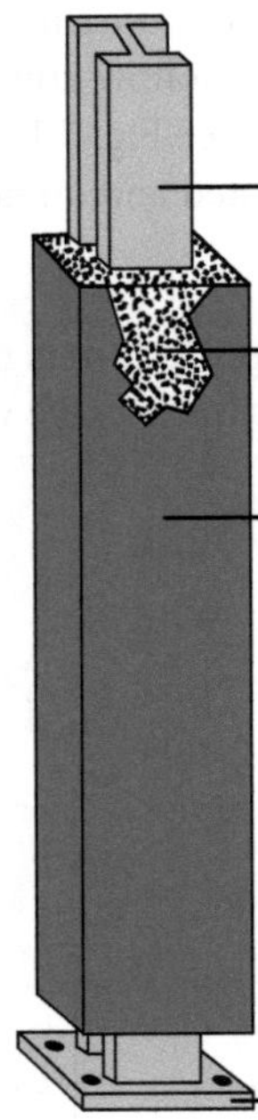

FIGURE 7.16 Fire-Trol column. (*Adapted from Division of Dean Lally L.P.*)

Table 7.12 provides a summary of some of the commonly used empirical equations for evaluating fire resistance of unprotected steel columns as well as hollow steel columns, protected with SFRM or board products as specified in the UL directory. Note that the listed expressions are developed in traditional units only.

Fire resistance of steel beams, similar to that of steel columns, is a function of the ratio of weight per lineal foot (W) of steel section and heated perimeter (D), i.e., (W/D) ratio. However, the main difference in the case of a steel beam arises from the fact that only three sides of a beam are typically exposed to fire, as the top flange of the beam is usually attached to a floor or slab. Hence, there is no direct (or negligible) heat exchange from fire surface between the floor or roof slab and the steel beam on the top surface. This is due to the fact that roof or slab normally absorbs some of the heat in the beam; the time to reach a critical (failure) temperature in steel beams is relatively longer than that in columns. Therefore, in calculation of heated perimeter, D, only three sides of the beam are to be included, as illustrated in Fig. 7.17.

For beams protected with spray-applied insulation materials, a scaling formula to evaluate required thickness of insulation to achieve a given fire resistance rating has been developed, utilizing data from standard fire tests conducted at UL (Ruddy et al., 2003). This formula can be applied to evaluate required thickness (t_1) to achieve a fire resistance in any given steel beam (referred to as beam 1) of a certain (W_1/D_1) ratio based on the thickness provided to achieve the same level of fire resistance in a tested beam (referred to as beam 2) having a W_2/D_2 ratio. This equation is incorporated in the International Committee for the Study and Development of Tubular Structures (1976) and CIB (1983) and equivalent insulation thickness (h_1) is given as

$$h_1 = \left[\frac{W_1/D_1 + 0.036}{W_2/D_2 + 0.036}\right] h_2 \qquad \text{(7.6a—SI units)}$$

Member/Protection Type	Expression*	Remarks
Unprotected steel column	$R = 10.3\ (W/D)0.7$, for $W/D < 10$ $R = 8.3\ (W/D)0.8$, for $W/D \geq 10$	Valid for critical temperature of 1000°F
Hollow steel columns protected with SFRM or board products	$R = C_1\ (A/P)h + C_2$	C_1 and C_2 are constants for specific insulation materials. A/P is the ratio of a circular pipe calculated as $A/P_{pipe} = t(d - -t)/d$ or rectangular or square tube calculated as $A/P_{tube} = t(a + b - 2t)/(a + b)$ where: D is the outer diameter t is the wall thickness a is the outer width of tube b is the outer length of tube

*These expressions are for use in traditional units.

Table 7.12 Empirical Relations for Evaluating Fire Resistance (R) of Steel Columns (UL, 2007)

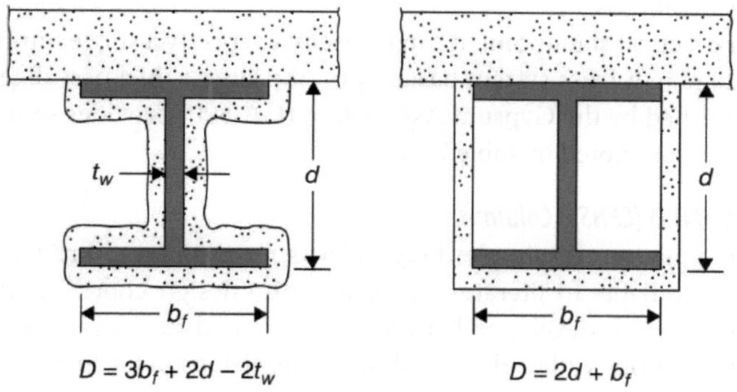

Figure 7.17 Heated perimeter calculation in insulated steel beams. (*AISI, 1985.*)

$$h_1 = \left[\frac{W_1/D_1 + 0.60}{W_2/D_2 + 0.60}\right] h_2 \qquad \text{(7.6b—FPS units)}$$

where h is the thickness of insulating spray in mm (in.).[12]

Knowing W_2/D_2 and h_2 of the tested beam for a given fire resistance rating, together with W_1/D_1 of the steel beam for which the resistance is required, insulation thickness to achieve a rating similar to that in the tested beam can be obtained.

Steel Trusses

Three categories of trusses are commonly used in buildings: transfer, staggered, and interstitial trusses. Because of the inherent features of each type of truss, certain insulation schemes are more appropriate than others (see Table 7.13). Fire protection schemes often used on trusses can be broadly grouped as membrane, envelope, and individual member protection. Membrane and envelope protection is installed through the use of a fire-resistant ceiling assembly where design parameters for such an assembly can be

[12]Subscript 1 refers to the desired beam size and required material thickness of the new member, while subscript 2 refers to the given beam size and material thickness shown on listings for the tested member. Further, this equation is limited to beams with $W/D > 0.022$ kg/m^2 (0.37 lb/ft·in.) and thickness of protection >9.5 mm (3/8 in.).

Truss Type	Membrane	Envelope	Individual (Single) Element
Transfer		✓	✓
Staggered		✓	✓
Interstitial	✓	✓	✓

Table 7.13 Typical Fire Protection Methods for Steel Trusses

Fire Rating (hour)	Gypsum Type X (mm)	SFRM Insulation Thickness (mm)
1	16	16
2	32	—
3	—	38

Table 7.14 Guidelines for Gypsum Wallboard Thickness in Steel Trusses (ASCE, 1992)

determined from listings of fire-rated designs. Spray-on-type insulation is applied on steel truss members with smaller cross sections and very slender shapes. In most cases, fire resistance of steel trusses, protected with certain insulation thickness, can be evaluated by applying similar empirical equations as that for steel columns.

From a practical standpoint, it can be useful to specify a single minimum thickness of protection for all truss members in a truss system based upon the largest thickness determined. The fire resistance design manual published by the Gypsum Association (GA-600) provides some practical rules of thumb based on test results and are noted in Table 7.14.

Concrete-Filled HSS (CFHSS) Columns

Correlation equations or simple design charts for evaluating fire resistance of HSS sections filled with concrete are available in literature as well as fire design codes. As these correlation equations rely on the composite action developed between concrete and steel and take applied loading into account, these correlation equations will be discussed in further details in Chap. 9.

7.6 Rational Approaches for Evaluating Fire Resistance

The Section 7.5.2 correlation equations for evaluating fire resistance, although simple and easy to apply, have numerous drawbacks and are limited in achieving innovative engineering design. An alternative to these prescriptive-based equations and tabulated data is through applying *rational engineering approaches* wherein many of the drawbacks, such as not accounting for realistic fire loading, restraint conditions, and failure limit states, can be overcome. Fire resistance analysis can be carried out by applying simplified or advanced analysis procedures. The level of complexity in rational engineering approaches increases with the type of analysis (simplified equations or advanced analysis) adopted and with the number of variables considered in such an analysis.

7.6.1 Simplified Approaches

Fire resistance analysis through simplified approaches involves applying simplified equations for estimating temperature or capacity degradation in a steel structural member, ideally for evaluating sectional-level response. Fire resistance evaluation can be carried out at a given fire exposure time (e.g., 60 minutes) or in several time increments (e.g., 5 minutes, 10 minutes, etc.) of fire exposure until failure occurs in a structural member.

General Procedure

Fire resistance evaluation through simplified approaches can be quite involved, compared to perspective-based approaches, since it involves evaluation of thermal and structural response parameters at a given time-step. The analysis is to be carried out in two stages: thermal and structural. Thermal analysis involves generating cross-sectional temperatures in a structural member at a given fire exposure time. These sectional temperatures can be compared against thermal failure limit state (i.e., critical temperature) to check if the steel member fails (e.g., when cross-sectional temperature exceeds *critical temperature*). The cross-sectional temperature can also be utilized to evaluate reduction in strength and modulus properties of steel at that fire exposure time. Knowing reduction in strength and modulus properties, reduced sectional capacity (i.e., moment, shear, or axial, etc.) or deflection of the steel structural member can be evaluated. Then, these response parameters are compared against applicable limiting values corresponding to different limiting states (strength or deflection) to evaluate failure. Failure occurs when capacity or deflection values at that time-step exceed the limiting criterion. If the analysis is carried out at a series of time-steps, the time to reach the preceding time-step, at which failure occurs, is taken as fire resistance of the member. If the failure is not reached at that time-step, the time is further incremented and the analysis is repeated. A flowchart illustrating steps associated with fire resistance analysis of a typical structural member is shown in Fig. 6.3.

Revisiting Chap. 6, strength limit state can also be applied to evaluate failure of a structural member. Depending on the member type, effects arising from external loading, U_{fire}, which can be axial force, bending moment, or shear force, or even a combination of these forces, are compared against reduced capacity, R_{fire}, of the fire-exposed structural member. The resistance of the member at any time-step, axial capacity, moment, and/or shear capacity can be evaluated by applying the same equations as that used in ambient temperature design, but with due consideration to degradation in strength and modulus properties in steel, and also consideration to other fire-induced effects (i.e., restraint force, local buckling).

Thus, in strength analysis, once loading effects, U_{fire} (i.e., axial, bending, or shear force), present on the member during fire exceed the corresponding reduced capacity, R_{fire}, of the fire-exposed structural member, failure is said to have attained such that

$$U_{\text{fire}} \leq R_{\text{fire}} \tag{7.7}$$

Approach for Evaluating Sectional Temperature

There are number of simplified approaches for estimating cross-sectional temperature in steel. Alternatively, advanced analysis methods such as finite difference and finite element can also be applied for evaluation of sectional temperature in steel member.

Best-Fit Method Temperatures in a steel section, in its crudest form, can be determined by applying the *best-fit equation*. Two sets of empirical equations, one for unprotected steel section and the other for protected steel section, were developed for estimating failure in a steel structural member exposed to a standard fire through correlation of temperature data generated in fire tests carried out on steel structural members. Accordingly, the time to failure (t) in an unprotected steel structural member exposed to fire is expressed in terms of surface area and volume of section such that

$$t = \frac{0.54(T_{\text{lim}} - 50)}{\left(F/V\right)^{0.6}} \tag{7.8}$$

where F is the surface area of unit length (m^2), V is the volume of steel in unit length (m^3), t is the time to reach a limit temperature,[13] and T_{lim} is the limiting temperature in steel as per ECCS (1985) (in the 400 to 600°C range). This equation is valid for F/V (section factor) value ranging from 10 to 300 m^{-1} or V/F from 3.3 to 100 mm and when in steel members results in temperatures in the 400 to 600°C range. The calculation of F/V ratios for various steel sections is illustrated in Fig. 7.18.

Equation 7.8 can be rearranged to express limiting temperature in terms of failure time and F/V ratio:

$$T_{\text{lim}} = \frac{t\left(F/V\right)^{0.6}}{0.54} + 50 \tag{7.9}$$

The above expression is for the case of unprotected steel sections. When steel members are provided with fire protection, insulation properties and their thickness will have significant influence on temperature rise in steel in addition to F/V values for steel sections. To account for this, Eq. 7.9 is modified for protected

[13] This method is valid for failure times for unprotected steel members of 10 to 80 minutes.

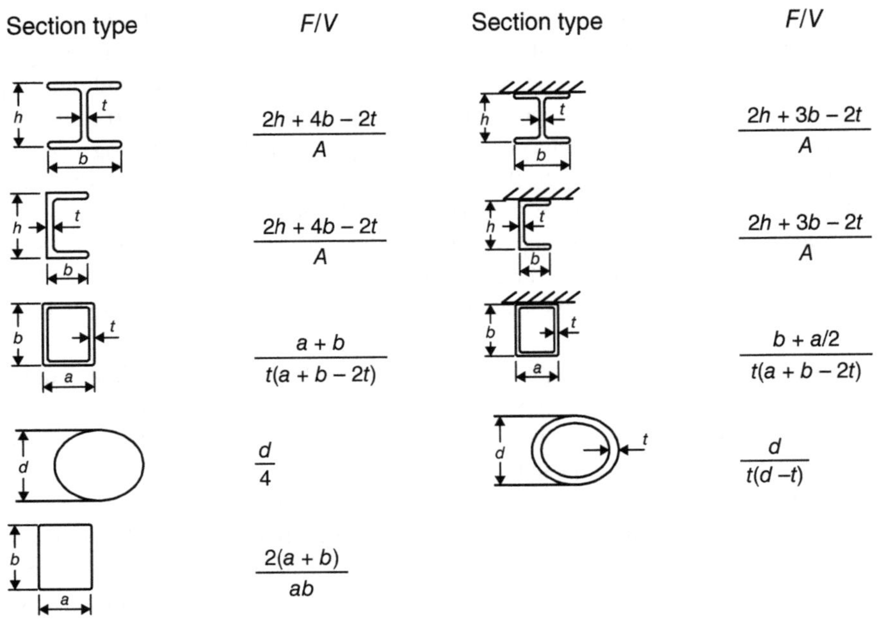

Figure 7.18 Calculations of *F*/*V* ratio for various steel sections.

steel sections, and the time to reach a limiting temperature in a steel member protected with certain insulation thickness is given as

$$t = 40(T_{\text{lim}} - 140)\left(\frac{d_i / k_i}{F/V}\right)^{0.77} \tag{7.10}$$

where k_i is the thermal conductivity of insulation (W/m·K) and d_i is the thickness of insulation (m). This equation is valid for fire exposure ranging from 30 to 240 minutes with limiting steel temperature of 400 to 600°C. *F*/*V* and d_i/k_i range from 10 to 300 m^{-1} and from 0.1 to 0.3 m^2·K/W, respectively.

In certain insulation types (such as SFRM), significant moisture can be present in insulation material throughout the service life of the structure. Such moisture will have a positive contribution in delaying temperature rise in steel and thus enhancing fire resistance of steel members. As a result, additional time delay (t_v), arising from moisture in insulation, to reaching a limiting temperature can be accounted for through adding t_v to t calculated in Eq. 7.10. t_v is evaluated as

$$t_v = m\rho_i d_i^2 / 5k_i \tag{7.11}$$

where ρ_i is the density of the insulation (kg/m^3) and m is the moisture content (%). The time delay t_v is to be added to time t from Eq. 7.10 to arrive at total failure time (fire resistance).

The *best-fit* equation is applicable for standard fire exposures only and cannot be used for evaluating steel temperatures when the member is subjected to a much different fire scenario than a standard fire. The reader should also note that the best-fit method was developed for estimating average steel temperature in proximity to critical temperature of steel, which corresponds to 50% strength loss in steel. Further, this method assumes uniform temperature across the depth of cross section and thus no estimation of temperature gradient can be made. Thus, the temperature predicted by best-fit method may not be reliable at temperatures beyond roughly ±25% of critical temperature of steel, which is 550°C.

Lumped Heat Capacity Approach Another simplified method for evaluating temperature rise in fire-exposed steel sections is through applying *lumped heat capacity approach.* In this approach, commonly referred to as *step-by-step method*, temperature in a steel section at a given time-step is evaluated by treating steel as lumped mass and assuming the temperature in the overall mass (of section) to be uniform. At each time-step, heat entering steel over an exposed surface area during a small time increment (Δt) is equal to the heat required to raise the temperature of the steel by ΔT_s (°C). This assumption of uniform steel temperature is a reasonable approximation, especially when calculating temperature of unprotected or protected steel sections exposed to fire from all four sides (due to high thermal conductivity of steel) such that heat entering a steel member equals to the heat utilized to raise temperature in a steel member.

Since steel is directly exposed to fire, as in the case of unprotected steel sections, heat transfer from fire zone to steel member is through radiation and convection. Therefore, heat balance equation can be expressed as

$$mc\frac{dT_s}{d_t} = \alpha A \Delta T \tag{7.12}$$

where $\Delta T = T_f - T_s$ is the temperature difference between fire and steel temperature. Dividing both sides by mc, the above equation transforms into[14]

$$\frac{\Delta T_s}{\Delta t} = \frac{\alpha A}{\rho c V}(T_f - T_s) \tag{7.13}$$

Knowing that D = perimeter of exposed steel, $A = DL$ and $V = A_c L$, where L is the linear length of the steel member, Eq. 7.12 can be rearranged as

$$\frac{\Delta T_s}{\Delta t} = \frac{\alpha}{\rho c \frac{A_c}{D}}(T_f - T_s) \tag{7.14}$$

Substituting for $W = \rho A_c$, the above equation becomes

$$\frac{\Delta T_s}{\Delta t} = \frac{\alpha}{c\frac{W}{D}}(T_f - T_s) \tag{7.15}$$

Thus

$$\Delta T_s = \frac{\alpha}{c\frac{W}{D}}(T_f - T_s)\Delta t \tag{7.16}$$

where α represents heat transfer to the steel from fire surface arising from convection (h_c) and radiation (h_r) components. After substituting for h_c and h_r, α translates to $h_c + h_r$.

$$\alpha = h_c + h_r \tag{7.17}$$

$$\alpha = h_c + \sigma\varepsilon(T_f^2 - T_s^2)T_f - T_s \tag{7.18}$$

where h_c is the convective heat transfer coefficient (25 and 50 W/m^2·K for standard and hydrocarbon fires), σ is the Stefan-Boltzmann constant (56.7×10^{-12} kW/m^2·K^4), ε is the resultant emissivity (0.5–0.67), T_f is the temperature in the fire environment (K), and T_s is the temperature of steel (K).

Substituting for and expressing W/D in terms of F/V, Eq. 7.18 takes the following form:

$$\Delta T_s = \frac{F}{V}\frac{k_{sh}}{\rho_s c_s}\left\{(h_c(T_f - T_s) + \sigma\varepsilon(T_f^4 - T_s^4))\right\}\Delta t \tag{7.19}$$

where k_{sh} is a correction factor to account for the shadow effect.

The above procedure can also be applied to deduce equations for temperature rise in steel members protected with fire insulation. In this scenario, the main difference is in the mechanism of heat transfer from fire zone to steel surface. In protected steel sections, heat transfer first occurs from fire source to insulation through radiation and convection. However, steel section experiences temperature rise resulting from heat transfer from fire insulation, and this occurs through conduction. For simplicity, it can be assumed that the outermost layer of insulation has a similar temperature to that of the fire source. The thickness of insulation, together with thermal properties, is the main parameter influencing the temperature

[14] $m = pV$

rise in steel. Thus, the use of radiative and convective heat transfer parameters may be ignored as seen in the following equation:

Since

$$V_s = AL, \text{ and } V_i = DhL \tag{7.20}$$

and

$$W = \rho A \tag{7.21}$$

and if

$$\frac{c_s W}{D} > 2h\rho_i c_i \tag{7.22}$$

Substituting the above parameters in Eq. 7.16 transforms into

$$\Delta T_s = \frac{F}{V}\frac{k_i}{h\rho_s c_s}\left\{\frac{\rho_s c_s}{\rho_s c_s + \frac{F}{V}\frac{\rho_i c_i h}{2}}\right\}(T_f - T_s)\Delta t \tag{7.23}$$

where h is the thickness of the insulation, ρ_i is the density of insulation (kg/m^3), and c_i is the specific heat of insulation (J/kg·J).

Knowing the temperature rise in steel at a given time-step ΔT_s, the resulting temperature in steel until that time-step (from the beginning of the fire) can be evaluated by adding ΔT_s to steel temperature T_{s-1} from the preceding time-step.

In the lumped heat capacity approach, unlike the *best-fit method*, temperature rise in steel can be evaluated in time increments throughout the fire exposure duration, thus making the temperature procedure more accurate. The calculation can be set up in a spreadsheet. Still, one of the main drawbacks of this method is that it assumes the temperature in the steel section to be uniform across for the lumped mass. This may lead to slightly conservative temperatures, especially in deep sections or those exposed to partial fire exposure, such as the presence of a concrete slab, since it is not able to account for thermal gradients.

Eurocode 3 Equation for Evaluating Temperature in Steel Members Eurocode 3 has adopted similar equations for unprotected and protected steel sections, similar to that in Eq. 7.19, and are given as

$$\Delta T_s = 0.9k_{sh}\frac{F/V}{\rho_s c_s}h_{net}\Delta t \qquad \text{for unprotected steel} \tag{7.24}$$

$$\Delta T_s = \frac{F}{V}\frac{k_i}{\rho_s}\frac{1}{\rho_s c_s}\frac{(T_f - T_s)}{(1+\phi/3)}\Delta t - (e^{\phi/10} - 1)\Delta T_a \qquad \text{for protected steel} \tag{7.25}$$

where h_{net} is the design value of the net heat flux per unit area (W/m^2) and $\phi = \frac{F}{V}\frac{d_i \rho_i c_i}{\rho_s c_s}$, where c_i is the specific heat of the insulation (J/kg·K), ρ_i is the density of the insulation (kg/m^3), k_i is the thermal conductivity of the insulation (W/m·K), d_i is the thickness of the insulation (m), and ΔT_a is the increase in the ambient gas temperature during the time interval Δt (K). It can be seen that ϕ is influenced by the thermal properties of insulation. In the case where fire insulation is of low mass and specific heat, heat capacity will tend to be small and then ϕ can be zero.

Eurocode 3 Approach for Evaluating Temperature in Steel Connections Eurocode 3 presents a simplified approach to evaluate the temperature in steel connections in beam-to-column and beam-to-beam connections such that

$$T = 0.88T_{bf}\left(1 - 0.3\left(\frac{h}{D}\right)\right) \tag{7.26}$$

where T_{bf} is the bottom flange temperature of the steel beam remote from the connection, h is the height of the component being considered above the bottom of the beam in mm, and D is the depth of the beam in mm. This equation is valid for beams of a depth <400 mm.

Dwaikat and Kodur Approach Recently, Dwaikat and Kodur (2012) proposed another simplified approach for evaluating cross-sectional temperature in unprotected or protected steel members subjected to either standard or design fire. This approach, which utilizes many of the assumptions used in the lumped heat capacity approach (LHCA), was derived by applying basic heat transfer principles to evaluate steel temperature at a specified time (in one time-step)—without the need for evaluating incrementing temperature at a series of time-steps. The equation for evaluating temperature in steel section is given as

$$T_s(t) = T_f(1 - e^{-st}) \tag{7.27}$$

where $T_s(t)$ is the steel temperature at a specified time (t), T_f is the fire temperature at that time (t), and s is a parameter calculated as

$$s = \frac{F_1}{n+1} = \frac{(F_p/A_s)}{c_s\rho_s(1/h + t_p/k_p)\left(1 + \frac{c_p\rho_p}{c_s\rho_s}\frac{F_p}{A_s}\frac{t_p}{m}\right)(n+1)} \tag{7.28}$$

where F_p is the heated perimeter of the cross section (or the steel plate), ρ_p is the density of insulation (kg/m^3), A_s is the cross-sectional area of the section (or the steel plate), h is the convective heat transfer coefficient (25 to 50 W/m^2·K for standard and hydrocarbon fires), m is the constant used for averaging temperature of insulation layer around steel section, n is the regression coefficient for fitting growth phase of fire scenarios ($n = 0.1677$ and 0.1478 for ASTM E119 and ISO 834, respectively), c_s, c_p are specific heat of steel and insulation (J/kg·J), respectively, and t_p is the insulation thickness.

This approach accounts for all critical factors influencing temperature rise in steel including thermal properties of insulation and steel, thickness of insulation, convection, and radiation heat transfer coefficients. This approach is capable of estimating steel temperature for the whole section or in parts of the section, which enables accounting for fire-induced thermal gradients arising in steel section exposed to fire from one, two, or three sides. Further, it has been shown that this approach better estimates steel temperature than the best-fit method and LHCA for the entire range of fire exposure time. This equation can be applied to all fire exposure scenarios including those with a decay phase. Perhaps one of the strengths of this approach is that steel temperature can be evaluated in time-steps, without the need for spreadsheet calculations, as it does not require iterative processes (i.e., spreadsheets).

Knowing temperature in steel (T_s), fire resistance of a steel structural member can be evaluated by applying the thermal failure limit state in which T_s can be checked against limiting temperature (e.g., 550°C). Alternatively, once T_s is determined from one of the above approaches, this temperature is then input into the strength analysis to examine the structural response of steel members, as detailed in the following section.

Approach for Evaluating Sectional Capacity

Strength limit state can be applied to evaluate failure of a steel member at sectional or member level. At the sectional level, reduced sectional capacity at a critical section at any given fire exposure time resulting from reduction in material properties due to temperature is evaluated. As an example, the reduced moment capacity (M_f) at a critical section of a fire-exposed steel beam can be evaluated by applying the room temperature capacity equation, but with due consideration to temperature-induced strength degradation in steel.

$$M_f = Zk_{y,T}f_y \tag{7.29}$$

where f_y is the yield strength of steel at room temperature in MPa, Z_x is the plastic section modulus of the section in mm^3, and $k_{y,T}$ is the reduction factor of yield strength of steel at temperature T (which can be evaluated as detailed in the previous section). Knowing the temperature in steel at any given fire exposure, the reduction factor for yield strength can be obtained. Then, the reduced moment capacity can be obtained by extending room temperature equations in design codes and standards. A similar procedure can be applied to evaluate fire-induced degradation in shear capacity of beams or axial capacity in compression members (columns). The reduced moment (or shear or axial) capacity is compared against the effect of actions (bending moment, shear force, or axial force) to determine if the fire-exposed member has attained failure or not at a given time-step.

Eurocode 3 Equations *Eurocode 3* presents a detailed procedure for fire resistance evaluation of structural members based on strength limit state. Relevant equations for evaluating the resistance of a member in bending or in shear mode, which is an extension of ambient (room temperature) equations, are given in Eurocode 3. The design moment resistance of a section with a uniform temperature is given by[15]

$$M_{fi,\theta,Rd} = k_{y,\theta}\left[f_y/\gamma_{M,fi}\right]W \tag{7.30}$$

where W is the plastic modulus of the section W_{pl} for a Class 1 or a Class 2 section or the elastic modulus of the section W_{el} for a Class 3 section. $\gamma_{M,fi}$ is the partial safety factor under fire conditions, which is taken as 1.0.

In the case of shear failure limit state, Eurocode 3 provides the following equation to evaluate reduced shear capacity at a given temperature in a steel beam subjected to fire conditions:

$$V_{fi,t,Rd} = k_{y,\theta,web} V_{RD}\left[\gamma_{M,0}/\gamma_{M,fi}\right] \tag{7.31}$$

where θ, web is the average temperature in the web of the steel section, $k_{y,\theta,web}$ is the reduction factor for the yield strength of steel at the average temperature of the web, and V_{RD} is the shear resistance of the gross cross section for normal temperature design according to Eurocode 3.[16]

In the case of a steel column, factors relating to buckling are to be considered in a fire situation. If the column is a continuous member that extends through several floors of a braced building and if each story forms a separate fire compartment, then the buckling length of a column exposed to fire in an intermediate story may be taken as $l_{fl} = 0.5L$ and in the top story as $l_{fl} = 0.7L$, where L is the system length in the story that is under fire. The design buckling resistance $N_{b,fi,t,Rd}$ of a member in compression with a uniform sectional temperature θ_a is given as

$$N_{b,fi,\theta,Rd} = \chi_{fi} A k_{y,\theta} f_y/\gamma_{M,fi} \tag{7.32}$$

where χ_{fi} is the buckling coefficient, A is the cross-sectional area of the compression member, $k_{y,\theta}$ is the reduction factor for yield strength of steel, and $\gamma_{M,fi}\lambda_{M,fi}$ is the partial safety factor for the relevant material property for the fire situation (taken as 1.0).

Eurocode 3 also provides guidance for evaluating fire resistance of bolted and welded connections. For instance, the reduced temperature-induced strength in bolts loaded is determined as

$$F_{v,t,Rd} = F_{v,Rd} k_b \frac{\gamma_{M2}}{\gamma_{M,fi}} \tag{7.33}$$

where k_b is the reduction factor determined for the appropriate bolt temperature. $F_{v,Rd}$ is the design shear resistance of the bolt per shear plane calculated assuming that the shear plane passes through the threads of the bolt, and γ_{M2} is the partial safety factor at normal temperatures.

Eurocode 3 also details fire resistance evaluation procedure for shapes made of stainless steel as well as for slender members in its Annexes C and E.

AISC Equations Similar to the Eurocode equations, the AISC also provides a design procedure to evaluate sectional capacity under fire conditions. This procedure allows extending ambient temperature design expressions into fire conditions given due consideration to temperature-induced degradation in steel material properties. Such that the reduced moment and shear capacity of steel beam under fire conditions can be evaluated as

$$M_{\text{fire}} = f_{y,T} \times Z \tag{7.34}$$

$$V_{\text{fire}} = 0.6 f_{y,T}\, d\, t_w C_v \tag{7.35}$$

[15]The AISC shares a similar strength-based approach to that of Eurocode 3. However, AISC only provides guidance to evaluating moment capacity under fire conditions.

[16]Eurocode 3 procedure can be applied to fire design as per other codes (AISC) following the same procedure. Refer to Numerical Example 6.2.

where Z is the plastic modulus, f_{yT} is the yield strength at temperature T. f_{yT} can be evaluated as, $f_{yT°C} = k_{fyT°C} \times f_{y,20°C}$, where $k_{fyT°C}$ is the reduction factors for yield strength of steel at temperature T. These reduction factors are plotted in Fig. 7.1a.

Further, the AISC manual provides other equations to account for lateral torsional buckling in steel beams under fire conditions. The AISC provides recommended reduction factors for yield strength and modulus of steel in the temperature range of 20 to 1200°C as well as temperature-dependent reduction factors for steel bolts (Groups A and B). It is worth noting that the AISC specifically states that temperature effects need to be accounted for when temperature in steel section exceeds 200°C.

7.6.2 Advanced Analysis

The simplified approaches discussed above for evaluating fire resistance, although one step above prescriptive-based approaches, do not yield realistic fire response since these approaches do not account for all critical factors influencing response of a structure exposed to fire. Further, at the sectional level, redistribution of stresses as well as fire-induced forces generated in a member cannot be fully accounted for. For accounting for such effects, *advanced analysis* is to be carried out. In advanced analysis, all influencing parameters including varying fire scenarios, buckling effects, composite action developed between members (such as beam-slab), and temperature-induced restraint forces can be fully accounted for.

Advanced analysis can be carried using finite element-based computer programs. Over the last three decades, there have been numerous works on applying general-purpose computer models such as ANSYS, ABAQUS, and ADINA for undertaking fire resistance analysis of steel structures. A number of special-purpose computer models such as SAFIR and VULCAN have also been developed. The application of ANSYS for undertaking advanced analysis for evaluating fire resistance of a typical fire-exposed member (beam), as that shown in Fig. 7.19, is briefly discussed below. Full details associated in such advanced analysis can be found in various publications (Ding and Wang, 2007).

Analysis Details

Discretization of Structure As discussed in Chap. 6, the first step in undertaking fire resistance analysis through advanced analysis is to discretize the structural member (beam) under consideration into elements

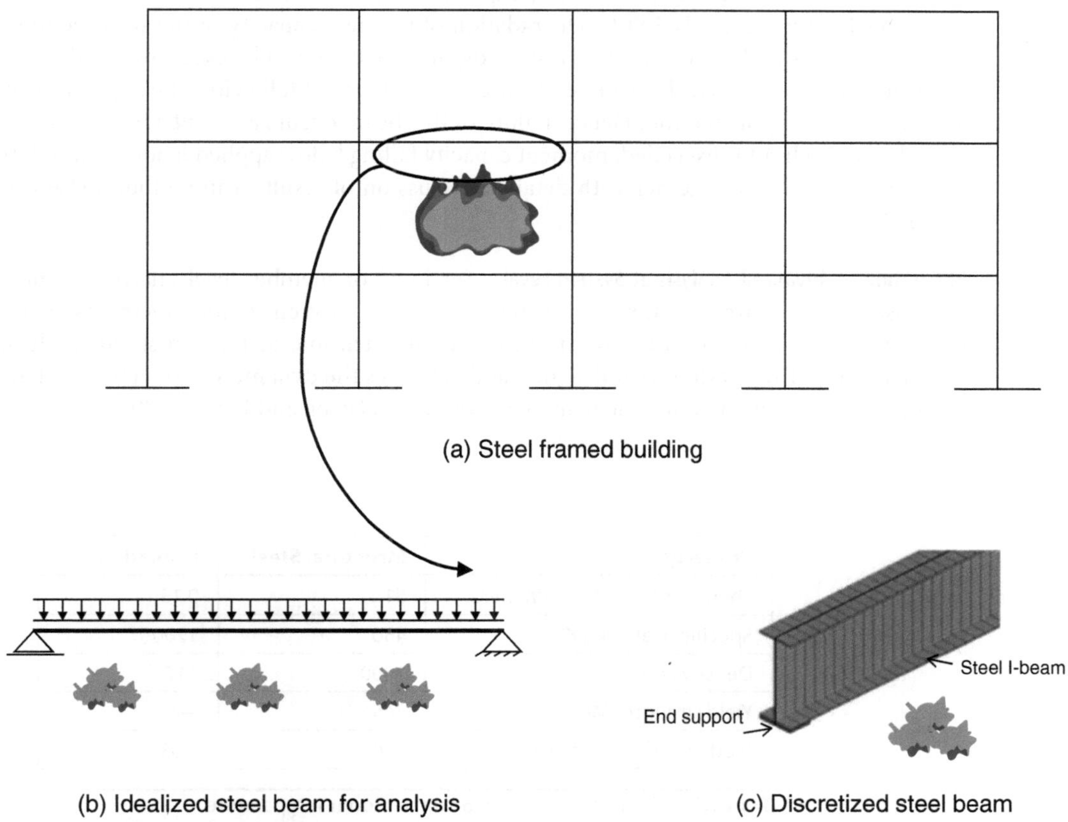

FIGURE 7.19 Finite element model of a steel beam exposed to fire.

through the use of various element libraries available in commercial programs. In this example, solid elements are used to discretize the beam shown in Fig. 7.19. Figure 7.19*c* shows discretization of a beam where SOLID70 and SURF152 elements are used as thermal elements to simulate heat transfer from fire source to fire-exposed beam. For structural analysis, SOLID185 is used for modeling a three-dimensional solid beam member, as this element can account for large deflections and large strain effects. It should be noted that shell elements such as SHELL131 and SHELL181 can also be used to discretize steel structural members for thermal and structural analysis, respectively.

Temperature-Dependent Material Properties For advanced analysis, temperature-dependent thermal and mechanical properties of steel and insulation materials are provided as input into ANSYS through property relations such as those provided in Sec. 7.2. The room temperature properties are listed in Table 7.15. The nonlinearity in steel material is accounted for through a bilinear elastoplastic constitutive material model based on the von Mises plasticity yielding criterion. The insulation material has weak mechanical properties and as such is neglected in the structural analysis.

Failure Limit State As part of advanced analysis, numerous thermal and structural response parameters were generated at incrementing time-steps. All critical limit states were applied to check failure at each incrementing time-step and derive final fire resistance. This includes consideration to sectional temperatures, strength (axial, moment, and shear capacity), and deflection, as well as local buckling limit states.

Illustration of Advanced Analysis at Member Level The aforementioned beam (shown in Fig. 7.19) is subjected to a fire scenario similar to that of a standard fire exposure. The beam is analyzed using the above procedure, and various output parameters generated in the analysis—namely temperatures, midspan deflection, and failure mode—are compared against measured data from tests. For instance, Fig. 7.20 shows that the temperature in the steel section rises steadily, but slowly due to the protection provided by fire insulation. The structural response in terms of midspan deflection progression with time is also plotted in Fig. 7.20*b*. In this figure, it can be seen that the beam undergoes only a small increase in deflection in the first 90 minutes. This is attributed to low temperature in the steel beam facilitated by the presence of fire insulation. Similarly, as can be seen in Fig. 7.20*c*,*d*, moment and shear capacity decrease slowly with fire exposure time at corresponding critical sections: midspan section for moment and support section for shear. The moment and shear capacity start to reduce at a higher pace when the temperature in the steel section exceeds 500°C. Degradation of moment capacity of the steel section continues until 130 minutes, at which point a rapid rise in midspan deflection produces runaway failure of the beam. Due to the low level of applied shear force, shear capacity does not fall below the applied shear force near the vicinity of the support section. Hence, failure of this beam occurs at about 130 minutes, very close to the time deflection limit is exceeded, moment capacity falling below applied loading limit. Further details on analysis of this beam, together with detailed discussion of results, can be found elsewhere (Kodur and Naser, 2017).

Illustration of Advanced Analysis at System Level Sectional or member-level analysis cannot capture global response of the entire structure, since the interaction between various members that are part of the structural frame (system) is not taken into account. Such interaction can significantly affect structural response under fire conditions and significantly changes the dynamics of structural behavior (and failure mechanisms) as compared to that at ambient conditions (Jiang and Usmani, 2017).

Property	Structural Steel	Insulation
Thermal conductivity (W/m·K)	51	0.15
Specific heat (J/kg·K)	450	1200
Density (kg/m³)	7900	217
Yield strength (MPa)	345	—
Modulus of elasticity (GPa)	210	0.03

Table 7.15 Properties of Steel (Beam) at Room Temperature

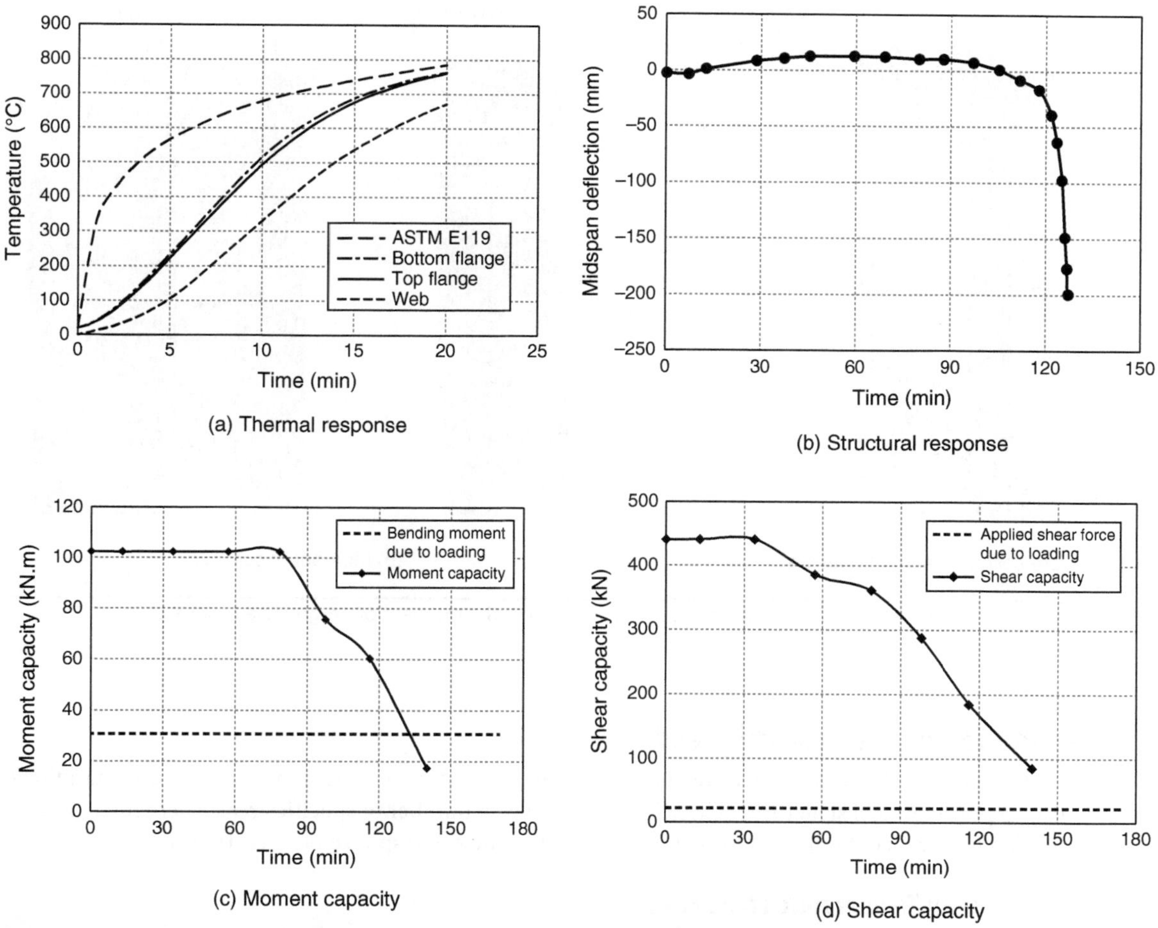

Figure 7.20 Predicted temperature and midspan deflection as a function of fire exposure time.

Although principles that are applicable to the use of advanced analysis are the same as those of isolated structural members, it is worth noting that system-level analysis is highly complex and requires numerous input parameters and boundary conditions. Advanced analysis at the system level can be carried out through modeling a steel structure using different subassemblies, some of which may be represented in 2D or using standard finite elements in one assembly, while the areas requiring more focused attention are modeled using more complex elements (3D elements) in another assembly. In this analysis, all steel structure subassemblies interact at every time-step using a middleware software, such as *OpenSees*.

One of the assemblies is selected to act as the master assembly, which solves the complete structure, while the other assemblies are selected to act as slave assemblies. The master assembly implements the boundary conditions on the slave assembly, and the slave program (or programs) returns the reaction forces to the master program. The boundary conditions that are transferred at the interface degrees-of-freedom from the master to the slave assembly can be defined as displacements and rotations.

For undertaking system-level analysis, a large number of input parameters—geometric and material features of all structural members, interaction of structural members, and so on—are to be input. A few researchers managed to apply advanced analysis on large-scale steel structures under fire conditions through simulating full steel structure such as that carried out by NIST as part of the 9/11 investigation and shown in Fig. 7.21. In this analysis, NIST examined both thermal and structural response of WTC7 to understand the fire-induced progressive collapse of this tower. Other researchers carried system-level analysis by simplifying the problem with various assumptions such as use of simple "fiber-based" finite element types and neglecting connections (Nguyen et al., 2016). Some of the recent research on system-level analysis for evaluating fire response of steel structures can be found in Khan et al. (2018), LaMalva et al. (2009), and NIST (2006), shown in Fig. 7.20.

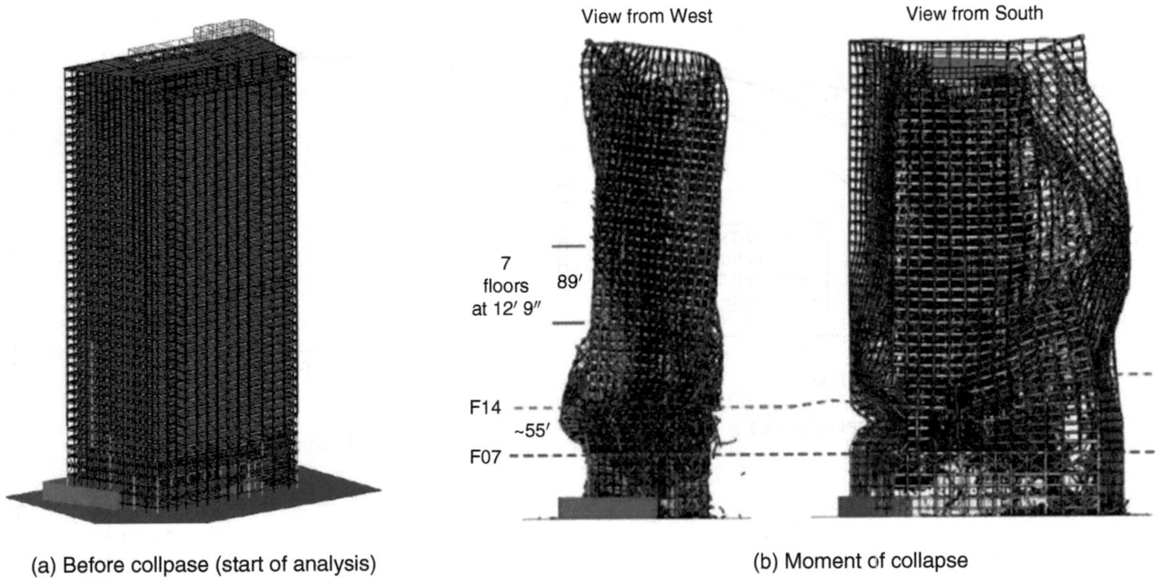

Figure 7.21 System-level analysis of WTC7 carried out by NIST that shows the moment of progressive collapse.

7.7 Emerging Trends and Research Needs

In the last three decades, there have been significant advances on developing further understanding on the fire response of steel structures. There is still, however, a number of specific areas where further research is needed to advance the state of the art with respect to the fire response of steel structures.

7.7.1 High-Temperature Creep Effects

Steel structures, under elevated temperature exposure, undergo permanent (plastic) deformations even when applied stress level is below that of yield stress. This deformation occurs due to high-temperature *creep* effects. Progression of creep in steel occurs at a given temperature and stress level in three stages: primary (initial, or Stage I) creep, secondary (steady state, or Stage II) creep, and tertiary (accelerating, or Stage III) creep, as shown in Fig. 7.22. In Stage I, creep strain starts at a relatively high rate, but this rate slows down eventually, depending on the stress level and temperature. Following that, in Stage II, the strain rate remains almost constant with time. However, in Stage III, creep strain increases at a rapid pace with an increase with stress level, and this results from the necking phenomenon and damage to the material (e.g., formation of microcracks). The effect of creep in Stage III gets accelerated under elevated temperature exposure (Harmathy, 1967).

Creep strains can be very high at elevated temperatures and high stress levels, and this can lead to significant creep-based deformations in structural members under fire conditions. Limited research has shown that creep effects in conventional carbon steel can significantly accelerate at temperatures above 40% of the melting temperature of the material ($T_{melting}$). At temperatures above 500°C, a large degradation in strength and modulus properties take place, along with diffusing dislocations and deformations, and this can lead to a large creep in steel structures. In spite of creep effects dominating structural response at elevated temperature, creep is often not fully accounted for in evaluating response of fire-exposed steel structures. Most published stress-strain material models for steel, including Eurocode 3, only account for partial creep effects. Creep effects, if not fully accounted for, can lead to underestimation of deflections as well as fire-induced forces, which in turn will influence failure time and mode, including onset of fire-induced progressive collapse in steel structures.

Recently there has been some work at the material and structure levels through incorporating high-temperature creep effects explicitly in tracing the response of steel structures under fire conditions (Morovat et al., 2012). Still, further research is needed to develop test setup and testing procedure for measuring creep effects with high reliability. Data collected from such creep tests can help develop temperature-dependent stress-strain models for different types of steel that fully account for creep effects. Another research need is to develop sensors that can capture creep effects and withstand effects of high temperature.

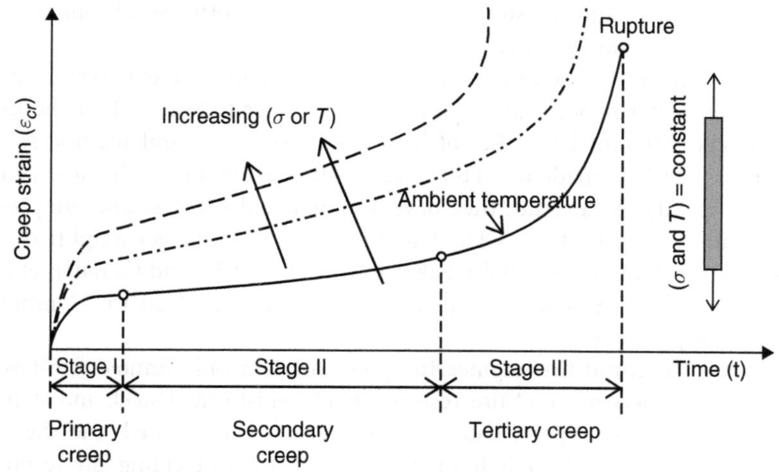

FIGURE 7.22 Creep of steel with increasing temperature.

7.7.2 Temperature-Induced Instability

Stability criterion is one of the key considerations in the design of steel structures at ambient conditions. As such, steel structural members are designed to satisfy stability limit states (i.e., flexural and shear limit states in beams and buckling requirements in columns). However, in current practice, stability effects are not specifically considered in design under fire conditions. Although stability limit state may not govern most building applications, stability effects can be dominant in a fire-exposed member subjected to eccentric loading, or when concentrated loads are placed near end supports, as in the case of beams connecting offset columns, transfer beams, coped beams (with notched ends), deep beams, and plate girders (Chung et al., 2003).

Temperature-induced instability occurs in steel structures when internal stress buildup approaches limiting yield strength. At this point, stiffness and strength properties of steel start deteriorating, which can induce sectional instability (local buckling). This is due to the fact that local buckling (slenderness) limits are a function of modulus and strength properties ($\sqrt{E/f_y}$) of steel. As a result, these limits can vary (reduce) with temperature levels in the section according to net loss of steel strength and modulus. Since the modulus of steel starts decreasing at 150°C, and at a faster pace than yield strength, local buckling in steel structures can get initiated at much lower temperatures than critical temperature (around 550°C), at which point the strength limit state becomes a governing factor. Therefore, the stability limit state is to be applied in evaluating fire resistance (failure) of steel structural members.

The effect of sectional instability in steel structures has been investigated recently by Kodur and Naser (2014), as well as by Couto et al. (2014) and Quiel and Garlock (2010). Results from these studies show that degradation in sectional capacity of steel members, specifically in Class 2, 3, or 4 sections (as per Eurocode 3 classifications), result from (1) temperature-dependent strength and modulus degradation and (2) the onset of temperature-induced instability. Such instability effects are more apparent in the case of flexural members subjected to high shear loading, due to the onset of shear web (local) buckling, which can significantly reduce shear capacity in slender steel beams or girders. Further research on this subject is needed to establish guidance for incorporating stability limit state as one of the governing failure criteria.

7.7.3 Properties of Specialized Steels

In recent years, a number of specialized steel structures such as cold-formed steel, fire-resistant steel, and stainless steel structures have been developed for use in residential and commercial buildings. With the increasing use of specialized steels, there is a need to address fire performance concerns of such steels. Cold-formed steel (CFS) is the common term for products made by rolling or pressing steel into semifinished or finished goods at relatively low temperature. CFS uses less energy than hot-rolled steel, has increased yield and tensile strengths, and has improved surface finish with fewer imperfections. Stainless steel, on the other hand, is a steel alloy with a minimum of 10.5% chromium content by mass and has high resistance to corrosion and fire.

The behavior of structural members made of stainless steel is very similar to that of hot-rolled steel with better performance under fire conditions (refer to Sec. 7.2). However, the structural behavior of CFS is very different from that of hot-rolled steel structural members. This is due to the fact that structural members made of CFS are very thin and slender with shear center not coinciding with the center of gravity. As a result, they have low torsional stiffness and are prone to buckling (Laím et al., 2009). This vulnerability of CFS structural members can get amplified under fire conditions. Tests carried out by Laím et al. (2009), Sakumoto et al. (2003), and Gunalan et al. (2013) reported adequate performance of CFS structural members in the range of 30 to 73 minutes, given that appropriate insulation is provided.

From a structural fire engineering point of view, only limited fire tests have been carried out on structural members made of fire-resistant steel, cold-formed steel, and stainless steel. The temperature-dependent material property data on these steels is very limited, and there is a wide scatter in limited published data. Similarly, only limited numerical studies tackling key response parameters of structural members made of these steels exist (Gunalan and Mahendran, 2013). Further fire tests, in conjunction with numerical studies, are needed to derive appropriate material models (i.e., temperature-dependent stress-strain relations, temperature-induced creep, etc.) for this specialty steel. Also, there is a need for developing fire resistance information on steel structural members made of these specialized steels through fire tests and numerical studies.

7.8 References

ACI. *Code Requirements for Determining Fire Resistance of Concrete and Masonry Construction Assemblies.* ACI 216.1-14/TMS-0216-14. Farmington Hills, MI: American Concrete Institute, 2014.

AISI. *Designing Fire Protection for Steel Beams.* Washington, DC: AISI, 1985.

AISI, *Load and Resistance Factor Design Specification for Cold-Formed Steel Structural Members*, Washington: American Iron and Steel Institute, 1991.

American Iron and Steel Institute (AISI). *Fire Resistant Steel Frame Construction.* Washington, DC: AISI, 1974.

Anderberg, Y. 1988. "Modeling Steel Behavior." *Fire Safety Journal* 13 (1988): 17–26.

ASCE. *Structural Fire Protection: Manual of Practice, No. 78.* New York: American Society of Civil Engineers, 1992.

ASCE/SFPE 29. *Standard Calculation Methods for Structural Fire Protection.* Reston, VA: American Society of Civil Engineers, 2005.

Bond, G. V. L. *Fire and Steel Construction: Water Cooled Hollow Columns.* London: Constrado, 1975.

Calculated Fire Resistance, Appendix "P." Standard Building Code, Southern Building Code Congress International, Incorporated, 1985.

CEN. *Eurocode 3: Design of Steel Structures—Part 1-2: General Rules—Structural Fire Design. EN 1993-1-2.* Brussels: European Committee for Standardization, 2005.

Chung, K. F., C. H. Liu, and A. C. H. Ko. "Steel Beams with Large Web Openings of Various Shapes and Sizes: An Empirical Design Method Using a Generalised Moment-Shear Interaction Curve." *Journal of Constructional Steel Research* 59, no. 9 (2003): 1177–1200.

CIB, A. "Conceptual Approach Towards a Probability Based Design Guide on Structural Fire Safety. In *CIB W14 Workshop Report.* Rotterdam, Netherlands: CIB, 1983.

Couto, Carlos, Paulo Vila Real, Nuno Lopes, and Bin Zhao. "Effective Width Method to Account for the Local Buckling of Steel Thin Plates at Elevated Temperatures." *Thin-Walled Structures* 84 (2014): 134–149.

Ding, J., and Y. C. Wang. "Experimental Study of Structural Fire Behaviour of Steel Beam to Concrete Filled Tubular Column Assemblies with Different Types of Joints." *Engineering Structures* 29, no. 12 (2007): 3485–3502.

Dorn, John Emil. "Some Fundamental Experiments on High Temperature Creep." *Journal of the Mechanics and Physics of Solids* 3, no. 2 (1955): 85–116.

Dwaikat, M. M. S., and V. K. R. Kodur. "A Simplified Approach for Predicting Temperature Profile in Steel Members with Locally Damaged Fire Protection." *Fire Technology* 48, no. 2 (2012): 493–512.

ECCS (1985). Design Manual on the European Recommendations for the Fire Safety of Steel Structures. European Commission for Constructional Steelwork, Brussels, Belgium.

Fellouh, Abdelkadir, Nourredine Benlakehal, Paulo Piloto, Ana Ramos, and Luís Mesquita. "Load Carrying Capacity of Partially Encased Columns for Different Fire Ratings." *Fire Research* 1, no. 1 (2017). https://doi.org/10.4081/fire.2017.23.

Fields, B. A., and R. J. Fields. *Elevated Temperature Deformation of Structural Steel: Report NISTIR 88-3899.* Gaithersburg, MD: NIST, 1989.

Gann, Richard G. *Final Report on the Collapse of World Trade Center Building 7, Federal Building and Fire Safety Investigation of the World Trade Center Disaster (NIST NCSTAR 1A).* National Construction Safety Team Act Reports (NIST NCSTAR), 2008.

Gunalan, Shanmuganathan, and Mahen Mahendran. "Finite Element Modelling of Load Bearing Cold-Formed Steel Wall Systems under Fire Conditions." *Engineering Structures* 56 (2013): 1007–1027.

Gunalan, Shanmuganathan, Prakash Kolarkar, and Mahen Mahendran. "Experimental Study of Load Bearing Cold-Formed Steel Wall Systems under Fire Conditions." *Thin-Walled Structures* 65 (2013): 72–92.

Harmathy, Tibor Zoltan. "A Comprehensive Creep Model." *Journal of Basic Engineering* 89, no. 3 (1967): 496–502.

International Conference of Building Officials (ICBO). *Uniform Building Code*. Lansing, MI: ICBO, 1997.

Jiang, Liming, and Asif Usmani. "Modelling Structures in Fire Using OpenSees: An Integrated Approach." *OpenSees Days Europe 2017* (2017): 43.

Khan, Mustesin Ali, Liming Jiang, Katherine A. Cashell, and Asif Usmani. "Analysis of Restrained Composite Beams Exposed to Fire Using a Hybrid Simulation Approach." *Engineering Structures* 172 (2018): 956–966.

Kirby, B. R., and R. R. Preston. "High Temperature Properties of Hot-Rolled, Structural Steels for Use in Fire Engineering Design Studies." *Fire Safety Journal* 13, no. 1 (1988): 27–37.

Kodur, Venkatesh, and Amir Arablouei. "Effective Properties of Spray-Applied Fire-Resistive Material for Resistance to Cracking and Delamination from Steel Structures." *Construction and Building Materials* 84 (2015): 367–376.

Kodur, Venkatesh K., and Esam M. Aziz. "Effect of Temperature on Creep in ASTM A572 High-Strength Low-Alloy Steels." *Materials and Structures* 48, no. 6 (2015): 1669–1677.

Kodur, V. K., and David H. MacKinnon. "Design of Concrete-Filled Hollow Structural Steel Columns for Fire Endurance." *Engineering Journal—American Institute of Steel Construction* 37, no. 1 (2000): 13–24.

Kodur, Venkatesh, and Mohannad Z. Naser. "Effect of Local Instability on Fire Response of Steel Beams." *PSU Research Review* 1, no. 2 (2017): 170–179.

Kodur, V. K. R., and A. M. Shakya. "Effect of Temperature on Thermal Properties of Spray Applied Fire Resistive Materials." *Fire Safety Journal* 61 (2013): 314–323.

Kodur, V. K. R., and M. Z. Naser. "Effect of local instability on capacity of steel beams exposed to fire." *Journal of Constructional Steel Research* 111 (2015): 31–42.

Kodur, Venkatesh, Mahmud Dwaikat, and Rustin Fike. "High-Temperature Properties of Steel for Fire Resistance Modeling of Structures." *Journal of Materials in Civil Engineering* 22, no. 5 (2010): 423–434.

Kodur, Venkatesh K.R., and T. T. Lie. "Fire performance of concrete-filled hollow steel columns." *Journal of Fire Protection Engineering* 7, no. 3 (1995): 89–97.

Laím, Luís, J. P. C. Rodrigues, and A. M. Correia. "Fire Resistance of a Cold Formed Steel Roof Beam." Nordic Steel Conference, Malmö, Sweden, 2009.

LaMalva, Kevin J., Jonathan R. Barnett, and Donald O. Dusenberry. "Failure Analysis of the World Trade Center 5 Building." *Journal of Fire Protection Engineering* 19, no. 4 (2009): 261–274.

Lawson, R. M., and D. Oshatogbe. "Design of Fabsec Beams in Non-composite and Composite Applications." The Steel Construction Institute, RT777 (1999).

Lie, T. T., ed. *Structural Fire Protection*. Reston, VA: American Society of Civil Engineers, 1992.

Lie, T. T., and V. K. R. Kodur. "Fire Resistance of Steel Columns Filled with Bar-Reinforced Concrete. *Journal of Structural Engineering* 122, no. 1 (1996): 30–36.

Lie, T. T., and W. W. Stanzak. "Fire Resistance of Protected Steel Columns." *Engineering Journal—American Institute of Steel Construction* 10, no. 4 (1973): 134–136.

Lie, Tiam Tjoan, and T. Z. Harmathy. "Fire Endurance of Concrete-protected Steel Columns." *Journal Proceedings* 71, no. 1 (1974): 29–32.

Liu, T. C. H. "Fire resistance of unprotected steel beams with moment connections." *Journal of Constructional Steel Research* 51, no. 1 (1999): 61–77.

Malhotra, H. L. "Spalling of Concrete in Fires." CIRIA. Technical Note 118, 1984.

Milke, James A. "Analytical Methods for Determining Fire Resistance of Steel Members." In *SFPE Handbook of Fire Protection Engineering*. New York: Springer, 2016: 1909–1948.

Morovat, Mohammed, Michael Engelhardt, Todd Helwig, and Eric Taleff. "High-Temperature Creep Buckling Phenomenon of Steel Columns Subjected to Fire." *Journal of Structural Fire Engineering* 5, no. 3 (2014): 189–202.

Morovat, Mohammed Ali, Jin Woo Lee, Michael D. Engelhardt, Eric M. Taleff, Todd A. Helwig, and Victoria A. Segrest. "Creep Properties of ASTM A992 Steel at Elevated Temperatures." In *Advanced Materials Research*. Vol. 446. Zurich: Trans Tech Publications, 2012: 786–792.

Muratov, A. N., Yu D. Morozov, O. N. Chevskaya, and G. A. Filippov. "Technology for the Commercial Production of Fire-resistant Steel for Building Structures." *Metallurgist* 51, no. 7–8 (2007): 446–453.

Nguyen, Ha, Ann E. Jeffers, and Venkatesh Kodur. "Computational Simulation of Steel Moment Frame to Resist Progressive Collapse in Fire." *Journal of Structural Fire Engineering* 7, no. 4 (2016): 286–305.

NIST, Final Report on the Collapse of World Trade Center Building 7, Federal Building and Fire Safety Investigation of the World Trade Center Disaster (NIST NCSTAR 1A), 2008. https://nvlpubs.nist.gov/nistpubs/Legacy/NCSTAR/ncstar1a.pdf

PABCO (PABCO Super Firetemp Fireproofing Board). *Fire Protection Guide*. Ruston, LA: PABCO, 1984.

Pakala, Purushotham, Venkatesh Kodur, Serdar Selamet, and Maria Garlock. "Fire Behavior of Shear Angle Connections in a Restrained Steel Frame." *Journal of Constructional Steel Research* 77 (2012): 119–130.

Poh, K. W. "Stress-Strain-Temperature Relationship for Structural Steel." *Journal of Materials in Civil Engineering* 13, no. 5 (2001): 371–379.

Quiel, Spencer E., and Maria EM Garlock. "Calculating the Buckling Strength of Steel Plates Exposed to Fire." *Thin-Walled Structures* 48, no. 9 (2010): 684–695.

Ruddy, John L., Joseph P. Marlo, Socrates A. Ioannides, and Farid Alfawakhiri. *Fire Resistance of Structural Steel Framing*. Chicago, IL: American Institute of Steel Construction, 2003.

Sakumoto, Y., T. Hirakawa, H. Masuda, and K. Nakamura. "Fire Resistance of Walls and Floors Using Light-Gauge Steel Shapes." *Journal of Structural Engineering* 129, no. 11 (2003): 1522–1530.

Seigel, L. G. "Designing for Fire Safety with Exposed Structural Steel." *Fire Technology* 6, no. 4 (1970): 269–278.

Sha, Wei. *Steels: From Materials Science to Structural Engineering*. Berlin: Springer Science & Business Media, 2013.

Stanzak, W. W., and Tiam Tjoan Lie. "Fire Resistance of Unprotected Steel Columns." *Journal of the Structural Division* 99, no. 5 (1973): 837–852.

Twilt, L., and J. Witteveen. "The Fire Resistance of Wood-Clad Steel Columns." *Fire Prevention Science and Technology* 11 (1974): 14–20.

Underwriters Laboratories of Canada (UL). CAN/ULCS101. "Standard Methods of Fire Endurance Tests of Building Construction and Materials." Toronto: Underwriters Laboratories of Canada, 2007.

Wang, Wei-yong, Bing Liu, and Venkatesh Kodur. "Effect of Temperature on Strength and Elastic Modulus of High-Strength Steel." *Journal of Materials in Civil Engineering* 25, no. 2 (2012): 174–182.

7.9 Nomenclature

A	Surface area of column (m^2)
A_s	Cross-sectional area of the steel column, mm^2 (in^2)
b_f	Flange width of the steel column, mm (in.)
C	Compressive force due to unfactored dead and live load, in kN (kips)
C_1 and C_2	Material-dependent constants for wide-flange columns (SFRM)
c_c	Ambient temperature specific heat of concrete or masonry, kJ/kg·K (Btu/ft·°F)
c_s	Specific heat of steel (J/kg·J)
D	Heated perimeter of structural steel column, mm (in.)
D	Heated perimeter of the structural steel beam or girder, mm (in.)
D	Outside diameter for circular columns, least outside dimension for square or rectangular columns, m (in.)
d	Depth of the steel column, mm (in.)
d_i	Thickness of insulation (m)
E_o	Modulus of elasticity at room temperature
F	Surface area of unit length (m^2)
$f'c$	Specified 28-day compressive strength of concrete, MPa (kips/in^2)
f_y	Yield strength at room temperature
f_y	Yield strength of steel at room temperature, MPa (ksi)
$f_{y,T}$	Yield strength at elevated temperatures
H	Ambient temperature thermal capacity of steel column = 0.46 W·kJ/m·K (0.11 W·Btu/ft·°F)
h	Thickness of concrete or equivalent thickness (for masonry), m (in.)
h	Thickness of spray, mm (in.)
h	Total thickness of gypsum wallboard, mm (in.)
h_c	Convective heat transfer coefficient (25–50 W/m^2·K for standard and hydrocarbon fires)
h_f and h_0	Heat transfer coefficient on fire side and the cold side, respectively
k	Conductivity matrix
k_c	Ambient temperature thermal conductivity of concrete, W/m·K (Btu/hr·ft·°F)
k_i	Thermal conductivity of insulation (W/m·K)
KL	Column effective length, m (ft)
$k_{y,T}$	Ratio of yield strength at elevated temperatures to the yield strength at room temperature
L	Interior dimension of one side of square concrete or masonry box protection, m (in.)
m	Equilibrium moisture content of concrete, volume (%)
n_y and n_z	Components of the vector normal to the boundary in the plane of the cross section
q	Heat transferred to the column during a fire test per unit surface area (kJ/m^2)

r	Decay rate of design fires according to Eurocode parametric fires
R	Fire resistance (hrs)
R_0	Fire resistance rating (hrs) at zero moisture content
S	Elastic section modulus, mm^3 (in^3)
T	Temperature (°C)
t	Time to reach a limit temperature
t_1	Time corresponding to maximum fire temperature in design fires
T_f	Temperature in the fire environment (K)
T_f and T_0	Temperature at fire-exposed and cold side, respectively
$T_{f,max}$	Maximum temperature of design fire scenario
T_s	Temperature of steel (K)
T_{s1}	Steel temperature at the time of maximum fire temperature (proposed approach)
V	Volume of steel in unit length (m^3)
V_w	Require external storage water (m^3)
W	Average weight of steel columns, kg/m (lbs/lin ft)
W	Average weight of the steel column, kg/m (lbs/ft)
W	Weight of the structural steel beam or girder, kg/m (lbs/lin ft)
W'	Total weight of steel column and gypsum wallboard protection, kg/m (lbs/lin ft)
Z_x	Plastic section modulus, mm^3 (in^3)
$\dot{q}''$	Heat transfer at the surface (W/m^2)
ΔT_s	Change in steel temperature in the time-step (°C or K)
ε	Resultant emissivity (0.5–0.67)
∇	Spatial gradient operator
ρ_c	Concrete or masonry density, kg/m^3 (lbs/cu ft)
ρc	Heat capacity
ρ_i	Density of the insulation (kg/m^3)
ρ_s	Density of steel (kg/m^3)
σ	Stefan-Boltzmann constant (56.7×10^{-12} $kW/m^2{\cdot}K^4$)
σ_{yo}	Yield strength at room temperature

7.10 Numerical Examples

7.1

UL Design X517 for a steel column of W10×49 with *W/D* ratio of 49.4 (0.84 imperial) and protected with two 12.7 mm (0.5 in.) layers of Type X gypsum wall board insulation meets 2-hour fire resistance rating. In a newly designed building, another steel column (W12×106) is being used. Check and determine if W12×106 column meets the 2-hour fire rating criteria.

W10×49

Solution

As per prescriptive-based solution, the substitution equation can be applied to evaluate fire resistance of a column that is similar to the tested column.

Step 1.

Determine W/D for W12×106 column:

$W = 106$ lb/ft $= 158$ kg/m

$D = 2d + 2b$

$D = 2(12.9 + 1 + 1) + 2(12.2 + 1 + 1) = 58.2$ in. $= 1.478$ m

Step 2.

As $W/D = 106.9$ metric (1.82 imperial) > 49.4 metric (0.84 imperial), then substitution is permitted, i.e., W12×106 column will have 2-hour fire rating.

7.2

Calculate the fire resistance of a square HSS column protected with GWB given that this section is of 254 × 254 × 12.7 mm size, has a weight, W, of 93 kg/m and heated perimeter of $D = 4 \times 254 = 1016$ mm.

Solution

As per discussion in the section "Membrane Protection":

Step 1.

Check if W/D ratio lies within the applicable limit of 0.215.

$W/D = 93/1016 = 0.092 \leq 0.215$

Step 2.

Total weight, $W' = W + 0.0008\,h\,D = 93 + 0.0008 \times 12.7 \times 1016 = 103.32$ kg/m

$$R = 1.60\left[\frac{h\left(W'/D\right)}{2}\right]^{0.75}$$

$$R = 1.60\left[\frac{12.7\left(103.3/1016\right)}{2}\right]^{0.75} = 1.15 \text{ hours}$$

7.3

Calculate the fire resistance of a circular steel column filled with reinforced concrete following discussion of the "Steel Columns" section in the chapter under Sec. 7.5.2. This column has a diameter of 324 mm, thickness of 6.4 mm, and has fixed boundary conditions. The concrete is of carbonate aggregate type and has a compressive strength of 40 MPa. The steel reinforcement in the column is 3% and rebars are provided at a concrete cover thickness of 25 mm.

Solution

For circular HSS column filled with RC filling of carbonate aggregate type, 3% reinforcement ratio, and 25-mm cover thickness, $a = 0.095$ (from Table 7.15). The fire resistance of this column using Eq. 7.7a is

$$R = a\frac{(f_c' + 20)}{(KL - 1000)}D^2\left(\frac{D}{C}\right)^{1/2}$$

$$R = 0.095\frac{(40 + 20)}{((0.67 \times 4800) - 1000)} \times 324^2\left(\frac{324}{1344}\right)^{1/2} = 132 \text{ minutes}$$

7.4

A steel column of section W200×250 mm is exposed to a standard fire. Assuming that flange thickness $T = 15$ mm and web thickness $t = 10$ mm, calculate the time required to reach limiting temperature to be 550°C through best-fit method for two cases:

(a) Consider section to be similar to that of box profile.

(b) Consider section to be similar to that contour profile.

Solution

(a) Box profile

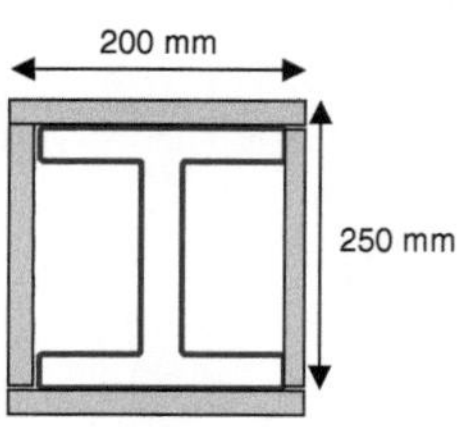

Area of cross section:

$$A = 2(b \times T) + t(h - 2T) = 2(200 \times 15) + 10(250 - 2 \times 15) = 8200\ \text{mm}^2 = 0.0082\ \text{m}^2$$

Volume of 1-m length:

$$V = A \times 1.0\ \text{m} = 0.00082\ \text{m}^3$$

Perimeter of section:

$$H_p = 2\,(b + h) = 900\ \text{mm} = 0.9\ \text{m}$$

Surface area of 1-m length:

$$F = H_p \times 1.0\ \text{m} = 0.9\ \text{m}^2$$

Section factor:

$$H_p/A = 0.9/0.0082 = 109.7\ \text{m}^{-1}$$

Section factor:

$$F/V = 10.9/0.0082 = 109.7\ \text{m}^{-1}$$

Effective thickness:

$$V/F = 1000/(F/V) = 9.1\ \text{mm}$$

Time to reach T_{lim}:

$$t = \frac{0.54(T_{\text{lim}} - 50)}{\left(F/V\right)^{0.6}}$$

$$t = \frac{0.54(550 - 50)}{(109.7)^{0.6}} = 16.1\ \text{minutes}$$

(b) Contour profile

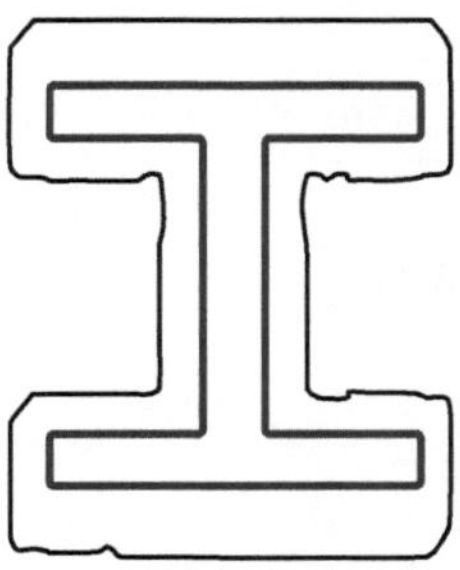

$$A = 2(b \times T) + t(h - 2T) = 2(200 \times 15) + 10(250 - 2 \times 15) = 8200\ \text{mm}^2 = 0.0082\ \text{m}^2$$

Volume of 1-m length:

$$V = A \times 1.0\ \text{m} = 0.0082\ \text{m}^3$$

Perimeter of section:

$$H_p = 2\,(b + h + b - t) = 1280 \text{ mm} = 1.28 \text{ m}$$

Surface area of 1-m length:

$$F = H_p \times 1.0 \text{ m} = 1.28 \text{ m}^2$$

Section factor:

$$H_p/A = 1.28/0.0082 = 156.1 \text{ m}^{-1}$$

Section factor:

$$F/V = 1.28/0.0082 = 156.1 \text{ m}^{-1}$$

Effective thickness:

$$V/F = 1000/(F/V) = 6.4 \text{ mm}$$

Time to reach T_{lim}:

$$t = \frac{0.54(T_{\text{lim}} - 50)}{\left(F/V\right)^{0.6}}$$

$$t = \frac{0.54(550 - 50)}{(156.1)^{0.6}} = 13 \text{ minutes}$$

7.5

An unprotected W18×46 steel beam with $F/V = 156.4\ \text{m}^{-1}$ is exposed to the ISO 834 standard fire. Use the step-by-step method to evaluate when the temperature in steel section would reach 500°C. Assume a convective heat transfer coefficient $h_c = 25\ \text{W/m}^2\cdot\text{K}$ and emissivity of 0.6. The density of steel is 7850 kg/m^3 and its specific heat is 600 J/kg·K. Use a time-step of 0.5 minutes.

Solution

Unprotected steel (three-sided exposure)
Surface area per unit length:

$$F = (2d + 3b_f - 2t_w) \times 1 = (2 \times 18.2 + 3 \times 6.06 - 2 \times 0.36) \times 1 = 53.66 \text{ in.}$$

Volume per unit length:

$$V = A \times 1 = 13.5 \times 1 = 13.5 \text{ in}^3$$

$$F/V = 3.97 \text{ in}^{-1} = 156.4 \text{ m}^{-1} \text{ (between 10 and 300 m}^{-1}\text{)}$$

For the step-by-step method, define time increment; use 0.5 minutes.
The first step at a half time-step of 0.25 minutes:

$$T_f = 345 \log_{10}(8t + 1) + T_0$$

$$T_f = 345 \log_{10}(8 \times 0.25 + 1) + 20 = 184.6°\text{C}$$

$$\Delta T_s = \frac{F}{V}\frac{k_{sh}}{\rho_s c_s}\left\{(h_c(T_f - T_s) + \sigma\varepsilon\,(T_f^{\,4} - T_s^{\,4}))\right\}\Delta t$$

$$\Delta T_s = (156.4)\frac{1}{7850(600)}\left\{(25(184.6 - 20) + 5.67 \times 10^{-7} \times 0.7\,((184.6 + 273)^4 - (20 + 273)^4))\right\} \times 0.25 = 5.542$$

Set up the above procedure in a spreadsheet package. Results from the temperature are shown in Table E7.1 and also plotted in the companion figure.

7.6

Calculate the temperature in a W10×49 steel column after 10 minutes of exposure to ASTM E119 fire

Time (min)	Time at Half Step (min)	Steel Temp. T_s by Step-by-Step, °C	E119 Fire Temp. at Half Step (T_f), °C	Difference in Temp. ($T_f - T_s$), °C	Change in Steel Temp. (ΔT_s), °C
0.00	0.25	20.00	184.60	164.60	5.54
0.50	0.75	25.54	311.55	286.01	11.42
1.00	1.25	36.96	379.28	342.31	15.31
1.50	1.75	52.28	425.75	373.46	18.28
2.00	2.25	70.56	461.17	390.60	20.66
…	…	…	…	…	…
…	…	…	…	…	…
10.00	10.25	494.15	682.08	187.92	23.88
10.50	10.75	518.04	689.13	171.09	22.66

Table E7.1 Temperature Rise in W18×46 Steel Beam due to ISO 834 Fire Exposure

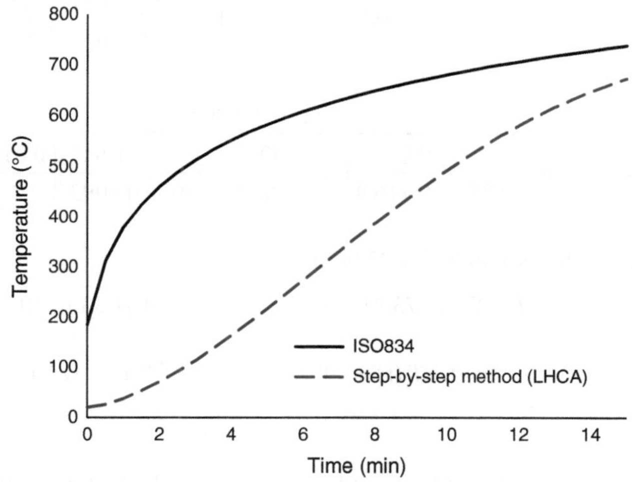

Figure E7.1 Temperature rise in W18 × 46 steel beam due to ISO 834 fire exposure.

conditions. Also, calculate the temperature for a protected W10×49 column after 80 minutes of exposure to standard fire. The insulation has 19-mm-thick protection with the properties indicated below. Use Dwaikat and Kodur approach (refer to Sec. 7.6.2).

Given:

$$c_s = 450 \text{ J/kg·K}, \rho_s = 7850 \text{ kg/m}^3, k_p = 0.116 \text{ W/m·K}, c_p = 1825 \text{ J/kg·K}, \rho_p = 240.5 \text{ kg/m}^3$$

Solution

$$F_p = 2\,(b + h + (b - t)) = 2\,(10 + 9.98 + (10 - 0.34)) = 59.28 \text{ in.} = 1.51 \text{ m}$$

$$A_s = 14.4 \text{ in}^2 = 0.0093 \text{ m}^2$$

(a) Temperature in unprotected column after 10 minutes of fire exposure:

$$s = \frac{F_1}{n+1} = \frac{\left(F_p/A_s\right)}{\rho_s c_s \left(\dfrac{1}{h_{con}} + \dfrac{t_p}{k_p}\right)\left(1 + \left(\dfrac{\rho_p c_p}{\rho_s c_s}\right)\left(F_p/A_s\right)\left(t_p/m\right)\right)(n+1)}$$

$$s = \frac{(1.51\ /\ 0.0093)}{7850 \times 450 \times 10^6 \left(\dfrac{1}{25} + 0\right)(1 + 0)(0.1478 + 1)} = 0.00115$$

At 10 minutes (600 seconds = 0.16 hour),

$$T(°\text{C}) = 750(1 - e^{-3.79\sqrt{t}} + 170.41\sqrt{t} + 20)$$

$$T_f(°\text{C}) = 750(1 - e^{-3.79\sqrt{0.16}}) + 170.4\sqrt{0.16} + 20 = 673.5°\text{C}$$

$$T_s(t) = T_f(1 - e^{-st})$$

$$T_s(t) = 673.5(1 - e^{-0.00115 \times 600}) = 335.69°\text{C}$$

(b) Temperature in protected column after 80 minutes of fire exposure.

$$s = \frac{F_1}{n+1} = \frac{\left(F_p / A_s\right)}{\rho_s c_s \left(\dfrac{1}{h_{con}} + \dfrac{t_p}{k_p}\right)\left(1 + \left(\dfrac{\rho_p c_p}{\rho_s c_s}\right)\left(F_p / A_s\right)\left(t_p / m\right)\right)(n+1)}$$

$$s = \frac{(1.51\ /\ 0.0093)}{7850 \times 450 \times \left(\dfrac{1}{25} + \dfrac{0.019}{0.116}\right)\left(1 + \left(\dfrac{240.5 \times 1825}{7850 \times 450}\right)\left(\dfrac{1.51}{0.0093}\right)\left(\dfrac{0.019}{2}\right)\right)(0.1478 + 1)} = 0.00016$$

At 80 minutes (4800 seconds = 1.33 hours),

$$T_f(°\text{C}) = 750(1 - e^{-3.79\sqrt{1.33}}) + 170.4\sqrt{1.33} + 20 = 957°\text{C}$$

$$T_s(t) = 957(1 - e^{-0.00016 \times 4800}) = 513°\text{C}$$

7.7

For a simply supported steel beam of known span, load, yield strength, and section properties, calculate the time to failure when exposed on three sides to the standard fire: (a) unprotected and (b) protected with insulation of known thickness and properties. Use an equivalent section to W12×170 (from AISC manual).

Given:

Dead load: $DL = 3.0$ kN/m (including self-weight)
Live load: $LL = 6$ kN/m
Beam span: $L = 12$ m
Beam size: 760 mm deep and 147 kg/m
This is a "compact" section (type 1), as per AISC manual.
Plastic section modulus: $Z_x = 4480 \times 10^3$ mm^3
Strength reduction factor: $\phi = 0.9$
Yield strength: $f_y = 250$ MPa
Area to volume ratio: $F/V = 119$ m^{-1} (Effective thickness: $V/F = 8.4$ mm)
Use a load combination of $DL + 0.5LL$ under fire conditions.

Solution

Cold Calculations

Design load (cold):

$$w_c = 1.2DL + 1.6LL = 13.2 \text{ kN/m}$$

Bending moment:

$$M_{\text{cold}} = \frac{w_c l^2}{8}$$

$$M_{\text{cold}} = 237.6\ \text{kN·m}$$

Bending strength:

$$M_n = Zf_y$$

$$\phi M_n = 0.9 \times 1120 = 1008\ \text{kN·m} > M_{\text{cold}}\ \text{(Design is safe)}$$

Fire Calculations

Strength reduction factor: $\phi_f = 1.0$
Design load (fire):

$$w_c = DL + 0.5LL = 6\ \text{kN/m}$$

Bending moment:

$$M_{\text{fire}} = \frac{w_c l^2}{8}$$

$$M_{\text{fire}} = 108\ \text{kN·m}$$

Load ratio:

$$r_{\text{load}} = \frac{M_{\text{fire}}}{M_{\text{cold}}} = \frac{108}{1120} = 0.10$$

then

$$T_{\text{lim}} = 905 - 690 r_{\text{load}}$$

$T_{\text{lim}} = 836°\text{C}$, failure will occur when steel temperature reaches 836°C.

(a) Unprotected steel (three-sided exposure):

Time to reach limiting temperature; using best-fit equation:

$$t = \frac{0.54(T_{\text{lim}} - 50)}{\left(F/V\right)^{0.6}}$$

$$t = \frac{0.54(836 - 50)}{(119)^{0.6}} = 24.1\ \text{minutes}$$

Design is OK if the equivalent fire severity is no more than 24.1 minutes.

(b) Protected steel

Thickness of insulation: $d_i = 0.020$ m (20 mm)
Thermal conductivity of insulation: $k_i = 0.10$ W/m·K
Time to reach limiting temperature:

$$t = 40(T_{\text{lim}} - 140)\left(\frac{d_i/k_i}{F/V}\right)^{0.77}$$

$$t = 40(836 - 140)\left(\frac{0.02/0.10}{119}\right)^{0.77} = 203.4\ \text{minutes}$$

Moisture content of insulation: $m = 15\%$
Density of insulation: $\rho_i = 800$ kg/m^3

$$t_v = m\rho_i d_i^2 / 5k_i$$

$$t_v = \frac{15 \times 800 \times 0.02^2}{5 \times 0.10} = 9.6 \text{ minutes}$$

Total time:

$$t + t_v = 213 \text{ minutes}$$

Design is adequate if the equivalent fire severity is less than 213 minutes.

CHAPTER 8

Concrete Structures

8.1 Overview

This chapter provides an overview of the response of concrete structures subjected to fire conditions. Temperature-dependent thermal and mechanical property relations for concrete, masonry, reinforcing, and prestressing steel are presented. The fire behavior of a typical concrete structures is illustrated, and the main strategies that can be incorporated for achieving required fire resistance in concrete structures are discussed. The prescriptive and rational engineering approaches that can be applied for evaluating the fire resistance of concrete structures are laid out. Solutions to overcome fire-induced spalling in concrete structures, together with strategies for enhancing fire resistance of high-strength concrete structures, are discussed.

8.2 Properties of Concrete and Steel Reinforcement at Elevated Temperature

8.2.1 General

Concrete, one of the most widely used building materials in the construction industry, possesses the best fire resistance properties of any building material. This is due to the composition of constituent materials (i.e., cement and aggregates) which, when chemically combined, form a material that is essentially inert. When subjected to high temperatures as encountered under fire exposure, cement paste in concrete undergoes an endothermic reaction which absorbs much of the heat that is applied to the concrete member, thus slowing the temperature rise in concrete. This reaction continues to occur throughout the depth and width of concrete member. However, due to the low thermal conductivity and high specific heat of concrete, interior layers (core) of the concrete member remain cooler much longer than the outer layer close to the fire (exposed surface). With increased sectional temperatures, the strength and stiffness properties of concrete gradually degrade. However, the rate of degradation is at a much lower rate than other construction materials. The lower thermal conductivity, higher specific heat, and slower degradation in the mechanical properties of concrete, combined with a larger thermal mass, enable concrete structures to continue to efficiently resist the applied loading under fire conditions for a longer duration.

Concrete is available in various forms and is grouped under different categories based on:

Weight: normal-weight concrete (NWC) and lightweight concrete (LWC)

Strength: normal-strength concrete (NSC), high-strength concrete (HSC), and ultrahigh performance concrete (UHPC)

Presence of fibers: steel or polypropylene fiber-reinforced concrete, or FRC

The definitions for classifying concrete into different groups vary in different codes and standards. Generally, concrete up to 60 MPa is considered to be NSC, and when the strength exceeds 60 MPa, it is considered to be HSC. Further, when strength exceeds 150 MPa, it is classified as UHPC. Fire safety practitioners further subdivide concretes into silicate (siliceous) and carbonate (limestone) aggregate concrete, according to the composition of the principal aggregate used in concrete mix. The ACI 216.1 (ACI, 2014) defines carbonate aggregate concrete as that made with a coarse aggregate consisting mainly of calcium carbonate or a combination of calcium and magnesium carbonate (e.g., limestone or dolomite). On the other hand, the same ACI code defines siliceous-aggregate concrete as concrete made with normal-density aggregates having constituents composed mainly of silica or silicates (such as quartz or granite). The newer types of concrete, although they share some of the basic attributes of normal-strength concrete, exhibit a significant variation in some of the properties especially under high-temperature exposure; this is due to the additional additives used in their mix, such as fibers (Khaliq and Kodur, 2012).

8.2.2 Effect of Temperature on Properties

Concrete, similar to other materials, experiences degradation in properties under elevated temperature exposure. This is mainly because thermal, mechanical, and deformational properties of concrete depend not only on the composition and characteristics of concrete but also on the temperature level reached. To overcome the weak tensile strength of concrete, steel reinforcement (either reinforcing or prestressing steel) is always provided in concrete structures. As discussed in Chap. 7, steel also experiences significant degradation in properties under elevated temperatures. Therefore, the variation in properties of concrete and reinforcement with temperature must be given due consideration in fire resistance evaluations of concrete structures.

The trends for variation in various properties of concrete with temperature are specified as temperature-dependent property relations or as graphs in fire codes and standards such as Eurocode 2 (CEN, 2004), the *ASCE Manual of Practice* (ASCE, 1992), and ACI 216.1 (ACI, 2014). While Eurocode 2 gives relations for properties of both NSC and HSC, the *ASCE Manual of Practice* gives relations only for NSC. In addition to relations in standards, a few researchers proposed high-temperature property relations for newer types of concrete (Cheng et al., 2004; Khaliq and Kodur, 2014). A closer look at these relations reveals that there are significant variations specified in different standards. These variations arise from differences associated with test setup and varying procedures including heating rate, sample size, moisture content, and concrete mix proportion, and conditions adopted to measure properties at elevated temperatures. Another major difference between the relations specified in the Eurocode and the ASCE standards relates to the effect of aggregate type on the high-temperature properties of concrete. While the ASCE manual accounts for the effects of the type of aggregate (siliceous, carbonate, expanded shale, and pure quartz) on thermal capacity and other thermal properties, Eurocode does not specifically account for the effect of aggregate type on the thermal capacity of concrete at high temperatures, and hence, heat capacity is considered to be the same for all aggregate types used in concrete. Some of these property relations that are needed for fire resistance modeling of concrete structures are discussed here.

8.2.3 High-Temperature Property Relations of Concrete

Thermal Properties

The thermal properties—thermal conductivity, specific heat (or heat capacity), and mass loss of concrete—are significantly influenced by the composition of concrete mix, aggregate type, and moisture content. Relations for high-temperature thermal properties of different concrete types are available in the literature (Harmathy, 1970; Kodur et al., 2013) and the widely used relations present in Eurocode, ASCE, and other published sources listed in Table 8.1. While temperature has an effect on thermal properties of concrete, published property data clearly shows that the effect of temperature on different properties of concrete (NSC, HSC, FRC, etc.) follows a similar trend overall. The thermal conductivity of concrete generally decreases with temperature rise (see Fig. 8.1*a*) due to a decrease of moisture content with a rise in temperature. It is interesting to note that the Eurocode proposes upper and lower bound limits for the thermal conductivity of concrete but without indicating which limit to use for a given aggregate type in concrete.

The variation in heat capacity of concrete with temperature, which is defined as the product of specific heat and density of concrete, as per ASCE and Eurocode 2 recommendations, is shown in Fig. 8.1*b*. ASCE relations take into account the effect of aggregate type in the evaluation of thermal capacity of concrete.[1] However, the Eurocode property relation for heat capacity does not distinguish between different aggregate types in concrete, and this can vary widely from one type of concrete to another and also with age and humidity conditions. The presence of steel fibers in concrete has a minor influence on the specific heat of normal- or high-strength concrete at room and elevated temperatures. Eurocode 2 also recommends the use of the same thermal property relations for both NSC and HSC.

Concrete generally undergoes expansion when subjected to elevated temperatures. This increase is due to high thermal expansion resulting from constituent aggregates and cement paste in concrete. Thermal expansion of concrete can reach about 1.3% at 700°C and then generally remains constant through 1000°C. The variation in thermal expansion depends on the type of aggregate as per Eurocode 2 recommendations. However, ASCE (1992) and Kodur et al. (2004) do not consider the effect of varying aggregate type on the expansion of concrete.

---

[1] Refer to Chap. 4 for more details on concrete material behavior at elevated temperatures.

(a) Thermal Properties of Normal-Strength and High-Strength Concrete

	Normal-Strength and High-Strength Concrete		
	ASCE Manual (ASCE, 1992) (NSC)	**Kodur et al. (2004) (HSC)**	**CEN (2004) (NSC and HSC)**
Thermal Capacity	Siliceous Aggregate Concrete $\rho c = \begin{cases} 0.005T + 1.7 & 20°C \leq T \leq 200°C \\ 2.7 & 200°C < T \leq 400°C \\ 0.013T - 2.5 & 400°C < T \leq 500°C \\ 10.5 - 0.013T & 500°C < T \leq 600°C \\ 2.7 & 600°C < T \end{cases}$ Carbonate Aggregate Concrete $\rho c = \begin{cases} 2.566 & 20°C \leq T \leq 400°C \\ 0.1765T - 68.034 & 400°C < T \leq 410°C \\ 25.00671 - 0.05043T & 410°C < T \leq 445°C \\ 2.566 & 445°C < T \leq 500°C \\ 0.01603T - 5.44881 & 500°C < T \leq 635°C \\ 0.16635T - 100.90225 & 635°C < T \leq 715°C \\ 176.07343 - 0.22103T & 715°C < T \leq 785°C \\ 2.566 & 785°C < T \end{cases}$	Siliceous Aggregate Concrete $\rho c = \begin{cases} 0.005T + 1.7 & 20°C \leq T \leq 200°C \\ 2.7 & 200°C < T \leq 400°C \\ 0.013T - 2.5 & 400°C < T \leq 500°C \\ -0.013T + 10.5 & 500°C < T \leq 600°C \\ 2.7 & 600°C < T \leq 635°C \end{cases}$ Carbonate Aggregate Concrete $\rho c = \begin{cases} 2.45 & 20°C \leq T \leq 400°C \\ 0.026T - 12.85 & 400°C < T \leq 475°C \\ 0.0143T - 6.295 & 475°C < T \leq 650°C \\ 0.1894T - 120.11 & 650°C < T \leq 735°C \\ -0.263T + 212.4 & 735°C < T \leq 800°C \\ 2 & 800°C < T \leq 1000°C \end{cases}$	Specific heat (J/kg ·°C) $c = 900$, for $20°C \leq T \leq 100°C$ $c = 900 + (T - 100)$, for $100°C < T \leq 200°C$ $c = 1000 + (T - 200)/2$, for $200°C < T \leq 400°C$ $c = 1100$, for $400°C < T \leq 1200°C$ Density change (kg/m^3) $\rho = \rho(20°C)$ = Reference density for $20°C \leq T \leq 115°C$ $\rho = \rho(20°C)\,[1 - 0.02\,(T - 115)/85]$ for $115°C < T \leq 200°C$ $\rho = \rho(20°C)\,[0.98 - 0.03(T - 200)/200]$ for $200°C < T \leq 400°C$ $\rho = \rho(20°C)\,[0.95 - 0.07(T - 400)/800]$ for $400°C < T \leq 1200°C$ Thermal capacity $= \rho \times c$

(b) Thermal Conductivity of Normal-Strength and High-Strength Steel Fiber-Reinforced Concrete

	Siliceous Aggregate Concrete		Carbonate Aggregate Concrete	
Normal-Strength Steel -Fiber-Reinforced Concrete	$20°C \leq T \leq 200°C$	$k = 3.22 - 0.007T$	$20°C \leq T \leq 500°C$	$k = 2.00 - 0.001775T$
	$200°C < T \leq 400°C$	$k = 2.24 - 0.0021T$	$500°C < T \leq 1000°C$	$k = 1.402 - 0.000579T$
	$400°C < T \leq 1000°C$	$k = 1.4$	-	
High-Strength Steel Fiber-Reinforced Concrete	$20°C \leq T \leq 200°C$	$k - 2.50 - 0.0034T$	$20°C \leq T \leq 500°C$	$k = 1.8 - 0.0016T$
	$200°C < T \leq 400°C$	$k = 2.24 - 0.0021T$	$500°C < T \leq 1000°C$	$k = 1.20 - 0.0004T$
	$400°C < T \leq 1000°C$	$k = 1.4$	-	

Table 8.1 Relations for Variation in Thermal Properties of Normal and High-Strength Concrete with Temperature

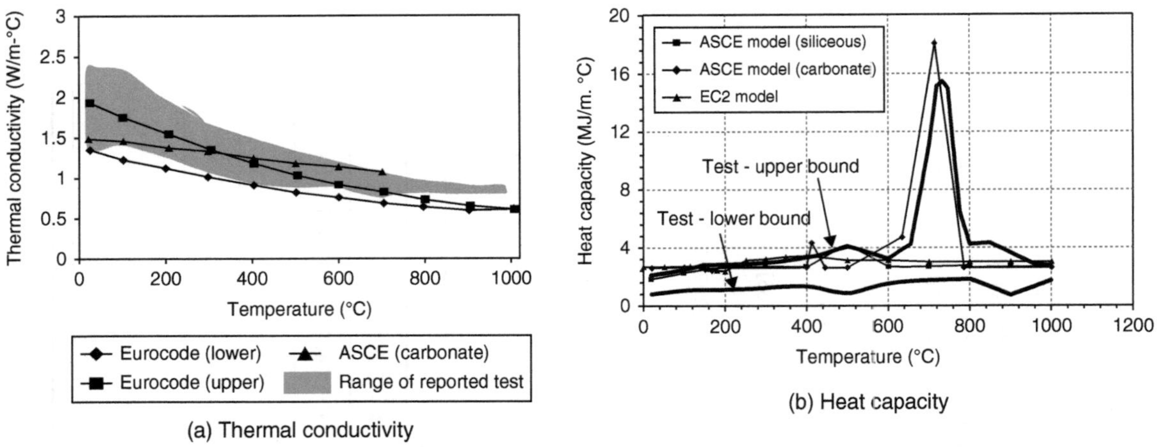

Figure 8.1 Variation in thermal properties of normal-strength concrete as a function of temperature. (*With permission from ASCE.*)

Mechanical Properties

The mechanical properties that are of primary interest in fire resistance evaluations of concrete structures are compressive strength, tensile strength, elastic modulus, and stress-strain response in compression. The degradation in mechanical properties of normal- and high-strength concrete types at elevated temperatures has been extensively studied in the literature. Data from these tests have been utilized to develop temperature-dependent property relations for normal- and high-strength concrete, and these relations are specified in ASCE (1992), Eurocode 2 (CEN, 2004), and Kodur et al. (2004), among others. ASCE trends for variation in properties are applicable for NSC, whereas Eurocode 2 has separate relations for NSC and HSC.

A closer look at such relations, plotted in Fig. 8.2, shows that the variation in mechanical properties is highly influenced by concrete strength, mainly arising from concrete batch mix proportions. The strength degradation in the case of higher strength concrete occurs at a faster pace than that in normal-strength concrete, and this is due to the dense microstructure of HSC. This can also be seen by comparing the stress-strain response of these concretes plotted in Fig. 8.2*e,f*. The reduction factors to represent variations in compressive strength of NSC and HSC with temperature are reproduced in Table 8.2 as per Eurocode 2. For generating stress-strain curves at elevated temperature, these reduction factors are to be used in conjunction with the stress-strain property relations listed in Table 8.3. Note that Eurocode 2 does not give strain values for HSC but rather refers designers to its country's National Annex for more information.

The tensile strength of concrete, due to its low magnitude, is often neglected in strength (capacity) calculations at room temperatures. However, from a fire resistance point of view, it is an important property, since cracks are induced when tensile stress in a member exceeds low tensile strength of concrete. Such cracking accelerates temperature rise in steel reinforcement during fire exposure, which in turn can significantly lower the strength of reinforcement. Further, tensile strength of concrete can induce fire-induced spalling in concrete members, with lower tensile strength of concrete leading to higher probability of spalling. The variation in splitting tensile strength of NSC and HSC with temperature, as per Eurocode 2 provisions and as reported in previous studies, is plotted in Fig. 8.2*c*. The tensile strength of concrete decreases with temperature at a higher pace than compressive strength. Eurocode 2 provides simplified relations for degradation in tensile strength of normal strength concrete as:

$$k_{c,t} = 1.0 \qquad \text{for } 20 < T < 100°\text{C} \tag{8.1}$$

$$k_{c,t} = 1.0 - 1.0\,(T - 100)/500 \qquad \text{for } 20 < T < 100°\text{C} \tag{8.2}$$

where T is temperature.

The addition of steel fibers to concrete enhances its tensile strength, but at room and elevated temperature, and the increase can be up to 50% higher at room temperature (Kodur, 1999). This increased tensile strength can delay the propagation of cracks in steel fiber-reinforced concrete structural members and is highly beneficial when the member is subjected to bending stresses.

The elastic modulus of concrete is mainly dependent on the water-cement ratio, the age of concrete, the method of conditioning, and the amount and nature of the aggregates. The variation in elastic modulus at

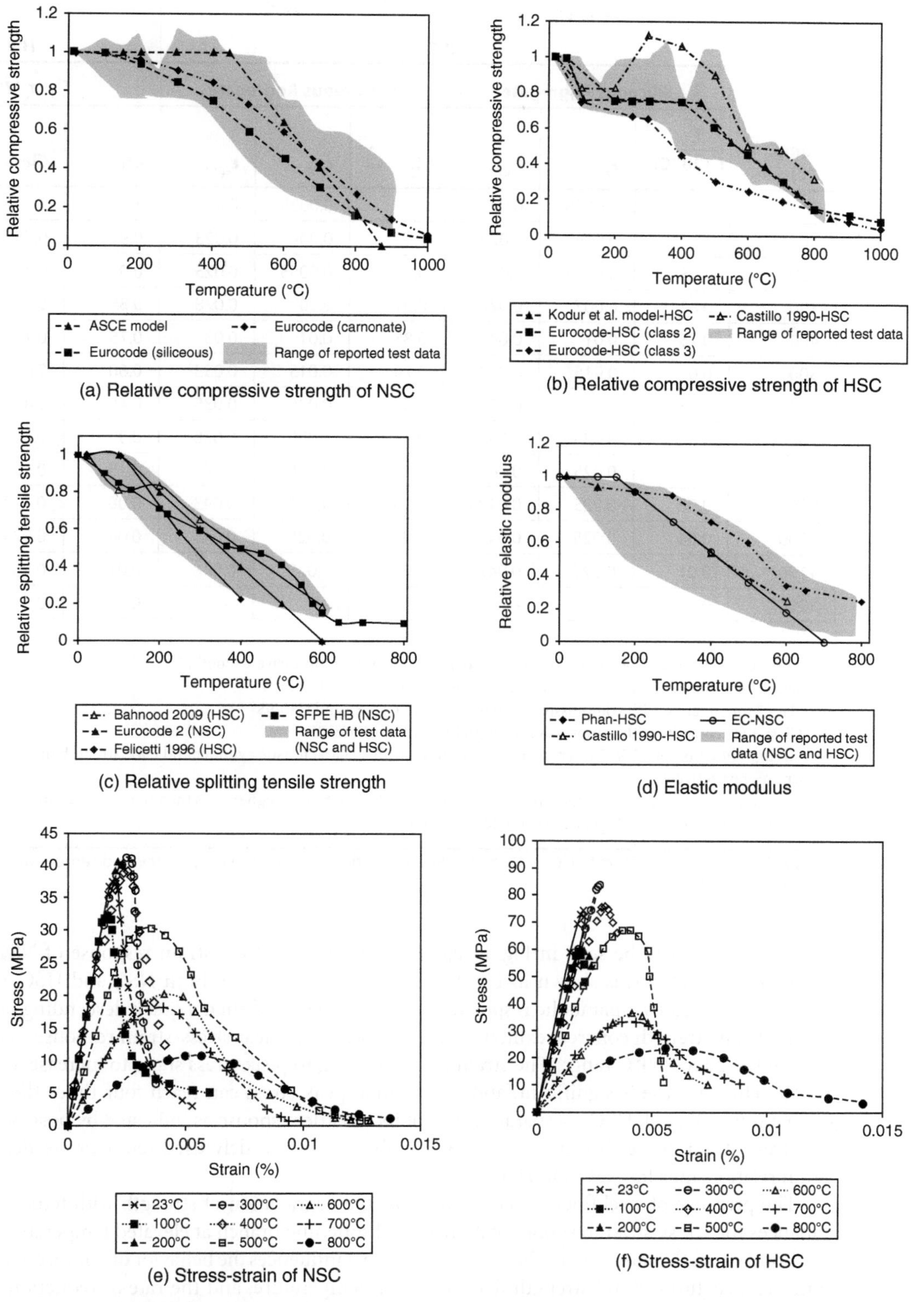

FIGURE 8.2 Variation in mechanical properties of concrete as a function of temperature. (*With permission from ASCE.*)

a given temperature to that at room temperature for NSC and HSC is shown in Fig. 8.2*d*. The degradation in modulus of both NSC and HSC can be attributed to excessive thermal stresses and physical and chemical changes in concrete microstructure. The elastic modulus of NSC and HSC starts deteriorating even at moderate temperatures (150 to 200°C). The modulus of normal-strength concrete decreases at a higher pace with the rise in temperature than that of lightweight aggregate concretes.

The compressive strength of concrete decreases with increase in temperature, and this is accompanied with increased softening of concrete. This results in decreasing slope of stress-strain response at elevated

Temp. (°C)	NSC						HSC		
	Siliceous Aggregate			Calcareous Aggregate			$f'_{c,T}/f'_c(20°C)$		
	$\frac{f'_{c,T}}{f'_c(20°C)}$	$\varepsilon_{c1,T}$	$\varepsilon_{cu1,T}$	$\frac{f'_{c,T}}{f'_c(20°C)}$	$\varepsilon_{c1,T}$	$\varepsilon_{cu1,T}$	Class 1	Class 2	Class 3
20	1	0.0025	0.02	1	0.0025	0.02	1	1	1
100	1	0.004	0.0225	1	0.004	0.023	0.9	0.75	0.75
200	0.95	0.0055	0.025	0.97	0.0055	0.025	0.9	0.75	0.70
300	0.85	0.007	0.0275	0.91	0.007	0.028	0.85	0.75	0.65
400	0.75	0.01	0.03	0.85	0.01	0.03	0.75	0.75	0.45
500	0.6	0.015	0.0325	0.74	0.015	0.033	0.60	0.60	0.30
600	0.45	0.025	0.035	0.6	0.025	0.035	0.45	0.45	0.25
700	0.3	0.025	0.0375	0.43	0.025	0.038	0.30	0.30	0.20
800	0.15	0.025	0.04	0.27	0.025	0.04	0.15	0.15	0.15
900	0.08	0.025	0.0425	0.15	0.025	0.043	0.08	0.113	0.08
1000	0.04	0.025	0.045	0.06	0.025	0.045	0.04	0.075	0.04
1100	0.01	0.025	0.0475	0.02	0.025	0.048	0.01	0.038	0.01
1200	0	—	—	0	—	—	0	0	0

* The Eurocode classifies HSC into three classes,† depending on its compressive strength:
(i) Class 1 for concrete with compressive strength between C55/67 and C60/75
(ii) Class 2 for concrete with compressive strength between C70/85 and C80/95
(iii) Class 3 for concrete with compressive strength higher than C90/105
The strength notation of C55/67 refers to a concrete grade with a characteristic cylinder and cube strength of 55 N/mm^2 and 67 N/mm^2, respectively.
†Note: Where the actual characteristic strength of concrete is likely to be of a higher class than that specified in the design, the relative reduction in strength for the higher class should be used for fire design.

Table 8.2 Reduction Factors to Generate Stress-Strain Relations of NSC and HSC at Elevated Temperatures (Eurocode 2*) *(Reproduced from CEN)*

temperatures, as can be seen in Fig. 8.2*e,f*. The variation in stress-strain response of NSC and HSC, at elevated temperatures, is shown in Fig. 8.2*e,f* (Cheng et al., 2004). Both NSC and HSC exhibit a linear response followed by a parabolic response until peak stress, and then a quick descending portion prior to failure. High-strength concrete exhibits a steeper and more linear stress-strain response in comparison to NSC in the 20 to 800°C range. The strain corresponding to peak stress starts to increase, especially above 500°C. This increase is significant, and the strain at peak stress can reach four times the strain at room temperature. Overall, HSC comparatively exhibits a brittle response as indicated by post-peak behavior of the stress-strain curve shown in Figure 8.2*e*. Table 8.3 lists widely used temperature-dependent stress-strain relationships for NSC and HSC.

The presence of steel fibers in concrete slows the degradation of strength with temperature and also enhances the stress-strain response with increased softening branch at elevated temperatures.

Bond strength is another mechanical property that influences the behavior of reinforced and prestressed concrete structures. Bond strength decreases with temperature, and the rate of reduction is higher than that of decrease in compressive strength with temperature. Bažant et al. (1996) showed that concrete, with carbonate aggregate, exhibits a slower rate of bond degradation than that of quartz aggregates at elevated temperatures. Due to the complexity involved in experimentally evaluating bond degradation at elevated temperatures, there is a significant variation in reported data on bond strength. Figure 8.3 shows bond strength degradation with temperature in the case of smooth and ribbed reinforcing rebars (Huang, 2010). The bond strength drops to 80% of room temperature strength at 400°C, beyond which the rate of degradation in bond strength is much more rapid. Bond degradation in the case of smooth rebars is almost twice as rapid as that in ribbed rebars, wherein it reduces to 40% of the room temperature capacity at 400°C. Recent research (Kodur and Agrawal, 2017) noted the increased bond strength degradation in HSC with temperature as compared to NSC.

ASCE Manual (ASCE, 1992) (NSC)	Kodur et al. (2004) (HSC)	CEN (2004) (NSC and HSC)*
$\sigma_c = \begin{cases} f'_{c,T}\left[1-\left(\frac{\varepsilon-\varepsilon_{\max,T}}{\varepsilon_{\max,T}}\right)^2\right], & \varepsilon \le \varepsilon_{\max,T} \\ f'_{c,T}\left[1-\left(\frac{\varepsilon_{\max,T}-\varepsilon}{3\varepsilon_{\max,T}}\right)^2\right], & \varepsilon > \varepsilon_{\max,T} \end{cases}$ $f'_{c,T} = \begin{cases} f'_c, & 20^\circ\text{C} \le T \le 450^\circ\text{C} \\ f'_c\left[2.011 - 2.353\left(\frac{T-20}{1000}\right)\right], & 450^\circ\text{C} < T \le 874^\circ\text{C} \\ 0, & 874^\circ\text{C} < T \end{cases}$ $\varepsilon_{\max,T} = 0.0025 + (6.0T + 0.04T^2) \times 10^{-6}$	$\sigma_c = \begin{cases} f'_{c,T}\left[1-\left(\frac{\varepsilon_{\max,T}-\varepsilon}{\varepsilon_{\max,T}}\right)^H\right] & \varepsilon \le \varepsilon_{\max,T} \\ f'_{c,T}\left[1-\left(\frac{30(\varepsilon-\varepsilon_{\max,T})}{(130-f'_c)\varepsilon_{\max,T}}\right)^2\right] & \varepsilon > \varepsilon_{\max,T} \end{cases}$ $f'_{c,T} = \begin{cases} f'_c[1.0 - 0.003125(T-20)] & T < 100^\circ \\ 0.75f'_c & 100^\circ\text{C} \le T \le 400^\circ \\ f'_c[1.33 - 0.00145T] & 400^\circ\text{C} < T \end{cases}$ $\varepsilon_{\max,T} = 0.0018 + (6.7f'_c + 6.0T + 0.03T^2) \times 10^{-6}$ $H = 2.28 - 0.012f'_c$	$\sigma_c = \frac{3\varepsilon f'_{c,T}}{\varepsilon_{c1,T}\left(2+\left(\frac{\varepsilon}{\varepsilon_{c1,T}}\right)^3\right)}, \varepsilon \le \varepsilon_{cu1,T}$ For $\varepsilon_{c1(T)} < \varepsilon \le \varepsilon_{cu1(T)}$, the Eurocode permits the use of linear as well as nonlinear descending branches in the numerical analysis. For the parameters in this equation, refer to Table 8.2.

*To be used in conjunction with reduction factors in Table 8.2.

Table 8.3 Stress-Strain Relations of Concrete at Elevated Temperature

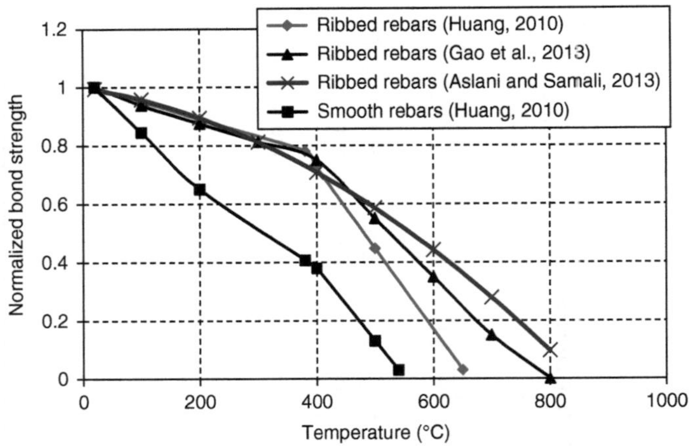

FIGURE 8.3 Bond strength degradation as a function of temperature for NSC.

Deformational Properties

Deformational properties of concrete that include creep and transient strains are highly variant with mix proportions, the type of aggregate present, and the chemical and physical reactions that occur in the concrete, which in turn depend on the rate of heating and cooling during fire exposure. Creep and transient strains become accelerated at elevated temperatures and higher stress levels. The creep phenomenon in concrete under high temperatures is attributed to moisture movement within the concrete core. Limited test data on the creep of concrete at elevated temperatures is available from the works of Cruz (1966), Gross (1975), Anderberg and Thelandersson (1976), and Schneider (1986).

Eurocode 2 implicitly accounts for partial creep in its temperature-dependent stress-strain relations, and these relations can take into account only primary and secondary creep strains that occur at elevated temperatures. To fully capture creep effects, creep strains need to be explicitly accounted for in fire resistance analysis. Kodur and Alogla (2017) proposed a numerical approach to explicitly account for creep through advanced fire resistance analysis.

In order to explicitly account for high-temperature creep effects in fire resistance analysis, temperature-dependent creep models of concrete are required. The classical steady-state creep strain (ε_{cr}) is expressed as a function of stress level, temperature, and time and is given as

$$\varepsilon_{cr}(\sigma, T, t) = -0.00053\left(\frac{\sigma}{f'_{cT}}\right)\left(\frac{t}{180}\right)^{0.5} e^{0.00304(T-20)} \tag{8.3}$$

Anderberg and Thelandersson (1976) proposed simplified creep and transient strain models at elevated temperatures, which are given as

$$\varepsilon_{cr} = \beta_1 \frac{\sigma}{f_{c,T}} \sqrt{t}\, e^{d(T-293)} \tag{8.4}$$

$$\varepsilon_{tr} = k_2 \frac{\sigma}{f_{c,20}} \varepsilon_{th} \tag{8.5}$$

where ε_{cr} = Creep strain
ε_{tr} = Transient strain
$\beta_1 = 6.28 \times 10^{-6}\ s^{-0.5}$
$d = 2.658 \times 10^{-3}\ K^{-1}$
T = Concrete temperature (K) at time t (s)
$f_{c,T}$ = Concrete strength at temperature T
σ = Stress in the concrete at the current temperature
k_2 = A constant range between 1.8 and 2.35
ε_{th} = Thermal strain
$f_{c,20}$ = Concrete strength at room temperature

Schneider's Transient Creep Model	$$(\varepsilon_{cr} + \varepsilon_{tr}) = \frac{\Phi}{g} \cdot \frac{\sigma}{E}$$ where Φ is an empirical function to account for stress and temperature effects through the function ϕ, as follows: $$\Phi(\sigma, T) = \begin{cases} g\phi + \frac{\sigma}{f_c'} \cdot \frac{T-20}{100}, & \frac{\sigma}{f_{cT}'} \le 0.3 \\ g\phi + \frac{0.3(T-20)}{100}, & \frac{\sigma}{f_{cT}'} > 0.3 \end{cases}$$ $$\phi = C_1 \cdot \tanh[\gamma_\omega \cdot (T-20)] + C_2 \cdot \tanh[\gamma_0 \cdot (T - T_g)] + C_3$$ The function g, accounts for the increase in initial tangent modulus due to preapplied stress, f_{ci}, and is given as $$g(f_{ci}, T) = \begin{cases} 1 + \frac{f_{ci}}{f_c'} \cdot \frac{T-20}{100}, & \frac{f_{ci}}{f_c'} \le 0.3 \\ 1 + \frac{0.3(T-20)}{100}, & \frac{f_{ci}}{f_c'} > 0.3 \end{cases}$$ and the symbols γ_0, C_1, C_2, and C_3 are constants for different types of concrete and aggregate. The symbol γ_ω represents a function which considers moisture content effect and is given as $$\gamma_\omega = (0.3\omega + 2.2) \times 10^{-3}$$ where ω represents moisture content in concrete.
Diederichs's Transient Creep Model	$$(\varepsilon_{cr} + \varepsilon_{tr}) = \frac{\sigma}{f_c'} \cdot f(T)$$ $$f(T) = 3.3 \times 10^{-10}(T-20)^3 - 1.72 \times 10^{-7}(T-20)^2 + 0.0412 \times 10^{-3}(T-20)$$ where $f(T)$ is an empirical function proposed by fitting experimental results.
Nielsen et al. Transient Creep Model	$$\varepsilon_{tr} = 0.000038 \cdot \left(\frac{\sigma}{f_c'}\right) \cdot T$$

TABLE 8.4 Creep Models of Concrete for Undertaking Advanced Fire Resistance Analysis

Other available creep models for normal-strength concrete are listed in Table 8.4. There is still a lack of test data and creep models to evaluate creep and transient strain in HSC and fiber-reinforced concrete types.

Spalling

Concrete structures under certain fire exposure conditions are susceptible to spalling. Such spalling results in the breakup of chunks of concrete from a concrete member, and this occurs due to the accumulation of pore pressure that develops in a concrete member at elevated temperatures. This fire-induced spalling is grouped under three stages: early spalling, intermediate spalling, and late spalling. In typical concrete structures, early spalling can start early into fire exposure (5 to 10 minutes) and can continue up to 30 to 60 minutes. This type of spalling is generally of an explosive nature and results from high pore pressure generated due to large thermal gradients developed within the concrete section. Early spalling can lead to the breakup of large chunks of concrete from the fire-exposed surface of the member and thus can have a detrimental effect on the overall fire resistance of a concrete member, particularly when the extent of spalling reaches the level of main reinforcing bars or prestressing strands.

Intermediate spalling occurs midway into fire exposure (about 30 to 90 minutes) and results in the form of surface scaling (peel-off of smaller pieces of concrete). This process can continue throughout the fire exposure duration. This type of spalling is less violent and does not significantly influence the fire resistance of a concrete member, since only thin layers (chunks) of concrete detach from the surface of the member. Late-stage spalling occurs just prior to failure of a structural member and primarily results from a significant loss of strength and stiffness properties in concrete (and reinforcement) due to prolonged fire exposure. Late-stage spalling is more due to load-induced phenomenon and can reduce overall fire resistance of a member to a certain extent.

Overall, fire-induced spalling leads to the loss of the concrete section, an accelerated temperature rise, and the rapid strength loss in rebars, thereby lowering the fire resistance of a concrete member. The main factors influencing fire-induced spalling in concrete are moisture content, concrete permeability, concrete strength, rate of heating, and stress levels (Kodur and Phan, 2007). Compared to normal-strength concrete members, high-strength concrete members are much more susceptible to fire-induced spalling, due to the low permeability and dense microstructure of HSC, which can prevent the escape of steam generated from temperature-induced pore pressure within a concrete section. In addition, if high-strength concrete members are subjected to higher stress levels, the probability of fire-induced spalling further increases. There are two broad theories for explaining the spalling phenomenon in concrete:

- *Pressure buildup.* Spalling is caused by the buildup of pore pressure during heating. HSC is believed to be more susceptible to this pressure buildup than NSC because of its low permeability. The extremely high water vapor pressure, generated during exposure to fire, cannot escape due to the high density (and low permeability) of HSC. When the effective pore pressure (porosity times pore pressure) exceeds the reduced tensile strength of concrete, chunks of concrete fall off from the structural member. This falling off concrete can often be explosive depending on the fire and concrete characteristics (Anderberg, 1997). This phenomenon is shown in Fig. 8.4.
- *Restrained thermal dilatation.* This hypothesis is based on the phenomenon that spalling results from restrained thermal dilatation close to the fire-exposed heated surface. Such thermal dilatation leads to the development of compressive stresses parallel to the heated surface as shown in Fig. 8.5. These

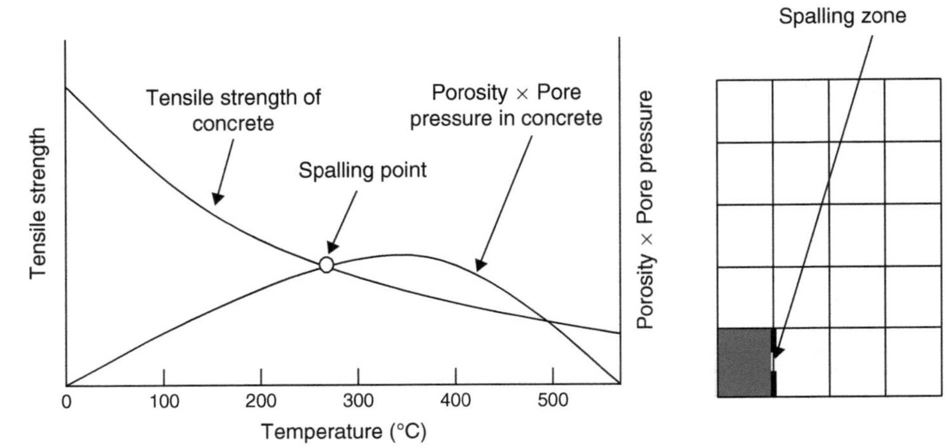

FIGURE 8.4 Spalling through pressure buildup mechanism.

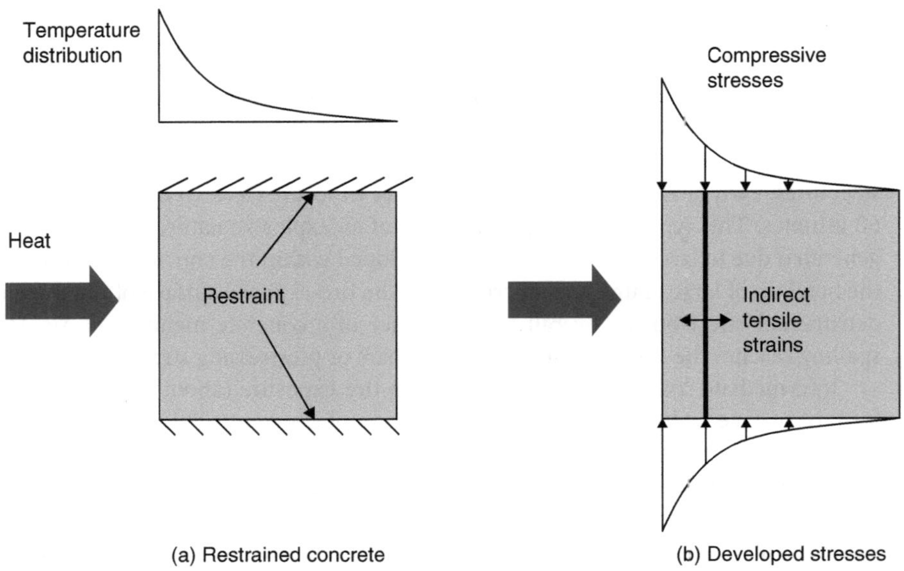

FIGURE 8.5 Spalling through restrained thermal dilatation mechanism.

compressive stresses are released by the brittle fracture of concrete referred to as spalling. The pore pressure developed within the concrete can play a significant role in the onset of instability in the form of the breakage of concrete chunks (Bažant, 1997).

Spalling of concrete is governed by its *hygral properties*, which include to a large extent the moisture diffusion, pore pressure, permeability, and porosity levels. Bažant et al. (1996) infer that moisture content of concrete at elevated temperatures is affected by the size and shape of the concrete member, rate and period of heating, and environmental conditions. At temperatures above 100°C, moisture transforms into vapor, which results in pore pressure developing internally. This pore pressure is a function of temperature and of the pore size of the specimen. An example of pore pressure generation in a heated concrete is shown in Fig. 8.6, as measured by Bažant et al. (1996) in concrete wall specimens under elevated temperature exposure.

Dwaikat and Kodur (2009) developed a hydrothermal-based numerical spalling model for evaluating temperature-induced pore pressure and spalling in concrete members by giving due consideration to most of the parameters influencing spalling. In this model, the governing equation that determines the water vapors in concrete and hence spalling is

$$A\frac{dP_V}{dt} = \nabla B \nabla P_V + C \tag{8.6}$$

where

$$A = \left[\left(1 - \frac{m_V}{V_V \rho_L}\right)\frac{dm_L}{dP_V} + \frac{V_V M}{RT}\right],\ B = m_V \frac{k_T}{\mu_V}$$

$$C = \left[\left(1 - \frac{m_V}{V_V \rho_L}\right)\left(-\frac{dm_L}{dT} + \frac{dm_D}{dT}\right) + \frac{m_V}{T} + \frac{m_V}{V_V \rho_L^2}\frac{d\rho_L}{dT}(m_D - m_L)\right]\frac{dT}{dt}$$

where m_V = Mass of water vapor
M = Molar mass of water
V_V = Volume fraction of water vapor
T = Temperature (K)
m_L = Mass of liquid water at any time, t
m_D = Mass of liquid water formed due to dehydration
T = Temperature
P_V = Vapor pressure
ρ_L = Density of liquid water
t = Time

Accounting for spalling involves a complex set of calculations. Further details on the derivation of Eq. 8.6 can be found elsewhere (Dwaikat and Kodur, 2009).

Permeability represents the rate of flow of a fluid/gas into a porous solid. The permeability depends on the porosity, as it increases with an increase in the average pore size, and on the type of aggregate. Also, permeability can increase with the formation of microcracks caused by the sudden change of the phase of liquid water into steam. Permeability is expressed through intrinsic permeability and relative permeability. Compared to normal-strength concrete, HSC or high-performance concrete has lower permeability. The permeability, k, can be evaluated as (Gawin et al., 1999):

$$k = k_0 10^{A_k (T - T_0)} (P_g / P_0)^{B_k} \tag{8.7}$$

where A_k and B_k are constants that depend on the concrete type and k_0 is the initial permeability at the reference temperature $T_0 = 293.15$ K and reference pressure $P_0 = 101325$ Pa. For silicate concrete, $A_k = 0.005$ and $B_k = 0.368$.

Porosity is the ratio of the pore volume to the total volume in concrete. Gradual heating of concrete causes changes in the distribution of pore structure, with the increase in total pore volume affecting the distribution of pore sizes. As can be seen in Fig. 8.6*b*, heating up concrete results in a further increase in pore volume, with limestone concrete achieving the highest porosity. Based on experimental results carried out

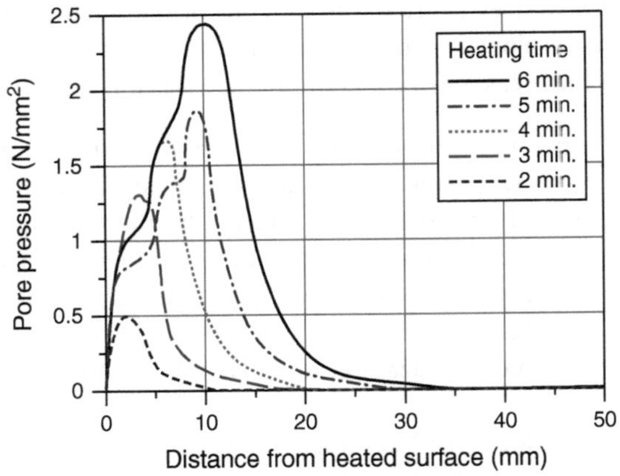

(a) Pore pressure in a concrete wall exposed to fire from one side

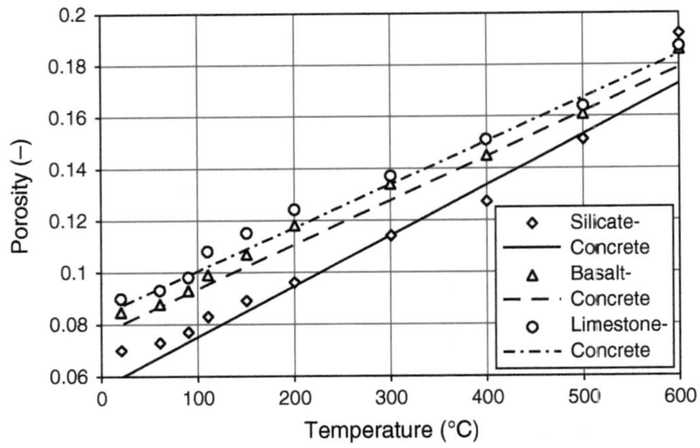

(b) Comparison of the experimental values of total porosity for three types of concrete

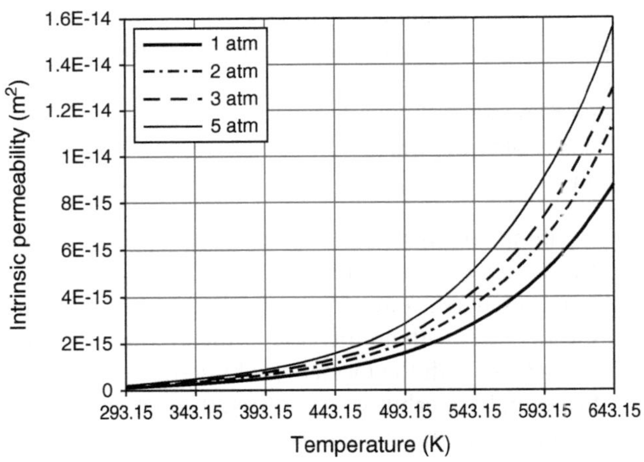

(c) Intrinsic permeability of concrete as a function of temperature

FIGURE 8.6 Spalling-related properties of concrete.

by Schneider et al. (1980), Schneider and Herbst (1989), and Bažant et al. (1996) on three types of concrete, the change in porosity, ϕ, with temperature was approximated by a linear relation (Gawin et al., 1999):

$$\phi(T) = \phi_0 + A_\phi(T - T_0) \tag{8.8}$$

where ϕ_0 can be assumed at 9×10^{-20} m^2 and A_ϕ is a constant that depends on the type of concrete. For limestone concrete $A_\phi = 0.000165$ K^{-1}.

(a) Reinforcing and Prestressing Steel as per Eurocode 3 Properties

Thermal Conductivity	$20°C \le T \le 800°C$	$k = 54 - 3.33 \times 10^{-2}T$
	$800°C \le T \le 1200°C$	27.3
Specific Heat	$20°C \le T \le 600°C$	$c = 425 + 7.73 \times 10^{-1}T - 1.69 \times 10^{-3}T^2 + 2.22 \times 10^{-6}T^3$
	$600°C \le T \le 735°C$	$c = 666 + 13002/738 - T$
	$735°C \le T \le 900°C$	$c = 545 + 17820/T - 731$
	$900°C \le T \le 1200°C$	$c = 650$

(b) Thermal Strain

Reinforcing Steel	ASCE	$20°C \le T \le 1000°C$	$\varepsilon_{ths} = 1.0 \times 10^{-5}T + 0.4 \times 10^{-8}T^2 - 2.016 \times 10^{-4}$
	Eurocode2	$20°C \le T \le 750°C$	$\varepsilon_{ths} = -2.416 \times 10^{-4} + 1.2 \times 10^{-5}T + 0.4 \times 10^{-8}T^2$
		$750°C \le T \le 860°C$	$\varepsilon_{ths} = 11 \times 10^{-3}$
		$860°C \le T \le 1200°C$	$\varepsilon_{ths} = -6.2 \times 10^{-3} + 2 \times 10^{-5}T$
Prestressing Steel	Eurocode2	$20°C \le T \le 1000°C$	$\varepsilon_{ths} = -2.016 \times 10^{-4} + 10^{-5}T + 0.4 \times 10^{-8}T^2$

Table 8.5 High-Temperature Thermal Properties of Reinforcing and Prestressing Steel

8.2.4 High-Temperature Property Relations of Steel Reinforcement

Thermal Properties

The temperature-dependent thermal properties of reinforcing steel, prestressing strands, and wires primarily vary with the composition of steel. The thermal conductivity of reinforcing steel is up to 50 times higher in comparison to concrete, and thus heat gets transmitted through steel rebars quite rapidly. However, in concrete members the area of reinforcing and prestressing steel is very small compared to the concrete area and, further, the reinforcement is embedded in concrete. For these reasons, in fire resistance calculations, steel reinforcement is generally assumed to be a perfect conductor, which implies that temperature is uniform within the steel area. Therefore, thermal conductivity and the specific heat of steel reinforcement are not specifically taken into consideration in evaluating cross-sectional temperature in a concrete member.

However, temperature-dependent thermal expansion of reinforcing (and prestressing) steel is an important property that needs to be considered in analysis, as it can influence the deformational response of fire-exposed reinforced concrete (RC) members, especially under restraint supports. Thermal expansion linearly increases with temperature rise. Different thermal properties of reinforcing and prestressing steel are provided in Eurocode 3 and the ASCE manual and are listed in Table 8.5.

Mechanical Properties

Mechanical properties of reinforcing and prestressing steel needed for fire resistance analysis include yield strength, modulus of elasticity, and stress-strain relations. These properties decrease more rapidly than those of concrete at elevated temperatures. The relations for reduction in yield strength of reinforcing and prestressing steel with temperature as specified in the ASCE (1992) manual and Eurocodes 2 and 3 are provided in Table 8.6.

(a) ASCE Manual

Reinforcing Steel	Yield strength	$20°C \le T \le 600°C$	$f_{y,T} = \left[1.0 + \frac{T}{900\, ln\left(\frac{T}{1750}\right)}\right] f_{yo}$
		$20°C \le T \le 1000°C$	$f_{y,T} = \left[\frac{340 - 0.34T}{t - 240}\right] f_{yo}$
	Modulus of elasticity	$20°C \le T \le 600°C$	$E_{y,T} = \left[1.0 + \frac{T}{2000\, ln\left(\frac{T}{1100}\right)}\right] E_{yo}$
		$20°C \le T \le 1000°C$	$E_{y,T} = \left[\frac{690 - 0.69T}{t - 53.5}\right] E_{yo}$

(b) Eurocode 2*

Reinforcing Steel	Yield strength	$20°C \le T \le 350°C$ $350°C \le T \le 500°C$ $500°C \le T \le 700°C$ $700°C \le T \le 1200°C$	1.0 $1.0 - 0.4(T - 350)/150$ $0.61 - 0.5(T - 500)/200$ $1.0 - 0.1(T - 700)/500$
Prestressing Steel	Yield strength	$20°C \le T \le 100°C$ $100°C \le T \le 350°C$ $350°C \le T \le 500°C$ $550°C \le T \le 1200°C$	1.0 $1.0\text{-}0.45(T - 100)/250$ $0.55 - 0.45(T - 350)/200$ $1.0 - 0.1(T - 550)/650$

*Eurocode provides reduction factors to evaluate degradation in modulus properties. These factors are similar to that of structural steel and are presented in Chap. 7.

Table 8.6 Relations for Temperature-Induced Reduction in Yield Strength and Modulus of Elasticity of Reinforcing and Prestressing Steel

For the case of prestressing strands, the variation in yield strength and elastic modulus of low relaxation (LR) prestressing strands with temperature is specified in Eurocode 2 for cold-worked wires and strands. For the case of quenched and tempered (hot-rolled bars) prestressing steel at elevated temperatures, the reduction factors are shown in Table 8.7.

Reinforcing and prestressing recover nearly all of the original (yield) strength upon cooling, as long as the temperature in steel during heating phase does not exceed 500°C (Neves, 1996). The temperature-dependent stress-strain relations for reinforcing and prestressing steel in Eurocode 2 are idealized into a uniform trilinear form, by truncating response at a stress level, typically taken as the yield stress and ignoring any strain hardening phase. This idealization can be modeled through relations presented in Sec. 7.2 for reinforcing steel. For prestressing steel, temperature-dependent relations for generating strain curves are given in Table 8.2, together with reduction factors needed to generate stress-strain curves.

Deformational Properties

Deformational properties of reinforcing and prestressing steel include creep strain. Creep can be a significant parameter in steel, which occurs due to the movement of dislocations in the slip plane within the microstructure of steel. Normally, the steel microstructure contains a variety of defects, for example, solute atoms, that act as obstacles to dislocation movement. At room temperature, the amount and

(a) Stress-Strain Relations

Stress-Strain Relations

$$\sigma_s = \begin{cases} \varepsilon_p E_{p,T} & \varepsilon_p \le \varepsilon_{pp,T} \\ f_{pp,T} - c + (b/a)(a^2 - (\varepsilon_{py,T} - \varepsilon_p)^2)^{0.5} & \varepsilon_{pp,T} < \varepsilon_p \le \varepsilon_{py,T} \\ f_{py,T} & \varepsilon_{py,T} < \varepsilon_p \le \varepsilon_{pt,T} \\ f_{py,T}\left(1 - \dfrac{\varepsilon_p - \varepsilon_{pt,T}}{\varepsilon_{pu,T} - \varepsilon_{pt,T}}\right) & \varepsilon_{pt,T} < \varepsilon_p \le \varepsilon_{pu,T} \\ 0.0 & \varepsilon_p > \varepsilon_{pu,T} \end{cases}$$

Parameters

$$\varepsilon_{pp,T} = \frac{f_{pp,T}}{E_{p,T}} \qquad \varepsilon_{py,T} = 0.02 \qquad \varepsilon_{pt,T} = 0.15 \qquad \varepsilon_{pu,T} = 0.2$$

Functions

$$a^2 = (\varepsilon_{py,T} - \varepsilon_{pp,T})\left(\varepsilon_{py,T} - \varepsilon_{pp,T} + \frac{c}{E_{p,T}}\right)$$

$$b^2 = c(\varepsilon_{py,T} - \varepsilon_{pp,T})E_{p,T} + c^2$$

$$c = \frac{(f_{py,T} - f_{pp,T})^2}{(\varepsilon_{py,T} - \varepsilon_{pp,T})E_{p,T} - (f_{py,T} - f_{pp,T})}$$

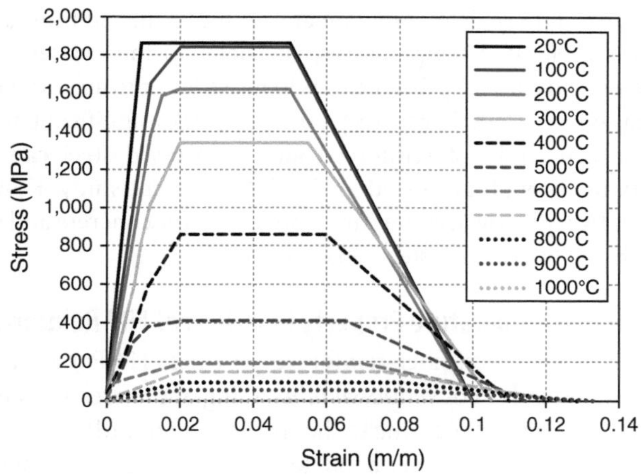

Stress-strain curves for prestressing strands under elevated temperature

(b) Reduction Factors (*With permission from Shakya, 2014*)

Steel Temp. T (°C)	$\frac{F_{py,T}}{\beta f_{yp}}$ cw			$\frac{F_{pp,T}}{\beta f_{yp}}$		$\frac{E_{p,T}}{E_p}$		$\varepsilon_{pt,T}[-]$	$\varepsilon_{pu,T}[-]$
	Class A	Class B	*q* and *t*	*cw*	*q* and *t*	*cw*	*q* and *t*	*cw*, *q* and *t*	*cw*, *q* and *t*
1	2a	2b	3	4	5	6	7	8	9
20	1	1	1	1	1	1	1	0.05	0.1
100	1	0.99	0.98	0.68	0.77	0.98	0.76	0.05	0.1
200	0.87	0.87	0.92	0.51	0.62	0.95	0.61	0.05	0.1
300	0.7	0.72	0.86	0.32	0.58	0.88	0.52	0.055	0.105

Table 8.7 Stress-Strain Relations and Associated Strength Reduction Factors for Prestressing Steel at Elevated Temperatures

400	0.5	0.46	0.69	0.13	0.52	0.81	0.41	0.06	0.11
500	0.3	0.22	0.26	0.07	0.14	0.54	0.2	0.065	0.115
600	0.14	0.1	0.21	0.05	0.11	0.41	0.15	0.07	0.12
700	0.06	0.08	0.15	0.03	0.09	0.1	0.1	0.075	0.125
800	0.04	0.05	0.09	0.02	0.06	0.07	0.06	0.08	0.13
900	0.02	0.03	0.04	0.01	0.03	0.03	0.03	0.085	0.135
1000	0	0	0	0	0	0	0	0.09	0.14
1100	0	0	0	0	0	0	0	0.095	0.145
1200	0	0	0	0	0	0	0	0.1	0.15

Note: For intermediate values of temperature, linear interpolation may be used.

Where:

cw = cold worked, q = quenched, and t = tempered

$$\beta = \begin{cases} \left(\dfrac{\varepsilon_{ud} - f_{po,1T} / E_p}{\varepsilon_{uT} - \varepsilon_{po,1T} / E_p}\right) \times \left(\dfrac{f_{pT} - f_{po,1T}}{f_{pT}}\right) + \dfrac{f_{po,1T}}{f_{pT}} & ClassA \\ 0.9 & ClassB \end{cases}$$

$\varepsilon_{ud}, \varepsilon_{uT}, f_{po,1T}, f_{pT}$, and E_p are material properties at room temperature as per EN1992-1-1.

Table 8.7 Stress-Strain Relations and Associated Strength Reduction Factors for Prestressing Steel at Elevated Temperatures (*Continued*)

distribution of these defects remain almost uniform, and thus creep strains occur at a very slow time pace. However, at elevated temperatures, vacancies in the crystal structure can diffuse to the locations of a dislocation and cause the dislocations to move faster to an adjacent slip plane, therefore allowing for further deformation to occur. Thus, creep strain becomes significant at elevated temperatures, particularly above 450°C. Creep strain of reinforced and prestressed steel can be explicitly included in fire resistance calculations through advanced analysis. In a recent study, Kodur and Alogla (2017) have developed an approach to explicitly account for creep strains for concrete and structural steel, presented in Chap. 7. This approach can be used for steel rebars as well.

8.2.5 High-Temperature Property Relations of FRP Reinforcement

In some specific cases, fiber-reinforced polymers (FRPs) are used as internal reinforcement in concrete structures, instead of conventional reinforcing or prestressing steel. In such cases, the thermal properties of FRP are not of importance due to the smaller size of FRP rebars as compared to the thermal mass provided by the concrete. On the other hand, temperature-dependent mechanical properties of FRP are to be input for fire resistance analysis. There is a good amount of data on these properties, and some of the temperature-dependent thermal and structural properties of FRPs are presented in Chap. 11 (Bisby, 2003).

8.3 Behavior of Concrete Structures under Fire Exposure

Concrete structures pose good inherent fire resistance, and this is mainly due to low thermal conductivity, high specific heat, and slower degradation of mechanical properties of concrete with temperature, combined with the larger mass of concrete members. The primary factors that influence the level of fire resistance in a reinforced (or prestressed) concrete structural member are cover thickness to steel reinforcement and the mass (size) of the concrete member. Other factors that influence the fire resistance of concrete members include the level of applied loading,[2] aggregate type, restraint (support) conditions, and moisture content.

Under fire conditions, the behavior of a concrete structural member can be quite different from that at room temperature. This is mainly due to the fact that properties of concrete as well as reinforcing and

[2] Heavily loaded reinforced concrete members generate higher internal stresses to counterbalance (resist) the applied loading. Thus, a reinforced concrete member subjected to a high level of loading can reach critical stage (i.e., yield limit of steel) much earlier than a member loaded with a lower load level, especially under fire conditions when strength properties degrade rapidly.

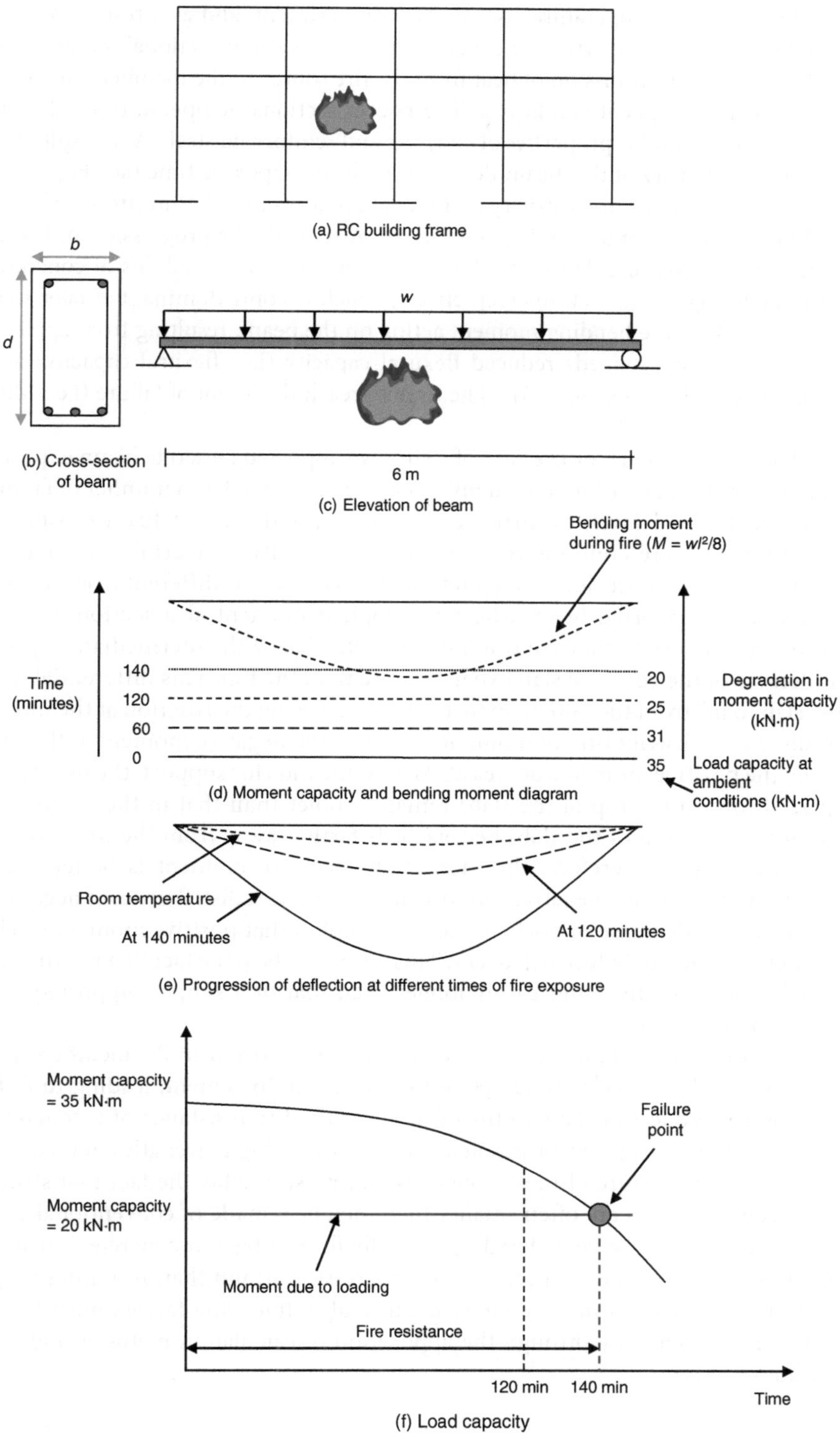

Figure 8.7 Behavior of a typical RC beam under fire exposure. (*With permission from Dwaikat, 2009.*)

prestressing steel degrade with a rise in temperature. Further, temperature-induced mechanisms, such as spalling, cracking, development of restraint forces, and so on, also contribute to member degradation. The temperature-induced reduction in properties of concrete and steel leads to a gradual decrease in the capacity and stiffness of the member, as well as leads to excessive cracking and large deflections. This will eventually lead to the failure of the member.

To illustrate the response of a typical reinforced concrete member under fire conditions, the behavior of a typical RC beam located in a concrete building is illustrated in Fig. 8.7. When fire breaks out in the building, the fire temperature rises in the compartment, and as a result, the RC beam (as well as other members in that compartment) experiences an increase in sectional temperatures through the radiative and convective transmission of heat from the fire source to the member. The beam has a flexural capacity of 35 kN·m at ambient conditions. The rise in sectional temperature results in gradual degradation in strength and modulus properties of concrete and reinforcing steel.[3] As a result, the load-carrying (moment and shear) capacity of this beam decreases with fire exposure time (see Fig. 8.7*d,f*). After 140 minutes of fire exposure, the flexural capacity of this beam reduces to 20 kN·m (from 35 kN·m at ambient conditions) in 2 hours and continues to degrade further. Also, with the progression of fire exposure time, the beam experiences increasing deflection due to a reduction in the modulus of concrete and reinforcement and also due to high-temperature creep effects, which become dominant at later stages of fire exposure (see Fig. 8.7*e*). Once the bending moment acting on the beam, resulting from applied loading present during the fire exposure, exceeds reduced flexural capacity (i.e., flexural capacity) at 140 minutes, the beam experiences failure (see Fig. 8.7*f*). The time to reach this point of failure (i.e., 140 minutes) is taken as the fire resistance of this beam.

The illustration is for the case of a simply supported concrete beam exposed to fire. In other cases, the fire resistance of a concrete member can be influenced by a number of factors such as level restraint at supports, load level, occurrence of spalling, and type of fire exposure. For instance, concrete structures that are continuous or otherwise statically indeterminate undergo changes in moments when subjected to fire due to a reduction of forces. This is different from that shown above for the case of a simply supported beam where the applied moment at a section remains constant during fire exposure. In a two-span continuous concrete slab with intermediate support at the midspan, the underside of the concrete slab expands more than the top. This differential heating causes the ends of the slab to lift from the outer supports, thus increasing the reaction at the interior support. This action results in a redistribution of moments—that is, the negative moment at the interior support increases while the positive moments decrease. Also, at the interior support, the negative moment reinforcement (placed toward the top of the slab) remains cooler than that in the positive moment region (placed toward the bottom of the slab) because it is farther away from the fire exposure zone (i.e., increased fire protection). Therefore, the increase in negative moment is better accommodated due to the combination of cooler negative reinforcement and the redistribution of negative and positive moments. The resulting decrease in positive moment implies that positive moment steel can withstand a higher temperature before failure will occur. This all amounts to the fact that the fire resistance of a continuous member is generally significantly longer than that of a simply supported member having the same cover and load intensity.

In other cases, and if spalling of concrete were to occur in PC members, it can lead to loss of cross section as well as accelerate temperature rise in reinforcement, leading to faster degradation of rebar mechanical properties. This in turn can lower the fire resistance of a concrete member. The effect of spalling can be a significant problem in the case of high-strength concrete and prestressed concrete, given the dense nature of these concretes and most notably the fact that structural members made of these concrete types are often smaller than members made of conventional concretes. Another critical point that needs to be considered, specifically for prestressed concrete structures, is faster degradation in strength and modulus properties of prestressing strand than in reinforcing steel. Therefore, in fire resistance evaluations of concrete structures, all influencing factors must be given due consideration. This can be done only through the application of calculation methods and not through standard fire resistance tests.

8.4 Strategies for Achieving Required Fire Resistance

Concrete structures generally possess a high inherent fire resistance (as compared to those made of steel or timber) and therefore in most cases can be designed to achieve required fire resistance even up to 4 hours, without any external fire insulation. The two primary factors that determine the level of fire resistance in a concrete structure are sectional size (mass of concrete) and cover thickness to steel reinforcement. Concrete structures with a larger mass experience slower temperature rise due to high thermal inertia, resulting from the lower thermal conductivity and higher heat capacity of concrete. Also, larger cover thickness to steel reinforcement slows down temperature rise in rebars or strands, which in turn slows down degradation

[3] In the case of a PC beam, a loss in the prestressing force will also occur due to temperature-induced relaxation.

of strength in reinforcement. Thus, slowing down temperature rise in steel reinforcement through larger concrete cover and also in the inner concrete layers, through larger cross-sectional size of the member, is the main strategy that can be adopted in achieving the required fire resistance in a concrete structure. Through the provision of sufficient sectional size and cover thickness to reinforcement, fire resistance up to 1 to 4 hours can be achieved in most concrete members.

In the case of concrete structures, concrete cover thickness provisions are to be met to satisfy corrosion requirements. These cover thickness requirements can be sufficient to achieve 1 to 2 hours of fire resistance in a member. However, if the required fire resistance is higher than 3 hours, then additional cover thickness may be needed.

If a concrete member is susceptible to fire-induced spalling, as in the case of high-strength concrete members, additional measures are to be incorporated to mitigate spalling effects, since such spalling can cause a loss of cross section, which in turn leads to the loss of sectional capacity at a rapid pace and thus lowers the fire resistance of the member. In such cases, specific strategies to mitigate spalling in a member are to be incorporated. Optimal strategies to enhance fire resistance of concrete structural members are discussed in Sec. 8.8.

Fire resistance of a concrete structure can also be enhanced through the provision of external fire insulation as in the case of steel structures. The provision of such insulation delays temperature rise in the section, thereby slowing down degradation in strength and stiffness properties of concrete and steel. This in turn will slow down capacity degradation and thus enhance the fire resistance of a reinforced or prestressed concrete structural member. This strategy is rarely used in practice, since concrete members can be designed to achieve the required fire resistance through the provision of sufficient sectional dimensions and cover thickness to reinforcement (i.e., rebars and strands). However, fire insulation strategies can be an effective strategy in the case of repair or retrofitting concrete members, which are also required to have higher fire resistance levels than the original (unstrengthened) member. Another application for insulation strategies is in the case of externally strengthened concrete members by incorporating fiber-reinforced polymers (FRPs). Strategies on enhancing fire resistance of FRP-strengthened structures are discussed further in Chap. 11.

8.5 Methods for Evaluating Fire Resistance

The fire resistance of concrete structural members, similar to other structural members, can be evaluated through full-scale fire tests or through calculation methods. In a standard fire test, a concrete member is exposed to a standard fire exposure for a certain duration, and if the member maintains load-bearing functionality, as well as satisfies insulation and integrity criteria, the member is said to attain fire resistance equivalent to exposure duration. Data obtained through such fire tests is utilized to assign fire resistance ratings for structural members, and these ratings are listed in standards as well as industry directories. An alternative to undertaking fire tests is to apply engineering approaches for evaluating the fire resistance of reinforced and prestressed concrete members. These approaches can range from simple hand calculation methods to sophisticated advanced analysis approaches through the use of finite element-based numerical models.

8.6 Prescriptive-Based Approaches for Evaluating Fire Resistance

Current provisions for fire resistance evaluation of concrete members in most codes and standards are prescriptive based. These provisions are derived from standard fire tests and as such are only applicable to standard fire exposure as applied in tests. Further, most of these provisions assume failure to occur when steel reinforcement attains a critical temperature, without due consideration to realistic failure limit state (strength or stability failure limit states). Prescriptive-based approaches can be grouped under tabulated data and correlation equations and are discussed below.

8.6.1 Tabulated Data

Tabulated data provides the simplest means to satisfy prescribed fire rating requirements in concrete members. Fire ratings for concrete structural members are specified in terms of minimum dimensions of a member, together with minimum concrete cover thickness to steel reinforcement. In some standards, this tabulated rating takes into consideration support conditions (simply supported or restrained conditions) as well as aggregate type in concrete and provides separate size and cover thickness requirements depending on aggregate type and restraint conditions. The size and cover thickness requirements for ratings are different for reinforced and prestressed concrete members, since prestressing steel loses its strength at a faster rate under elevated temperature exposure. These fire ratings, specified as tabulated data, are applicable only under standard fire exposure conditions. Further, these ratings do not fully take into account the level of applied load or restraint conditions present in the member that can significantly influence the fire resistance of the member. Although the main criterion used in compiling tabulated fire ratings is mostly based on data from standard fire tests, the format of such tabulated data, as well as some of the assumptions, varies from one standard to another. The tabulated data in some of the standards such as BS 8110, Eurocode 2, AS 3600, and ACI 216.1 are briefly discussed here.

British Standard (BS 8110)

The British Standard (BS 8110, 2010) provides the simplest and most direct table for deriving fire ratings in a reinforced or prestressed concrete member. The table, reproduced here in as Table 8.8, specifies minimum width (or depth) and minimum cover thickness requirements to achieve fire ratings of 0.5, 1, 1.5, 2, 3, and 4 hours in RC beams, columns, and floors made of concrete. Concrete cover is defined as the distance from the exterior surface of concrete member to the surface (edge) of reinforcing steel. BS 8110 also provides a similar table for prestressed concrete members. This table lists minimum width and cover thickness requirements to achieve fire ratings in the range of 0.5, 1, 1.5, 2, 3, and 4 hours and is applicable only to standard fire exposure (ISO 834 or equivalent). Table 8.8 does not take into account geometric features (width and height) or level of applied loading. BS 8110, however, provides

Reinforced Concrete Members									
Fire Rating (hours)	Beams			Floors			Columns	
	Minimum Width	Simply Supported	Continuous	Minimum Thickness	Simply Supported	Continuous	Width	Cover
0.5	200	20	20	75	20	20	150	20
1.0	200	20	20	95	20	20	200	20
1.5	200	20	20	110	25	20	250	20
2.0	200	40	30	125	35	25	300	25
3.0	240	60	40	150	45	35	400	25
4.0	280	70	50	170	55	45	450	25
Prestressed Concrete Members**								
Fire Rating (hours)	Beams		Floors					
	Simply Supported	Continuous	Simply Supported	Continuous				
0.5	20	20	20	20				
1.0	20	20	20	20				
1.5	35	20	30	25				
2.0	60	35	40	35				
3.0	70	60	55	45				
4.0	80	70	65	55				

*For the purpose of assessing a nominal cover for beams and columns, the cover to main bars has been reduced by a notional allowance for stirrups of 10 mm to cover the range 8 to 12 mm.

** BS 8110 states that rapid rates of heating, large compressive stresses, or high moisture content (over 5% by volume or 2% to 3% by mass of dense concrete) can lead to spalling of concrete cover at elevated temperatures, particularly for thicknesses exceeding 40 to 50 mm. Strategies to mitigate fire-induced spalling are discussed in Sec. 8.8.

Table 8.8 Minimum Sectional Dimensions (mm) and Minimum Cover Thickness (mm) Requirement for Achieving Fire Ratings in Reinforced and Prestressed Concrete Structural Members* (*Authorized Reprint from ACI, 2014*)

supplementary tables for columns, and this table accounts for fire exposure conditions (full exposure, 50% exposure, or one-face exposure), as well as the nature of aggregate type (dense vs. lightweight) and restraint conditions in the case of beams and slabs. This type of prescriptive fire rating table is adopted in many other codes and standards, including the National Building Code of India and United Arab Emirates Code.

Eurocode 2 (EC2)

Eurocode 2 (CEN, 2004), *Part 1—2: Structural Fire Design*, lists fire ratings for reinforced and prestressed concrete members of common structural shapes and sizes in a tabulated format. The basis for this data is through fire tests carried out as per ISO 834 standard. A set of tables composed of minimum distance from the center of steel bars to the heated edge of cross section along with minimum sectional dimensions required for a certain fire rating of a specific concrete member type is listed. Eurocode 2 fire rating tables are derived based on a limiting (critical) temperature in steel reinforcement: 500°C for reinforcing steel, 400°C for prestressed bars, and 350°C for prestressed wires and strands. These values assume failure to occur under a loading level of 70% of initial capacity in a fire situation—meaning that, in the case the other factors stay equal, the steel is only loaded with 70% of the maximum allowed stress at normal temperature (cold situation).

The tabulated data in Eurocode 2 applies to normal-weight concrete (density of 2000 to 2600 kg/m^3) made with siliceous aggregates, but the specifications state that when calcareous aggregates or lightweight aggregates are used in beams or slabs, the minimum dimension of the cross section may be reduced by 10%. The same tables can also be applied for high-strength concrete if the minimum cross-section dimension is increased by

- $(k - 1)a$ for walls and slabs exposed on one side only
- $2(k - 1)a$ for all other structural members and the axis distance is factored by k

where k is the factor given of a value equal to 1.1 for Class 1 and 1.3 for Class 2 and a is the axis distance required.

The codal provisions further state that when tabulated data is used in fire design, no further checks are required concerning shear or torsion capacity, or for anchorage details. Also, no further checks are required concerning spalling, except when cover to surface reinforcement exceeds 70 mm. Eurocode 2 provides a separate set of tables for fire ratings of different member types (i.e., separate tables for walls, beams, columns, slabs, etc.). A sample of such a table for RC and PC beams is shown in Table 8.9.

Eurocode 2 also provides tables for fire ratings for non-load-bearing and load-bearing walls, in terms of minimum sectional dimensions. For load-bearing walls, the fire ratings are for exposure to one or two sides and under two load levels of 35% and 70%[4] (see Table 8.10). Fire walls made of normal-weight concrete need to comply with both fire rating and impact criteria. The minimum thickness requirement for unreinforced, reinforced load-bearing, and reinforced non-load-bearing fire walls are 200, 140, and 120 mm, respectively. If minimum dimensions are provided as per the tables, then the walls will automatically satisfy separating functions [Criterion E (integrity) and I (insulation)]. To avoid excessive thermal deformation and subsequent integrity breach between wall and slab, the ratio of clear height of wall to wall thickness should not exceed 40.

For assessing the fire resistance of RC and PC columns, Eurocode 2 specifies two tabulated methods: *Method A* and *Method B*. Both methods are only applicable for reinforced and prestressed concrete columns under compression loading and when the column is part of a braced frame (structure). The table, as part of Method A, provides the required minimum dimension and the axis distance to the main reinforcement to achieve the specified fire rating in columns.

The alternative method, Method B, is a subset of Annex C of Eurocode 2-1, whose provisions are derived based on buckling analysis. It compares reasonably well with the published results of fire tests and can be more conservative compared with Method A. The two methods mainly vary with regard to allowance of eccentricity, load level, and effective length of column. For instance, Method A assumes effective length of a column under fire conditions, $l_{o,fi}$, to be equal to effective length l_o at normal temperature in all cases (and to be $\leq$3 m), while Method B lays out a more realistic assumption on

[4] It is worth noting that cover thickness requirements for room temperature design will govern over those provided in such tables when the required fire rating is 30, 60, or 90 minutes.

(a) Simply Supported Beams

Standard Fire Resistance	Minimum Dimensions (mm)			
	Possible Combinations of a and b_{min} Where a Is the Average Axis Distance and b_{min} Is the Width of Beam			
R 30	$b_{min} = 80$	120	160	200
	$a = 25$	20	15	15
R 60	$b_{min} = 120$	160	200	300
	$a = 40$	35	30	25
R 90	$b_{min} = 150$	200	300	400
	$a = 55$	45	40	35
R 120	$b_{min} = 200$	240	300	500
	$a = 65$	60	55	50
R 180	$b_{min} = 240$	240	400	600
	$a = 80$	80	65	60
R 240	$b_{min} = 280$	280	500	700
	$a = 90$	90	75	70

(b) Continuous Beams

Standard Fire Resistance	Minimum Dimensions (mm)			
	Possible Combinations of a and b_{min} Where a Is the Average Axis Distance and b_{min} Is the Width of Beam			
R 30	$b_{min} = 80$	160	—	—
	$a = 15^*$	12^*	—	—
R 60	$b_{min} = 120$	200	—	—
	$a = 25$	12^*	—	—
R 90	$b_{min} = 150$	250	—	—
	$a = 35$	25	—	—
R 120	$b_{min} = 200$	300	450	500
	$a = 45$	35	35	30
R 180	$b_{min} = 240$	400	550	600
	$a = 60$	50	50	40
R 240	$b_{min} = 280$	500	650	700
	$a = 75$	60	60	50

*Normally the cover required by EN 1992-1-1 for other effects (i.e., corrosion) will control.

TABLE 8.9 Eurocode Specifications for Fire Resistance Rating of Simply Supported and Continuous RC and PC Beams (*Reproduced from CEN*)

Standard Fire Resistance	Minimum Dimensions (mm) Wall Thickness/Axis Distance			
	Load Ratio = 35%		Load Ratio = 70%	
	Wall Exposed on One Side	Wall Exposed on Two Sides	Wall Exposed on One Side	Wall Exposed on Two Sides
R 30	100/10	120/10	120/10	120/10
R 60	110/10	120/10	130/10	140/10
R 90	120/20	140/10	140/25	170/25
R 120	150/25	160/25	160/35	220/35
R 180	180/40	200/45	210/50	270/55
R 240	230/55	250/55	270/60	350/60

TABLE 8.10 Minimum Dimensions and Axis Distances for Load-Bearing Reinforced Concrete Walls According to EC2 Provisions (*Reproduced from CEN*)

(a) Method A

Standard Fire Resistance	Possible Combinations of b_{min}/Axis Distance a to Main Bars			
	Column Exposed on More Than One Side			Exposed on One Side
	Load Ratio = 20%	Load Ratio = 50%	Load Ratio = 70%	Load Ratio = 70%
R 30	200/25	200/25	200/32	155/25
			300/27	
R 60	200/25	200/36	250/46	155/25
		300/31	350/40	
R 90	200/31	300/45	350/53	155/25
	300/25	400/38	450/40*	
R 120	250/40	350/45*	350/57*	175/35
	350/35	450/40*	450/51*	
R 180	350/45*	350/63*	450/70*	230/55
R 240	350/61*	450/75*	-	295/70

* Requires a minimum of 8 bars for prestressed concrete columns the increase of axis distance according to Eurocode 2.

(b) Method B

Standard Fire Resistance	Ratio of Steel Reinforcement	Possible Combinations of b_{min}/Axis Distance a of Main Bars			
		Load Ratio = 15%	Load Ratio = 30%	Load Ratio = 50%	Load Ratio = 70%
R 30	0.1	150/25	150/25	200/25: 250/25	300/30: 350/25
	1.0	150/25	150/25	150/25	200/25: 300/25
R 60	0.1	150/30:200/25	200/40:300/25	300/40:500/25	500/25
	1.0	150/25	150/30:200/25	200/40:400/25	300/50:600/30
R 90	0.1	200/40:250/25	300/40:400/25	500/50:550/25	550/40:600/25
	1.0	200/25	200/40:300/25	250/40:550/25	500/50:600/45
R 120	0.1	250/50:350/25	400/50:550/25	550/25	550/60:600/45
	1.0	200/40:250/25	250/50:400/25	450/45:600/30	600/60
R 180	0.1	400/50:500/25	500/60:550/25	550/60:600/30	—
	1.0	300/35:400/25	450/50:550/25	500/60:600/45	—
R 240	0.1	500/60:550/25	550/40:600/25	600/75	—
	1.0	400/45:500/25	400/40:600/30	600/60	—

TABLE 8.11 Minimum Column Dimensions and Axis Distances for Reinforced Concrete Columns with a Rectangular or Circular Section as per EC2 (*Reproduced from CEN*)

effective length that relies on column position in a building (i.e., intermediate floor, higher floor, etc.). Method A also assumes a maximum eccentricity to be equal to 0.15 to 0.4 times the width of the column, b (as oppose to Method B, which assumes a maximum eccentricity of $0.25b$ and not exceeding 100 mm). Finally, Method A is applicable for load levels of 0.2, 0.5, and 0.7, while Method B applies to load levels of 0.15, 0.3, 0.5, and 0.7, and for various reinforcement ratios in a column. Table 8.11 shows a sample of tables listed in Eurocode 2.

Australian Code (AS 3600)

The Australian code, AS 3600, provides tables, graphs, and equations that can be used to meet specific fire ratings in concrete members. For example, the graphs for simply supported or continuous beams are given in terms of minimum beam width and cover thickness to achieve a certain fire rating. If the minimum width of the beam and the cover thickness to the main rebar are known, the fire resistance period exhibited by that beam can be read from the graph. This code also provides simplified equations and tables for evaluating the fire ratings of columns, slabs, and load-bearing and non-load-bearing walls.

American Standards

The ACI 216.1 standard (ACI/TMS, 2014), as well as ASCE/SFPE29 (2010), PCI (MNL 124, 2011), PTI (2006), and lists prescriptive-based specifications for evaluating fire resistance ratings of concrete (both reinforced and prestressed) and masonry structures, and these specifications are mainly a function of minimum width and concrete cover thickness. These specifications are derived mostly based on fire tests and cover members with dimensions commonly used in design situations. The data used for deriving these tables was from fire tests wherein concrete members were tested under *unfactored* full-service load (i.e., load factor of 1.0 for both dead and live loads) or per ASTM E119 requirements. The required concrete cover thickness and minimum dimensions in a member are mostly based on insulation and structural endpoint criterion. The structural endpoint criterion is based on limiting temperature rise of 593°C in main reinforcing steel bars and 427°C for prestressing steel strands/bars so as to limit the temperature-induced strength loss in main reinforcement and prestressing strands to 50%. The concrete cover is measured from the exterior surface of concrete to the surface of steel reinforcement.

In the case of beams, ACI 216.1 (ACI, 2014) provisions give consideration to aggregate type and support conditions. Separate fire rating tables are given for restrained and unrestrained beams made of siliceous and carbonate aggregate. These ratings are provided for 1, 1½, 2, 3, and 4 hours (unlike Eurocode 2, which provides ratings at 30-minute intervals in the range of 30 to 240 minutes). The fire ratings are a function of minimum width and minimum cover thickness. The recent edition of ACI 216.1 provides guidance on what constitutes restraint support conditions, and this is reproduced in Table 8.12.

For achieving fire ratings in prestressed concrete beams, ACI 216.1 (and PCI's MNL 124, 2011) provides a separate set of tables (see Table 8.13). Values in Table 8.13 apply to beams of carbonate, siliceous, or semi-lightweight aggregate and with widths not less than 200 mm. In addition, separate tables for prestressed concrete members made of any aggregate types and widths that have cross-sectional areas not less than 0.026 m^2 are given. The technical basis for these tables is fire tests carried out at the National Research Council in Canada and originally adopted by the National Building Code of Canada (Galbreath, 1966; Lin, 1981). As a result, these provisions are similar to those of National Building Code of Canada (NBCC) provisions. Similar tables for post-tensioned concrete structures are also available in the IBC as well as PTI (2006) codes. These tables were derived based on a thermal limit state determined by the point at which tendons reach a critical temperature of 427°C. It is worth noting that the cover to the prestressing steel at the anchor should be at least 12.5 mm greater than that specified by tables for use in post-tensioned concrete structures.

ACI 216.1 (ACI, 2014) also provides separate tables for evaluating fire ratings of reinforced concrete columns in the range of 1 to 4 hours.[5] These ratings, as shown in Table 8.14, are derived based on fire tests originally undertaken on building columns by Hull and Ingberg (1925) in the 1920s. The minimum column dimensions for each fire rating are specified for columns made of three aggregate types: carbonate, silicate, and semi-lightweight. These five ratings are applicable for RC columns made of compressive strength not exceeding 83 MPa. In the case of columns made of HSC (>83 MPa), ACI 216.1 provides additional guidance for overcoming fire-induced spalling in concrete. This guidance requires ties in HSC columns to be formed with hooks that have six-diameter extensions and engage the longitudinal reinforcement into the interior of the hoop. Rectangular hoops are to be formed with minimum 135-degree bends, while circular hoops are to be formed with minimum 90-degree bends.

For slabs, ACI 216.1 (as well as PCI's MNL 124, 2011) provides minimum sectional depth for achieving specific fire resistance in the range of 1 to 4 hours. The provisions are for concrete slabs made of various types of concrete as well as constructions such as one- or two-way cast-in-place beam/slab systems or precast solid or hollow-core slabs with flat undersurfaces for a range. In addition, ACI 216.1 states that fire resistance of up to 4 hours in concrete bearing or nonbearing walls, floor slabs, and roof slabs can be obtained using minimum equivalent thickness (T_e) values provided in Table 8.15. ACI 216.1

[5] ACI 216.1 also lays out certain guidance for minimizing fire-induced spalling only in RC columns made of high-strength concrete.

Reinsured Concrete Structures [ACI 216.1 (ACI, 2014)]	
Unrestrained	
Wall bearing*	Single spans and simply supported end spans of multiple bays such as concrete or precast units†
Restrained	
Wall bearing*	Interior spans of multiple bays: 1. Cast-in-place concrete slab systems 2. Precast concrete where the potential thermal expansion is resisted by adjacent construction‡
Concrete framing	1. Beams fastened securely to the framing numbers 2. Cast-in-place floor or roof systems (such as beam/slab systems, flat slabs, pan joists, and waffle slabs) where the floor or roof system is cast with the framing members 3. Interior and exterior spans of precast systems with cast-in-place joints resulting in restraint equivalent to that of Condition 1, Concrete framing 4. Prefabricated floor or roof systems where the structural members are secured to such systems and the potential thermal expansion of the floor or roof systems is resisted by the framing system of the adjoining floor or roof construction‡
Post-Tensioned Concrete Structures (PTI, 2006)	
1. In a continuous post-tensioned concrete slab, end spans are typically considered to be unrestrained and interior spans are considered restrained. 2. In the case of two-way slabs, this criterion applies independently to the reinforcement in each orthogonal direction.	

*Wall bearing relates to elements bearing on walls.

†Consider floor and roof systems restrained when they are tied into walls with or without tie beams, provided the walls are designed and detailed to resist thermal thrust from the floor or roof system.

‡For example, resistance to potential thermal expansion is considered to be achieved when:

c) Continuous concrete structural topping is used.

d) The space between the ends of precast units or between the ends of units and the vertical face of supports is filled with concrete or mortar.

e) The space between the ends of the precast units and the vertical face of supports, or between the ends of solid or hollow-core slab units, does not exceed 0.25% of the length for normal-weight concrete members or 0.1% of the length for structural lightweight concrete members.

Table 8.12 Classification for Restraint Conditions as per ACI 216.1 (ACI, 2014) and PTI (2006) (*Authorized Reprint from ACI, 2014*)

	Condition	Minimum Width (mm)	Minimum Cover (mm)				
			1 hour	1.5 hours	2 hours	3 hours	4 hours
RC	Restrained	127	19.05	19.05	19.05	25.4	31.75
		178	19.05	19.05	19.05	19.05	19.05
		254	19.05	19.05	19.05	19.05	19.05
	Unrestrained	127	19.05	25.4	31.75	–	–
		178	19.05	19.05	19.05	44.45	76.2
		254	19.05	19.05	19.05	25.4	44.45
PC	Restrained	203.2	38.1	38.1	38.1	44.45	63.5
		>304.8	38.1	38.1	38.1	38.1	47.63
	Unrestrained	203.2	38.1	44.45	63.5	127	–
		>304.8	38.1	38.1	47.63	63.5	76.2

Table 8.13 Minimum Width and Cover Thickness Requirements for Achieving Fire Resistance in RC and PC Beams (*Authorized Reprint from ACI, 2014*)

Aggregate Type	Minimum Column Dimension (mm)				
	1 hour	1.5 hours	2 hours	3 hours	4 hours
Minimum Column Size (mm)—Fire Exposure on Four Sides					
Carbonate	203.2	228.6	254	279.4	304.8
Siliceous	203.2	228.6	254	304.8	355.6
Semi-lightweight	203.2	215.9	228.6	266.7	304.8
Minimum Column Size—Fire Exposure on Two Parallel Sides					
Carbonate	203.2	203.2	203.2	203.2	254
Siliceous	203.2	203.2	203.2	203.2	254
Semi-lightweight	203.2	203.2	203.2	203.2	254

Table 8.14 Minimum Sectional Size for Concrete Columns (*Authorized Reprint from ACI, 2014*)

Aggregate Type	Minimum Equivalent Thickness (mm)				
	1 hour	1.5 hours	2 hours	3 hours	4 hours
Carbonate	88.9	109.2	127	157.5	177.8
Siliceous	81.28	101.6	116.8	144.8	167.6
Semi-lightweight	68.58	83.82	96.5	116.8	137.2
Lightweight	63.5	78.74	91.4	111.8	129.5

Table 8.15 Fire Resistance of Single-Layer Concrete Walls, Floors, and Roofs as per ACI 216.1 (*Authorized Reprint from ACI, 2014*)

standard provides guidance to calculate requirements of thickness on solid walls and hollow cross sections.

For solid walls and slabs with flat surfaces, the actual thickness shall be the equivalent thickness of the member. For walls and slabs constructed with precast concrete hollow-core panels with constant core cross section throughout their length, the equivalent thickness is calculated by dividing the net cross-sectional area by the panel width[6] (see Fig. 8.8). Thus, for hollow-core panel walls, $T_e = A/w$. Where all of the core spaces are filled with grout or loose fill material, such as perlite, vermiculite, sand or expanded clay, shale, slag, or slate, the fire resistance of the wall or slab shall be the same as that of a solid wall or slab of the same type of concrete.

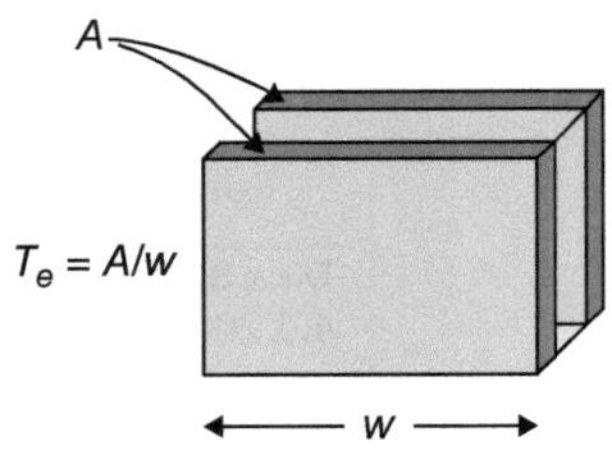

Figure 8.8 Equivalent thickness calculation in a concrete hollow-core wall.

Prescriptive-based fire ratings for concrete walls and floor assemblies are available in various codes and standards such as NBCC. A sample of an NBCC listing of fire ratings for these assemblies is shown in Table 8.16. For example, a wall assembly, *B2a*, made of 140-mm concrete blocks and protected with a 12.7-mm gypsum-sand plaster on each side can achieve a 2-hour fire rating under standard fire exposure. Similarly, a floor assembly, *F2a*, made of 50-mm-thick concrete deck (on top), with open steel joists spaced at 400 mm o.c. and furring channels spaced not more than 600 mm o.c., and protected with a 15.9-mm Type X gypsum board on the ceiling side can achieve a 45-minute fire rating under standard fire exposure.

8.6.2 Correlation Equations

Although tabulated data offers an easy and convenient way to determine required cover thickness and sectional dimensions for satisfying a fire rating in a member, such data are available only for certain member dimensions and may not be applicable for members with unique configuration, or exposed to fire

[6] ACI 216.1 also provides guidance as to calculating the equivalent thickness of flanged, ribbed, and undulating panels as well as for walls, floors, and roofs consisting of two or more layers of different types of concrete, masonry, or both.

Type of Wall	Wall Number	Description	Fire-Resistance Rating: Load-bearing	Fire-Resistance Rating: Non-loadbearing	Typical Sound Transmission Class
	B2	• 140 mm or 190 mm concrete block • no absorptive material • 1 layer of gypsum-sand plaster or gypsum board on each side	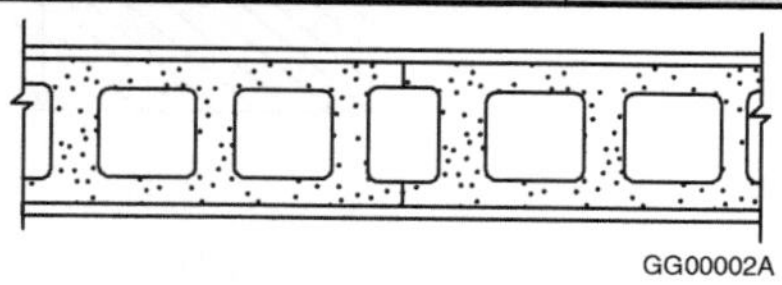		
	B2a	B2 with • 140 mm concrete block • 12.7 mm gypsum-sand plaster	2 h	2 h	50
	B2b	B2 with • 140 mm concrete block • 12.7 mm Type X gypsum board or 15.9 mm Type X gypsum board	2 h	2 h	47
	B2c	B2 with • 140 mm concrete block • 12.7 mm regular gypsum board	1.5 h	1.5 h	46

Type of Assembly	Assembly Number	Description[(2)(3)(4)]	Fire-Resistance Rating	Typical Sound Transmission Class (STC)	Typical Impact Insulation Class (IIC)
Floors and Ceilings					
Open Web Steel Joists	F2	• open web steel joists with concrete floor	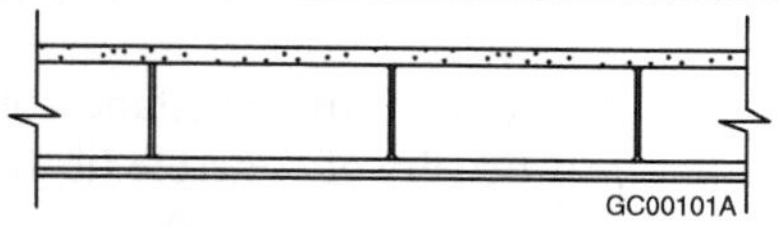		
	F2a	• 50 mm thick concrete deck • on open web steel joists spaced 400 mm o.c. • furring channels spaced not more than 600 mm o.c wired to underside of joists • 1 layer of 15.9 mm Type X gypsum board on ceiling side	45 min	53	27
	F2b	• 60 mm regular concrete minimum 155 kg/m² • on composite steel joists spaced 1250 mm o.c. • furring channels spaced not more than 600 mm o.c. wired to underside of joists • 1 layer of 12.7 mm or 15.9 mm Type X gypsum board on ceiling side	1.5 h	53	28
	F1e	152 × 152 × MW3.8 × MW3.8 wire mesh • resilient metal channels 400 mm or 600 mm o.c. • 2 layers of 12.7 mm Type X gypsum board or 2 layers of 15.9 mm Type X gypsum board	1.5 h	57	36

Table 8.16 Sample of Fire Rating for Lightweight Concrete Walls and Floors in NBCC (*Reproduced with Permission of the National Research Council of Canada, Copyright Holder*)

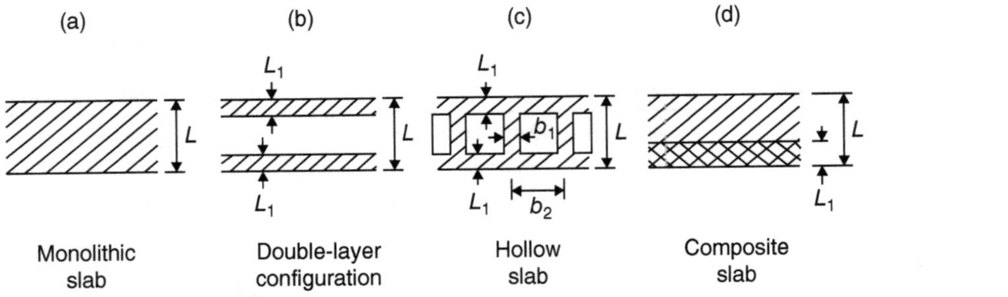

FIGURE 8.9 Slabs and walls. (*With permission from ASCE.*)

scenarios other than standard fire conditions. In these situations, correlation methods can come in handy. A number of simplified equations (or correlations) are available for specific concrete members—beams, columns, slabs, and walls—and can be found in trade association manuals, building codes (NBCC, Eurocode 2, ASCE 29), and Underwriters Laboratories (UL) Fire Resistance Directory.

Slabs and Walls

ASCE (1992)[7] manual provides empirical equations for evaluating the fire resistance of different concrete slabs with different configurations, such as monolithic, double-layer, hollow, and composite, to be evaluated using a number of simplified (empirical) equations. In the case of dry monolithic normal-weight concrete slabs (see Fig. 8.9), fire resistance is expressed through the following semiempirical formula (in FPS units):

$$R_1 = 0.205\frac{(\rho c)^{1.2} L^{1.85}}{k^{0.65}} \tag{8.9}$$

where R_1 = fire resistance of slab based on heat transmission criterion (hr), L = thickness of slab (ft), p = density of concrete (lb/ft³), c = specific heat of concrete (Btu/lb·°F), and k = thermal conductivity of concrete (Btu/ft·h·°F).

In cases where specific data on the thermal properties of the specific concrete used in the walls/slabs is not available, ASCE (1992) recommends the following conservative values to be used for normal- and lightweight concrete:

k = 1.0 Btu/ft·h·°F for normal-weight concrete, k = 0.45 Btu/ft·h·°F for lightweight concrete, and c = 0.20 Btu/lb·°F for both concretes.

Applying recommended thermal property values for normal-weight concrete, Eq. 8.9 for fire resistance becomes

$$R_1 = 0.03\rho^{1.2} L^{1.85} \tag{8.10}$$

Equations for fire resistance of lightweight concrete becomes

$$R_1 = 0.05\rho^{1.2} L^{1.85} \tag{8.11}$$

Note that the above two equations (Eqs. 8.10 and 8.11) are in FPS units.

For concrete slabs with two layers of concrete, separated by a continuous air gap of any thickness (Fig. 8.9*b*), an analogous formula as that of the monolithic slab is applicable. The fire resistance of double-layered slabs in dry condition, based on obtaining a failure temperature rise of 139°C at the unexposed surface, is given by the following semiempirical formula (in FPS units):

$$R_2 = 0.75\frac{(\rho c)^{1.1} L_1^{\,1.6}}{k^{0.5}} \tag{8.12}$$

where R_2 = the thermal fire resistance of the slab (hr), L_1 = thickness of one layer of the slab (ft), p = density of the concrete (lb/ft³), c = specific heat of the concrete (Btu/lb·°F), and k = thermal conductivity of the concrete (Btu/ft·h·°F).

[7] Some of the correlation equations adopted by the ASCE fire protection handbook were derived for use in traditional units.

Using the above recommended material properties, Eq. 8.12 for the fire resistance of normal-weight concrete gets transformed to

$$R_1 = 0.13\rho^{1.1}L_1^{1.6} \tag{8.13}$$

and for lightweight concrete

$$R_1 = 0.216\rho^{1.1}L_1^{1.6} \tag{8.14}$$

The equations given for the monolithic concrete slabs and for the double-layer slabs can be extended to evaluate the fire resistance of hollow-core slabs (Fig. 8.9*c*). In general, a hollow slab is regarded as a monolithic slab at the locations of the webs and as a double-layer slab at the location of the cavity. The fire resistance of this type of slab can be evaluated based on attaining a temperature rise of 139°C (250°F) at the side away from the fire (top surface) and is given by the following semiempirical formula:

$$R = \left[\frac{1}{\dfrac{b_1/b_2}{R_1^{0.5}} + \dfrac{1-\left(b_1/b_2\right)}{R_2^{0.5}}} \right]^2$$

where R = the thermal fire resistance of the hollow slab (hr) (see Fig. 8.9*c*), R_1 = the thermal fire resistance of the monolithic slab (hr), R_2 = the thermal fire resistance of the double layer slab (hr), b_1 = thickness of a web (ft), and b_2 = distance between the centerlines of two webs (ft).

For the case of multilayered concrete slabs composed of a layer of normal-weight concrete and a layer of lightweight concrete (see Fig. 8.9*d*), ACI 216.1 provides two equations. These equations were developed by Abrams and Gustaferro (1969) and have been derived for the calculation of the fire resistance of composite slabs that have a relative humidity in the range of 50% to 75%.

For composite multilayered slabs made of normal-weight concrete

$$R = 0.057\left(2l^2 - dl + \frac{6}{l}\right) \tag{8.15}$$

and when the base slab consists of lightweight concrete

$$R = 0.063\left(l^2 + dl + d^2 + \frac{4}{l}\right) \tag{8.16}$$

where R = fire resistance of slab (hr), l = total thickness of slab (in.), d = thickness of base slab (in.), and $l - d$ is not less than 25 in.

A few correlation equations are also available for determining minimum thickness required for achieving certain fire resistance in composite slabs, floors, and roofs, made of lightweight topping. The required thickness for the slab made of normal-weight concrete as a base layer and semi-lightweight concrete as a top (finishing) layer can be evaluated from Eqs. 8.15 and 8.16. For floor and roof slabs and walls made of one layer of normal-weight concrete and one layer of semi-lightweight or lightweight concrete, where each layer is 25.4 mm (1 in.) or greater in thickness, the combined fire resistance of the assembly can also be determined using the following expression (in FPS units):

(a) When the fire-exposed base layer is of normal-weight concrete

$$R = 0.057\left(2t^2 - dt + \frac{6}{t}\right) \tag{8.17}$$

(b) When the fire-exposed layer is of lightweight or semi-lightweight concrete

$$R = 0.063\left(2t^2 - 2dt - 2d^2 + \frac{4}{t}\right) \tag{8.18}$$

where R is the fire endurance (hr), d is the thickness of base slab (in.), and t is the overall thickness of entire slab (in.); $t \geq 1 + d$.

For multiple-wythe walls, the ACI 216.1 standard utilizes a graphical solution in the form of charts. Through these charts, the fire resistance of wythe walls made of carbonate and siliceous aggregates can be determined for various wall thicknesses.

In lieu of the above equations, ACI 216.1 also provides an empirical equation for calculating fire resistance of multiple layers of wythe. The fire resistance of multilayered wythe walls and floors is the sum of individual values for each layer and is given as

$$R = (R_1^{0.59} + R_2^{0.59} + \ldots + R_n^{0.59})^{1.7} \tag{8.19}$$

where R_i values for wythes of concrete are given in Fig. 8.10 for various concrete types. The assumptions and limitations of this empirical equation are that this formula does not account for the orientation of the assembly—walls, floors, and roofs with better insulating material on the exposed side will have greater fire endurance than if better insulating material is on the unexposed side. Further, exponents present in the equations are average values and are not really equal for all materials.

RC Columns

A number of correlation (empirical) equations are available for evaluating the fire resistance of concrete columns. These equations are generally applicable for columns subjected to four-side heating in a standard fire exposure.

ASCE Manual ASCE (1992) manual provides empirical formulas to determine the minimum sectional dimensions and minimum concrete cover to vertical steel reinforcement in an RC column to achieve specific fire resistance in the column. The formulas take into account the type of concrete, the effective

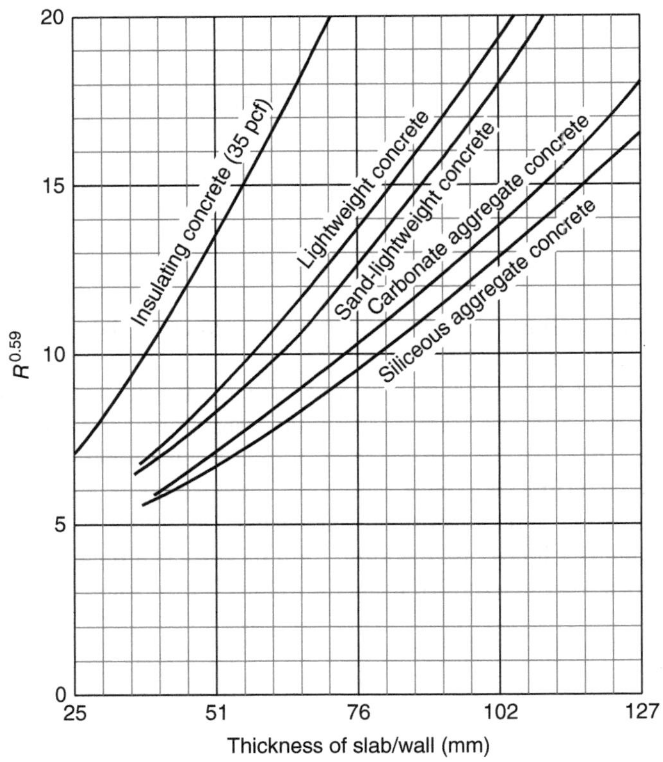

FIGURE 8.10 R_i values based on slab thickness and aggregate type for deriving the fire resistance of concrete slabs based on a 250°F rise in temperature on an unexposed surface. (*With permission from ASCE.*)

length of the column, and the area of the vertical reinforcement. Accordingly, the minimum dimension of a rectangular column t_{min} (in inches) for obtaining a fire resistance $R(h)$ is given as

$$t_{min} = 3.2/(R + 1) \tag{8.20}$$

For normal-weight siliceous aggregate concrete

$$t_{min} = 3.2f(R + 0.75) \tag{8.21}$$

For normal-weight carbonate aggregate concrete

$$t_{min} = 4f(R + 1) \tag{8.22}$$

where f is the over design factor and depends on the effective length of the column and reinforcement in the column. The ASCE manual recommends this factor to be taken as 1.0 for all cases. Other values for f are given in ASCE (1992).

In addition to a minimum dimension of the column, there is also a minimum cover requirement to prevent the main longitudinal steel reinforcement from reaching critical temperature. Following is the minimum cover C_{min} (in inches) to the vertical reinforcement of columns made of all concrete types:[8]

$$\text{for } R \leq 3 \text{ hr, } C_{min} = R \text{ or 2 in. (50 mm), whichever is less}$$

$$\text{for } R > 3 \text{ hr, } C_{min} = 112\,(R - 3) + 2$$

Eurocode 2 (EC2) Eurocode 2 provides the following simplified equation that can be used to evaluate the fire resistance of RC columns which can substitute the use of tabular data for Method A described earlier. This equation was originally developed by Dotreppe and Franssen (1985) and has the same assumptions and limitations as Method A. Accordingly, fire resistance of an RC column is given as

$$R = 120\left(\frac{R_{\eta fi} + R_a + R_l + R_b + R_n}{120}\right)^{1.8} \tag{8.23}$$

where

$$R_{\eta fi} = 83 < \left[1 - \mu_{fi}\frac{(1+\omega)}{\left(\left(\dfrac{0.85}{\alpha_{cc}}\right) + \omega\right)}\right],\ R_a = 1.6(a - 30),\ R_l = 9.6(5 - l_{o,fi}),\ R_b = 0.09b',\ R_n = 0,\ \text{for}$$

$n = 4$ (corner rebars only), $R_n = 12$, for $n > 4$

where μ_{fi} is the ratio of the applied load to the load-carrying capacity of the column at room temperature, α_{cc} is the coefficient for compressive strength as per EN 1992-1-1, a is the shortest distance between the centroid of the longitudinal reinforcement and the nearest concrete surface exposed to fire (mm), 25 mm $\leq a \leq$ 80 mm, $l_{o,fi}$ is the effective length of the column under fire conditions, 2000 mm $\leq l_{o,fi} \leq$ 6000 mm, $b' = 2A_c/(b + h)$ and 200 mm $\leq b' \leq$ 450 mm, $h \leq 1.5b$, h and b are the longer and shorter dimensions of the column, respectively, ω is the longitudinal reinforcement ratio at room temperature conditions $= A_s f_{yd}/A_c f_{cd}$ and A_s is the area of steel reinforcement, f_{yd} is the yield strength of reinforcing steel, A_c is the characteristic strength of concrete, and f_{cd} is the area of concrete.

Australian Code (AS 3600) The Australian code (AS 3600) provides an equation for evaluating the fire resistance of RC columns exposed to a standard fire. This equation was developed based on computer simulations of the fire resistance of columns and were validated against fire tests undertaken at the

[8] These conditions are quite similar to those required by ACI 216.1 and are shown in Sec. 8.6.1.

Building Research Association of New Zealand (BRANZ). The results of both the fire tests and the model were analyzed to fit a conservative expression (Eq. 8.24) using a multiple linear regression technique.

$$R = \frac{(k \times f_c'^{1.3} \times D_c^{3.3} \times D_g^{1.8})}{10^5 \times N^{*1.5} \times L_e^{0.9}} \tag{8.24}$$

where R is the fire resistance in minutes; k is a coefficient dependent on steel ratio = 1.5 when (area of reinforcing steel, A_s)/(gross area of cross section, A_g) < 0.0025, otherwise $k = 1.7$, f_c' is the characteristic compressive strength of concrete (MPa); D_c is the smaller column dimension of a rectangular column (mm); D_g is the greater column dimension of a rectangular column (mm); N^* is the design axial force for the fire limit state (kN); and L_e is the effective length of the column (mm).

Kodur and Raut Approach To overcome some of the limitations in the current ACI 216.1 provision, Raut and Kodur (2010) proposed an empirical equation for evaluating the fire resistance of RC columns. This equation is applicable under ASTM E119 or ISO 834 fire exposure and accounts for the effects of one-, two-, three-, or four-sided fire exposure, bi-eccentric loading, and design fire scenarios. Fire resistance of a reinforced concrete column is evaluated as

$$R = C_t \left[8 \times k \times ((S_R + 5) \times (0.2 - L_R) + 30)\right]^{0.94} \tag{8.25}$$

where R is the fire resistance in minutes, S_R is the slenderness ratio of the column = $\frac{L_e}{r} = \frac{\text{effective length (depends on end conditions)}}{\text{radius of gyration}}$, L_R is the load ratio of the column, C_t is the constant based on aggregate type used in concrete, C_t is 1.0 for siliceous aggregate and C_t is 1.1 for carbonate aggregate, k is the constant based on the cover thickness and the percentage steel:

$$k = \frac{\left[(C_e - 82) \times (S_p + 10.5) + 870\right]}{390}$$

where C_e is the effective concrete cover thickness (mm) (clear cover), and S_p is the percentage of steel in the column.

8.7 Rational Approaches for Evaluating Fire Resistance

In lieu of prescriptive-based approaches, rational design approaches can be applied for evaluating the fire resistance of concrete structures. As in the case of steel structures, fire resistance can be evaluated by either applying the simplified analysis procedure or undertaking advanced analysis. In the case of simplified analysis, failure of a concrete member under fire conditions can be evaluated either based on attaining critical temperature in reinforced or prestressed reinforcement or based on attaining sectional capacity or deflection limit state.

8.7.1 General Procedure

As part of a rational approach, fire resistance of an RC or PC member can be evaluated at a specified time-step or in a series of cumulative time-steps. If the thermal limit state is applied to determine the failure of the member, only cross-section temperature, T_{sec} (in reinforcement or in some cases layers of concrete), in a fire-exposed structural member, is to be evaluated. The cross-sectional temperatures can be determined through the use of simple graphs, charts, or equations or through undertaking detailed numerical analysis. The temperature in reinforcing or prestressing steel and concrete layers (including the unexposed side) is then compared against a limiting temperature specified in codes and standards. For example, as per Eurocode 2 recommendations, critical temperature in reinforcing steel (T_{ren}) is 500°C or in prestressing steel (T_{pre}) is 350°C. Thus, failure is said to occur once

$$T_{ren} \text{ or } T_{pre} \leq T_{sec} \tag{8.26}$$

In the case that strength or deflection limit state is to be applied, temperature rise in reinforcement and concrete layers within the fire-exposed concrete member is to be evaluated first. Knowing sectional temperature as a given time-step, the reduced strength in concrete and reinforcement can be determined, and then the decreased sectional capacity (axial, shear, or moment) can be calculated. Also, other response parameters such as deflection in the member of the member can be evaluated at any given time-step. Then failure is checked by comparing sectional capacity with applicable loading effect (i.e., moment, shear, axial) or by comparing maximum deflection to limiting deflection criteria such that

$$U_{fire} \leq R_{fire} \tag{8.27}$$

$$\Delta_{fire} \leq \delta_{fire} \tag{8.28}$$

where U^*_{fire} is the resulting actions from applied loading due to the fire event and R_{fire} is the reduced capacity of the fire-exposed member. U^*_{fire} can be axial force N^*_{fire}, bending moment M^*_{fire}, or shear force V^*_{fire}, and correspondingly R_{fire} is the reduced axial force N_f, bending moment M_f, or shear force V_f at a given time-step. For deflection limit state, Δ_{fire} is the deflection (deformation) limit specified in applicable standards under fire conditions, and δ_{fire} is the actual deflection (deformation) in the concrete member at a given fire exposure time. The steps for applying rational approaches are detailed in the flowchart given in Fig. 6.5.

8.7.2 Simplified Approaches

In simplified approaches, fire resistance can be evaluated by undertaking heat transfer and/or strength calculations at a specified time or a series of time-steps until the failure of a structural member. The main response parameters needed for evaluating fire resistance are cross-sectional temperature, especially rebars or strands temperatures, and sectional capacity of the member. Simplified equations or charts have been developed to determine sectional temperatures in a member at any given fire exposure time. Also, simplified equations for evaluating the reduced capacity of a member are available in the literature.

Approaches for Evaluating Sectional Temperature

In concrete members, significant thermal gradients develop during fire exposure, and these have to be accounted for in cross-sectional temperature calculations. These gradients arise from the low thermal conductivity and high specific heat of concrete, which make temperature propagation slow within the concrete. Simplified charts or equations are available for evaluating cross-sectional temperature across different concrete cross-sectional layers. Design guides such as the FIP/CEB report (de la Precontrainte, Federation Internationale, 1978) and Eurocode 2 (CEN, 2004), as well as ACI 216.1 (ACI, 2014), provide simple charts or equations for different evaluating sectional temperatures in concrete members (beams, slabs, floors, etc.) exposed to standard fire exposure. These graphs take into consideration the type of concrete (aggregate type, concrete density, etc.) in estimating cross-sectional temperature.

Eurocode 2 provides temperature profiles for standard fire resistance periods of slabs and beam/column sections cast from normal-weight concrete. The temperature profiles given in Annex A can be used to determine temperature rise in cross sections with siliceous aggregate (with a moisture content of 1.5%) exposed to a standard fire, which can be given conservative temperature predictions in Annex A of Eurocode 2, shown in Fig. 8.11. It should be noted

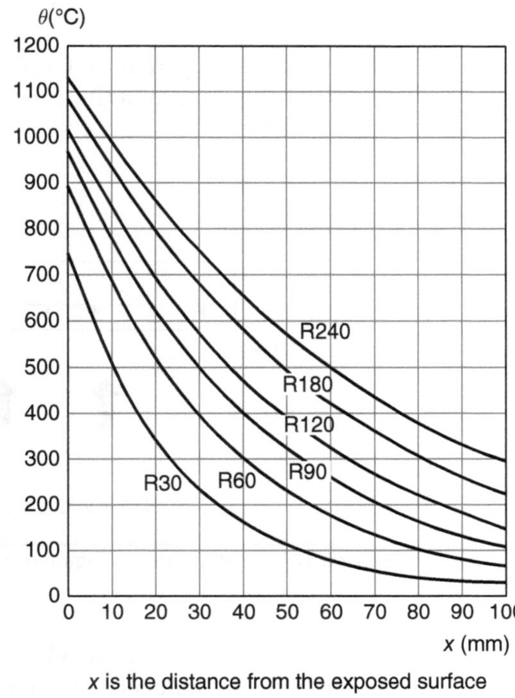

FIGURE 8.11 Eurocode 2 temperature profiles in a concrete slab at depth *x* from a fire-exposed surface. (*Reproduced from CEN.*)

that temperatures obtained from these charts may not yield exact values but can be a close approximation to actual values that can be generated with detailed heat transfer analysis. Similar figures/charts are also available in the ACI 216.1 as well as PCI's MNL 124 for reinforced and prestressed concrete members.

In lieu of the above-discussed charts and graphs, simplified equations for evaluating cross-sectional temperature are also available in Eurocodes and published in the literature. These temperature equations are derived analytically based on the application of heat transfer principles or simply developed as a fit to cross-sectional data obtained in furnace tests, or through the superposition of simple solutions to the Fourier heat transfer equations. Two examples of available simplified equations for evaluating cross-sectional temperature in a beam, column, or slab are the empirical expressions proposed by Wickström (1985, 1986) and by Kodur et al. (2013).

Wickström equation for evaluating temperature, T_c, at a given point in a fire-exposed concrete member with unidimensional heat flow and at time t is given as

$$T_c = \eta_x \eta_w T_f \tag{8.29}$$

where

$$\eta_x = 0.18\ln\left(\frac{t_h}{x^2}\right) - 0.81 \tag{8.30}$$

$$\eta_w = 1 - 0.0616 t_h^{-0.88} \tag{8.31}$$

where T_f is the fire temperature (°C), t_h is the time (hours), and x is the distance across concrete section in meters.

The above equation is for 1D heat transfer. However, heat flow is bidirectional (two-sided) in a section in the case of a corner of a beam or column. In that case, the above equation for T_c is modified to take into consideration heat transfer occurring from two sides of the member and is given by (see Fig. 8.12):

$$T_c = [T_w(\eta_x + \eta_y - 2\eta_x\eta_y) + \eta_x\eta_y]T_f \tag{8.32}$$

where T_f is fire temperature (°C), T_w is the temperature at fire-exposed surface, x and y are distance from fire-exposed surface in m (in.), and η_y is calculated similar to η_x, but in the x-direction.

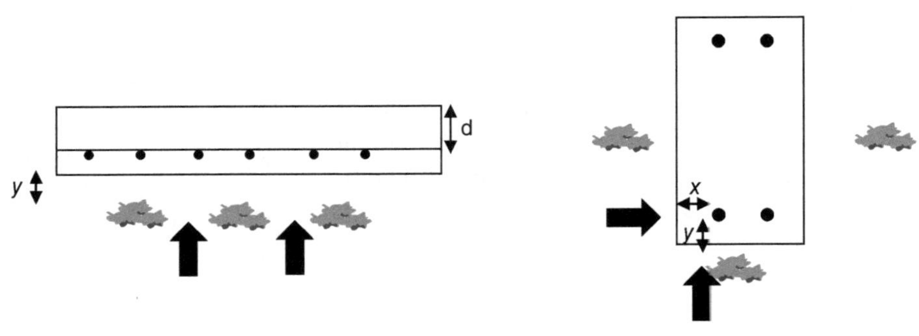

FIGURE 8.12 Temperature propagation in fire-exposed concrete members.

The above equation, although easy to apply, has some limitations. This equation does not directly[9] take into account the influence of aggregate type (siliceous or carbonate), newer concrete types (such as

[9] Wickström's equation can be modified to account for different types of concrete through derived parameters as provided in Wickström (1985).

high-strength concrete), and temperature-dependent variations in thermal properties of concrete (Kodur et al., 2013).

To overcome some of the limitations in Wickström's equation, Kodur et al. (2013) proposed a slightly modified equation for evaluating cross-sectional temperature in fire-exposed concrete members. The equations are modified by introducing a number of correction factors to take into consideration the effect of aggregate type (silicate or carbonate), concrete type (NSC or HSC), and type of exposure (ASTM E119 or ISO 834). The modified equations can be applied for both unidimensional and bidimensional heat transfer situations.

For one-dimensional heat transfer

$$T_c = c_1 \cdot \eta_z \cdot (at^n) \geq 20°\text{C} \tag{8.33}$$

where

$$\eta_z = 0.155\ln\left(\frac{t_h}{z^{1.5}}\right) - 0.348\sqrt{z} - 0.371 \tag{8.34}$$

For 2D heat transfer

$$T_c = c_2 \cdot (-1.481 \cdot (\eta_z \cdot \eta_y) + 0.985 \cdot (\eta_z + \eta_y) + 0.0017\,(at^n)) \geq 20°\text{C} \tag{8.35}$$

where

$$\eta_z = 0.155\ln\left(\frac{t_h}{z^{1.5}}\right) - 0.348\sqrt{z} - 0.371 \tag{8.36}$$

where t_h is the fire exposure time in hours, z and y are the distance from the point in concrete section to fire exposure surface in meters, and at^n is the temperature values under standard fire exposure (Kodur et al., 2013). The factors a, n, c_1, and c_2 needed for applying the above equations can be found in Table 8.17. These equations are validated against measured temperatures in fire tests and also predicted temperature from advanced analysis. Good correlation in cross-sectional temperatures was obtained for different cross-sectional shapes as well as different concrete types used in RC and PC members.

In lieu of graphs or simplified equations, temperature rise within a concrete member can also be carried out by applying finite difference or finite element-based numerical methods (Lie and Allen, 1972). Details for undertaking such an analysis is provided in Sec. 8.7.3. Knowing the cross-sectional temperature, the fire resistance of a RC or PC member can be evaluated by applying thermal limit criterion, as discussed in Sec. 8.7.1.

Approaches for Evaluating Sectional Capacity

General Procedure In the case of of reinforced or prestressed concrete members, fire resistance can also be evaluated based on strength limit state. At the sectional level, reduced sectional capacity at a critical section is calculated by considering mostly reduction in properties due to temperature. As an example, the reduced sectional capacity at a critical section of a fire-exposed concrete beam can be evaluated by applying room temperature capacity equations, but with due consideration to temperature-induced

Factor	ASTM E119		ISO 834	
a	910		935	
n	0.148		0.168	
	NSC-carbonate	NSC-silicate	HSC-carbonate	HSC-silicate
c_1	1.0	1.01	1.12	1.12
c_2	1.0	1.06	1.12	1.20

Table 8.17 Factors Proposed by Kodur and Shakya (2017) for Use in Temperature Evaluation in Concrete Members

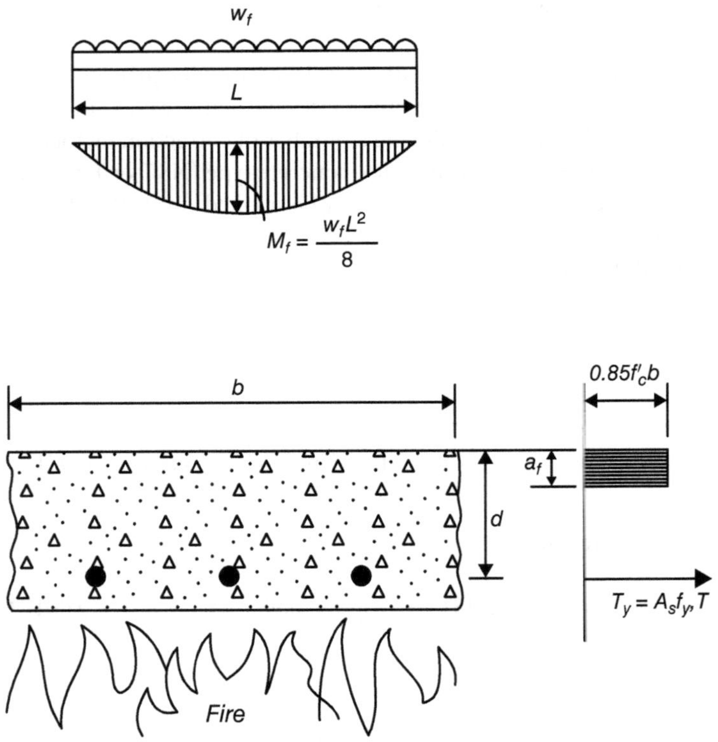

Figure 8.13 Concrete beam under fire conditions (where f'_c is the compressive strength of concrete, b is the width of concrete beam, A_s is the area of steel reinforcement, and $f_{y,T}$ is the yield strength of steel reinforcement at temperature T).

reduction in strength of concrete and reinforcement. This reduced capacity can be moment (bending), shear, or axial capacity.

In the case of a simply supported beam or slab, without consideration to any moment redistribution and any fire-induced forces, the failure at a midspan critical section occurs when

$$M_f \leq M_{\text{fire}} \tag{8.37}$$

where M_{fire} is the flexural capacity at the critical section under fire conditions and M_f is the bending moment.

In the case of a simply supported beam exposed to fire and subjected to a bending moment, M_{fire}, as shown in Fig. 8.13, the top portion of the concrete member is under compression, while the bottom portion is under tension. The compressive force developed in the concrete section can be evaluated by taking equivalent stress value of a rectangular stress block of concrete, as in room temperature calculations, and are equal to $0.85f'_{c,T}\,b$, where $f'_{c,T}$ is the compressive strength of concrete. The tensile forces developed can be determined as $A_s f_{y,T}$, where $f_{y,T}$ is the reduced tensile strength of reinforcement and A_s is the area of steel reinforcement. The tensile strength of concrete can be neglected in a capacity evaluation at any given time-step, as in room temperature calculations.

Evaluating the forces at a section, the depth of the rectangular stress block can be determined as

$$a_f = \frac{A_s f_{y,T}}{0.85 f'_{c,T}\, b} \tag{8.38}$$

Then the reduced flexural strength M_f can be evaluated as

$$M_f = A_s f_{y,T}\left(d - \frac{a_f}{2}\right) \tag{8.39}$$

where A_s is the area of reinforcing steel, $f_{y,T}$ is the yield stress of reinforcing steel at temperature T, d is the effective depth of the section, and a_f is the depth of the rectangular stress block reduced due to thermal effects and can be evaluated by applying equilibrium principles as shown in Fig. 8.13.

The above equations get further simplified if the temperature rise in the compressive zone of the concrete beam (or slab) is not beyond a certain temperature (≈ 300°C), causing any notable reduction in compressive strength. This scenario is likely in many situations, specifically in slabs and beams when fire exposure is two hours or less and for sections with higher sectional dimensions. However, under severe fire exposure (and at higher fire resistance times), temperature rise in compressive zones in a concrete slab or beam can go beyond 300°C and can lead to a reduction in the compressive strength of concrete. In this case, the reduced compressive strength of concrete $f'_{c,T}$ at temperature T is to be taken into account. This can be done through a layered approach or through the 500°C isotherm approach as described in the following section.

The above procedure can be extended to evaluate the reduced shear capacity in beam or axial capacity in columns at a given fire exposure time. The reduced shear or axial capacity can be compared against applied shear or axial force, and this way, failure can be checked under these limit states. Shear is not usually a governing criterion in fire-exposed concrete beams, especially in typical beams used in most building applications. Shear capacity at any given time-step can be evaluated by extending the room temperature shear capacity equation, but duly accounting for shear strength degradation in concrete and steel. The shear capacity of a concrete beam under fire conditions can be evaluated as

$$V_f = 2\sqrt{f'_{c,T}}\, bd \tag{8.40}$$

where $f'_{c,T}$ is the concrete compressive strength at a given temperature T, and b and d are width and depth of concrete beam, respectively, as shown in Fig. 8.14.

In the case of RC columns or walls, the reduced axial capacity under fire conditions can be evaluated in a similar manner to that of moment and shear capacity. With the knowledge of the reduced strength properties of concrete and reinforcement at a fire exposure time, the axial capacity (P_f) is given as

$$P_f = \omega f'_{c,T}(A_g - A_{st}) + f_{y,T} A_{st} \tag{8.41}$$

where A_g and A_{st} are the gross area of concrete column and area of reinforcing steel, respectively. ω is a factor that accounts for reinforcement configuration (0.80 for tied and 0.85 for spiral configuration).

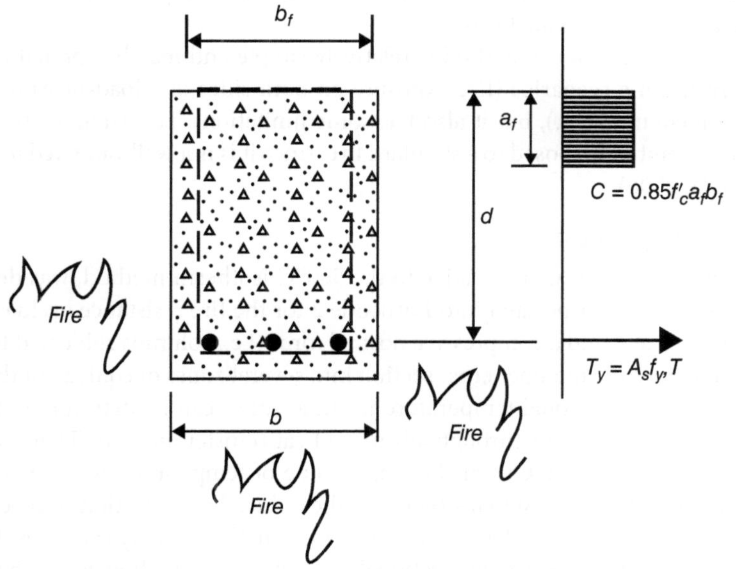

Figure 8.14 Depth of the rectangular stress block.

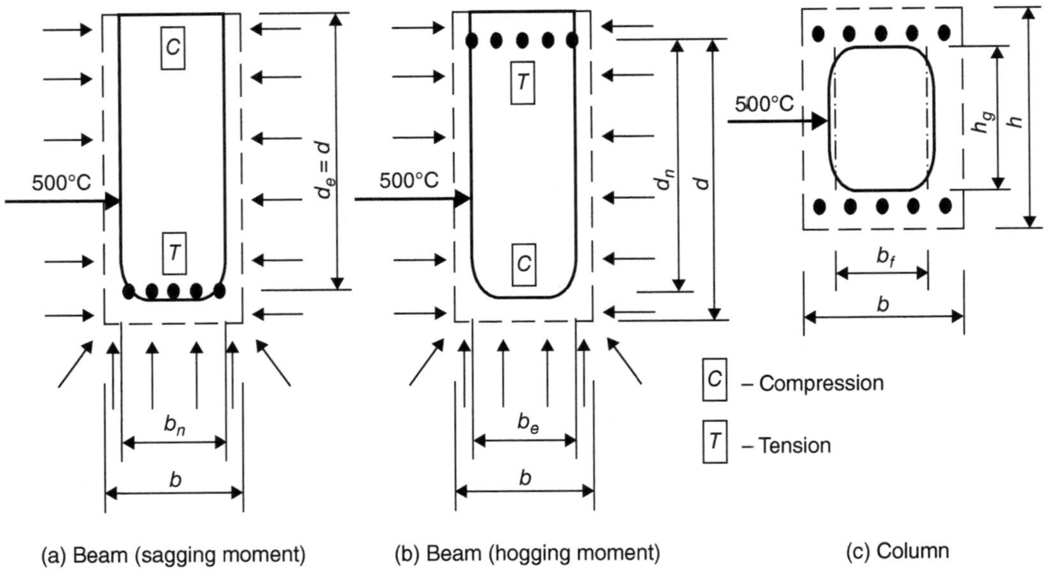

FIGURE 8.15 Reduced cross section of reinforced concrete beam or column in the 500°C Isotherm method.

Eurocode 2 provides two simple calculation methods to evaluate the sectional capacity of any given fire exposure time: the 500°C Isotherm method and the Zone method.

Eurocode 2 500°C Isotherm Method

The *500°C isotherm method* was developed in the 1970s by Anderberg (1978) and was introduced in the 1990s into Eurocode 2 (CEN, 2004) and the CEB-FIP Bulletin (1991). This method is based on the assumption that the reduction in compressive strength of concrete is not significant when the concrete temperatures remain below 500°C. Once concrete temperature exceeds 500°C, the compressive strength of concrete decreases at a rapid rate, reaching to about 30% at 700°C. The sectional temperatures in a member (or location of 500°C isotherm) can be determined through the use of charts in Eurocode 2 or equations by Wickström (1985, 1986) and by Kodur et al. (2013). Layers of concrete having a temperature above 500°C are to be removed from the overall cross section as shown in Fig. 8.15. In the case of reinforcing or prestressing reinforcement, all bars contribute to sectional capacity, based on reduced strength in these bars as per the temperature reached in each rebar at that time. Once the reduced cross-sectional dimensions (after removal of concrete layers) and the reduced level of yield stress for steel are determined, the reduced sectional capacity (flexural, shear, or axial) at a given fire exposure time can be evaluated by applying room temperature strength equations.

The 500°C Isotherm method is relatively simple and may be applied for all cross-section shapes and different heating scenarios (fire exposure on both sides of a load-bearing wall, fire exposure on all four sides of a column, etc.), but it also has some limitations. For example, this method is only applicable to concrete members exposed to a standard fire, since it is not well validated for other fire scenarios (Bamonte and Meda, 2006).

Eurocode 2 Zone Method

The *Zone method*, as an alternative to the 500°C Isotherm method, was developed in the 1980s by Hertz (2003) and was incorporated into Eurocode 2 for the fire resistance evaluation of members subjected to a bending moment and a compressive normal force (i.e., columns subjected to eccentric loading). The Zone method divides the member cross section into several zones of equivalent thickness (based on temperature isotherms). The sectional temperature in these zones can be determined through simplified analysis, as discussed in Sec. 8.7.2 or through advanced heat transfer analysis. Then, the average reduced strength of each zone is calculated. Out of the single value of temperature in each zone (i.e., concrete or rebars), a damaged zone and the residual strength of the reduced cross section is to be determined. With the reduced strength in concrete and rebar in each zone known, the capacity can be evaluated from each zone and this can be summed up to get the sectional capacity. This method yields better accuracy than the 500°C Isotherm method, but it is more laborious.

8.7.3 Advanced Analysis

Through advanced analysis, the behavior of reinforced and prestressed concrete structures under fire conditions can be simulated with a higher degree of accuracy, since many influencing factors such as temperature-induced spalling, restraint forces, and bond degradation between concrete and rebars can be duly incorporated in the analysis. Also, additional response parameters such as member deflections (deformations), extent of spalling, and stresses and strains in concrete and steel can be generated in tracing the fire response of such structures.

Advanced analysis can be carried out either at the macroscopic or microscopic level through finite element-based software/packages. In the macroscopic finite element-based methods, a sectional analysis is carried out at a number of cross sections along the length of a structural member and a series of sectional responses are then utilized to predict the overall member response (Kodur et al., 2009). Alternatively, microscopic finite element-based conventional computer packages such as ANSYS, ABAQUS, and SAFIR can be utilized for analysis, wherein a structural member or whole structure is discretized into a three-dimensional meshed model to undertake coupled (or uncoupled) thermal and structural analyses to evaluate fire response (Kodur and Agrawal, 2017). The steps needed to undertake fire resistance analysis through microscopic and macroscopic approaches on fire exposure concrete structures is briefly discussed here.

Microscopic Finite Element Approach

Fire resistance analysis of concrete structures is carried out at incrementing time-steps until the failure of the structure or the end of fire exposure. Commercial software packages (i.e., ABAQUS, ANSYS) or special-purpose programs (e.g., SAFIR) can be applied to undertake this analysis.

The first step in undertaking analysis is to idealize a concrete structure for thermal and structural analysis. This is to be done by discretizing the member into finite elements (mesh) available within the selected computer package. Different temperature-dependent thermal and mechanical properties of concrete and reinforcing (or prestressing) steel, discussed in Sec. 8.2, form input into such finite element models. Further, various temperature-induced phenomena such as the spread of plasticity in concrete—in both compression and tension regimes—cracking, and crushing can be accounted for through constitutive material models such as those proposed by William and Warnke (1975), Drucker and Prager (1952), and others. In addition, factors such as temperature-induced bond degradation, spalling, and high-temperature creep can be accounted for in the analysis.

The tensile behavior of concrete can be modeled using a trilinear model as recommended by William and Warnke (1975). In this model, the concrete tensile strength is taken as $0.3\text{–}0.62\sqrt{(f_c')}$; where f_c' is the compressive strength of concrete. Once the concrete reaches its peak tensile strength, a tensile stiffness multiplier of 0.4 to 0.6 is used to simulate a sudden drop of tensile stress to 40% (or 60%) of the initial rupture stress. Then the drop is followed by a linearly descending response to zero at a strain value of six times the rupture strain. The William and Warnke (1975) constitutive material model requires additional parameters—identified as the open and close crack shear transfer coefficients (β_t and β_c). Shear transfer coefficients are taken as zero when there is a total loss of shear transfer (representing a smooth crack) and 1.0 when there is no loss of shear transfer (representing a rough crack). The values of β_t and β_c in the developed model can be fairly assumed to be 0.2 and 0.7, respectively.

The response parameters generated in finite element analysis (i.e., cross-sectional temperatures, deformations, sectional capacity) are compared against corresponding limiting values (criteria) at each step to evaluate the failure state of the member. Failure occurs once any of these limiting states is exceeded. For instance, when a concrete beam attains a deflection of $L/20$ or rate of deflection reaches $L^2/9000d$, where L and d are the span and depth of the flexural member, respectively, the beam is said to fail. Additional failure criteria arising from spalling can also be applied to evaluate the failure of a member, if the extent of spalling at a given time-step reaches a specified spalling limit.

The above steps can be applied for analyzing a structure at the sectional, member, or system level. The main difference is in the discretization of the structure and in supplying relevant boundary conditions for thermal and structural analysis. Microscopic finite element models have been successfully applied to evaluate fire behavior of concrete structures at sectional and member levels (Huang et al., 2003). However, only limited studies have demonstrated the applicability of fire resistance evaluation of concrete structures at the system level (Ulm et al., 1999).

Although microscopic finite element models can model the behavior of a reinforced concrete member with better accuracy, such an analysis is quite complex, requiring significant effort and expertise for analysis, as well as for interpreting output response parameters. Some of the sophistication and complexity that can be adopted in three-dimensional modeling of structures can be limited, since currently available high-temperature property relations for concrete and reinforcement are mostly based on uniaxial strength properties. Further, most commercial finite element programs are not fully capable of modeling moisture transport or evaporation phenomena that have a significant influence on parameters such as temperature-induced cracking and spalling.

Macroscopic Finite Element Approach

To overcome the complexities associated with microscopic finite element-based fire resistance analysis for concrete structures, researchers have proposed the use of macroscopic finite element analysis in recent years. This type of analysis is suited for fire resistance evaluation based on sectional or member-level response in a fire-exposed concrete structure. In this macroscopic finite element formulation, specific high-temperature properties of concrete or reinforcing (or prestressing) steel can be duly accounted for in the analysis. Factors such as fire-induced spalling and high-temperature creep in concrete members can also be fully accounted for in the analysis (Kodur and Alogla, 2017). As in the case of the microscopic finite element approach, the analysis in the microscopic finite element approach is carried out at incrementing time-steps.

The analysis involves the discretization of a concrete member into segments (see Fig. 8.16). These segments are then discretized into a number of beam elements. Temperature rise in each segment is then calculated by applying heat balance equations. For evaluating such cross-sectional temperatures, temperature-dependent thermal properties of concrete are to be specified as input.

Within each beam element, the calculation of pore pressure, and associated spalling evolution, is carried out through a simplified hydrothermal model. This hydrothermal model uses principles of mechanics and thermodynamics, including the conservation of the mass of liquid water and water vapor, to compute time-dependent pore pressure development within concrete.

Following cross-sectional calculation, pore pressure in concrete in each element is calculated at each time-step, and this pressure is then compared against the temperature-dependent tensile strength of concrete. When the pore pressure exceeds the tensile strength of concrete, spalling is assumed to occur in that element. Once spalling occurs in an element, that element is removed from the overall concrete section, and the reduced section, with new boundary surface, is considered in the subsequent time-step. In this way, spalling and thermal analyses are coupled with pore pressure calculations.

Both thermal and spalling analyses are coupled and lead to a strength analysis to evaluate the structural response of the reinforced (or prestressed) concrete member. Through strength analysis, time-dependent sectional moment curvature (M-κ) relations are generated at each segment of the discretized beam. These M-κ relations form the basis to trace the response of a concrete member in the entire range of loading, up to collapse, under fire exposure.

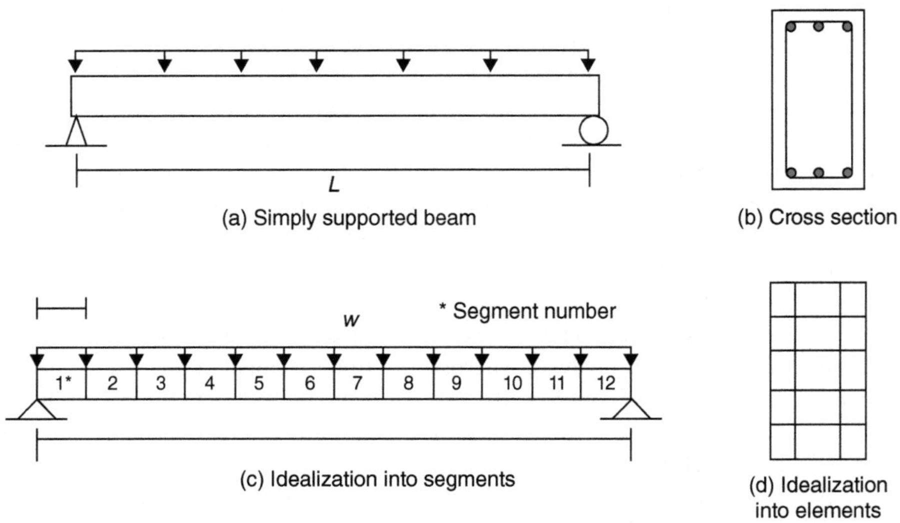

Figure 8.16 Layout of a typical RC beam for moment-curvature analysis and its idealization into FE analysis.

A flowchart showing the steps associated in the fire resistance analysis of an RC member through macroscopic analysis is presented in Fig. 8.17. The fire resistance analysis is carried out in a series of time increments until the failure of the member or the end of fire exposure. At each time interval, a coupled heat transfer-spalling analysis to predict cross-sectional temperature and spalling is carried out. The generated sectional temperatures are then supplied to perform strength analysis wherein M-κ relations are generated for each of the segments. These time-dependent M-κ relations form the basis for structural analysis of the overall concrete member to compute deflections and internal forces. In macroscopic FE analysis, a number of response parameters, such as cross-sectional temperatures, axial restraint force, deflections, and moment capacities, can be generated at each fire exposure time.

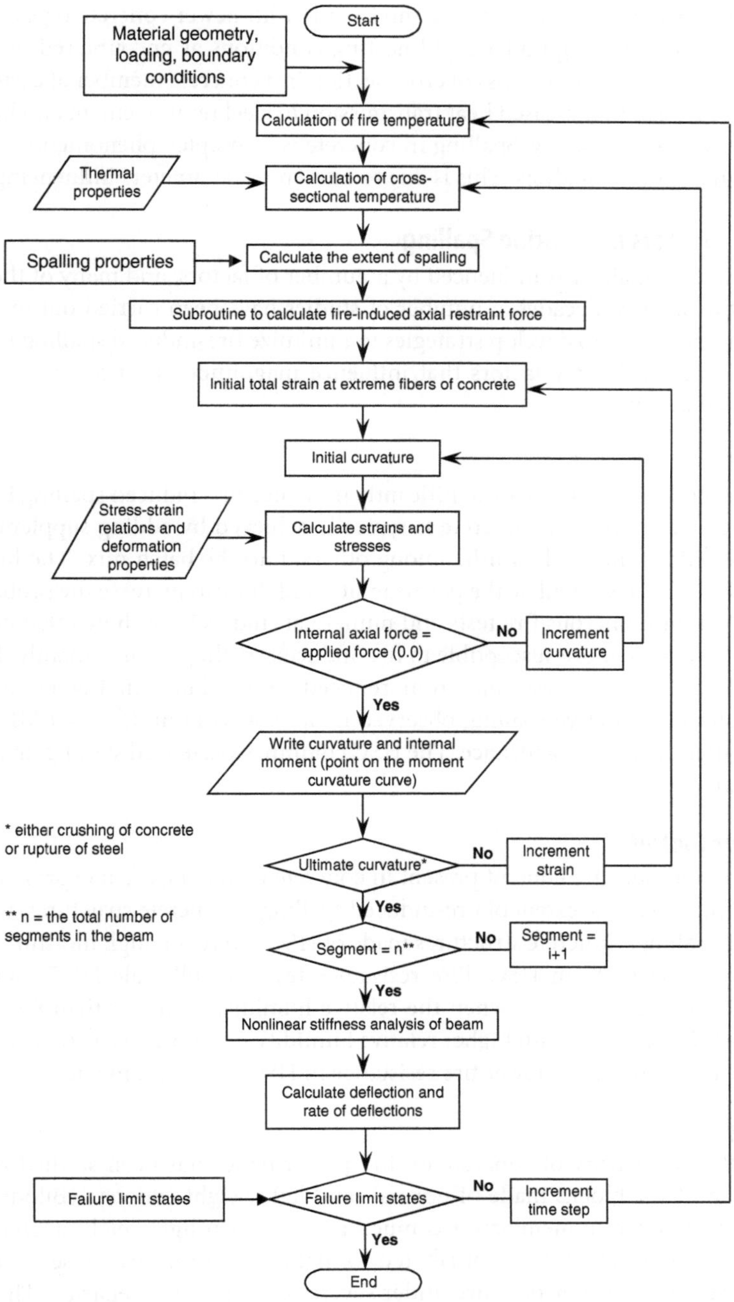

FIGURE 8.17 Flowchart showing the steps associated with fire resistance analysis of an RC member (beam) through macroscopic analysis. (*With permission from Dwaikat, 2009.*)

The generated temperatures and capacity at each segment, or computed deflections (deformations) in the member, are used to check failure state of the concrete member at each time-step. At every time-step, each segment is checked against predetermined failure criteria, which can include the prescriptive-based thermal criterion, or performance-based strength and deflection considerations. Once any of these criteria is exceeded (e.g., midspan deflection exceeding maximum deflection limit of $L/30$), failure is said to occur and the duration to reach the preceding time-step is taken as the fire resistance of the member. Further details of the analysis procedure for macroscopic finite element-based analysis can be found in Kodur et al. (2009).

8.8 Strategies for Enhancing Fire Resistance of High-Strength Concrete Members

As discussed in Sec. 8.2.3, concrete structures, under certain conditions, are susceptible to fire-induced spalling. Specifically, concrete structures made with newer concrete types (higher-strength concretes) are prone to such spalling under rapid heating conditions as encountered in petrochemical or tunnel fires. Such spalling can lead to a loss of cross section in a concrete member at early stages of fire exposure, which in turn can lead to increased heat transmission to steel reinforcement and has the potential to significantly reduce sectional capacity. Spalling in concrete is a complex phenomenon and is not easy to predict even through advanced analysis. This is due to numerous parameters influencing the spalling phenomenon.

8.8.1 Factors Influencing Spalling

Fire-induced spalling is influenced by a number of factors, and many of these factors are interdependent. Over the last two decades, a number of studies have been carried out to understand the mechanism of spalling, as well as to develop strategies to minimize fire-induced spalling in concrete structures. Based on recent studies, the key factors that influence magnitude and occurrence of spalling (Kodur, 2018) are compiled below.

Strength

The strength of concrete can significantly influence fire-induced spalling in concrete structures. A higher compressive strength in concrete is typically achieved by adding supplementary cementitious materials such as silica fume and fly ash, among others, into the batch mix. The higher the silica fume (or other additives), the lower will be the permeability, and this can increase the probability of fire-induced spalling. Results from numerous fire tests and numerical studies have shown that concretes with strengths higher than 70 MPa are more susceptible to fire-induced spalling. This is mainly due to buildup of pore pressure within the concrete resulting from reduced permeability in higher-strength concretes. Figure 8.18 illustrates comparative spalling observed in an NSC column ($f'_c \approx 40$ MPa) and HSC column ($f'_c \approx 100$ MPa) after fire tests in a furnace. The HSC column experienced significant spalling under ASTM E119 fire exposure.

Moisture Content

The level of moisture content present in concrete members, often expressed in terms of relative humidity (RH), influences the extent of fire-induced spalling in concrete structures, with higher RH levels leading to greater spalling. Concrete structures made of HSC can retain high moisture content for longer periods due to low permeability in HSC. Fire resistance tests on full-scale HSC columns have clearly shown that significant spalling occurs when the relative humidity is higher than 80% (Kodur and McGrath, 2003). Thus, HSC members, with higher relative humidity, are more susceptible to spalling under fire conditions, which in turn results in lower fire resistance in HSC structural members.

Density

The effect of density of concrete on fire performance has been studied through fire tests on columns, slabs, and large blocks made of normal- and lightweight concrete (Bilodeau et al., 2004). The extent of spalling in concrete members was much greater when lightweight aggregate was incorporated in HSC blocks. This could be partly attributed to higher free moisture present in lightweight aggregate, which generates higher vapor pressure under severe fire exposure scenarios. Thus, the density of concrete can influence the fire performance of concrete structures and should be duly accounted for in evaluating fire resistance.

(a) NSC column (b) HSC column

Figure 8.18 The state of four columns made of different concrete types after fire resistance tests. (*With permission from Khaliq, 2014.*)

Fire Intensity

Fire severity, specifically the rate of heating, affects the extent of spalling and thus can indirectly influence fire resistance of high-strength concrete members. HSC members subjected to severe fires, as characterized by rapid heating rates or high fire intensities (typical of hydrocarbon fires), are prone to severe spalling (Kodur and Phan, 2007).

Specimen Dimensions

Generally, the fire resistance of a concrete member made of NSC increases with member dimensions, due to larger thermal mass, increased sectional capacity, and nonoccurrence of spalling in concrete. However, the risk of explosive thermal spalling increases with member size specifically when the structure is made of HSC. This is due to the fact that the specimen size is directly related to heat and moisture transport through the member, as well as the capacity of larger structures to store more energy. When appropriate spalling mitigation measures are incorporated, the risk of explosive spalling decreases and the fire resistance increases with the size of the members.

Fiber Reinforcement

The presence of steel or polypropylene fibers in concrete minimizes the extent of fire-induced spalling and thus enhances the fire resistance of concrete members (see Fig. 8.19). Results from previous fire tests show that the addition of polypropylene fibers to concrete minimizes spalling in HSC structural members and also enhances its fire resistance. This improved performance is due to the fact that polypropylene fibers melt at about 160 to 170°C and create pores (channels) within concrete, which help in dissipating temperature-induced vapor pressure in concrete. Alternatively, steel fibers in concrete can also minimize spalling and enhance fire resistance of HSC structural members (Kodur and Sultan, 2003). This strategy works through the enhancement of tensile strength of concrete, facilitated by the presence of steel fibers in concrete. Further, the presence of steel fibers also slows down the degradation of tensile strength with temperature. This increased tensile strength of concrete helps to overcome temperature-induced tensile stresses generated by the pore pressure and thus minimize the occurrence of spalling, which in turn leads to higher fire resistance in a member. Recently, the combination of steel and polypropylene fibers for minimizing spalling in concrete has been explored and found to be very effective, specifically in tunnels and petrochemical facilities (Kodur and Naser, 2020).

Figure 8.19 illustrates the state of four columns made of different concrete types after fire tests. As can be seen, while the concrete columns made of plain HSC (HSC-3) experience severe spalling, columns made of concrete with embedded fibers underwent slight spalling (in the case of polypropylene fibers—HSCP-2) and no spalling in the case of columns made of concrete with steel (HSC-S) and hybrid (polypropylene and steel fibers—HSC-H). The optimum fiber content for minimizing spalling is 0.1% to

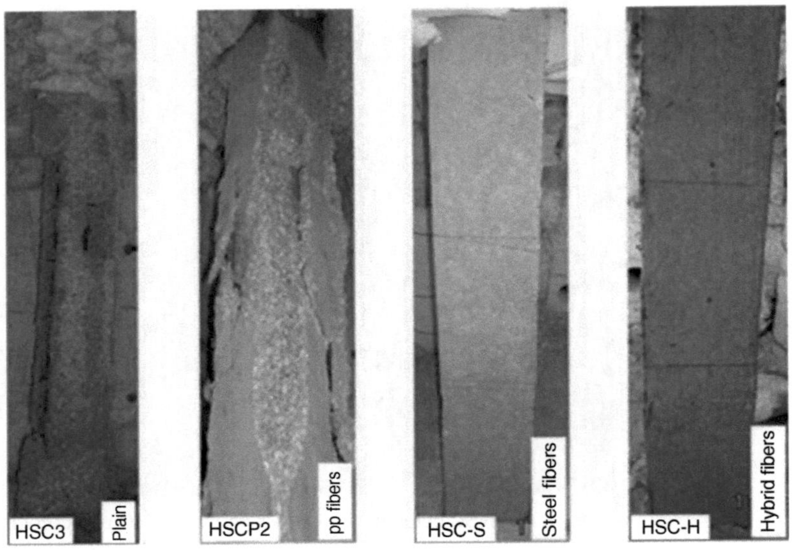

Figure 8.19 The state of four columns made of different concrete types after fire resistance tests. (*With permission from Khaliq, 2014.*)

0.15% in the case of polypropylene fibers, and about 1.75% to 2.0% in the case of steel fibers (Kodur, 2018).

Load Intensity and Type

The type of load and its intensity have a significant influence on fire-induced spalling and the resulting fire resistance in a concrete member. Higher load intensity leads to lower fire resistance. The effect is more pronounced in HSC structural members, since the strength loss with temperature occurs at a faster pace in HSC than in NSC. A loaded HSC structural member is prone to higher fire-induced spalling than an unloaded member. This is due to the fact that a loaded structural member is subjected to stresses because of structural (mechanical) loading, in addition to stresses arising from pressure generated by temperature. Further, the extent of spalling is higher if the load is of an eccentric (or bending) type, as in the case of columns, due to the development of higher tensile stress in concrete (resulting from pore pressure and eccentric loading) under fire exposure, specifically when the member is subjected to nonsymmetrical heating. Once such tensile stress exceeds the tensile strength of concrete, spalling will occur, and this phenomenon can lower fire resistance.

Type of Aggregate

Of the two commonly used aggregates in structural concrete, carbonate aggregate (predominantly composed of limestone) provides higher fire resistance in concrete members than siliceous aggregate (predominantly composed of quartz) concrete members. This is due to an endothermic reaction, arising from a reaction in limestone that occurs around 700°C (Kodur and Harmathy, 2012). This will result in lower thermal conductivity and higher specific heat of carbonate aggregate, which lowers the rate of heat increase in concrete members and, in turn, fire-induced spalling. In general, the fire resistance of a concrete member made with carbonate aggregate concrete is about 10% higher than a concrete member made with siliceous aggregate concrete (Kodur et al., 2013). This trend is applicable for both NSC and HSC structural members (columns, beams, slabs, etc.).

Sectional Shape/Configuration

Sectional configuration and sectional mass have a significant influence on the fire resistance of a concrete member. As an illustration, a hollow-core slab of an equivalent cross section exhibits lower fire resistance, as compared to a traditional solid slab, specifically when the slabs are made of HSC. In a solid slab, the temperature transmission from a fire-exposed surface to an unexposed surface is mainly governed by the conduction of heat, whereas such temperature transmission through a hollow-core slab occurs via conduction in the solid concrete portion, and convection and radiation in the hollow cores. This can cause significant variation in the cross-sectional temperature profile between the two slabs. Also, a hollow-core slab experiences faster transmission of heat to the unexposed side of the slab, as compared to a solid slab, due to lower mass of concrete. In addition, shear capacity in hollow-core slabs can be

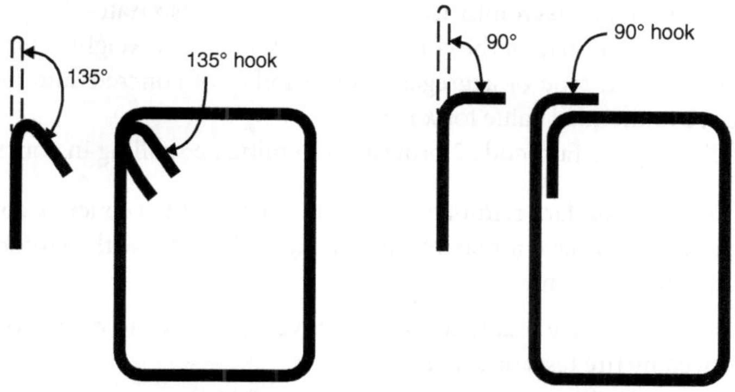

FIGURE 8.20 Bent tie configurations in concrete members (right: modified tie configuration at 135°; left: conventional tie configuration at 90°).

much lower than in solid slabs due to a significant loss of cross section around mid-depth of the slab, as a result of voids (Kodur and Shakya, 2014). Further, the occurrence of even minor spalling in a hollow-core slab or in a beam section with slender stems can have an adverse impact on the fire resistance of a structural member.

Along similar lines, the fire resistance of concrete columns is influenced by the shape of the cross section. An RC column with a circular cross section will have higher fire resistance than that of an RC column with a square or rectangular cross section, of equivalent cross-sectional area. The lower fire resistance in square or rectangular columns is due to faster heating at the corners arising from the edge effect.

Reinforcement

Concrete structures, depending on the type of reinforcement, are broadly grouped under reinforced concrete and prestressed concrete structures. Prestressed concrete members possess lower fire resistance compared to reinforced concrete members. This is due to the fact that the prestressing strand experiences a higher rate of strength loss than that of reinforcing bars when subjected to elevated temperatures. If spalling were to occur in prestressed concrete members, then the fire resistance can be much lower due to faster temperature transmission to strands.

Lateral Reinforcement

The layout of lateral reinforcement in columns has an influence on the fire performance of reinforced concrete columns, especially made of HSC (Khaliq and Kodur, 2012). Higher fire resistance in HSC columns can be achieved by providing improved tie configuration through the provision of bent ties, bent at 135° back into the core of the column section, and closely spaced ties (at 0.75 times than required for NSC columns). Such improved lateral reinforcement (see Fig. 8.20) also reduces the extent of late-stage fire-induced spalling in concrete columns due to ties holding the longitudinal rebar firmly in place under fire exposure conditions. This is possible due to lower temperature developing in tie hooks (since ties or stirrups are bent into the cooler core of concrete), and thus ties will retain higher strength and stiffness properties for a longer fire duration. This mechanism helps minimize the movement and buckling of longitudinal bars, which in turn reduces thermal and load-induced strains in concrete. The theoretical basis for this mechanism is discussed by Kodur et al. (2013).

8.8.2 Guidance for Minimizing Spalling

Based on detailed studies, researchers have developed solutions to overcome fire-induced spalling in concrete structures. These solutions include modifications that can be incorporated at the material and structural levels, when spalling is likely to occur in a concrete structure. Most researchers concur spalling to be a key issue in the case of higher strength concrete (above 70 MPa) and also when structural members are subjected to severe fire exposure scenarios.

Although many research articles have been published on solutions to fire-induced spalling in concrete, there is still limited guidance in codes and standards for developing strategies to mitigate spalling in concrete structures. Most codes ignore any reference to spalling, and only a few codes and standards include broad general guidelines for mitigating spalling effects, without giving specific detailing provisions. At the outset, Eurocode 2 provides a few clauses that specifically address spalling issue. Eurocode 2 states, "Spalling shall be avoided by appropriate measures or the influence of spalling on performance requirements

(R and/or EI) shall be taken into account." Eurocode 2 also states, "Explosive spalling is unlikely to occur when the moisture content of the concrete is less than *k*% by weight. Above *k*% a more accurate assessment of moisture content, type of aggregate, permeability of concrete and heating rate should be considered where the recommended value for *k* is 3."

The following are Eurocode 2 provisions to mitigate spalling in concrete structures.

- Use of a secondary reinforcement (wire) mesh with a concrete cover of 15 mm. The reinforcement mesh should have a maximum spacing of 50 mm, with reinforcement bars having a minimum diameter of 2.0 mm.
- Use of a concrete mix that does not have a tendency to explosive spalling at high temperatures, as proved by fire tests or experience.
- Limiting maximum content of silica fume to less than 6% by weight of cement.
- Use of a thermal barrier (between fire zone and structural member) so as to ensure no spalling occurs.
- Use of 2.0 kg/m^3 monofilament polypropylene fibers in concrete batch mix.

The most recent edition of ACI 216.1 lists some guidelines for mitigating spalling in high-strength concrete; accordingly:

- Installation of bent ties in columns (when the ties are bent at 135° into the concrete core) can minimize spalling and increases fire resistance (see Fig. 8.20).

A few researchers (Khoury, 2008; Kodur, 2018; Liu et al., 2018) also proposed the use of some solutions such as the following:

- Addition of polypropylene fibers, about 0.1% to 0.15% by volume, to the concrete batch mix reduces spalling.
- Addition of steel fibers enhances tensile strength and reduces spalling.
- Use of carbonate aggregate, instead of siliceous aggregate, reduces spalling and enhances fire endurance.

It should be noted that the above-cited guidelines for minimizing spalling are mainly derived based on observations in furnace tests (mostly standard fire exposure) and do not give due consideration to critical factors that influence this phenomenon. These guidelines are qualitative and not specific enough. Further, there are no predictive models or guidance to determine the likely occurrence of spalling in any given concrete structure under specific fire exposure conditions. Further research is needed to develop specific codal guidance for mitigating fire-induced spalling in concrete structures.

8.9 Masonry Structures

Masonry is one of the most widely used materials in building construction, specifically in low-rise buildings. Masonry as a material generally exhibits similar variation in properties as that of concrete at elevated temperatures. Therefore, much of the property data and fire resistance approaches discussed for concrete at the material and structure level generally holds well for evaluating fire resistance of masonry structures. There have been only limited efforts in applying rational engineering approaches for evaluating resistance of masonry structures. This is due to the fact that the main use of masonry is mostly in non-load-bearing fire-separating walls in low-rise buildings to form isolated compartments within a large open-plan structure, and not as a primary structural member (e.g., columns and beams). Fire ratings for masonry structures, including masonry walls, are available as listings and tabulated data in Eurocode 6 and ACI 216.1, National Building Code (IBC, NBC, etc.), as well as the TEK design manual (which cites ACI 216.1 as the main source for fire ratings).

In general, the level of fire resistance achieved by a clay masonry wall is a function of mass and thickness of the wall. This fact is well established based on the results of many fire resistance tests conducted on walls of solid and hollow clay units. Typically, the insulation criterion (i.e., temperature rise on unexposed surface of the wall) governs failure of a masonry wall. For satisfying the compartmentation criterion, the temperature on the unexposed face should not exceed the limit

Standard Fire Resistance (min)	Minimum Wall Thickness (mm)				
	Non-Load-Bearing	Group 1 Units*		Group 2 Units**	
		Load Level = 60%	Load Level = 100%	Load Level = 60%	Load Level = 100%
30	60	100	100	90	90
45	90	170	200	90	90
60	90	170	200	90	90
90	100	200	200	100	100
120	100	200	200	190	140
180	160	200	200	190	190
240	190	300	300	190	190

*For use when compressive strength of mortar is between 5 and 25 MPa.
**For use when compressive strength of mortar is between 5 and 35 MPa.

Table 8.18 Minimum Thickness for Load-Bearing Walls Made of Masonry as Listed in EC 6 (*Reproduced from CEN*)

prescribed in the standard fire test of 139°C as per ASTM E119 or similar temperatures given in other tests' standards (140°C in EC6).

Eurocode 6 provides three methods to evaluate fire resistance of masonry members: fire tests, tabulated data, and calculation methods (including advanced analysis).[10] In current practice, the fire resistance of conventional masonry structures is mostly evaluated through the use of tabulated methods. The tabulated data provided in Eurocode 6 are compiled from a large set of fire tests on load-bearing masonry walls conducted in Belgium, Germany, and England.[11] Eurocode 6 broadly groups masonry structures under six categories and specifies which limiting (failure) criterion is relevant for each category (E-integrity, I-insulation, and R-load bearing):

- Non-load-bearing separating walls (Criterion EI)
- Load-bearing separating walls (Criterion REI)
- Load-bearing non-separating walls (Criterion R)
- Short load-bearing non-separating walls (Criterion R)
- Fire walls (load-bearing or not, single or double leaf, Criterion REI-M, EI-M[12])
- Load-bearing cavity walls with one leaf loaded (REI)

An example of a typical table in Eurocode 6 for evaluating fire resistance requirement in load-bearing separating walls is shown in Table 8.18. The minimum wall thickness for achieving a specified fire rating in a masonry wall can be evaluated from Table 8.18. The thickness depends on the compressive strength of mortar and also on load level on the wall.

The ACI 216.1 standard (as well as the National Concrete Masonry Association guide, TEK 7-1C) also provides tables for determining fire resistance of masonry construction assemblies. These tables provide minimum equivalent thickness (see Table 8.19) needed to achieve a certain fire rating in plain and reinforced concrete masonry walls, floors, and roof slabs. A sample table from ACI 216.1 (ACI, 2014) giving fire resistance ratings of clay masonry walls is shown in Table 8.19. It is worth noting that the same table also appears in the TEK 7-1C guide document (2009). The TEK 7-1C guide also lists minimum longitudinal cover to reinforcement in reinforced concrete masonry lintels as shown in Table 8.20.

[10] Calculation methods are available in Eurocode 6; however, they are more complicated and tedious compared to those for steel and concrete structures.
[11] It is worth noting many of these tests were often not carried out until failure but up to a specific time of resistance meeting a required fire rating (i.e., requirement aimed to be met).
[12] M indicates a fire wall.

Masonry Type	Minimum Equivalent Thickness for Fire Resistance (mm)			
	1 hour	2 hours	3 hours	4 hours
Solid brick of clay or shale	68.6	96.52	124.5	152.4
Hollow brick of tile of clay or shale, unfilled	58.4	86.36	109.2	139.7
Hollow brick of tile of clay or shale, filled	76.2	111.8	139.7	167.4

Table 8.19 Fire Resistance of Clay Masonry Walls as per ACI 216.1 (ACI, 2014) (*Authorized Reprint from ACI, 2014*)

Nominal Lintel Width (mm)	Fire Resistance Rating/Cover (mm)			
	1 hour	2 hours	3 hours	4 hours
152	38	51	—	—
203	38	38	44	76
254 or more	38	38	38	44

Table 8.20 Minimum Cover (mm) to Longitudinal Reinforcement for Lintels as per TEK 7-1C Guide Document (*Adapted from TEK7-1C, 2009*)

8.10 Emerging Trends and Research Needs

Concrete is an excellent material from a fire resistance consideration and continues to be a primary material in building construction, where fire safety is a primary design issue. Recent advances in concrete technology have led to the development of newer concrete types (HSC, UHPC, FRC) with superior mechanical, durability, and sustainability properties. Further, a number of additives (silica fumes, superplasticizers, admixtures) are added to the batch mix to enhance different properties, even in the case of traditional concretes. The addition of such ingredients has an effect on properties of concrete at elevated temperatures. Further, recent research has clearly shown that certain types of concretes (HSC, UHPC) are more prone to fire-induced spalling, with faster deterioration in mechanical properties at elevated temperatures. To address these problems, needed research relating to fire performance of modern concrete and masonry structures is discussed below.

8.10.1 Predictive Spalling Models

Spalling in concrete can take several different forms and can adversely impact the overall fire resistance of a concrete structure. Such spalling can lead to the reduction of cross section, which results in an overall loss of the thermal protection to reinforcing or prestressing steel, thereby accelerating strength loss in the reinforcement. Spalling is a complex phenomenon and is influenced by numerous parameters including temperature-dependent properties, differential thermal stresses, moisture movement, and physiochemical changes that occur in concrete during fire exposure.

Current engineering approaches specified in codes and standards for fire design of concrete structures rely on the assumption that the concrete will not spall during fire. This assumption is based largely on data from furnace tests on normal-weight/normal-strength concrete members tested in the 1970s and 1980s. However, recent research, both at the material and the structural level, has shown that the higher strength of concretes is more susceptible to spalling under severe fire exposure scenarios. While some guidance on the means by which spalling can be minimized was developed in recent years, still there is no comprehensive understanding on the mechanisms through which spalling occurs. Furthermore, there is a lack of numerical models that can predict when spalling is likely to occur in a given concrete structure and under what conditions.

The lack of such numerical models and approaches is perhaps due to the complexity in modeling spalling phenomenon, as well as limited equipment and test methods available to generate needed data for validation of models (Kodur, 2014). To improve the state of the art, there is an urgent need for developing reliable numerical models for predicting spalling in concrete structures. Such models should account for specific temperature-dependent material properties such as porosity, permeability, and tensile strength of concrete in predicting spalling. The validity of these models is to be established through validation against data, both at material-level and structural-level fire tests. Once validated, such spalling models can be integrated into overall fire resistance numerical models for evaluating realistic fire resistance of concrete structures.

8.10.2 Temperature-Dependent Property Data for Newer Concrete Types

The temperature-dependent properties of concrete are critically important for quantifying the fire resistance of reinforced and prestressed structures. As shown in Sec. 8.2, there is a large variability in currently available thermal and mechanical properties of conventional concretes at elevated temperatures. Further, there is very minimal data on some of the temperature-dependent properties such as porosity, permeability creep, transient strain, and so on, particularly under rapid heating regimes or in the cooling phase of fire, that are critical for evaluating the fire resistance of concrete structures. There is very limited information on these properties for newer concrete types such as HSC, FRC, and UHPC, as well emerging alternatives such as geopolymer based concretes and mortars (Zhang et al., 2014; Zhang et al., 2016).

To overcome the current limitation in property data, a series of property tests relating to high-temperature properties of concrete are needed. A combination of experimental and analytical studies is needed to gain insights into high-temperature material properties, especially those related to fire-induced spalling and creep. Property data such as temperature-dependent permeability, tensile strength, porosity, and moisture movement are key parameters where there is a serious lack of test data. Experimental data which include measurement of internal pore pressures and moisture distribution are also needed for the development of spalling models and validation of analytical approaches.

In such tests, adequate consideration should be given to the rate of heating, testing regime, and specimen conditions (moisture content, size) so as to be applicable for a range of conditions encountered in practice. For measuring these properties, there is a need to develop relevant testing procedure instrumentation to measure pore pressure and associated sensors that can work in a high-temperature range of 20 to 800°C. Data compiled from these tests is used to derive constitutive models and property relations for advanced analysis to evaluate the fire resistance of concrete structures. Enhancement of such constitutive models for material properties will help improve the accuracy of numerical model predictions under different fire exposure scenarios (Kodur and Agrawal, 2016).

8.11 References

Abrams, Melvin S., and A. H. Gustaferro. "Fire Endurance of Two-course Floors and Roofs." *Journal Proceedings* 66, no. 2 (1969): 92–102.

American Concrete Institute (ACI). *Code Requirements for Determining Fire resistance of Concrete and Masonry Construction Assemblies.* ACI 216.1-07/TMS-0216-14. Farmington Hills, MI: ACI, 2014.

Anderberg, Y. "Spalling Phenomena of HPC and OC." In *NIST Workshop on Fire Performance of High Strength Concrete in Gaithersburg.* Gaithersburg, MD, February 1997, pp. 69–73.

Anderberg, Y., S. E. and S. Magnusson Thelanders-son, and U. Wickström. "An Analytical Approach to Fire Engineering Design of Concrete Structures." *Special Publication* 55 (1978): 409–438.

Anderberg, Yngve, and Sven Thelandersson. "Stress and Deformation Characteristics of Concrete at High Temperatures. 2. Experimental Investigation and Material Behaviour Model." *Bulletin of Division of Structural Mechanics and Concrete Construction, Bulletin 54* (1976).

ASCE. *Structural Fire Protection: Manual of Practice, No. 78.* New York: American Society of Civil Engineers, 1992.

Aslani, F., and B. Samali. "Predicting the Bond Between Concrete and Reinforcing Steel at Elevated Temperatures." *Structural Engineering and Mechanics* 48, no. 5 (2013): 643–660.

Bamonte, Patrick, and Alberto Meda. *Towards a Simplified Approach for the Sectional Analysis of R/C Members in Fire.* Aisberg, an Archive and Research Registry. Bergamo, Italy: University of Bergamo, 2006. https://aisberg.unibg.it/handle/10446/19969#.XgZzhEdKjD4.

Bažant, Z. P. "B10. Analysis of Pore Pressure, Thermal Stress and Fracture in Rapidly Heated Concrete." In *International Workshop on Fire Performance of High-Strength Concrete.* Gaithersburg, MD, February 1997.

Bažant, Z. P., M. F. Kaplan, and Zdeněk P. Bazant. *Concrete at High Temperatures: Material Properties and Mathematical Models.* Harlow, Essex: Longman, 1996.

Behnood, A. and M. Ghandehari, "Comparison of Compressive and Splitting Tensile Strength of High-Strength Concrete With and Without Polypropylene Fibers Heated to High Temperatures." *Fire Safety Journal* 44, no. 8 (2009): 1015–1022.

Bilodeau, A., V. K. R. Kodur, and G. C. Hoff. "Optimization of the Type and Amount of Polypropylene Fibres for Preventing the Spalling of Lightweight Concrete Subjected to Hydrocarbon Fire." *Cement and Concrete Composites* 26, no. 2 (2004): 163–174.

Bisby, L. A. *Fire Behaviour of Fibre-Reinforced Polymer (FRP) Reinforced or Confined Concrete.* Belfast: Queen's University, 2003: pp. 2520.

BS 8110-1:2010. *Structural Use of Concrete. Code of Practice for Design and Construction.* London: British Standards Institution, 2010.BSI. *Eurocode 6—Design of Masonry Structures.* London: British Standard Institution, 2005.

Castillo, C., Durrani, A. J., "Effect of Transient High-temperature on High-Strength Concrete." *ACI Materials Journal* 87, no. 1 (1990): 47–53.

CEB-FIP, M. C. "Fire Design of Concrete Structures in Accordance with CEB-FIP Model Code 90-Final Draft." Bulletin d'Information du CEB 208 (1991).

CEN *Eurocode 2: Design of Concrete Structures—Part 1–2: General Rules—Structural Fire Design. EN 1992-1-2.* Brussels: European Committee for Standardization, 2004.

Cheng, Fu-Ping, V. K. R. Kodur, and T.-C. Wang. "Stress-Strain Curves for High Strength Concrete at Elevated Temperatures." *Journal of Materials in Civil Engineering* 16, no. 1 (2004): 84–90.

Cruz, C. R. *Elastic Properties of Concrete at High Temperatures*. No. 191. Skokie, IL: Portland Cement Association,1966.

de la Precontrainte, Federation Internationale. "FIP/CEB Report on Methods of Assessment of the Fire Resistance of Concrete Structural Members." Wexham Springs: Cement and Concrete Association, 1978.

Dotreppe, J-C., and J-M. Franssen. "The Use of Numerical Models for the Fire Analysis of Reinforced Concrete and Composite Structures." *Engineering Analysis* 2, no. 2 (1985): 67–74.

Drucker, Daniel Charles, and William Prager. "Soil Mechanics and Plastic Analysis or Limit Design." *Quarterly of Applied Mathematics* 10, no. 2 (1952): 157–165.

Dwaikat, Monther BM. Flexural Response of Reinforced Concrete Beams Exposed to Fire. PhD thesis. Michigan State University, 2009.

Dwaikat, M. B., and V. K. R. Kodur. "Hydrothermal Model for Predicting Fire-Induced Spalling in Concrete Structural Systems." *Fire Safety Journal* 44, no. 3 (2009): 425–434.

Felicetti, R., P. G. Gambarova, G. P. Rosati, F. Corsi, and G. Giannuzzi, "Residual Mechanical Properties of High-Strength Concretes Subjected to High-Temperature Cycles." In *Proceedings of the International Symposium of Utilization of High-Strength/High-Performance Concrete*, Paris, France, 1996, pp. 579–588.

Galbreath, Murdoch. *Fire Endurance of Light-Framed and Miscellaneous Assemblies: A Compilation of Published Information on Fire Endurance of a Variety of Light-Frame Walls, Partitions, and Floors*. Ottawa: National Research Council of Canada, 1966.

Gao, W. Y., J-G. Dai, J. G. Teng, and G. M. Chen. "Finite Element Modeling of Reinforced Concrete Beams Exposed to Fire." *Engineering Structures* 52 (2013): 488–501.

Gawin, D., C. E. Majorana, and B. A. Schrefler. "Numerical Analysis of Hydro-thermal Behaviour and Damage of Concrete at High Temperature." *Mechanics of Cohesive-Frictional Materials: An International Journal on Experiments, Modelling and Computation of Materials and Structures* 4, no. 1 (1999): 37–74.

Gross, H. "High-Temperature Creep of Concrete." *Nuclear Engineering and Design* 32, no. 1 (1975): 129–147.

Harmathy, T. Z. "Thermal Properties of Concrete at Elevated Temperatures." *ASTM Journal of Materials* 5, no. 1 (1970): 47–74.

Hertz, Kristian Dahl. "Limits of Spalling of Fire-exposed Concrete." *Fire Safety Journal* 38, no. 2 (2003): 103–116.

Huang, Zhaohui. "Modelling the Bond between Concrete and Reinforcing Steel in a Fire." *Engineering Structures* 32, no. 11 (2010): 3660–3669.

Huang, Zhaohui, Ian W. Burgess, and Roger J. Plank. "Modeling Membrane Action of Concrete Slabs in Composite Buildings in Fire. II: Validations." *Journal of Structural Engineering* 129, no. 8 (2003): 1103–1112.

Hull, Walter Austin, and Simon H. Ingberg. "Fire Resistance of Concrete Columns." *Journal of the Franklin Institute* 200, no. 3 (1925): 379–381.

Jin-Cheng Liu, Kang Hai Tan, and Yao Yao. "A New Perspective on Nature of Fire-Induced Spalling in Concrete." *Construction and Building Materials* 184 (2018): 581–590.

Khaliq, Wasim, and Venkatesh Kodur. "High Temperature Mechanical Properties of High-Strength Fly Ash Concrete with and without Fibers." *ACI Materials Journal* 109, no. 6 (2012).

Khoury, G. A. "Concrete Spalling Assessment Methodologies and Polypropylene Fibre Toxicity Analysis in Tunnel Fires." *Structural Concrete* 9, no. 1 (2008): 11–18.

Kodur, Venkatesh. "Fiber Reinforced Concrete for Enhancing Structural Fire Resistance of Columns." *Special Publication* 182 (1999): 215–234.

Kodur, Venkatesh. "Properties of Concrete at Elevated Temperatures." *ISRN Civil Engineering* 2014 (2014):1–15.

Kodur, Venkatesh. "Innovative Strategies for Enhancing Fire Performance of High-strength Concrete Structures. *Advances in Structural Engineering* 21 (2018): 1369433218754335.

Kodur, V. K., and N. K. Raut. "Design Equation for Predicting Fire Resistance of Reinforced Concrete Columns." *Structural Concrete* 10, no. 2 (2009): 73–86.

Kodur, V. K. R., and A. M. Shakya. "Modeling the Response of Precast Prestressed Concrete Hollowcore Slabs Exposed to Fire." *PCI Journal* 59, no. 3 (2014): 78–94.

Kodur, V. K. R., and S. M. Alogla. "Effect of High-Temperature Transient Creep on Response of Reinforced Concrete Columns in Fire." *Materials and Structures* 50, no. 1 (2017): 27.

Kodur, V. K. R., and Ankit Agrawal. "An Approach for Evaluating Residual Capacity of Reinforced Concrete Beams Exposed to Fire." *Engineering Structures* 110 (2016): 293–306.

Kodur, V. K. R., and Ankit Agrawal. "Effect of Temperature Induced Bond Degradation on Fire Response of Reinforced Concrete Beams." *Engineering Structures* 142 (2017): 98–109.

Kodur, V. K. R., and T. Z. Harmathy. "Properties of Building Materials. Section 1, Chapter 10." *SFPE Handbook of Fire Protection Engineering*, 2012.

Kodur, V. K. R., Baolin Yu, and M. M. S. Dwaikat. "A Simplified Approach for Predicting Temperature in Reinforced Concrete Members Exposed to Standard Fire." *Fire Safety Journal* 56 (2013): 39–51.

Kodur, V. K. R, M. M. S. Dwaikat, and M. B. Dwaikat. "High-temperature Properties of Concrete for Fire Resistance Modeling of Structures." *ACI Materials Journal* 105, no. 5 (2008): 517.

Kodur, Venkatesh, Monther Dwaikat, and Nikhil Raut. "Macroscopic FE Model for Tracing the Fire Response of Reinforced Concrete Structures." *Engineering Structures* 31, no. 10 (2009): 2368–2379.

Kodur, Venkatesh, and Richard Mcgrath. "Fire Endurance of High Strength Concrete Columns." *Fire Technology* 39, no. 1 (2003): 73–87.

Kodur, Venkatesh, and M. Z. Naser. "Frontiers of Structural and Civil Engineering." *International Journal of Critical Infrastructure Protection* (2018).

Kodur, Venkatesh, and M. Z. Naser. "Fire Hazard in Transportation Infrastructure: Review, Assessment and Mitigation Strategies." *Frontiers of Structural and Civil Engineering* (2020).

Kodur, V. K. R., and L. Phan. "Critical Factors Governing the Fire Performance of High Strength Concrete Systems." *Fire Safety Journal* 42, no. 6–7 (2007): 482–488.

Kodur, V. K. R., and A. M. Shakya. "Factors Governing the Shear Response of Prestressed Concrete Hollowcore Slabs under Fire Conditions." *Fire Safety Journal* 88 (2017): 67–88.

Kodur, V. K. R., and M. A. Sultan. "Effect of Temperature on Thermal Properties of High-Strength Concrete." *Journal of Materials in Civil Engineering* 15, no. 2 (2003): 101–107.

Kodur, V. K. R., T. C. Wang, and F. P. Cheng. "Predicting the Fire Resistance Behaviour of High Strength Concrete Columns." *Cement and Concrete Composites* 26, no. 2 (2004): 141–153.

Lie, T. T., ed. *Structural Fire Protection*. Reston, VA: American Society of Civil Engineers, 1992.

Lie, T. T., and David E. Allen. "Calculation of the Fire Resistance of Reinforced Concrete Columns." Technical Paper. *Division of Building Research, National Research Council Canada* 378 (1972): 44.

Lin, T. D., A. H. Gustaferro, and M. S. Abrams. "Fire Endurance of Continuous Reinforced Concrete Beams." (1981): 23. https://trid.trb.org/view/170962

NCMA Tek 7-1C (2009) Fire Resistance Ratings of Concrete Masonry Assemblies. National Concrete Masonry Association. ncma.org/etek/Pages/Manualviewer.aspx?filename=TEK%2007-01C.pdf

Neves, I. Cabrita, Joao Paulo C. Rodrigues, and António de Pádua Loureiro. "Mechanical Properties of Reinforcing and Prestressing Steels after Heating." *Journal of Materials in Civil Engineering* 8, no. 4 (1996): 189–194.

PCI. *Design for Fire Resistance of Precast/Prestressed Concrete, 3rd Edition*, (MNL-124-11EH), 2011. https://www.pci.org/ItemDetail?iProductCode=MNL-124-11EH

Post Tensioning Institute. PTI TAB.1-06: *Post-Tensioning Manual*, 6th ed. Farmington Hills, MI: Post Tensioning Institute, 2006.

PTI. *Post-Tensioning Manual, 6th Edition*, Post-Tensioning Institute: Farmington Hills, MI, 2006.

Raut, N. Response of High Strength Concrete Columns under Fire-Induced Biaxial Bending. 2011. PhD Thesis, Michigan State University.

Raut, N. K., and V. K. R. Kodur. "Response of High-strength Concrete Columns under Design Fire Exposure." *Journal of Structural Engineering* 137, no. 1 (2010): 69–79.

Schneider, U. "Modelling of Concrete Behaviour at High Temperatures." *Design of Structures against Fire* (1986): 53–69.

Schneider, U., U. Diedrichs, W. Rosenberger, and R. Weiss. *Sonderforschungsbereich* 148, Arbeitsbericht 1978–1980, Teil II, B 3, Technical University of Braunschweig, Germany, 1980.

Schneider, U., and H. J. Herbst. "Permeabilitaet und Porositaet von Beton bei hohen Temperaturen." *Deutscher Ausschuss für Stahlbeton* 403 (1989): 23–52.

Ulm, Franz-Josef, Paul Acker, and Michel Lévy. "The 'Chunnel' Fire. II: Analysis of Concrete Damage." *Journal of Engineering Mechanics* 125, no. 3 (1999): 283–289.

Wickström, Ulf. "Temperature Analysis of Heavily-insulated Steel Structures Exposed to Fire." *Fire Safety Journal* 9, no. 3 (1985): 281–285.

Wickström, Ulf. *A Very Simple Method for Estimating Temperature in Fire Exposed Concrete Structures*. Sweden: Statens provningsanst., 1986.

William, K. J., and E. D. Warnke. *Constitutive Model for the Triaxial Behavior of Concrete. Proceedings*. International Association for Bridge and Structural Engineering. Vol. 19. Bergamo, Italy: ISMES, 1975: p. 174.

Zhang, H., V. K. R. Kodur, S. Qi, L. Cao, and B. Wu. "Development of Metakaolin-Fly-Ash Based Geoplymers for Fire Resistance Applications." *Journal of Construction and Building Materials* 55 (2014): 38–45.

Zhang, H., V. K. R. Kodur, B. Wu, L. Cao, and F. Wang. "Thermal Behavior and Mechanical Properties of Geopolymer Mortar After Exposure to Elevated Temperatures." *Journal of Construction and Building Materials* 109 (2016): 17–24.

8.12 Nomenclature

w_c	Weight of concrete, kg/m^3 (lb/ft^3)
f'_c	Compressive strength of concrete, MPa (psi)
s	Spacing of ribs or undulations, mm (in.)
t_{min}	Minimum thickness, mm (in.)
t_e	Equivalent thickness, mm (in.)
R	Fire endurance (hr)
d	Thickness of base slab, mm (in.)
t	Overall thickness of entire slab, mm (in.)
μ_{fi}	Ratio of the applied load to the load carrying capacity at room temperature
α_{cc}	Coefficient for compressive strength as per EN 1992-1-1
$l_{o,fi}$	Effective length of the column under fire conditions
ω	Longitudinal reinforcement ratio at room temperature conditions

A_s	Area of steel reinforcement
f_{yd}	Yield strength of reinforcing steel
A_c	Characteristic strength of concrete
f_{cd}	Area of concrete
FRP	Fire resistance (min)
k	Coefficient dependent on steel ratio
D_c	Smaller column dimension of a rectangular column (mm)
D_g	Greater column dimension of a rectangular (mm)
N^*	Design axial force for the fire limit state (kN)
L_e	Effective length of the column (mm)
R	Fire resistance (min)
S_R	Slenderness ratio of the column
L_R	Load ratio of the column
C_t	Constant based on aggregate type used in concrete
k	Constant based on the cover thickness and the percentage steel
T_f	Fire temperature (°C)
t_h	Time (hours)
t	Fire exposure time (hr)
U^*_{fire}	Design force from applied loads (code)
R_{fire}	Load-bearing capacity
A_s	Area of reinforcing steel
$f_{y,T}$	Yield stress of reinforcing steel at temperature T
d	Effective depth of the section
a_f	Depth of the rectangular stress block reduced due to thermal effects and given
b_f	Fire-reduced effective width of the beam
a	Shortest distance between the centroid of the longitudinal reinforcement and the nearest concrete surface exposed to fire (mm)
h and b	Longer and shorter dimension of the column, respectively
z and y	Distance from the point in concrete section to fire exposure surface (m)
$\varepsilon_{th}(T)$	Thermal strain
$\varepsilon_{\sigma}(\sigma,T)$	Stress-related strain
$\varepsilon_{cr}(\sigma,T,t)$	Creep strain
$\varepsilon_{tr}(\sigma,T)$	Transient strain

8.13 Numerical Examples

8.1

What is the minimum width and cover thickness required to achieve a 2- and 4-hour fire rating each in a reinforced concrete beam and a column as per (a) BS 8110 provisions and (b) ACI 216.1 provisions?

Solution

(a) Applying BS 8110 provisions (see Table 8.8), minimum width and cover thickness can be evaluated.

For RC beam:

For 2-hour fire rating: minimum beam width is 200 mm, with minimum cover of 40 mm.
For 4-hour fire rating: minimum beam width is 280 mm, with minimum cover of 70 mm.

For RC column:

For 2-hour fire rating: minimum width of column is to be 300 mm, with minimum cover of 25 mm.

For 4-hour fire rating: minimum width of column is to be 450 mm, with minimum cover of 25 mm.

(b) As per ACI 216.1 provisions (see Tables 8.13 and 8.14), minimum width and cover thickness can be evaluated.

For RC (unrestrained) beam:

Use Table 8.13 to determine minimum size and cover thickness.

For 2-hour fire rating: minimum width is 127 mm, with minimum cover of 19 mm.
For 4-hour fire rating: minimum width is 127 mm, with minimum cover of 32 mm.

For RC column:

Use Table 8.14 to determine minimum size and cover thickness.

Assume the columns to be of carbonate aggregate concrete.
For 2-hour fire rating: minimum column dimension is to be 254 mm.
For 4-hour fire rating: minimum column dimension is to be 304.8 mm.

In all cases:

The minimum thickness of concrete cover to main longitudinal reinforcement in columns, regardless of the type of aggregate used in the concrete and the specified compressive strength of the concrete, shall not be less than 25.4 mm times the number of hours of required fire resistance, or 50.8 mm, whichever is less.

Thus:

1. The required cover for fire resistance of 2 hours—$25.4 \times 2 = 50.8 \leq 50.8$ mm; use 50.8 mm.
2. The required cover for fire resistance of 2 hours—$25.4 \times 4 = 101.6 \geq 50.8$ mm; use 50.8 mm.

8.2

An RC column fabricated with siliceous aggregate concrete of compressive strength of 31 MPa and reinforcing steel of 444 MPa yield strength is subjected to an applied load of 1178 kN and has fixed ends, as shown in Fig. E8.1. Calculate the fire resistance of the column using the provisions of ACI 216.1, Eurocode 2, Australian Code 3600, and the approach proposed by Kodur and Raut (2009).

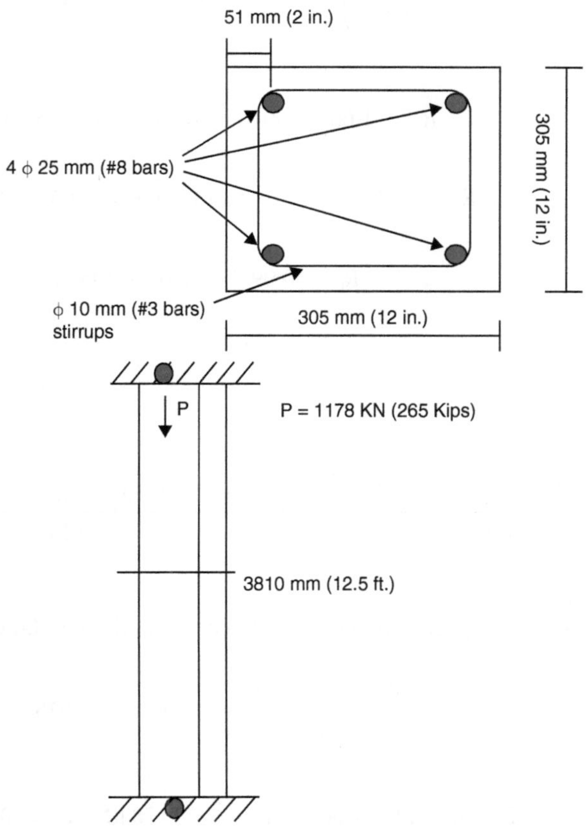

Figure E8.1 Elevation and cross sections of an RC column.

Solution

ACI 216.1

The fire resistance of the column as per ACI 216.1 provisions can be evaluated using Table 8.14. For a 305-mm column made of siliceous aggregate concrete, the fire rating can be taken similar to that of 304.8 mm, which is 3 hours.

Eurocode 2 Empirical Equation

Applying Eq. 8.23 and the procedure laid out in Sec. 8.6.2, fire resistance can be evaluated for this column.

$$\omega = \frac{A_s \times f_y}{A_s \times f_c'}$$

$$\omega = \frac{1964 \times 444}{((305 \times 305) - 1964) \times \dfrac{31}{0.85 \times 1.5}} = 0.39$$

$$b' = 305 \text{ mm}, a = 51 \text{ mm}$$

$$l_{o,fi} 0.5 \times 3.81 = 1.905 \text{ m}$$

$$\mu_{fi} = \frac{P}{\left(A_s \dfrac{f_y}{1.15} + A_c \dfrac{f_c'/0.85}{1.5}\right)} = \frac{1178}{\left(A_s \dfrac{f_y}{1.15} + A_c \dfrac{f_c'/0.85}{1.5}\right)} = 0.32$$

$$R_{nfi} = 83\left[1 - \mu_{fi}\frac{(1+\omega)}{\left(\left(\dfrac{0.85}{\alpha_{cc}}\right) + \omega\right)}\right] = 83\left[1 - 0.32\frac{(1+0.39)}{\left(\left(\dfrac{0.85}{1}\right) + 0.39\right)}\right] = 53.23$$

$$R_a = 1.6(a - 30) = 1.6(51 - 30) = 33.6$$

$$R_l = 9.6(5 - l_{o,fi}) = 9.6(5 - 0.5 \times 3.81) = 29.7$$

$$R_b = 0.09b' = 0.09 \times 305$$

$$R_n = 0$$

Fire resistance:

$$R = 120\left(\frac{R_{nfi} + R_a + R_l + R_b + R_n}{120}\right)^{1.8} = 120\left(\frac{53.23 + 33.6 + 29.7 + 27.4 + 0}{120}\right)^{1.8} = 166 \text{ minutes}$$

Australian Method (AS 3600)

Applying Eq. 8.24 and the procedure laid out in Sec. 8.6.2., fire resistance can be evaluated for this column.

$$\frac{A_s}{A_g} = \frac{1964}{305 \times 305} = 0.021 < 0.025$$

$$k = 1.5, D_c = 305 \text{ mm}, D_g = 305 \text{ mm}, N^* = 1178 \text{ kN}, L_e = 0.5 \times 3810 = 1905 \text{ mm}$$

For fire resistance:

$$R = \frac{(k \times f_c^{'1.3} \times D_c^{3.3} \times D_g^{1.8})}{10^5 \times N^{*1.5} \times L_e^{0.9}}$$

$$R = \frac{(1.5 \times 31^{1.3} \times 305^{1.3} \times 305^{1.8})}{10^5 \times 1178^{1.5} \times 1905^{0.9}} = 168 \text{ minutes}$$

Kodur and Raut Approach

Applying Eq. 8.25 and procedure laid out in Sec. 8.6.2, fire resistance can be evaluated for this column.

The clear cover thickness to surface of steel rebars = distance to center or rebar − diameter of stirrup − half the rebar diameter.

Thus, thickness of clear cover:

$$C_e = 51 - 10 - 25/2 = 28.5 \text{ mm}$$

$$S_p = \frac{A_s}{A} \times 100 = \frac{4 \times \frac{\pi \times 1^2}{4}}{12 \times 12} \times 100 = 2.18$$

$$k = \frac{[(C_e - 82) \times (S_p + 10.5) + 870]}{390} = \frac{[(28.5 - 82) \times (2.18 + 10.5) + 870]}{390} = 0.49$$

Radius of gyration:

$$r = \sqrt{\frac{I}{A}} = \sqrt{\frac{bd^3}{12bd}} = \sqrt{\frac{b^2}{12}} = \sqrt{\frac{305^2}{12}} = 88.05 \text{ mm}$$

Then:

$$S_R = \frac{l_e}{r} = \frac{0.5 \times 3810}{88.05} = 21.64$$

Load-carrying capacity:

$$P_{cr} = 0.8(0.85A_c f_c' + A_s f_y)$$

$$P_{cr} = 0.8[0.85((305 \times 305) - 1964) \times 31 + 1964 \times 444] = 2617 \text{ kN}$$

Load ratio:

$$L_R = \frac{P}{P_{cr}} = \frac{1178}{2617} = 0.45$$

For siliceous aggregate concrete:

$$C_t = 1.0$$

Fire resistance:

$$R = C_t[8 \times k \times ((S_R + 5) \times (0.2 - L_R) + 30)]^{0.94}$$

$$R = 1.0[8 \times 0.49 \times (30 - (21.64 + 5) \times (0.2 - 0.45) + 30)]^{0.94} = 187 \text{ minutes}$$

It can be seen that there are some discrepancies in the estimated fire resistance following perspective approaches of different codes of practice.

	ACI 216.1	Eurocode 2	Australian Code 3600	Kodur and Raut Approach
Fire resistance (min)	8 180	9 166	10 168	11 187

8.3

A simply supported RC slab of 200-mm thickness is subjected to standard fire exposure ISO 834 (see Fig. E8.2). The provided concrete cover and area of steel reinforcement is 15 mm spaced at 125 mm. Use (a) Wickström's formula and (b) Kodur and Raut Approach to calculate temperature rise in reinforcing steel after one hour of fire exposure (at a height of 23 mm from soffit of the slab).

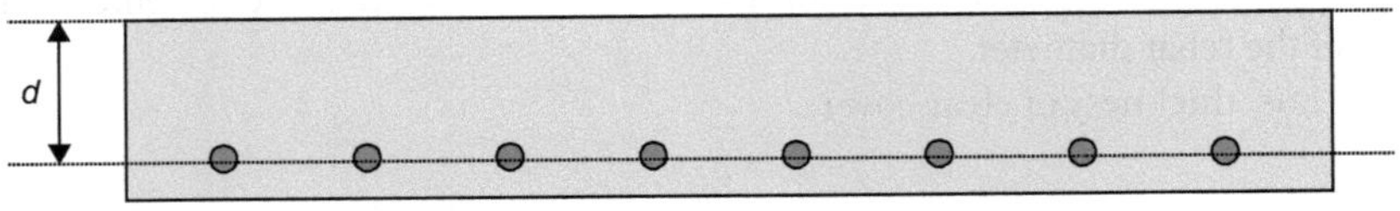

FIGURE E8.2 Concrete slab.

Solution

(a) Wickström's Formula

After 60 minutes of standard fire exposure:

$$t = 60 \text{ minutes } (t_h = 1.0 \text{ hour})$$

Fire temperature as per ISO 834:

$$T_f = 20 + 345\log(8t + 1) = 945°\text{C}$$

Surface temperature on the slab:

$$T_w = \left[1 - 0.0616 t_h^{-0.88}\right] T_f$$

$$T_w = \left[1 - 0.0616 \times 1^{-0.88}\right] \times 945 = 887°\text{C}$$

Concrete temperature at a given depth of 23 mm:

$$T_c = \left[0.18\ln\left(\frac{t_h}{x^2}\right) - 0.81\right] T_w$$

$$T_c = \left[0.18\ln\left(\frac{1}{0.023^2}\right) - 0.81\right] \times 887 = 486°\text{C}$$

Steel temperature:

$$T_s = T_c = 486°\text{C}$$

(b) Kodur and Raut Approach

After 60 minutes of standard fire exposure (ISO 834), $t = 60$ minutes ($t_h = 1.0$ hour)

$$\eta_z = 0.155\ln\left(\frac{t_h}{z^{1.5}}\right) - 0.348\sqrt{z} - 0.371$$

$$\eta_z = 0.155\ln\left(\frac{1}{0.023^{1.5}}\right) - 0.348\sqrt{0.023}) - 0.371 = 0.453$$

$$T_c = c_1 \cdot \eta_z \cdot (at^n)$$

$$T_c = 1.01 \times 0.453 \times (935 \times 1^{0.168}) = 428°C$$

Concrete temperature (at 0.023 m) = Steel rebar temperature:

$$T_s = T_c = 428°C$$

8.4

An RC beam of rectangular cross section of 400 mm × 600 mm and span length of 5 m is subjected to a standard fire exposure (see Fig. E8.3). The provided concrete cover to main rebars and area of steel reinforcement are 50 mm and 1885 mm², respectively. This beam is made of 40 MPa concrete and 400 MPa grade steel reinforcement. During fire, the beam is subjected to a uniformly distributed dead load of 20 kN/m and a live load of 40 kN/m. After 2 hours of fire exposure, assume that temperature in steel reinforcement reaches 550°C and temperature rise in the compression block of concrete is minor (below 300°C). Does the beam have sufficient capacity to withstand 2 hours of fire exposure?

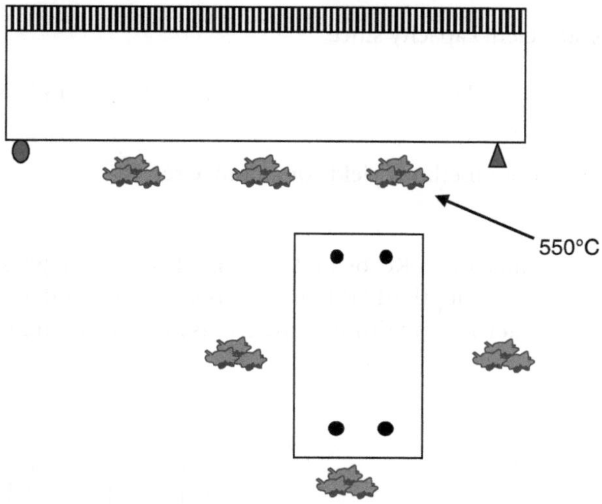

Figure E8.3 Concrete beam.

Solution

(a) General Approach

Following ASCE 7 load combination under fire conditions:

$$w_u = 1.2 \text{ DL} + 0.5 \text{ LL} = 1.2 \times 20 + 0.5 \times 40 = 44 \text{ kN/m}$$

$$M_u = w_u l^2/8 = 44 \times 5^2/8 = 137.5 \text{ kN·m}$$

At 550°C, the yield strength of steel reinforcement reduces according to the following expression (see Sec. 8.2.3):

$$f_{y,T} = \left[1.0 + \frac{T}{900 \ln\left(\frac{T}{1750}\right)}\right] f_{yo} = \left[1.0 + \frac{550}{900 \ln\left(\frac{500}{1750}\right)}\right](400) = 222.6 \text{ MPa}$$

Thus, nominal moment capacity of this beam after 2-hour fire exposure is

$$M_{n,\text{at 2 hours}} = A_s f_{y,550°C}\left(d - \frac{A_s f_{y,550°C}}{1.7 f_c' b}\right) = 1885 \times 222.6\left((600 - 50) - \frac{(1885)(222.6)}{1.7(40)(400)}\right) = 224.3 \text{ kN·m}$$

The beam has sufficient capacity since

$$M_{n,\text{at 2 hours}} = 224.3\text{ kN}\cdot\text{m} > M_{\text{fire}} = 137.5\text{ kN}\cdot\text{m}$$

(b) Eurocode 2 (500°C Isotherm Method)

The first step is to remove layers of concrete that exceed 500°C (around steel reinforcement). The 500°C isotherm in the section is at 50 mm from the outer surface of the column. The new cross section of concrete is

$$b = 400 - 50 - 50 = 300\text{ mm}$$

$$d = 600 - 50 = 550\text{ mm}$$

Concrete below 500°C is assumed to have full compressive strength, while steel rebars at 550°C have reduced strength of 222.6 MPa as computed above.

Thus:

$$M_{n,\text{at 2 hours}} = A_s f_{y,550^\circ\text{C}}\left(d - \frac{A_s f_{y,550^\circ\text{C}}}{1.7 f_c' b}\right) = 1885 \times 222.6\left((550) - \frac{(1885)(222.6)}{1.7(40)(300)}\right) = 222.1\text{ kN}\cdot\text{m}$$

The beam has sufficient capacity since

$$M_{n,\text{at 2 hours}} = 222.1\text{ kN}\cdot\text{m} > M_{\text{fire}} = 137.5\text{ kN}\cdot\text{m}$$

It can be seen that both methods yield comparative results.

8.5

Evaluate fire resistance of an RC beam through advanced analysis. This beam has a span of 5 m and a width of 300 mm and a depth of 500 mm. A cross section and an elevation of this beam are shown in Fig. E8.4. Use (a) macroscopic finite element-based analysis and (b) microscopic-based finite element model.

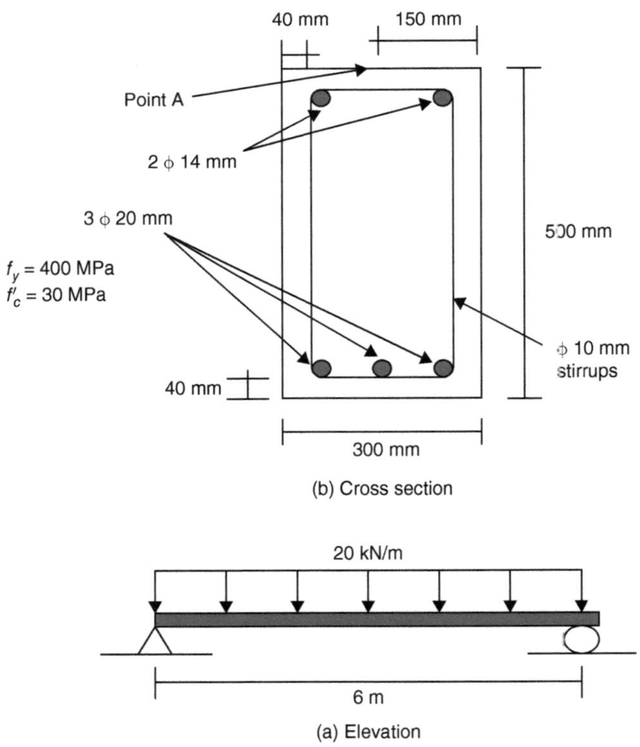

Figure E8.4 Elevation and cross section of RC beam.

Solution

(a) Macroscopic Finite Element-Based Analysis

Step 1. The RC beam is divided into a number of segments along its length (as shown in Fig. E8.5), and the midsection of the segment is assumed to represent the behavior of the whole segment.

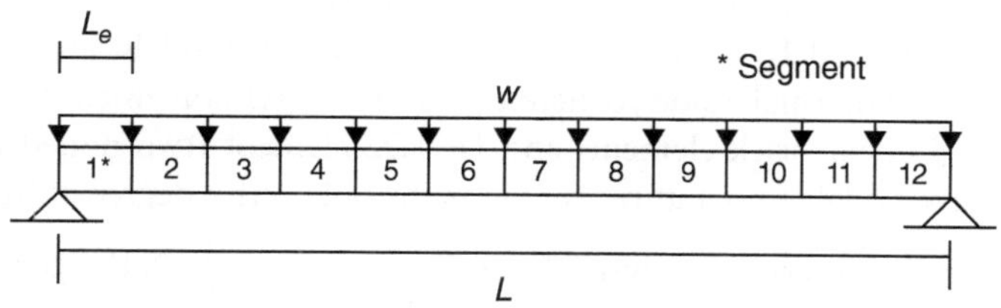

Figure E8.5 Discretization of beam.

Step 2. Evaluate temperatures due to fire exposure through thermal analysis as shown in Fig. E8.6.

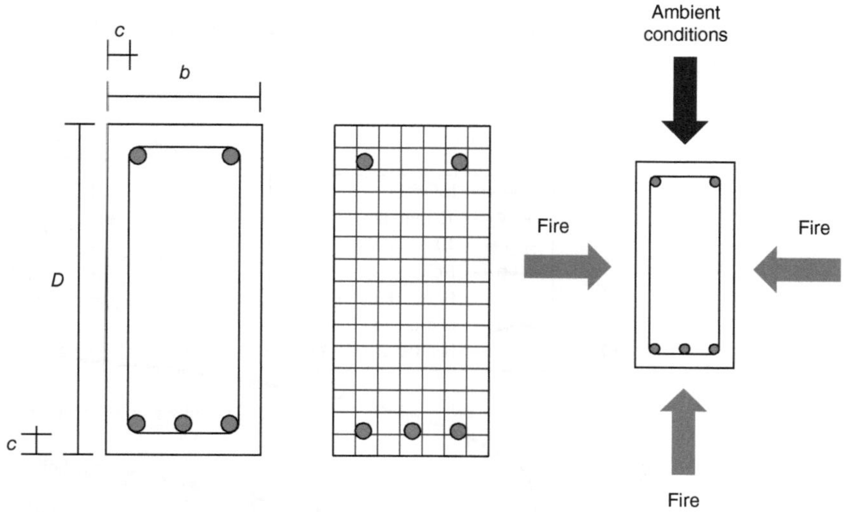

Figure E8.6 Discretization of RC beam cross section. (*With permission from Dwaikat, 2009.*)

Step 3. Following this, time-dependent sectional moment-curvature (M-κ) relations are generated for each longitudinal segment in the beam, and these M-κ relations are in turn used to trace the response of the beam under fire conditions (see Fig. E8.7). The M-κ relations, at various time-steps, are generated using the changing properties of constituent materials, namely concrete and reinforcement.

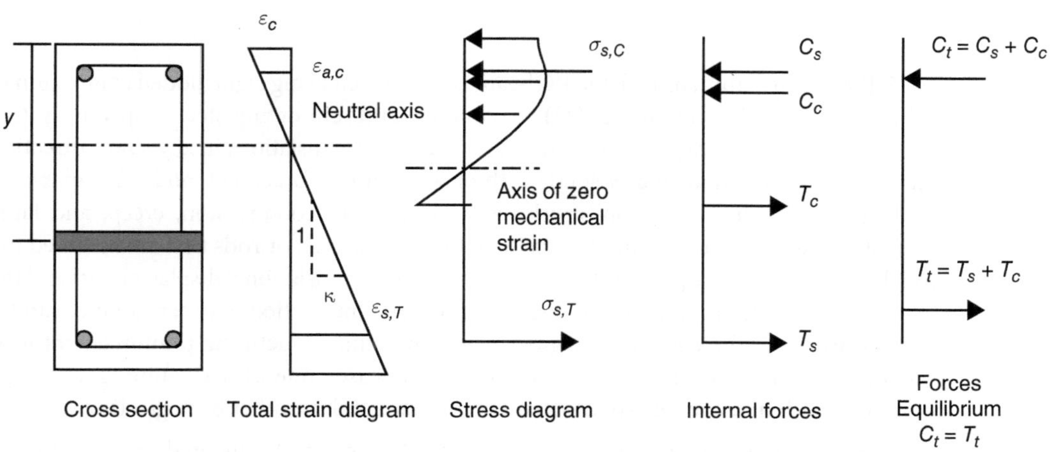

Figure E8.7 Stress and strain distribution within a concrete beam.

Follow the discussion in Sec. 8.7.3 and Fig. 8.17, along with the procedure laid out in following publication:

Kodur, V. K. R., and M. Dwaikat. "A Numerical Model for Predicting the Fire Resistance of Reinforced Concrete Beams." *Cement and Concrete Composites* 30, no. 5 (2008): 431—443.

(b) Microscopic-Based Finite Element Model (Using ABAQUS Program)

Step 1. Two submodels, namely thermal and structural models, are needed to carry out three stages of analysis. For the thermal model, concrete portion and reinforcement are discretized using DC3D8 element (eight-noded linear brick element) and DC1D2 element (two-noded link element), respectively. These elements have nodal temperature (NT11) as the only active degree of freedom.

Step 2. Carry out thermal analysis. According to Eurocode 2, the convective heat transfer coefficient is taken to be 25 W/m^2·°C on fire-exposed surface. For the unexposed surfaces, a convective coefficient of 9 W/m^2·°C is used to account for the effects of heat transfer through radiation. The emissivity for radiative heat transfer at the exposed surfaces of the concrete member is taken as 0.8.

Step 3. Input temperature-dependent thermal and mechanical properties (including bond degradation if it is to be accounted) of reinforcing steel and concrete as per Eurocode 2 and 3, or any other standard.

Step 4. Evaluate cross-sectional temperature obtained from thermal analysis and link these temperatures to structural analysis.

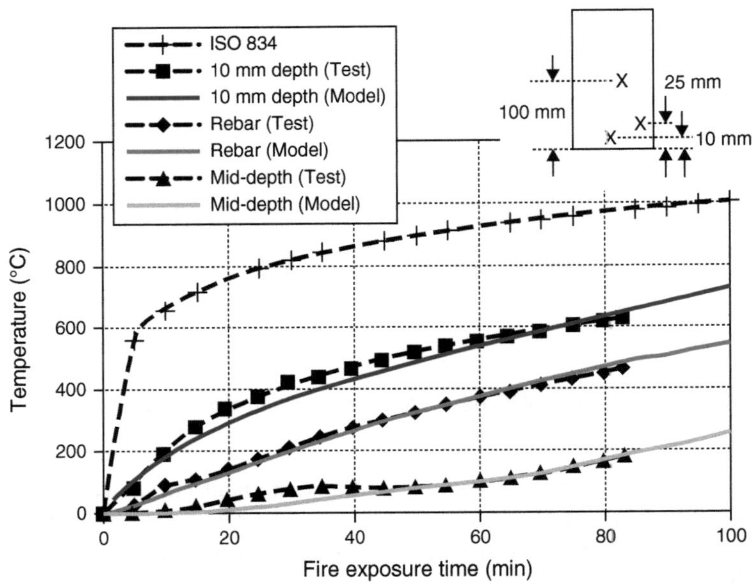

Figure E8.8 Cross-sectional temperature in an analyzed beam.

Step 5. For the structural model, the RC beam is discretized using eight-noded continuum elements (C3D8) and two-noded link elements (T3D2) for concrete and reinforcing steel, respectively. C3D8 element has eight nodes with three degrees of freedom, namely three translations in *x*, *y*, and *z* directions. This element can be used for 3D modeling of solids with or without reinforcement, and it is capable of accounting for cracking of concrete in tension, crushing of concrete in compression, creep, and large strains. T3D2 elements are used to model one-dimensional reinforcing bars or rods that are assumed to deform by axial stretching only. They are pin-jointed at their nodes; only translational displacements and the initial position vector at each node are used in the discretization. The interaction between concrete and reinforcement is achieved by using the embedded region constraint—that is, defining reinforcement to be embedded in concrete. Note that the analysis is carried out for two cases, one with accounting for temperature-induced bond degradation in concrete-reinforcement interface and the other by neglecting such bond degradation.

Step 6. Evaluate structural response (i.e., midspan deflection) and compared against fire exposure time (min) shown in Fig. E8.9.

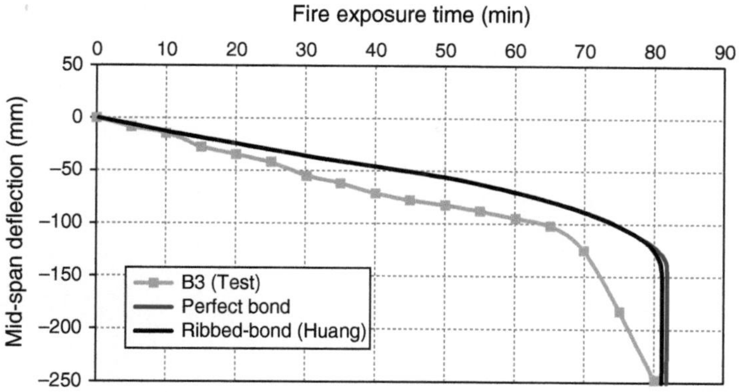

Figure E8.9 Midspan deflection in analyzed beam.

Follow the discussion in Sec. 8.7.3 and Fig. 8.17, as well as the procedure laid out in the following publication. Please note that finite element predictions are also plotted in Figs. E8.8 and E8.9.

Kodur, V. K. R., and Ankit Agrawal. "Effect of Temperature Induced Bond Degradation on Fire Response of Reinforced Concrete Beams." *Engineering Structures* 142 (2017): 98–109.

CHAPTER 9

Composite Structures

9.1 Overview

This chapter presents strategies and associated calculation methodologies for achieving the required fire resistance in steel structures through composite construction. Temperature-dependent properties of structural steel, concrete, and shear connecters, needed for evaluating fire response of composite steel members, are reviewed. Strategies that can be developed for achieving required fire resistance in steel-concrete composite structural members are discussed. Different calculation approaches for evaluating fire resistance of composite steel structures, including tabulated data, simplified approaches, and advanced analysis procedure, are presented, together with relevant numerical examples.

9.2 Composite Construction

Steel-framed structures often utilize concrete to enhance structural performance; this combination of concrete and structural steel is referred to as composite construction. A composite structural member is fabricated utilizing both concrete and steel and is integrally connected to deflect as a single unit. Such a composite member offers numerous advantages over other types of construction (i.e., isolated steel or concrete member only) including high load-bearing capacity, high stiffness, and excellent durability characteristics. Examples of composite construction include metal deck (floor) with concrete infill, steel girder attached to concrete slab through shear studs (or connectors), steel hollow structural section column filled with concrete, and shear wall made of steel plates with concrete infill. Figure 9.1 illustrates commonly used composite structural members in building and bridge applications.

For effective load transfer between steel and concrete components, connectors (studs) are to be provided in composite structural members, and this can significantly enhance structural capacity under ambient conditions. Shear connectors should possess sufficient strength and stiffness properties to enable the two components to develop full capacity and act as a single structural member until failure. That means a mechanical shear connection does not rely on bond or adhesion at the interface between steel and concrete. Composite action, with rigid connectors, typically enables a 100% to 200% increase in the moment capacity of a beam slab assembly relative to the steel beam alone, depending on the degree of shear connection provided and on the size of the cross section.

This composite action, which develops between concrete and steel, can also improve performance during fire exposure. The concrete component of the member, due to its excellent fire resistance properties, delays temperature propagation in steel components either through insulation or a heat sink mechanism and also can carry part of the load during fire exposure conditions. This mechanism can enhance fire resistance of a composite structure significantly. However, the extent of fire resistance enhancement depends on a number of factors, including the level of applied loading, connector properties, and structural characteristics. This chapter discusses strategies and associated calculation methods that can be applied to enhance fire resistance of steel structures through composite construction.

9.3 Material Properties at Elevated Temperatures

A composite structural member is primarily constructed from two materials: steel and concrete. As such, the fire resistance of a composite structural member depends on the temperature-dependent properties of steel, concrete, and any connecting components (e.g., steel studs). Most of the high-temperature property relations for steel and concrete, discussed in Chaps. 7 and 8, respectively, are applicable for the fire resistance evaluation of composite structures. As such, it is not repeated here and the reader is encouraged to revisit

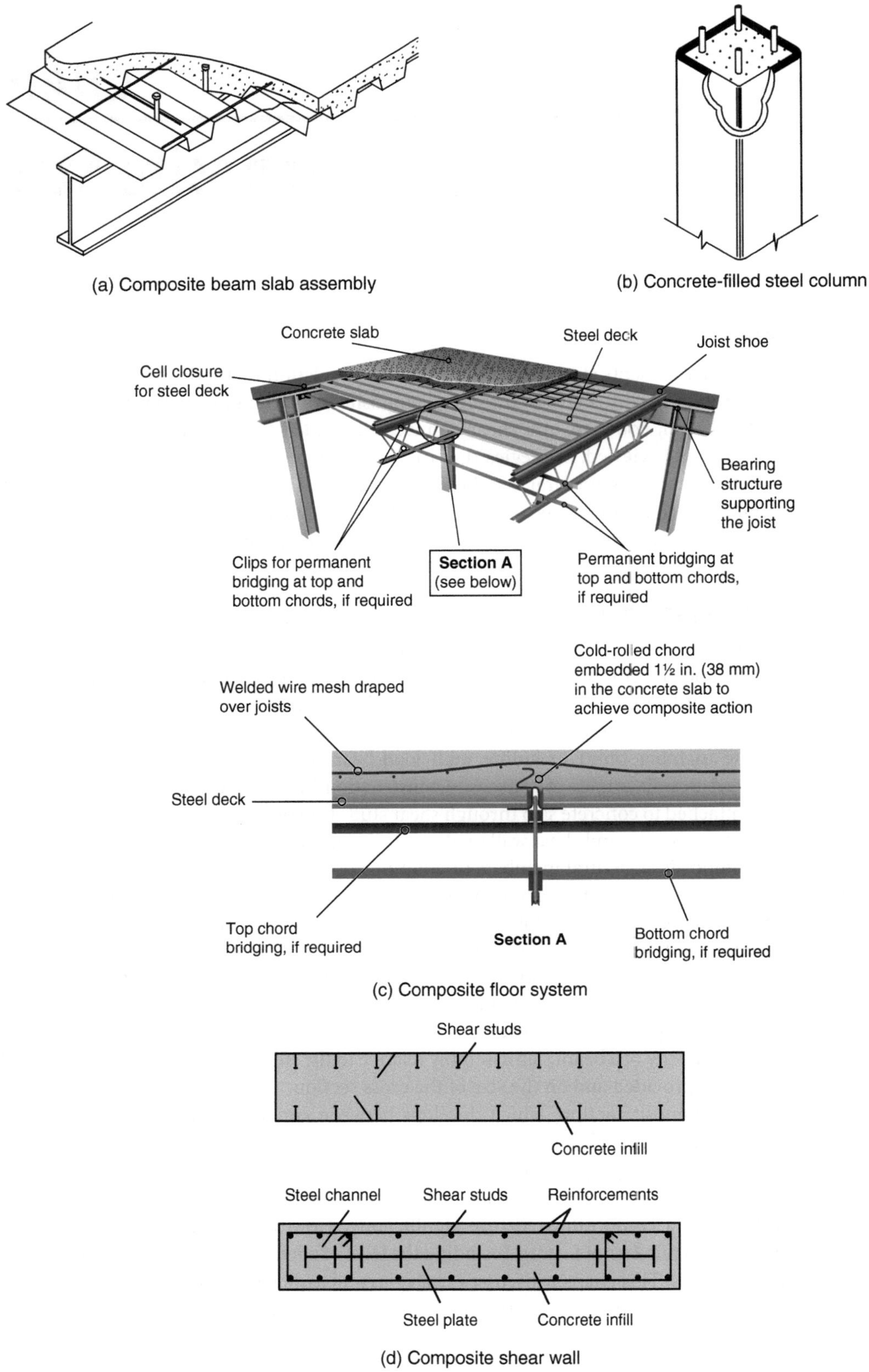

FIGURE 9.1 Various composite systems used in buildings and bridges.

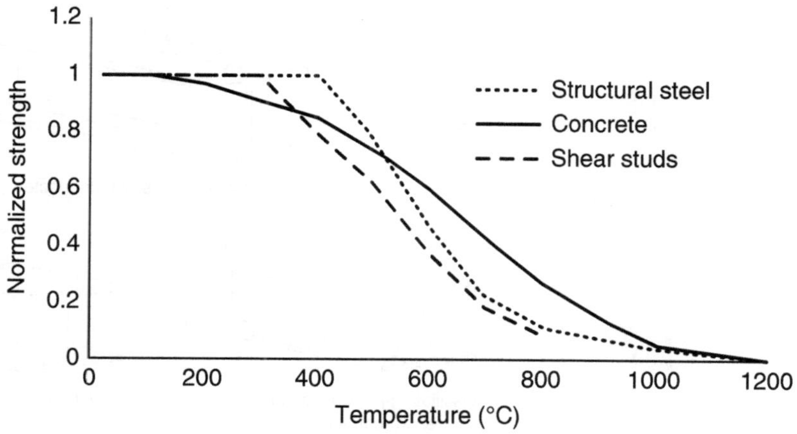

Figure 9.2 Degradation in strength properties of steel as well as concrete at elevated temperatures as per Eurocode 4.

Secs. 7.2 and 8.2 for detailed information regarding elevated temperature property relations for steel and concrete.

For illustrative purpose, the variation in the compressive strength of concrete and the tensile strength of structural steel and steel in shear studs is shown in Fig. 9.2. It can be seen that the strength (both compression and tension) of steel deteriorates at a much more rapid pace than the compressive strength of concrete. The strength of steel reduces to about 50% at 550°C, while concrete maintains about 70% to 75% of its initial strength at the same temperature. Thus, in a composite structural member, such as a steel column filled with concrete, unprotected exterior steel loses most of its initial strength rather quickly, while the interior cooler concrete will still possess sufficient strength to support the load due to the slower rate of strength reduction in concrete. Figure 9.2 also shows that the connectors (shear studs) also lose their strength with a rise in temperature and the rate of strength loss is rapid beyond 300°C. Eurocode 4 recommends that strength reduction of shear connectors follow a similar trend as that of yield stress of structural (carbon) steel as given in Chap. 7. This can be attributed to the fact that shear connectors are made of steel with similar chemical and physical composition as structural steel.

In fire-exposed composite structural members, in addition to a loss of strength and stiffness properties of steel and concrete, the bond that develops at the interface of steel and concrete is critical, since the extent of bond degradation determines the level of force transfer from steel to concrete and vice versa. This bond is facilitated through shear studs (connectors) made of steel with different shapes (perfobond ribs, oscillating perfobond strips, and channel shear ribs). Such studs are welded to the steel section, often to the top flange of a steel beam or to the interior surface of steel shear walls or columns (see Fig. 9.3). As a result, these studs are assumed to possess rigid behavior (i.e., no slip to occur between concrete and stud/beam) once concrete attains its design strength.

However, when composite members undergo high deformations under high load levels, especially as in the case of beams and slabs, concrete cracks and shear studs can bend, causing slip to occur between steel and concrete. The possibility of such slip occurring is even higher when the composite member is exposed to fire conditions, mainly due to degradation in strength and stiffness properties of steel with a rise in temperature. Once this slip occurs, the extent of horizontal shear transfer (i.e., composite interaction) between steel and concrete components decreases. As a result, overall flexural, shear, or axial capacity of a composite assembly reduces. Such reduction in composite action is to be duly accounted for in evaluating fire resistance of composite structures (Kodur and Naser, 2017).[1]

For incorporating the appropriate level of composite action developed in a structural member during fire exposure, temperature-dependent force-slip relations, which depend on properties of steel, concrete, and connectors, are required. The horizontal force-slip that develops between shear studs and concrete slab

[1] While composite columns rarely incorporate shear studs, the natural bond that occurs at the steel column and concrete filling interface might reduce as well. Little information has been discussed on the behavior and development of composite interaction in the case of fire-exposed composite columns.

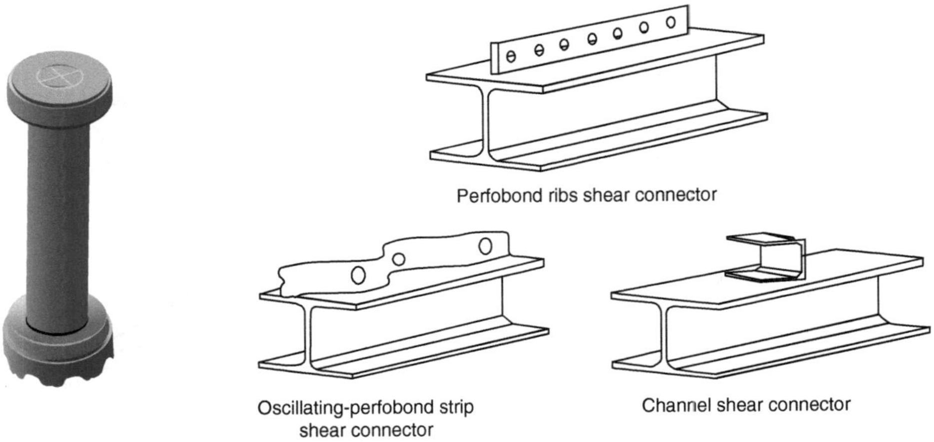

Figure 9.3 Different types of shear studs used as connectors.

can be described under ambient temperature conditions through the following force-slip relation based on a nonlinear constitutive relation (proposed by Ollgaard et al., 1971):

$$V_h = Q_u[1 - e^{-4.75S}] \tag{9.1}$$

where V_h is the shear force and Q_u is the strength of the studs. Q_u is calculated as $Q_u = 0.43A_s\sqrt{E_c f_c'} \leq 0.7A_s f_u$ where A_s is the cross-sectional area of the shear studs, E_c is the elastic modulus of the concrete taken as $4600\sqrt{f_c'}$ (in MPa), f_c' is the compressive strength of the concrete, and f_u is the ultimate strength of the studs. S is the slip length, and the maximum slip length is specified in CEB-FIP Model Code (1993) or other design standards. The above force-slip expression derived by Ollgaard et al. (1971) can be extended to fire exposure conditions through substituting temperature-induced degradation in strength (f_c') and modulus properties (E_c) of concrete in the calculation of Q_u.

Eurocode 4 also specifies the temperature-dependent force-slip relation for accounting composite action in fire resistance evaluation of composite structures based on the work of Zhao and Kruppa (1997). This relation is composed of three parts; the first part represents the linear behavior of connectors, the second part represents the nonlinear behavior of connectors until reaching its maximum (peak) resistance, and the third part is the descending branch to take into account a limited softening that is present in a connector. The expression is given as

$$\frac{Q}{Q_{u,T}} = \begin{Bmatrix} \dfrac{k\gamma}{\gamma_0} & \text{for } 0 \leq \gamma \leq \gamma_0 \\ e^{\frac{1}{a}\left(\frac{\gamma}{\gamma_0}\right)^a} & \text{for } \gamma_0 \leq \gamma \leq \gamma_u \\ Q_{u,T} & \text{for } \gamma > \gamma_u \end{Bmatrix} \tag{9.2}$$

where $a = 1/\log k$, $\gamma =$ slip, $\gamma_o =$ initial slip, $\gamma_u =$ slip at failure, $Q =$ shear force, and $Q_{u,T} =$ shear force at temperature T.

The associated parameters needed to develop force-slip relations in Eq. 9.2 are given in Table 9.1. The temperature-dependent force-slip relations for a shear stud connector casted in solid concrete slab, as per Eq. 9.2., are illustrated in Fig. 9.4.

9.4 Behavior of Composite Structures under Fire Conditions

The response of a typical composite structural member under fire conditions can be illustrated through tracing the behavior of a concrete-filled hollow structural steel (CFHSS) column (as shown in Fig. 9.5*a*) under combined effects of thermal and structural loading. At ambient conditions, a CFHSS column behaves as a composite member wherein steel and concrete components jointly contribute to load-carrying capacity. When this column is exposed to fire conditions, the temperature in the exterior steel section

ϕ (°C)	$Q_u/Q_{u(20°C)}$	k	λ_o	λ_u	λ_{max}
≤100	1.0	0.46	0.36	5.20	11
200	1.0	0.52	0.65	5.70	12
300	1.0	0.55	0.90	6.25	13
400	0.8	0.68	1.30	6.70	14
500	0.624	0.70	1.45	7.10	15
600	0.376	0.72	1.48	7.50	16
700	0.184	0.74	1.50	7.90	17
800	0.088	0.75	1.50	8.30	18

$k = Q_{u(T)}/Q_{u(20°C)}$

TABLE 9.1 Parameters for Deriving Temperature-Dependent Force-Slip Relations of Shear Stud Connector with Flat Concrete Slab (*Adapted from Zhao and Kruppa, 1997*)

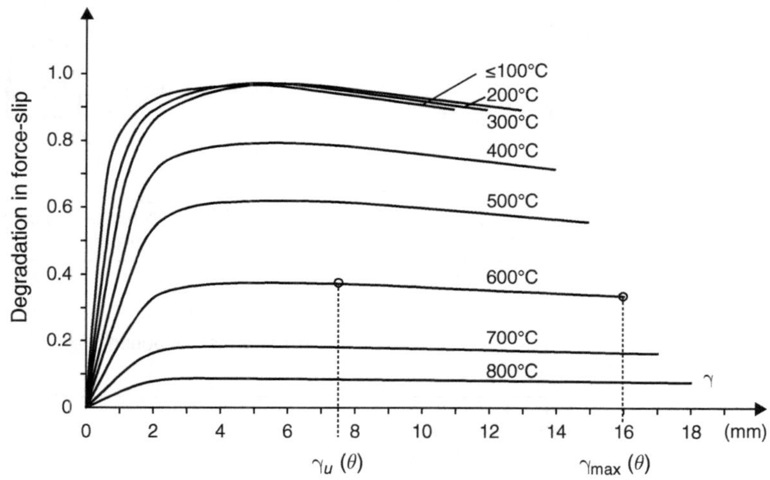

FIGURE 9.4 Temperature-dependent force-slip relations of shear stud connector attached to a flat concrete slab. (*Adapted from Zhao and Kruppa, 1997.*)

raises rapidly, as there is no fire insulation on steel and also because steel has high conductivity and low specific heat. However, due to better thermal properties of concrete—that is, low thermal conductivity and high specific heat—heat from steel (HSS) gets transferred, at a slow pace, into the interior concrete filling. As a result of concrete absorbing much of the heat from steel, temperature rise in steel section slows down to some extent. Typical temperature distribution at various points in a CFHSS column section under standard fire exposure, as measured in fire tests, is shown in Fig. 9.5*b*.

The above-illustrated temperature progression has a significant influence on the resulting structural response of a CFHSS column during fire exposure, and this is shown in Fig. 9.5*c* through axial deformation-fire exposure time plot. At the early stages of fire exposure, the steel section carries most of the load. This is because the steel section (HSS) expands more rapidly than the interior concrete core. With a further increase in steel section temperature, the strength (and modulus) of steel decreases, and as a result, the HSS alone cannot support the axial load acting on the column, leading to gradual yielding of steel. Because steel cannot support the load, the column starts contracting. This occurs in about 20 to 30 minutes of fire exposure (see Fig. 9.5*c*), and this contraction is still often accompanied by local bulging in the HSS section. At this stage, since the interior concrete is at a very low temperatures, the concrete core begins to support the load acting on the column. The concrete core carries an increasing portion of the load with increasing fire exposure time. Ultimately, when the concrete core can no longer support much of the load acting on the column, failure occurs in the column through either buckling or compression mode. The time to reach this failure point in a CFHSS column is much longer than that of an unfilled HSS column and can be anywhere from 1 hour to 3 hours. This duration to failure depends on a number of factors, such as fire severity, load level on the column, cross-sectional dimensions, support conditions, concrete strength, type

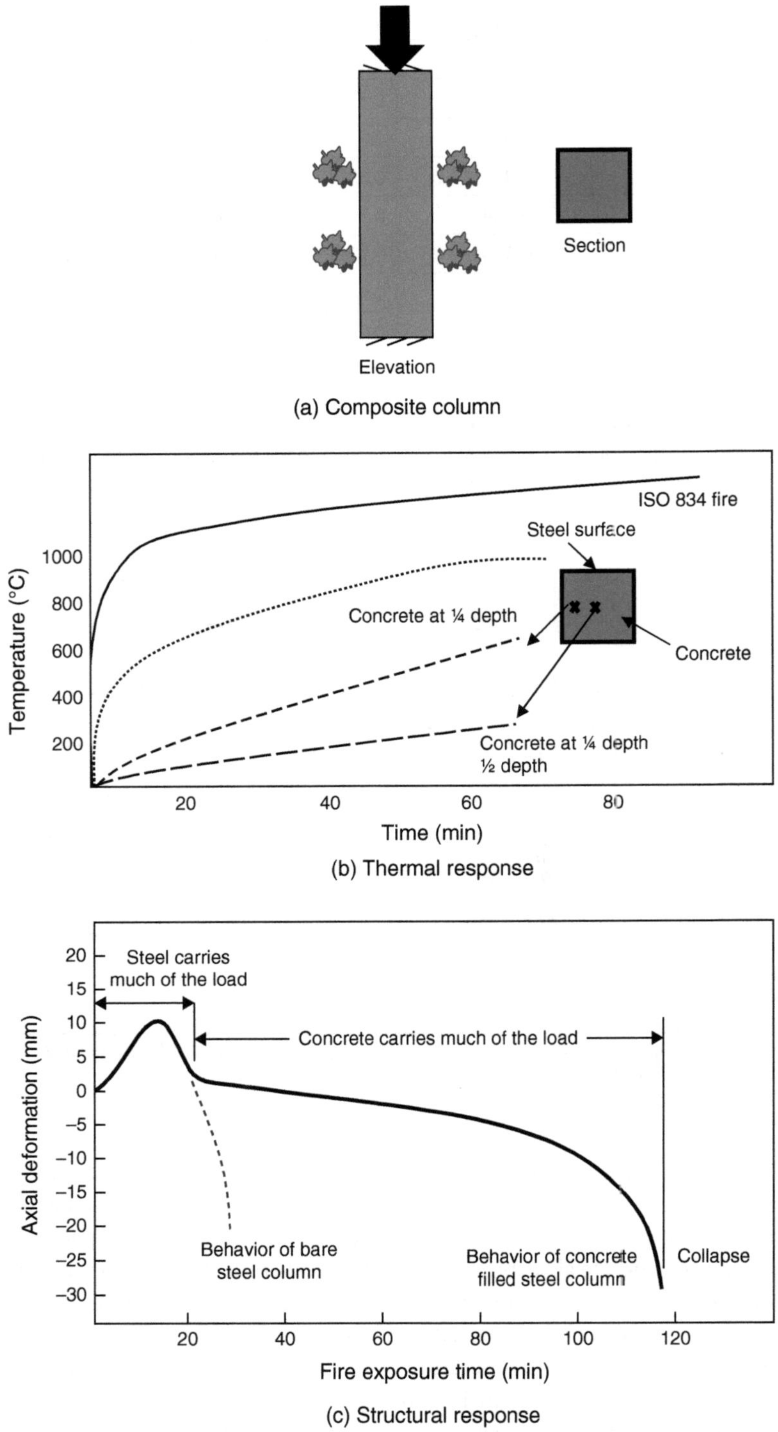

Figure 9.5 Behavior of the concrete-filled HSS column during fire exposure.

of concrete filling, and level of composite action that develops between steel and concrete components (Kodur and Lie, 1995). However, as can be seen in the figure, a bare HSS column (unprotected, i.e., without any concrete filling) can experience failure in about 20 to 30 minutes due to exposed steel column attaining high temperature at a rapid pace.

To illustrate the effect of the type of concrete filling on the fire behavior of a composite column, the response of HSS column filled with three types of concrete—plain concrete (PC), reinforced concrete (RC), and steel fiber-reinforced concrete (FC)—is shown in Fig. 9.6. This comparative data is generated through fire resistance tests on structural columns with different concrete fillings (Kodur and Lie, 1995). In all three

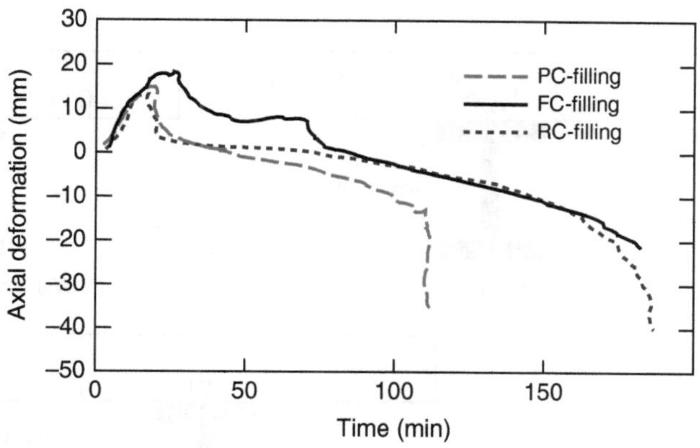

FIGURE 9.6 Comparative fire behavior of PC-, RC-, and FC-filled HSS columns exposed to fire.

cases, the steel section gradually yields, and the column rapidly contracts at some point between 20 and 30 minutes into fire. Until this point, the behavior of all three columns is similar since the load is mainly carried by the HSS section. After this point in time, the concrete core starts carrying more and more of the load, and as such the behavior of these three columns is mainly dependent on the type of concrete filling.

In the case of a PC-filled HSS column, when the concrete core loses much of its compressive strength, resulting from a temperature rise in interior core, the composite column cannot support the applied loading and thus experiences failure. In the case of RC-filled HSS columns, rebars, together with concrete, contribute to the load-carrying capacity and thus RC-filled column attains higher fire resistance. Also, these rebars experience a slower temperature rise, since they are embedded within the concrete mass, thus experiencing a slower loss of strength with time. Along the same lines, in the case of an FC-filled HSS column, the presence of steel fibers in concrete leads to slower degradation in compressive strength of fiber-reinforced concrete with temperature rise. Thus, the FC-filled HSS column attains failure at a longer duration as compared to a PC-filled HSS column. Overall, the fire resistance of RC-filled HSS columns is higher than that of FC-filled HSS columns, which in turn is higher than PC-filled HSS columns. Through a series of fire tests and numeral studies, Kodur (1997) has shown that FC filling in HSS columns is very effective and can provide fire resistance comparable to that of RC-filled HSS columns.

9.5 Strategies for Achieving Fire Resistance

With all things being equal, fire resistance of a well-designed composite structural member is higher than that of an unprotected steel structural member but lower than that of a comparative reinforced concrete member. This is due to better fire resistance properties (i.e., thermal and mechanical) of concrete as compared to steel. Thus, when steel structural members are used in conjunction with concrete, fire resistance of such a composite structural member can be significantly enhanced, often eliminating the need of fire insulation on steel.

Although protecting steel structures with fire insulation offers a convenient solution to increase fire resistance, such an approach has a number of drawbacks including problems such as durability and stickability of fire insulation to steel, poor aesthetics of steel structure, and high costs of insulation application and its maintenance. As an alternative, a composite construction strategy can be effectively utilized to achieve required fire resistance in steel structures. Since modern construction often utilizes both structural steel and concrete, through proper design and careful construction detailing, it is often possible to achieve required fire resistance through composite construction. For optimizing the fire resistance levels that can be achieved in a composite member, relevant strategies should be developed, specific to each type of structural member (e.g., beam, slab, wall, or column), so as to facilitate effective load transfer between steel and concrete components under high-temperature exposure. The various strategies that can be developed in different member types are illustrated in Fig. 9.7.

In the case of a steel beam-concrete slab composite assembly, provision of shear studs is required to enhance transfer of forces (shear) between concrete slab and steel beam (or vice versa), so as to enhance fire resistance (Fig. 9.7). Facilitating such interaction between steel beam and concrete slab through studs enable concrete slab to carry a portion of applied load throughout the service life of the composite beam

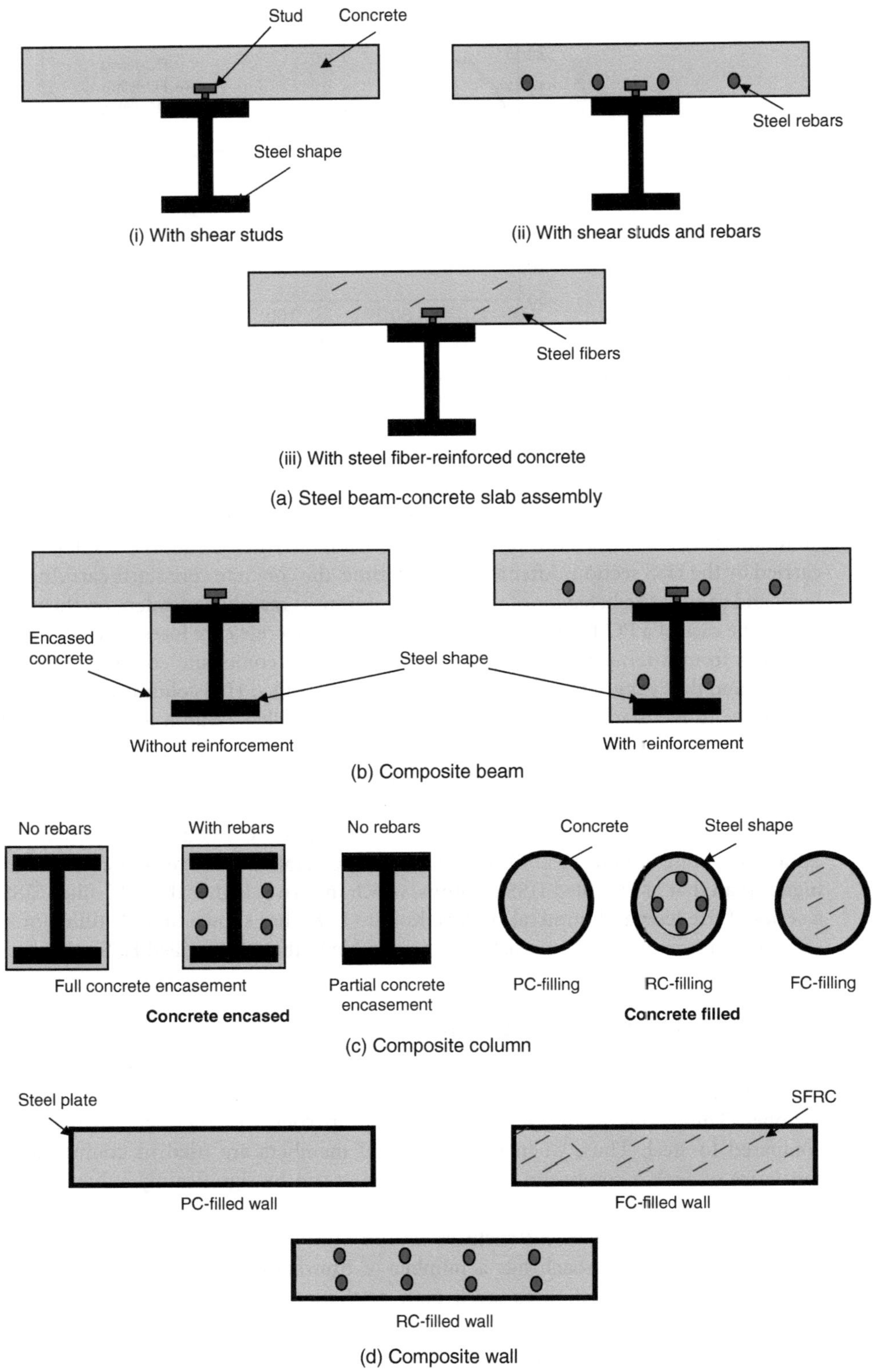

FIGURE 9.7 Different strategies for enhancing fire resistance of structural members through composite construction.

under ambient conditions. When a fire-exposed steel beam loses much of its capacity due to high temperatures developing in steel sections, a portion of the forces carried by steel can get transferred to the cooler concrete slab. Also, the sectional temperatures in steel sections rise at a slower rate, as concrete absorbs heat from steel through the *heat sink effect*. Further, a slower rise in temperatures in concrete section, due to its high thermal inertia, leads to the gradual degradation in strength properties of concrete. Thus, the contribution of concrete slab to load-carrying capacity increases with fire exposure time. This can enable the composite beam-slab assembly to sustain service loads for up to 1 to 2 hours without the need for any external fire protection to the steel beam. The fire resistance in such a composite beam-slab assembly can be further improved through adding discrete steel reinforcement (rebars) in the concrete slab, mainly for resisting tensile forces during fire exposure. Alternatively, the use of steel fibers in concrete can improve both the tensile and compressive strength of concrete under elevated temperatures and limit crack progression. This can enhance fire resistance significantly (Fike and Kodur, 2011). Both of these options have been shown to improve fire resistance significantly in a composite beam-slab assembly (see Fig. 9.7*a*).

In the case of a steel beam, concrete encasement can provide thermal insulation to steel sections during fire exposure and also can carry a portion of the total load, thus enhancing fire resistance of a composite beam (see Fig. 9.7*b*). Adding reinforcement (rebars) in concrete can further enhance fire resistance.

Through concrete encasement or concrete filling of steel columns, fire resistance can be significantly enhanced. Different types of concrete filling can be used, as discussed in Sec. 9.4 and as plotted in Fig. 9.7*c*, to arrive at the optimum level of fire resistance for a given steel shape. The use of shear connectors may not be easily applied in the case of CFHSS columns due to complexities in installing studs into the interior of steel shapes. Alternatively, W-shaped columns can be encased in concrete and specifically designed such that steel columns and concrete encasement work compositely.

In addition, steel-plated shear walls can be filled between concrete to enhance load-bearing capacity, as well as shear resistance of walls during fire exposure, thus enhancing fire resistance (as well as limit temperature progression) (see Fig. 9.7*d*). The addition of rebars and/or steel fibers can further improve their structural performance and fire response.

The use of the above strategies can be quite effective in enhancing fire resistance in different types of structural members. However, detailed calculations are necessary to optimize the level of fire resistance that can be achieved in a specific member type. In such calculations, due consideration must be given to various factors, including thermal inertia of concrete, force transfer between concrete and steel components, and extent of composite action developed during high-temperature exposure. In the simplest form, prescriptive calculation approaches can be applied for evaluating fire resistance wherein only limited influencing factors are taken into account. Alternatively, advanced analysis approaches can be applied wherein all critical factors, including the level of composite action, can be taken into account to evaluate realistic fire resistance of a composite structure.

9.6 Prescriptive Approaches for Evaluating Fire Resistance

In the case of a composite member, failure is to be evaluated with due consideration to the overall composite system, rather than steel or concrete components alone. Therefore, failure of the member cannot be based on simply looking at the critical temperature reached in the steel member alone, as done in the case of isolated steel structural members. The prescriptive approaches that can be applied to evaluate fire resistance of different composite structural members are discussed here.

9.6.1 Testing

Fire resistance of a composite member can be evaluated through a standard fire resistance test wherein a composite member, subjected to a specified structural loading, is exposed to standard fire in a specially built fire test furnace. Following the procedures outlined in applicable standards such as ASTM E-119 and ISO 834, failure in the case of a composite member occurs once the insulation, deflection, or strength criterion is breached as outlined in Chap. 5.[2] Integrity failure is not an issue in most types of composite members, if the member is designed as per room temperature specifications outlined in applicable design standards. Based on these fire tests, a number of listings are available for different composite members in directories such as Underwriters Laboratories (UL). The major difference in evaluating failure in composite members is that the critical temperature criterion applied to evaluate failure in isolated steel members or concrete members (temperature in reinforcing or prestressing steel) cannot be applied. This is because the concrete portion of the composite member can continue to sustain loading, well beyond steel member attains critical temperature of limit, acting on the member.

[2] The critical temperature limit applied to evaluate failure used in isolated steel members (see Chap. 7) may not be valid for evaluating failure of a composite member or assembly.

9.6.2 Tabulated Data

Listings are available in directories, such as UL, for evaluating minimum sectional requirements to achieve certain fire rating in a composite structural member. Codes and standards, such as Eurocode 4 and ACI 216.1, present tabular data in the form of tables or graphs for evaluating fire resistance of certain composite members. The tables and procedures listed in Eurocode 4 are applicable for partially encased W-shaped steel columns and beams, in which the concrete has only an insulating function. In addition, in Eurocode 4, tabular data/listings are also available for fully or partially encased steel columns, as well as for CFHSS columns. Similarly, ACI 216.1 provides tabulated data for satisfying fire resistance criterion in concrete encased, as well as concrete-filled, steel columns. This type of tabulated data is only applicable for composite structural members under standard fire exposure (i.e., ISO 834 or ASTM E119).

Eurocode 4 Provisions

Eurocode 4—Part 2 (EC4-2) provides a number of tables for deriving fire resistance of different composite structural members. The provisions specified in these tables are derived with due consideration to strength contribution from both concrete and steel in resisting loading under fire conditions. For evaluating fire resistance of a composite beam, a five-step procedure is to be applied. In Step 1, maximum bending moment (M_{sd}) that is to be carried by the composite beam at ambient conditions is evaluated as per load combinations listed in EC1. In this step, loading is evaluated, taking into account both permanent (dead) and live (variable) loads acting on the member. In Step 2, room temperature properties of concrete, steel, and rebars (modulus of elasticity and yield stress for structural steel and reinforcement, f_y, E_a, f_s, E_s, as well as compressive strength and modulus of elasticity of concrete, f_c, E_{cm}) that constitute composite section are to be compiled. In Step 3, the geometrical sectional properties (sectional area, sectional modulus, and moment of inertia) of individual components of composite section—steel, concrete, and rebars components—as well as that of the overall composite section, are to be determined. Once material (from Step 2) and geometric (from Step 3) properties are known, sectional moment capacity ($M_{b,Rd}$) can then be evaluated, as per Eurocode 4—Part 1 equations in Step 4.

In Step 5, load level ($\eta_{fi,t}$), defined as the ratio of maximum bending moment ($M_{fi,d}$) to be carried by the beam under fire conditions to that of bending moment at ambient conditions, is determined as

$$\eta_{fi,t} = \frac{M_{fi,d}}{M_{b,Rd}} \tag{9.3}$$

where $M_{fi,d}$ is the design effect of applied loading (e.g., bending moment) during fire exposure (also referred to as $E_{fi,d}$, for general effect of actions) and $M_{b,Rd}$ is the moment capacity at ambient conditions.

Knowing load level $\eta_{fi,t}$ and type of composite beam (which depends on steel shape), the minimum width and reinforcement area (of concrete core) needed to achieve a certain fire rating in a composite beam can be read off the tables listed in Eurocode 4—Part 2. These tables list minimum sectional dimensions required for load levels of 0.3, 0.5, and 0.7 corresponding to fire resistance ratings of 30, 60, 90, 120, and 180 minutes. Eurocode 4 tables are only applicable for composite beams made of structural steel of Grade S355 and reinforcing steel of Grade S500. Further, fire ratings given in these tables may be used for beams connected to composite floors with profiled steel sheets, if at least 85% of the upper side of the steel profile is directly covered by the steel sheet. If not, void fillers have to be applied on top of the beam to achieve specified ratings. A sample table from Eurocode 4 showing minimum sectional requirements for achieving a certain fire rating in a composite beam is reproduced in Table 9.2.

Eurocode 4 also provides a tabulated approach for evaluating the fire resistance of concrete-encased and concrete-filled steel columns. These tables require a five-step calculation procedure, as in the case of composite beams, for evaluating minimum dimensions corresponding to a desired fire rating in a composite column. In Step 1, the applied design load (N_{sd}) to be carried by a composite column at ambient conditions is determined based on load combinations specified in Eurocode 1. Then, in Step 2, mechanical properties of structural steel, steel reinforcement (rebars), and concrete that form the composite column are to be compiled. In Step 3, sectional properties, including area, moment of inertia, and radius of gyration of individual components and of the overall composite column, are to be determined. In Step 4, sectional axial capacity ($N_{b,Rd}$) of the composite column at ambient condition is calculated using room temperature expressions as per the procedure outlined in EC4-1. Then, in Step 5, the reduced load ($N_{fi,d}$) acting on the column during fire exposure is determined and a load level factor ($\eta_{fi,t}$), defined as the ratio of applied axial load at ambient conditions to that under fire conditions, is evaluated as in the case of composite beams discussed above (see Eq. 9.3). Knowing $\eta_{fi,t}$, minimum dimension (h_c and b_c) required in the column for

	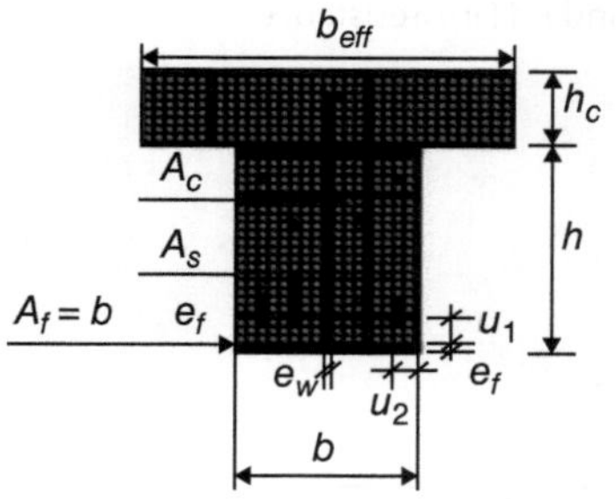 Condition for application: slab: $h_c \geq 120$ mm, $b_{eff} \leq 5$ m; steel section: $b/e_w \geq 15$, $e_f/e_w \leq 2$; additional reinforcement area, related to total area between the flanges: $A_s/A_c + A_s \leq 5\%$	Standard Fire Resistance				
1		R30	R60	R90	R120	R180
	Minimum cross-sectional dimensions for load level $\eta_{fi,t} \leq 0.3$					
	min b [mm] and additional reinforcement A_s in relation to the area of flange A_s/A_f					
1.1	$h \geq 0.9 \times \min b$	70/0.0	100/0.0	170/0.0	200/0.0	260/0.0
1.2	$h \geq 1.5 \times \min b$	60/0.0	100/0.0	150/0.0	180/0.0	240/0.0
1.3	$h \geq 2.0 \times \min b$	60/0.0	100/0.0	150/0.0	180/0.0	240/0.0
2	Minimum cross-sectional dimensions for load level $\eta_{fi,t} \leq 0.5$					
	min b [mm] and additional reinforcement A_s in relation to the area of flange A_s/A_f					
2.1	$h \geq 0.9 \times \min b$	80/0.0	170/0.0	250/0.4	270/0.5	–
2.2	$h \geq 1.5 \times \min b$	80/0.0	150/0.0	200/0.2	240/0.3	300/0.5
2.3	$h \geq 2.0 \times \min b$	70/0.0	120/0.0	180/0.2	220/0.3	280/0.3
2.4	$h \geq 3.0 \times \min b$	60/0.0	100/0.0	170/0.2	200/0.3	250/0.3
3	Minimum cross-sectional dimensions for load level $\eta_{fi,t} \leq 0.7$					
	min b [mm] and additional reinforcement A_s in relation to the area of flange A_s/A_f					
2.1	$h \geq 0.9 \times \min b$	80/0.0	270/0.4	300/0.6	–	–
2.2	$h \geq 1.5 \times \min b$	80/0.0	240/0.3	270/0.4	300/0.6	–
2.3	$h \geq 2.0 \times \min b$	70/0.0	190/0.3	210/0.4	270/0.5	320/1.0
2.4	$h \geq 3.0 \times \min b$	70/0.0	170/0.2	190/0.4	270/0.5	300/0.8

Table 9.2 Minimum Sectional Dimensions for Evaluating Fire Resistance Tables in a Composite Beam (*Reproduced from CEN*)

achieving a certain fire ratings of 30, 60, 90, 120, 180, and 240 minutes can be read off in tables provided in Eurocode 4. A series of tables covering various composite column configurations—totally encased steel sections, partially encased steel sections, and concrete-filled hollow sections—are listed in Eurocode 4. A sample table from Eurocode 4 for the case of concrete-filled round and square steel columns is shown in Table 9.3.

It should be noted that tabular data listed in Eurocode 4 are for isolated composite columns assuming pin-ended boundary conditions, and that column ends are rotationally restrained under fire conditions. These tables are also applicable to composite columns that are continuous and restrained under fire conditions. Tabular data is valid both for concentric or eccentric axial loads in the range of 0.28 to 0.66 of ultimate axial capacity and only to columns with a maximum length of 30 times the minimum external dimension (breadth or depth) of the cross section. Eurocode 4 further states that a totally encased steel column, without any fire insulation, can achieve a fire rating of 30 minutes, without the need for any calculation. Eurocode 4 does not provide any tabular data for composite slabs.

Much of the tabulated data adopted in Eurocode 4 are from CIDECT (International Committee for the Development and Study of Tubular Structures) design guide (Twilt et al., 1996). This design guide gives a number of tables and associated calculation procedures for determining fire resistance of composite steel members, as well as connections. A number of tables for fire resistance rating of HSS columns with three types of fire protection—through external insulating materials or special paints; through water cooling, either statically or with circulation; and through concrete filling—are given in CIDECT design girder, together with numerical examples illustrating fire resistance evaluation in design situations.

	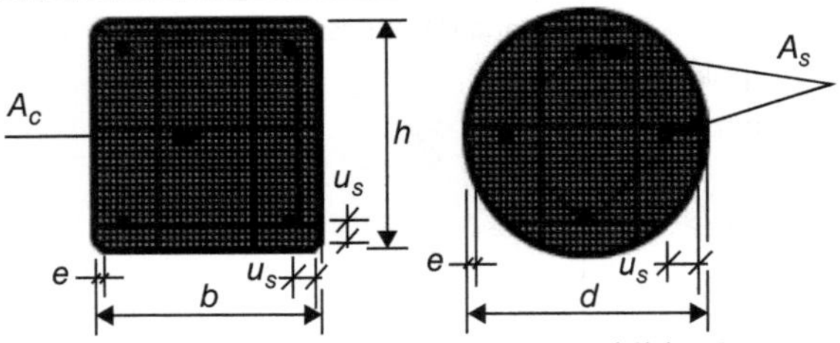 stell section: $(b/e) \geq 25$ or $(d/e) \geq 25$	Standard Fire Resistance				
		R30	R60	R90	R120	R180
1	Minimum cross-sectional dimensions for load level $\eta_{fi,t} \leq 0.28$					
1.1	Minimum dimensions h and b or minimum diameter d (mm)	160	200	220	260	400
1.2	Minimum ratio of reinforcement $A_s/(A_c + A_s)$ in (%)	0	1.5	3.0	6.0	6.0
1.3	Minimum axis distance of reinforcing bars u_s (mm)	–	30	40	50	60
2	Minimum cross-sectional dimensions for load level $\eta_{fi,t} \leq 0.47$					
2.1	Minimum dimensions h and b or minimum diameter d (mm)	260	260	400	450	500
2.2	Minimum ratio of reinforcement $A_s/(A_c + A_s)$ in (%)	0	3.0	6.0	6.0	6.0
2.3	Minimum axis distance of reinforcing bars u_s (mm)	–	30	40	50	60
3	Minimum cross-sectional dimensions for load level $\eta_{fi,t} \leq 0.66$					
2.1	Minimum dimensions h and b or minimum diameter d (mm)	260	450	550	–	–
2.2	Minimum ratio of reinforcement $A_s/(A_c + A_s)$ in (%)	3.0	6.0	6.0	–	–
2.3	Minimum axis distance of reinforcing bars u_s (mm)	25	30	40	–	–

Table 9.3 Eurocode 4 Table for Evaluating Fire Resistance of Concrete-Filled Steel Columns (*Reproduced from CEN*)

ACI 216.1 Provisions

ACI 216.1 (ACI, 2014) specifies a tabular approach for determining fire resistance of fully encased W-shaped steel columns—that is, steel shape protected with normal- or lightweight concrete, or precast concrete. The listed tables are applicable for W-shaped columns under a combination of section size, concrete type, and concrete cover thickness. The ACI 216.1 approach for determining fire ratings is much simpler than that of Eurocode 4, as it does not require calculation of ambient sectional capacity and involves simply reading off cover thickness requirements for a specified fire rating from the listed tables. Tabulated listings in AIC 216.1 is only applicable to certain W-shaped steel sections that vary from W150×13.5 (equivalent to W6×9) to W360×347 (equivalent to W14×223). A sample table from ACI 216.1 is reproduced as Table 9.4.

9.6.3 Empirical Equations

In lieu of listings, *correlation equations* are available in some codes and design standards, as well as in published articles, for evaluating fire resistance of different types of composite members. These correlations are developed by applying extension rules as discussed in Chaps. 5 and 6 or through the curve-fitting technique to fire resistance data obtained from standard fire resistance tests or numerical models. An example of the correlation equation used in practice for evaluating fire resistance of a composite columns is discussed herein.

Based on a large set of fire test data and results from numerical studies on CFHSS columns, Lie and Kodur (1995) and Kodur (1999) proposed a unified equation for evaluating fire resistance of CFHSS columns. This equation has been incorporated into many North American codes and standards (ACI, 2014; NBCC, 2015). This equation accounts for a range of parameters, including load level, section size, column length, type of concrete filling and its strength, and any steel reinforcement that is present, in evaluating fire resistance of CFHSS columns. The equation is applicable for three types of concrete filling: plain concrete (PC), steel fiber-reinforced concrete (FC), and reinforced concrete (RC) filling. The fire resistance of an HSS column filled with PC, FC, or RC is given as

$$R = a\frac{(f_c' + 20)}{60(KL - 1000)}D^2\left(\frac{D}{C}\right)^{1/2} \qquad \text{(9.4a—SI units)}$$

<table>
<tr><th rowspan="2">Structural Shape</th><th colspan="5">Fire Resistance Rating (hr)</th></tr>
<tr><th>1</th><th>1.5</th><th>2</th><th>3</th><th>4</th></tr>
<tr><td>W360×347</td><td rowspan="8">1-1/2</td><td rowspan="8">1-1/2</td><td rowspan="5">1-1/2</td><td rowspan="3">2</td><td rowspan="2">2-1/2</td></tr>
<tr><td>W360×262</td></tr>
<tr><td>W360×216</td><td rowspan="4">3</td></tr>
<tr><td>W360×196</td><td rowspan="4">2-1/2</td></tr>
<tr><td>W360×162</td></tr>
<tr><td>W360×147</td><td rowspan="3">2</td></tr>
<tr><td>W360×101</td><td rowspan="2">3-1/2</td></tr>
<tr><td>W360×64</td><td>3</td></tr>
<tr><td>W310×283</td><td rowspan="8">1-1/2</td><td rowspan="8">1-1/2</td><td rowspan="5">1-1/2</td><td rowspan="3">2</td><td rowspan="2">2-1/2</td></tr>
<tr><td>W310×226</td></tr>
<tr><td>W310×202</td><td rowspan="2">3</td></tr>
<tr><td>W310×158</td><td rowspan="4">2-1/2</td></tr>
<tr><td>W310×143</td><td rowspan="4">3-1/2</td></tr>
<tr><td>W310×129</td><td rowspan="3">2</td></tr>
<tr><td>W310×97</td></tr>
<tr><td>W310×60</td><td>3</td></tr>
<tr><td>W250×167</td><td rowspan="7">1-1/2</td><td rowspan="6">1-1/2</td><td rowspan="3">1-1/2</td><td>2</td><td rowspan="4">3</td></tr>
<tr><td>W250×149</td><td rowspan="4">2-1/2</td></tr>
<tr><td>W250×131</td></tr>
<tr><td>W250×115</td><td rowspan="4">2</td></tr>
<tr><td>W250×89</td><td rowspan="3">3-1/2</td></tr>
<tr><td>W250×58</td><td rowspan="2">3</td></tr>
<tr><td>W250×49.1</td><td>2</td></tr>
<tr><td>W200×100</td><td rowspan="5">1-1/2</td><td rowspan="3">1-1/2</td><td>1-1/2</td><td rowspan="2">2-1*2</td><td rowspan="2">3</td></tr>
<tr><td>W200×71</td><td rowspan="3">2</td></tr>
<tr><td>W200×52</td><td rowspan="3">3</td><td rowspan="2">3-1/2</td></tr>
<tr><td>W200×41.7</td><td rowspan="2">2</td></tr>
<tr><td>W200×26.6</td><td>2-1/2</td><td>4</td></tr>
<tr><td>W150×37.1</td><td rowspan="3">1-1/2</td><td rowspan="3">2</td><td>2</td><td rowspan="2">3</td><td>3-1/2</td></tr>
<tr><td>W150×22.5</td><td rowspan="2">2-1/2</td><td rowspan="2">4</td></tr>
<tr><td>W150×13.5</td><td>3-1/2</td></tr>
</table>

Table 9.4 Minimum Concrete Cover Thickness to Steel for Achieving Certain Fire Rating in Encased Steel Columns (in Inches) (*Authorized Reprint from ACI, 2014*)

$$R = 0.58a\frac{\left(f_c' + 2.9\right)}{\left(KL - 3.28\right)}D^2\left(\frac{D}{C}\right)^{1/2} \qquad \text{(9.4b—FPS units)}$$

where R is the fire resistance in hours, a is a factor specified in Table 9.5, f_c' is the specified 28-day compressive strength of concrete in MPa (kips/in^2), KL is the column effective length in m (ft), D is the outside diameter for circular columns, least outside dimension for square or rectangular columns in m (in.), and C is the compressive force due to unfactored dead and live load in kN (kips). In the above equation, a is to account for aggregate type in concrete and shape of HSS (rectangular and circular), and values of a to be used in this equation are given in Table 9.5. In the case when plain concrete is used as an

Filling Type	**Plain Concrete (PC)**		**Bar-Reinforced Concrete (RC)**								**Fiber-Reinforced Concrete (FC)**	
Aggregate Type	S*	C**	S				C				S*	C**
% of Steel Reinforcement	N/A		<3%		≥3%		<3%		≥3%		—	—
Thickness of Concrete Cover	N/A		<25 mm	≥25 mm	<25 mm	≥25 mm	<25 mm	≥25 mm	<25 mm	≥25 mm	—	
Circular HSS	0.07	0.08	0.075	0.08	0.08	0.085	0.085	0.09	0.09	0.095	0.075	0.085
Square HSS	0.06	0.07	0.065	0.07	0.07	0.075	0.075	0.08	0.08	0.085	0.065	0.075

* Siliceous.
** Carbonate.

TABLE 9.5 Value of Parameter *a* in Eq. 9.4 for Evaluating Fire Resistance of CFHSS Columns

in-fill, the fire resistance of an HSS columns is limited to a maximum of 2 hours, while for the case of FC or RC filling, up to 3 hours of fire resistance can be obtained.

The above equation can be rearranged to evaluate the maximum load (C_{max}) a CFHSS column can sustain for a required fire rating:

$$C_{max} = \left(\frac{a(f_c' + 20)D^{2.5}}{R(KL - 1000)} \right)^2 \qquad \text{(9.5a—SI units)}$$

$$C_{max} = \left(\frac{0.58a(f_c' + 2.9)D^{2.5}}{R(KL - 3.28)} \right)^2 \qquad \text{(9.5b— FPS units)}$$

The applicability of the above equation is limited to a range of parameters, as specified in Table 9.6. Further discussion on the test data utilized in the development of the equation and main assumptions adopted in the equation are given by Kodur (1999).

9.7 Rational Approaches for Evaluating Fire Resistance

A number of rational engineering approaches are available for evaluating fire resistance of composite members. In these approaches, many of the critical parameters governing fire resistance, including the level of composite action available at a specified fire exposure time, can be incorporated. The major difference in the application of these rational methods is that the critical temperature concept, as often used in prescriptive approaches for steel structures, cannot be applied to evaluate the failure time of a composite member, but strength or deflection limit state is to be considered. Eurocode 4 lists detailed steps, as part of simplified calculations methods, for evaluating fire resistance of columns, beams, and slabs.

9.7.1 Simplified Approaches

Fire resistance of a composite member can be evaluated through simplified approaches by undertaking heat transfer and strength calculations at a specified fire exposure time or in a series of time increments until failure of the given composite member. The main response parameters needed for deriving fire resistance are cross-sectional temperatures, in steel sections, rebars, and concrete and on unexposed side of member, and sectional capacity of the member. The procedure for evaluating these response parameters is discussed herein.

Parameters	PC	FC	RC
f_c' (MPa)	20 to 40	20 to 55	20 to 55
D (round) (mm)	140 to 410	140 to 410	165 to 410
D (square) (mm)	140 to 305	102 to 305	175 to 305
Reinforcement (%)	N/A	~2% of the concrete mix by mass	1.5% to 5%, with limits on size, number and spacing of bars and ties
R (min)	120	180	180
KL (mm)	2000–4000	2000–4500	2000–4500
Class	1, 2, or 3	1, 2, or 3	1, 2, or 3
C (AISC)	$\leq C_r'$	£1.1 C_r'	£1.7 C_r'

Table 9.6 Range of Applicability of Fire Resistance Equation (Eq. 9.4) for CFHSS Columns

Standard Fire Resistance	Minimum Effective Thickness, h_{eff} (mm)
R30	60
R60	80
R90	100
R120	120
R180	150
R240	175

Table 9.7 Minimum Effective Thickness for Achieving Required Fire Resistance in a Composite Deck Slab Made of Normal-Weight Concrete (*Reproduced from CEN*)

Eurocode 4 Approach for Composite Slabs

Eurocode 4 provides a separate set of simplified equations for evaluating fire resistance of composite slabs, columns, and beams. Although the underlying principles and procedures for evaluating fire resistance are the same for different member types, there are subtle differences in accounting for specific parameters in each member type.

The Eurocode 4 method for evaluating fire resistance of a composite slab takes into account all three failure limit states: *integrity* (E), *insulation* (I), and *load-bearing* (R) criteria. The integrity criterion, E, in a composite slab is said to be automatically satisfied if the composite slab has continuous steel decking and meets room temperature design provisions as per Eurocode 4—Part 1. As per Eurocode 4 provisions, a composite slab with profiled steel sheets, with or without any additional reinforcement (rebars), will provide a minimum fire resistance of 30 minutes without any need for calculation or verification. Further, the load-bearing criterion, R, in a composite slab is satisfied as long as the temperature of the steel deck of the composite slab is lower or equal to 350°C under standard fire exposure. Once temperature in the steel deck exceeds this level, the sectional capacity of composite slab starts to reduce, and this reduction in sectional capacity is to be evaluated through an outlined procedure as discussed below. There are two methods for checking insulation criteria, either through the *effective thickness approach* or through *unexposed temperature calculation*. The strength-based failure criterion is to be checked through additional strength calculations and is discussed in detail.

Effective Thickness Approach The insulation criterion can be satisfied by checking whether the minimum effective slab thickness meets the limits given in Eurocode 4 (see Table 9.7). Accordingly, for achieving a certain level of fire resistance rating, minimum effective thickness (h_{eff}) of the slab, taken as $h_2/h_1 \leq 1.5$ and

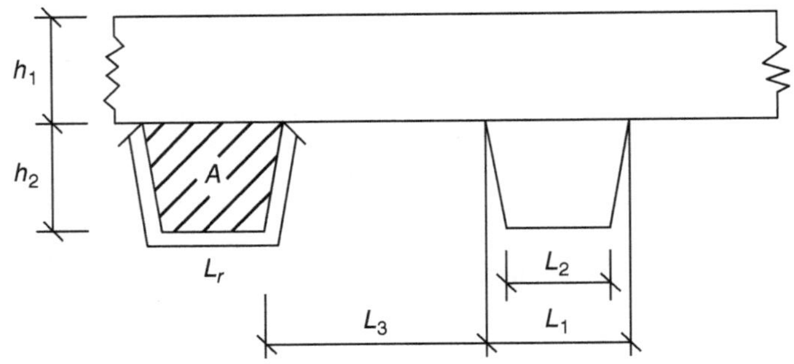

Figure 9.8 Rib geometry calculations in a composite slab according to Eurocode 4 procedure. (*Reproduced from CEN.*)

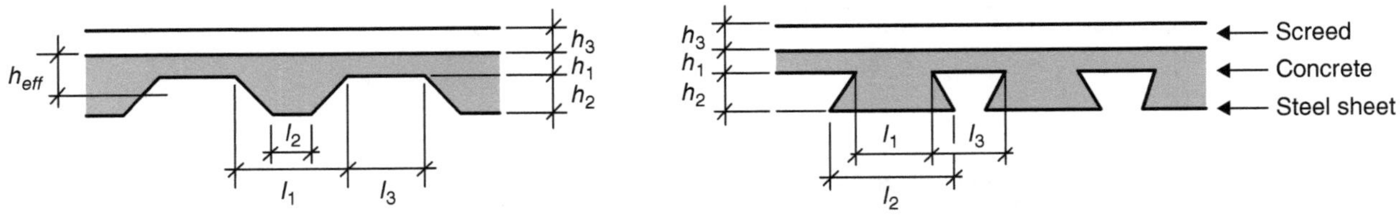

Figure 9.9 Parameters needed to determine unexposed temperature in a deck slab for checking insulation criteria as per Eurocode 4 provisions. (*Reproduced from CEN.*)

$h_1 > 40$ mm, is to be below the limiting thickness (h_{eff}) specified in Table 9.7. h_{eff} in a given slab is computed based on the slab geometry (see Fig. 9.8) as

$$h_{eff} = h_1 + 0.5h_2 \frac{l_1 + l_2}{l_1 + l_3} \tag{9.6}$$

However, when $h_2/h_1 > 1.5$ and $h_1 > 40$ mm for trapezoidal and/or dovetail sections, h_{eff} is evaluated as

$$h_{eff} = h_1 \left(1 + 0.75 \frac{l_1 + l_2}{l_1 + l_3}\right) \tag{9.7}$$

where h_1, h_2, l_1, l_2, and l_3 are sectional dimensions as represented in Fig. 9.9. If $l_3 > 2l_1$, $h_{eff} = h_1$. The limiting values of thickness, h_{eff}, given in Table 9.7 are for normal-weight concrete (NWC). For lightweight concrete slabs, the limiting values of h_{eff} are to be reduced by 10% to account for lower thermal conductivity of lightweight concrete (LWC).

Unexposed Temperature Calculation As an alternative to the effective thickness method, Eurocode 4 also lays out a set of simplified equations for evaluating unexposed surface temperature that can be applied to check the failure based on insulation (I) criterion in unprotected composite deck slabs. The empirical equation to evaluate time to failure (t_i) in a deck slab exposed to standard fire, based on insulation limit state, is given as

$$t_i = a_0 + a_1 h_1 + a_2 \Phi + a_3 \frac{A}{L_r} + a_4 \frac{1}{l_3} + a_5 \frac{A}{L_r}\frac{1}{l_3} \tag{9.8}$$

where t_i is the fire resistance based on the insulation (I) criterion, A is the concrete volume of the rib per meter rib length, L_r is exposed area of the rib per meter rib length, A/L_r is the rib geometry factor, Φ is the view factor of the upper flange, and l_3 is the width of the upper flange. a_0, a_1, a_2, a_3, a_4, and a_5 are regression coefficients given in a table in Eurocode 4—Part 2, for a concrete slab made of NWC or LWC (as shown in Table 9.8).

	a_0 (min)	a_1 (min/mm)	a_2 (min)	a_3 (min/mm)	a_4 (mm·min)	a_5 (min)
Normal- weight concrete	−28.8	1.55	−12.6	0.33	−735	48
Lightweight concrete	−79.2	2.18	−2.44	0.56	−542	52.3

TABLE 9.8 Regression Coefficients for Evaluation of Unexposed Surface Temperature in a Deck Slab (*Reproduced from CEN*)

The rib geometry factor (A/L_r) is to be evaluated based on the geometry of the composite deck (see Fig. 9.10) as

$$\frac{A}{L_r} = \frac{h_2 \frac{l_1 + l_2}{2}}{l_2 + 2\sqrt{h_2^2 + \left(\frac{l_1 - l_2}{2}\right)^2}} \tag{9.9}$$

and Φ is a view (shadow) factor and can be evaluated as

$$f = \left(\sqrt{h_2^2 + \left(l_3 + \frac{l_1 - l_2}{2}\right)^2} - \sqrt{h_2^2 + \left(\frac{l_1 - l_2}{2}\right)^2}\right) \Big/ l_3 \tag{9.10}$$

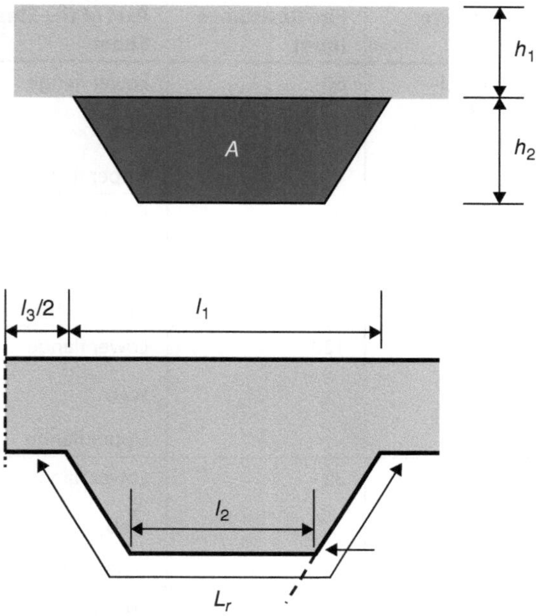

FIGURE 9.10 Geometry of steel deck in a composite slab. (*Reproduced from CEN.*)

To satisfy strength limit state (R criteria) of a composite slab under fire exposure, Eurocode 4 lays out a series of simplified equations to evaluate sectional temperatures first, and then the reduced sectional capacity of a slab. The first step is to compute sectional temperatures in various components (concrete, rebar, deck) of a composite slab (see Fig. 9.11). With the knowledge of these sectional temperatures, the sectional capacity of the slab at a given exposure time can be evaluated by applying a plastic global analysis wherein the composite slab is analyzed assuming that plastic hinges form at critical midspan and support sections. It should be ensured that the slab has sufficient rotational capacity and tensile reinforcement with sufficient deformation capacity. For end spans, where continuity exists at one support only, Eurocode 4 recommends to assume the plastic hinge to occur at midspan section alone as per Eurocode 4—Part 1 (CEN, 2005) provisions.

Separate expressions for evaluating temperatures in critical components (deck, steel, rebars, concrete) of the slab are given in Eurocode 4 at fire resistance times of 30, 60, 90, and 120 minutes and for the case of two types of concrete. The equation for calculating temperature θ_a in the bottom flange, web, or top flange of the decking is

$$\theta_a = b_0 + b_1 \frac{1}{l_3} + b_2 \frac{A}{L_r} + b_3 \Phi + b_4 \Phi^2 \tag{9.11}$$

where b_0 to b_4 are coefficients for determining temperature in parts of a steel deck at 60, 90, or 120 minutes, and these are given for two types of concrete, namely NSC and LWC. A sample table of coefficients from Eurocode 4—Part 2 is shown in Table 9.9. Φ is the shadow factor and can be calculated as in Eq. 9.12.

Similarly, for evaluating temperature in steel (rebars) reinforcement, θ_s, present in the slab, the associated empirical equation is given as

$$\theta_s = c_0 + c_1 \frac{u_3}{h_2} + c_2 z + c_3 \frac{A}{L_r} + c_4 \alpha + c_5 \frac{1}{l_3} \tag{9.12}$$

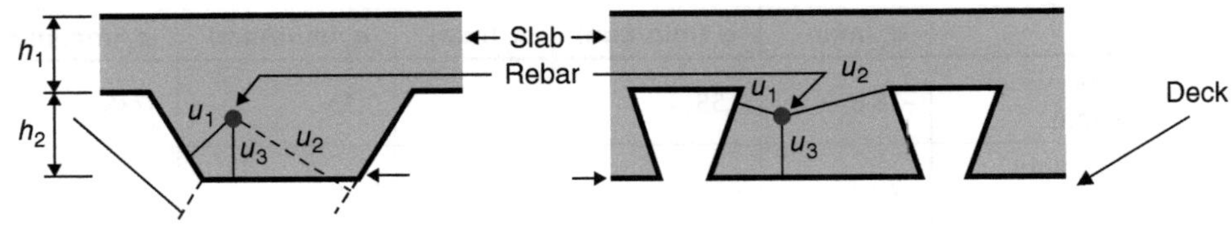

FIGURE 9.11 Typical composite slab with various components of parameters u_1, u_2, and u_3. (*Reproduced from CEN.*)

Concrete	Fire Resistance (min)	Part of the Steel Sheet	b_0 (°C)	b_1 (°C)·mm	b_2 (°C)·mm	b_3 (°C)	b_4 (°C)
Normal-weight concrete	60	Lower flange	951	−1197	−2.32	86.4	−150.7
		Web	661	−833	−2.96	537.7	−351.9
		Upper flange	340	−3269	−2.62	1148.4	−679.8
	90	Lower flange	1018	−839	−1.55	65.1	−108.1
		Web	816	−959	−2.21	464.9	−340.2
		Upper flange	618	−2786	−1.79	767.9	−472.0
	120	Lower flange	1063	−679	−1.13	46.7	−82.8
		Web	925	−949	−1.82	344.2	−267.4
		Upper flange	770	−2460	−1.67	592.6	−379.0
Lightweight concrete	30	Lower flange	800	−1326	−2.65	114.5	−181.2
		Web	483	−286	−2.26	439.6	−244.0
		Upper flange	331	−2284	−1.54	488.8	−131.7
	60	Lower flange	955	−622	−1.32	47.7	−81.1
		Web	761	−558	−1.67	426.5	−303.0
		Upper flange	607	−2261	−1.02	664.5	−410.0
	90	Lower flange	1019	−478	−0.91	32.7	−60.8
		Web	906	−654	−1.36	287.8	−230.3
		Upper flange	789	−1847	−0.99	469.5	−313.0
	120	Lower flange	1062	−399	−0.65	19.8	−43.7
		Web	989	−629	−1.07	186.1	−152.6
		Upper flange	903	−1561	−0.92	305.2	−197.2

TABLE 9.9 Coefficients for the Determination of Temperature in a Steel Deck of a Composite Slab (*Reproduced from CEN*)

where c_0 to c_5 are temperature coefficients as listed in Table 9.10 and which depend on the duration at which fire resistance is to be evaluated (60, 90, or 120 minutes). A/L_r and L_3 are sectional parameters as explained earlier (see Eq. 9.11 and Fig. 9.10).

The parameters u_3 and z (in Eq. 9.13) characterize the position of the reinforcing bar and α is the angle of the web (in degrees) as shown in Fig. 9.11. The parameter z is evaluated as

$$\frac{1}{z} = \frac{1}{\sqrt{u_1}} + \frac{1}{\sqrt{u_2}} + \frac{1}{\sqrt{u_3}} \tag{9.13}$$

where u_1, u_2 are the shortest distances of the center of the reinforcement bar to any point of the webs of the steel sheet; u_3 is the distance of the center of the reinforcement bar to the lower flange of the steel sheet; l_1,

Concrete	Fire Resistance (min)	c_0 (°C)	c_1 (°C)	c_2 (°C)·mm$^{0.5}$	c_3 (°C)·mm	c_4 (°C/°)	c_5 (°C)·mm
Normal-weight concrete	60	1191	-250	-240	-5.01	1.04	-925
	90	1342	-256	-235	-530	1.39	-1267
	120	1387	-238	-227	-4.79	1.68	-1326
Lightweight concrete	30	809	-135	-243	-0.70	0.48	-315
	60	1336	-242	-292	-6.11	1.63	-900
	90	1381	-240	-269	-5.46	2.24	-918
	120	1397	-230	-253	-4.44	2.47	-906

*For the mechanical analysis it may be assumed that for concrete temperatures below 250°C, Eurocode 4 assumes no strength reduction of concrete is considered.

Table 9.10 Coefficients for Determination of Temperature in Steel Reinforcement in Ribs of Composite Deck* (*Reproduced from CEN*)

l_2, and l_3 are the specific dimensions of the reentrant steel sheet profile or the trapezoidal steel profile; h_1 is the height of the upper concrete part situated on the reentrant steel sheet profile or on the trapezoidal steel profile; h_2 is the height of the bottom concrete part situated on the reentrant steel sheet profile or on the trapezoidal steel profile; and z is the indication of the position in the rib.

Knowing temperatures in various components, and thus temperature-induced reduction in strength properties of steel deck and rebars, the reduced moment capacity of the composite slab can be evaluated by applying strength equations. No reduction to compressive strength of concrete is to be applied, since temperature rise in the concrete (in compression zone layers) will be small enough, for fire exposure time of 2 hours or less. The rationale for this is the fact that upper compression layers of slab are unlikely to experience temperature rise beyond 300 to 400°C and thus any loss in compressive strength of concrete is minimal. As a result, the sagging moment resistance at the midspan section depends mainly on the amount of tensile reinforcement and its temperature level. With the knowledge of temperature in steel rebars, reduction in strength properties of steel can be arrived at using temperature-strength relations discussed in Chaps. 7 and 8.[3] Thus, flexural capacity at a given exposure duration (M_f) is given as

$$M_{fi,t,Rd} = \sum_{i=1}^{n} A_i z_i k_{y,\theta,i}\left(\frac{f_{y,i}}{\gamma_{M,fi}}\right) + \alpha_{slab}\sum_{j=1}^{m} A_j z_j k_{c,\theta,j}\left(\frac{f_{c,j}}{\gamma_{M,fi,c}}\right) \tag{9.14}$$

where $f_{y,i}$ is the nominal yield strength for the elemental (i.e., bottom or top flanges and web) steel area, A_i, $f_{c,j}$ is the design strength for the elemental concrete area A_j at 20°C (and concrete parts tension is ignored), z is the distance from the plastic neutral axis to the centroid of the elemental area A_i, or A_j. $k_{y,\theta,i}$ and $k_{c,\theta,j}$ are strength reduction factors for concrete and steel at a given temperature as per Eurocode 2 and Eurocode 3, respectively. $\gamma_{M,fi,c}$ is the partial factor for the strength of concrete in the fire situation. A comparison between the above reduced moment capacity against the bending moment due to applied loading under fire conditions will determine the failure states of the composite slab.

Composite Slabs with Hogging Moment When composite slabs are subjected to hogging moment, as in the case of support regions, slight modifications to the above equations are needed. In such a case, at the hogging moment zone the bottom portion of the deck which is directly exposed to fire will be in compression, while the top layers of concrete slab will be in tension. Eurocode 4 states that for evaluating the hogging moment capacity at the support section of a composite slab, the contribution of steel deck to compressive force can be ignored, as the temperature in deck will reach high values quickly. The hogging moment resistance of the slab is calculated by considering a reduced cross section. The parts of the bottom

[3] In the case of calculation of sagging moment, Eurocode 4 does not specify if steel deck contributes to the sectional capacity of a composite slab. In general, the contribution of the steel deck is about 5% of total moment capacity.

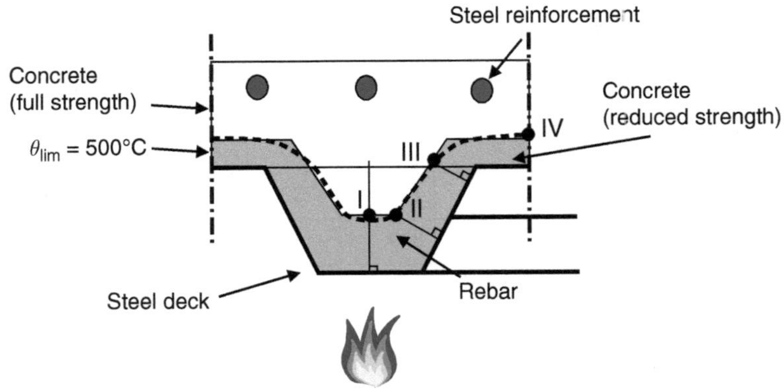

FIGURE 9.12 Isotherms in the hogging moment region of a composite deck slab. *(Reproduced from CEN.)*

portion concrete section, with temperatures beyond a certain limiting temperature, θ_{lim}, are neglected. The remaining bottom portion of concrete section is considered to be fully effective as in the case of room temperature conditions. Similar to the isotherm method discussed in Chap. 8, bottom layers of concrete section with temperatures less than a limiting value (θ_{lim}) is considered to be contributing fully (at room temperature strength) to the moment capacity. As illustrated in Fig. 9.12, concrete outside (below) the limiting temperature isotherm, approximated by four points (I, II, III, and IV), can be neglected and the remaining concrete is considered to be contributing full strength to compressive force as at 20°C. The geometric points I, II, III, and IV are as defined in Eurocode 4.[4] The tension forces in this case are resisted by reinforcement present in the upper layers of slab.

The equation for evaluating limiting temperature isotherm points, θ_{lim}, in concrete slab is given as

$$\theta_{lim} = d_0 + d_1 N_s + d_2 \frac{A}{L_r} + d_3 \Phi + d_4 \frac{1}{l_3} \tag{9.15}$$

where N_s is the normal force in the hogging reinforcement and d_i are coefficients for the determination of the limiting temperature and are as given in Table 9.11. Φ is the shadow factor and can be calculated as in Eq. 9.12.

By knowing the location of I, II, III, and IV points, effective sectional area, the hogging moment capacity at end supports of a composite (profiled) slab can be evaluated in a similar manner to that of a flat slab having an effective depth equal to h_{eff}, measured from the top of the slab. The temperature profile in concrete is given by the isotherm values as in Table 9.11, where the depth is measured from the bottom of the equivalent deck. Where appropriate, strength of steel reinforcement should be reduced and the rebars temperature can be taken equal to concrete temperature at the level of the reinforcement. The moment capacity can then be calculated using the conventional design approach, but with an iterative procedure to calculate the depth of the neutral axis at each step similar to that shown in Chap. 8 for a concrete slab. Then, the failure of the slab, as per the strength limit state, can be evaluated by comparing the reduced moment capacity against the bending moment arising from applied loading present during fire exposure.

Eurocode 4 Approach for Composite Beams

Composite beams, similar to composite slabs, need to satisfy three criteria to achieve a certain level of fire resistance: integrity, insulation, and strength. The integrity criterion is said to be automatically satisfied if the slab is designed as per Eurocode 4—Part 1 procedure. For satisfying the strength criterion, Eurocode 4 lays out two approaches, one based on limiting temperature criteria and the other based on moment capacity limit state. In both cases, for simplicity, sectional temperatures are evaluated by considering only

[4] Point I is situated at the central line of the rib, at a distance from the lower flange of the steel sheet and calculated as a function of the limiting temperature.

Point II is situated on a line through point I, parallel to the lower flange of the steel sheet, at a distance from the web of the steel sheet, equal to that from the lower flange.

Point III is situated on a line through the upper flange of the steel sheet, at a distance from the web of the steel sheet, equal to the distance of point IV to the upper flange.

Point IV is situated at the central line between two ribs, at a distance from the upper flange of the steel sheet, calculated as a function of the limiting temperature.

Concrete	Fire Resistance (min)	d_0 (°C)	d_1 (°C) · N	d_2 (°C) · mm	d_3 (°C)	d_4 (°C) · mm
Normal-weight concrete	60	867	$-1.9 \cdot 10^{-4}$	−8.75	−123	−1378
	90	1055	$-2.2 \cdot 10^{-4}$	−9.91	−154	−1990
	120	1144	$-2.2 \cdot 10^{-4}$	−9.71	−166	−2155
Lightweight concrete	30	524	$-1.6 \cdot 10^{-4}$	−3.43	−80	−392
	60	1030	$-2.6 \cdot 10^{-4}$	−10.95	−181	−1834
	90	1159	$-2.5 \cdot 10^{-4}$	−10.88	−208	−2233
	120	1213	$-2.5 \cdot 10^{-4}$	−10.09	−214	−2320

TABLE 9.11 Coefficients for Determining Limiting Temperature in Hogging Moment Region of Normal-Weight and Lightweight Concrete Slab (*Reproduced from CEN*)

isolated steel beams and without due consideration of thermal resistance of concrete, and thus development of thermal gradients is neglected.

Critical Temperature Method This simplified method is applicable for simply supported composite beams subjected to sagging moment and when the beam depth is less than 500 mm and the concrete slab depth h_c is greater than 120 mm. This method assumes uniform steel temperature across the beam depth and is allowed for simply supported beams made of hot-rolled sections. Thus, the strength failure limit state is said to be attained when moment capacity ($R_{fi,d,t}$) at the critical section of the beam decreases to the level of applied bending moment, $E_{fi,d,t}$, which is said to occur when steel (shape) reaches its critical temperature. For simplicity, temperature rise in the entire steel section is assumed to be the same as the temperature at lower flange.

The allowable critical temperature in steel, θ_{cr}, is determined as a function of level of applied loading, $\eta_{fi,t}$, and available steel strength, $f_{ay,\theta cr}$. This critical temperature for simply supported beams can be reduced by 10% if fire exposure (standard fire) is expected to be less than 30 minutes.

$$0.9\eta_{fi,t} = f_{ay,\theta cr}/f_{ay} \quad \text{(for } R = 30 \text{ minutes)} \tag{9.16}$$

and if the beam is subjected to standard fire exceeding 30 minutes, then the critical temperature is calculated as

$$\eta_{fi,t} = f_{ay,\theta cr}/f_{ay} \tag{9.17}$$

where f_{ay} is the ambient yield or characteristic strength, and the load level $\eta_{fi,t}$ is given as the ratio of bending moment present during fire exposure ($E_{fi,d,t}$) to that of room temperature moment capacity (R_d) of beam.

$$\eta_{fi,t} = \frac{E_{fi,d,t}}{R_d} \tag{9.18}$$

Moment Resistance Method This method evaluates reduced sectional moment capacity of the beam at a given fire exposure time and then compares it against applied bending moment present during fire exposure to evaluate failure time. At a given fire exposure time, the reduced tensile force, T, is equated to reduced compressive force, C, shown in Fig. 9.13, to evaluate neutral axis depth (h_u) and the associated lever arm ($y_F - y_T$). Then, the reduced moment capacity is calculated as

$$M_{fi,Rd^+} = T(y_F - y_T) \tag{9.19}$$

The above moment capacity calculation is made assuming the web and both flanges of steel profile are at uniform temperatures and the compression zone of the concrete slab to be unaffected by any minimal temperature raise. Eurocode 4 gives simple equations for evaluating overall steel sectional temperatures, as well as temperature in individual flanges and web under both cases of protected and unprotected sections. To determine temperature, $\Delta\theta_{a,t}$, for the whole section, without any fire protection, or with contour fire protection, the section factor, A_m/V, is to be evaluated first. Then the equations for determining steel temperature are given as

For unprotected sections:

$$\Delta\theta_{a,t} = k_{sh}\frac{1}{c_a\rho_a}\frac{A_m}{V}h_{net,d}\Delta t \tag{9.20}$$

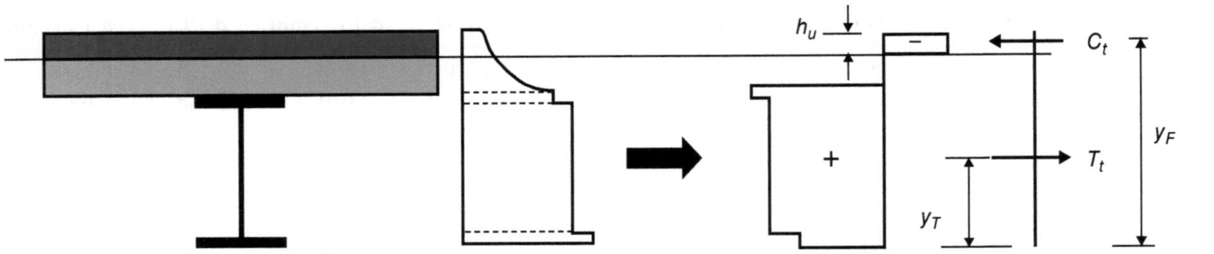

FIGURE 9.13 Calculation of moment capacity in composite beams at a given fire exposure time. (*Peproduced from CEN.*)

where k_{sh} = Correction factor for the shadow effect

$\frac{A_m}{V}$ = Section factor for unprotected steel members (1/m)

A_m = Surface area of the member per unit length (m^2/m)

V = Volume of the member per unit length (m^3/m)

c_α = Specific heat of steel (J/kg·K)

ρ_α = Unit mass of steel (kg/m^3)

$h_{net,d}$ = Design value of the net heat flux per unit area (W/m^2)

Δ_t = Time interval (seconds)

For protected sections:

$$\Delta\theta_{a.t} = \frac{\lambda_p / d_p \, A_p}{c_a \rho_a \quad V}\left(\frac{1}{1+\phi/3}\right)\left(\theta_{g.t} - \theta_{a.t}\right)\Delta t - \left(e^{\phi/10} - 1\right)\Delta\theta_{g.t} \tag{9.21}$$

with $\phi = \frac{c_p \rho_p}{c_s \rho_s} d_p \frac{A_p}{V}$

where c_p = Specific heat of fire protection material (J/kg·K)

ρ_p = Unit mass of fire protection material (kg/m^3)

λ_p = Thermal conductivity of fire protection material (W/m·K)

$\theta_{a.t}$ = Steel temperature at time t (°C)

$\Delta\theta_{g.t}$ = Ambient gas temperature at time t (°C)

In case of individual steel plates (flanges and web), temperature rise can be calculated by modifying the A_m/V factor of the whole section with that of individual plates, and then the reduced strength of steel in each steel component (plate) at that temperature level can be evaluated. Knowing the reduced strength and area, the tensile force, T_f, developed in the steel beam, as shown in Fig. 9.13, can be calculated as

$$T_f = (1/\gamma_{M,fi,a})\,(f_{ay,\theta 1}\, b_1\, e_1 + f_{ay,\theta w}\, h_w\, e_w + f_{ay,\theta 2}\, b_2\, e_2) \tag{9.22}$$

where $f_{ay,\theta 2}$, $f_{ay,\theta w}$, and $f_{ay,\theta 1}$ are the temperature-dependent stresses in steel at the top flange, web, and bottom flange, respectively, $\gamma_{M,fi,a}$ is the materials partial safety factor applied to the steel section, h_w and e_w are the height and thickness of the web, respectively, and e_1 and e_2 are thicknesses of the bottom and top flanges, respectively.

The tensile force T_t acts at a depth y_T (as shown in Fig. 9.13) from the bottom of the beam, which is calculated as

$$y_T = \frac{f_{ay,\theta 1}\dfrac{b_1 e_1^2}{2} + f_{ay,\theta w} b_w e_w\left(e_1 + \dfrac{h_w}{2}\right) + f_{ay,\theta 2} b_2 e_2\left(h - \dfrac{e_2}{2}\right)}{T_{\gamma M,fi,a}} \tag{9.23}$$

where h is the overall depth of the beam.

However, development of tensile force T_f is limited to the condition when

$$T_f \leq NP_{fi,Rd} \tag{9.24}$$

Depth x (mm)	Temperature θ_c(°C) after a Fire Duration in Minutes of					
	30 minutes	60 minutes	90 minutes	120 minutes	180 minutes	240 minutes
5	535	705				
10	470	642	738			
15	415	581	681	754		
20	350	525	627	697		
25	300	469	571	642	738	
30	250	421	519	591	689	740
35	210	374	473	542	635	700
40	180	327	428	493	590	670
45	160	289	387	454	549	645
50	140	250	345	415	508	550
55	125	200	294	369	469	520
60	110	175	271	342	430	495
80	80	140	220	270	330	395
100	60	100	210	210	260	305

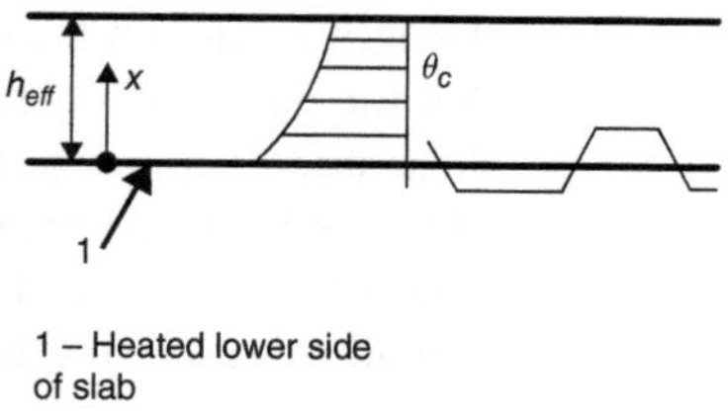

Table 9.12 Temperature Distribution in a Solid Slab of 100-mm Thickness Composed of Normal-Weight Concrete (*Reproduced from CEN*)

where N is the number of shear connectors in any critical length of the beam and $P_{fi,Rd}$ is the temperature-reduced capacity of the shear connectors.

The tensile force T_f is resisted by an equal force developed in the compression block of concrete, where the depth of the concrete stress block h_u is calculated as

$$h_u = T_f/(b_{eff} f_{ck}/\gamma_{M,fi,a}) \tag{9.25}$$

where b_{eff} is the effective width of the concrete slab taken as the ambient design condition and f_{ck} is the ambient strength of the concrete.

- If $h_c - h_u \geq h_{cr}$, temperature of concrete is below 250°C and concrete may be assumed to have full strength.
- If $h_c - h_u < h_{cr}$, then some layers of the concrete are at temperatures greater than 250°C.

T_f is calculated using iterative procedure using 10-mm-thick layers in the temperature-affected zone, and T_f is then given by

$$T_f = (1/\gamma_{M,fi,a})(b_{eff}(h_c - h_{cr}) f_{ck} + \sum_{i=1}^{n-1} 10 b_{eff} f_{c,\theta i} + h_{u,n}\, b_{eff} f_{ck,\theta n} \tag{9.26}$$

where $h_u = (h_c - h_{cr}) + 10\,(n - 2) + h_{u,n}$
in which n is the total number of layers of concrete including the top concrete layer with a temperature less than 250°C. The position of h_{cr} and the elemental temperatures is then taken from tables in Eurocode 4 (see a sample as in Table 9.12). The point of application of the compression force y_F will be at

$$y_F = h + h_c - \frac{h_u}{2} \tag{9.27}$$

and the reduced moment capacity of the section $M_{fi,Rd}$ is calculated as

$$M_{fi,Rd} = T\left(h + h_c - y_T - \frac{h_u}{2}\right) \tag{9.28}$$

This reduced moment capacity at a given fire exposure time is to be compared against applied bending moment under fire conditions to check if the failure has occurred at that time and then to evaluate fire resistance of the composite beam.

The horizontal shear force is also to be checked against shear stud capacity ($P_{fi,Rd}$), and this stud capacity is to be taken as the lesser of

$$P_{fi,Rd} = \begin{cases} 0.8k_{u,\theta}P_{Rd}Rd \\ k_{c,\theta}P_{Rd}Rd \end{cases} \tag{9.29}$$

where $k_{u,\theta}$ is the temperature-induced reduction in the ultimate strength of steel and $k_{c,\theta}$ is the concrete strength reduction factor. The values of P_{Rd} should be calculated in accordance with Eurocode 4, except that the partial safety factor γ_v should be replaced by $\gamma_{M,fi,v}$. Eurocode 4 states that temperature θ_v in the shear studs, Q_v, may be taken as 0.8 of the temperature in the top flange of the steel beam. Eurocode 4 also allows estimating temperature in the concrete, θ_c, as 0.4 of that in the top flange. It is worth noting that Annex F to Eurocode 4 provides a procedure for calculating sagging and hogging moment resistances of a partially encased steel beam connected to a concrete slab under fire exposure conditions.

Eurocode 4 Approach for Composite Columns

Eurocode 4 provides a separate methodology for the calculation of fire resistance of two types of composite columns: steel columns with partial or full concrete encasement and concrete-filled hollow structural section columns. In both cases, fire resistance is evaluated by comparing reduced sectional capacity, with buckling assumed to occur about the minor axis, against applied loading present during fire exposure. This method is only applicable to braced frames when fire exposure is occurring over a single story and when fire-exposed columns are fully connected to the colder columns located below and above fire-affected stories as shown in Fig. 9.14.

Concrete-Filled HSS Columns For calculating axial load capacity from various components, the cross section of a composite column is divided into various parts. These are denoted "a" for the steel profile, "s" for the reinforcing bars, and "c" for the concrete. Thus, the reduced design value of the plastic resistance to axial compression at a given fire exposure time is given as

$$N_{fi,pl.Rd} = \sum_j \left(A_{a,\theta} f_{a\max,\theta}\right) / \gamma_{M,fi,a} + \sum_k \left(A_{c,\theta} f_{c,\theta}\right) / \gamma_{M,fi,c} + \sum_m \left(A_{s,\theta} f_{s\max,\theta}\right) / \gamma_{M,fi,s} \tag{9.30}$$

where $A_{i,\theta}$ is the area of each element, i, of the cross section to which is attributed a certain temperature θ.

Then, effective flexural stiffness of the composite column is calculated as

$$(EI)_{fi,\text{eff}} = \sum_j \left(\phi_{a,\theta} E_{a,\theta} I_{a,\theta}\right) + \sum_k \left(\phi_{s,\theta} E_{s,\theta} I_{s,\theta}\right) + \sum_m \left(\phi_{c,\theta} E_{c,\sec,\theta} I_{c,\theta}\right) \tag{9.31}$$

where I_a, I_c, and I_s are second moments of area of the steel section, concrete, and reinforcement, and $E_{a,\theta}$, $E_{c,sec,\theta}$, and $E_{s,\theta}$ the tangent moduli calculated at the stress level σ and temperature level θ for each of the constituent materials, and $\phi_{i,\theta}$ is the reduction coefficient depending on the effect of thermal stresses.

The critical buckling capacity of the composite column is evaluated by applying the buckling equations as

$$N_{fi,cr} = \pi^2 (EI)_{fi,eff} / \ell_\theta^2 \tag{9.32}$$

where l_θ is the buckling length of the column in the fire situation.

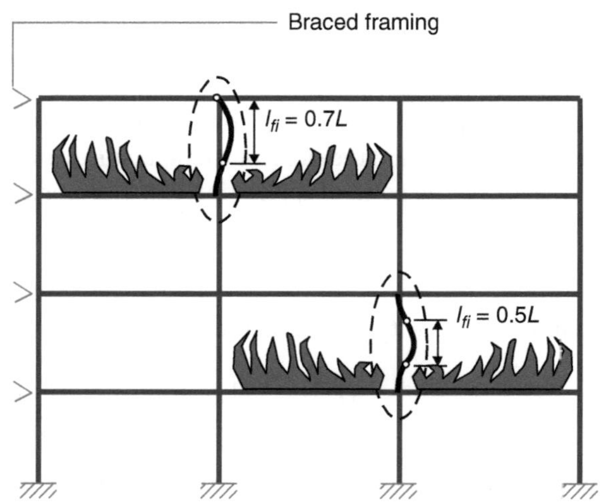

FIGURE 9.14 Effective length calculations in fire-exposed columns at different stories of a braced steel frame.

The buckling capacity of the composite column is then compared against reduced plastic axial capacity, $N_{fi,pl,Rd}$, which can be evaluated as

$$N_{fi,pl,Rd} = A_a \frac{\sigma_{c,\theta}}{\gamma_{M,fi,a}} + A_c \frac{\sigma_{c,\theta}}{\gamma_{M,fi,c}} + A_s \frac{\sigma_{s,\theta}}{\gamma_{M,fi,s}} \tag{9.33}$$

where A_a, A_c, and A_s are the respective areas of the steel section, concrete, and reinforcement, and $\sigma_{a,\theta}$, $\sigma_{c,\theta}$, and $\sigma_{s,\theta}$ are the reduced stresses in the steel, concrete, and reinforcement, respectively, due to a rise in temperature. If the loading is eccentric, the eccentricity should not exceed half of the least dimension of the section.

The smaller of these two values, Euler buckling capacity ($N_{fi,cr}$) or design value of the plastic resistance ($N_{fi,pl,Rd}$), is considered to be the axial capacity of the composite column, and this is to be compared against the applied load acting on the column to evaluate the failure state of the column.

In the case of a continuous column, over multiple stories, where local buckling does not occur, the buckling length may be taken as 0.55 times the actual length of the column (Newman et al., 2000). In case local buckling is anticipated, then its effect needs to be included through a buckling factor of 0.75. This buckling factor is to be increased to 0.8 at the topmost story. In the case of a composite column subjected to an eccentric loading, then an *equivalent axial capacity*, N_{equ}, is computed as

$$N_{equ} = \frac{N_{fi,Sd}}{\phi_s \phi_\delta} \tag{9.34}$$

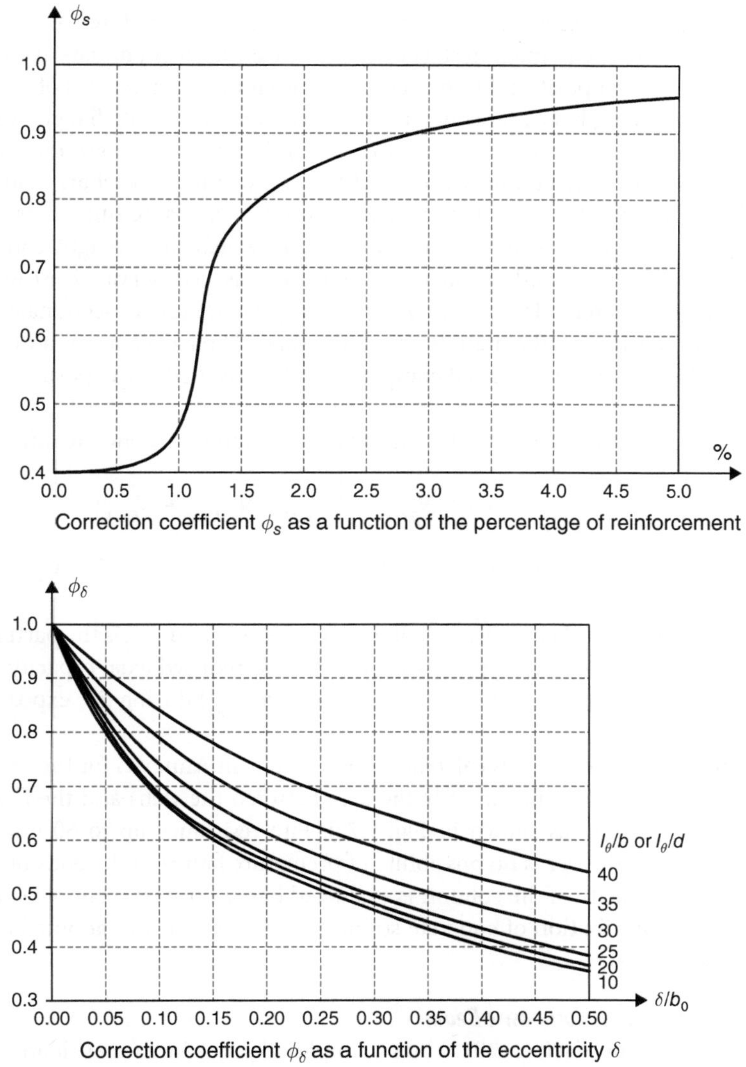

Figure 9.15 Charts with correlation coefficients ϕ_s and ϕ_δ for evaluating equivalent axial capacity in a composite column subjected to eccentric loading. (*Reproduced from CEN.*)

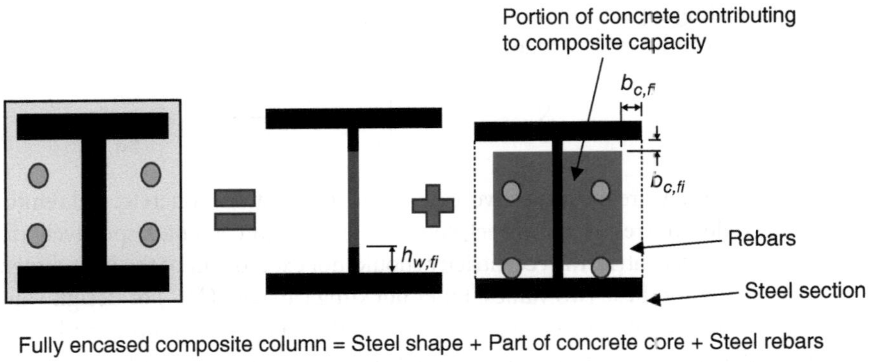

FIGURE 9.16 Configuration of various zones contributing to capacity in a fully encased steel column. (*Reproduced from CEN.*)

where $N_{fi,Sd}$ is the design value of the axial load in the fire situation, ϕ_s accounts for the effect of reinforcement (rebars) within the concrete section (with no reinforcement $\phi_s = 0.4$), and ϕ_δ allows for the effect of eccentricity. Values of these two correction factors, ϕ_s and ϕ_δ, can be obtained from Fig. 9.15 (as given in Eurocode 4), respectively.

Concrete-Encased Steel Columns Eurocode 4 lists detailed procedures for evaluating fire resistance of two types of concrete encasement in steel columns: full encasement or partial encasement. The cross section of a fully encased composite column is grouped into four zones: flanges of the steel section, web of the steel section, reinforcing bars, and concrete infill (see Fig. 9.16). Only flanges and web of steel section, rebars, and parts of concrete core (as shown in Fig. 9.16) contribute to sectional capacity during fire exposure. The temperature distribution within steel flanges, within steel rebars, and within inner sides of web up to a distance of $h_{w,fi}$ (see Fig. 9.16*b*), at any given fire exposure time, is assumed to be uniform. Another assumption is made relating to concrete infill, in which the strength contributions from outer parts of concrete, $b_{c,fi}$, are neglected for capacity calculations as a conservative estimate. This is due to the fact that outer parts of concrete, closer to fire zone, will experience higher sectional temperatures. The strength and modulus properties in all materials within the shaded area decrease as a function of temperature.

The plastic resistance to axial compression at a specified fire exposure time is taken as

$$\text{Plastic capacity} = \text{Capacity of \{Steel section + Concrete infill + Rebars\}}$$

$$N_{fi,pl.Rd} = N_{fi,pl.Rd} + N_{fi,pl.Rd,c} + N_{fi,pl.Rd,s} \tag{9.35}$$

$$N_{fi,pl.Rd} = \Sigma_j(A_{a,\theta} f_{amax,\theta})/\gamma_{M,fi,a} + \Sigma_m(A_{c,\theta} f_{c,\theta})/\gamma_{M,fi,c} + \Sigma_k(A_{s,\theta} f_{smax,\theta})/\gamma_{M,fi,s} \tag{9.36}$$

where $A_{i,\theta}$ is the area of each element of the cross section and $\gamma_{M,fi}$ is the partial factor for respective material strength under elevated temperature exposure. This reduced axial capacity $N_{fi,pl.Rd}$ is to be checked against applied axial load acting on the composite column (N_{Rd}) during fire exposure and failure is said to occur when $N_{fi,pl.Rd} < N_{Rd}$.

Eurocode 4 also provides calculation procedures in Annex G for the case of steel columns with partial concrete encasement, wherein only the area between the webs and the flanges of a W-shaped section is filled with concrete (as shown in Fig. 9.7*c*). Fire resistance up to 60 minutes can be achieved in such columns provided there is no possibility of premature failure at the ends of the column and that the room temperature capacity of the column is evaluated based on steel section capacity alone without taking into account the contribution of concrete strength. This method was adopted based on the work of Newman and Lawson (1991).

AISC Methodology for Composite Beams

The recent edition of AISC (2017) lists a simplified procedure for evaluating fire resistance of a composite beam (or composite beam-slab assembly). The reduced flexural capacity at a given fire exposure time is taken as the sum of the strength contributions from steel beams and concrete slabs. This flexural capacity depends on the level of composite action that develops between concrete slab and steel beam, which is

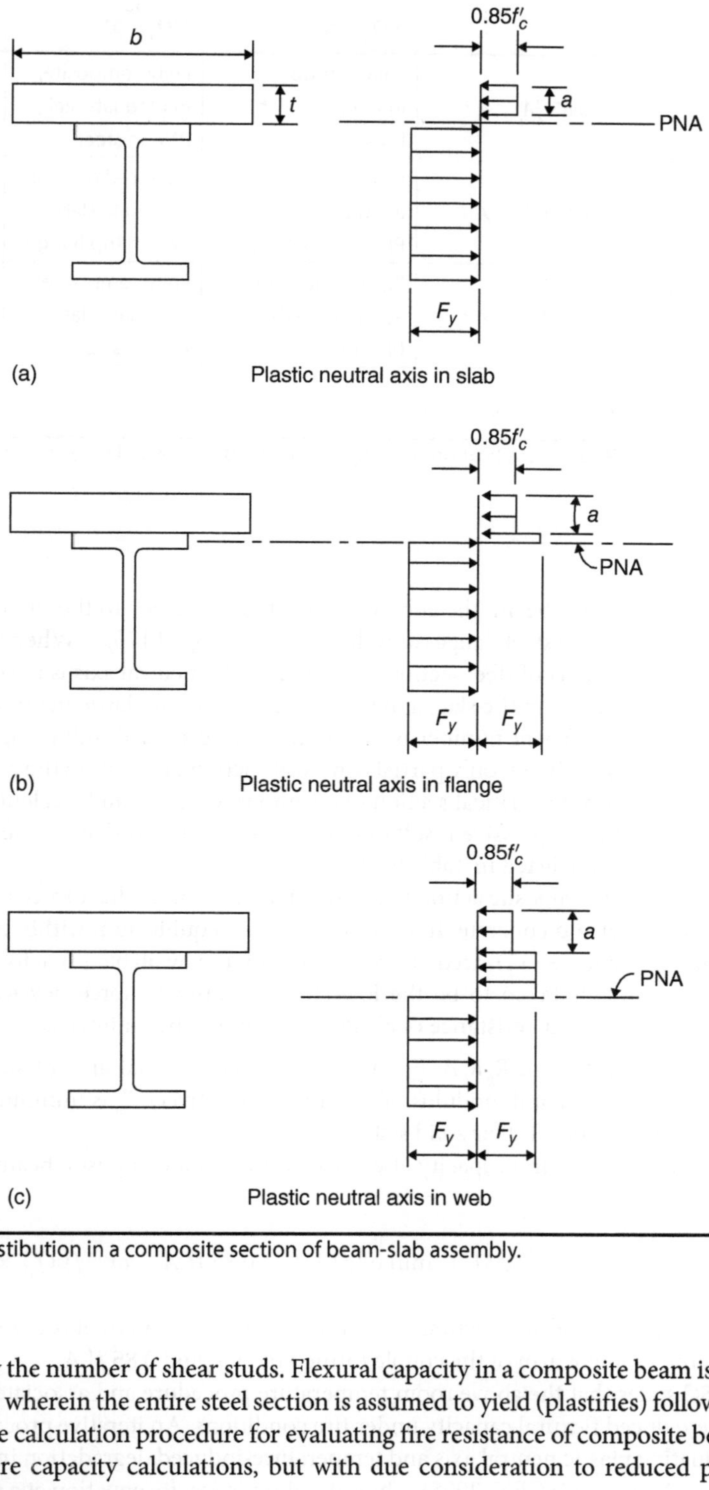

FIGURE 9.17 Stress distibution in a composite section of beam-slab assembly.

mainly governed by the number of shear studs. Flexural capacity in a composite beam is derived, similar to that in a steel beam, wherein the entire steel section is assumed to yield (plastifies) followed by the crushing of concrete slab. The calculation procedure for evaluating fire resistance of composite beam is an extension of room temperature capacity calculations, but with due consideration to reduced properties resulting from elevated temperature exposure.

The distribution of internal stresses at critical sections of a composite beam under fire exposure can follow one of the three distribution profiles depending on the extent of composite action developing at the concrete slab-steel beam interface as shown in Fig. 9.17 (and Table 9.12). In these three cases, concrete is assumed to develop a uniform compressive stress distribution at 0.85 f_c', where f_c' is the compressive strength of concrete. This stress distribution, referred to as the Whitney equivalent stress block, is a simplified version of average stress block. The Whitney stress block has a resultant that matches actual stress distribution and extends from the top of the slab to a depth equal to or less than the thickness of slab.

For the case of a composite beam-slab assembly shown in Fig. 9.17*a*, the plastic neutral axis lies in the concrete slab and the corresponding stress distribution resembles tensile yielding in the steel section (beam) and compression resistance developing in the concrete slab. If the plastic neutral axis lies in the

		$\Sigma Q_n > S^*$	$\Sigma Q_n = S^*$	$\Sigma Q_n < S^*$
I.	$0.85f'_cA_c < A_sf_y$	Fully composite; inadequate slab; PNA in steel	Fully composite; inadequate slab; PNA in steel	Partially composite; inadequate slab; PNA in steel
II.	$0.85f'_cA_c = A_sf_y$	Fully composite; adequate slab; PNA in top flange	Balanced design; adequate slab; PNA in top flange	Partially composite; adequate slab; PNA in steel
III.	$0.85f'_cA_c > A_sf_y$	Fully composite; adequate slab; PNA in slab	Fully composite; adequate slab; PNA in slab	Partially composite; adequate slab; PNA in steel

$S^* = Q_n$ is capacity of shear studs.

Table 9.13 Classification of Composite Beam Based on Relative Strength of Its Components

flange or web of steel beam, the concrete stress block extends to the full depth of the slab and the plastic neutral axis lies in the steel flange (or web as shown in Fig. 9.17*b*,*c*). When the plastic neutral axis lies in the flange or web, the part of steel section above the plastic neutral axis is in compression, which supplements the compression force in the slab, and the portion below all plastic neutral axis resists tension force. In all three cases, if a sufficient number of shear studs are provided, full composite action can develop in the beam-slab assembly. If not, only partial composite action develops in the composite assembly. The nominal moment capacity at the critical section of a composite beam can be calculated by computing the moment of the tensile and compressive resultants. The three scenarios that can develop depending on the level of composite action are listed in Table 9.13.

The extent of composite action is a function of horizontal shear force (V_h) that develops at the interface between the steel and concrete. In order to maintain equilibrium within the composite cross section, this horizontal shear force (V_h) needs to be resisted by the available capacity in shear studs. The capacity of these shear studs is taken to be the lowest of compressive force developed in the slab (either A_sf_y or $0.85 f'_c A_c$) or the shear resistance of all studs at the slab-beam interface (ΣQ_n), where ΣQ_n is taken as the lesser of $0.5A_{sc}\sqrt{f'_cE_c}$ or $R_gR_bA_{sc}F_u$, A_{sc} is the cross-sectional area of shear stud, and f'_c and E_c are the compressive strength and modulus of concrete, respectively. F_u is minimum tensile strength of studs, and $R_g R_b$ are slab factors = 1.0 for solid slab.

The nominal flexural capacity at critical section of a composite beam, at ambient conditions, can be evaluated as

$$\phi M_n = \text{min of } \phi\{A_sf_y,\ \ 0.85 f'_c A_c \ \ or \ \ f'_c Q_n\} \times y \tag{9.37}$$

where A_s is the area of steel section, f'_c is the compressive strength of concrete, A_c is the area of concrete slab, and y is the moment arm of the couple formed by A_sf_y and $0.85 f'_c A_c$.

AISC states that the above room temperature procedure and associated equations can be extended to evaluate reduced flexural capacity under fire conditions. An iterative procedure is to be applied to account for evaluating plastic neutral axis and temperature-induced degradation in strength properties of steel and concrete (AISC, 2011; CEN, 2005). Thus, the above capacity equation at a given time, *t*, can be rewritten as

$$\phi_f M_{n,t} = \text{min of } \phi\{A_sf_{y,Ts},\ \ 0.85 f'_{c,Tc} A_c \ \ or \ \ \sum Q_{n,TQ}\} \times y_T \tag{9.38}$$

where $f'_{c,T}$ is the reduced compressive strength of concrete at concrete temperature, T_c, $f_{y,Ts}$ is the yield strength of steel section at temperature, T_s, and $Q_{n,TQ}$ is the shear strength of shear studs at temperature, T_{stud}.

The reduced sectional capacity is to be compared against the bending moment present during fire exposure to determine failure state of a beam. No specific equations are given in AISC to calculate sectional temperature at a fire exposure time. Such cross-sectional temperature in a composite beam (or beam-slab assembly) can be calculated through simplified approaches prescribed in Eurocode 4 or through detailed heat transfer analysis by applying finite element-based software (Kodur and Naser, 2015).

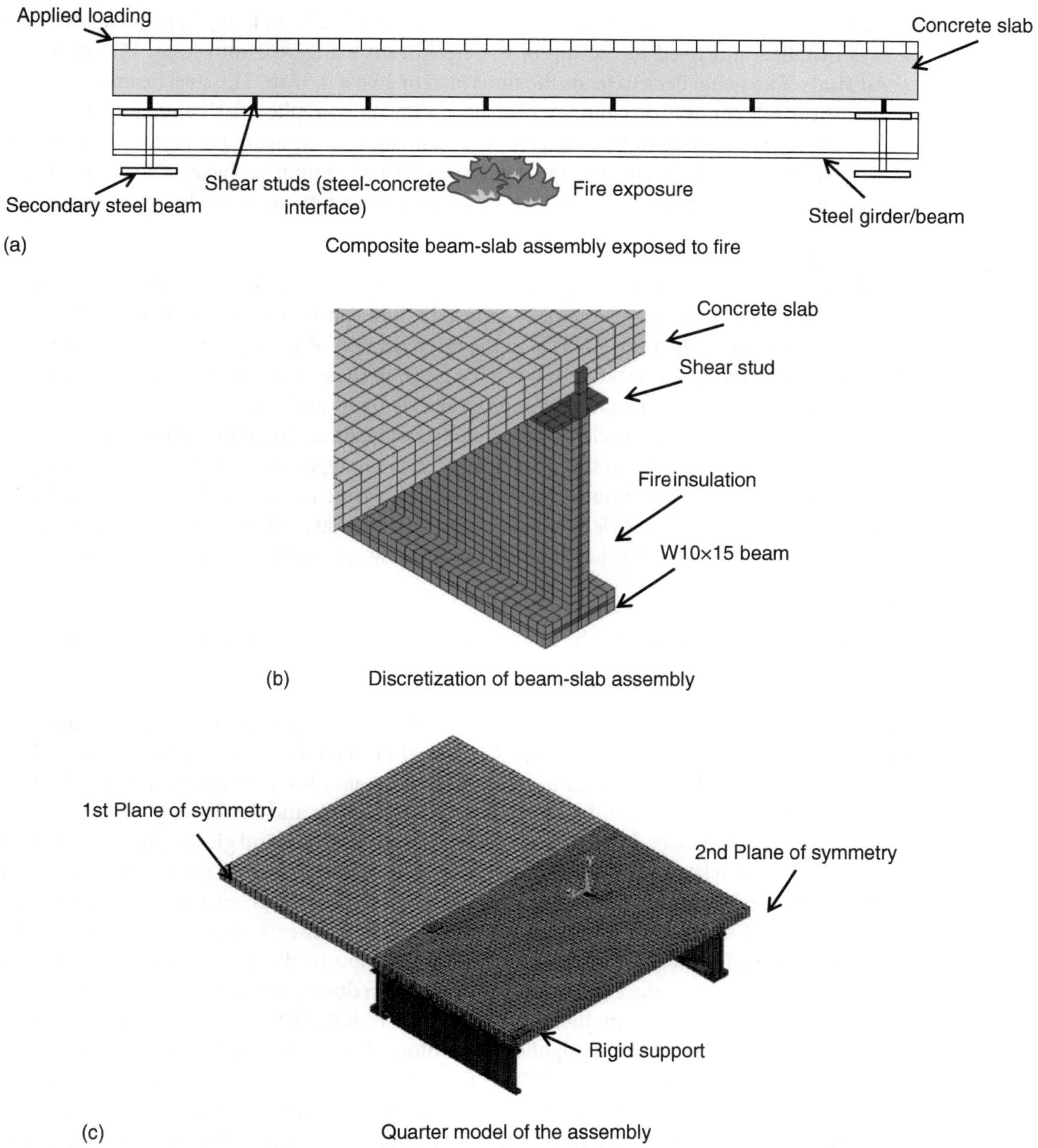

Figure 9.18 Discretization of typical composite beam-slab assembly for finite element analysis using ANSYS. (*With permission from Elsevier.*)

9.7.2 Advanced Analysis

The simplified approaches discussed in Sec. 9.7.1 can yield mostly fire resistance of composite structural members. For tracing detailed response parameters in a fire-exposed composite structure, advanced analysis is to be carried out. In such an analysis, carried out at incremental fire exposure times, all critical factors including extent of composite action developed at elevated temperatures, temperature-induced instability effects, and specific reduction in properties of shear connectors can be duly accounted for. Application of such advanced analysis for evaluating detailed fire response parameters in a composite structure is illustrated below.

Case Study

To illustrate the application of advanced analysis, fire resistance analysis of a composite beam-slab assembly system carried out using finite element-based software ANSYS is discussed here. This beam-slab assembly, shown in Fig. 9.18, is composed of metal deck-concrete slab placed on the top of two A992 steel W12×16 girders (3.96 m long) and three A992 steel W10×15 beams (2.13 m long).[5] The metal deck is 38.1 mm deep and is installed and fixed to the top of the steel beam using 76.2-mm-long, 15.9-mm-diameter headed shear studs. The metal deck extends the floor plan to 3.96 × 4.57 m. The steel beams are connected through bolted shear-tab connections and are protected with spray-applied fire insulation. This composite beam-slab assembly is assumed to be subjected to a standard fire exposure for 90 minutes and then left to cool down naturally. Full details on the analysis of this composite beam-slab assembly, together with comprehensive presentation of generated response parameters, are given by Kodur et al. (2013).

Computer Programs

Finite element-based computer programs such as ANSYS, ABAQUS, and ADINA can be applied for tracing the fire response of a composite member exposed to fire conditions. Kodur et al. (2013) carried out analysis on a beam-slab assembly using ANSYS. In the discretization of a composite structural member for advanced fire resistance analysis, special attention is to be given to shear connectors, deck profile, and boundary layers between steel and concrete to simulate appropriate elements and nodes. Also, tensile membrane action is an important mechanism that should be taken into consideration. To capture the stress transfer at the interface of steel and concrete, specific "contact" or "spring" element types are to be used in developing a finite element model. These elements can simulate the bond action between shear studs and concrete slab as well (Kodur et al., 2013). The availability of element types as well as material models varies from one finite element package to the other, and this should be kept in mind in selecting appropriate finite element program.

Analysis Details

Specific details on modeling of the above selected composite beam-slab assembly using ANSYS software are outlined in this section.

Geometry and Discretization The composite beam-slab assembly is discretized into different element types for advanced analysis as depicted in Fig. 9.18. For thermal analysis, the beam-slab assembly is discretized with SOLID70 and LINK33 elements, with thermal capabilities. For structural analysis, this assembly was discretized using SOLID65, SOLID45, LINK8, CONTA174, and TARGE170. As various components of composite beam-slab assembly, namely concrete slab, steel beams and girders, shear studs, and connections, are to be modeled, a large number of elements are needed to discretize a composite structure. In the current problem, the total number of elements required to discretize this beam-slab assembly was 89,000.

The thermal elements were converted to structural elements upon completion of thermal analysis. Such conversion (i.e., elements) should be able to capture the various high-temperature phenomena including temperature-induced bond-slip phenomenon during stress analysis. The conversion of elements was carried out as follows: for the concrete material, SOLID70 (3D thermal solid) is converted to the structural element SOLID65 (3D eight-node reinforced concrete solid). For the steel girders, connections, bolts, and insulation material, SOLID70 (3D thermal solid) element is converted to structural element SOLID45 (3D eight-node structural solid). Steel mesh reinforcement, LINK33 thermal element (3D conduction bar), is converted to LINK8 (3D two-node structural bar) for structural analysis.

In analyzing a composite member under fire conditions, the contact behavior at the bolt-connection interface is to be given special consideration since large fire-induced forces develop at these connection zones. The interface at the bolted connection was modeled using CONTA174 and TARGE170 elements. The contact regions were defined as surface-to-surface areas that only permit sliding of the adjunct faces. In addition, Coulomb's frictional law was applied to govern the amount of sliding in the connected components. The frictional model uses a coefficient of friction of 0.3 to represent the amount of friction present in the connections.

High-Temperature Material Properties and Constitutive Laws For analysis, high-temperature thermal and mechanical property relations of structural steel, concrete, and shear connectors are to be provided as input to ANSYS. The room temperature thermal and mechanical properties of concrete, steel, and fire insulation at room temperature are listed in Table 9.14. The nonlinearity in steel material is accounted for through a bilinear elastoplastic constitutive material model based on the von Mises plasticity yielding criterion. Similarly, the compressive plastic behavior of concrete is defined using an isotropic multilinear compressive stress-strain curve that varies with temperature (Kodur et al., 2013). Tensile strength of concrete, often neglected in simplified analysis, can be given due consideration

[5]Steel shapes are given in the AISC design manual.

Property	Value at Ambient Temperature		
	Steel/Shear Studs	Concrete	Insulation
Thermal conductivity (W/m·K)	51	27.3	2.5
Specific heat (J/kg·K)	450	725	1652
Density (kg/m³)	7,900	1762	832
Strength (MPa)	345/450	45.3	–
Modulus of elasticity (GPa)	210	32.3	–

Table 9.14 Properties of Various Materials That Constitute Composite Beam-Slab Assembly

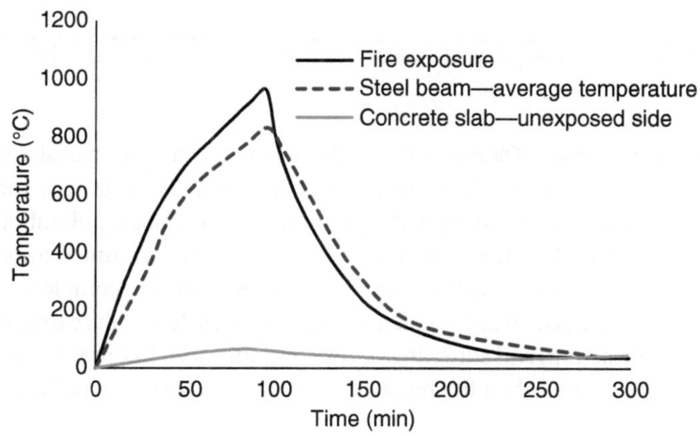

Figure 9.19 Temperature rise in beam-slab assembly as a function of fire exposure time. (*With permission from Elsevier.*)

in advanced analysis. Tensile strength of concrete is taken as $0.62\sqrt{(f_c')}$, where f_c' is the compressive strength of concrete. Once the concrete reaches its tensile rupture stress, a tensile stiffness multiplier of 0.6 is used to simulate a sudden drop of the tensile strength to 60% of the initial rupture stress. Then, the drop is followed by a linearly descending response to zero stress at a strain value of six times the rupture strain. The high-temperature property relations of concrete and steel for this analysis are as per Eurocode 4 provisions (see Chaps. 7 and 8 for these relations).

The bond-slip phenomenon between concrete slab and steel beam, modeled through spring elements, can be accounted for in advanced analysis through bond property relations prescribed in Sec. 9.3. Since fire insulation has significantly lower mechanical strength and stiffness, the strength contribution to capacity of the composite member from the fire insulation is neglected. However, thermal properties of insulation play a significant role in temperature development, and these are accounted for in the analysis. The thermal properties of insulation can be obtained from manufacturers' product data sheets.

Loading and Boundary Conditions Under fire conditions, a composite structure resists applied loading through contribution from concrete and steel components of the assembly. Thus, both thermal and structural loadings are applied on the composite assembly until failure of the assembly or end of fire exposure. The design fire applied in this case had a 90-minute growth phase analogous to the ASTM E119 standard fire, followed by rapid cooling in the decay phase (see Fig. 9.19 for fire exposure). The beam-slab assembly has simply supported end conditions.

Utilizing symmetry is highly important in a composite structure, as a large number of elements are needed for discretizing a structure due to the presence of different material types, as well as various components (i.e., beam, slab, studs, connections) that form the assembly. In this problem, due to the symmetric geometry, loading, and boundary conditions present in this beam-slab assembly, a quarter model of this assembly is analyzed. The main advantage of building a quarter model is the tremendous reduction in the total number of elements and thus reduced computational time needed for the analysis. Symmetrical boundary conditions were simulated by inserting vertical restraints perpendicular to the nodes of elements present in the outermost surface of the two planes of symmetry.

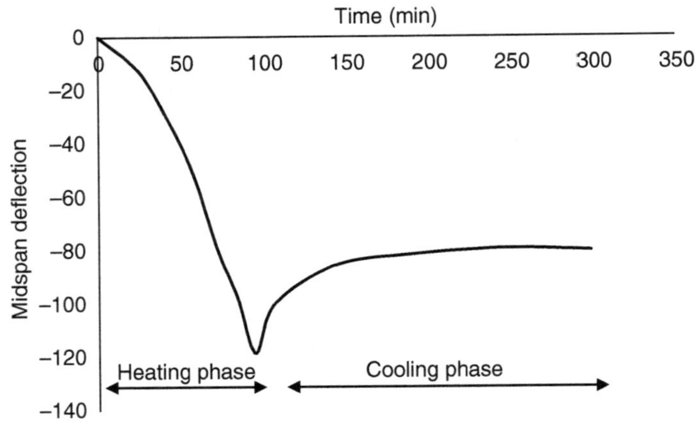

Figure 9.20 Midspan deflections of interior beam in the composite beam-slab assembly. (*With permission from Elsevier.*)

Failure Criteria and Numerical Convergence In the advanced analysis, realistic failure limit states can be applied to check the failure of the assembly at any fire exposure time. These criteria include strength, deflection, cracking and crushing of concrete, yielding of steel, failure of connections, pullout of shear studs, or detachment of slab from the steel beam. Thus, failure in a composite assembly is assumed to occur when one or a combination of the limiting criteria, as discussed in Chap. 5, is violated. Another complexity in the advanced analysis is relating to convergence, which can be a big challenge especially close to failure time due to softening effects arising from loss of composite action, and softening of concrete and steel. In order to avoid such issues, the use of elemental crack softening can smoothen fire resistance analysis in later stages of fire exposure (Kodur et al., 2013).

Results

Advanced analysis can generate large sets of response parameters at each time-step of analysis. Sectional temperatures, generated from ANSYS analysis, is plotted in Fig. 9.19 to illustrate the typical response of a composite beam-slab assembly under entire fire exposure, including cooling phase. As expected, temperature in steel beam rapidly increases within the first few minutes of fire and reaches about 800°C after 100 minutes of exposure. This is due to high thermal conductivity and low specific heat of steel, as well as due to the fact that this interior steel beam was not protected with any fire insulation. On the other hand, the unexposed surface of concrete slab experienced relatively low peak temperatures (65 to 160°C) due to high thermal inertia of concrete.

The progression of midspan deflection in the middle interior beam, as obtained in ANSYS, is plotted in Fig. 9.20. The deflection increases steadily with fire exposure time. Heating from the bottom and side surfaces of the beams and girders produced increasing deflections as a result of high-temperature gradients and temperature-induced degradation in stiffness and strength of steel. The reduction of strength and stiffness in the tension zone of the steel beam (bottom layers) causes the neutral axis to shift upward, resulting in additional deflection. It should be noted that concrete did not experience much loss of strength or stiffness, since the slab did not experience a significant rise in temperature (remained below 200°C).

Prior to failure—that is, in the last stage of fire exposure—and due to significant loss of strength and stiffness properties in steel beams, much of the forces in the steel beams get transferred into the cooler concrete slab through tensile membrane action, and this increases midspan deflection. The midspan deflection stabilizes once the cooling phase starts and the slab regains some levels of deflection (as can be seen in Fig. 9.20). Tensile membrane action is an important phenomenon in beam-slab assemblies through which an alternative load transfer mechanism develops between concrete slab and steel, and thus the load is mostly being carried by the concrete slab, especially when the composite member undergoes large deflections.

Another benefit of applying advanced analysis is that various phenomena occurring in a composite assembly under fire conditions such as degradation in composite action, connection failure, crack progression along the concrete slab, and so on can be captured. To illustrate the usefulness of such results from advanced analysis, crack progression across the concrete slab is shown in Fig. 9.21, which gives an indication of failure modes developing under fire exposure. Analysis results from ANSYS simulated a major diagonal shear crack developing in the slab, as observed in fire tests on this composite beam-slab assembly. The main shear crack spreads from the middle of the concrete slab to end supports. Another phenomenon that the advanced analysis was able to capture is failure of bolted connection. Figure 9.21*b* shows the state of the bolts after the fire test on the composite assembly; this implied the presence of

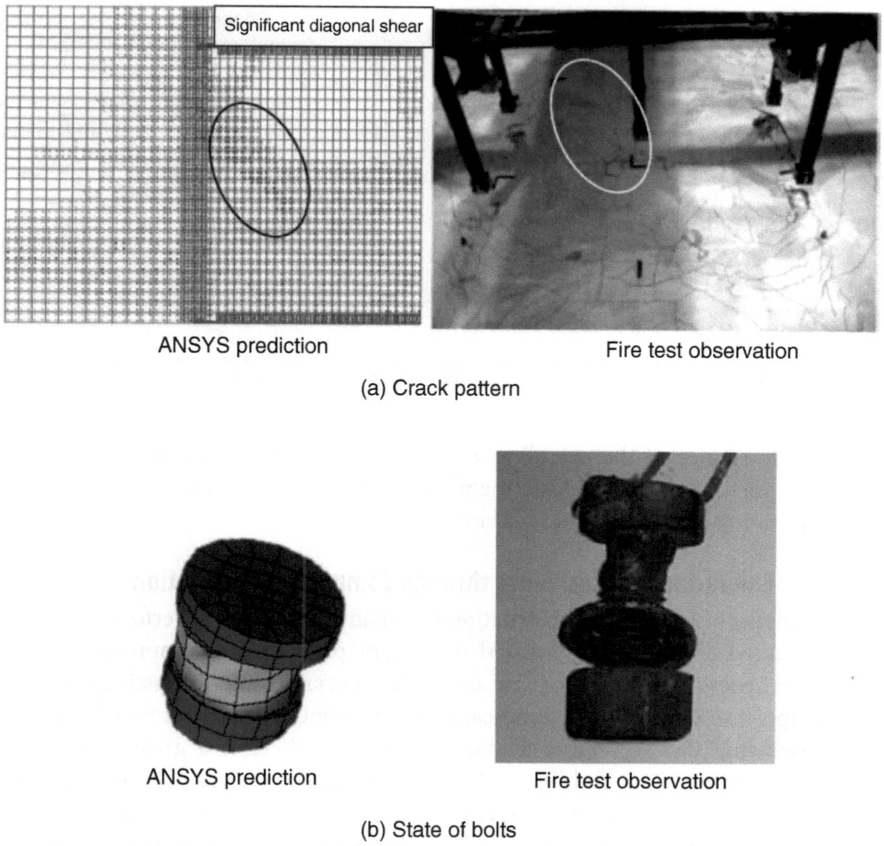

Figure 9.21 Comparison of observed and predicted structural response parameters in beam-slab assembly (left: ANSYS; right: fire test). (*With permission from Elsevier.*)

permanent deformations, an indication of reaching the yield stress in the bolts. This was also simulated in the output generated from ANSYS.

9.8 Emerging Trends and Research Needs

In the last three decades, there have been significant research efforts to effectively utilize the beneficial effects of composite construction for enhancing fire resistance of steel structures. While some of this research has translated into practice, there are still certain areas where further research and development is needed in realizing the full benefits of composite construction, especially under fire conditions.

9.8.1 Tensile Membrane Action

When a concrete slab, which is part of a composite assembly, undergoes large deflections under fire exposure, tensile membrane action can develop within the floor slab if the slab perimeter is vertically supported and horizontally restrained. The expansion and bowing that occurs under later stages of fire exposure can be restrained by surrounding areas of the cooler structure, leading to the development of large membrane forces in the slab. In this case, the slab can support additional load through tensile membrane action developing in the center of the slab and compressive membrane action, forming a supporting "ring" around the perimeter of the slab, as shown in Fig. 9.22. The development of such compressive membrane action can significantly enhance the load-carrying capacity of the slab and thus its fire resistance. Still, this enhancement in fire resistance is not accounted for in most codal provisions.

Recent studies have evaluated the behavior of composite slabs under fire conditions, specifically quantifying the extent of tensile membrane action developed in such a slab (Bailey et al., 2000). Although most of these studies were able to successfully demonstrate the positive effects of membrane action in enhancing fire resistance of composite slabs, codal provisions still do not give credence to tensile membrane action, and as a result, in most design situations, steel beams/girders that are part of a beam-slab assembly are designed with conventional fire insulation. Some of the factors that need further research include level of horizontal restraint at the slab perimeter, quantifying the level of composite action available between the steel beam and slab interface, effects of thickness of a concrete slab, and possible spalling in a concrete slab

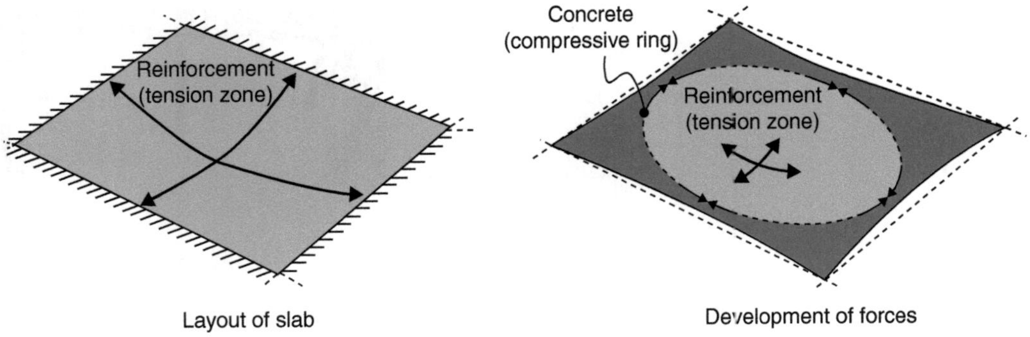

Figure 9.22 Development of tensile membrane action in a composite slab under fire conditions.

on fire resistance of the overall assembly (Bailey, 2001, 2003). Also, simplified calculation methods to account for the effect of tensile membrane action in fire resistance calculations must be developed for incorporating into codes and standards.

9.8.2 Enhancing Fire Resiliency through Composite Construction

The benefits of composite construction in enhancing structural performance at ambient conditions is well documented and is widely utilized in current practice. This improved performance is mainly due to inherent synergy between concrete and steel, especially under inelastic range of response. Another benefit of composite construction is significant improvement in fire resistance of a structural member (Fike, 2014).

Following the September 11 terrorist attacks, there is a great emphasis on enhancing structural resiliency in important buildings and critical infrastructure under extreme loading conditions. Composite construction can provide the most effective solution to achieve structural resiliency under extreme loading events, including fire and blast effects. Although composite construction is often adopted in buildings, and infrastructure, its beneficial contribution to fire resistance is not taken into consideration. Thus, steel structures that are part of composite member (e.g., concrete-filled steel columns) continue to be protected with fire insulation. This is mainly due to a lack of calculation of methods and design approaches to take into account beneficial effects of composite action in enhancing fire resistance. Therefore, there is a strong need for advancing calculation approaches that can be incorporated in design standards. Specifically, there is a need for simplified methods for evaluating fire resistance of composite shear walls, partially enchased composite beams and columns, and beam-slab assemblies.

9.9 References

ACI. *Code Requirements for Determining Fire Resistance of Concrete and Masonry Construction Assemblies.* ACI 216.1-14/TMS-0216-14. Farmington Hills, MI: American Concrete Institute, 2014.

AISC. *Steel Construction Manual.* 15th ed. Chicago, IL: 2017. https://www.aisc.org/products/publication/manuals/steel-construction-manual-15th-ed-print/

Bailey, C. G. "Efficient Arrangement of Reinforcement for Membrane Behaviour of Composite Floor Slabs in Fire Conditions." *Journal of Constructional Steel Research* 59, no. 7 (2003): 931–949.

Bailey, C. G., D. S. White, and D. B. Moore. "The Tensile Membrane Action of Unrestrained Composite Slabs Simulated under Fire Conditions." *Engineering Structures* 22, no. 12 (2000): 1583–1595.

Bailey, Colin G. "Membrane Action of Unrestrained Lightly Reinforced Concrete Slabs at Large Displacements." *Engineering Structures* 23, no. 5 (2001): 470–483.

CEN. *Eurocode 3: Design of Steel Structures—Part 1-2: General Rules—Structural Fire Design. EN 1993-1-2.* Brussels: European Committee for Standardization, 2005.

CEN. *Eurocode 4: Design of Composite Steel and Concrete Structures—Part 1-2: General Rules—Structural Fire Design. EN 1994-1-2.* Brussels: European Committee for Standardization, 2005.

CEP-FIP Model Code. *90. 1993. Model Code for Concrete Structures.* Luasanne, Switzerland: Comité Euro-International du Béton (CEB)–Fédération Internationale de la Précontrainte (FIP).

Fike, Rustin. Strategies for Enhancing the Fire Resistance of Steel Framed Structures Through Composite Construction. Michigan State University. Civil Engineering, 2010

Fike, Rustin, and Venkatesh Kodur. "Enhancing the Fire Resistance of Composite Floor Assemblies through the Use of Steel Fiber Reinforced Concrete." *Engineering Structures* 33, no. 10 (2011): 2870-2878.

Kodur, V. K. R. "Studies on the Fire Resistance of High-Strength Concrete at The National Research Council of Canada." In *International Workshop on Fire Performance of High-Strength Concrete.* Gaitherburg, MD: *NIST*, 1997, p. 75.

Kodur, V. K. R. "Performance-Based Fire Resistance Design of Concrete-Filled Steel Columns." *Journal of Constructional Steel Research* 51, no. 1 (1999): 21–36.

Kodur, V. K. R., M. Naser, P. Pakala, and A. Varma. "Modeling the Response of Composite Beam-Slab Assemblies Exposed to Fire." *Journal of Constructional Steel Research* 80 (2013): 163–173.

Kodur, V. K. R., and M. Z. Naser. "Effect of Local Instability on Capacity of Steel Beams Exposed to Fire." *Journal of Constructional Steel Research* 111 (2015): 31–42.

Kodur, Venkatesh, and Mohannad Z. Naser. "Effect of Local Instability on Fire Response of Steel Beams." *PSU Research Review* 1, no. 2 (2017): 170–179.

Kodur, Venkatesh K. R., and T. T. Lie. "Fire Performance of Concrete-filled Hollow Steel Columns." *Journal of Fire Protection Engineering* 7, no. 3 (1995): 89–97.

Lie, T. T., and Kodur, V. K. R. "Fire Protection of Hollow Steel Columns through Concrete Filling." In *Proceedings of the First CSCE Construction Speciality Conference*. Ottawa, 1995, pp. 215–224.Naser, M. Z., and V. K. R. Kodur. "Comparative Fire Behavior of Composite Girders under Flexural and Shear Loading." *Thin-Walled Structures* 116 (2017): 82–90.

NBCC. *National Building Codes of Canada*. Ottawa: National Research Council of Canada, 2015.

Newman, Gerald, M., and Robert Mark Lawson. *Fire Resistance of Composite Beams*. Berkshire, UK: Steel Construction Institute, 1991.

Newman, Gerald M., Jef T. Robinson, and Colin G. Bailey. *Fire Safe Design: A New Approach to Multi-Storey Steel-Framed Buildings*. Berkshire, UK: Steel Construction Institute, 2000.

Ollgaard, J. G., R. G. Slutter, and J. W. Fisher. "Shear Strength of Stud Connectors in Lightweight and Normal Weight Concrete." *AISC Engineering Journal* (April 1971): 71–10.

Twilt, L., R. Hass, W. Klingsch, M. Edwards, and D. Dutta. *Design Guide 4—For Structural Hollow Section Columns Exposed to Fire*. 1st ed. TÜV Verlag, ed. AISC, 1996. https://www.aisc.org/technical-resources/cidect-design-guides/.

Zhao, B., and J. Kruppa. *Fire Resistance of Composite Slabs with Profiled Steel Sheet and of Composite Steel Concrete Beams (Part 2, Composite Beams)*. Luxembourg: Directorate-General for Research and Innovation (European Commission), 1997. https://cordis.europa.eu/publication/rcn/199711697_en.html.

9.10 Numerical Examples

9.1

A circular steel column (350 × 6.4 mm) filled with concrete is exposed to standard fire. The concrete filling is reinforced with steel bars of 3% area with a 25-mm cover. The column is 5000 mm long with fixed-fixed boundary conditions and subjected to 1344 kN during fire exposure. The carbonate aggregate concrete infill has a design compressive strength of 40 MPa. Calculate the fire resistance of this column using Lie and Kodur's equation adopted by ACI 216.1.

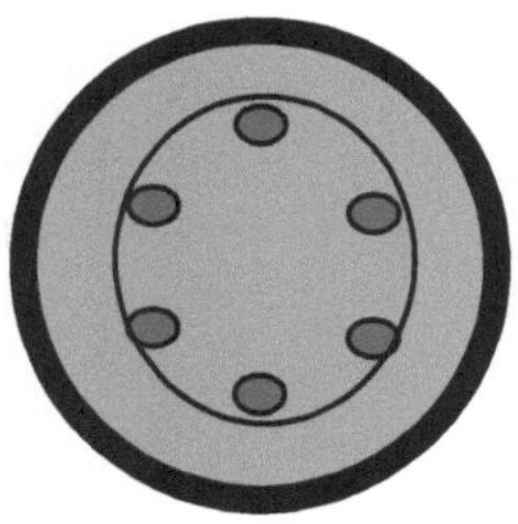

Solution

For fixed-fixed boundary conditions, $kl = 0.67 \times 5000 = 3350$ m

Using AIC 216.1 expression (as per Eq. 9.4a)

$$R = a\frac{(f_c' + 20)}{60(KL - 1000)}D^2\left(\frac{D}{C}\right)^{1/2}$$

where $a = 0.08$ for a circular column filled with carbonate aggregate concrete (as per Table 9.5).

Thus

$$R = a\frac{(f_c' + 20)}{60(KL - 1000)}D^2\left(\frac{D}{C}\right)^{1/2} = R = 0.08\frac{(40 + 20)}{60(3350 - 1000)}350^2\left(\frac{350}{1344}\right)^{1/2} = 2.13 \text{ hours}$$

9.2

Evaluate the fire resistance rating for a composite column (324 × 6.4 mm) filled with concrete and exposed to standard fire. The concrete filling is reinforced with steel bars of 3% area with a 25-mm cover. The column is 4800 mm long and subjected to 1100 kN during fire exposure. The steel section is made of grade 355 MPa, steel rebars are of grade 500 MPa, and the compressive strength of concrete is 25 MPa. Assume the concrete cover is equal to 25 mm and the diameter of rebars is 25 mm.

Solution

Applied loading under fire conditions = 1100 kN

Following the discussion in Sec. 9.6.2, the buckling capacity of composite column = 4363.3 kN

Level of applied loading = 1100/4363.3 = 0.252 < 0.28

Thus, using Table 9.3, the fire resistance of this column equals 60 minutes, given that concrete cover is less than 30 mm.

9.3

Evaluate the fire resistance of a composite slab that is part of a shopping center, and the span between supports is 4.8 m. The slab is designed as a series of simply supported beams. The required standard fire resistance class for the slab is 90 minutes based on thermal criteria. Given that:

Material Properties:

- Steel sheet:
  - Yield stress, f_{yp} = 350 MPa
  - Cross-sectional area = 1562 mm²/m
- Concrete:
  - Strength category, C = 25/30
  - Comprehensive strength, f_c = 25 MPa
  - Height, h = 140 mm
  - Cross-sectional area = 131,600 mm²/m

Loads:

- Permanent loads:
  - Steel sheet, $g_{p,k}$ = 0.13 kN/m²
  - Concrete, $g_{c,k}$ = 3.29 kN/m²
  - Finishes, $g_{c,k}$ = 1.2 kN/m²
- Variable loads:
  - Live load, P_k = 5.0 kN/m²
  - Assume the design bending moment under fire, M_{fid}, is 55% of that at ambient conditions.

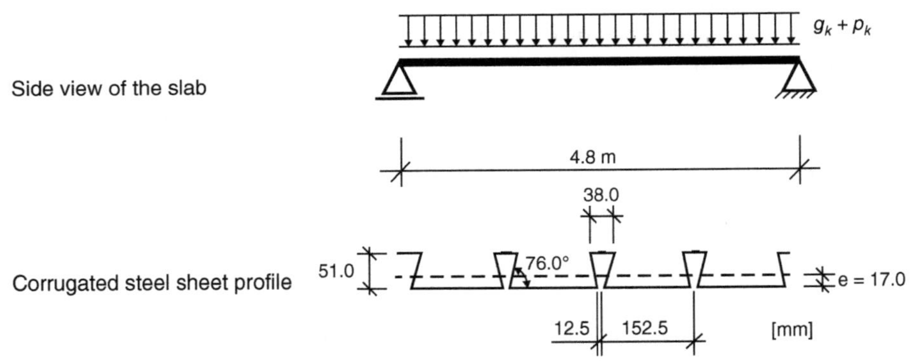

Solution

According to Eurocode 4,

Factored load = $\gamma_G G_k + \gamma_{Q,1} Q_{k,1}$

$$= 1.35 \times (0.13 + 3.29 + 1.2) + 1.5 \times 5.0 = 13.74 \text{ kN/m}$$

Design sagging moment at ambient temperatures:

$M_{s,d} = wl^2/8$ = 39.56 kN·m/m, thus $M_{fi,d}$ = 0.55 × 39.56 = 21.94 kN·m/m

For thermal insulation criterion:

- The thermal insulation criterion "I" limits the permitted temperature rise on the unexposed face of a separating element. In accordance with the insulation criteria, the temperature rise on the top surface of the slab should not exceed an average value of 140°C or a maximum of 180°C.
- The verification is carried out in the time domain as follows:

$$t_i = a_0 + a_1 h_1 + a_2 \phi + a_3 \frac{A}{L_r} + a_5 \frac{1}{l_3} + a_3 \frac{A}{L_r}\frac{1}{l_3}$$

$$\frac{A}{L_r} = \frac{h_2\left(\frac{l_1 + l_1}{2}\right)}{l_2 + 2\sqrt{h_2^2 + \left(\frac{l_1 - l_2}{2}\right)^2}} = \frac{52\left(\frac{115 + 140}{2}\right)}{140 + 2\sqrt{51^2 + \left(\frac{115 - 140}{2}\right)^2}} = 27 \text{ mm}$$

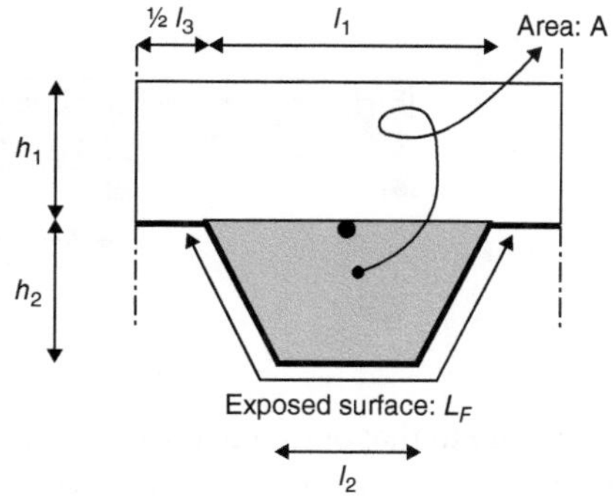

The view factor ϕ considers the shadow effect of the rib on the upper flange:

$$\phi = \frac{\left[\sqrt{h_2^2 + \left(l_3 + \frac{l_1 - l_2}{2}\right)^2} - \sqrt{h_2^2 + \left(\frac{l_1 - l_2}{2}\right)^2}\right]}{l_3}$$

$$= \frac{\left[\sqrt{51^2 + \left(38 + \frac{115 - 140}{2}\right)^2} - \sqrt{51^2 + \left(\frac{115 - 140}{2}\right)^2}\right]}{38} = 0.119$$

The coefficients a_i for normal-weight concrete corresponding to 90 minutes of fire exposure are given in Table 9.5:

$$t_i = (-28.8) + 1.55(89) + (-12.6)0.119 + 0.33(27) + (-735)\frac{1}{38} + 48(27)\frac{1}{38}$$

$t_i = 131.48 > 90$ minutes, the section is adequate

Following Eurocode 4 procedure, the moment capacity of this composite slab can be evaluated such that

$$\theta_s = c_o + c_1\frac{u_3}{h_2} + c_2 z + c_3\frac{A}{L_r} + c_4\alpha + c_5\frac{1}{l_3}$$

$$\theta_s = 1342 + (-256)\frac{61}{51} + (-235)(2.72) + (-5.3)(27) + (1.39)(104) + (1267)\frac{1}{38} = 364.12°\text{C}$$

and comparing this value against Table 9.8 shows that this slab has a fire resistance of 90 minutes.

The moment capacity of this slab can be evaluated as

$$M_{fi,t,Rd} = \sum A_i z_i k_{y,\theta,i}\left(\frac{f_{y,i}}{\gamma_{M,fi}}\right) + \alpha_{\text{slab}}\sum A_j z_j k_{c,\theta,j}\left(\frac{f_{c,j}}{\gamma_{M,fi,c}}\right) = 22.51 \text{ kN}\cdot\text{m/m}$$

$M_{fi,t,Rd} = \frac{3.42}{0.152} = 22.51$ kN·m/m < $M_{s,d} = wl^2/8 = 39.56$ kN·m/m, thus M_{fid}

$= 0.55 \times 39.56 = 21.94$ kN·m/m

Therefore, the fire resistance of this slab is 90 minutes.

9.4

Using advanced analysis, evaluate fire resistance when the composite girder made of W24×62 section is subjected to high shear loading similar to that shown below. Each of the applied loading is 244.6 kN. Use similar material properties to that used in Example 6.3 for steel and concrete.

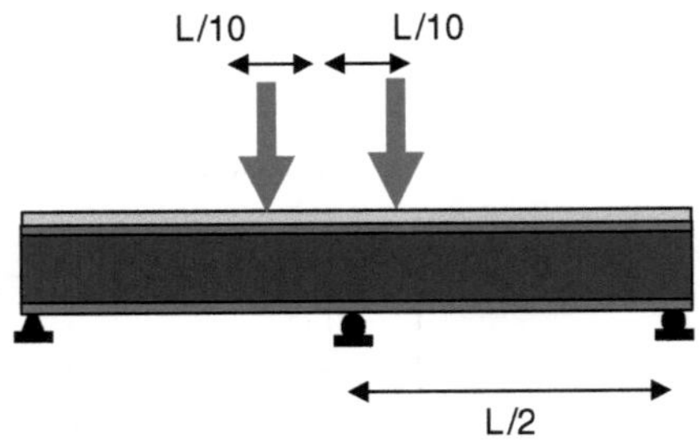

Solution

Following a similar procedure to that outlined in Chap. 6 and Example 6.3, the first four steps are identical.

Step 1. Heat transfer analysis of the composite steel-concrete girder is carried out using two types of elements available in ANSYS, namely SOLID70 and SURF152. For the discretization of girder, slab, and the stiffeners, SOLID70 elements are used.

Step 2. In the structural analysis, the bottom flange, web, top flange, and stiffeners of the steel girder were modeled with SHELL181 elements, and the concrete slab was modeled with SOLID185 elements.

Step 3. High-temperature property relations of concrete and structural steel are input to the analysis as per Eurocode 2 and Eurocode 3, respectively.

Step 4. Heat transfer analysis gives temperature rise across the depth of the beam, as shown in Example 6.3.

Step 5. Structural analysis gives deflection time history as a function of fire exposure. Due to the different loading configuration (high shear loading as opposed to shear loading), the structural response of the composite is expected to be different. Hence, only the variation in midspan deflection of this beam (noted as CB2 with fire exposure time), as compared to that shown in Example 6.3 (in beam CB1), is shown here:

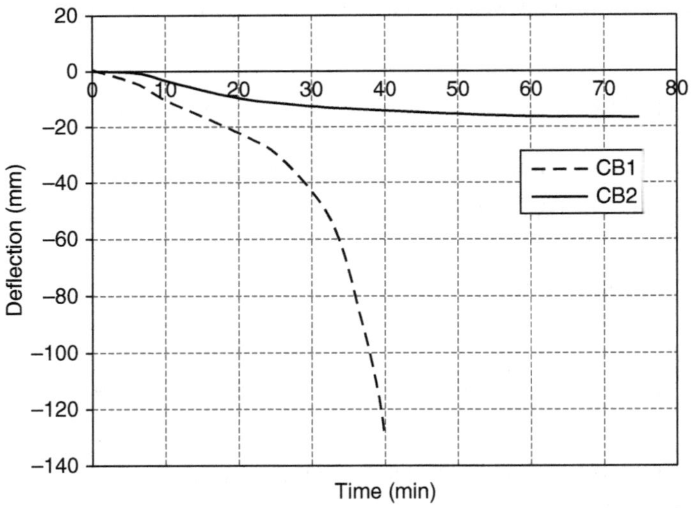

Note: CB1 and CB2 are composite beams 1 and 2.

Refer to the following published work and compare your solution:

Naser, M. Z., and V. K. R. Kodur. "Comparative Fire Behavior of Composite Girders under Flexural and Shear Loading." *Thin-Walled Structures* 116 (2017): 82–90.

CHAPTER 10

Timber Structures

10.1 Overview

Historically, until the early 1900s, wood was the primary choice of material in residential construction in most parts of the world. However, due to declining forest cover and rapid advancements in other construction materials, the use of wood has decreased significantly in many countries in Asia (such as China, India, and the Middle East) and some parts of Europe over the last 50 years. But in the last few years, wood construction is coming back due to the recent focus on sustainability and the significant advances in wood technology. This is especially true in North America, many parts of Europe, Japan, South America, Australia, and New Zealand. Since wood is used predominantly in building construction, fire resistance is one of the key considerations in the design of wood structural members.

This chapter presents various approaches that can be applied for evaluating the fire resistance of wood structures. The variation in thermal and mechanical properties of timber with temperature is discussed. The typical fire behavior of structural members made of wood is illustrated, and the relevant strategies that can be developed for enhancing fire resistance of wood structural members are explained. Both prescriptive and rational calculation methods for fire resistance analysis and design of wood structural member are laid out.

10.2 Properties of Timber

Timber, unlike steel, concrete, or masonry, is a combustible material, and this aspect limits the use of timber to low-rise buildings in most countries. The mechanical properties of timber, at room temperature, vary significantly depending on the species, moisture content, orientation of grain, and state of stress. These mechanical properties deteriorate at relatively lower temperatures, as compared to concrete, steel, or masonry. Thus, fire resistance is one of the key considerations in the case of wood structural members.

Structural members made of wood are broadly grouped into two categories: heavy timber and light timber construction. Heavy timber construction refers to timber framing members with larger dimensions, such as beams, columns, decks, or trusses, that are primary load-bearing members in a structure. Heavy timber construction utilizes either *natural wood, glue laminated timber (glulam),* or *laminated veneer lumber (LVL).* Glulam wood is manufactured by face-bonding layers of kiln-dried laminations, typically 2 × 4 in.[1] or 2 × 6 in. size, together with waterproof adhesives (such as resorcinol or casein) to form a timber section. On the other hand, LVL is manufactured from veneers laid into long, parallel strands and bonded together. A benefit of LVL is that it can be fabricated into narrower sectional sizes (1½ in., 1¾ in. width), and multiple plies can be nail-laminated together to form a larger beam.

Light timber construction refers to members built with smaller-size wood sections, and most common applications are in walls, floors, decking, and the like. These members typically carry lower levels of load and are constructed utilizing studs, joists, and so on, together with layers of oriented strand board, plywood panels, or gypsum wallboard. Both heavy and light timber constructions are widely utilized in residential housing, low-rise commercial constructions, and specialized structures such as religious places, museums, and libraries.

10.2.1 Properties at Ambient Conditions

Since timber cannot be grown or fabricated to certain specifications as accurately as can be done with alternative construction materials such as steel and concrete, it is important to give due consideration to generalizing the properties of wood over a large range of species. Depending on the wood species of the timber and its mechanical properties, wood is often classified under two broad classes: softwood and

[1] FPS units.

hardwood. Softwood is not necessarily a softer material (less dense) than hardwood. Softwood and hardwood are distinguished botanically in terms of their reproduction process, not by their end use or appearance. In general, hardwood comes from deciduous trees, which lose their leaves annually, and softwood comes from coniferous trees, which usually remain evergreen. Hardwood trees tend to grow at a slower rate and are therefore usually denser. Softwood trees are known as a gymnosperm. Gymnosperms reproduce by forming cones which emit pollen to be spread by the wind. Thus, softwood trees grow from seeds that are dropped to the ground or spread through wind. Some examples of softwood include pine, redwood, Douglas fir, cypress, and larch. A hardwood tree is referred to as an angiosperm, a plant that produces seeds with some sort of covering such as a shell or a fruit. Examples of hardwood trees include eucalyptus, beech, and blackwood.

Wood, from a structural property standpoint, is an orthotropic material; that is, it has unique and independent mechanical properties in the directions of three mutually perpendicular axes: longitudinal, radial, and tangential. The longitudinal axis, *L*, is parallel to the fiber (grain); the radial axis, *R*, is normal to the growth rings (perpendicular to the grain in the radial direction); and the tangential axis, *T*, is perpendicular to the grain but tangent to the growth rings (see Fig. 10.1). Wood experiences a large variability, or variation, in its properties due to its natural origin—in other words, trees are subject to many constantly changing influences such as moisture, soil conditions, and growing space, as well as the presence of knots. As a result, wood properties vary considerably, even in clear wood (i.e., without any knots).

The strength of wood is fundamentally affected by the direction in which it is loaded (stressed) in relation to the orientation of grain. In the direction of the grain, the bending strength is directly proportional to the density of the wood. In uniform flawless wood, the bending tensile strength is as high as the direct tensile strength. Tensile strength in the direction of the grain is usually 10 to 20 times more than its strength perpendicular to the grain. Tensile strength also depends on the density of the wood: for example, the tensile strength of spring wood in a pine is only 1/6 of that of summer wood. The compression strength of air-dried wood is about half of the corresponding tensile strength. The shear strength of wood is 10% to 15% of its tensile strength in the direction of the grain. Shear strength is weakened by knots, faults, and cracks that appear in the wood. The elasticity and durability of wood increase as its density increases. The modulus of elasticity (MOE) of wood in the direction of the grain may be up to 100 times more than its elasticity perpendicular to the grain. In the radial direction, the MOE is about twice as great as the same parameter in a tangential direction. Wood properties vary within and between species and are primarily affected by levels of moisture content, knots, and specific gravity (or density).

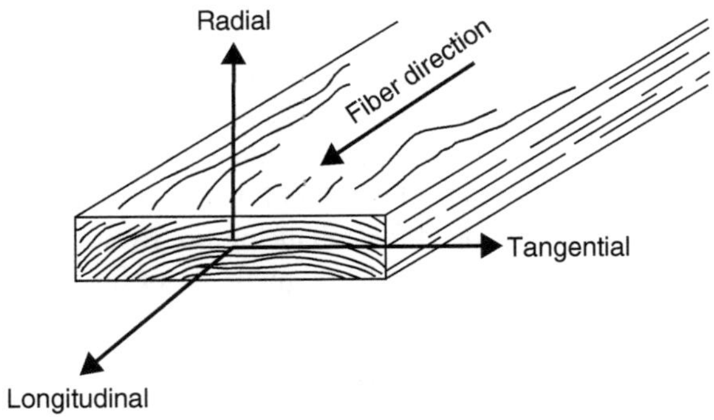

FIGURE 10.1 Three principal axes of wood with respect to grain direction and growth rings.

Wood is a hygroscopic material, which gains or loses moisture depending on the temperature and relative humidity of the surrounding air. Under ambient conditions, wood has significant moisture in its composition. Moisture content is expressed as a percentage of ratio of weight of water present in a piece of wood to its oven-dry weight. Fresh-cut trees can have a wood moisture of more than 90%, while completely dried wood will have a wood moisture content of 0%. Typically, seasoned wood, used in construction, can have up to 20% of moisture. Even in dry conditions, wood can have up to 16% of moisture. From a material property point of view, most properties are affected by any changes in moisture content, below the fiber saturation point. An excess presence of moisture in wood can prevent adhesives from making a secure bond, promote growth of mold in the wood, and cause shrinkage as the excess moisture leaves the wood. Hence, mechanical properties of wood generally increase with the decrease in moisture content.

A knot is an imperfection developing in wood (a tree) while growing, producing defects from a structural properties consideration. The influence of a knot on the mechanical properties of a wood member is due to the interruption of continuity and change in the direction of wood fibers associated with the knot. The extent of influence of a knot on properties of wood depends on its size, location, shape, soundness, and type of stress to which the wood structural member is subjected. Most mechanical properties are lower in sections containing knots than in clear straight-grained wood because (a) the clear wood is displaced by the knot, (b) the fibers around the knot are distorted, and (c) the discontinuity of wood fiber leads to stress concentrations. In general, knots have a much greater effect on strength in axial tension than in axial compression, and the effects on bending are somewhat less than those in axial tension.

Wood is a porous material made up of cells of various kinds. The amount of solid wood substance for a given volume determines wood density, which is the simplest index of the strength of wood. With increasing density, the strength of wood also increases. Higher density results from a greater proportion of cells with thick walls and small cavities, and this in turn leads to a higher strength in wood. For this reason, the density of wood varies with species type. The oven-dry density of wood can range from 160 to over 1000 kg/m^3, but most wood species used in construction applications are in the 300 to 700 kg/m^3 range (USDA, 2010). For example, the density of Douglas fir (a commonly used wood in interior construction in North America that accounts for 25% of all wood produced) is about 530 kg/m^3, which is relatively lower than the 740 kg/m^3 density of American red oak, which is extremely durable and resistant to insect damage or rot. Since density of wood is affected by the amount of moisture it contains, it is important to also know the moisture level in wood.

10.2.2 Properties at Elevated Temperature

The properties of wood, as in the case of other materials, deteriorate at elevated temperatures. Thus, the fire resistance of a timber structure depends on the temperature-dependent properties of wood. As sectional temperatures in a wood member rise with fire exposure time, the properties of wood deteriorate due to changes in the cellular structure of wood at the molecular level. From a fire resistance consideration, the property variations that occur in wood in the range of 20 to 300°C are of primary interest. The variation in properties in this temperature range is mainly influenced by density, moisture content, species of wood, and any adhesives (or glue) used in fabrication (as in engineering wood). While a general discussion on the variation in properties of wood with temperature is laid out in Chap. 4, this section mostly focuses on available relations of variation in temperature-dependent properties of wood.

A review of the reported thermal, mechanical, and deformational properties of wood at elevated temperatures indicates significant variation across the published works. In fact, this variation is much higher in the case of wood, as compared to concrete and steel, since reported data reflects tests carried out on different wood species, at different moisture content, of differing specimen sizes, and under different test conditions. From this data, a few researchers have proposed generalized temperature-dependent property relations (or trends) for wood, suitable for undertaking fire resistance calculations. The well-accepted relations are adopted in Eurocode 5 and will be featured herein. Other property relations for wood can be found elsewhere (Knudson and Schniewind, 1975; Gerhards, 1982). It is worth noting that limited research has been carried out to deduce temperature-dependent material properties for woods glued together using bamboo scrimber (Young and Tsao, 2014). A common trend in all of the proposed property relations, including those in Eurocode 5, is that the mechanical properties start to deteriorate at a relatively low temperature, close to 100°C. These properties continue to deteriorate, reaching 50% and 10% of initial strength at around 200 and 300°C, respectively.

The density of wood at room temperature varies greatly between different species and reduces 90% of its room temperature value at about 100°C, and then 50% of its original value about 300°C. The reduction in the density of typical wood with temperature is shown in Fig. 10.2, which shows that a sharp change in wood density occurs in the temperature range of 100 to 120°C due to the evaporation of moisture present in wood. The density continues to decrease, at a much more rapid pace, between 200 and 400°C, reaching to about 30%.

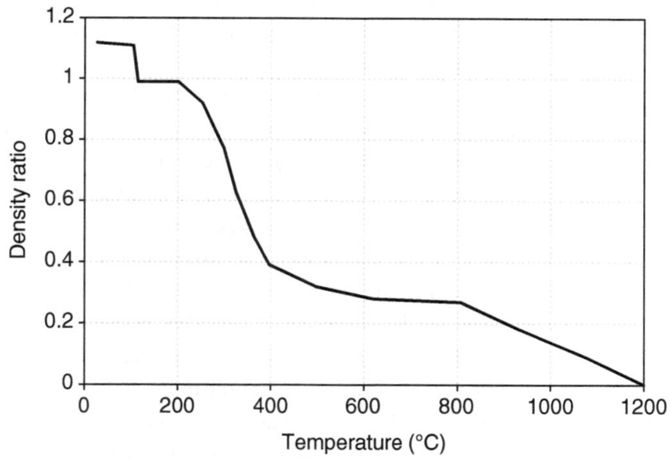

FIGURE 10.2 Variation in density with temperature in softwood with an initial moisture content of 12%. (*Reproduced from CEN.*)

The variation in thermal conductivity of typical wood at elevated temperatures is shown in Fig. 10.3. The thermal conductivity of wood, which is about 0.11 W/m·K at room temperature, slightly increases with temperature up to about 200°C, and then decreases linearly between 200 and 350°C, and then increases again beyond 350°C when the wood turns to char. Shrinkage cracks in the charcoal above 525°C further increase heat transfer through wood, thus increasing thermal conductivity.

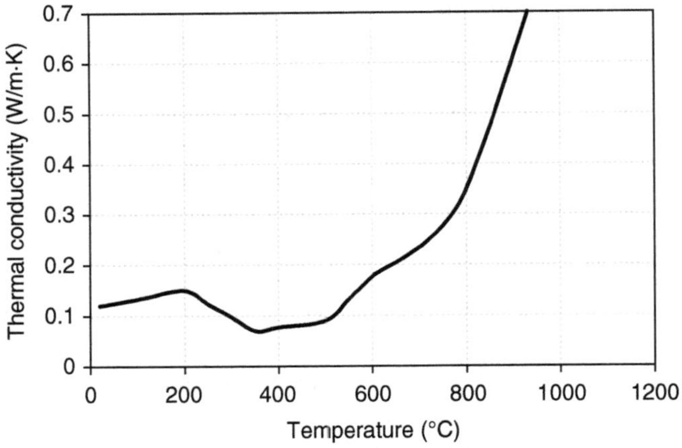

Figure 10.3 Thermal conductivity of wood as a function of temperature. (*Reproduced from CEN.*)

The specific heat of wood, which is about 1.5 to 2 kJ/kg·K at ambient conditions, increases significantly around 100°C. This spike represents a surge in thermal energy (latent heat) required to evaporate the free water present within the wood. Beyond this temperature, the specific heat reduces to a similar level to that at ambient conditions. The variation in specific heat is illustrated in Fig. 10.4. Wood has very low coefficient of thermal expansion (1×10^{-4} °C^{-1}) at ambient temperature, and as such this property is often neglected in fire resistance calculations (Frangi and Fontana, 2003).

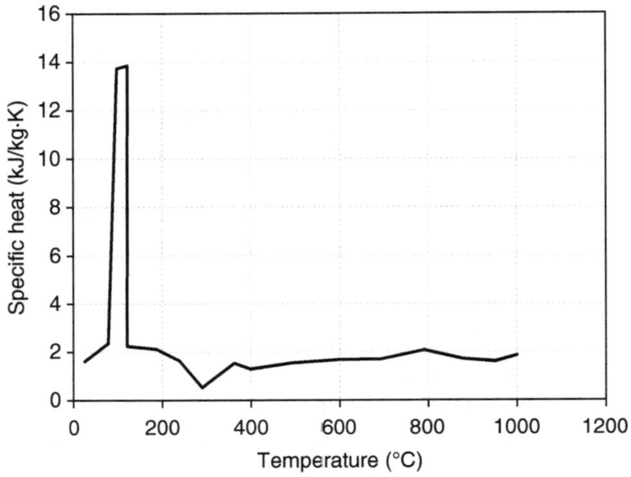

Figure 10.4 Specific heat of wood as a function of temperature. (*Reproduced from CEN.*)

The mechanical properties of timber degrade with a rise in temperature, as in the case of other construction materials. However, due to the anisotropic nature of wood, the extent of degradation in its mechanical properties varies in different directions (i.e., parallel or perpendicular to the grain). For example, Fig. 10.5 shows the temperature-dependent variation in the elastic MOE and strength properties of wood oriented parallel to the grain. The degradation in the modulus is linear up to 100°C, reaching about 35% and 50% of tensile and compression modulus to that at ambient conditions, respectively. The moduli continue to degrade at a lower rate until diminishing to almost zero at about 300°C. The strength properties of wood also deteriorate with a rise in temperature. Eurocode 5 depicts this deterioration in tensile, compressive, and shear strength of softwood, as shown in Fig. 10.5*b*. The degradation in compressive strength of wood is more rapid than in shear or tensile strength, primarily due to higher creep effects that wood undergoes under compressive loading. Eurocode 5 states that for compression perpendicular to the grain, the same reduction of strength may be applied as for compression parallel to the grain.

Eurocode 5[2] gives idealized stress-strain relations for timber at elevated temperatures, as shown in Fig. 10.6. These relations are based on the work of Konig and Walleij (2000), through fitting data to a series of bending tests carried out at elevated temperatures. As can be seen in Fig. 10.6, the stress-strain response is very brittle in tension but highly ductile in compression. These relations assume linear elastic behavior in the tension and elastoplastic behavior in the compression and account for partial creep effects that are generated at elevated temperatures.

Unlike other construction materials, wood undergoes shrinkage at elevated temperatures, as moisture levels in wood significantly reduce at temperatures above 100°C. The extent of shrinkage in wood is a function of the initial moisture content, the type of species, and the grain orientation. In most cases, shrinkage will be about 12% and 8% in the tangential and radial directions, respectively. In the longitudinal direction, the average values for shrinkage from wet to oven-dry conditions are between 0.1% and 0.2% for most species of wood (USDA, 2010).

[2] It is worth noting that Eurocode 5 only provides a chart and does not provide relations to temperature-dependent material properties of wood.

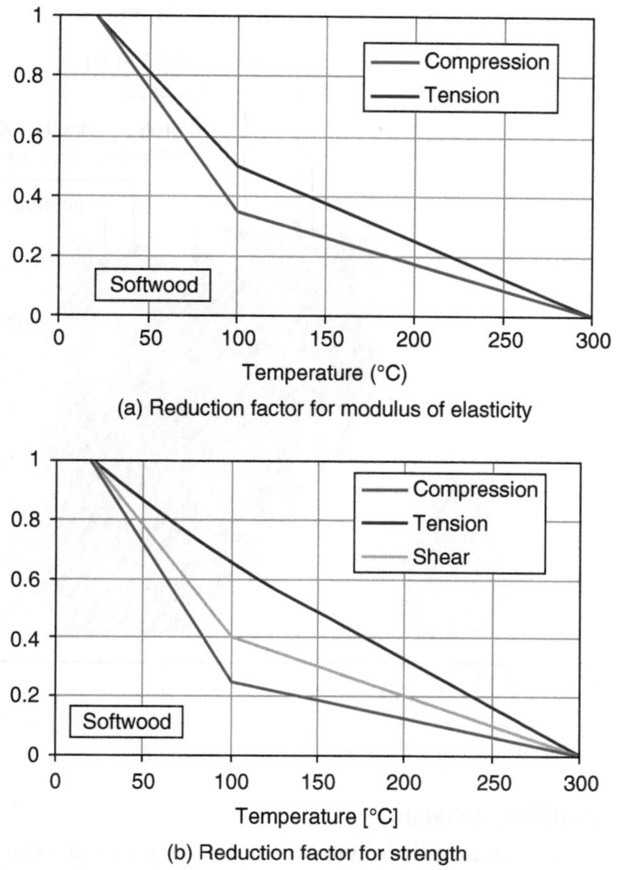

FIGURE 10.5 Temperature-dependent degradation of mechanical properties of wood (parallel to the grain) as per Eurocode 5. *(Reproduced from CEN.)*

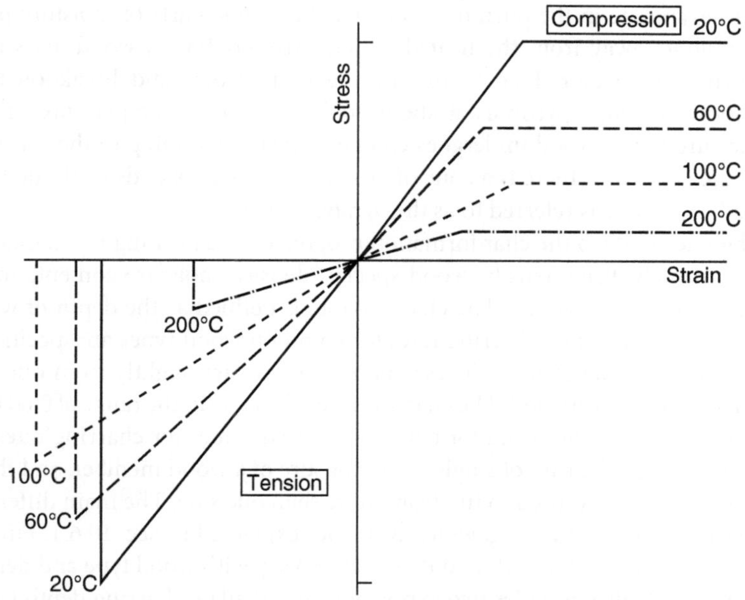

FIGURE 10.6 Stress-strain response of wood at elevated temperature. *(Reproduced from CEN.)*

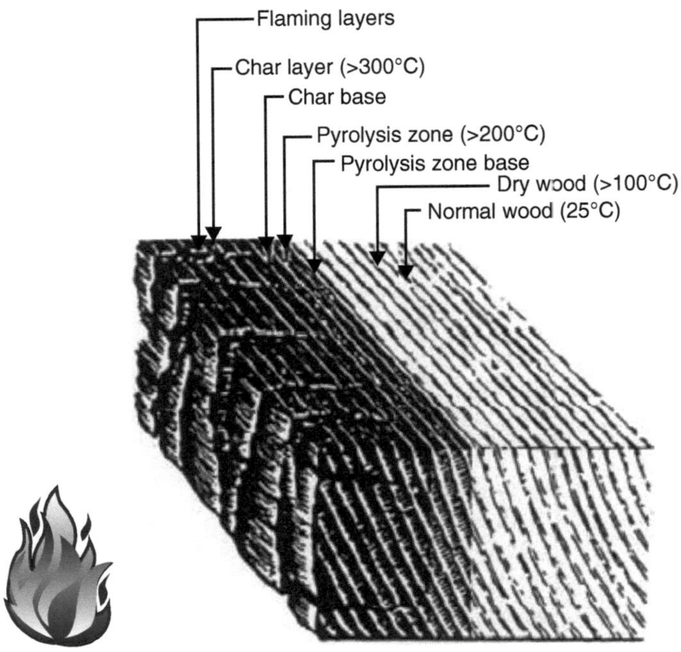

Figure 10.7 Burning wood.

10.2.3 Charring Phenomenon

Another factor that significantly influences the fire resistance of a wood structural member is the charring phenomenon that occurs in timber under fire conditions. When a wood structural member is exposed to fire, it undergoes thermal degradation, referred to as *pyrolysis*. The pyrolysis of wood is a complex interplay of chemical and physical reactions involving heat and mass transfer, and transforms the solid wood to char and gases (such as carbon dioxide and hydrocarbons). This pyrolysis process consequently changes the properties of wood significantly. The progression of char during the burning of wood is illustrated in Fig. 10.7, through a piece of wood exposed to fire from one side.

When exposed to fire, outer layers of wood start to combust with a rise in temperature, and moisture starts to evaporate in the burning layers. Further, some parts of moisture present in adjoining (beneath) layers migrate away from the heated region. Around 100°C, wood loses much of the moisture, which transforms into vapor. Due to this drying of moisture and breakdown of cellular structure, wood decomposes through pyrolysis at about 200°C. When the temperature in wood is in the 200 to 300°C temperature range, wood undergoes complex reactions leading to the formation of char at about 300°C. Below the char layer, there is a zone of heated wood that extends to about 35 mm toward the inner layers of wood. This zone is referred to as the *pyrolysis zone*.

The rate at which the char formation progresses in a burning wood member depends on a number of factors—namely, fire intensity, wood species, density, moisture content, and grain orientation (parallel/perpendicular to the grain). The charring rate is defined as the depth of wood transforming to char per minute during burning. Charring rates for common wood types are specified in design standards for use in fire resistance calculations. These charring rates vary slightly from one standard to another and also from hardwood to softwood. The charring rate is broadly in the range of 0.60 to 0.75 mm/min for softwoods and about 0.5 to 0.6 mm/min for hardwoods. Two values for charring rates are given in Eurocode 5; the first one (β_0) is in the case of single-face exposure of a wood member, and the second one (β_n) is for when a wood member is exposed to fire from more than one side. The main difference between β_0 and β_n is that β_n allows for corner rounding, which is further explained in Sec. 10.6.1. Eurocode 5-specified values of β_0 and β_n are given in Table 10.1, and these values vary with wood type and density (ρ).

Eurocode 5 also provides two expressions to calculate charring depth (d_{char}) based on exposure under standard furnace conditions or under real (parametric) fire conditions. These expressions can be applied to calculate the charring depth in a wood member exposed to one- or two-directional fire.

$$\text{For one-directional exposure: } d_{char,0} = \beta_0 t \qquad (10.1)$$

Timber Type	Charring Rate (mm/min)	
	β_o	β_n
Softwood and beech		
Glulam ($\rho \geq 290$ kg/m^3)	0.65	0.7
Solid ($\rho \geq 290$ kg/m^3)	0.65	0.8
Hardwood		
Solid or glulam ($\rho = 290$ kg/m^3)	0.65	0.7
Solid or glulam ($\rho > 450$ kg/m^3)	0.5	0.55

Table 10.1 Charring Rates for Timber as per Eurocode 5

$$\text{For two-directional exposure: } d_{char,n} = \beta_n t \tag{10.2}$$

where t is the exposure time.

In North America, recommendations for charring rate are given in the American Forest and Paper Association (AFPA) handbook (White, 1988). As per these recommendations, the average charring rate, β (mm/min), in a wood member exposed to standard fire exposure is given by

$$\beta = 2.58\ \beta_n/t^{0.187} \tag{10.3}$$

where β_n is a nominal charring rate and equals to 0.635 mm/min, and t is the time (min).

The above equation was also adopted by the American Wood Council (AWC). In its recent edition, NDS (2018) specifies an effective char rate, β_{eff} (mm/min), which accounts for the corner rounding effect; β_{eff} is computed as

$$\beta_{eff} = 1.2\ \beta_n/t^{0.187} \tag{10.4}$$

where β_n = nominal char rate (in./hr) and t = exposure time (hr). A nominal char rate, β_n, of 1.5 in./hr (0.63 mm/min) is recommended for sawn lumber and timbers, glued laminated timbers, laminated veneer lumber, parallel strand lumber, laminated strand lumber, and cross-laminated timber.

In lieu of Eq. 10.4, the NDS manual (AWC manual) provides additional guidance for calculating the charring rate and depth of char for other types of wood including sawn lumber and timber, glued laminated timbers, laminated veneer lumber, parallel strand lumber, and laminated strand lumber. For example, the effective char rate for 60, 90, and 120 minutes of burning for these timber types are 1.8, 1.67, and 1.58 in./hr (or 0.76, 0.71, and 0.67 mm/min), respectively. It is worth noting that in spite of some differences in charring rate specified in different standard codal provisions, charring rates measured on different wood species, in various parts of the world, are broadly in the range of 0.6 to 0.8 mm/min.

Adhesives

With the recent move to engineered wood products, adhesives play a critical role in the fire performance of engineered wood, including the charring rate and degradation of mechanical properties at elevated temperatures. Modern adhesives [such as polyurethane (1C-PUR)] have proven to provide a reliable alternative to commonly used commercial adhesives, such as water-based aminoplast (UF, MF, MUF) or phenoplast-based (PR, PRF) adhesives. Modern adhesives cure at a faster rate than traditional adhesives, do not release chemicals/smell like formaldehyde, and do not require complex installation procedures. As a result, these modern adhesives are finding increasing acceptance in the field of structural timber gluing. From the point of view of fire resistance, contemporary adhesives have a moderately positive performance under fire conditions. In fact, PUR adhesives can achieve a good performance up to 170°C (Frangi et al., 2012). Results from fire tests indicate that the performance of 1C-PUR adhesives under fire exposure depends strongly on chemical composition and filler materials in their formulations.

One of the major limitations of adhesives is their vulnerability to high-temperature creep. In order to limit the development of such creep, it is advisable to use a higher initial strength (i.e., with high concentrations of isocyanate) to reduce creep effects. Other chemical parameters that have an impact on the thermal stability of adhesives include the relative proportion of polyurethane, the degree of polymerization of the prepolymer, and the rate of reaction (Frangi et al., 2012). Eurocode 5 clearly states that the softening temperature of some adhesives is considerably below the charring temperature of the wood, and hence relevant measures to avoid delamination of engineering wood under fire conditions need to be incorporated.

10.3 Behavior of Timber Structures under Fire

When a large-sized timber member is exposed to fire, the outer layers of wood initially ignite and start to combust slowly. With gradual burning on the outer surface, a layer of char develops beneath the burning layers, and this char slows down the burning rate of the remaining wood beneath the char layers. The char layer is composed of CO and CO_2 gases, and as such the layer below char does not burn due to an insufficient oxygen supply for oxidation to occur. Above 100°C, the moisture in the wood evaporates. Some of this moisture travels out to the burning face of the member, and the remaining moisture travels toward the unburnt side (interior) of the section, which increases its moisture content. This moisture movement and its evaporation slow down rate of heating, as well as the charring of wood. One should note that with extended fire exposure time—that is, when the residual cross section becomes thinner—the charring rate increases (NDS, 2018).

The response of a typical wood structural member can be illustrated by considering a wood column subjected to the combined effects of axial loading and fire exposure as shown in Fig. 10.8*a*. Outer layers in the wood column start to combust slowly, and with increasing fire exposure time, this combustion process gradually spreads into the inner layer of the section of the column (see Fig. 10.8*b*,*c*). This will result in the formation of charring at layers beneath the burning surface as shown in Fig. 10.7. Below the layers of char is a pyrolysis zone where moisture starts to migrate toward the inner layers of wood. Therefore, with increasing fire exposure time, the overall combustion process, composed of burning, charring, pyrolysis, and moisture vaporization, moves deeper into the interior layers of the wood section. This process will lead to the loss of the effective cross section, as well as the temperature-induced reduction in mechanical properties of the unburnt portion of the wood.

Figure 10.8*d* shows how axial deflection in a fire-exposed column increases with time as a result of degradation in the strength and modulus properties of wood, as well as a reduction in the cross section. Due to the gradual loss of the cross section, the stress level on the column increases as a result of load bearing (see Fig. 10.8*f*). This is quite different from that in steel and concrete members, where there is no loss in the effective cross section throughout the fire exposure time (assuming there is no spalling in the concrete). The combination of increasingly higher stress levels and deterioration in the mechanical properties of wood leads to a rapid rise in axial deformation and a loss of axial capacity (see Fig. 10.8*d*,*e*). When the reducing axial capacity or increasing axial deformation exceeds the liming criterion, failure occurs, and this point in time is taken as the fire resistance of the column.

10.4 Strategies for Achieving Required Fire Resistance

The behavior of wood structural members under fire conditions, as illustrated in Fig. 10.8, is quite different from that of concrete, steel, or masonry structural members. This is due to the fact that wood, unlike steel, concrete, or masonry, combusts at elevated temperatures, leading to the formation of char. This burning and associated charring is a continuous process and grows into interior layers with fire exposure time. Charring after the initial period slows down the burning process in the interior wood layers. This phenomenon slows the cross-sectional temperature rise in the inner layer of a wood member. Therefore, the fire resistance of wood structural members depends on numerous factors including the rate of charring, the extent of loss of cross section, and the loss of strength/stiffness properties of wood.

In low-rise building applications, where wood is predominantly used, structural members need to have 1 to 2 hours of fire resistance. Therefore, appropriate strategies must be developed to achieve this level. Three strategies are commonly applied to enhance the fire resistance of timber structures: the use of larger sectional size, the provision of fire insulation, and the application of fire retardants.

The most common strategy that can be adopted for achieving the required fire resistance is through the provision of a larger cross section in a wood member than that required to meet strength design at ambient conditions. In other words, if a wood column of sectional size $b \times d$ is needed to resist a specified loading at room temperature and if this column is to meet a 1-hour fire rating, then selecting a larger section size (say, $b' \times d'$, where b' and d' are larger than b and d) can help in achieving such fire resistance. When such

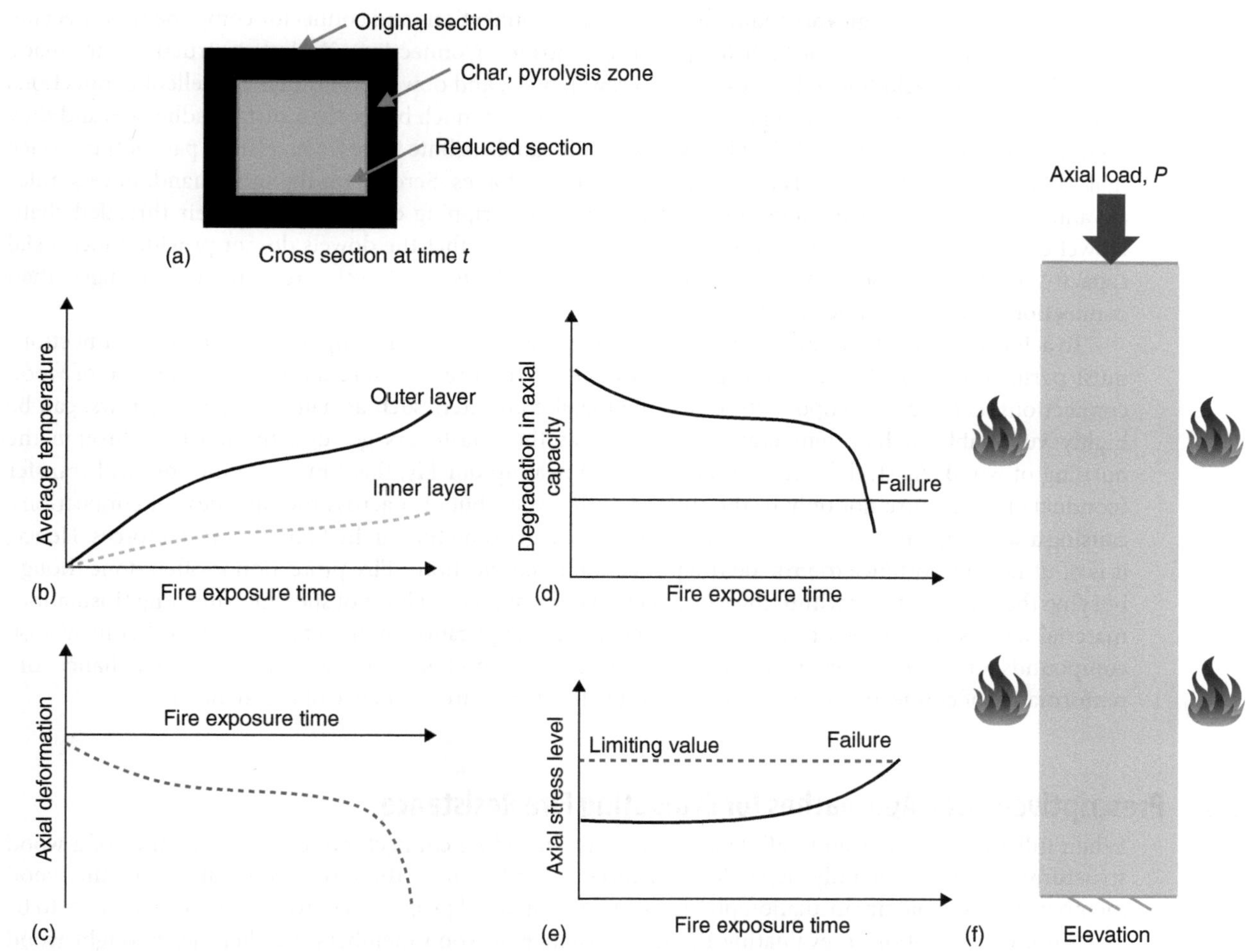

Figure 10.8 Behavior of a typical wood column under fire exposure.

a column is subjected to fire, the outer layers of the enhanced wood section act as sacrificial layers and at 1 hour the effective (uncharred) section reduces to a smaller size of $b'' \times d''$ due to burning/charring. This reduced section should be sufficient to withstand the axial load acting on the column during fire exposure.

Alternatively, the fire resistance of wood structural members can be enhanced through the provision of external fire insulation, as in the case of steel structures. The provision of such external fire insulation delays the overall combustion process in the wood, thereby slowing down the development of sectional temperatures, which in turn prolongs the failure time of a wood member. The commonly used fire protection materials for insulating timber structures are gypsum board, mineral fiber, glass fiber, or rock fiber insulation. The use of spray-applied fire resistive material (SFRM) is not that common in timber structures, possibly due to poor bond (stickability) and application issues (Wang et al., 2008).

Another strategy that can be adopted is to apply fire-retardant chemicals to reduce combustibility and flame spread across the wood surface. The application of fire retardants such as paints or coatings delays the combustion process; however, they do not significantly improve the fire resistance of wood members, since their effect is mainly in delaying initial combustion (Östman and Tsantaridis, 1995). Other limitations associated with the use of fire retardants include possible corrosion to metal fasteners (arising from chemicals in retardants), long-term durability (arising from weathering effects), and their effectiveness. Overall, the use of fire retardants in enhancing fire resistance of timber structures is still not proven to be very effective.

10.4.1 Connections

Connections of joints play a critical role in transferring loads from different components within a structural member or from one member to the other. In the case of wood structures, numerous connections are to be made to join not only different structural components of a member but also from one member to another.

Therefore, specific strategies are required in the case of connections and connector components, especially to resist the adverse effects of high-temperature exposure. Connections in timber structures are made through fasteners including adhesives, nails (screws), bolts, and dowels. Nails make excellent connections in timber structures since they can penetrate deep into wood much better than surface adhesives and they do not weaken the wood with drilled holes. Also, nails can distribute forces over a larger part of the surface than bolts, as a number of nails are required to transfer forces. Screws, on the other hand, have similar advantages as nails, with the additional benefit of better gripping capacity due to their threaded shaft. Dowel connections are similar to bolted connections, except that the dowels do not provide much axial capacity. Overall, connections made with many small bolts or dowels are generally stronger than connections with a few large fasteners.

To achieve fire resistance in a wood member, both the members (or components) and their connections must perform to provide required capacity throughout the fire exposure duration. In the case of wood connections, connector components, especially metal connecters such as nails, bolts, and screws, can be highly vulnerable to high-temperature exposure, as these fasteners get directly attacked through the burning of wood. As steel fasteners (nails or bolts) heat up quickly, these metal connectors will transfer (conduct) temperature not only to the surface of the wood but also across the thickness of connections, causing a softening in the connector, leading to a loss of grip and thus an inability to transfer forces. Hence, it is of utmost importance to provide fire protection to connections. Fire protection is often done though burying the metal fastener within the wood section or by applying a layer of sheeting/shielding (insulation) material such as solid wood or gypsum plasterboard. The application of fire retardants and other insulation compounds on joints/fasteners (connected surfaces) is another strategy that can help enhance fire performance of connection components and thus the overall fire resistance of the member.

10.5 Prescriptive-Based Approaches for Evaluating Fire Resistance

When full attention is given to all detailing aspects (including connectors), the fire resistance of a wood structural member is mainly dependent on three critical factors: the cross-sectional size of the wood member, any external fire insulation provided, and the material properties of wood. These factors are to be given full consideration in evaluating the fire resistance of wood members. Further, heavyweight wood members, such as columns and beams, derive their fire resistance primarily through cross-sectional size, while members that are part of lightweight construction, such as walls and floors, derive their fire resistance through supplemental fire insulation. Therefore, fire resistance calculation approaches are prescribed under two different categories: lightweight construction and heavyweight construction. Although both prescriptive and rational approaches can be applied for evaluating the fire resistance of wood members, prescriptive-based approaches are more common and practical for lightweight construction. Such prescriptive-based approaches for evaluating fire resistance at three levels, namely, testing, listing, and correlation methods, are discussed here.

10.5.1 Heavyweight Construction

Heavyweight wood members comprise load-bearing structural members such as columns and beams. These members are typically required to have a fire resistance rating of 30 minutes to 2 hours in most building applications.

Testing and Tabulated Data

The fire resistance rating of a wood member can be evaluated, as in the case of steel or concrete members, through undertaking standard fire tests in a furnace. However, since wood is a combustible material, unlike other types of construction materials, appropriate measures must be taken during testing to ensure that the burning of wood does not increase thermal energy, causing fire exposure temperatures in the furnace to exceed that to be applied as per the specified standard fire time-temperature curve. In addition, a simplified failure criterion, such as the critical temperature concept as in case of steel or concrete members, cannot be applied for evaluating the failure of wood members. Failure is to be determined based on the member-sustaining applied loading during fire exposure or limiting deflection to a specified limit. Based on the failure time attained in a test, fire ratings are assigned to columns, beams, and so on as 30, 45 (in some countries such as Canada), 60, 90, and 120 minutes. Data obtained from such fire resistance tests is utilized to develop listings for referencing in directories such as the Underwriters Laboratories (UL) directory and design standards and manuals (NDS, 2018).

Correlation Methods

In the last four decades, a number of sample *correlations* have been developed for evaluating the fire resistance of heavy wood members. The recent edition of the NDS (2018) incorporates some of these correlations for wood members stressed in one principal direction (under bending or axial loading) under 1 to 2 hours of exposure to a standard fire scenario. The fire resistance is expressed in terms of load ratio, R_s, for a given dimension of the wood member, where a value of unity or less indicates satisfying the specified fire rating. The value of load ratio, R_s, can be evaluated using simple expressions; one such expression for deriving fire resistance in a beam is given as

$$R_s = 2.85\ S_f\ C_{L\text{-}F} / S_s\ C_{L\text{-}s}\ C_D\ C_M\ C_t \tag{10.5}$$

where S_s = Section modulus using full (original) cross-sectional dimensions
S_f = Section modulus of reduced cross section as a result of burning
$C_{L\text{-}s}$ = Beam stability factor using full cross-sectional dimensions
$C_{L\text{-}f}$ = Beam stability factor using cross-sectional dimensions reduced from fire exposure
C_D = Load duration factor
C_M = Wet service factor
C_t = Temperature factor for in-service moisture conditions

Values of these factors are provided in the NDS wood design manual.

A number of tables for evaluating the fire resistance of different types of wood members are given in the NDS manual. A sample from NDS tables is shown in Table 10.2 for beams subjected to fire exposure from three sides. This table can be used to check if a wood beam of a certain sectional size ($b \times d$) and design load ratio (R_s) can achieve a 1-hour fire rating. Similar tables for 1.5- and 2-hour fire ratings for beams, columns, and decks under three and four fire side exposures are provided in the NDS (2018) manual.

The American Wood Council (AWC) manual and Canadian Wood Council (CWC) design manual provide empirical equations for evaluating the fire resistance of heavy wood members. These equations, adopted from the work by Lie (1977), are specific to fire resistance evaluation in heavy wood beams and columns. The failure of a wood beam or column is said to occur when the reduced sectional capacity exceeds the loading effect (bending moment or axial force) in the member at a given fire exposure time. Empirical equations are given to calculate the reduced cross-sectional capacity of a wood member under standard fire exposure of one hour or less. By knowing the mechanical and physical properties of the wood, the reduced sectional capacity of the member can be directly calculated at a required fire exposure duration. Sectional properties are to be computed assuming an effective char rate, β_{eff}, at a given time, t. The effective char depth is to be taken as 45.7, 42.4, and 40.1 mm at 60, 90, and 120 minutes of exposure to a standard fire. To account for reduction in strength and stiffness properties of wood directly, beneath the char layer, an additional 20% of the section is said to be not effective with respect to strength contribution.

As an illustration, in a wood beam exposed to standard fire exposure, failure is said to occur when the reduction in the cross section, and thus reduction in the original section modulus, reaches a critical value. Given the initial dimensions B (width) and D (depth), the fire resistance time can be calculated simply by using the following equations:

Beams heated on four sides (see Fig. 10.9):

$$R = 2.54fB\ [4 - 2(B/D)] \tag{10.6}$$

Beams heated on three sides (see Fig. 10.9):

$$R = 2.54\ fB(4 - B/D) \tag{10.7}$$

where $f = 1.3$ for $R < 0.5$ and $f = 0.7 + 0.3/R$ for $R \geq 0.5$.

Columns heated on four sides:

$$R = 2.54\ fB(3 - B/D) \tag{10.8}$$

Columns heated on three sides:

$$R = 2.54\ fB(3 - B/20) \tag{10.9}$$

Width (*b*) / Depth (*d*)	5-½	6	6-¾	6-⅞	7-¼	7-½	8-¼	8-½	8-¾	9	9-¼	9-⅝	10-½	10-¾	11	11-¼	12	12-¼	12-2/8	13-¼	13-½	13-¾	15
Design Load Ratio, R_s																							
5-½	0.45	0.52	0.60	0.61	0.65	0.67	0.73	0.74	0.76	0.77	0.79	0.81	0.85	0.86	0.87	0.88	0.90	0.91	0.91	0.94	0.95	0.95	0.98
6	0.48	0.56	0.65	0.67	0.70	0.73	0.79	0.81	0.82	0.84	0.85	0.87	0.95	0.93	0.94	0.95	0.98	0.99	0.99	1.00	1.00	1.00	1.00
6-¾	0.53	0.61	0.72	0.73	0.77	0.80	0.86	0.88	0.90	0.92	0.94	0.96	1.00	1.00	1.00	1.00	1.00	1.00	1.00	1.00	1.00	1.00	1.00
6-⅞	0.54	0.62	0.72	0.74	0.78	0.81	0.88	0.90	0.91	0.93	0.95	0.97	1.00	1.00	1.00	1.00	1.00	1.00	1.00	1.00	1.00	1.00	1.00
7-¼	0.56	0.64	0.75	0.77	0.81	0.84	0.91	0.93	0.95	0.97	0.98	1.00	1.00	1.00	1.00	1.00	1.00	1.00	1.00	1.00	1.00	1.00	1.00
7-½	0.57	0.66	0.77	0.78	0.83	0.86	0.93	0.95	0.97	0.99	1.00	1.00	1.00	1.00	1.00	1.00	1.00	1.00	1.00	1.00	1.00	1.00	1.00
8-¼	060	0.70	0.81	0.83	0.88	0.91	0.98	1.00	1.00	1.00	1.00	1.00	1.00	1.00	1.00	1.00	1.00	1.00	1.00	1.00	1.00	1.00	1.00
8-½	0.61	0.71	0.83	0.84	0.89	0.92	1.00	1.00	1.00	1.00	1.00	1.00	1.00	1.00	1.00	1.00	1.00	1.00	1.00	1.00	1.00	1.00	1.00
8-¾	0.62	0.72	0.84	0.86	0.91	0.93	1.00	1.00	1.00	1.00	1.00	1.00	1.00	1.00	1.00	1.00	1.00	1.00	1.00	1.00	1.00	1.00	1.00
9	0.63	0.73	0.85	0.87	0.92	0.95	1.00	1.00	1.00	1.00	1.00	1.00	1.00	1.00	1.00	1.00	1.00	1.00	1.00	1.00	1.00	1.00	1.00
9-¼	0.64	0.74	0.86	0.88	0.93	0.96	1.00	1.00	1.00	1.00	1.00	1.00	1.00	1.00	1.00	1.00	1.00	1.00	1.00	1.00	1.00	1.00	1.00
9 ⅝	0.65	0.75	0.88	0.90	0.95	0.98	1.00	1.00	1.00	1.00	1.00	1.00	1.00	1.00	1.00	1.00	1.00	1.00	1.00	1.00	1.00	1.00	1.00
10-½	0.68	0.78	0.91	0.93	0.99	1.00	1.00	1.00	1.00	1.00	1.00	1.00	1.00	1.00	1.00	1.00	1.00	1.00	1.00	1.00	1.00	1.00	1.00
10-¾	0.68	0.79	0.92	0.94	0.99	1.00	1.00	1.00	1.00	1.00	1.00	1.00	1.00	1.00	1.00	1.00	1.00	1.00	1.00	1.00	1.00	1.00	1.00
11	0.69	0.80	0.93	0.95	1.00	1.00	1.00	1.00	1.00	1.00	1.00	1.00	1.00	1.00	1.00	1.00	1.00	1.00	1.00	1.00	1.00	1.00	1.00
11-¼	0.69	0.80	0.94	0.96	1.00	1.00	1.00	1.00	1.00	1.00	1.00	1.00	1.00	1.00	1.00	1.00	1.00	1.00	1.00	1.00	1.00	1.00	1.00
12	0.71	0.82	0.96	0.98	1.00	1.00	1.00	1.00	1.00	1.00	1.00	1.00	1.00	1.00	1.00	1.00	1.00	1.00	1.00	1.00	1.00	1.00	1.00
12-¼	0.72	0.83	0.97	0.99	1.00	1.00	1.00	1.00	1.00	1.00	1.00	1.00	1.00	1.00	1.00	1.00	1.00	1.00	1.00	1.00	1.00	1.00	1.00
12-2/8	0.72	0.83	0.97	0.99	1.00	1.00	1.00	1.00	1.00	1.00	1.00	1.00	1.00	1.00	1.00	1.00	1.00	1.00	1.00	1.00	1.00	1.00	1.00

Table 10.2 Design Load Ratio, R_s, for One-Hour Fire Rating for Wood Flexural Member Exposed to Fire on Three Sides (*Adapted from NDS, 2018*)

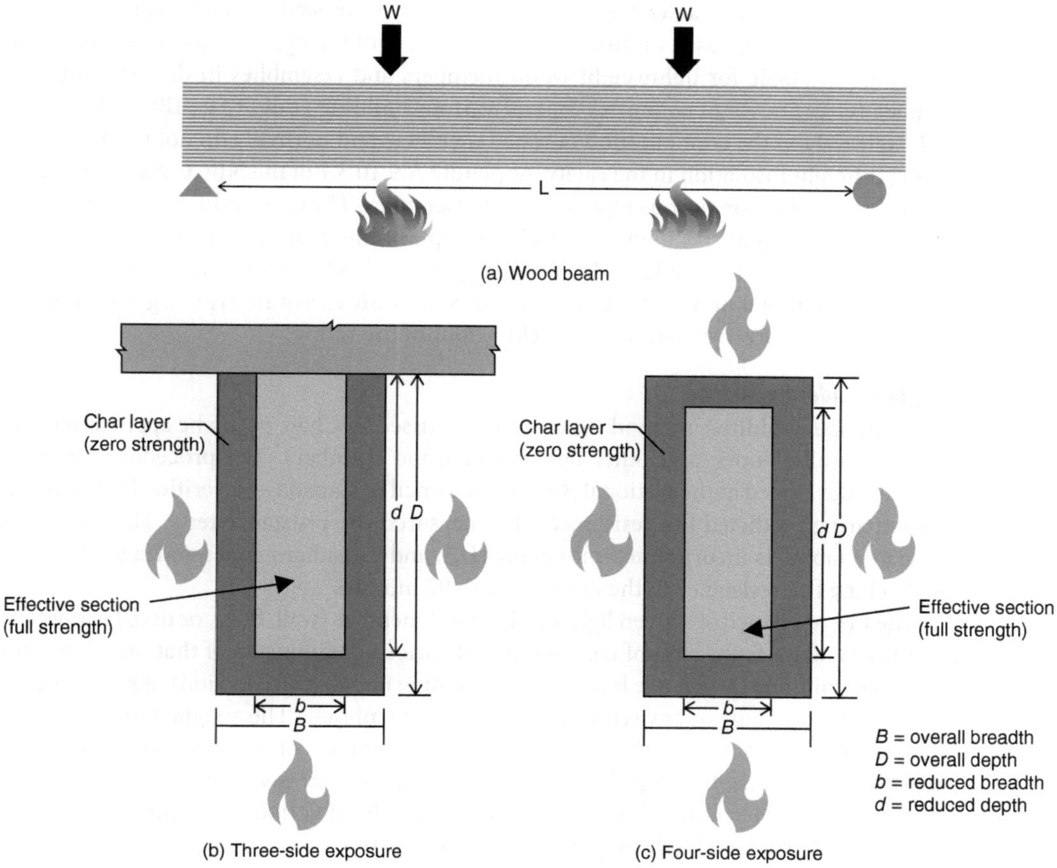

FIGURE 10.9 Cross-sectional reduction in a wood member exposed to fire.

where R = fire resistance of the member (min), B and D are the smaller and larger side of the member before exposure to fire (mm), f = a factor to take into account the load and effective length in the case of columns and can be obtained from graphs as shown in Fig. 10.10, K = effective length factor, and L = unsupported length of column (mm).

The above empirical formulas are applicable for a wood beam or a column with minimum nominal dimension (B or D) of 152 mm. In the case of a glued laminated member, the net finish width is to be a minimum of 130 mm. These equations are applicable up to one hour of standard fire exposure only.

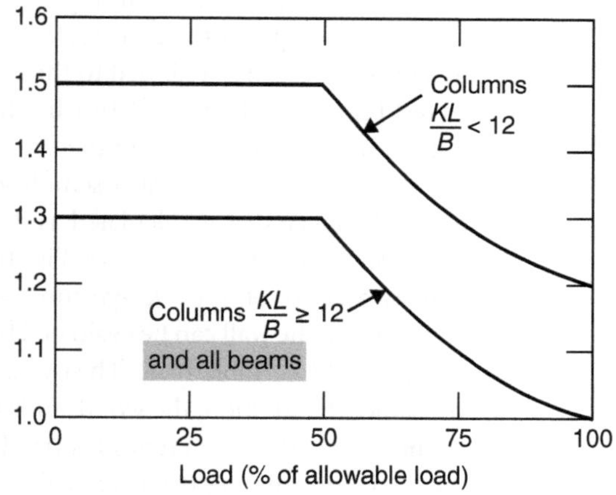

FIGURE 10.10 Factor, Z, for timber columns and beams. *(With permission from ASCE.)*

10.5.2 Lightweight Construction

Lightweight construction comprises walls, floors, and decks fabricated using wood studs or joists, together with layers of sheathing made of plywood, gypsum wallboard (GWP), oriented strand board (OSB), and additional insulation in the cavity. The required fire resistance ratings for walls, floors, and decks used in building applications are in the range of 30 minutes to 2 hours. The fire resistance of such members depends on the overall configuration of the assembly and contribution of each component of the assembly.

Testing and Tabulated Data

Fire resistance of walls, floors, and decks is mostly evaluated through standard fire tests. As in the case of wood columns and beams, special attention is to be paid to ensure that the specified time-temperature curve of the standard fire is followed during a fire test, since the wood member itself adds to the fuel.

The failure of the tested assembly (wall or floor) is said to occur when insulation, integrity, and stability criteria are exceeded. Based on fire test data, a number of listings, or associated tables, with fire resistance ratings are available for lightweight wood members and assemblies in directories (such as UL), design manuals (Gypsum Association, NDS), and various building codes (e.g., IBC, NBCC). These listings are applicable only to the configuration of tested members and normally do not permit modifications such as adding different insulation in the cavity. Appendix A 9.10.3.1 of the NBCC contains fire rating (and sound transmission class) for various types of wood assemblies. These assemblies can only be used for dwellings and small buildings as specified in NBCC. A sample of these ratings for wood walls and floors is shown in Table 10.3. As per these tables, a load-bearing or non-load-bearing wood wall made of 3889-mm studs spaced at 400 or 600 mm center to center protected with 15.9-mm Type X gypsum board and filled with 89-mm-thick absorptive material can yield 1 hour of fire rating.

Component Additive Method

The component additive method (CAM),[3] as discussed in Chap. 6, can be applied to derive fire resistance ratings of walls, floors, and other lightweight wood members. The procedure for this assessment was originally developed at the National Research Council of Canada and verified for various wood assemblies by comparing predicted fire resistance with measured fire resistance tests. The procedure, together with associated tables, is incorporated into many (U.S. and Canadian) codes and standards and can be applied for checking fire resistance in the range of 20 to 90 minutes.

The fire resistance of a given lightweight wood member (wall, floor, or deck) exposed to a standard fire exposure is taken as the sum of contribution of various components of that member, namely (1) the fire-exposed membrane (protective layer), (2) the framing members (wood joist or stud), and (3) any additional insulation in the cavity or any reinforcement of the membrane. The assigned times are based on the ability of the membrane to remain in place for a specified duration of fire exposure. The times assigned to the protective membrane, framing elements, and other factors are given in codes, standards, and handbooks. These times are determined through fire resistance tests on various configurations of assemblies. The fire resistance, *FR*, of a wood lightweight member is given as

$$FR = t_{m,layer\ 1} + t_{m,layer\ 2} + \cdots + t_{m,layer\ n} \tag{10.10}$$

where *FR* is fire resistance and $t_{m,layer\ n}$ is fire resistance of particular layer (component).

The times given in Table 10.4[4] are based on the membrane's contribution to the total fire rating of the assembly. The times assigned to the protective membranes are not the "finish ratings" of the material cited in test reports or listings. (A finish rating is defined as the time for an average temperature rise of 139°C or a maximum rise of 181°C on the fire-exposed side of the wood framing.)

As an illustration, a wall assembly with two or more protective membranes as that shown in Fig. 10.11 has a fire resistance that is calculated as the sum of the times, t_m, assigned for the individual components. As an example, for a wall assembly (see Fig. 10.11) made of 2 × 4-in. wood studs placed at 406 mm o.c. and profiled with two layers of Type X gypsum wallboard, together with insulation in the cavity on each side, the fire resistance of the wall can be evaluated by adding the contribution of each component as tabulated in the ASCE 29 standard (see Table 10.4). The two layers of Type X gypsum wallboard will give 25 × 2 = 50 minutes of fire resistance. In addition, the wood studs (of 5/8 in.) will yield 20 minutes, and the based slag mineral insulation in the cavity adds 15 minutes (as per Table 10.4). Thus, the total fire resistance of this wall assembly is 50 + 20 + 15 = 85 minutes (as no contribution can be taken from the unexposed layer, as per ASCE guidelines).

Provisions for Connections

The performance of connections in a wood member is critical during a fire event, and thus appropriate strategies, as outlined in Sec. 10.4, are to be adopted for achieving required fire performance. This is especially critical in the case of lightweight wood members, as a number of connections are required to join not only the member but also various components of the member. Prescriptive provisions are laid out for detailing connections in codes/standards and design manuals for achieving a certain level of fire resistance. These provisions are in terms of minimum dimension of attachments (screws, nails, etc.), spacing distances between connectors, type of fasteners, and any fire protection to be applied on these connectors.

---

[3] A similar procedure is also described in Eurocode 5 but is only applicable to evaluate the fire separation function of wall and floor assemblies. It is calculated as the sum of the contribution to fire resistance from each layer of material.

[4] It is worth noting that Eurocode 5 does not present explicit fire ratings for wood structural members and directs designers to adhere to national fire regulations.

(a) Walls

Type of Wall	Wall Number	Description	Fire-Resistance Rating		Typical Sound Transmission Class
			Loadbearing	Non-Loadbearing	
• Wood Studs	W1	• 38 mm × 89 mm studs spaced 400 mm or 600 mm o.c. • with or without absorptive material • 1 layer of gypsum board on each side	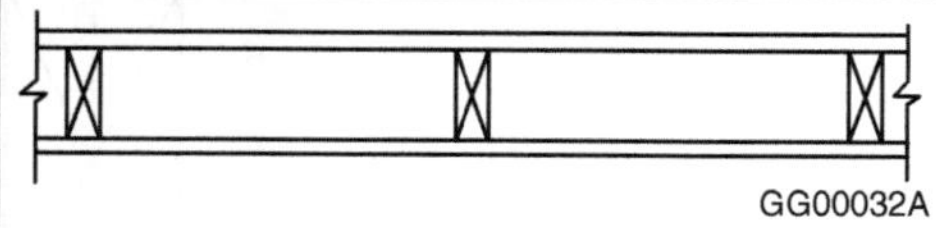		
• Single Row	W1a	W1 with • 89 mm thick absorptive material • 15.9 mm Type X Gypsum board	1 h	1 h	36
• Loadbearing or Non-Loadbearing	W1b	W1 with • 89 mm thick absorptive material • 12.7 mm Type X gypsum board	45 min [1 h]	45 min [1 h]	34
	W1c	W1 with • 89 mm thick absorptive material • 12.7 mm regular gypsum board	30 min	30 min [45 min]	32
	W1d	W1 with • no absorptive material • 15.9 mm Type X gypsum board	1 h	1 h	32
	W1e	W1 with • no absorptive material • 12.7 mm Type X gypsum board	45 min	45 min	32

(b) Floors

Type of Assembly	Assembly Number	Description	Fire-Resistance Rating	Typical Sound Transmission Class(STC)	Typical Impact Insulation Class (IIC)
	F8	• subfloor of 15.5 mm plywood, OSB or waterboard, or 17 mm tongue and groove lumber • on wood joists or wood I-joists spaced not more than 600 mm o.c. • with or without absorptive material in cavity • resilient metal channels spaced 400 mm or 600 mm o.c. • 1 layer of gypsum board on ceiling side	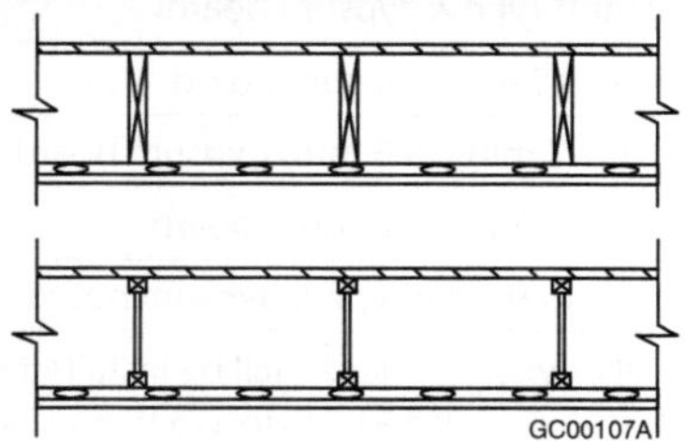		
	F8a	F8 with • no absorptive material in cavity • resilient metal channels spaced 400 mm o.c. • 15.9 mm Type X gypsum board	30 min	41	33
	F8b	F8 with • no absorptive material in cavity • resilient metal channels spaced 600 mm o.c. • 15.9 mm Type X gypsum board	30 min	43	36
	F8c	F8 with • absorptive material in cavity • resilient metal channels spaced 400 mm o.c. • 15.9 mm Type X gypsum board	30 min [45 min]	48	41
	F8d	F8 with • absorptive material in cavity • resilient metal channels spaced 600 mm o.c. • 15.9 mm Type X gypsum board	30 min	50	44

Table 10.3 Sample of Fire Ratings for Light Wood Construction (*Reproduced with Permission of the National Research Council of Canada, Copyright Holder*)

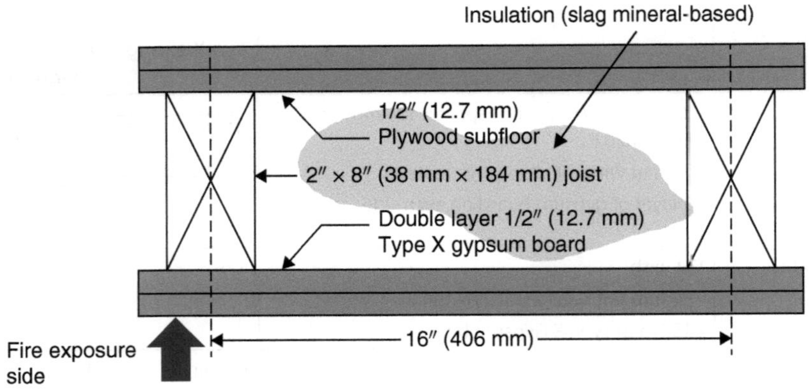

Figure 10.11 Illustration of fire resistance calculation in a wood wall.

(a) Protective layers

Description of Finish	Time (min)
3/8 in. (9.5 mm) Douglas fir plywood phenolic bonded	5
1/2 in. (12.7 mm) Douglas fir plywood phenolic bonded	10
5/8 in. (15.9 mm) Douglas fir plywood phenolic bonded	15
3/8 in. (9.5 mm) gypsum board	10
1/2 in. (12.7 mm) gypsum board	15
5/8 in. (15.9 mm) gypsum board	20
1/2 in. (12.7 mm) Type X gypsum board	25
5/8 in. (15.9 mm) Type X gypsum board	40
Double 3/8 in. (9.5 mm) gypsum board	25
1/2 ~ 3/8 in. (12.7 mm ~ 9.5 mm) gypsum board	35
Double 1/2 in. (12.7 mm) gypsum board	40
Double 5/8 in. (15.9 mm) Type X gypsum board	55
Note: On walls, gypsum board shall be installed with the long dimension (edge) parallel to framing members with all face layer joints and fasteners finished. 5/8 in. (15.9 mm) Type X gypsum board is permitted to be installed horizontally with the horizontal joints unsupported. On floor/ceiling or roof/ceiling assemblies, gypsum board shall be installed with the long dimension at right angles to framing members and shall have all face layer joints and fasteners finished.	

(b) Wood frame components

Description of Frames	Time (min)
Wood studs minimum 2 in. nominal (38 mm), 16 in. (406 mm) on center	20
Wood joists minimum 2 in. nominal (38 mm), 16 in. (406 mm) on center	10
Wood roof and floor truss assemblies, 24 in. (610 mm) on center	5

(c) Insulation or cavity components

Description of Additional Protection	Time (min)
Add to the fire endurance rating of wood stud walls if the spaces between the studs are filled with rockwool or slag mineral wool batts weighing not less than 2.5 nef of wall surface 40 kg/m^3	15
Add to the fire endurance rating of non-load-bearing wood stud walls if the spaces between the studs are filled with glass fiber batts weighing not less than 2 nef of wall surface 32 kg/m^3	5

Table 10.4 Fire Resistance Times for Various Components of a Wood Wall or Floor (*With Permission from ASCE*)

Component	Fire Resistance (min)	Provisions*
Nails	15	$d \geq 2.8$ mm
Screws	15	$d \geq 3.5$ mm
Bolts	15	$t \geq 45$ mm
Dowels	20	$t \geq 45$ mm

*d is the diameter of fastener; t is the thickness of side member.

Table 10.5 Fire Resistance of Unprotected Connections (*Reproduced from CEN*)

Eurocode 5 (CEN, 2004) lists a simple table for estimating the fire resistance of unprotected connections with side members of wood; this is reproduced in Table 10.5. The listed fire resistance rating is deemed to be satisfied in the case of unprotected wood-to-wood connections when specified spacing, edge-to-edge distances, and side member dimensions comply with the minimum requirements as per the room temperature design given in Eurocode 5–Part 1. For example, a nailed wood connection in a wood assembly designed according to Eurocode 5–Part 1, where nails have a diameter larger than 2.8 mm, would have a fire rating of 15 minutes. Eurocode 5 allows the fire resistance values specified in this table to be increased to 30 minutes by increasing the thickness or width of the wood side members, or by increasing the edge-to-edge distance to the fasteners.

10.6 Rational Approaches for Evaluating Fire Resistance

Rational engineering approaches can be applied for evaluating the fire resistance of wood structural members. In these approaches, degradation of temperature-dependent properties of materials, loss of cross section resulting from charring, and associated corner rounding effects can be taken into consideration in calculating the fire resistance of a wood member. However, the application of advanced analysis procedures for evaluating wood members' fire resistance is much more complex than that of steel or concrete members due to complexities associated with anisotropic properties of wood, presence of numerous layers of fire protection, charring phenomenon, and modeling connection performance.

10.6.1 General Analysis Procedure

The structural design of timber structures, even at ambient conditions, is quite different from that of concrete and steel structures due to the high variation in properties of wood, the anisotropic nature of these properties, and the need to include the effect of variations in environmental, geometric, and duration factors in strength calculations. Environmental (moisture), geometrical (e.g., flaws/cavities in wood), and durability (loading duration, etc.) properties can influence the capacity of a member. This complexity is overcome in the design process by incorporating specific simplified design values for each factor in strength calculations. Further, due to varying strength properties in different directions, the type of loading (i.e., bending, shear versus axial), member type (beam versus column), exposure type (dry versus moist state), and quality of wood (e.g., with/without knots), among others, the assigned design value for these factors also varies under different design limit states. For example, under limit state design at ambient conditions, the duration-of-load factor is 1.0 for short-duration loads acting on a wood member, and it is 0.8 or 0.6 for medium- or long-duration loads. In working stress design, the duration-of-load factor is 1.0 for long-duration loads and 1.25 or 1.6 for medium- or short-duration loads. A thorough discussion on such factors, together with guidance on the factors to be used in strength design calculations, can be found in NDS (2018) as well as in Eurocode 5 (CEN, 2004). Separate tables for each of these factors are provided, and a sample table from NDS for duration factor is shown in Table 10.6. These factors are to be given due consideration in evaluating the fire resistance of wood members as well.

The primary factors that influence the fire resistance of a wood member are reduction in sectional size from burning and charring and degradation in mechanical properties resulting from temperature rise. In addition, the secondary factors such as various geometric and environmental factors also influence resistance, as in the case of ambient strength design of wood members.

The application of rational calculation procedures for evaluating the fire resistance of a typical wood member can be illustrated by considering a typical wood beam of initial breadth, B, and depth, D (as shown in Fig. 10.12*a*) to be exposed to fire from three sides. Due to fire exposure, at a time t, the effective cross section (breadth and depth) of the beam reduces to b_f and d_f, respectively.

Load Duration	C_D	Typical Design Loads
Permanent	0.9	Dead Load
Ten years	1.0	Occupancy Live Load
Two months	1.15	Snow Load
Seven days	1.25	Construction Load
Ten minutes	1.6	Wind/Earthquake Load
Impact[2]	2.0	Impact Load

[1]Load duration factors shall not apply to reference modulus of elasticity, E, reference modulus of elasticity for beam and column stability, E_{min}, nor to reference compression perpendicular to the grain design values, $F_{c\perp}$, based on a deformation limit.
[2]Load duration factors greater than 1.6 shall not apply to structural members pressure-treated with water-borne preservatives AWPA Book of Standards (2011), or fire retardant chemicals. The impact load duration factor shall not apply to connections.

Table 10.6 Load Duration Factor, C_D^1, for Strength Calculation in Wood Members (*Adapted from NDS, 2018*)

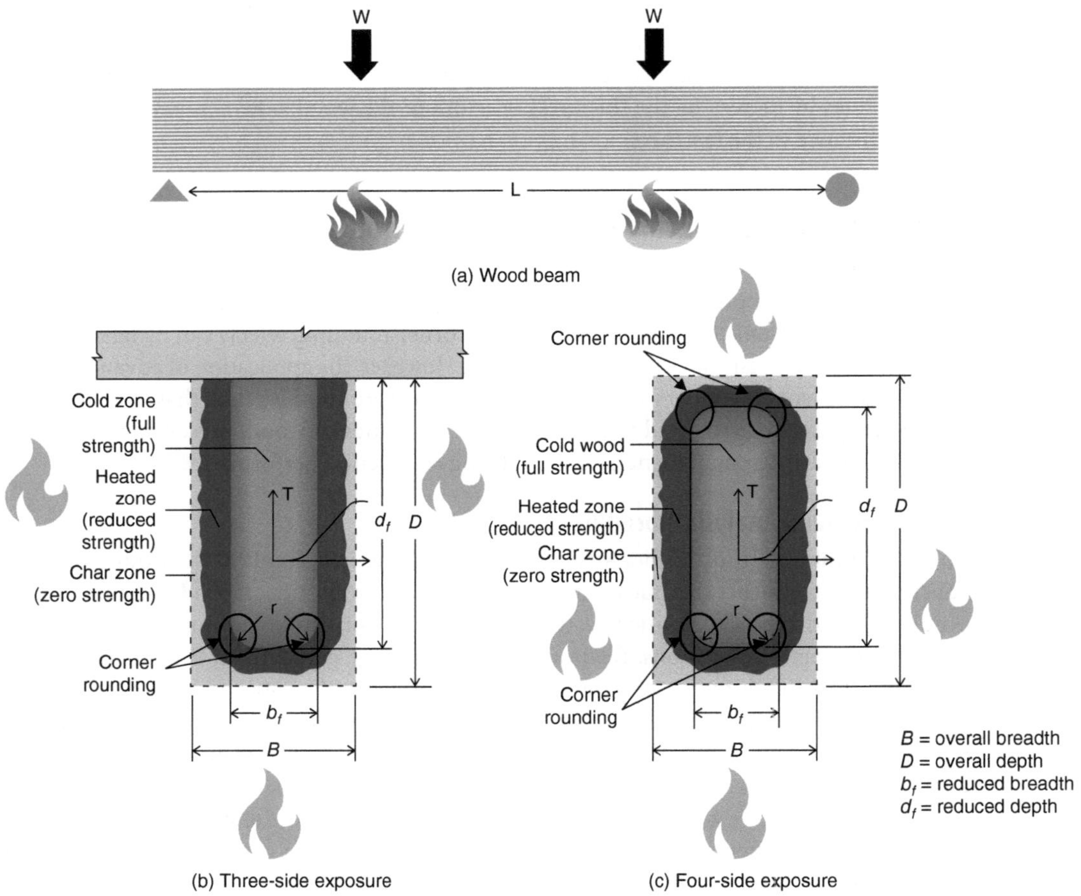

Figure 10.12 Reduction in member breadth and depth over time t. (*Adapted from NDS.*)

The reduced section at time t, resulting from burning and charring, can be calculated by knowing depth of charring, which is given as

$$c = \beta \times t \tag{10.11}$$

where c is the charred depth over time t, and β is the charring rate (mm/min) of wood as per the discussion in Sec. 10.2.3.

Knowing char depth, the effective breadth and depth of the section that can contribute to structural capacity at time t reduce to b_f and d_f, where

$$b_f = b - 2c \quad \text{and} \quad d_f = d - c \tag{10.12}$$

In addition to this, since the bottom two corners of the beam are subjected to heat flow from two directions, these corners get much more heated, leading to a larger temperature rise and thus a higher loss of section in corners, making the remaining cross section no longer rectangular but rather round around the corners. This will lead to a higher loss of section at the corners than the zones away from corners (see Fig. 10.12). The extent of such corner rounding, referred to as *arris,* varies with a number of factors such as heating rate, sectional size, and so on, and is expressed in terms of radius of arris (r). Although some researchers reported radius of the rounding of corners to be in 80% of the char depth (Majamaa, 1991), Eurocode 5 assumes a bilinear relation for calculating the radius of arris as a function of fire exposure duration (see Fig. 10.13). It is worth noting that most design codes assume the radius of the rounding to be equal to the depth of the char.[5]

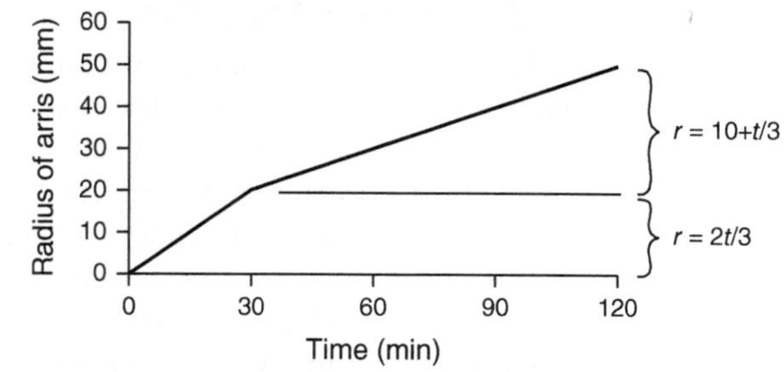

Figure 10.13 Radius of arris as a function of fire exposure time. *(Reproduced from CEN.)*

This corner rounding effectively reduces the overall cross section of a fire-exposed wood member and is to be accounted for in determining the reduced section modulus $Z_{f,r}$ at a given fire exposure time. Therefore, the reduced section modulus for a beam with three-side exposure, at time t, can be calculated knowing the reduced breadth and depth and radius of arris:

Without corner rounding:

$$Z_f = b_f d_f^2/6 \tag{10.13a}$$

With corner rounding:

$$Z_f = b_f d_f^2/6 - 0.215\, r^2 d_f \tag{10.13b}$$

where b_f is the beam residual width, d_f is the beam residual depth, and r is the radius of arris (see Fig. 10.14). The radius of arris rounding, r, equals the calculated depth of charring. The area of the section lost due to rounding will be $A = 0.215r^2$ and the center of gravity of this area will lie at a distance from either side of $y = 0.223r$.

By knowing the reduced sectional modulus and reduced strength properties of wood, the reduced moment capacity at time t can then be calculated as $M_f = Z_f \times f_{b,t}$, where $f_{b,t}$ is the reduced bending strength of the remaining wood that can be evaluated through temperature-dependent strength relations such as those given by Eurocode 5 and as discussed in Sec. 10.2.2. This sectional capacity is then compared against the applied bending moment present on the beam during fire exposure. Once the level of the reduced moment capacity falls below the applied bending moment, failure occurs and the duration to this point is taken as the fire resistance of the member.

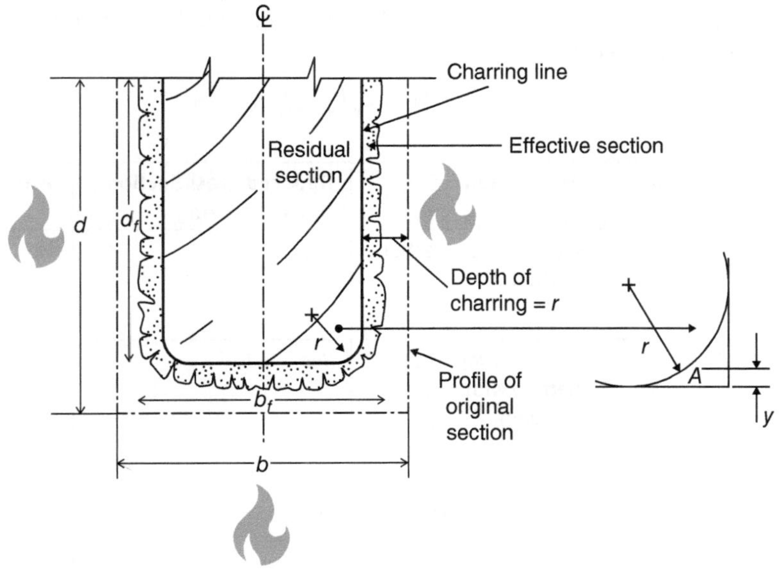

Figure 10.14 Corner rounding effect in a fire-exposed wood member.

[5] To account for rounding at the corners and reduction of strength and stiffness of the heated zone, NDS assumes that the nominal char rate values, β_n, are increased by 20%. The NDS does not provide specific expressions that account for the rounding effect.

The above discussion is for three-side exposure in beams. If a wood member (beam or column) is exposed to fire from four sides (see Fig. 10.12*c*), then the reduction in the cross section (on all four sides, together with the associated corner rounding effects) is to be accounted for in calculating reduced sectional modulus (Z_f) and thus reduced moment/axial capacity.

10.6.2 Simplified Approaches

Kersken-Bradley (1993) applied simplified approaches for evaluating the cross-sectional capacity of wood members based on the above-discussed general procedure. Eurocode 5 and European standard BS-5268 adopted these approaches for evaluating fire resistance in wood beams and columns.

Eurocode 5 recommends three methods for evaluating fire resistance of wood structural members, as tabulated in Table 10.7. The first two methods are based on simplified equations, while the third method requires the application of the advanced analysis procedure. Method I, referred to as the *effective cross-section method,* evaluates reduced sectional capacity by taking into account the loss of cross section of wood due to charring (with no consideration of corner rounding), and indirectly taking into account the effect of reduced strength properties in the uncharred portion of the wood section. On the other hand, Method II, referred to as the *reduced properties method with corner rounding,* takes into account the loss of cross section including corner rounding effects (arris) as well as reduced strength properties of the uncharred portion of the wood section. Method III, the *advanced analysis approach*, requires the finite difference/element method where all influencing parameters can be accounted for in tracing detailed response (including capacity, stresses, deflections, etc.) of fire-exposed wood members.

Reduced Cross-Section Method

The reduced cross-section method is the most direct method for evaluating fire resistance of a given wood member. In this method, the charred layer of wood, together with an additional layer of wood below the charred layer, is assumed not to contribute to reduced capacity of the member. The depth of charring is to be calculated as per the charring rate specified in Eurocode 5. The additional layer, referred to as the *zero-strength layer* (ranging anywhere from 0 to 7 mm), is said to possess no strength. For fire exposure duration exceeding 20 minutes, the zero-strength layer is taken as 7 mm, while for fire exposure between 0 to 20 minutes, it is taken as a fraction of 7 mm. Knowing the char depth, the effective depth and width of section are evaluated. The total reduced depth, d_{ef}, is given as the sum of two components, $d_{char,n}$, and an additional zero layer calculated as $k_0 d_0$, such that

$$d_{ef} = d_{char,n} + k_0 d_0 \tag{10.14}$$

where d_0 is taken as 7 mm and k_0 to be taken as equal to $t/20$ for $t < 20$ minutes, and equals 1.0 for $t > 20$ minutes. Knowing d_{ef}, the reduced breadth and width of the section is evaluated as $B\text{-}d_{ef}$ and $D\text{-}d_{ef}$. The remaining portion of the section is taken to have full room temperature strength. Then, the reduced section modulus of effective uncharred section Z_f can be calculated as in Eq. 10.13a. So, the flexural capacity of a wood member is calculated as

$$M_f = k_{fi}\, k_{mod,fi}\, f_b\, Z_f \tag{10.15}$$

	Method	Designation	Reduced Cross Section due to Charring	Arris	Reduced Properties
Simplified approaches	I	Reduced cross-section method	Yes	No	No
Simplified approaches	II	Reduced properties method with corner rounding	Yes	Yes	Yes
Advanced approach	III	Advanced analysis	Yes	Yes	Yes

Table 10.7 Eurocode Methods for Fire Resistance Analysis in Wood Members

where M_f = moment capacity in kN·m, $k_{mod,fi}$ = 1.0, and k_{fi} = 1.25 to account for property reduction due to knots, etc. for solid timber (1.15 for glulam, and 1.1 for laminated veneer lumber, or LVL), f_b = ambient temperature bending strength (MPa) of wood (and equal to f_{20} = 20% fractile of "cold" strength), Z_f = section modulus of the reduced section (m³) as in Eq. 10.13a.

Reduced Properties Method with Corner Rounding

The reduced properties method with corner rounding (referred to as Method II) gives consideration to reduction in breadth and depth as well as additional reduction at the corners due to corner rounding effect (arris). Further, the reduced strength properties in uncharred wood sections are accounted for in evaluating the reduced capacity of a wood member. The reduced cross-sectional dimensions (b_f, d_f) at a given time are evaluated by subtracting the charred width and depth using Eq. 10.12. Then, the loss of section at corners due to the corner rounding effect (radius of rounding, r) is evaluated as described in Sec. 10.6.1. Knowing reduced dimensions (b_f, d_f) and radius of rounding (r), the reduced section modulus Z_f can be evaluated as per Eq. 10.13.

Then, the reduced flexural capacity at time t of the beam is evaluated using the following equation, keeping in mind that the corner rounding effect is accounted for by modifying the plastic section modulus (Z_f) through Eq. 10.13, such that

$$M_f = k_{fi}\, k_{mod,fi}\, f_b\, Z_f \tag{10.16}$$

where Z_f includes corner rounding and is calculated as per Eq. 10.13b, k_{fi} = 1.25 to account for property reduction due to knots, etc. for solid timber (1.15 for glulam and 1.1 for laminated veneer lumber, or LVL), $k_{mod,fi}$ = 1.0, which is the strength reduction factor in residual sections, and it depends on the temperature and moisture content of the heated section. This is approximated as $k_{mod,fi} = 1.0 - 1/g\,(A_r/p)$, where p is the perimeter of the fire-exposed residual cross section (m), A_r is the area of the effective residual cross section (m²), and g is a property modification factor depending on the loading type, with the value of 200 for bending, 125 for compression, and 330 for tensile strength and MOE (see Fig. 10.15).

The reduced sectional capacity (moment or axial) evaluated in Method II is to be compared against corresponding actions (bending moment or axial force) during fire to evaluate the failure time or fire resistance. It should be noted that Method II yields slightly better fire resistance predictions, since the corner rounding effect is accounted for.

In the case of compression members (short columns), the reduced axial capacity at a given fire exposure time (t) can be evaluated similarly to the above procedure of evaluating the moment capacity in beams. The equation for reduced axial capacity in wood columns is given as

$$F_f = k_{fi}\, k_{mod,fi}\, f_b\, A_f \tag{10.17}$$

where A_f is the reduced cross-sectional area. In the case of slender (long) columns, appropriate consideration should be given to buckling effects as well.

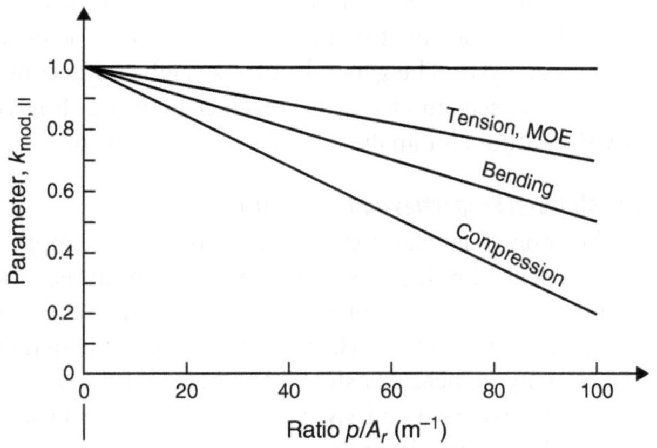

FIGURE 10.15 Parameter, k, as a function of ratio p/A_r. (*Reproduced from CEN.*)

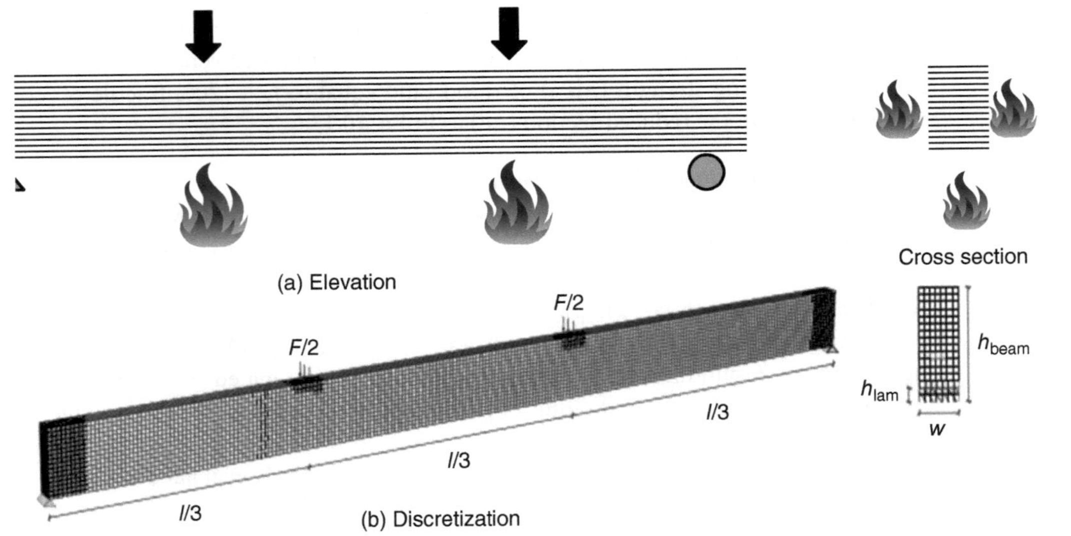

FIGURE 10.16 Discretization of a typical wood beam for finite element analysis. (*With permission from ASCE.*)

10.6.3 Advanced Analysis

The advanced analysis procedure can also be applied for evaluating detailed response parameters in a fire-exposed wood structure. However, such an analysis can be extremely tedious in the case of wood structures, and the accuracy of predicted response parameters may not be as close to realistic values as that in steel or concrete structures due to a high variation in the properties of wood and variations arising from charring, corner rounding, and so on. In such an advanced analysis, critical parameters specific to a wood structure, namely charring, rate of property degradation, moisture migration, and corner rounding effect, as well as different failure modes (i.e., buckling, delamination, crushing, etc.), are to be given appropriate consideration. Due to the complexity arising from anisotropic material properties of wood and the charring phenomenon, as well as a lack of reliable property data for many types of wood and a lack of validated computer models, rarely has system-level analysis been undertaken on wood structures. However, in recent years significant research and development efforts are underway to apply advanced analysis procedures at the sectional and member levels (Klippel and Frangi, 2017).

For example, in a recent study, Klippel and Frangi (2017) applied a finite element approach for modeling the fire behavior of glulam beams. A detailed response of glulam beams was simulated in ABAQUS through two stages of analysis: thermal and structural. Critical parameters that influence fire response of glulam beams, including geometric flaws in wood, charring phenomenon, temperature-dependent properties in wood and adhesive, and any variability in properties across wood lamella, are accounted for in this analysis. A summary of the analysis details is provided below.

Geometry and Discretization

Klippel and Frangi (2017) analyzed a glulam beam of 320-mm depth and 140-mm width, and spanning over 5760 mm. The wood beam was discretized into various layers, with each lamella composed of two elements across its depth, and this fine mesh was selected to achieve a high accuracy at both room and elevated temperature analysis. The general-purpose eight-node continuum element of type C3D8R with reduced integration was used, and the size of each element was limited to 3 mm. The discretization of the beam adopted by the authors for analysis is shown in Fig. 10.16.

High-Temperature Material Properties and Constitutive Laws

For tracing the thermomechanical response of wood beams, properties of wood at room and elevated temperatures and associated reduction factors are to be input. For thermal modeling, density of wood of 450 kg/m^3, with initial moisture content of 12%, was assumed. The temperature-dependent variation in the density and specific heat of wood and charring rate were implemented according to Eurocode 5 relations. As part of thermal analysis, heat transfer to the beam surfaces was evaluated using values specified in Eurocode 1 (CEN, 2002) wherein emissivity and convective heat transfer coefficient were taken as 0.8 and 25 W/(m^2·K), respectively. Charring of wood was taken into account by gradually changing the thermal properties of wood into those of char with increasing wood temperatures.

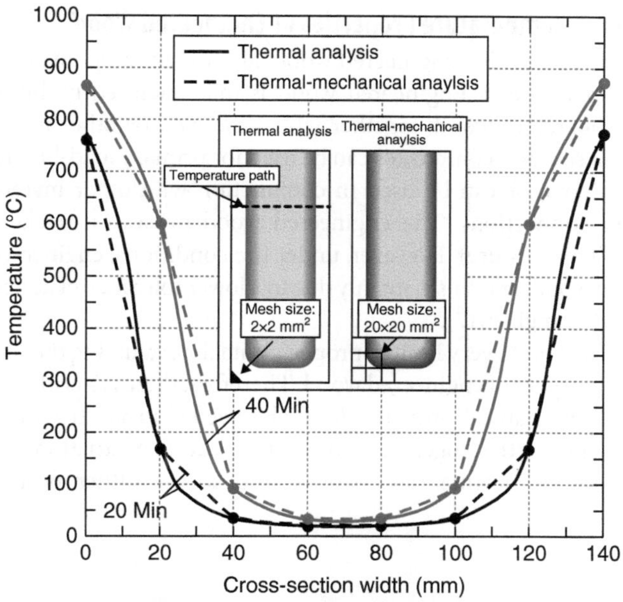

Figure 10.17 Temperature rise in analyzed glulam beam. (*With permission from ASCE.*)

In the case of mechanical (structural) analysis, timber was considered as a homogeneous solid with elastic material properties and a timber strength class of GL24. For this analysis, only the strength properties in the direction parallel to the grain are input. The mechanical properties in compression and tension of both timber and adhesive were defined as elastoplastic and elastobrittle stress-strain relations for different temperatures. Thereby, timber was modeled using the relative strength and stiffness reduction values as given in Eurocode 5 for the compressive and tensile strength and the MOE in tension.

Further, the mechanical behavior of timber at elevated temperatures was modeled by making some modifications to the constitutive model named "Concrete Damaged Plasticity" (CDP), as this material model is capable of modeling not only concrete but also other quasi-brittle materials. Timber, as in the case of concrete, has a quasi-brittle behavior in the tension zone and a ductile behavior in compression. The different behaviors in compression and tension in wood can be implemented in the CDP model by considering, in addition, the influence of temperature on the mechanical properties.

Failure Criteria

The failure of the beam in the analysis was evaluated by comparing the elemental stresses generated in the lengthwise direction of the beam in the tension zone with the ultimate strength allocated to this specific element. Klippel and Frangi (2017) state that the analysis was aborted when they were faced with numerical convergence problems. At this stage, the tensile strength in some elements of the beam exceeds the limiting stress and no further stress redistribution can occur. For the mechanical analysis at room temperature, the maximum load applied on the member was determined and thus the bending strength of the beam. In the case of a thermomechanical analysis, the time to reach the final time-step, at which failure occurs, represents the fire resistance of the beam. The authors only reported the visual outcome of temperature propagation across the analyzed beam (as shown in Fig. 10.17). Full details of the analysis including additional results can be found in Klippel and Frangi (2017).

10.7 Emerging Trends and Research Needs

In recent years, wood has been strongly promoted as a viable material for building construction due to today's focus on sustainability, as well as significant research and development efforts in advancing design and construction aspects relating to wood structures. However, the issue of resistance continues to be one of the main hindrances for the wider use of wood in low- to mid-rise buildings. Two critical areas where further research and development effort relating to the fire resistance of wood needed are with respect to establishing reliable property data for different types of wood and on the development of rational engineering approaches for fire resistance evaluation.

10.7.1 High-Temperature Properties of Engineered Wood

In the last three decades, engineered wood, also called *composite wood,* is being introduced in a variety of structural applications. Engineered wood is manufactured by binding or gluing the strands, fibers, or veneers or boards of wood, together with adhesives, or other methods of fixation to form a composite material. These wood composites can be used for primary load-bearing members, such as columns, beams, walls, and floors, or can be used in conjunction with other materials in building components such as trusses and connections. The engineered wood possesses excellent mechanical properties at ambient conditions (temperature). However, under fire conditions, engineered wood performs poorly, often worse than natural wood, and this is mainly due to a lower tolerance of adhesives and glue to elevated temperature (Klippel and Frangi, 2017).

Currently there is very limited property data for evaluating the fire resistance of specific types of woods, especially new types of engineered wood. Therefore, there is a critical need for developing high-temperature thermal and mechanical property data. The needed properties include temperature-dependent strength and stiffness properties, together with charring rate, and variations in the properties with different levels of moisture content, heating rate, density, and so on. Other properties that are needed include bond

Figure 10.18 Fourteen-story wood building (Brock Commons student residence) at the University of British Columbia (UBC), Vancouver, Canada.

degradation and delamination between wood layers, together with properties of adhesives suitable to apply on engineered wood products to enhance fire resistance. Further, properties of nails and screws often used to fasten connections are also needed. Other pressing research needs include the development of reliable fire insulation materials that can be used on composite wood structures with the potential of significantly enhancing fire resistance. In this direction, the effectiveness of some of the fire insulation materials (such as SFRM) typically used in steel structures should be explored for wood members.

10.7.2 Calculation Approaches for Fire Resistance Evaluation

Currently, wood buildings in most parts of the world are restricted to four to six stories in height, and one of the main reasons is the vulnerability of wood structures to fire. In the last few years, there is a strong push toward permitting wood buildings more than four stories in height in some of the countries. As an example, Canada is moving toward allowing wood construction for buildings up to six stories in general. In a unique case, a high-rise wood building, named Brock Commons student residence (see Fig.10.18), was built in 2016 at the University of British Columbia (UBC) in Vancouver, Canada. This building is 18 stories and 53 m tall and is the tallest wood structure in the world (Poirier et al., 2016). This building is made of concrete core but with many load-bearing members made of wood.

In order to move toward the use of wood in low- to mid-rise buildings, fire resistance issues must be addressed. For this purpose, the development of engineering approaches for evaluating the fire resistance of wood structures is warranted. These approaches, whether simplified or advanced analysis, should be able to account for all critical factors including realistic fire scenarios (incorporating effects of sprinklers), loading and restraint scenarios, and associated unique phenomena in wood under fire conditions such as charring, delamination, and specific property variations at elevated temperatures. There is a need to develop validated numerical models for tracing the response of wood structures under fire exposure to establish realistic fire performance. Case studies demonstrating the applicability of such models for tracing fire behavior on large-scale wood structures can enhance the acceptability of wood in low-rise construction. In parallel, research with the aim of improving protection systems to overcome vulnerability of wood structures is critical.

10.8 References

Audebert, M., D. Dhima, M. Taazount, and A. Bouchaïr. "Numerical investigations on the thermo-mechanical behavior of steel-to-timber joints exposed to fire." *Engineering Structures* 33, no. 12 (2011): 3257–3268.

AWPA Book of Standards, American Wood Preservers' Association, Selma, AL, 2011.

CEN. *Eurocode 1: Actions on Structures–Part 1-2: General Actions–Actions on Structures Exposed to Fire*. Brussels: European Committee for Standardization, 2002.

CEN. *Eurocode 5: Design of Timber Structures Part 1-2: General--Structural Fire Design*. Brussels: European Committee for Standardization, 2004. https://doi.org/10.1680/cien.2001.144.6.39.

Frangi, Andrea, Marco Bertocchi, Sebastian Clauß, and Peter Niemz. "Mechanical Behaviour of Finger Joints at Elevated Temperatures." *Wood Science and Technology* 46, no. 5 (2012): 793–812. https://doi.org/10.1007/s00226-011-0444-9.

Frangi, Andrea, and Mario Fontana. "Thermal Expansion of Wood and Timber-Concrete Composite Members under ISO-Fire Exposure." In *IAFSS Symposiums*. 2003. https://www.iafss.org/publications/fss/7/1111/view/fss_7-1111.pdf.

Gerhards, C. C. "Effect of Moisture Content and Temperature on the Mechanical Properties of Wood: An Analysis of Immediate Effects." *Wood and Fiber Science* 14, no. 1 (2007): 4–36. https://wfs.swst.org/index.php/wfs/article/view/501.

Kersken-Bradley, M. "EuroCode 5, Part 10 Structural Fire Design." In *Proceedings of the Oxford Fire Conference*. Oxford, July 1–2, 1993. Timber Research and Development Association, Princess Risborough, England.

Klippel, Michael, and Andrea Frangi. "Fire Safety of Glued-Laminated Timber Beams in Bending." *Journal of Structural Engineering* 143, no. 7 (2017): 04017052. https://doi.org/10.1061/(ASCE)ST.1943-541X.0001781.

Knudson, RM, and AP Schniewind. "Performance of Structural Wood Members Exposed to Fire." *Forest Products Journal*. 1975. http://agris.fao.org/agris-search/search.do?recordID=US7501067.

Konig, J, and L Walleij. "Timber Frame Assemblies Exposed to Standard and Parametric Fires, Part 2: A Design Model for Standard Fire Exposure." Rapport I 0001001. 2000. Retrieved from http://urn.kb.se/resolve?urn=urn:nbn:se:ri:diva-28624.

Lie, T.T. "A Method for Assessing the Fire Resistance of Laminated Timber Beams and Columns." *Canadian Journal of Civil Engineering* 4, no. 2 (1977): 161–169. https://doi.org/10.1139/l77-021.

Majamaa, J. *Calculation Models of Wooden Beams Exposed to Fire*. Technical Report. Espoo, Finland: Technical Research Centre of Finland,1991. https://www.osti.gov/etdeweb/biblio/10141486.

NDS. *Technical Report No. 10: Calculating the Fire Resistance of Exposed Wood Members and Assemblies*. Leesburg, VA: American Wood Council, 2018.

Östman, Birgit A.-L., and Lazarous D. Tsantaridis. "Heat Release and Classification of Fire Retardant Wood Products." *Fire and Materials* 19, no. 6 (1995): 253–258. https://doi.org/10.1002/fam.810190603.

Poirier, Erik, Manu Moudgil, Azadeh Fallahi, Sheryl Staub-French, and Thomas Tannert. "Design and Construction of a 53-Meter-Tall Timber Building at the University Of British Columbia." In *World Conference on Timber Engineering*, August 22–25, Vienna, Austria, 2016. http://www.proholz.at/fileadmin/proholz/media/documents/Thomas-Tannert.pdf.

USDA. *Wood Handbook*. Forest Product Agriculture Laboratory. Washington, DC: USDA, 2010. https://www.fs.usda.gov/treesearch/pubs/37440.

Wang, Song-Yung, Te-Hsin Yang, Li-Ting Lin, Cheng-Jung Lin, and Ming-Jer Tsai. "Fire-Retardant-Treated Low-Formaldehyde-Emission Particleboard Made from Recycled Wood-Waste." *Bioresource Technology* 99, no. 6 (2008): 2072–2077. https://doi.org/10.1016/J.BIORTECH.2007.03.047.
White, RH. *Charring Rates of Different Wood Species*. Ph.D. dissertation. Madison, WI: University of Wisconsin, 1988.
Young, Wen Bin, and Yu Chi Tsao. "The Mechanical and Fire Safety Properties of Bamboo Fiber Reinforced Polylactide Biocomposites Fabricated by Injection Molding." *Journal of Composite Materials*. 2014. https://doi.org/10.1177/0021998314554437.

10.9 Numerical Examples

10.1

Calculate the fire resistance of an interior floor assembly, shown below, by applying the component additive method. Use the fire resistance values listed in Table 10.3 for different components.

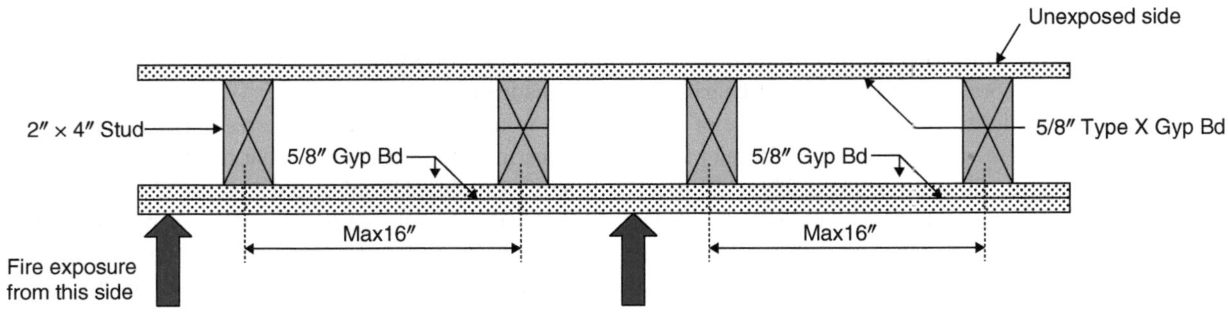

Solution

Fire resistance of this floor assembly is the sum of fire resistance contributions of different components:

Contribution to fire resistance from two Type X gypsum board layers of 5/8 in. on fire-exposed side = $2 \times 20 = 40$ minutes

Contribution to fire resistance from 2 in. × 4 in. wood studs = 20 minutes

No contribution is allowed for GWB located on fire unexposed side = 0.

Thus, total fire resistance of wall assembly = 60 minutes.

10.2

Calculate the maximum design load ratio limit as per NDS provisions for a simply-supported Douglas fir glulam beam with a cross section of 6¾ in. × 13½ in. The beam is exposed to fire from three sides for 60 minutes. Assume a charring rate of 0.03 in./min as per NDS. Also assume various wood design factors as $C_D = 1.0$, $C_M = 1.0$, $C_t = 1.0$, $C_{L\text{-}s} = 1.0$, $C_{L\text{-}F} = 1.0$.

Solution

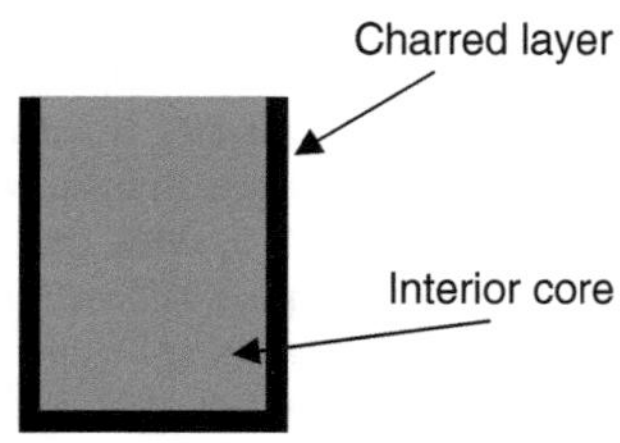

Original section modulus (S_s) of wood beam:

$$S_s = bd^2/6 = (6.75)(13.5)2/6 = 205 \text{ in}^3$$

Charred depth, z, at 60 minutes $= 0.03 \times 60 = 1.8$ in.

Reduced section modulus of beam after charring of 1.8 in. depth on three sides:

$$S_f = (b - 2z)(d - z)^2/6 = (6.75 - 3.6)(13.5 - 1.8)^2/6 = 71.9 \text{ in}^3$$

Applying NDS equation (Eq. 10.7); load ratio (R_s)[6] of the beam:

$$R_s = 2.85\, S_f\, C_{L\text{-}F} / S_s\, C_{L\text{-}s}\, C_D\, C_M\, C_t$$

$$R_s = 2.85\, (71.9)\, (1)/(205)\, (1)\, (1)\, (1)\, (1)\, (1) = 1$$

Since the load ratio equals 1.0, the beam meets the 1-hour fire rating requirement.

[6] These factors are given in NDS (2016) for typical loading conditions.

10.3

Determine the fire resistance of a wood column made of glulam. The effective length (KL) of the column is 20 ft and the cross section is 8¾ in. (B) and 10½ in. deep (D). The beam as per room temperature has an axial load capacity of 78.7 kips. The column to a concentrated total floor load (DL + LL) is 50 kips.

Solution

The column dimensions are $B = 8.75$ in., $D = 10.5$ in., $KL = 20$ ft.

Load ratio of the column, i.e., applied load/axial capacity under ambient conditions = 50/78.7 = 0.635.

For a slenderness ($KL/B = 27.4$), and load ratio = 0.635, the f factor for the column (as per Fig. 10.10 from NDS) = 1.16.

Applying Lie's approach (Eq. 10.13), fire resistance of this column

$$R = 2.54\, fB(3 - B/D)$$
$$R = 2.54\,(1.16)\,(8.75)(3 - 8.75/10.5) = 56 \text{ minutes}$$

10.4

Evaluate whether the following simply supported timber beam has 60 minutes of fire resistance as per simplified methods specified in Eurocode 5. This beam is of a rectangular cross section of 720 × 130 mm and has a span of length 7.5 m. The beam is subjected to a dead load (D) and live load (L) of 4 and 3.5 kN/m under fire conditions. Assume three-sided exposure (top is protected by supporting deck) and ambient temperature bending strength of wood (f_b) to be 17.7 MPa. Assume k_{fi} to be 1.15 as given in Eurocode 5.

Solution

Calculate maximum applied (M_{app}) moment on the beam under fire conditions:

$$w = D + 0.5L$$
$$w = 4 + 3.5 = 7.5 \text{ kN/m}$$
$$M_{app} = wL^2/8 = 52.7 \text{ kN·m}$$

Fire resistance for this beam can be evaluated by applying either Method I or Method II (as per Eurocode 5) as discussed in Sec. 10.6.2. Given: $f_b = 17.7$ MPa and $k_{fi} = 1.15$

Reduced Cross-Section Method (Method I)

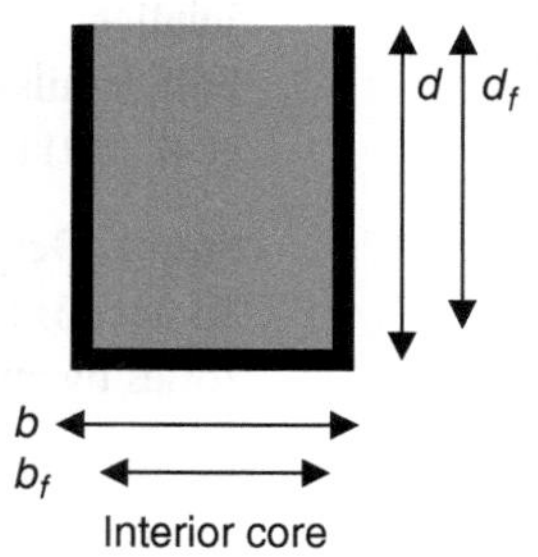

Rate of charring = 0.65 mm/min

Depth of char at 60 minutes = 60 × 0.65 = 39 mm

Thickness of zero strength layer: 7 mm

Effective depth of char = 46 mm

Reduced breadth, $b_f = 130 - 2 \times 46 = 38$ mm

Reduced depth, $d_f = 720 - 46 = 674$ mm

Reduced section modulus, $Z_f = b_f \times d_f^2/6 = 2.87 \times 10^6 \text{ mm}^3 = 0.00287 \text{ m}^3$

Reduced moment capacity at 60 minutes = $M_f = k_{mod,fi}\, k_{fi}\, f_b\, Z_f = 1.0 \times 1.15 \times 17.7 \times 10^3 \times 0.00287 =$ 58.46 kN·m

$M_{cap} > M_{app}$ Therefore, the beam will not fail at 60 minutes and will meet 1-hour fire rating requirement.

Reduced Properties Method (Method II)

Rate of charring = 0.65 mm/min

Depth of char at 60 minutes = 60 × 0.65 = 39 mm

Reduced breadth, $b_f = 130 - 2 \times 39 = 52$ mm

Reduced depth, $d_f = 720 - 39 = 681$ mm

Reduced beam area = $b_f \times d_f \approx 36.3 \times 10^3 \text{ mm}^2 = 36.3 \times 10^{-3} \text{ m}^2$

Reduced beam perimeter = $(2d_f + b_f) = 1414$ mm = 1.414 m

$p/A_r = 36.3 \times 10^{-3}/1.414 = 39.04\ \text{m}^{-1}$

Reduction property factor, $k_{mod,fi} = 1.0 - 1/g\,(A_r/p)$, where $g = 200$ for bending

$k_{mod.fi} = 1 - 39.04/200 = 0.805$

Reduced section modulus $(Z_f) = b_{fd}\,{}_f^2/6 = 53.2 \times 681^2/6 = 4.12 \times 10^6\ \text{mm}^3 = .00412\ \text{m}^3$

Reduced sectional moment capacity at 60 minutes $= M_{cap} = k_{mod.fi}\, k_{fi} f_b\, Z_f$

$Z_f = 53.2 \times 681^2/6 - 0.215 \times 39^2 \times 681 = 3.88 \times 10^6\ \text{mm}^3$

Reduced moment capacity at 60 minutes $= 0.805 \times 1.15 \times 17.7 \times 0.00412 \times 10^3 = 67.5$ kN·m

$M_{cap} > M_{app}$ Therefore, the beam will not fail at 60 minutes and will meet 1-hour fire rating requirement.

10.5

Analyze the response of a fire-exposed metal-wood connection, shown in the figure below, by applying the advanced analysis procedure (MSC-Marc software). This connection is composed of a wood member connected to a steel plate via steel dowels as shown below.

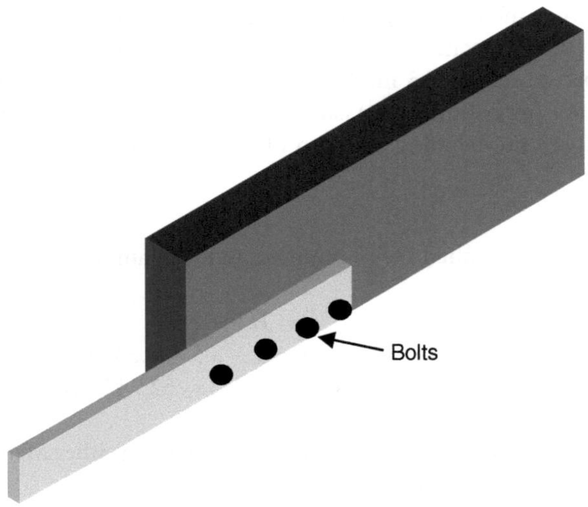

Solution

Full details on the steps for undertaking advanced analysis, together with results, are provided by Audebert et al. (2011). The following steps are to be applied.

Step 1. Develop a geometric model of the connection and mesh it using HEX20 elements (as shown below). To account for the complex nature of timber, the mesh of the model can be discontinuous at the contact zones by applying friction between the components in contact using the Coulomb criterion.

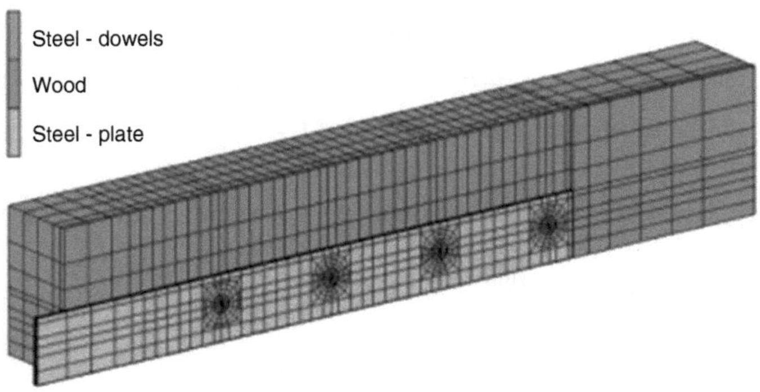

Step 2. Input temperature-dependent properties of wood and steel dowels as per Eurocode 5 (for wood) and Eurocode 3 (for steel) recommendations.

Step 3. Heat transfer analysis will output temperature rise across the depth of the connection.

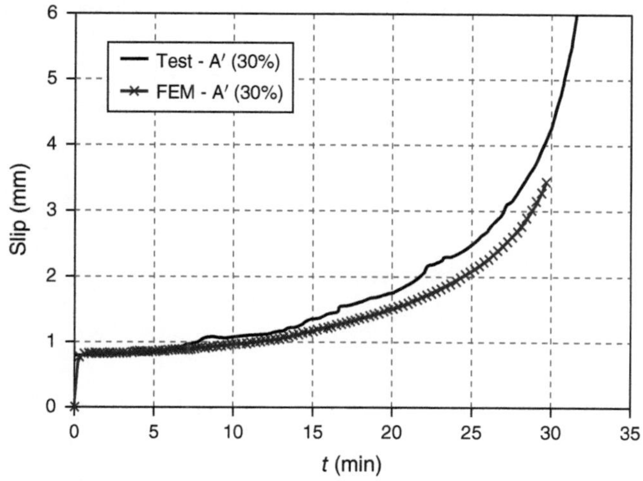

FIGURE E10.1 Time–slip progression of connection subjected to a load ratio of 30%.

Step 4. Strength analysis will output slip within the wood-metal connection.
Step 5. The failure of this connection under fire conditions can occur in different modes. In the analysis, Audebert et al. (2011) determined failure based on the rate of increase in slip between two successive time increments. When this rate of slip exceeded 50%, failure was said to occur. A sample of time-slip history of modeled connection, loaded at 30% of its initial capacity, is shown in Fig. E10.1.

CHAPTER 11

Concrete Structures Strengthened with Fiber-Reinforced Polymer Composites

11.1 Overview

This chapter provides an overview of the fire behavior of concrete structural members strengthened with fiber-reinforced polymers (FRPs). The variation in thermal and mechanical properties of FRP with temperature, as well as that of commonly used insulation materials, is reviewed. Strategies that can be developed for achieving the required fire resistance in FRP-strengthened RC structures are outlined. The calculation methods, both simplified and advanced approaches, for evaluating fire resistance of FRP-strengthened concrete structures are presented. Finally, research needs for overcoming some of the shortcomings relating to the fire performance of FRP-strengthened concrete structures are discussed.

11.2 Strengthening of Structures

11.2.1 General

Concrete structures usually have decades of service life. During this long service time, they can experience severe deterioration and damage due to poor maintenance, corrosion in steel reinforcement, and aging of concrete. Therefore, retrofitting deteriorated and damaged concrete structures is becoming a routine exercise in the civil engineering field to keep civil infrastructure in workable condition. In addition, structures are often strengthened to facilitate increased load-carrying capacity to withstand extreme loading events arising from earthquake, blast, and impact. Traditionally, retrofitting and strengthening has been accomplished using conventional materials and techniques, such as extending the size of the concrete section, bonding steel plate at the outer surface of the member, or applying external post-tensioned steel tendons.

In recent years, fiber-reinforced polymers have emerged as an attractive alternative to conventional strengthening techniques in retrofitting and strengthening applications due to the numerous advantages they offer over traditional construction materials such as concrete and steel. This is because FRP composites have high strength- and stiffness-to-weight ratio, corrosion resistance, environmental durability, and inherent tailorability, and can be easily adopted for use in the repair and rehabilitation of existing infrastructure. Applications for FRP include seismic retrofitting of bridges and building columns; repair and strengthening of beams, girders, and slabs; and the rehabilitation of special structures such as chimneys and reinforced and unreinforced masonry walls. A number of research studies and practical applications have demonstrated that FRP strengthening can offer an effective solution for enhancing the axial, flexural, and shear capacity of RC structures (Teng et al., 2002; Rasheed, 2014).

11.2.2 Type of Strengthening

Two types of techniques can be applied for strengthening existing concrete structures—either as externally bonded (EB) FRP reinforcement or as near-surface mounted (NSM) internal reinforcement (Fig. 11.1). In the case of EB strengthening, FRP plates, strips, or sheets are applied to the surface of a concrete member. The external surface of concrete is prepared through grinding, sandblasting, or high-pressure water jet to remove

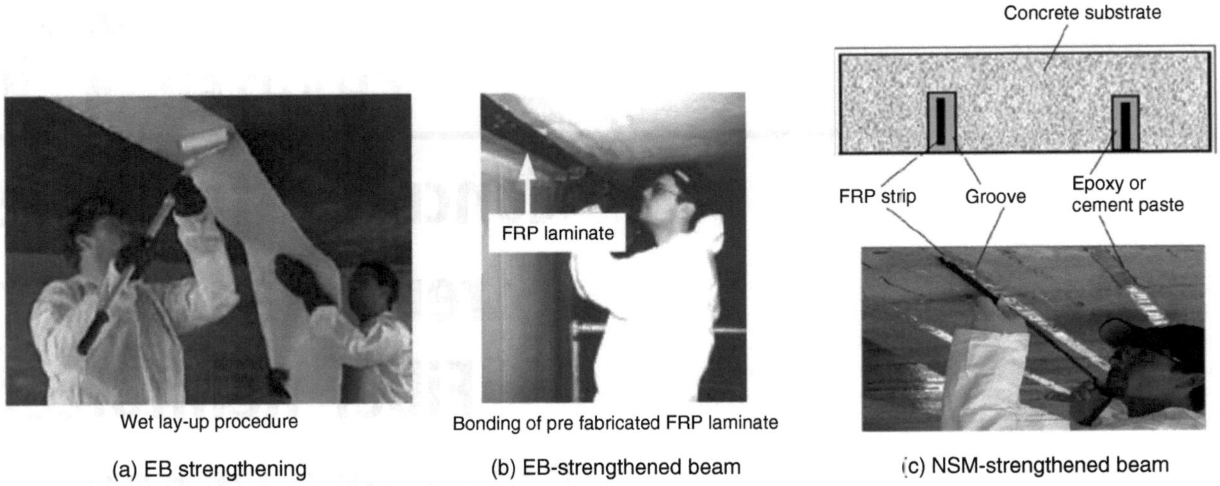

Figure 11.1 Application of FRP strengthening on concrete structures in the field.

the weak concrete surfaces. For RC columns, FRP sheets are wrapped around the exterior sides of a column. In the case of beams or slabs, the FRP plates or strips are attached to the soffit (and sides) of the beam/slab. The application of this type of strengthening technique is direct and involves less effort and time. However, one of the major drawbacks is the fact that externally applied FRP cannot develop full bonding with the concrete surface, as only one side of FRP (attachment) is in contact with the concrete surface. Also, low levels of axial strain get developed in FRP. Thus, in the EB method of strengthening, the full tensile strength potential of FRP cannot effectively be utilized (Lorenzis and Teng, 2007). In order to overcome such limitations, an alternative technique referred to as near-surface-mounted FRP strengthening has recently emerged.

In the NSM strengthening technique, FRP strips or rods are inserted into precut grooves on the concrete cover of an RC member, and then the grooves are filled with an epoxy adhesive or cementitious grout. The adhesive or grout ensures that the NSM FRP strip or rod is well anchored to the concrete inside and functions as an effective tensile reinforcement in resisting loading on a concrete member. NSM-strengthened members, compared to EB strengthening, can develop better bond adherence, as all sides of the strips/rods are attached to the adhesive (Barros et al., 2016).

There are a few disadvantages with FRP as a strengthening material, and these have limited its wider use in certain applications. One such major disadvantage is its poor performance under fire exposure. Under elevated temperature exposure, the polymer matrix of FRP composite decomposes, releasing heat, smoke, soot, and toxic volatiles. From a structural point of view, the strength and stiffness properties of the FRP composite decrease rapidly at elevated temperatures. Specifically, because FRP is bonded to a concrete structure with epoxy, the bond between the FRP and the concrete degrades significantly when the temperature in the FRP reinforcement exceeds the glass transition temperature (T_g) of the epoxy, thus affecting the stress transfer between the concrete member and the FRP reinforcement. Due to these problems, potential fire exposure is a major concern for the wider use of FRP strengthening in RC structures, especially in buildings, where fire resistance is a key requirement of design.

11.2.3 FRP Materials

Fiber-reinforced polymers belong to a class of materials commonly referred to as composites. Composite materials are manufactured through the combination of two or more individual materials to form an enhanced material with improved properties that are superior to those of its constituents. FRP composite materials are composed of high-strength continuous fibers embedded in a polymer matrix. The embedded fibers are the main reinforcing elements, while the polymer matrix acts as a binder, and thus protects the fibers and transfers stresses to and between the fibers. Different fiber types—glass, carbon, and aramid—are utilized in the fabrication of FRPs commonly used in strengthening civil engineering structures. The matrix material that can be used for fabricating FRP is made of polymers or resins, metals, or ceramics. The polymers are the most common among matrix materials used in FRP for civil engineering applications because of its ease of manufacturing and relatively inexpensive tooling. The combination of different fiber types with polymer matrices has led to the development of various FRP materials such as carbon FRP (CFRP), glass FRP (GFRP), aramid FRP (AFRP), basalt FRP (BFRP), and the newly developed polyethylene naphthalate (PEN) and polyethylene terephthalate (PET) fiber composites (Teng et al., 2002; Bai et al., 2014).

The aforementioned FRP composites are manufactured in a variety of products, namely prefabricated plates, bars, sheets, and anchorages. Prefabricated FRP elements are typically stiff and cannot be bent on-site to wrap around columns or beams but can be used for internal (or NSM) reinforcement (bars or stirrups). FRP fabric, on the other hand, is available as continuous unidirectional or bidirectional sheets, supplied on rolls, that can be easily tailored to fit any geometry and can be wrapped around complex profiles of structural members to be strengthened. FRP fabrics can be adhered to the tension side of flexural members as additional tension reinforcement to increase flexural capacity or can be wrapped around the sides of beams (webs of joists) to enhance shear capacity. In the case of columns, FRP sheets can be wrapped around the columns to increase axial capacity, as well as to improve ductility and energy dissipation capacity.

The resins and fibers forming FRP are usually developed as one system, based on material testing and structural testing. Mixing or replacing a component of one FRP system with a component from another system is not acceptable and can adversely affect the properties of the cured system. The bond between the FRP system and the existing concrete structure is critical, and as such, proper surface preparation of the concrete member to be strengthened is essential. The adhesive systems used to bond FRP to the concrete substrate is made of two parts: a primer that is used to penetrate the concrete substrate and improve the bonding of FRP to the member and an epoxy putty to fill small surface voids in the substrate and provide a smooth surface to which the FRP system is bonded. Most epoxies used in FRP strengthening systems are adversely affected by exposure to ultraviolet light, but can be protected using acrylic coatings, cementitious coatings, and other types of coatings (Czarnecki et al., 2013).

11.2.4 Need for Fire Resistance

Typically, conventional reinforced or prestressed concrete structural members provide the required fire resistance, without any external fire protection, as long as appropriate concrete cover is maintained to the steel reinforcing (or prestressing) bars. However, when the same concrete member is strengthened with FRP reinforcement, its fire resistance (or basic load-carrying capacity at elevated temperature exposure) can be lower than that of the prestrengthened concrete member. This is due to the fact that the fire resistance of an FRP-strengthened RC member depends not only on the properties of the original concrete member but also on the properties of the FRP strengthening system. FRP experiences more rapid strength, stiffness, and bond degradation than steel rebars or concrete at elevated temperatures. In addition, FRP is highly flammable and susceptible to rapid degradation in bond properties at elevated temperatures, and thus, the flexural/axial/shear capacity of an FRP-strengthened member decreases at a faster pace under fire exposure than a conventional RC member (Kodur et al., 2007). Failure of a strengthened member occurs when the decreasing capacity at the critical section drops below the effect of actions (moment, shear, or axial force), resulting from applied loading. Therefore, under similar levels of applied loading, the fire resistance of an FRP-strengthened RC member can be much lower than that of an unstrengthened RC member.

Repaired or strengthened structural members are required to satisfy fire resistance requirements as specified in building codes and standards (ACI, 2017; ICC, 2018). Several researchers have conducted experimental and numerical studies on the response of fire-exposed FRP-strengthened RC beams, slabs, and columns, and have quantified critical factors influencing the fire resistance of these members. Results from these studies indicate that an FRP-strengthened structural member is highly susceptible to damage from fire exposure, and therefore external fire insulation needs to be applied to achieve required fire resistance ratings in most cases (Kodur et al., 2007; Palmieri et al., 2012). The resulting fire resistance in such an FRP-strengthened member will depend on the properties of the original concrete member, as well as on the properties of the added FRP system, together with any applied fire insulation. In order to evaluate the actual fire resistance of FRP-strengthened concrete members, calculations should be carried out, and for such calculations, temperature-dependent properties of FRP and fire insulation, in addition to concrete and reinforcing (or prestressing) steel, are needed.

11.3 Property Relations for FRP at Elevated Temperatures

An FRP-strengthened concrete member is composed of a number of constituent materials, namely concrete, reinforcing (or prestressing) steel, FRP, and fire insulation. As such, the fire resistance of an FRP-strengthened structural member depends on temperature-dependent properties of these constituent materials. Temperature-dependent property relations and their trends for steel and concrete, as well as fire insulation, are discussed in Chaps. 7 and 8 and are applicable for fire resistance evaluation of FRP-strengthened structures. As such, this topic is not repeated in this chapter. Only temperature-dependent properties of FRP are discussed in detail.

The properties of fiber-reinforced polymers[1] at room and elevated temperatures are highly influenced not only by the properties of fibers and matrix but also by the orientation of the fibers. Further, FRP materials are highly combustible and burn when exposed to fire. Flame spread and toxic smoke generation are the two major concerns with FRP materials, and the extent of smoke largely depends on the type of FRP formulation (composition). Thermal, mechanical, and deformational properties of FRP deteriorate at a rapid rate with temperature rise, leading to a loss of capacity in an FRP-strengthened concrete member. Therefore, a variation in thermal, mechanical, and deformational properties of FRP with temperature is to be taken into consideration in evaluating the fire resistance of an FRP-strengthened concrete member. The properties of various FRP materials at ambient conditions can be obtained from the literature (Teng et al., 2002; Rasheed, 2014) or from the manufacturer, and typical properties of some FRP types are listed in Tables 4.3 and 4.4. The variation in thermal, mechanical, and deformational properties of FRP at elevated temperature is discussed in detail herein.

11.3.1 Thermal Properties

The extent to which thermal properties of FRP influence temperature rise in an FRP-strengthened RC member depends on the type and amount of FRP applied on a structural member. In the case of columns or walls wrapped with an external FRP sheet, a large amount of FRP might cover much of the surface of a concrete member. When exposed to fire directly, FRP laminate essentially burns and loses much of its strength and modulus as a result of temperature rise. Therefore, thermal properties of FRP can significantly affect heat propagation and sectional temperature rise within a concrete column or wall. In the case of concrete beams or slabs, only small plates or strips of FRP are applied for strengthening. Due to their small size, the effect of FRP burning on temperature rise in concrete will be not that high. In cases where FRP is applied as internal or NSM reinforcement, the influence of thermal properties of FRP on temperature rise within the concrete member is negligible, as the reinforcement is embedded within the concrete section and also because the cross-sectional area of FRP is much smaller compared to the overall area of the concrete section.

The thermal properties of FRPs commonly used in strengthening applications are listed in Table 4.3. There is limited information on the variation in thermal properties of FRP at elevated temperatures in the range of 20 to 600°C. The following relations express thermal conductivity, k_T (kJ/kg·°C), of carbon and glass FRP as a function of temperature (T) (Ahmed, 2010):

$$k_T = 1.4 + \frac{-1.1}{500}T: 0 \le T \le 500 \tag{11.1}$$

$$k_T = 1.4 + \frac{-0.1}{150}(T - 500): 500 \le T \le 650 \tag{11.2}$$

$$k_T = 0.2: T \ge 650 \tag{11.3}$$

Specific heat, c_T (W/m·°C), is another property that influences temperature rise in a member. The temperature-dependent relations for the specific heat of carbon and glass FRP are shown below (Ahmed, 2010):

$$c_T = 1.25 + \frac{0.95}{325}T: 0 \le T \le 325 \tag{11.4}$$

$$c_T = 2.2 + \frac{2.8}{18}(T - 325): 325 \le T \le 343 \tag{11.5}$$

$$c_T = 5.0 + \frac{-0.15}{167}(T - 343): 343 \le T \le 510 \tag{11.6}$$

$$c_T = 4.85 + \frac{-3.59}{28}(T - 510): 510 \le T \le 538 \tag{11.7}$$

[1]Thermal and mechanical properties of FRP at ambient temperatures are provided in Chap. 3.

$$c_T = 1.265 + \frac{1.385}{2778}(T - 538): 538 \le T \le 3316 \quad (11.8)$$

$$c_T = 0: T \ge 3316 \quad (11.9)$$

The variation in density, ρ_T (g/cm^3), of carbon and glass FRP with temperature can be evaluated as per the relations given herein (Ahmed, 2010):

$$\rho_T = 1.60: \le T \le 510 \quad (11.10)$$

$$\rho_T = 1.6 + \frac{-0.35}{28}(T - 510): 510 \le T \le 538 \quad (11.11)$$

$$\rho_T = 1.25: 538 \le T \le 120 \quad (11.12)$$

It is worth noting that very limited information is available on the thermal expansion of FRPs at elevated temperatures.

11.3.2 Mechanical Properties

Although FRP is highly susceptible to a loss of mechanical properties at elevated temperatures, compared to resin, the embedded fibers can retain significant strength until very high temperatures (as high as 600°C) are reached. It is the polymer matrix component of FRP composite that is highly vulnerable to decomposition even at low to moderate temperatures, in the range of 65 to 140°C (Kodur and Harmathy, 2016). Due to this, the FRP system as a whole starts to experience a loss of strength, stiffness, and bond properties at temperatures beyond 100°C. In fact, tensile strength tests have shown that CFRP and GFRP lose about 50% of their room temperature strength at 250 and 325°C, respectively (Wang and Kodur, 2005).

Firmo et al. (2015) compiled temperature-dependent strength and modulus variations of FRP from a number of studies and proposed the following empirical relations to express the strength ($f_{f,T}$) and stiffness ($E_{f,T}$) degradation with temperature (T):

$$f_{f,T} = f_{20°C}\left(\frac{1-a_\sigma}{2}\right)\tanh\left(-b_\sigma(T - c_\sigma) + \frac{1+a_\sigma}{2}\right) \quad (11.13)$$

$$f_{f,T} = f_{20°C}\left(\frac{1-a_E}{2}\right)\tanh\left(-b_E(T - c_E) + \frac{1+a_E}{2}\right) \quad (11.14)$$

where $f_{20°C}$ and $E_{20°C}$ are the strength and elastic modulus of FRP at room temperature, respectively. a_σ, b_σ, c_σ, a_E, b_E, and c_E are the coefficients obtained from the curve-fitting of test data for different types of FRP commonly used in externally bonded strengthening applications (Fig. 11.2).[2]

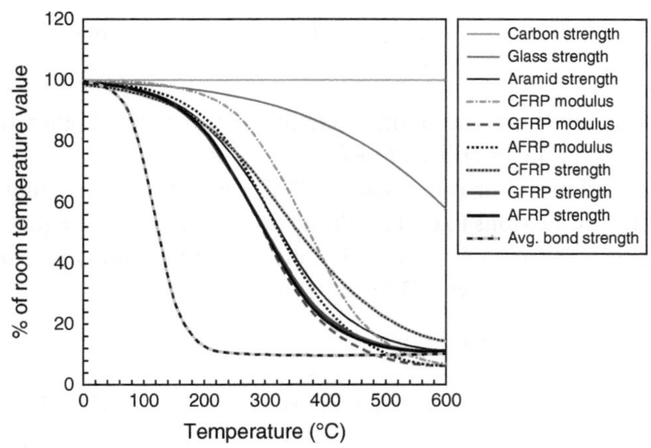

FIGURE 11.2 Variation in strength, elastic modulus, and bond strength of FRP with temperature. (*With permission from Bisby, 2003.*)

[2]CFRP: $a_\sigma = 0.1$, $b_\sigma = 5.83e{-}3$, $c_\sigma = 339.54$
$a_E = 0.05$, $b_E = 8.68e{-}3$, $c_E = 367.41$
GFRP: $a_\sigma = 0.1$, $b_\sigma = 8.10e{-}3$, $c_\sigma = 289.14$
$a_E = 0.05$, $b_E = 7.91e{-}3$, $c_E = 320.35$
AFRP: $a_\sigma = 0.1$, $b_\sigma = 8.48e{-}3$, $c_\sigma = 287.65$
$a_E = 0.05$, $b_E = 7.93e{-}3$, $c_E = 290.49$

Yu and Kodur (2014) proposed a set of relations based on test data for degradation of tensile strength and elastic modulus for NSM FRP at elevated temperatures. These property reduction factors are applicable for CFRP strips and rebars mainly used in NSM strengthening applications:

For CFRP strips:

$$\text{Strength: } f(T) = 0.56 - 0.44\tanh[0.0052(T - 305)] \tag{11.15}$$

$$\text{Modulus: } E(T) = 0.51 - 0.49\tanh[0.0035(T - 340)] \tag{11.16}$$

For CFRP rebars:

$$\text{Strength: } f(T) = 0.54 - 0.46\tanh[0.0064(T - 330)] \tag{11.17}$$

$$\text{Modulus: } E(T) = 0.51 - 0.49\tanh[0.0033(T - 320)] \tag{11.18}$$

where T is temperature of FRP in °C, and $f(T)$ and $E(T)$ are reduction factors for tensile strength and elastic modulus, respectively.

11.3.3 Bond Properties

As discussed in Sec. 11.2, the development of an effective bond between FRP and concrete is critical for transferring stresses from the FRP reinforcement to the concrete and vice versa. Such load transfer occurs through shear stresses that develop in the polymer matrix or adhesive layer. Due to temperature-induced changes in the composition of the matrix material (resin), the bond property of FRP is affected even at a moderate rise in temperature (as low as 80°C). Thus, temperature rise in FRP leads to a loss of interaction (bond) between the FRP and concrete surface, which in turn can reduce the load-carrying capacity of an FRP-strengthened concrete member.

The extent of bond retained at the FRP-concrete interface at elevated temperature is influenced by many factors including the type of FRP reinforcement, geometry of the surface, glass transition temperature of the adhesive, and temperature level reached in the FRP. The glass transition temperature of most commercially available FRPs used in structural strengthening applications is in the range of 60 to 85°C.

There is somewhat limited data in the literature on variation in bond (strength) with temperature rise. The following relation, proposed by Ahmed and Kodur (2011), can be used to express temperature-induced reduction variation in the bond strength of externally bonded FRP and concrete:

$$f_T = f_{20} (\text{if } T \leq 40^\circ\text{C}) \tag{11.19}$$

$$\frac{f_T}{f_{20}} = 1 - \left(\frac{1}{80}\right)(T - 40) \quad (\text{if } 40^\circ C < T \leq 120^\circ \text{C}) \tag{11.20}$$

where f_{20} and f_T are the bond strengths at room and higher temperatures, respectively, and T is the temperature at the interface of FRP and concrete.

For FRP rebars or rods used in NSM strengthening technique, Yu and Kodur (2014) recommended two sets of expressions to evaluate bond degradation. These equations were arrived at through a number of pullout tests carried out on a wide range of temperatures (20 to 400°C) for two types of commercially available FRP rebars: Tyfo T300 and Tyfo S.

For Tyfo T300 epoxy:

$$\text{Bond strength: } \tau(T) = 0.55 - 0.45\tanh[0.011(T - 119)] \tag{11.21}$$

$$\text{Bond modulus: } E(T) = 0.6 - 0.4\tanh[0.009(T - 143)] \tag{11.22}$$

For Tyfo S epoxy:

$$\text{Bond strength: } \tau(T) = 0.55 - 0.45\tanh[0.012(T - 129)] \tag{11.23}$$

$$\text{Bond modulus: } E(T) = 0.6 - 0.4\tanh[0.009(T - 143)] \tag{11.24}$$

where T is the NSM epoxy temperature in °C and $\tau(T)$ and $E(T)$ are reduction factors for bond strength and bond modulus, respectively.

11.3.4 Deformational Properties

The deformational properties of FRP depend on the type and orientation of fibers, as well as on the type of polymer used as the matrix and volume fraction of fibers. Like thermal and mechanical properties, the coefficient of thermal expansion (CTE) of FRP in the longitudinal direction is dominated by fibers, while the transverse direction of CTE depends on the polymer matrix (Bank, 1993). The CTE of polymer matrix is considerably higher than the CTE of fibers and increases significantly with the increase in temperature (Silverman, 1983). Therefore, the CTE of FRP in the transverse direction is higher than the CTE in the longitudinal direction.

Limited information is available on the variation in the CTE of FRP at elevated temperatures. In the temperature range of 20 to 200°C, the CTE of carbon FRP (CFRP) is quite small and fluctuates around zero, while the CTE of glass FRP (GFRP) reaches around 15×10^{-6}/K. This is attributed to the fact that glass fibers experience much higher expansion as compared to carbon fibers. There is a lack of test data on CTE of FRP in the temperature range of 200 to 800°C, as the polymer matrix starts melting beyond 200°C and it is difficult to measure the CTE of FRP as one whole piece. Based on the theoretical studies by Schapery (1968) and Nomura and Ball (1993), CTE values of CFRP and GFRP in the temperature range of 20 to 1000°C can be assumed to be 5×10^{-6}/K and 15×10^{-6}/K, respectively.

Creep is defined as the time-dependent plastic strain under constant stress and temperature. At room temperature and under service loads, creep deformations of FRP are insignificant. However, at elevated temperatures, creep deformations accelerate and may affect the global response of structures. In general, creep strain in FRP increases with an increase in temperature and is largely governed by the type of polymer used in FRP. For instance, FRPs incorporating thermoset polymers, which have high T_g due to heavy cross-linking, exhibit less creep than FRPs incorporating thermoplastic polymeric materials (Mallick, 2008). Creep strain in FRP also depends on the type and orientation of fibers. For instance, when fibers are in the loading direction, the creep in fibers governs the creep in composite. However, it has been reported that commercially available fibers (except aramid fibers) do not creep significantly (Mallick, 2008). Therefore, thermally accelerated creep at elevated temperature exposure can be neglected for unidirectional FRP sheets and laminates used in strengthening applications (Rahman et al., 1993). Further, since the FRP area is much smaller than that of concrete in the overall strengthened member, creep of FRP can be neglected in fire resistance analysis (Kodur and Yu, 2013).

Owing to the softening and melting of polymer at elevated temperatures, it is extremely difficult to measure creep strain of FRP at high temperatures. Therefore, very limited information is available on high-temperature creep strain of FRP. The limited information available is based on the tests carried out by Raghavan and Meshii (1998) on CFRP sheets at various stress levels in a temperature range of 20 to 150°C. The study showed that at the same stress levels, the CFRP composite experienced twice the creep effect at 150°C as that at room temperature. Based on the test data, Raghavan and Meshii (1998) also proposed a relation to predict creep strain of FRP at elevated temperatures, given as

$$\begin{aligned} \varepsilon_{crf} &= \int_0^t B\sigma^{0.01} e^{H/kT} \sinh\left(\frac{\sigma}{kT}\right) dt \\ B &= 2.03 \times 10^{-4} T^{1.55} t^{0.25} \end{aligned} \tag{11.25}$$

where ε_{crf} is the creep strain of FRP, T is the FRP temperature (K), σ is the stress, t is the fire exposure time (s), k is the Boltzmann's constant, and H is the activation energy which depends on experimental data.

11.4 Behavior of FRP-Strengthened Concrete Members under Fire Exposure

The behavior of an FRP-strengthened concrete member at ambient conditions is influenced by the properties of concrete, steel rebars, and FRP, specifically the extent of the bond that develops at the interface of FRP and concrete. When such an FRP-RC concrete member is exposed to fire, the behavior is much more dominated by the properties of FRP than that of concrete or steel rebars in the early stages of fire exposure. This is mainly due to the fact that FRP reinforcement has a low tolerance to elevated temperature exposure and thus experiences significant degradation in properties at early stages of fire exposure, which in turn influences the overall fire performance of a strengthened concrete member.

Typically, a conventional concrete member can achieve required fire resistance with the provision of sufficient concrete cover to steel rebars. However, when the same RC member is strengthened, FRP reinforcement undergoes faster strength and stiffness degradation with temperature than those of steel rebar and concrete, and thus the capacity of an FRP-strengthened member decreases at a much faster pace in early stages of fire exposure. Once the effectiveness of FRP is gone, the capacity degradation is mainly controlled by the properties of concrete and steel reinforcement, and thus the behavior of the strengthened member at later stages of fire exposure follows that of an RC member.

Another critical issue that influences the fire response of an FRP-strengthened member is bond deterioration between FRP and concrete. In the case of an EB-strengthened member, FRP laminates are bonded to the external surface of a concrete member through epoxy-based adhesives. These adhesives are capable of providing somewhat good adhesion at ambient conditions and thus are stable. However, epoxy-based adhesives soften at relatively low temperatures (as low as 65°C), as discussed earlier, and thus the bond (strength) between FRP and concrete decreases significantly. When the glass temperature of FRP (T_g) is reached, the bond strength of epoxy can be lower than the shear stress that develop at the FRP-concrete interface, leading to the debonding of FRP laminates. Once complete debonding occurs, FRP reinforcement hardly contributes to any flexural or shear capacity of a concrete member. In fire tests on FRP-strengthened RC beams (Ahmed and Kodur, 2011), part of the external FRP laminate debonded from the original RC beam at about 20 minutes into fire exposure. So, much of the strength contribution from FRP to flexural member capacity is lost once full debonding occurs. It is worth mentioning that current provisions in design standards do not account for the strength contribution of FRP in evaluating sectional capacity under fire conditions (ACI, 2017; FiB, 2007a).

Unlike externally bonded FRP, NSM FRP reinforcement is embedded into concrete substrate, and thus concrete cover provides a certain level of thermal protection to NSM FRP in the event of fire. Therefore, during fire exposure, NSM FRP experiences a slower temperature rise than that of externally bonded FRP, and hence the strength (as well as bond) degradation in NSM FRP is at a much lower pace with fire exposure time. Based on recent studies (Kodur and Yu, 2013), the temperature developed in NSM FRP-strengthened RC beams can be up to 300°C lower than that in EB FRP reinforcement in similar strengthened beams throughout fire exposure duration.

To illustrate typical behavior under fire conditions, the response of EB and NSM FRP-strengthened RC beams is compared to that of a conventional (unstrengthened) RC beam in Fig. 11.3. These three beams, namely, unstrengthened conventional RC beam (Beam I-1), EB FRP-strengthened RC beam (Beam I-2), and the NSM FRP-strengthened RC beam (Beam I-3), are of similar geometry, with the main difference being the presence of EB FRP or NSM FRP in strengthened beams, as shown in Fig. 11.3*b*. No external fire insulation is present in any of these beams. The unstrengthened RC beam (Beam I-1) has an initial (prior to fire exposure) flexural capacity of 250 kN·m, while the ambient sectional capacity of two strengthened beams, Beam I-2 (with EB FRP) and Beam I-3 (with NSM FRP), is increased to 400 kN·m due to the presence of FRP. When these beams are exposed to a fire, cross-sectional temperatures rise with fire exposure time. The temperatures reached in steel rebar and FRP mainly dictate the level of moment (or shear) capacity in these beams at any given fire exposure time.

In the case of Beam I-2 (the beam with EB FRP), FRP temperature increases at a rapid pace initially with fire exposure time (see Fig. 11.3*c*). However, in the case of Beam I-3, the temperature in NSM FRP increases at a much slower rate due to the thermal protection provided by the concrete cover. Due to different temperature levels reached in FRP reinforcement (and in steel rebars to a lower extent), the respective reductions in the moment capacity in Beams I-2 and I-3 with fire exposure time vary depending on the type of strengthening (EB or NSM) as plotted in Fig. 11.3*d* for these three beams.

In the case of an unstrengthened RC beam (Beam I-1), the moment capacity decreases gradually with time and is mainly dependent on the loss of strength in reinforcing steel bars with a rise in temperature. In the case of an externally bonded FRP-strengthened RC beam (Beam I-2), the moment capacity degradation is initially rapid due to high temperatures developing in FRP laminate, and after FRP burns out and loses much of its strength and effectiveness, the degradation of the moment capacity follows that of the unstrengthened RC beam. In the case of the NSM FRP-strengthened RC beam (Beam I-3), the moment capacity decreases at a slower rate throughout the fire exposure time. This can be attributed to a lower strength loss in NSM FRP due to a slower temperature rise facilitated by the presence of concrete cover over the FRP reinforcement.

Due to the variation in moment capacity degradation, these three beams experience failure at different times. If the loading present on each beam during fire exposure is taken to be 50% of the room temperature sectional capacity, the loads to be carried by FRP-strengthened RC beams (Beams I-2 and I-3) during fire

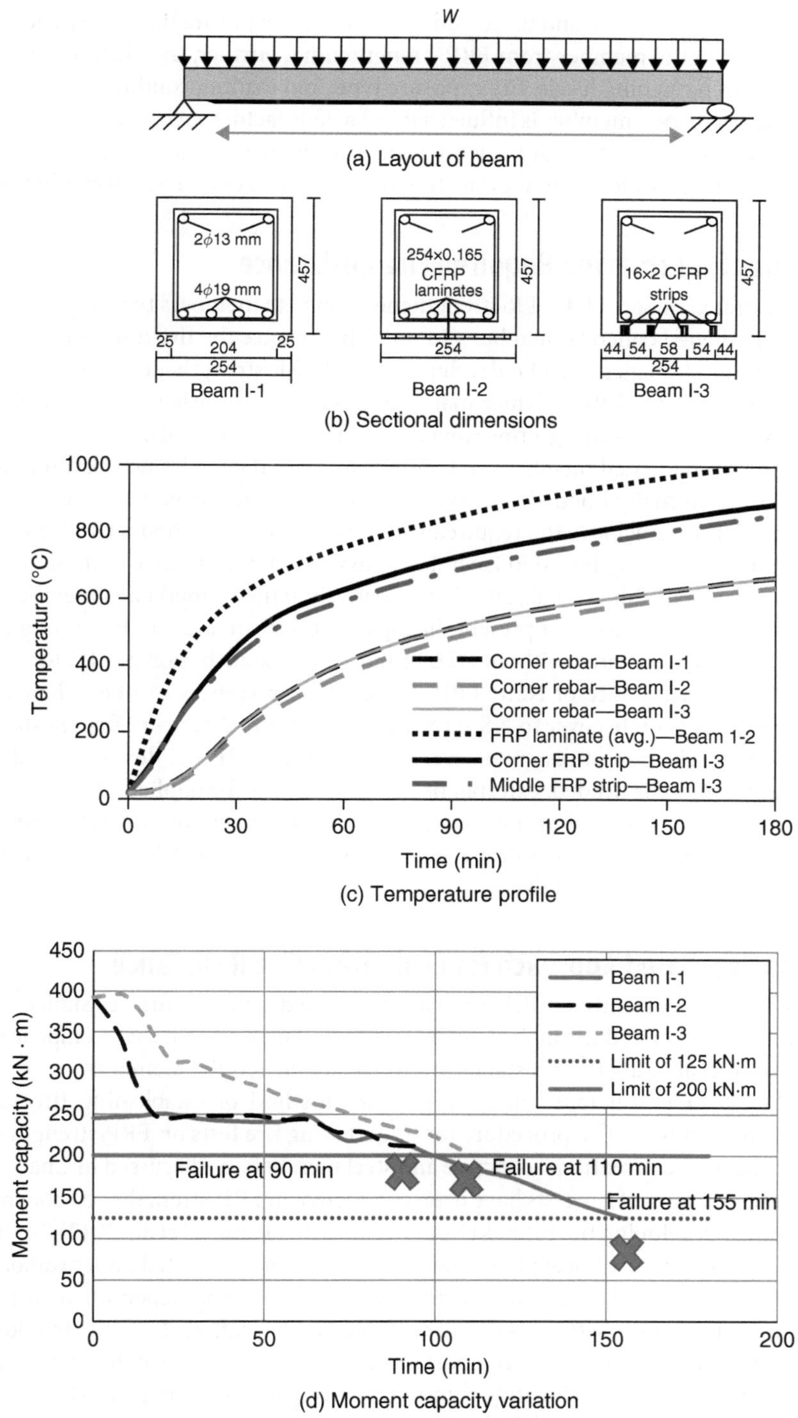

Figure 11.3 Typical response of an EB and NSM FRP-strengthened concrete beam under fire exposure. (*With permission from Yu, 2013.*)

are obviously higher than that by an unstrengthened RC beam (in this case 50% higher). The EB FRP-strengthened RC beam (Beam I-2) experiences failure at an earlier time than the unstrengthened RC Beam I-1, because this EB-strengthened beam must carry increased service loading under fire conditions, and also because the exposed FRP loses much of its strength and stiffness at a rapid pace, due to increased temperature in FRP. However, the NSM FRP-strengthened RC beam (Beam I-3) experiences failure at a much longer fire exposure time, as sectional capacity reduces at a slower pace than in Beams I-1 and I-2 due to a slower degradation of moment capacity with fire exposure time, facilitated by a much slower temperature rise in NSM FRP.

The typical fire behavior of FRP-strengthened compression members such as columns and walls is similar to the behavior illustrated above for beams. One major difference is the level of axial capacity developed in a column, and its degradation with fire exposure time is dependent on the confinement effect developed in the member from FRP strengthening. Further, the illustration in Fig. 11.3 is for beams with similar strengthening levels, fire exposure type, and loading conditions. In reality, the fire resistance of an FRP-strengthened member is influenced by various factors such as the level of strengthening, fire exposure scenario, FRP properties, and any fire insulation applied to the strengthened member. These factors are to be given due consideration in evaluating fire resistance of an FRP-strengthened RC member.

11.5 Strategies for Achieving Required Fire Resistance

As illustrated in Sec. 11.4, FPR-strengthened concrete members typically possess lower fire resistance than unstrengthened concrete members. In order to enhance the fire resistance of such strengthened members, appropriate strategies must be developed. In EB FRP-strengthened concrete members, the glass transition temperature of the FRP polymer matrix is reached at an early stage of fire exposure (20 minutes or less). Reaching T_g of FRP during a fire event does not lead to failure of an FRP-strengthened concrete member. In fact, the strengthened member can continue to carry the load much beyond the point where T_g is reached. However, externally bonded FRP-strengthened concrete members, in most cases, require supplementary fire insulation to achieve the required fire resistance of 1 to 4 hours (Kodur et al., 2007). Without such fire insulation, EB-strengthened RC members can yield a fire resistance of about 1 hour in most design scenarios (Ahmed, 2010; Palmieri et al., 2012). Alternatively, if the strengthened member is designed for lower service loads (i.e., reduced loading present during a possible fire event), then it might be possible to achieve 1 to 2 hours of fire resistance without accounting for the contribution of EB FRP during fire exposure.

The fire resistance of NSM FRP-strengthened members can be enhanced through the provision of appropriate concrete cover to NSM rebars/strips.[3] Up to 2 hours of fire resistance can be achieved through the proper selection of concrete cover in most NSM-FRP RC flexural members. When fire resistance of 2 hours or more is needed, external fire insulation must be applied.

In order to arrive at an optimum insulation thickness or concrete cover to satisfy the required fire resistance, prescriptive and rational approaches can be applied. These approaches are discussed in detail in Secs. 11.6 and 11.7, respectively.

11.6 Prescriptive-Based Approach for Evaluating Fire Resistance

Much of the currently available perspective-based data for fire resistance of an FRP-strengthened RC member is generated through standard fire tests. There are not many empirical-based equations or design charts for evaluating fire resistance of FRP-strengthened RC members.

Fire testing still represents the primary method of establishing fire ratings in FRP-strengthened concrete members. The procedure for undertaking fire tests on FRP-strengthened RC members is similar to fire tests on conventional concrete and steel members, as discussed in Chap. 5. Over the last two decades, a limited number of fire tests have been conducted on FRP-strengthened concrete beams and columns with the aim of developing fire ratings (Kodur et al., 2007; Palmieri et al., 2012). In these tests, FRP-strengthened concrete members with supplementary fire insulation were tested under standard fire exposure conditions.

A summary of early fire tests carried out on FRP-strengthened RC members at the National Research Council of Canada (NRCC) is shown in Table 11.1 (Kodur et al., 2007). In addition to these standard tests, a number of additional (mostly nonstandard) tests have been carried out on EB- and NSM-strengthened RC members (beams, slabs, and columns) at Michigan State University (USA), Ghent University (Belgium), South China University of Technology, and elsewhere. These tests were on typical concrete members strengthened with EB or NSM FRP, with applied supplementary fire insulation, and were tested under service loads and under standard fire exposure. In many cases, fire resistance up to 4 hours was achieved, as these members were protected with supplementary fire insulation.

The testing undertaken thus far clearly indicates that FRP-strengthened RC columns, beams, and slabs protected with supplementary fire protection are capable of achieving required fire ratings (up to 4 hours) under standard fire exposure and full retrofitted service load levels. Even though the glass transition temperature of the polymer matrix was exceeded relatively early into fire exposure, the insulated FRP-strengthened concrete members exhibited fire resistance up to 4 hours. It is clear from the test observations and recorded data that this level of fire performance was due to supplementary thermal protection to FRP, as well as to internal steel rebars. The thermal protection provided by the insulation system led to a

[3] Another strategy often used in the case of FRP-strengthened RC members is to design the strengthened member such that the FRP contribution is not accounted for in the capacity evaluation during a fire incident.

Test #	FRP Type	Number of Layers of FRP	Insulation System	Insulation Thickness (mm)	Fire Test Load Ratio[g]	Matrix T_g (°C)	Time to Exceed T_g (min)[h]	Fire Resistance (min)
Intermediate-scale FRP-strengthened RC slabs								
1	CFRP[a]	2	System 1[d]	19	0.0	93	42	147
2	CFRP[a]	2	System 1[d]	38	0.0	93	104	>240
3	CFRP[b]	1	System 2[e]	38	0.0	71	36	>240
4	CFRP[b]	1	System 3[e]	38	0.0	71	43	>240
Full-scale FRP-strengthened RC columns								
1	CFRP[a]	1	System 1[d]	57	0.5	93	118	>300
2	CFRP[a]	1	System 1[d]	32	0.5	93	81	>300
3	GFRP[c]	3	System 1.1[f]	38	0.69	93	51	>240
4	CFRP[b]	2	System 2[e]	0	0.56	71	3	210
5	CFRP[b]	2	System 2[e]	53	0.56	71	31	>300
Full-scale FRP-strengthened RC beam-slab assemblies								
1	CFRP[a]	1	System 1.1[f]	25	0.53	93	33	>240
2	CFRP[a]	1	System 1.1[f]	38	0.53	93	52	>240
3	CFRP[b]	1	System 2[e]	30	0.50	71	30	>240
4	CFRP[b]	1	System 2[e]	28	0.50	71	30	>240

[a] Tyfo® SCH System—additional information available from http://www.fyfeco.com.
[b] MBrace® CF 130 System—additional information available from http://www.mbrace.com.
[c] Tyfo® SEH System—additional information available from http://www.fyfeco.com.
[d] Tyfo® EI/VG System—additional information available from http://www.fyfeco.com.
[e] Systems 2 and 3 are proprietary MBrace® systems and remain under development.
[f] Tyfo® VG/EI-R System—additional information available from http://www.fyfeco.com.
[g] Ratio of load applied during the fire test to the design ultimate load of the strengthened column (with design ultimate load calculated according to ACI 440.2R-02).
[h] Based on manufacturer-specified T_g values.

Table 11.1 Summary of Early Fire Resistance Tests Carried out at NRCC on FRP-Strengthened Members (*Adapted from Kodur et al., 2011*)

slowdown in sectional temperature rise (in FRP, concrete, and especially rebars) during fire. These observations and results validated the intuition that the fire resistance (and load-carrying capacity) of insulated FRP-strengthened concrete members does not depend on the temperatures reached in FRP alone, but rather on the temperatures attained in reinforcing steel and concrete as well.

Based on the above standard fire tests, listings with fire ratings on FRP-strengthened concrete structural member have been incorporated in fire rating directories. The UL directory includes a few listings: seven configurations for insulated and externally strengthened concrete T-beams and four configurations for FRP-strengthened columns (BXUV.GuideInfo—Fire-Resistance Ratings—ANSI/UL 263, n.d.). The listings are for beams under restrained and unrestrained conditions. These listings are for specific strengthened material (FRP) and specific (proprietary) spray-applied fire-resistive materials (SFRM). Two such UL listings, one for T-beams (UL Design No. N814) and the other for columns (UL Design No. X826), are shown in Fig. 11.4. As per Design No. N814, a reinforced concrete beam with a 30-mm web width, a 250-mm web depth, a 150-mm flange thickness, and a 1220-mm flange width; reinforced with a steel configuration (as shown in Fig. 11.4a); and insulated with 50-mm SFRM insulation around two layers of Sikadur 300 SikaWrap 103C (FRP) system can achieve a 4-hour fire rating under a UL263 (equivalent to ASTM E119) standard fire. Similarly, as per UL listing X826, a concrete column of 305 mm × 305 mm cross section, strengthened with Sikadur 300 SikaWrap 103C system and protected with 40 mm of spray-applied fire insulation system, can yield a 4-hour fire rating under Ul263 (equivalent to ASTM E119) standard fire exposure.

11.7 Rational Approaches for Evaluating Fire Resistance

The fire resistance of FRP-strengthened concrete structural members can be evaluated through rational approaches, as in the case of concrete and steel members. However, due consideration should be given to

Design No. N814

July 27, 2010

Restrained Beam Rating-4 h

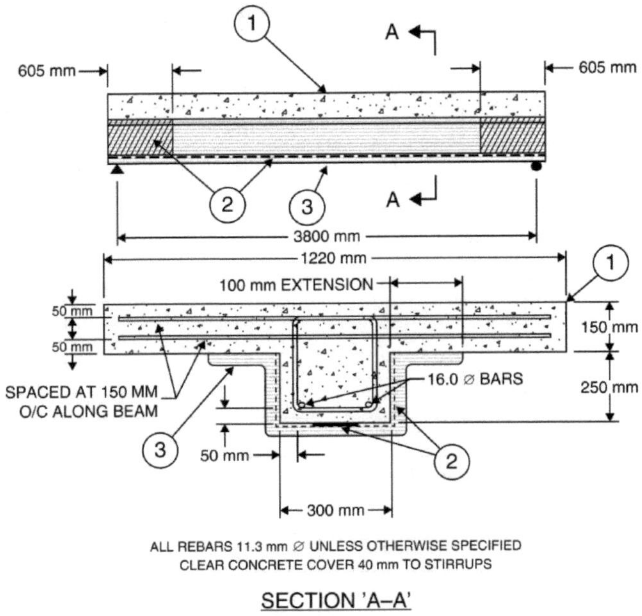

1. **Concrete beam** – Concrete tee beam with 300 mm web width, 250 mm web depth, 150 mm flange thickness, 1220 mm flange width, Reinforced with 2 No. 20M bottom longitudinal reinforcing steel in the web. Flange (slab) short direction transverse reinforcement No. 10M at 150 mm OC top and bottom, top longitudinal reinforcement 10M at 200 mm OC. Stirrup 10M at 150 mm OC. The minimum 28 day compressive strength of concrete shall be 28 Mpa. The external edges of the concrete that are to receive the composite material (Item 2) shall be rounded. Sections of the concrete that are to receive the composite material shall be mechanically abraded to expose the concrete aggregate. The applied load for the beam assembly shall be calculated in accordance with the Minimum Acceptable Design Criteria published in CSA S806-2, Design and Construction of Building Components with Fibre-Reinforced Polymers.

- 2. **Structural Concrete Fiber-Reinforced Composite System –** (CIZQC). Installed in accordance with the manufacturer's instructions, and shall include the following construction:

  (a) Concrete surface shall be cleaned with acetone or vacuumed and primed with Sikadur® 330 Epoxy Primer applied to sections of concrete that will be covered with Item 2b. and 2c. The primer shall also be applied to fill pittings, inundations and other surface defects in the concrete.

  (b) SikaWrap® 103C carbon fiber fabric, 200 mm wide, saturated with Sikadur® 300 epoxy shall be applied over the primed bottom web of the concrete tee beam (Item 1) and pressed-in with a roller to remove excess epoxy.

  (c) Two layers of 605 mm wide SikaWrap® 100G glass fiber fabric presaturated on both sides with Sikadur® 300 Epoxy or four layers of 605 mm wide SikaWrap® 430G glass fibre fabric each layer saturated in-place with Sikadur® 330 Epoxy, shall be applied as U wrap at the ends of the beam to anchor SikaWrap® 103C carbon fiber fabric or for shear reinforcement. Entrapped air shall be removed with a serrated roller.

  SIKA CANADA INC – Sikadur® 300, Sikadur® 330, SikaWrap® 103C, SikaWrap® 100G, SikaWrap® 430G

- 3. **Spray-Applied Fire Resistive Materials –** (CHPXC). Sikadur® 300 Epoxy Primer applied over the composite system. Before the primer is dry silica sand #28 applied over the primer. SikaCrete® 213F mixed with water and spray applied in one or more coats to minimum average thickness of 40 mm to concrete and to the Structural Concrete Fiber-Reinforced Composite System surfaces. Min avg and min individual dry density of 810/800 kg/m^3, respectively. For method of density determination, refer to General Information Section under heading "Fire Resistance Ratings". For extent of application of Spray-Applied Fire Resistive Material see illustration above.

SIKA CANADA INC – SikaCrete® 213F

(a) T-beam

Design No. X826

July 27, 2010

Rating-4 h

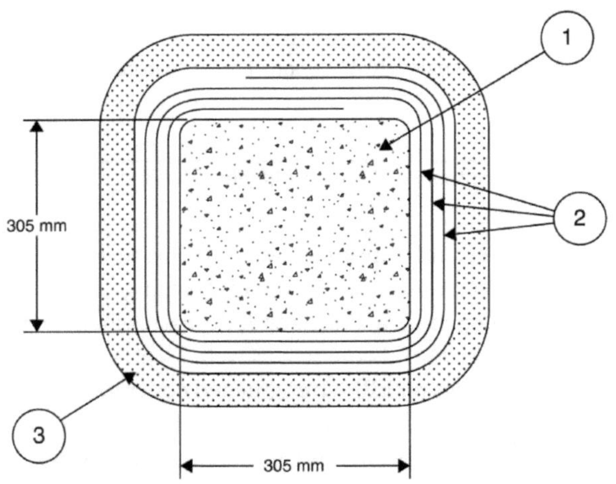

1. **Concrete Column** – 305 mm × 305 mm concrete column reinforced with 4-24 M vertical bars and 10 M at 406 mm c/c ties. Normal density 2400 kg/m^3 concrete with minimum 28 day strength 28 Mpa incorporating Type 10 Portland cement. Minimum concrete cover to main vertical bars shall be 50 mm and minimum concrete cover to ties shall be 40 mm. The external edges of the concrete shall be rounded and the concrete surface shall be mechanically prepared to expose the aggregate. The applied load for the column shall be calculated in accordance with the Minimum Acceptable Design Criteria published in CSA S806-2, Design and Construction of Building Components with Fibre-Reinforced Polymers.

- 2. **Structural Concrete Fiber-Reinforced Composite System –** (CIZQC). Installed in accordance with the manufacturer's instructions, and shall include the following construction:

  (a) Concrete surface shall be cleaned with acetone or vacuumed to removed dust and primed with Sikadur® 330 Epoxy Primer. The primer shall also be applied to fill pittings, inundations and other surface defects in the concrete.

  (b) Three layers of SikaWrap® 103 C Carbon Fibre Fabric, 600 mm wide, each layer saturated on both sides with Sikadur® 300 Epoxy shall be applied around the column. The saturated carbon fiber fabric shall be wrapped around the column with a 150 mm vertical overlap. All entrapped air shall be removed by a serrated roller.

  SIKA CANADA INC – Sikadur® 300, SikaWrap® 103C, Sikadur® 300

- 3. **Spray-Applied Fire Resistive Materials –** (CHPXC). Sikadur® 330 Epoxy Primer shall be applied over the composite system. Before the primer is dry Silica Sand #28 shall be applied over the primer. SikaCrete® 213F mixed with water shall be spray applied in one or more coats to minimum average thickness of 40 mm. Min avg and min individual dry density of 810/800 kg/m^3, respectively. For method of density determination, refer to General Information Section under heading "Fire Resistance Ratings". For extent of application of Spray-Applied Fire Resistive Material see illustration above.

SIKA CANADA INC – SikaCrete®213F

(b) Column

FIGURE 11.4 Sample of UL listings for FRP-strengthened RC members. (*With permission from Underwriters Laboratories.*)

the properties of FRP, including loss of strength, stiffness, and bond between the FRP strengthening system, reinforcing steel, insulation, and concrete. Current codes of practice recommend neglecting the strength contribution of FRP reinforcement in calculating the reduced capacity of an FRP-strengthened member under fire exposure (FiB, 2007b). Such an approach, although it leads to conservative design, may not be realistic and produce economical designs (Palmieri et al., 2012). Therefore, from the perspective of fire design, a rational calculation methodology must be applied to assess the realistic fire resistance of FRP-strengthened concrete members, taking into consideration the strength contribution of FRP.

In such rational calculation approaches, a number of parameters including severity of fire exposure, load level, temperature-induced property degradation in FRP, and different failure modes can be considered. Simplified and advanced analysis approaches can be applied for evaluating fire resistance of FRP-strengthened members.

11.7.1 Simplified Approaches

In the last two decades, researchers have proposed a few calculation approaches for evaluating fire resistance of FRP-strengthened concrete members (Gao et al., 2016; Kodur and Yu, 2013). The unified approach both for EB- and NSM-strengthened concrete members proposed by Kodur and Yu (2013) is an extension of the current approach used for evaluating the fire resistance of RC members, but it incorporates the effect of FRP and fire insulation into the fire resistance calculations. This strength-based approach involves three main steps: (1) evaluating fire temperature at a given time, (2) calculating sectional temperatures in concrete, FRP, and steel rebar of strengthened member, and (3) evaluating reduced sectional capacity of FRP-strengthened RC member at that time-step. The failure is said to occur if the reduced capacity drops below the loading effect (shear, flexure, or axial). A flowchart illustrating the steps associated in evaluating fire resistance of an FRP-strengthened concrete member is shown in Fig. 11.5.

Fire resistance analysis can be carried out in a series of time increments. At each time increment, the resulting cross-sectional temperatures from fire exposure is evaluated using specified time-temperature relations. The cross-sectional temperature in FRP, steel, and concrete can be calculated at a given time-step resulting from fire exposure, and these temperatures are then used to evaluate the reduced cross-sectional capacity of an FRP-strengthened RC member. Calculating such sectional temperatures can be complex. To simplify the sectional temperature calculations, empirical relations have been proposed and are discussed in this section (Kodur and Yu, 2013).

Evaluating Cross-Sectional Temperature

For undertaking a fire resistance analysis, the sectional temperature at a given time-step in a fire-exposed member must be evaluated. There are two possible scenarios in the case of FRP-strengthened members, one without any supplementary fire insulation and the other with supplementary fire insulation on the strengthened member. In the case of a fire-exposed FRP-strengthened RC member without any fire insulation, temperature can be evaluated as in the case of the original unstrengthened concrete member, since the area of applied FRP reinforcement (either NSM FRP or EB FRP) is too small, when compared to

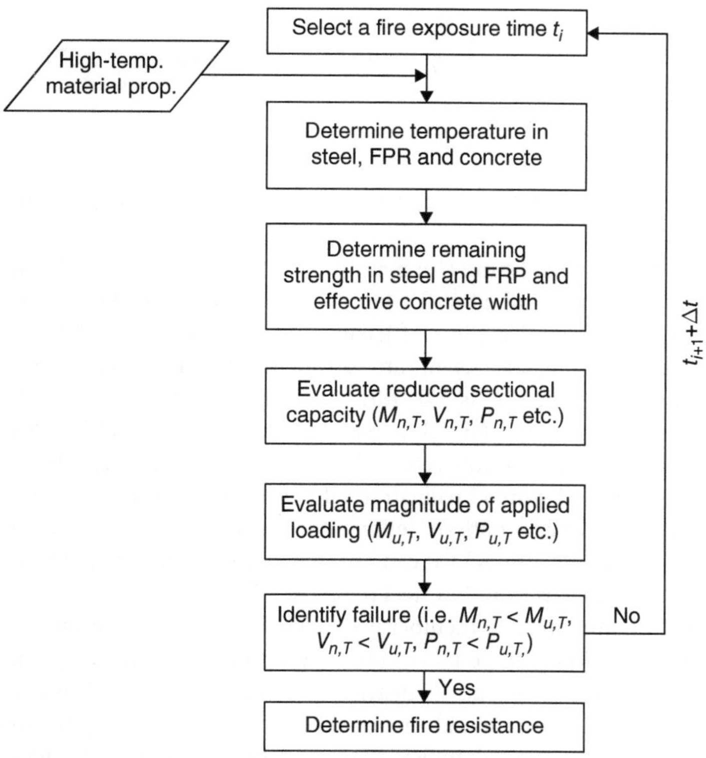

Figure 11.5 Flowchart illustrating a rational calculation approach for evaluating fire resistance of FRP-strengthened concrete members.[4]

[4]The above procedure can also be applied to determine fire resistance in FRP-strengthened columns, slabs, and walls by giving due consideration to proper evaluation of reduced cross-sectional capacity (i.e., axial and flexural capacity, etc.), as well as magnitude of applied loading ($M_{u,T}$, $P_{u,T}$).

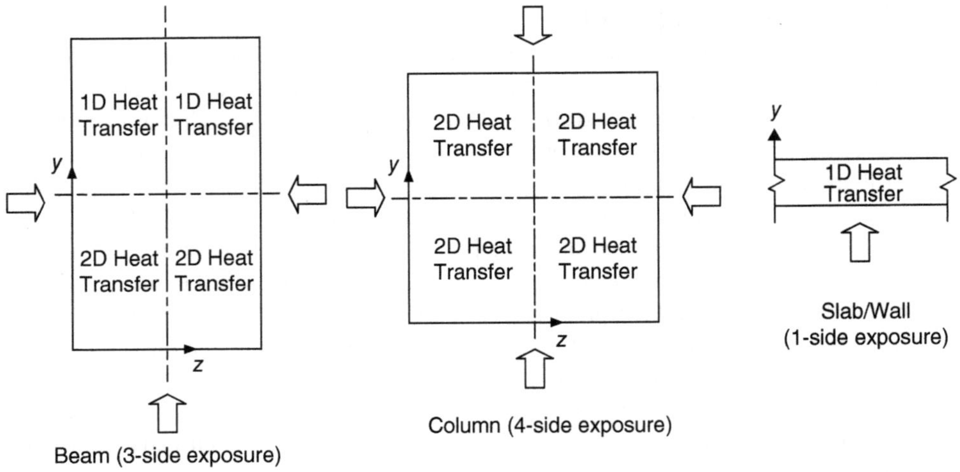

FIGURE 11.6 Different possible fire exposure scenarios on a concrete section.

the overall concrete area (or volume), to influence heat transfer within a strengthened RC member. In this case, temperature (T_c) at any point (at coordinates y and z in Fig. 11.6) in the cross section of a concrete member at a given fire exposure time can be obtained using the same procedure as outlined in Chap. 8.

Kodur and Yu (2013) derived a simple set of equations for calculating temperatures in a fire-exposed concrete member, shown in Fig. 11.6. For instance, for one-dimensional (1D) heat transfer, T_c (i.e., when the section is predominantly heated from one side as in a slab), the temperature in concrete at a given point in the section is given as shown in Chap. 8 and summarized herein.

$$T_c = c_1 \cdot \eta_z \cdot (at^n) \tag{11.26}$$

where η_z is given by

$$\eta_z = 0.155\ln\frac{t}{z^{1.5}} - 0.348\sqrt{z} - 0.371 \tag{11.27}$$

For 2D heat transfer

$$T_c = c_2[-1.481(\eta_z \cdot \eta_y) + 0.985(\eta_z + \eta_y) + 0.017](at^n) \tag{11.28}$$

where t = fire exposure time (h); z and y = distance from the point in the concrete section to the fire exposure surface (m); and at^n = fire temperature under ISO 834 or ASTM E119 standard fire exposure at time t. For an ISO 834 fire, the coefficients of a and n are 935 and 0.168, respectively; for an ASTM E119 fire, the corresponding coefficients are 910 and 0.148. The values c_1 and c_2 are the coefficients that account for concrete type, where c_1 is 1.0, 1.01, 1.12, and 1.12 for NSC-CA, HSC-CA, NSC-SA, and HSC-SA, respectively, and c_2 is 1.0, 1.06, 1.12, and 1.20 for NSC-CA, HSC-CA, NSC-SA, and HSC-SA, respectively. (NSC is normal-strength concrete; CA is carbonate aggregate; HSC is high-strength concrete; SA is siliceous aggregate.) Figure 11.6 illustrates the cases where 1D or 2D heat transfer calculations are applicable.

Through the above equations, temperature in concrete across different depths, including at the level of steel reinforcement as well as FRP, can be evaluated. Once the temperature is known, the strength reduction in various concrete layers and in steel rebars and FRP can be determined.

If the FRP-strengthened member is protected with fire insulation, then the effect of thermal protection offered by insulation is to be taken into account in evaluating temperature rise in different layers of the section. Temperature rise and its distribution in an insulated RC member is significantly influenced by properties of both concrete and fire insulation. Thus, predicting sectional temperatures in an insulated concrete member becomes complex, as thermal properties of both concrete and insulation are to be given due consideration.[5] In order to simplify sectional temperature evaluation in an insulated FRP-strengthened member, Kodur et al. (2017) extended the above set of simplified equations (Eqs. 11.26 to 11.28) to account

[5]The thermal properties of steel rebar and FRP reinforcement (laminate, strip, or rod added to the surface or placed internally) do not significantly affect temperature rise in the section due to their small sectional areas as compared to concrete and insulation sectional area.

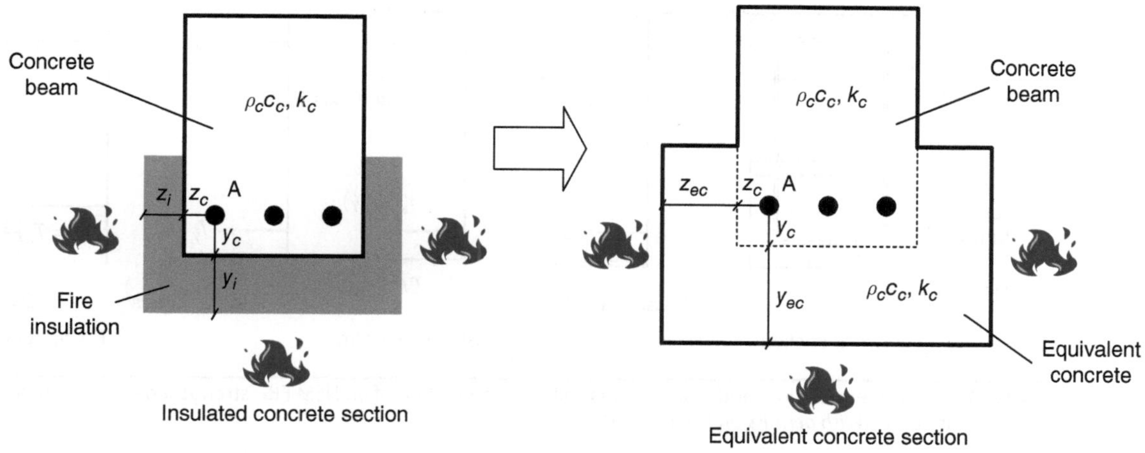

FIGURE 11.7 Equivalent concrete depth in an insulated concrete section.

for supplementary fire protection wherein insulation is converted to an equivalent layer of concrete (see Fig. 11.7). As a result, fire insulation is replaced with an additional concrete layer by calculating equivalent concrete thickness. For this purpose, they derived an approximate expression for transforming insulation into an equivalent concrete thickness.

Figure 11.7 shows a typical RC beam protected by a U-shaped supplementary fire insulation, and the thickness of insulation on the sides and bottom of the beam are z_i and y_i, respectively. If the fire insulation layer is replaced by an equivalent concrete layer (with the thickness z_{ec} of y_{ec} and on the sides and bottom) which has the same thermal resistance (inertia) as insulation layer, the temperature profiles remain the same within the beam cross section after this transformation. To illustrate this, the temperature calculation at a point (Point A) located in an insulated RC beam is considered (see Fig. 11.7). The distances from Point A to the side and bottom concrete surface of the beam are z_c and y_c, respectively, and the fire insulation thicknesses on the side and bottom surface of the beam are z_i and y_i, respectively. Thus, the equivalent (thermal) depth from Point A (z' or y') to the fire-exposed surface, after transforming fire insulation layer into equivalent concrete depth, can be expressed as (Kodur and Yu, 2013):

$$z' = z_c + z_{ec} = z_c + z_i \sqrt[\alpha]{t} \sqrt[\beta]{\frac{k_c}{(\rho c)_c} \frac{(\rho c)_i}{k_i}} \tag{11.29}$$

$$y' = y_c + y_{ec} = y_c + y_i \sqrt[\alpha]{t} \sqrt[\beta]{\frac{k_c}{(\rho c)_c} \frac{(\rho c)_i}{k_i}} \tag{11.30}$$

where k_i and $(\rho c)_i$ are thermal conductivity and heat capacity of fire insulation, respectively, and k_c and $(\rho c)_c$ are thermal conductivity and heat capacity of concrete, respectively.

Based on regression analysis, Kodur and Yu (2013) obtained correlation coefficients for α and β as 4.5 and 1.75, respectively, and thus the relation for expressing SFRM insulation thickness in terms of equivalent concrete thickness (z_{ec}) can be obtained by substituting for α and β in Eq. 11.29 as

$$z_{ec} = z_i \sqrt[4.5]{t} \sqrt[1.75]{\frac{k_c}{(\rho c)_c} \frac{(\rho c)_i}{k_i}} \tag{11.31}$$

Through the above set of equations, cross-sectional temperatures in an FRP-strengthened RC member can be evaluated for both cases of unprotected and protected members with fire insulation. In lieu of the above set of equations, other empirical equations such as Wickström's method (Wickström, 1986) or detailed finite element analysis can be applied to determine sectional temperature in FRP-strengthened RC members. The procedure laid out in the case of concrete members in Chap. 8 can be utilized for undertaking thermal analysis, with due consideration given to the presence of fire insulation.

Evaluating Sectional Capacity

Knowing cross-sectional temperatures (in FRP, steel bars, and concrete layers), the temperature-induced reduction in strength properties of steel rebar ($f_{y,T}$), concrete ($f_{c,T}$), and FRP ($f_{f,T}$, $E_{f,T}$), as well as effective concrete width (b_e) in the compression zone can be evaluated by applying property (or width) reduction factors. Then, the reduced sectional (moment, shear, axial) capacity of an FRP-strengthened RC member

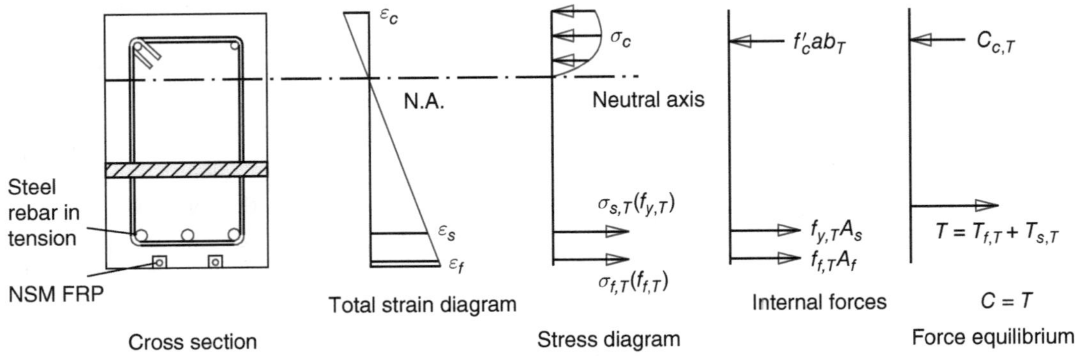

FIGURE 11.8 Force equilibrium and strain compatibility in the section of an NSM FRP-strengthened RC beam at a given fire exposure time. (*With permission from Yu, 2013.*)

at any given fire exposure time can be calculated. In the case of an FRP-strengthened RC beam, ACI 440.2 provisions state that two primary failure modes can occur, either through crushing of the top concrete before FRP debonding or through the rupture of FRP laminate after steel yielding. Thus, the reduced moment capacity in fire-exposed FRP strengthened beam is to be evaluated using different formulae for these two failure modes. Figure 11.8 illustrates stresses and internal forces developed in a cross section of a fire-exposed NSM FRP-strengthened RC beam.

At a given fire exposure time, if the failure is due to the crushing of the top concrete, the strain in the compression layer of the top concrete is considered to reach its limit ($\varepsilon_c = \varepsilon_{cu} = 0.003$), and strains in steel and FRP reinforcement can be calculated by applying strain compatibility principles as follows:

$$\frac{\varepsilon_s}{d-c} = \frac{\varepsilon_{cu}}{c} \tag{11.32}$$

$$\frac{\varepsilon_f}{h-c} = \frac{\varepsilon_{cu}}{c} \tag{11.33}$$

where ε_s is the strain in steel, d is the depth to steel rebars, h is the height of cross section, c is the neutral axis depth, and ε_f is the strain in FRP.

Since the strain hardening effect is usually neglected in strength calculations, as per design codes, the stress in reinforcing steel is assumed to be that of yield strength as long as $\varepsilon_s \geq \varepsilon_{sy}$. The stress in FRP reinforcement can be calculated as

$$f_{f,T} = E_{f,T}\varepsilon_f \tag{11.34}$$

Applying force equilibrium principles leads to the following strength equation:

$$0.85 f_c' a b_T = A_s f_{y,T} + A_f f_{f,T} \tag{11.35}$$

where A_s and A_f are the area of steel and FRP; $f_{y,T}$ and $f_{f,T}$ are the strength in rebars and FRP at respective temperatures, T; and f', a, and b_T are the compressive strength of concrete, depth of compression block, and effective (compressive) width of concrete, respectively (b_T being the width of the concrete—not at 500°C).

After solving for a, the reduced moment capacity at a given fire exposure time can be calculated as

$$M_{n,T} = A_s f_{y,T}(d - 0.5a) + A_f f_{f,T}(d - 0.5a) \tag{11.36}$$

If failure of the beam occurs due to a rupture of FRP reinforcement, FRP reaches its strain limit [$\varepsilon_f = \varepsilon_{fe} = \kappa_m \varepsilon_{fu}$ in which κ_m is the strain reduction factor, as specified in ACI 440.2 (ACI, 2017) to prevent the debonding failure of FRP reinforcement]. Equating tension and compression forces, at the critical section:

$$0.85 f_c' a b_T = A_s f_{y,T} + A_f (E_{f,T}\varepsilon_{fe}) \tag{11.37}$$

Solving Eq. 11.37, the depth of the equivalent rectangular stress block a can be evaluated as

$$a = \frac{A_s f_{y,T} + A_f (E_{f,T} \varepsilon_{fe})}{0.85 f_c' b_T} \tag{11.38}$$

Then, the reduced sectional moment capacity in the FRP-strengthened beam can be obtained as

$$M_{n,T} = A_s f_{y,T}(d - 0.5a) + A_f (E_{f,T} \varepsilon_{fe})(h - 0.5a) \tag{11.39}$$

In an FRP-strengthened RC beam exposed to fire, the reduced strength levels in constituent materials is known for any given fire exposure time, based on the above-described procedure. However, a preliminary calculation is needed to check whether the failure and thus moment capacity is governed by the crushing of top concrete or rupture of FRP. As part of this preliminary calculation, Eq. 11.35 is to be applied first to obtain the equivalent stress block a. Then the strain level in FRP can be evaluated using

$$\varepsilon_f = \frac{\varepsilon_{cu}}{c}(h - c) \tag{11.40}$$

If $\varepsilon_f > \varepsilon_{fe}$, FRP reinforcement will reach its strain limit before the crushing of top concrete, and thus the rupture of FRP governs the failure mode, and the moment capacity of the beam can be calculated using Eqs. 11.37 to 11.39. However, if $\varepsilon_f < \varepsilon_{fe}$, the crushing of top concrete governs the failure mode, and the reduced moment capacity of the beam can be calculated using Eqs. 11.32 to 11.36. As a result, the reduced sectional capacity (e.g., moment capacity in a beam) can be evaluated at a given fire exposure time with consideration to both failure modes.

Evaluating Fire Resistance

This reduced sectional capacity, computed above, is compared against the applied bending moment present during fire exposure. Failure is said to occur once the applied bending moment exceeds the level of moment capacity. The time to reach this failure point (time) is to be taken as the fire resistance of the FRP-strengthened RC member.

While the above approach is specifically illustrated for evaluating reduced moment capacity in an EB FRP-strengthened RC beam, the same procedure holds well for other FRP-strengthened concrete members including column, slab, or wall. Further, the same procedure can be applied for evaluating shear or axial capacity of a member, provided due consideration is given to applicable failure limit states. Due to simplification, limitations associated with above approach are discussed elsewhere (Kodur and Yu, 2016).

11.7.2 Advanced Analysis

The simplified approaches discussed in Sec. 11.7.1 can trace the fire response at the "section level," and as such, critical factors influencing the response of FRP-strengthened RC members such as restraint conditions, debonding of FRP across the member length, confinement effect, softening of concrete, among others, cannot be fully taken into consideration in the analysis. In order to account for these factors, the advanced analysis procedure must be applied. As in the case of concrete structures, microscopic or macroscopic finite element-based models can be applied to account for the aforementioned factors through advanced analysis. The application of such an advanced analysis approach for evaluating fire resistance of typical FRP-strengthened slab is illustrated in Fig. 11.9.

General

Finite-element-based numerical models for evaluating the fire response of FRP-strengthened concrete members have been applied by a number of researchers (Chen et al., 2011; Domenico, 2015; Kodur and Bhatt, 2018). One such model, proposed by Kodur and Bhatt (2018), for fire resistance analysis of FRP-strengthened concrete slabs is discussed here. This numerical model is based on the macroscopic finite element approach wherein time-dependent sectional moment curvature (M-κ) relations are utilized to trace the overall response of an FRP-strengthened RC slab, under combined effects of fire exposure and structural loading. This model specifically accounts for softening of concrete (in tension and compression), temperature-induced reduction in properties of materials, including bond degradation between FRP and concrete, and all applicable failure limit states governing fire response of an FRP-strengthened RC slab.

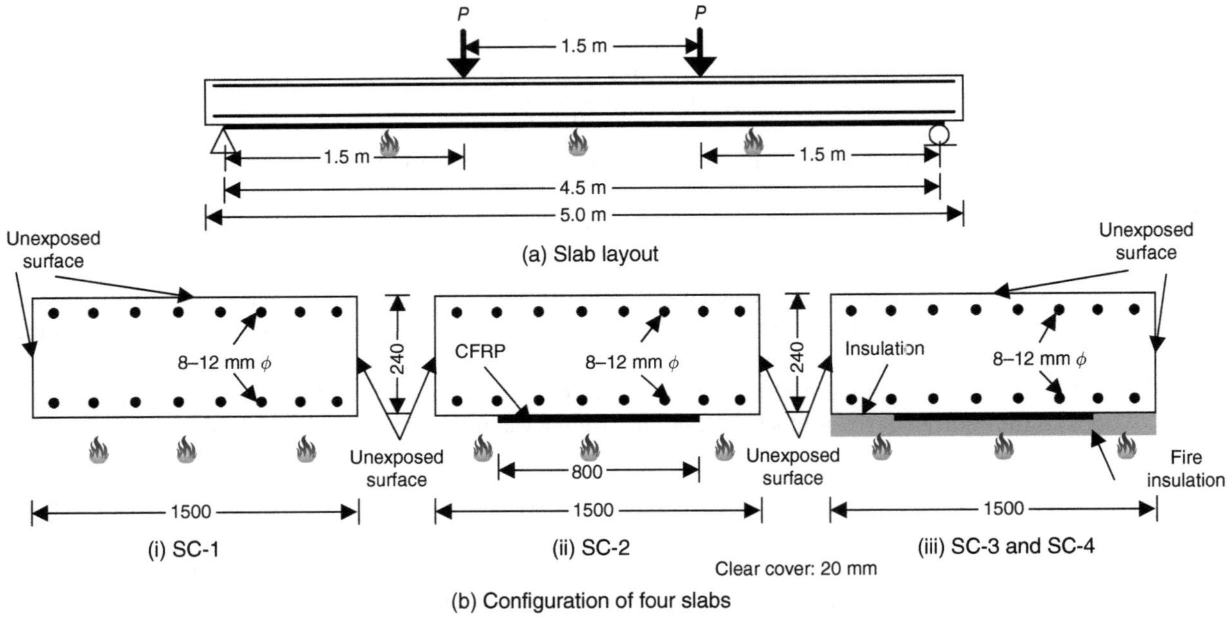

FIGURE 11.9 Layout of a typical EB FRP-strengthened RC slab and its discretization for fire resistance analysis. (*With permission from Elsevier.*)

In the analysis, the FRP-strengthened RC slab is first divided into a number of segments along its length and the midsection of each segment is further discretized into a number of elements as shown in Fig. 11.9. This midsection of the segment is assumed to represent the overall behavior of the segment. The fire resistance analysis is carried out in small time increments. At each time increment, thermal analysis followed by a structural analysis is carried out to trace the response of the slab.

Selected Slab for Analysis

For illustrating the application of advanced analysis, an FRP-strengthened RC slab tested in the laboratory under fire conditions is selected. The selected slab (shown in Fig. 11.9) is of 1500 mm width × 240 mm thickness and has an overall span of 5 m with an effective span of 4.5 m. The slab is reinforced with eight 12-mm-diameter steel rebars in tension and compression zone at a clear concrete cover of 20 mm. This discretized slab is strengthened with an 800-mm-wide and 0.165-mm-thick CFRP sheet at the soffit to enhance its flexural capacity by about 50% of its unstrengthened capacity at ambient conditions. The strengthened slab is supplemented with 12-mm-thick CAFCO (SFRM) fire insulation to enhance fire resistance (Fig. 11.10).

To illustrate the application of advanced analysis, as well as the importance of considering critical factors in the analysis, four configurations of this slab were analyzed, namely, SC-1, SC-2, SC-3, and SC-4. Of these configurations, SC-1 represents a conventional RC slab without any strengthening and without any fire insulation, SC-2 represents an RC slab strengthened with EB CFRP but without any fire insulation, and SC-3 and SC-4 represent EB-strengthened RC slabs insulated with 12-mm-thick CAFCO (SFRM) insulation wherein only the SC-3 slab considers the effect of bond-slip degradation in the FRP system. Such temperature-induced bond degradation is not specifically considered in the analysis of slab SC-4. All the slabs are analyzed under a four-point loading system developing a bending moment from service loads under fire conditions (equivalent to 50% of ambient moment capacity). The slab was analyzed with simply supported end conditions and with ASTM E119 standard fire exposure on the bottom surface. The unstrengthened slab (SC-1) has a room temperature capacity of 220 kN·m, while the corresponding moment capacity of strengthened slabs (SC-2, SC-3, SC-4) are 410 kN·m.

Material Properties

The properties at ambient temperature of various constituent materials composing this slab are listed in Table 11.2. The thermal and mechanical properties of concrete and reinforcing steel assumed to vary at elevated temperatures are as per Eurocode 2 (CEN, 2004) relations, whereas for CFRP, the high-temperature strength and elastic modulus property variation are assumed to follow the semiempirical relations proposed by Firmo et al. (2015) (see Sec. 11.3; Eqs. 11.13 and 11.14). For fire insulation, the temperature-dependent thermal conductivity and thermal capacity are assumed to follow published relations (Kodur and Shakya, 2013). Further details on temperature-dependent properties as well as discretization and analysis parameters are given by Kodur and Bhatt (2018).

Property	Value
Concrete compressive strength (f'_c, MPa)	38.8
Steel yield strength (f_y, MPa)	557
Modulus of elasticity of CFRP (E_f, MPa)	230,000
Ultimate tensile strength of CFRP (f_u, MPa)	3,500
Rupture strain of CFRP (ε_u, %)	0.85
Epoxy glass transition temperature (T_g, °C)	82
Fire insulation density (kg/m³)	298
Fire insulation thermal conductivity (W/m·K)	0.076
Fire insulation (CAFCO) specific heat (J/kg·K)	3,340

Table 11.2 Material Properties Used in Analysis of FRP-Strengthened RC Slabs (SC-1 to SC-4)

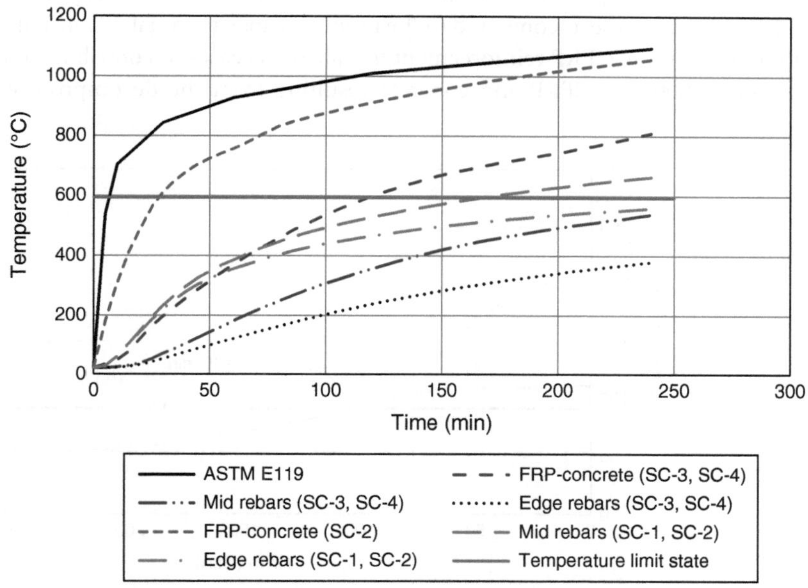

Figure 11.10 Temperature rise at the steel rebar and FRP-concrete interface of insulated FRP-strengthened RC slabs.

Results

Each slab was analyzed by exposing it to a standard fire (ASTM E119) while being subjected to a structural loading equivalent to 50% of room temperature capacity. As part of thermal analysis, fire exposure on this slab is input through specified standard fire time-temperature relations. Then cross-sectional temperatures in various elements of concrete, FRP, rebar, and fire insulation are calculated through thermal analysis. Brief results from thermal analysis is illustrated in Fig. 11.10 by plotting temperatures at CFRP-concrete interface, steel reinforcement, and various concrete depths as a function of fire exposure time. For example, the temperature at the FRP-concrete interface in slabs SC-1 and SC-2 exceeds 600°C within 30 minutes of fire exposure, as both slabs are not provided with any fire insulation. The provision of 12-mm fire insulation in slab SC-3 significantly reduces rise in sectional temperatures (at respective locations) throughout the fire exposure duration. The temperature at the CFRP-concrete interface exceeds the critical temperature of epoxy T_g in 5 minutes in slabs SC-1 and SC-2, while it is about 16 minutes of fire exposure in slabs SC-3 and SC-4. In addition, the temperature in rebars exceeds 593°C at about 175 minutes in slabs SC-1 and SC-2, while the temperature in steel reinforcement remains well below this limit for the entire fire exposure duration (4 hours) in slabs SC-3 and SC-4.

To illustrate the variation in structural response in all four slabs (SC-1 to SC-4), reduction in moment capacity with fire exposure time is plotted in Fig. 11.11. It can be seen that there is no decrease in moment capacity of the unstrengthened RC slab SC-1 until about 65 minutes of fire exposure. This is due to the fact that there is no loss of strength in steel rebars until 65 minutes, as the temperature rise in steel rebars is very slow and remains below 400°C until about 65 minutes. This slow temperature rise in the slab is attributed to the thermal protection provided by 20-mm concrete cover to rebars. With an increase in fire exposure time, the temperature in mid-rebars exceeds 400°C at about 70 minutes, and thus degradation of strength in steel rebars starts. As a result, the moment capacity of the slab gradually decreases and drops below the moment due to applied loading at 210 minutes, leading to the occurrence of strength failure in this slab.

For slab SC-2 (strengthened but uninsulated RC slab), the moment capacity decreases rapidly in the first 15 minutes of fire exposure. This is attributed to a rapid loss of strength in CFRP, combined with a degradation in bond between CFRP and concrete, due to a rapid rise in CFRP temperature (reaching much beyond T_g) resulting from direct exposure to fire. As a result, the sectional moment capacity in this slab decreases rapidly and falls below the moment due to applied loading at about 85 minutes. It should be noted that slabs SC-2, SC-3, and SC4 are designed to carry a higher service load (equivalent to a moment of 92 kN · m) during fire exposure, as a result of strengthening as opposed to slab SC-1 carrying 54 kN · m.

In the case of strengthened slabs SC-3 and SC-4, the reduction in moment capacity is at a much slower rate as compared to that in slab SC-2, due to thermal protection provided to the entire slab from supplementary fire insulation. The moment capacity of slabs SC-3 and SC-4 does not experience any decrease until 15 minutes of fire exposure, after which it starts decreasing gradually at a constant rate until 60 minutes of fire exposure. After about 65 minutes, the moment capacity in slab SC-3 experiences a slightly higher decrease as compared to the moment capacity in slab SC-4. This is attributed to the fact that after 65 minutes, the CFRP reinforcement in slab SC-3 ceases to contribute to the capacity of slab due to a loss of bond between CFRP and concrete resulting from the decomposition of the bonding adhesive.

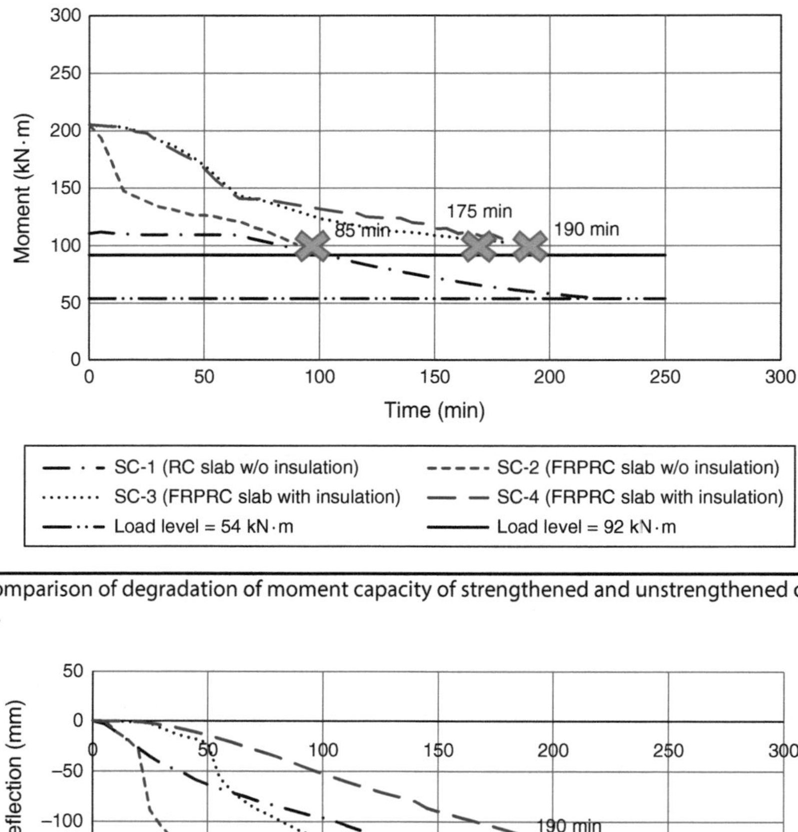

FIGURE 11.11 Comparison of degradation of moment capacity of strengthened and unstrengthened concrete slab subjected to fire.

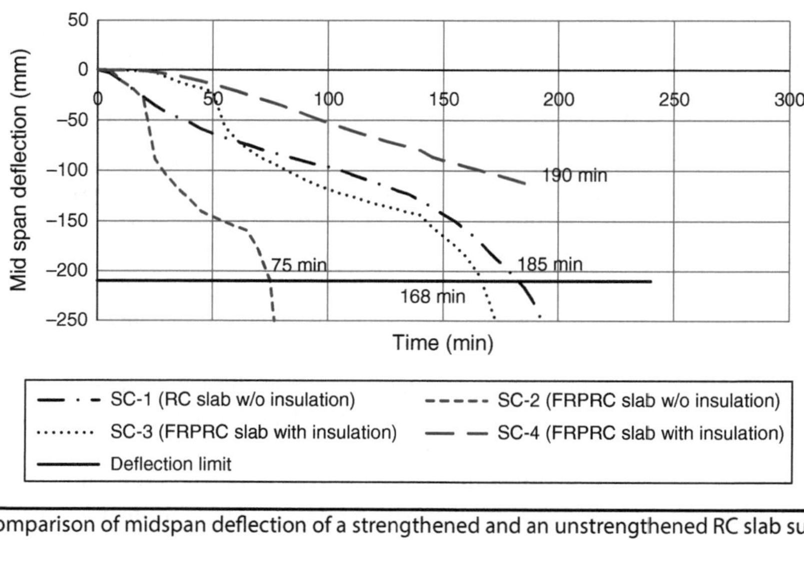

FIGURE 11.12 Comparison of midspan deflection of a strengthened and an unstrengthened RC slab subjected to fire.

Hence, after 65 minutes, slab SC-3 behaves as an insulated RC slab (without any contribution from CFRP), and the moment capacity of the slab decreases gradually, until failure. The failure in slab SC-3 occurs at 175 minutes, when the capacity of the slab falls below the bending moment due to applied loading, indicating strength failure of the slab. In the case of slab SC-4, since perfect bonding between CFRP and concrete is assumed in the analysis, there is no influence of bond degradation on the moment capacity of slab SC-4. Thus, the moment capacity of slab SC-4 decreases at a slower rate and remains slightly higher than the moment capacity of SC-3 for the major duration of fire exposure. The moment capacity of slab SC-4 degrades gradually with the increase in temperature of FRP and steel and ultimately drops below the bending moment due to applied loading at 190 minutes, leading to failure in the slab.

To further illustrate the variation in structural performance in different configurations of slabs (SC-1 to SC-4) under fire conditions, midspan deflection in all four slabs is plotted as a function of fire exposure time in Fig. 11.12. As per ASTM E119 criteria, the failure in the slab is said to occur when the deflection in the slab exceeds a limiting deflection of 210 mm (*L*/30). It can be seen from the figure that the deflection in RC slab SC-1 increases gradually, mostly due to a gradual reduction in strength and modulus of steel rebars as a result of temperature rise. After 180 minutes, the temperature in rebars of slab SC-1 increases to above 600°C. As a result, the strength and modulus properties of rebars drop below 50% of their room temperature value, which in turn leads to a rapid increase in midspan deflection of the slab beyond 180 minutes, until failure occurs at 185 minutes. The strengthened but uninsulated slab SC-2 initially (up to 20 minutes) deflects at a rate similar to slab SC-1, due to the application of the same load level and fire exposure conditions. However, after 20 minutes of fire exposure, slab SC-2 starts deflecting at an accelerated rate as the temperature at the CFRP-concrete interface increases beyond 500°C, which leads to a significant reduction in strength and modulus of CFRP (below 25% of their room temperature value). Further, the effective stress transfer from concrete to FRP (composite interaction) reduces significantly due to the temperature-induced bond degradation between FRP and concrete. After 70 minutes, the temperature in steel rebars increases beyond 400°C, causing strength and modulus properties of rebars to also degrade. As a result, deflection increases rapidly and exceeds the deflection limit at 75 minutes, indicating failure.

The presence of insulation in slabs SC-3 and SC-4 enhances the structural response of the slabs due to a slower sectional temperature rise. Thus, deflection in slabs SC-3 and SC-4 is significantly lower than deflection in slab SC-2, for the entire duration of fire exposure. Initially, the deflection in slabs SC-3 and SC-4 is negligible; however, after 25 minutes of fire exposure, slab SC-3 starts deflecting at a slightly faster rate compared to slab SC-4, due to the reduction in the contribution of CFRP to the stiffness of slab SC-3, resulting from temperature-induced bond degradation at the CFRP-concrete interface. At about 60 minutes, deflection of slab SC-3 increases abruptly, indicating a loss of composite action between CFRP and concrete, and as a result, slab SC-3 behaves as an insulated RC slab (without any contribution from CFRP). With an increase in fire exposure time, deflection in the slab increases gradually until about 160 minutes, at which point the temperature in steel rebars raises to above 400°C, leading to a rapid increase in deflection (in slab SC-3) followed by failure at 168 minutes.

In the case of slab SC-4, as a perfect bond is assumed between the CFRP and concrete, the overall structural response (deflection) primarily depends only on the degradation of strength and modulus properties of CFRP and steel rebars (no loss of bond). As the temperature rise at CFRP and steel rebars is relatively slow due to the presence of fire insulation, the degradation in strength properties of CFRP and steel rebars is also slow, resulting in a stiffer response. On an average slab, SC-4 exhibits about 50% stiffer response, as compared to slab SC-3, during the entire fire exposure duration. The deflection in slab SC-4 increases gradually and remains below the serviceability limits throughout fire exposure duration (no failure until 190 minutes). Thus, the assumption of a perfect bond between concrete and CFRP leads to unconservative fire resistance values (10% to 15% higher) in FRP-strengthened RC slabs. Hence, the effect of temperature-induced bond degradation is to be given due consideration in evaluating fire resistance of an FRP-strengthened members. A summary of results, including failure times, in all slabs is listed in Table 11.3.

11.8 Guidelines for Enhancing Fire Resistance of FRP-Strengthened Concrete Members

The use of FRP as EB or NSM reinforcement offers a quick, feasible, and effective solution to strengthen deteriorating structural members in buildings and other infrastructure. One of the key limitations that continues to hinder the use of FRP in building applications is its vulnerability under fire conditions. Recent research has established key factors influencing fire performance and has proposed guidelines that can be adopted for enhancing fire performance of FRP-strengthened members (Pacheco-Torgal et al., n.d.; Bisby et al., 2005; Kodur, 2016).

Property	Name of Case			
	SC-1	SC-2	SC-3	SC-4
Type of slab	RC slab (unstrengthened)	FRP-strengthened RC slab		
Moment due to applied loading (k·Nm)	54	92		
Applied load (%)	50			
Fire scenario	ASTM E119			
Fire exposure time (minutes)	240	240	240	240
Insulation material type	None	None	SFRM (CAFCO)	
Insulation thickness (mm)	NA	NA	12	
FRP-concrete bond degradation	NA	Considered	Considered	Neglected
Deflection (δ, mm)	453	305	400	390
Moment capacity (k·Nm) at end of analysis	53	89	90	91
Fire resistance (minutes)	185	75	175	190
Failure limit state attained	Strength	Deflection	Strength	
Failure mode	Steel yielding			

Table 11.3 Details of Parameters Considered in the Case Study

11.8.1 Factors Governing Fire Resistance

The main factors that influence fire resistance of FRP-strengthened RC members are the type of strengthening, extent of strengthening, load level, fire severity, restraint conditions, bond degradation, and properties of fire insulation, including thermal properties, thickness, and its geometric configuration (Blontrock et al., 2000; Bisby, 2003; Ahmed and Kodur, 2011; Kodur and Yu, 2013). The extent of influence of each of these factors is discussed below.

- *Type of strengthening.* The type of strengthening (EB or NSM) influences the fire response of FRP-strengthened RC members. In the case of flexural members, a concrete member strengthened with NSM FRP reinforcement yields fire resistance that is lower than a conventional concrete beam (or slab) member, but higher than that of a similar concrete member strengthened with EB FRP laminate. This is due to the fact that an FRP-strengthened concrete member is expected to carry a higher service load (say around 50% of strengthened capacity) under fire conditions, as compared to that in an unstrengthened member (50% of unstrengthened capacity). The same inference can also be made in the case of compression members (i.e., walls and columns).
- *Location of FRP reinforcement.* Unlike in EB FRP-strengthened RC beams (or slabs), the location of NSM FRP rebars on the soffit of the beam has an influence on the fire resistance of concrete beams strengthened with NSM FRP. As an example, the provision of NSM FRP rebars close to the center of the beam soffit yields slightly higher fire resistance than when NSM FRP rebars are located closer to the (bottom) corners of the beam [as shown in Fig. 11.13*b*(ii),*d*(ii) in rectangular and T-section beams, respectively].
- *Extent of strengthening.* The extent of strengthening or the level of contribution of EB or NSM FRP reinforcement to the overall sectional capacity of the strengthened member has an influence on the fire resistance of a concrete member strengthened with EB or NSM FRP. In general, members with a relatively higher strengthening level (in proportion to overall capacity) will have lower fire resistance than the case of members with a lower level of strengthening, since the FRP reinforcement experiences higher strength degradation rapidly with a rise in temperature.
- *Load level.* An EB or NSM FRP-strengthened member is to carry a higher load level after EB or NSM FRP ruptures (or delaminates) under elevated temperature exposure in a fire situation. This can lead to early strength failure of the member under fire conditions. Therefore, as in the case of concrete and steel members, a higher level of applied loading on an EB NSM FRP-strengthened member results in lower fire resistance.
- *Axial restraint.* The development of fire-induced axial restraint in the case of a flexurally strengthened RC member (i.e., beams or slabs) enhances fire resistance. This is attributed to arch action generated in the beam or slab as a result of fire-induced restraint force.

- *Fire severity.* Under most design (realistic) fire scenarios, representative of building fires, an NSM EB-strengthened concrete member achieves higher fire resistance than under standard fire exposure (ASTM E119). This is mainly due to the beneficial effect of the cooling phase of fire exposure (decay phase).
- *Bond degradation.* The bond between concrete and EB or NSM FRP reinforcement deteriorates considerably at temperatures above the glass transition temperature (T_g). This can lead to debonding of an EB NSM FRP, thus decreasing sectional capacity of a strengthened member and leading to lower fire resistance.
- *Fire insulation.* External fire insulation is an effective mechanism to delay temperature rise in FRP beyond a certain limiting temperature, and thus enhance fire resistance of an EB or NSM FRP-strengthened concrete member. The thickness of fire insulation and its layout, together with thermal properties of insulation, are the key parameters that govern the fire resistance of EB or NSM FRP-strengthened concrete members. In the case of NSM or EB FRP-strengthened concrete beams, the provision of U-shaped fire insulation provides a much higher fire resistance than that of providing insulation only at the beam soffit (Yu, 2013) (see Fig. 11.13 for effective insulation schemes). Similar insulation configuration can be developed for other structural members (column, wall, etc.) through proper fire resistance calculations.
- *Sectional configuration.* In the case of FRP-strengthened concrete columns, a circular column provides higher fire resistance than a column with a square/rectangular cross section, as temperature propagates at much faster rate on the edges of a square or rectangular section due to bidirectional heat transfer.

11.8.2 Guidelines for Enhancing Fire Resistance

Based on a series of experiments and numerical studies, a set of guidelines has been developed for enhancing fire resistance in FRP-strengthened concrete members (Kodur et al., 2007; Ahmed and Kodur, 2011; Kodur and Yu, 2013). The following strategies can be effective in enhancing the fire performance of FRP-strengthened RC members:

- In most cases, EB FRP-strengthened concrete members, unlike conventional RC members, require supplementary fire insulation to achieve the required fire resistance under increased service loads. The fire resistance of such fire-insulated FRP-strengthened concrete members can be similar to, or higher than, conventional RC members. While it is possible to get 1 to 2 hours of fire resistance without any external fire insulation in the case of NSM FRP-strengthened RC members, EB FRP-strengthened RC members require supplementing fire protection in most cases to achieve fire resistance of 1 hour or more.
- Reaching the glass transition temperature in the matrix (epoxy) of an FRP during fire exposure does not lead to failure of an FRP-strengthened concrete member. FRP-strengthened concrete members can continue to support external loading beyond when T_g of the FRP polymer matrix is reached during fire exposure.
- Proper detailing of fire insulation is needed to keep temperatures low not only in FRP but also in steel reinforcement to achieve optimum fire resistance. This is especially true in the case of strengthened beams (or slabs) where tensile rebars are close to the soffit of the beam (or slab). To develop optimum insulation detailing schemes, rational calculations have to be carried out (Kodur and Yu, 2013).
- The five possible insulation configurations that can be adopted in EB or NSM FRP-strengthened RC beams, with rectangular or T cross sections, are shown in Fig. 11.13. These are for five cases: (i) without any supplementary fire insulation, (ii) applied fire insulation at the bottom, (iii) insulation extending to the concrete cover (C_c) depth on the sides, (iv) insulation extending up to a depth of twice the concrete cover on the sides, and (v) insulation extending on the entire depth of the bottom and sides. The first option, without any supplementary fire protection, is not viable for EB FRP members in most cases, as some level of fire supplementary insulation is needed to achieve the required fire resistance ratings (1 hour or more) in building applications. However, for NSM FRP-strengthened concrete members, the first option may be workable provided that the required fire resistance is in 1- to 2-hour range. The second insulation configuration, shown in Fig. 11.13*b*, of

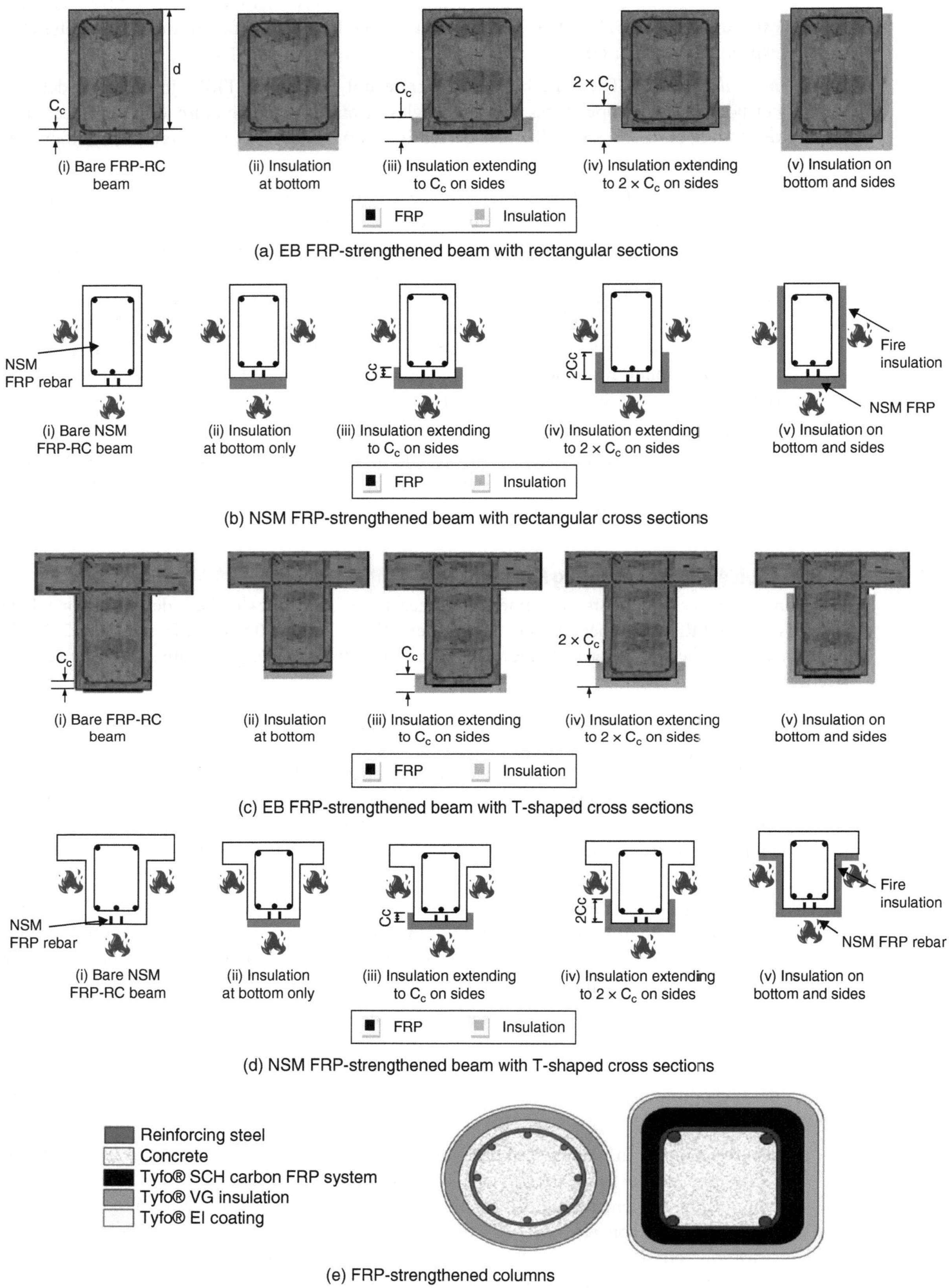

FIGURE 11.13 Geometric configuration schemes for fire insulation in FRP-strengthened concrete beams. (*With permission from Springer.*)

protecting the beam at the bottom soffit only may not be effective in achieving optimal fire resistance. In this configuration, corner rebar temperatures (in a beam) increase after a certain fire exposure time due to heat transfer to the corners from two sides, and thus any increase in fire resistance is only marginal, as compared to uninsulated FRP-strengthened concrete beams. In the case of configuration 3 [refer to Fig. 11.13*a*(iii),*c*(iii)], the needed fire insulation is provided at the beam soffit and extends to a depth equal to the concrete cover thickness on both sides of an FRP-strengthened RC beam (where C_c is the cover thickness to rebar as per applicable design standards such as ACI 318). In this case, the resulting temperatures in the steel rebars for a given insulation thickness are lower than in configuration 2. However, there is only a slight improvement in the fire resistance of the FRP-strengthened member (beam). In the case of configuration 4 [refer to Fig. 11.13*a*(iv),*c*(v)], the strengthened concrete beam is protected on three sides (bottom and two sides), where the depth of the insulation on two sides of the beam is held equal to twice the concrete cover thickness ($2 \times C_c$) to rebars. This configuration is most optimal, as it provides effective thermal insulation to the overall beam cross section by minimizing temperature rise at the FRP-concrete interface, as well as in tensile steel reinforcement for a considerable time, thus enhancing the fire resistance of the beam. Finally, in configuration 5 [refer to Fig. 11.13*a*(iv),*c*(v)], insulation is provided along the entire two side faces and beam soffit. This insulation scheme is not practically viable and may not be cost effective as well (Ahmed and Kodur, 2011).

- For a typical commercially available fire insulation, such as VG-EI, an optimum insulation thickness of 40 mm is required to achieve a fire resistance of 3 hours, while a 20-mm insulation thickness is needed to achieve a fire resistance of 2 hours in EB FRP-strengthened RC beams. In the case of NSM FRP RC beams, up to 2 hours of fire resistance can be achieved without any fire insulation through proper fire design calculations. An optimum insulation thickness of 20 mm is required to achieve a fire resistance of 3 hours, while a 40-mm thickness is needed to achieve a fire resistance of 4 hours in NSM FRP-strengthened RC beams. VG-EI fire insulation is a proprietary insulation material and is composed of vermiculite-gypsum spray-on insulation material with a top coating of EI intumescent paint. For other available fire insulations, the optimum fire protection scheme can be developed using rational engineering calculations (Kodur and Yu, 2016).
- To protect against the sudden and complete loss of effectiveness of EB FRP wrap (in columns of beams) or NSM rebars during fire, the service load on the strengthened (with EB or NSM) RC member should not exceed the design capacity of the unstrengthened (preexisting) RC members (see Fig. 11.13*e*). This requirement is similar to a strengthening limit requirement currently recommended by ACI Committee 440 (ACI, 2017) and also provides a measure of protection against poor installation practices or vandalism.

11.9 Emerging Trends and Research Needs

While significant progress has been made over the past three decades in developing efficient and cost-effective FRP strengthening systems, the poor fire performance of FRP materials continue to hinder their wide acceptance and full utilization in rehabilitation and strengthening applications. Two critical areas where further research and development efforts are needed with respect to the fire resistance of FRP-strengthened concrete members are discussed herein.

11.9.1 Fire-Resistant FRP Materials

One of the main limitations of fully incorporating FRPs into building applications is their vulnerability to fire hazard. As a result, externally strengthened FRP RC members need to be insulated in most cases in building applications to achieve required fire ratings. This is mainly due to the fact that FRP has a low tolerance to high-temperature exposure, mainly because of a low glass transition temperature of the matrix, which is in the range of 60 to 80°C. An effective strategy to enhance the fire performance of FRP systems is to increase the glass transition temperature of the matrix used in the composite.

In a recent study, Ji et al. (2013) explored the use of advanced FRP types, known as *fire-resistant FRP*, which are developed using nanoclay polymer-based nanocomposites reinforced with organically modified montmorillonite (MMT). These nanocomposites have high modulus properties, a large surface area, and a high aspect ratio. At relatively moderate temperatures, the morphology of nanoclay changes from discontinued intercalated status to well-dispersed and continued exfoliated status and forms insulating layers. These layers form thick char that could slow down the combustion of FRP, as well as the thermal diffusivity of the FRP composite. Data and observations from limited high-temperature exposure tests show that the addition of nanoclay improved fire resistance and increased residual strength of FRP-strengthened RC beams after exposure to elevated temperatures, up to a temperature of 1100°C (Ji et al., 2013). However, detailed fire tests and associated numerical studies at the material and structural levels are still lacking to validate this concept for enhancing fire resistance.

Another means to enhance fire resistance of FRPs is to incorporate inorganic matrices. Among the many types of inorganic matrices is a new class of material referred to as geopolymers (Zhang et al., 2016), which are synthesized by alkaline solutions (such as sodium silicate and potassium silicate) activating aluminosilicate source materials (such as metakaolin, fly ash, and slag). Compared to organic matrices, inorganic geopolymers possess better resistance to high-temperature exposure and UV radiation and produce minimal toxic smoke under fire exposure (Zhang et al., 2016). However, there is very limited property data and fire test data on concrete members strengthened with geopolymer-based FRP (Zhang et al., 2014). Additional material-level property tests, supplemented with numerical modeling, are needed to develop a better understanding of the properties and performance of FRP systems incorporating geopolymers under elevated temperatures. Further, full-scale fire tests and associated numerical studies on geopolymer FRP-strengthened concrete structural members are needed to validate the fire performance at the structural level (Zhang et al., 2014, 2015, 2018).

11.9.2 Simplified Approaches for Fire Resistance Evaluation

There is still a lack of simplified calculation methods and design guidance for evaluating the fire resistance of FRP-strengthened concrete members in codes and design standards. This lack of design guidance stems from the limited work in this area, especially with regard to developing temperature-dependent property data (i.e., on thermal, mechanical, and bond properties), reliable material models (specifically bond-slip degradation at elevated temperatures), suitable effective fire insulation materials or FRP-strengthened applications, and rational calculation methods for evaluating fire resistance. Further research and development efforts in the aforementioned area leading to the development of simplified approaches for evaluating fire resistance are urgently needed. Such an exercise can lead to the wider use of FRP in the retrofitting and strengthening of structures in building applications.

11.10 References

ACI. *Code Requirements for Determining Fire Resistance of Concrete and Masonry Construction Assemblies*. ACI 216.1-14/TMS-0216-14. Farmington Hills, MI: American Concrete Institute, 2014.

ACI. *Guide for the Design and Construction of Externally Bonded FRP Systems for Strengthening Concrete Structures: ACI 440*. Farmington Hills, MI: American Concrete Institute Committee 440, 2017.

Ahmed, A., and V. K. R. Kodur. "Effect of Bond Degradation on Fire Resistance of FRP-Strengthened Reinforced Concrete Beams." *Composites Part B: Engineering* 42, no. 2 (2011): 226–237. https://doi.org/10.1016/j.compositesb.2010.11.004.

Ahmed, Aqeel. *Behavior of FRP-Strengthened Reinforced Concrete Beams under Fire Conditions*. Ph.D. thesis. East Lansing, MI: Michigan State University, 2010.

Ahmed, Aqeel, and Venkatesh Kodur. "The Experimental Behavior of FRP-Strengthened RC Beams Subjected to Design Fire Exposure." *Engineering Structures* 33, no. 7 (2011): 2201–2211. https://doi.org/10.1016/J.ENGSTRUCT.2011.03.010.

ASCE. *Minimum Design Loads for Buildings and Other Structures*. ASCE 7-10. Reston, VA: ASCE, 2010.

Bai, Yu-Lei, Jian-Guo Dai, and J. G. Teng. "Cyclic Compressive Behavior of Concrete Confined with Large Rupture Strain FRP Composites." *Journal of Composites for Construction* 18, no. 1 (2014): 04013025. https://doi.org/10.1061/(ASCE)CC.1943-5614.0000386.

Bank, Lawrence C. "Properties of FRP Reinforcements for Concrete." In *Fiber-Reinforced-Plastic (FRP) Reinforcement for Concrete Structures*. Elsevier, 1993: pp. 59-86.

Barros, Joaquim A.O., Salvador J. E. Dias, Hadi Baghi, and A. Ventura-Gouveia. "New Shear Strengthening Configurations of Near-Surface-Mounted CFRP Laminates for RC Beams." *ACI Structural Journal* 113, no. 6 (2016). https://doi.org/10.14359/51689029.

Bisby, L.A. *Fire Behaviour of Fibre-Reinforced Polymer (FRP) Reinforced or Confined Concrete*. Ph.D. thesis. Ontario: Queen's University, 2003. https://www.researchgate.net/profile/Luke_Bisby/publication/36189430_Fire_behaviour_of_fibre-reinforced_polymer_FRP_reinforced_or_confined_concrete_microform/links/55cf69d808ae502646aa440e/Fire-behaviour-of-fibre-reinforced-polymer-FRP-reinforced-or-confined-concrete-microform.pdf.

Bisby, Luke A., Mark F. Green, and Venkatesh K. R. Kodur. "Response to Fire of Concrete Structures That Incorporate FRP." *Progress in Structural Engineering and Materials* 7, no. 3 (2005): 136–149. https://doi.org/10.1002/pse.198.

Blontrock, Hendrik, Luc Taerwe, and Paul Vandevelde. "Fire Tests on Concrete Beams Strengthened with Fibre Composite Laminates." In *Proceedings of the International PhD Symposium in Civil Engineering, Vienna (Austria), 5-7 October 2000*. Konrad Bergmeister, ed. Vol. 2, pp. 151–161. 2000.

"BXUV.GuideInfo—Fire-Resistance Ratings—ANSI/UL 263." n.d. Accessed January 18, 2019. https://database.ul.com/cgi-bin/XYV/template/LISEXT/1FRAME/showpage.html?name=BXUV.GuideInfo&ccnshorttitle=Fire-resistance+Ratings+-+ANSI/UL+263&objid=1074327030&cfgid=1073741824&version=versionless&parent_id=1073984818&sequence=1.

CEN. *Eurocode 2: Design of Concrete Structures—Part 1-2: General Rules—Structural Fire Design. EN 1992-1-2.* Brussels: European Committee for Standardization, 2004.

Chen, G. M., J. G. Teng, and J. F. Chen. "Finite-Element Modeling of Intermediate Crack Debonding in FRP-Plated RC Beams." *Journal of Composites for Construction* 15, no. 3 (2011): 339–353. https://doi.org/10.1061/(ASCE)CC.1943-5614.0000157.

Czarnecki, L., M. Kaproń, and D. Van Gemert. "Sustainable Construction: Challenges, Contribution of Polymers, Research Arena." *Restoration of Buildings and Monuments* 19. (2013). https://doi.org/10.1515/rbm-2013-6583.

Domenico, D. De.. "RC Members Strengthened with Externally Bonded FRP Plates: A FE-Based Limit Analysis Approach." *Composites Part B: Engineering* 71 (March 2015): 159–174. https://doi.org/10.1016/J.COMPOSITESB.2014.11.013.

FiB. *FRP Reinforcement in RC Structures: Technical Report.* Lausanne: International Federation for Structural Concrete, 2007a. https://www.fib-international.org/publications/fib-bulletins/frp-reinforcement-in-rc-structures-115-detail.html.

FiB. *FRP Reinforcement in RC Structures.* Bulletin No. 40. 2007b. https://doi.org/10.1371/journal.pntd.0001792.

Firmo, João P., João R. Correia, and Luke A. Bisby. "Fire Behaviour of FRP-Strengthened Reinforced Concrete Structural Elements: A State-of-the-Art Review." *Composites Part B: Engineering* 80 (October 2015): 198–216. https://doi.org/10.1016/J.COMPOSITESB.2015.05.045.

Gao, Wan-Yang, Jian-Guo Dai, and J. G. Teng. "Fire Resistance Design of Unprotected FRP-Strengthened RC Beams." *Materials and Structures* 49, no. 12 (2016): 5357–5371. https://doi.org/10.1617/s11527-016-0865-x.

ICC. *2018 International Building Code.* ICC, 2018. https://codes.iccsafe.org/content/IBC2018?site_type=public.

Ji, Gefu, Guoqiang Li, and Walid Alaywan. "A New Fire Resistant FRP for Externally Bonded Concrete Repair." *Construction and Building Materials* 42 (2013): 87–96. https://doi.org/10.1016/j.conbuildmat.2013.01.008.

Kodur, V. K. R., and P. P. Bhatt. "A Numerical Approach for Modeling Response of Fiber Reinforced Polymer Strengthened Concrete Slabs Exposed to Fire." *Composite Structures* 187 (March 2018.): 226–240. https://doi.org/10.1016/J.COMPSTRUCT.2017.12.051.

Kodur, V. K. R., and T. Z. Harmathy. "Properties of Building Materials." In *SFPE Handbook of Fire Protection Engineering.* New York: Springer, 2016: pp. 277–324. https://doi.org/10.1007/978-1-4939-2565-0_9.

Kodur, V. K. R., and A. M. Shakya. "Effect of Temperature on Thermal Properties of Spray Applied Fire Resistive Materials." Fire Safety Journal 61 (October 2013): 314–323. https://doi.org/10.1016/J.FIRESAF.2013.09.011.

Kodur, V. K. R., and Baolin Yu. "Evaluating the Fire Response of Concrete Beams Strengthened with Near-Surface-Mounted FRP Reinforcement." *Journal of Composites for Construction* 17, no. 4 (2013): 517–529. https://doi.org/10.1061/(ASCE)CC.1943-5614.0000348.

Kodur, V. K. R., and Baolin Yu. "Rational Approach for Evaluating Fire Resistance of FRP-Strengthened Concrete Beams." *Journal of Composites for Construction* 20, no. 6 (2016): 04016041. https://doi.org/10.1061/(ASCE)CC.1943-5614.0000697.

Kodur, V. K. R., L. A. Bisby, and M. F. Green. "Preliminary Guidance for the Design of FRP-Strengthened Concrete Members Exposed to Fire." *Journal of Fire Protection Engineering.* 17, no. 1 (2007): 5–26.

Kodur, V. K. R., B. Yu, and R. Solhmirzaei. "A Simplified Approach for Predicting Temperatures in Insulated RC Members Exposed to Standard Fire." *Fire Safety Journal* 92(2017): 80–90. https://doi.org/10.1016/j.firesaf.2017.05.018.

Kodur, Venkatesh K. R. "Strategies for Enhancing Fire Performance of NSM FRP-Strengthened Reinforced Concrete Beams." In *Multi-Hazard Approaches to Civil Infrastructure Engineering,* Gardoni, Paolo, James LaFave, eds. Cham: Springer, 2016: pp. 253–264. https://doi.org/10.1007/978-3-319-29713-2_12.

Lie, T. T., ed. *Structural Fire Protection.* New York: American Society of Civil Engineers, 1992. https://doi.org/10.1061/9780872628885.

Lorenzis, L. De, and J. G. Teng. "Near-Surface Mounted FRP Reinforcement: An Emerging Technique for Strengthening Structures." *Composites Part B: Engineering* 38, no. 2 (2007): 119–143. https://doi.org/10.1016/j.compositesb.2006.08.003.

Mallick, P. K. *Fiber-Reinforced Composites : Materials, Manufacturing, and Design.* Boca Raton, FL: CRC Press, 2008. https://www.crcpress.com/Fiber-Reinforced-Composites-Materials-Manufacturing-and-Design-Third/Mallick/p/book/9780849342059.

Nomura, Sechi, and Dale L. Ball. "Micromechanical Formulation of Effective Thermal Expansion Coefficients of Unidirectional Fiber Composites." *Advanced Composite Materials* 3, no. 2 (1993): 143–151. https://doi.org/10.1163/156855193X00142.

Pacheco-Torgal, Fernando, Robert Melchers, Nele de Belie, Xianming Shi, Kim Van Tittelboom, and Andrés Sáez Perez. *Eco-Efficient Repair and Rehabilitation of Concrete Infrastructures.* Duxford: UK: Woodhead Publishing, 2017.

Palmieri, Aniello, Stijn Matthys, and Luc Taerwe. "Experimental Investigation on Fire Endurance of Insulated Concrete Beams Strengthened with Near Surface Mounted FRP Bar Reinforcement." *Composites Part B: Engineering* 43, no. 3 (2012): 885–895. https://doi.org/10.1016/j.compositesb.2011.11.061.

Raghavan, J., and M. Meshii. "Creep of Polymer Composites." *Composites Science and Technology* 57, no. 12 (1998): 1673–1688. https://doi.org/10.1016/S0266-3538(97)00104-8.

Rahman, A. H., D. A. Taylor, and C. Y. Kingsley. "Evaluation of FRP as Reinforcement for Concrete Bridges." *Concrete. Org Special Publication* 138 (1993): 71–86. https://www.concrete.org/publications/internationalconcreteabstractsportal/m/details/id/3919.

Rasheed, H. A. *Strengthening Design of Reinforced Concrete with FRP.* Boca Raton, FL: CRC Press, 2014.

Schapery, R. A. "Thermal Expansion Coefficients of Composite Materials Based on Energy Principles." *Journal of Composite Materials* 2, no. 3 (1968): 380–404. https://doi.org/10.1177/002199836800200308.

Silverman, E. M. "Elevated Temperature Testing for Comparison of Glass/Resin Composites." *Polymer Composites* 4, no. 4 (1983): 214–218. https://doi.org/10.1002/pc.750040404.

Teng, J. G., J. Chen, S. Smith, and L Lam. *FRP-Strengthened RC Structures.* Hoboken, NJ: Wiley, 2002. https://www.wiley.com/en-us/FRP%3A+Strengthened+RC+Structures-p-9780471487067.

Wang, Y. C., and V. Kodur. "Variation of Strength and Stiffness of Fibre Reinforced Polymer Reinforcing Bars with Temperature." *Cement and Concrete Composites* 27, no. 9–10 (2005): 864–874. https://doi.org/https://doi.org/10.1016/j.cemconcomp.2005.03.012.

Wickström, U. "A Very Simple Method for Estimating Temperature in Fire Exposed Concrete Structures." In *Proceedings of New Technology to Reduce Fire Losses and Costs*. Grayson, S. J., and D. A. Smith, eds. New York: Elsevier, 1986. http://www.diva-portal.org/smash/record.jsf?pid=diva2%3A961620&dswid=-6040

Xu, X-P., and Alan Needleman. "Numerical Simulations of Fast Crack Growth in Brittle Solids." *Journal of the Mechanics and Physics of Solids* 42, no. 9 (1994): 1397–1434.

Yu, Baolin. Fire Response of Reinforced Concrete Beams Strengthened with Near-Surface Mounted FRP Reinforcement. Michigan State University, 2013.

Yu, Baolin, and Venkatesh Kodur. "Effect of Temperature on Strength and Stiffness Properties of Near-Surface Mounted FRP Reinforcement." *Composites Part B: Engineering* 58 (2014): 510–517. https://doi.org/10.1016/J.COMPOSITESB.2013.10.055.

Zhang, Hai-yan, Xu Hao, and Wang Fan. "Experimental Study on High Temperature Properties of Carbon Fiber Sheets Strengthened Concrete Cylinders Using Geopolymer as Adhesive." *Procedia Engineering* 135 (2016): 47–55. https://doi.org/10.1016/J.PROENG.2016.01.078.

Zhang, Hai Yan, Venkatesh Kodur, Shu Liang Qi, Liang Cao, and Bo Wu. "Development of Metakaolin–Fly Ash Based Geopolymers for Fire Resistance Applications." *Construction and Building Materials* 55 (2014): 38–45. https://doi.org/10.1016/J.CONBUILDMAT.2014.01.040.

Zhang, Hai Yan, Venkatesh Kodur, Shu Liang Qi, and Bo Wu. "Characterizing the Bond Strength of Geopolymers at Ambient and Elevated Temperatures." *Cement and Concrete Composites* 58 (2015): 40–49. https://doi.org/10.1016/J.CEMCONCOMP.2015.01.006.

Zhang, Hai Yan, Hao Ran Lv, Venkatesh Kodur, and Shu Liang Qi. "Performance Comparison of Fiber Sheet Strengthened RC Beams Bonded with Geopolymer and Epoxy Resin under Ambient and Fire Conditions." *Journal of Structural Fire Engineering* 9, no. 3 (2018): 174–188. https://doi.org/10.1108/JSFE-01-2017-0023.

11.11 Nomenclature

ε_{crf}	Creep strain of FRP
z_{ec}	Equivalent concrete thickness
a and n	Coefficients for fitting ISO 834 and E119 fire
a_σ, b_σ, c_σ, a_E, b_E, and c_E	Coefficients obtained from curve-fitting of test data for different types of FRP
b_T	Width of the concrete
c	Neutral axis depth
d	Depth to steel rebars
$E_{f,T}$	Stiffness of FRP degradation with temperature T
f_{20}	Bond strength at ambient conditions
$f_{f,T}$	Strength of FRP degradation with temperature T
f_T	Bond strength at elevated temperatures
H	Activation energy which depends on experimental data
h	Height of cross section
t	Fire exposure time
T	Temperature (°C)
T_g	Glass transition temperature T
z and y	Distance from the point in the concrete section to the fire exposure surface
ε_f	Strain in FRP
ε_s	Strain in steel

11.12 Numerical Examples

11.1

Develop a prescriptive-based fire insulation solution to satisfy a 4-hour fire rating for an EB FRP-strengthened RC column and an RC T-beam being used in a retrofitting application in a building. The T-beam and column have design specifications similar to the ones in UL listings N811 and X826.

Solution

FRP-Strengthened RC T-Beam

UL listing N811 forms the basis for developing a fire rating solution to this problem. VG-EI insulation scheme, outlined in UL N811 listing, can be applied to achieve a 4-hour fire rating. A fire insulation thickness of 34 mm is to be applied to the beam's soffit (covering the EB FRP system) and should extend to the sides (the full depth of the beam's web) and 125 mm under the beam's flange. This fire insulation is to have a minimum average dry density of 371 kg/m³ (see Fig. 11.4*a*).

FRP-Strengthened RC Column

In this case, UL listing X826 forms the basis for developing fire rating solutions. A minimum thickness of 40 mm of VG-EI (as in listing X826) insulation spray is to be applied uniformly around the column. This fire insulation is to have a minimum average dry density of 800 kg/m³ (see Fig. 11.4*b*).

11.2

An external FRP-strengthened RC beam (made of carbonate aggregate, normal strength) of span 6 m is exposed to a standard ASTM E119 fire and subjected to uniformly distributed loads. Dead load and live load are 12.5 and 14.4 kN/m, respectively. The beam is of 250 mm × 450 mm cross section and is in situ with the concrete floor slab. The beam has four 19-mm-diameter tension reinforcing bars at the bottom, two 13-mm-diameter compression bars at the top, 8-mm stirrups at a space of 150 mm, and the clear cover to the reinforcing bars 38 mm. One layer of carbon fiber-reinforced polymer (CFRP) laminate (250 × 1 mm) is applied as an external FRP strengthening. The beam is protected by U-shaped VG fire insulation with a thickness of 25 mm. The beam configuration and material properties are tabulated in Table E11.1 and Fig. E11.1. Calculate the moment capacity of the fire-exposed beam as a function of fire exposure time, and evaluate the fire resistance of the beam under the specified standard fire exposure and loading conditions. The thermal properties of Tyfo VG insulation and concrete are

$$k_{VG} = 0.116 \text{ W/m}\cdot\text{K},\ (\rho c)_{VG} = 413 \text{ kJ/m}^3\cdot\text{K},\ k_c = 1.355 \text{ W/m}\cdot\text{K},\ (\rho c)_c = 2566 \text{ kJ/m}^3\cdot\text{K}$$

Evaluate fire resistance of this FRP-strengthened concrete beam after 60 minutes of ASTM E119 standard fire.

Property	PROMATECT®-H
Density (Kg/m³)	870
Thermal conductivity (W/m·K)	@ 20°C → 0.17 @ 100°C → 0.19 @ 200°C → 0.21
Specific heat (kJ/Kg·K)	0.92
Coefficient of expansion (1/K)	–6.4 × 10⁻⁶

Table E11.1 Thermal Properties of PROMATECT-H Insulation

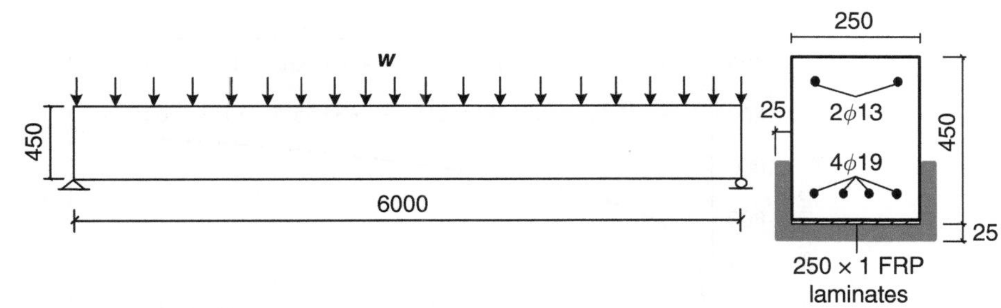

Figure E11.1 Layout and cross section of an external FRP-strengthened RC beam.

Solution

Step 1. Calculate external moment applied on the beam.

Based on the ASCE 7 (ASCE, 2010) recommendation, the external loads under fire condition can be calculated as

$$w_{\text{fire}} = 1.2DL + 0.5LL = 1.2 \times 12.5 + 14.4 \times 0.5 = 22.2 \text{ kN/m} \tag{E11.1}$$

Thus, the maximum external moment applied at the midspan of beam is

$$M_{\text{fire}} = w_{\text{fire}} L^2/8 = 22.2 \times 6^2/8 = 100 \text{ kN·m} \tag{E11.2}$$

Step 2. Calculate temperatures in steel rebars and external FRP laminates.
Apply proposed 1D and 2D heat transfer equations to calculate the temperatures in the steel rebar and external FRP, respectively.

Since the beam is protected by fire insulation, the insulation layer needs to be converted to an equivalent concrete layer. The thermal properties of concrete and VG insulation are taken from the literature (Bisby, 2003; Lie, 1992).

$$k_{VG} = 0.116 \text{ W/m·K}$$

$$(\rho c)_{VG} = 413 \text{ kJ/m}^3\text{·K}$$

$$k_c = 1.355 \text{ W/m·K}$$

$$(\rho c)_c = 2566 \text{ kJ/m}^3\text{·K}$$

With the thermal properties known, the thickness of an equivalent concrete layer is calculated using Eq. 11.31.

$$z_{ec} = z_i \sqrt[4.5]{t}\, \sqrt[1.75]{\frac{k_c}{(\rho c)_c}\frac{(\rho c)_i}{k_i}} = 25\sqrt[4.5]{1}\, \sqrt[1.75]{\frac{1.355}{2.566} \times \frac{0.413}{0.116}} = 36\text{mm} \tag{E11.3}$$

The distance from the center of the corner rebar to two fire exposure sides is 36 + 38 + 8 + 19/2 = 92 mm, so $y = z = 92$ mm. As an example, the temperature at 1-hour fire exposure time for corner rebar works out to be

$$T_c = 1.0 \times [-1.481 \times (0.08 \times 0.08) + 0.985 \times (0.08 + 0.08) + 0.017](910 \times 1^{0.148}) = 150°\text{C} \tag{E11.4}$$

For middle rebar,

$$z_{ec} = z_i \sqrt[4.5]{t}\, \sqrt[1.75]{\frac{k_c}{(\rho c)_c}\frac{(\rho c)_i}{k_i}} = 25 \times \sqrt[4.5]{1}\, \sqrt[1.75]{\frac{1.355}{2.566} \times \frac{0.413}{0.116}} = 36\text{mm} \tag{E11.5}$$

$$\eta_Z = -\left(0.155\ln\frac{1}{0.138^{1.5}} - 0.348\sqrt{0.138} - 0.371\right) = -0.04 \tag{E11.6}$$

$$T_c = 1.0 \times [-1.481 \times (0.04 \times 0.08) + 0.985 \times (0.04 + 0.08) + 0.017](910 \times 1^{0.148}) = 150°\text{C} \tag{E11.7}$$

For external FRP, the average temperature of FRP laminates is evaluated using the point at quarter FRP length from a fire-exposed surface. Accordingly, FRP laminate temperature is calculated using Eqs. 11.7 to 11.9.

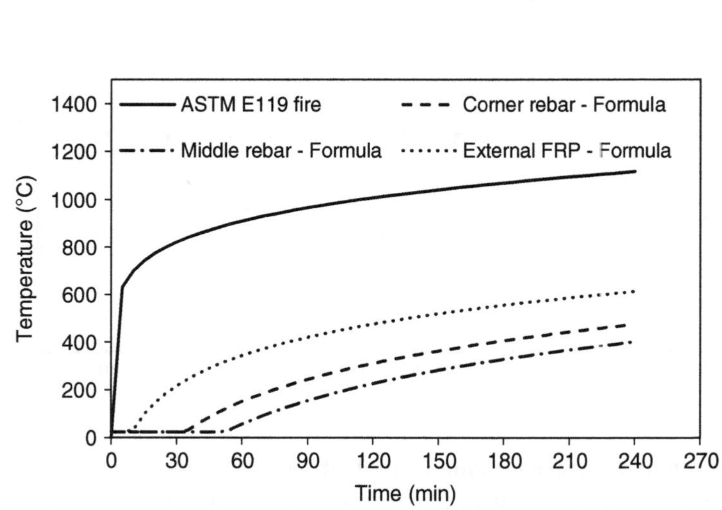

FIGURE E11.2 Prediction of temperatures in steel rebar and external FRP and comparison to FEA results.

$$\eta_Z = \left|0.155\ln\frac{1}{0.0985^{1.5}} - 0.348\sqrt{0.0985} - 0.371\right| = -0.06$$

$$T_c = 1.0 \times [-1.481 \times (0.06 \times 0.33) + 0.985 \times (0.06 + 0.33) + 0.017](910 \times 1^{0.148}) = 339°\text{C} \quad \text{(E11.8)}$$

By using a spreadsheet, the temperatures at other fire exposure times can be also calculated using the same procedure above. The resulting time-temperature progression in rebars and external FRP is plotted in Fig. E11.2.

Step 3. Evaluating the strength in steel and FRP and effective concrete depth.

By knowing the temperature in steel rebars and external FRP, the residual strength in steel rebar can be evaluated using temperature-dependent properties specified in Eurocode 2 (CEN, 2004), and the residual strength in external FRP laminate can be obtained utilizing empirical relations proposed by Bisby et al. (2005).

After 1 hour of fire exposure, the residual strength in steel is 100% of their room temperature strength for both corner and middle steel rebars, since their temperature is lower than 400°C.

The residual strength $f(T)$ and modulus $E(T)$ in external FRP are

$$f(T) = 0.55 - 0.45\tanh[0.00583(T - 339.54)]$$

$$= 0.55 - 0.45\tanh[0.00583(339 - 339.54)] = 0.55$$

$$E(T) = 0.525 - 0.475\tanh[0.00868(T - 367.41)]$$

$$= 0.525 - 0.475\tanh[0.00868(339 - 367.41)] = 0.64$$

To evaluate the effective concrete width in compression zones, the information given in Table E11.2 is needed. The effective concrete width for a section of 250 × 400 can be used in this calculation. For 1-hour fire exposure, the effective concrete width is 92.8% of original width.

Time (min)	Effective Width Factors (%)				
	150 × 200	200 × 300	250 × 400	300 × 500	400 × 600
0	100	100	100	100	100
15	99.7	99.9	99.9	99.9	99.9
30	91.1	97.2	98.2	97.4	98.2
45	83.0	95.6	96.0	94.0	95.6
60	75.5	91.9	92.8	91.2	93.3
75	69.1	88.8	90.7	88.1	91.4
90	61.7	86.4	88.0	86.1	88.9
105	63.3	83.5	86.2	83.5	87.7
120	68.2	80.6	83.8	82.2	86.3
135	69.5	78.5	82.3	80.2	84.7
150	66.6	75.2	80.1	78.8	83.5
165	61.7	72.1	78.0	77.2	82.4
180	56.4	69.8	76.8	75.7	81.2
195	52.4	66.7	74.6	74.6	80.4
210	49.8	63.0	72.8	73.3	79.3
225	48.3	59.7	70.6	71.9	78.3
240	47.1	56.8	68.9	70.6	77.8

Table E11.2 Factors for Calculating Effective Concrete Width for Various RC Beams Exposed to ASTM E119 Standard Fire

Step 4. Calculate moment capacity of external FRP-strengthened beam.
Based on the force equilibrium principle, apply Eq. 11.35 to find out the depth of effective concrete.

$$\begin{aligned} A_S \cdot f_{y,T} + A_f f_{f,T} &= 0.85 f_c' \cdot b_T \cdot (\beta_1 \cdot c) \\ 1134\ \text{mm}^2 \times 414\ \text{MPa} + 250\ \text{mm}^2 &\times (0.0045 \times 0.64 \times 75600\ \text{MPa}) \\ &= 0.85 \times 41.4\ \text{MPa} \times 0.928 \times 250\ \text{mm} \times (0.75c) \end{aligned} \tag{E11.9}$$

Thus, the neutral area depth, c, at 1-hour fire exposure is $c = 85.6$ mm.
The effective strain of FRP is

$$\varepsilon_{fd} = 0.41\sqrt{\frac{f_c'}{nE_f t_f}}\varepsilon_{fu} = 0.41\sqrt{\frac{41.4}{75600 \times 1.0}} \times 0.015 = 0.0045 \tag{E11.10}$$

Based on the strain compatibility principles

$$\begin{aligned} \frac{\varepsilon_c}{c} &= \frac{\varepsilon_s}{d_s - c} = \frac{\varepsilon_f}{d_f - c} \\ \frac{0.003}{85.6} &= \frac{\varepsilon_S}{394 - 85.6} = \frac{\varepsilon_f}{450.5 - 85.6} \end{aligned} \tag{E11.11}$$

$\varepsilon_s = 0.0108 > 414\ \text{MPa}/201{,}000\ \text{MPa} = 0.0021$, and thus steel yields.

Since $\varepsilon_s = 0.0128 > 0.0045$, FRP failure governs and thus Eq. E11.11 can be used.

Then the moment capacity of an EB FRP-strengthened beam at 1-hour fire exposure time is given as

$$\begin{aligned} M_n &= A_S f_{y,T} \cdot \left(d_s - \frac{a}{2}\right) + \psi_f A_f f_{f,T} \cdot \left(d_f - \frac{a}{2}\right) \\ &= 1134 \times 414 \times (394 - 0.5 \times 0.75 \times 85.6) / 10^6 \\ &\quad + 0.85 \times 250 \times (0.0045 \times 0.64 \times 75{,}600) \times (450.5 - 0.5 \times 0.75 \times 85.6) / 10^6 \\ &= 189.3\ \text{kN} \cdot \text{m} \end{aligned} \tag{E11.12}$$

By using a spreadsheet, the moment capacity at other fire exposure times can easily be calculated using the same procedure as above. The variation in moment capacity with fire exposure time for external FRP-strengthened RC beam is plotted in Fig. E11.3.

Step 5. Estimate failure time.
The reduced moment capacity of EB-strengthened RC beam at 4 hours is 154 kN·m, as shown on the time-moment capacity curve plotted in Fig. E11.3. Since this capacity is higher than the applied moment under fire conditions (100 kN·m from Step 1), the beam does not fail for 4 hours of fire exposure. Thus, the fire resistance of the beam is at least 4 hours.

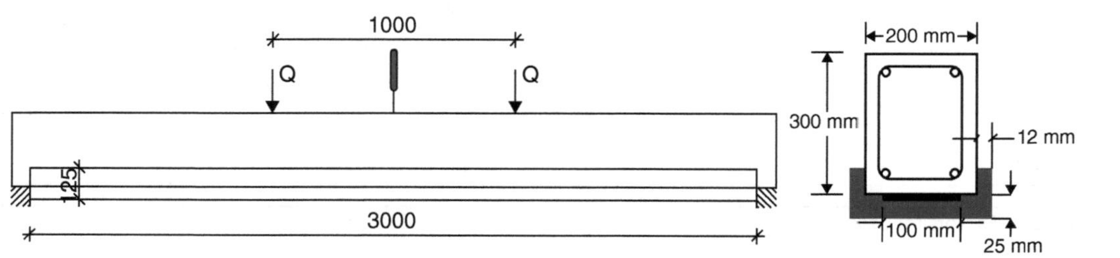

FIGURE E11.3 Details of an FRP-strengthened RC beam.

11.3

Evaluate the fire behavior of Beam I; that is, 200 × 300 mm in cross section (see Fig. E11.3) and 3 m in span length and subjected to a maximum service loading at two points, each having 40.6 kN. The beam has two 10-mm and 16-mm rebars as compression and tensile reinforcement, respectively. The beam was strengthened with 1.2-mm-thick and 100-mm-wide CFRP laminates and was insulated with Promatect-H type fire protection. The beam had 25 mm of insulation thickness at the beam soffit, and an additional 12-mm insulation thickness extending on both sides of the beam up to a height of 105 mm (measured from the soffit insulation thickness).

Solution

(Refer to Sec. 11.7.2 as well as Chap. 8 for advanced analysis procedure and representative results.)

Step 1. Develop the geometric model of the beam and mesh it using thermal elements: SOLID70 and LINK33, and structural elements: SOLID65, SOLID46, SOLID45, LINK8, and INTER205, respectively.

Step 2. Input the temperature-dependent material properties of concrete and reinforcing steel as per Eurocode 2 provisions. The properties for FRP can be taken from those presented in Sec. 11.2. The temperature-dependent bond-slip model proposed by Xu and Needleman (1994) (i.e., exponential cohesive zone) can be used at the FRP-concrete interface.

Step 3. Carry out a heat transfer analysis to get the temperature rise across the depth of the beam.

Step 4. Input cross-sectional temperatures obtained from Step 3 to the second stage of analysis (shown in Fig. E11.4). Carry out structural analysis to arrive at the loss in moment capacity and midspan deflection as a function of fire exposure time. A sample of these results is shown in Figs. E11.5 and E11.6, respectively.

Step 5. Apply strength/deflection limit state to evaluate failure of the beam.

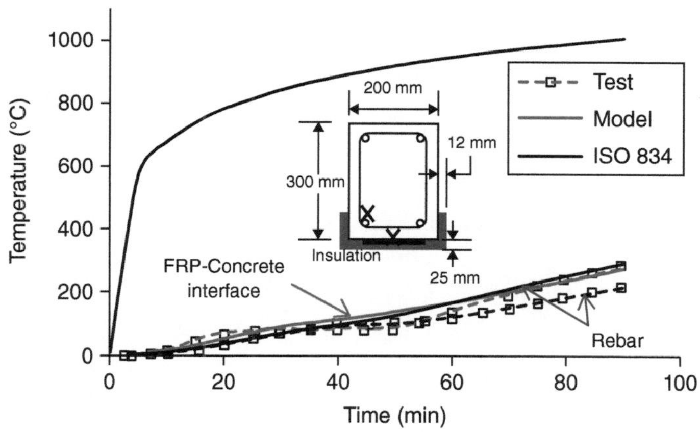

Figure E11.4 Temperature rise in RC beam.

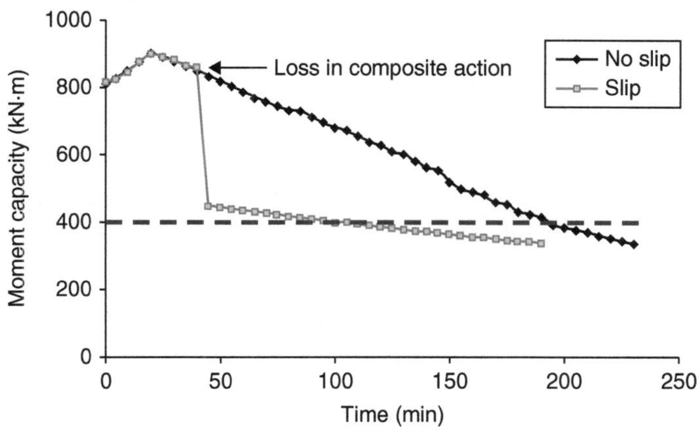

Figure E11.5 Loss in moment capacity.

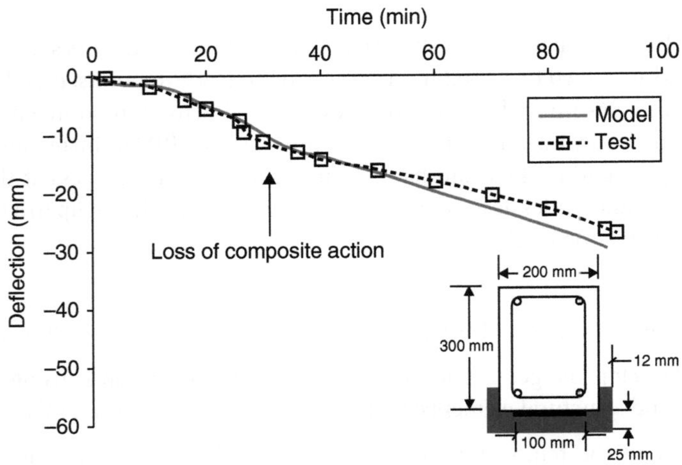

Figure E11.6 Midspan deflection.

CHAPTER 12

Transportation Structures

12.1 Overview

This chapter presents an overview of fire problems in critical transportation structures such as bridges and tunnels. The magnitude of such problems is illustrated, and the recent spurt of fires in bridges and tunnels is highlighted. Recent research undertaken to address fire problems in transportation structures is reviewed, and critical factors governing fire performance of these structures are discussed. Calculation approaches for evaluating the fire resistance of bridges, including the application of advanced analysis techniques, are laid out, and various strategies for mitigating fire hazard in bridges and tunnels are presented. Further, emerging trends and research needs for advancing the state of the art in this critical area are also discussed.

12.2 Fire Problems in Bridges and Tunnels

In the last two decades, there has been an increase in the number of fire incidents in transportation structures, and some of these fire incidents (as illustrated in Chap. 1) have led to destructive damage to structures and have also resulted in prolonged interruptions to traffic flow, leading to significant economic losses (Battelle, 2004). In this section, the magnitude of fire problems in bridges and tunnels is further highlighted and the need for fire resistance in these structures is laid out.

12.2.1 General

Transportation structures,[1] such as bridges and tunnels, are often built to last for several decades and serve a variety of commuters and modes of transportation. During their long service life, bridges, tunnels, and ports are exposed to various loading scenarios and hazards. Damage to structures under any hazardous conditions can adversely affect the overall traffic flow in the transportation grid and thus impacts travel, goods delivery, and the economy of the region. Therefore, transportation structures should be designed to withstand risks arising from natural (e.g., earthquake, scour, wind) or accidental disasters (e.g., fire, blast, etc.) (Garlock et al., 2012).

Despite the fact that fire is a growing concern, current codes of practice do not explicitly require specific fire safety provisions for bridges and only specify limited active fire protection measures for tunnels. In fact, in the case of bridges, and to some extent tunnels, structural members are often designed without any specific fire resistance provisions. Thus, in the occurrence of fire, transportation structures can be highly vulnerable to fire-induced damage, including partial or complete collapse.

12.2.2 Causes of Fires

A number of recent studies have documented fire incidents and associated losses in transportation structures (Bai et al., 2006; NYDOT, 2008; Kodur et al., 2010). Findings in these studies have shown that the majority of fires are caused by the burning of combustibles resulting from collisions of vehicles in the vicinity of bridges or in tunnels, such as fuel tankers/freight trucks colliding either with other vehicles or against bridge/tunnel components (e.g., piers, sidings). Another common reason for fires on bridges located in rural areas is from natural fires such as wildfires. Due to the location of a transportation infrastructure in open spaces, there is an unlimited supply of oxygen. Combined with a large amount of flammable fuel that can be available when fuel tankers are involved, fire in a transportation infrastructure

[1] Transportation infrastructure covers a broad range of structures including bridges, tunnels, ports, and metro stations. Fire safety is one of the major design considerations in most of these transportation structures. However, this chapter mostly focuses on fire resistance issues associated with bridges and tunnels.

can be *explosive* and *intense* in nature. For example, peak temperature in bridge and tunnel fires can reach in the range of 700 to 1000°C within the first few minutes of fire breakout. This rapid rise in fire temperature can induce high thermal gradients in structural members which could lead to temperature-induced instability (local or buckling) in steel structural members. Also, such high thermal gradients, along with high moisture conditions that can be present in concrete members in the outside environment, can produce ideal conditions for fire-induced spalling of concrete in tunnels linings, concrete girders, or piers/columns.

In many cases, fires in bridges (and tunnels to some extent) burn out quickly or are extinguished through active firefighting. However, in the case of high fire severity, a rapid rise in sectional temperatures can induce significant degradation in strength and modulus properties of constituent materials, leading to the rapid loss of load-bearing capacity in structural members. Such loss of capacity can trigger the partial or complete collapse of structural members. Even in the case of moderate fire incidents, where no collapse occurs, proper investigation, inspection, and maintenance are needed before the affected bridge/tunnel is opened to traffic. Shutting down a bridge or a tunnel for inspection or repair would require traffic detours to nearby routes. Eventually, this could affect the flow of traffic and impose significant traffic delays to the surrounding transportation network, which in turn can affect economic activity in the region.

12.2.3 Magnitude of Fire Problems

A review of the literature clearly indicates that fire incidents in bridges and tunnels have significantly increased in recent years (Battelle, 2004; NYDOT, 2008). In surveys compiled by Battelle (2004) and Schütz (2014), the average fire losses in bridges (in the United States alone) and tunnels (worldwide) are estimated to be at $1.28 and $1.06 billion, respectively, over the past few years.

While the common perception among engineers and the general public is that it is very unlikely that a bridge or tunnel would collapse due to fire, in reality intense fires can produce major damage to structural members. If there is significant degradation in the load-bearing capacity of main structural members, it can lead to partial or complete collapse of fire-exposed members that are a part of a bridge or a tunnel. Some of the notable fire incidents on bridges and tunnels are listed in Tables 12.1 and 12.2, respectively. For each fire incident, the cause of the fire together with the type of structural members and the extent of the damage are summarized in these tables. A closer look into compiled fire incidents reveals that fire hazard is a universal problem that can occur due to a variety of causes. While the majority of fires occurred as a result of the collision of fuel tankers (e.g., the I-75 bridge near Hazel Park, Michigan, and the Gudvanga Tunnel in Norway), a few bridge fires occurred due to arson (e.g., the I-85 bridge near Atlanta, Georgia) and from natural hazards such as wildfires (e.g., the Green Spot Bridge, California) and even from lightning (e.g., the Rio-Antirrio Bridge, Greece). There were a few mystery fires resulting from unknown causes (as in the case of the Zhuoshui Fengyu Bridge fire in China).

A United States-wide survey was commissioned by the New York State Department of Transportation in 2008 to compile the damage and collapse of bridges due to various hazards (NYDOT, 2008; United States Department of Transportation, 1995–2010). This survey of bridge failures, carried out across 18 states, classified the damage to bridges based on the type of bridge, material type, and type of hazard that produced the collapse. Results from this survey revealed that nearly three times more bridges have collapsed in the 1990 to 2005 period due to fire than earthquakes (Transportation, 2010). As per the survey, a total of 1746 bridges collapsed for various reasons between 1990 and 2005. A vast majority of bridges (1006) collapsed due to hydraulic reasons (scour, flood). Another 515 bridges collapsed due to collision, overload, or deterioration, and a total of 52 bridges collapsed due to fire. Only 19 bridges collapsed due to earthquake (seismic states like California were part of this survey). About half of the bridges that collapsed due to fire were timber bridges and in many cases were from wildfires.

In the case of tunnels, Schütz (2014) undertook a worldwide survey of fires in tunnels. This survey revealed that there were 177 tunnel fire incidents in 29 countries between 1866 and 2008. Of these 177 tunnel fire incidents, 28 major tunnel fires were recorded between 1949 and 2008. These tunnel fires led to more than 700 deaths, 1000 injuries, and the loss of more than 500 vehicles.

Still it is worth noting that the occurrence of a fire on a bridge or in a tunnel is a rare event, and even in such cases, only few of these fires grow into large-size fires that can affect the integrity and performance of main structural members. A recent study (Naser and Kodur, 2015), through comparative statistical analysis of fires in bridges, tunnels, and buildings in the United States, has shown that the annual probability of fire breaking out in a bridge is about 5%, which is slightly higher than that in tunnels (about 3%), but much lower than that in buildings (about 12%). Despite the small probability of fire occurrence, the fire incidents listed in Tables 12.1 and 12.2 clearly show the devastating effect of high-intensity fires in transportation structures, from the standpoint of structural damage, traffic disruption, and economic

Bridge Location	Date of Fire Incident	Material Type Used in Structural Members	Cause of Fire	Damage Description
I-85 bridge in Atlanta, GA, USA	March 30, 2017	Concrete structural members	Fire started due to vandalism. Led to burning of large polyvinyl chloride (PVC) pipes stored under the bridge	One span collapsed after 30 minutes of fire breakout
I-375 bridge over I-75 in Detroit, MI, USA	May 24, 2015	Composite deck (steel girders + reinforced concrete slab)	A gasoline tanker carrying 9000 gallons crashed over the bridge and caught fire	Concrete deck was significantly damaged by fire. Also, steel girders experienced some level of damage
I-15 bridge at Cajon, Hesperia, CA, USA	May 5, 2014	Composite deck (steel girders + reinforced concrete slab)	Workers cutting rebar with blowtorches spread the fire into the "formwork" of the bridge	Bridge collapsed due to burning of formwork
Bridge over Freeway 60, Los Angeles, CA, USA	December 14, 2011	Concrete deck (precast prestressed concrete I girders + cast in place reinforced concrete slab)	A tanker truck carrying 128 m^3 of gasoline caught fire and burned underneath the bridge	Concrete girders were significantly damaged, and bridge was demolished and replaced
Zhuoshui Fengyu Bridge, China	November 29, 2013	Wood girders on top of brick piers	Unknown	Bridge collapsed after burning of wooden superstructure
Bridge over I-75 near Hazel Park, MI, USA	July 15, 2009	Composite deck (steel girders + reinforced concrete slab)	A gasoline tanker struck an overpass on I-75	Complete collapse of the bridge to the freeway below
Big Four Bridge, Louisville, KY, USA	May 7, 2008	Steel truss bridge	Electrical problem of the lighting system	Minor structural damage resulting in large amount of debris on the bridge
Bill Williams River Bridge, AZ, USA	June 20, 2007	Concrete deck (precast prestressed concrete I girders + cast in place reinforced concrete slab)	A gasoline tanker overturned near the bridge	Concrete girders were severely damaged
MacArthur Maze I-80/880 interchange in Oakland, CA, USA	April 29, 2007	Composed of six plate girders supporting a reinforced concrete slab	A tanker truck carrying 8600 gallons of gasoline overturned underneath the bridge	Collapse occurred in 22 minutes. Post-fire analysis of the bridge revealed that the failure was due to overstressing of connections under high-temperature exposure effects
Rio-Antirrio Bridge, Greece	January 25, 2005	Cable stay composite bridge	One of the cable links of the bridge snapped after a lightning strike in one of the cables	Cable link failed after 40 minutes into fire. The bridge was reopened to traffic after replacement of cables
Wiehltalbrücke Bridge, Germany	August 26, 2004	Steel girder bridge	Collision between car and fuel tanker transporting 33,000 L of fuel	Major damages that cost €7.2 million
I-95 Howard Avenue Overpass in Bridgeport, CT, USA	March 26, 2003	Composite deck (steel girders + reinforced concrete slab)	A car struck a truck carrying 8000 gallons of heating oil near the bridge	Full collapse of girders in southbound lanes and partial collapse of girders in northbound lanes

Table 12.1 Notable Fire Incidents in Bridges in Recent Years

Tunnel Location	Date of Fire Incident	Material Type Used in Structural Members/ Linings	Cause of Fire	Damage Description
Weihai City Tunnel, China	May 9, 2017	Concrete lining	Bus collision	There were 12 deaths (11 children). Minor structural damage
Hachihonmatsu Tunnel, Japan	March 17, 2016	Concrete lining	A truck crashed into several vehicles stopped in a pile up. Fire burned for 3 hours	Two deaths and 70 injuries. Minor structural damage
Sangju Tunnel, South Korea	October 26, 2015	Concrete tiles	Truck carrying flammable paint thinner exploded after crashing into tunnel siding	Fire caused 21 injuries. Minor structural damage to lining
Shanxi Tunnel, China	March 1, 2014	N/A	Two methanol tanker trucks collided	There were 31 deaths. Major damages to tunnel
Gudvanga Tunnel, Norway	August 5, 2013	Concrete lining	Heavy goods vehicle (HGV) on fire due to engine breakdown	67 injuries. No report of structural damage
Melbourne Burnley Tunnel, Australia	March 23, 2007	Concrete lining	Collision of a truck and a car in the tunnel	Three deaths. Minor structural damage
Los Angeles Interstate 5 Tunnel, CA, USA	October 12, 2007	Concrete lining	Crashing of truck leading to explosion	No data on economic loss. Severe structural damage reported
Howard Street Tunnel, Maryland, USA	July 18, 2001	Concrete lining	Derailment of the train car	Losses estimated at $12 million
Tauern Tunnel, Austria	May 29, 1999	Concrete lining	Collision of cars and HGVs and lasted to 53 hours	Losses estimated at $32 million (and 12 deaths)
Channel Rail Tunnel, France–UK	November 18, 1996	Concrete lining	HGV caught fire and burned for 10 hours	Due to fire-induced spalling, tunnel lining along a 50-m length was reduced from 40 cm thick to 17 cm thick. Over a 240-m-long section damage to the concrete extended as far back as the first set of reinforcement bars. The total loss from fire was estimated at $278 million
Summit Tunnel fire, UK	December 20, 1984	Brick lining	Train carrying more than 260,000 gallons of gasoline derailed. The fire burnt for 3 days	The damage from fire was minimal. Approximately 800 m of pavement track had to be replaced, as did all the electrical services and signaling devices. Tunnel was shut down for nine months for repairs
Salang Tunnel fire, Afghanistan	November 3, 1982	Concrete lining	Two military convoys of Soviet Army collided, causing a traffic jam	176 deaths. Not much information on extent of structural damage
Caldecott tunnel, CA, USA	April 7, 1982	Concrete lining	Collision between gasoline tanker, car, and bus inside the tunnel. Fire lasted for 2 hours	Seven deaths and losses of $3 million
Holland Tunnel fire, NY, USA	May 13, 1949	Tiles	Truck carrying 4400 gallons of carbon disulfide malfunctioned and thus fire started	66 people were injured. Tunnel walls significantly spalled and concrete ceiling collapsed

Table 12.2 Notable Fire Incidents in Tunnels

losses. Even in the case of fires of low to moderate intensity, a rapid rise of fire temperature can lead to high thermal gradients within a structural member, which in turn can produce "unconventional" failures, such as fire-induced spalling in concrete, temperature-induced local buckling in steel members, or charring in wood members.

12.2.4 Illustrations of Typical Bridge and Tunnel Fires

To further illustrate the devastating effect of fire on transportation structures, two recent fire incidents that occurred in a bridge and a tunnel are detailed in this section. On July 15, 2009, a major fire occurred at a 9-mile road steel girder bridge overpass on the I-75 expressway near Hazel Park, Michigan. The fire started when a fuel tanker transporting 50,000 L of flammable fuel crashed into a truck close to the vicinity of the bridge. This collision resulted in the quick ignition and burning of fuel, with fire temperatures exceeding 1000°C in the first few minutes. This resulted in a rapid rise in sectional temperatures, which in turn led to the degradation of strength and stiffness properties of the unprotected steel girders. As a result, there was a significant loss of capacity in the steel girders, and this led to the partial collapse of bridge girders about 20 minutes after fire exposure. It took about 105 minutes for the firefighters to extinguish the fire. The collapse of this overpass caused major delays to traffic flow and also economic losses estimated at $2 million (Garlock et al., 2012). Following post-fire inspection, the entire 9-mile bridge over I-75 was rebuilt and the freeway was repaired. Repair of the bridge took several months.

Compared to bridge fires, tunnel fires can be much more severe and deadly due to intensive burning that occurs in an enclosed space. Further, from a firefighting perspective, tunnel fires are hard to fight and the quick evacuation of trapped people is not easy to facilitate. A fire that occurred in the 16.9-km-long Gotthard Road Tunnel pass in Switzerland on October 24, 2001 illustrates the growth of a typical fire in a tunnel and its devastating effect. This tunnel was constructed in 1980 and was equipped with various fire detection systems and ventilation systems, and the linings on the tunnel were of concrete. This fire started when two heavy-goods vehicles (HGVs) collided inside the tunnel. This collision ignited gasoline fuel present in one of the HGVs and then rapidly spread to the second HGV, along with five other vehicles. The active fire protection system was not effective in controlling fire growth and spread in the tunnel. As a result, seven vehicles continued to burn in the tunnel, releasing intense heat and toxic smoke. The burning inside the tunnel lasted for more than 24 hours, reaching temperatures of about 1200°C. Harsh conditions slowed down firefighting activities. There were 11 deaths and a number of injuries. From a structural point of view, the intense heat arising from the severe fire led to explosive spalling of concrete in tunnel linings, which in turn led to the collapse of a portion of the tunnel ceiling over a length of 250 m. Due to extensive damage to the linings and other fire protection systems, the tunnel had to be closed for 2 months to carry out necessary repair work.

12.2.5 The Need for Fire Resistance

Fire in a bridge or a tunnel, if it grows to full size, represents one of the most severe conditions a structural member has to withstand during such a fire event. As fire accidents are not easy to prevent, appropriate strategies must be developed to mitigate the adverse effects of fire on the structure, from a life safety standpoint (in the case of tunnels) and from a property protection standpoint (in the case of bridges, tunnels, or other transportation structures). Currently, the fire resistance of bridge or tunnel structures is lower than that in buildings due to the fact that the provision of fire resistance is not a design objective or consideration in most transportation infrastructure. Unlike buildings, only limited active and passive fire protection features are provided in tunnels, and in most cases no active or passive fire protection measures are installed in bridges. This is mainly due to the fact that it is not economical or practical to install active fire protection features in transportation structures, specifically bridges, to overcome the adverse effects of fire hazard. Therefore, the key fire safety measure is to incorporate appropriate passive fire protection features (i.e., fire resistance) to structural members. Ensuring the integrity and stability of structures through fire resistance provisions can minimize fire-induced damage or the possibility of the partial or complete collapse of structural members in bridges and tunnels. Therefore, provisions of fire resistance to structural members can overcome fire hazard in bridges, tunnels, and other transportation structures.

12.2.6 Current Fire Mitigation Provisions for Bridges and Tunnels

Although there is a growing realization that fire can threaten the structural integrity of a bridge or a tunnel and cause unexpected delays and substantial property losses, there is a lack of provisions and design approaches for overcoming fire hazards in transportation infrastructure. This is mainly due to the common thinking that the probability of fire occurrence in bridges and tunnels is quite low and does not justify the cost associated with passive fire resistance provisions. Also, in the case of bridges, it is due to the notion that they are open structures and hence people (commuters) can get out to safety from fire zones quickly. Other issues related to the durability of fire protection under harsher weather conditions (i.e., cycles of freezing/thawing, high humidity, etc.) also add to the complexity of installing and properly maintaining fire insulation.

Fire hazard in bridges and tunnels can be overcome to a certain extent through the provision of appropriate fire resistance to critical members, such as girders, piers, and tunnel linings. However, there is currently no guidance on minimum fire resistance provisions to structural members in bridges and tunnels. NFPA 502, "Standard for Road Tunnels, Bridges, and Other Limited Access Highways" (NFPA, 2017b), is one of the few documents that contains some general statements relating to fire hazard in bridges. The recommendations state that fire is to be considered as a possible hazard in bridges more than 300 m long. However, it does not have any specifics on fire loads that should be considered or the level of fire resistance needed in different structural members or types of bridges. Further, there is no guidance on how the structure should be protected or how the bridge should be analyzed following a fire with regard to restoration and traffic flow.

NFPA 502 also contains some recommendations for the installation of suppression systems in tunnels to mitigate the effects of fire. These fire provisions are recommended based on the length of the tunnel as follows:

1. Where tunnel length is less than 90 m, no suppression systems need to be applied.
2. Where tunnel length is 90 m or greater, standpipe systems and traffic control systems shall be installed.
3. Where tunnel length equals or exceeds 240 m and where the maximum distance from any point within the tunnel to a location of safety (for egress) exceeds 120 m, all provisions of this standard shall be installed.

NFPA 15, "Standard for Water Spray Fixed Systems for Fire Protection" (NFPA, 2017a), also provides some guidance on installing water spray fixed systems (sprinklers) in tunnels. European equivalent standards, "Fire and Smoke Control in Road Tunnels" (PIARC, 2007), contain similar provisions as that in NFPA 502 and NFPA 15. However, a closer review of the provisions in these standards leads to some ambiguity on the usefulness of sprinklers. These documents have statements such as "water spray will cause explosions in petrol," "vaporized steam can hurt people," and "visibility is reduced due to de-stratification of the smoke layer." Further, Appendix 4 of NFPA 502 also states that "sprinklers definitely not be installed in tunnels, except possibly for tunnels involving dangerous or hazardous cargo."

The future editions of NFPA 502 (NFPA, 2017), American Society of Heating, Refrigerating and Air-Conditioning Engineers (ASHRAE) Technical Committee 5.06, are likely to include new provisions directed toward the installation of passive fire protection systems (fire resistance) in transportation structures (ASHRAE, 2019).

12.2.7 Complexities in Applying Conventional Fire Resistance Provisions

In current practice, fire hazard in buildings is overcome through the provision of active and passive fire protection systems prescribed in building codes and standards (IBC, 2018; NBCC, 2015). However, there are no specific fire-related provisions in standards for structural members in bridges and only limited provisions are available for fire safety measures in tunnels. Even in some special cases, where fire protection to transportation structures is considered, designers tend to extend prescriptive-based fire requirements used in buildings to transportation structures despite major differences in key design features. As a result, these prescriptive fire provisions may not yield optimum fire safety measures in transportation structures.

In the absence of effective active and passive protection systems, bridges and tunnels can be highly vulnerable to the adverse effects of fire. The fire resistance provisions and evaluation approaches specified for building elements (members) may not be applicable for structural members in transportation infrastructure due to large differences in key factors such as fire severity, member characteristics, and design objectives (Peris-Sayol et al., 2017). The key differences associated with bridge fires and tunnel fires from those in buildings as well as structural features are summarized in Table 12.3. They include characteristics of fire (fuel load and type), ventilation conditions, geometrical features, and type of structural members/systems. For example, fire growth characteristics in a bridge or a tunnel are quite different from that in a building. As discussed in Chap. 3, combustible materials in buildings typically are

Scenario		Building	Bridge	Tunnel
Fuel source		Wood/plastic-based material	Hydrocarbon based	Hydrocarbon based
Ventilation		Restricted supply of oxygen	Unlimited supply of oxygen	Unlimited supply of oxygen
Fire severity		ASTM E119/ISO 834/ natural fire	Hydrocarbon fire/ ASTM E1529	Hydrocarbon fire/ASTM E1529
Enclosure		Compartmentation	Open area	Large space
Fire protection features		Presence of active and passive systems	None	Some (i.e., smoke control) active systems
Structural members	Failure limit state	Flexural mode	Flexural/shear mode	Failure in lining materials
	Connections (steel members)	Web and/or the flange	Bearing of the bottom flange	—
	Sectional slenderness (steel members)	Web slenderness (50)	Web slenderness (150 with no stiffeners)	—
	Likely loading during fire	Dead load and percent of live load	Dead load + (very little live load)	Mostly dead load
	Exposure conditions	Interior environment (lower humidity)	Outdoor environment (i.e., high humidity, hot temperature, etc.)	Outdoor environment (i.e., high humidity, high moisture content, etc.)

Table 12.3 Key Differences in Characteristics of Buildings, Bridges, and Tunnels

composed of cellulose-based fuels (e.g., wood, fabric). When ignited, these fuels gradually burn and hence produce less intense fires. However, gasoline-based or chemical-based fuels, commonly associated with bridge or tunnel fires, burn at rapid rates, producing high-intensity heating. Cellulose-based fuels can somewhat be represented through the ASTM E119 standard fire curve where fire temperature rises from 20 to 548°C after the first 5 minutes and reaches about 1000°C over a period of 120 minutes. Gasoline fires, on the other hand, can be described through hydrocarbon fire curves (such as ASTM E1529) or tunnel fires. The hydrocarbon fires can reach a temperature of 1140°C within the first 10 minutes of ignition.

Another key difference in fire growth arises due to varying ventilation characteristics between buildings, bridges, and tunnels. For instance, buildings are designed with compartmentation features, and as a result, there is a limited supply of fuel and oxygen for burning in each compartment. However, bridges and tunnels are somewhat open spaces which provide an unlimited supply of oxygen. When combined with a large amount of fuel available (whether in fuel tankers, chemical spills, or wildfires), this provides ideal conditions for rapid burning and faster fire spread. This effect is amplified in the case of a tunnel fire, as smoke spread and radiative heat transfer coming from tunnel linings can produce ideal conditions for a rapid rise in temperature.

The problem of fire hazard in buildings is overcome to a great extent through the implementation of active fire protection systems, such as alert and suppression systems like smoke detectors and water sprinklers. These systems help detect and mitigate fire growth while fire is still in early stages. While limited active systems (specifically ventilation/air circulation-related systems) are being used in tunnels, bridges still do not incorporate any active fire protection systems mainly due to major installation (practical) challenges and cost implications. It should be noted that active fire protection systems in tunnels are primarily installed to control the spread of smoke and to facilitate evacuation of commuters. In general, bridges and shorter tunnels do not incorporate any passive fire protection systems to mitigate fire growth and/or fire spread.

In addition to fire characteristics, there are major differences in structural features of buildings and transportation structures. For example, in the case of buildings, steel structural members (used for beams and columns) are made of compact (Class 1 or Class 2) sections. However, in bridges, structural systems often utilize slender (Class 3 or Class 4) sections (e.g., plate girders, T-beams, and box girders). This is mainly due to differences in performance requirements and cost considerations. Although they can satisfy load capacity requirements under ambient conditions, slender members can be vulnerable to fire-induced instabilities (such as local buckling and web shear buckling in steel members and fire-induced spalling in concrete members) as noted in recent fire incidents and laboratory fire tests (Aziz et al., 2015; Naser and

Kodur, 2017). From a construction material point of view, newer concrete types often used in bridges and tunnels have been shown to exhibit poor fire resistance properties as compared to conventional concretes. For example, high-performance (or high-strength) concrete (widely used in bridge and tunnel construction) is susceptible to fire-induced spalling and also experiences faster degradation in mechanical properties, specifically under severe (intense) fire conditions encountered in tunnels and bridges. Added to this are the environmental conditions in tunnels and bridges such as high levels of moisture content, which are ideal features in promoting spalling in concrete or corrosion in steel.

Therefore, the fire resistance provisions are used when buildings cannot be directly utilized for structures. Further, the calculation methods applied for evaluating fire resistance of structural members in buildings cannot be directly applied unless due consideration is given to various influencing factors. Specific strategies are to be developed with due consideration given to various factors influencing fire resistance of bridges and tunnels as outlined in Table 12.3.

12.3 Strategies for Enhancing Fire Safety

Although fire represents a significant hazard to bridges and other transportation structures, it is still a rare occurrence in the life span of a typical structure. As a result, it is not economical or practical to design all transportation structures and specifically all bridges and tunnels for fire hazard, and only structures that are highly vulnerable (at higher risk) to fire should be designed for fire safety. In such cases, identifying key factors governing the vulnerability of a bridge or a tunnel to fire is the first step. The second step is to develop a number of cost-effective strategies for enhancing the fire resistance of structural members to minimize the adverse consequences of fire hazard. These two steps are detailed in the following sections.

12.3.1 Identifying Factors Governing Vulnerability to Fire Hazard

The key factors that contribute to the vulnerability of a bridge or a tunnel to fire hazards are the geometrical features of structural members, materials used in their construction, loading and support (restraint) conditions of structural members, expected fire severity, critical nature of the bridge/tunnel, expected economic (property) losses, and presence of fire mitigation strategies. Following is a breakdown of these features.

Geometrical features. The geometry and dimensions of structural members can have a significant influence on their structural performance under fire conditions. For example, in the case of a steel bridge, factors such as slenderness or lateral restraint to girders can affect local or torsional buckling of girders under fire conditions. In the case of a concrete bridge, concrete cover thickness for internal steel reinforcement or sectional dimensions has a direct bearing on the resulting fire resistance of reinforced concrete structural members. Similarly, in the case of wood structures, the size of the member has a direct bearing on ensuing fire resistance.

Materials used in construction. The fire performance of a bridge structure is highly dependent on the thermophysical and mechanical properties of constituent materials that form a structural member. For instance, the temperature-induced loss of strength and stiffness properties in concrete is at a much slower pace than that of steel. Hence, concrete members generally exhibit higher fire resistance as compared to unprotected steel structural members. In comparison to concrete and steel, timber is a combustible material and also loses its cross section with fire exposure time. Thus, the type of material structural members are made of has a direct bearing on the vulnerability of that bridge under fire.

Loading and support conditions. Structural members under static loading or at lower load levels achieve higher fire resistance than similar members subjected to dynamic loading or at higher load levels. Further, restrained support conditions can significantly enhance fire resistance of flexural members due to the development of fire-induced restraint forces that can counterbalance the load-induced moments. However, in some scenarios, such restraint forces could also trigger temperature-induced phenomena such as spalling of concrete or buckling in steel web, thus initiating premature failure.

Fire intensity. The severity of fire has a direct bearing on the fire performance of a structure. Fire intensity and its duration depend on the fuel type and quantity, as well as on ventilation characteristics. Building fires tend to burn at a lower intensity and at a slower rate than bridge fires, due to cellulose-based fuels and the presence of active and passive protection systems. On the contrary, bridges (and tunnels to a certain extent) are open structures with an unlimited oxygen supply. The lack of active fire protection measures and the possible presence of highly flammable hydrocarbon products in bridges of tunnels can accelerate the rate of growth of a fire, producing high-intensity fires.

Critical nature of facility. In addition to the above factors, the criticality or importance of the transportation structure has a direct bearing on the impact of fire on a bridge or a tunnel. The factors that determine the critical nature of the structure are bridge location and traffic density. The importance of a transportation structure is directly related to its location in the traffic network grid. If the bridge/tunnel is located in a route connecting natural obstacles (such as valleys or rivers) and if there are no alternative routes for traffic detours, then any closure of that bridge/tunnel due to fire damage will significantly slow down or shut down the traffic movement in the region.

Similarly, traffic density can determine the critical nature of a transportation structure. For example, if a bridge is located on a condensed highway or in the surroundings of an urban area that serves a large number of vehicles daily, loss of operation of such a bridge will cause significant traffic disruptions in the region. These factors must be considered in evaluating the importance of a bridge.

Presence of fire mitigation features. When implemented to minimize fire hazard in a bridge, fire mitigation strategies can enhance the fire performance of a bridge or tunnel and reduce the adverse effect of fire. These measures are usually planned in the design stage of a bridge or implemented after regular maintenance or inspection. Examples of systems that can be added to minimize the state of a fire of adverse effects are installation of security guards and trained personnel, monitoring systems for fire safety, limits on cargo size and operating time, availability of on-site fire equipment/foam deluge systems, use of fire-rated insulation material (on main load-carrying structural members), and provisions of fire resistance to structural components of the bridge or tunnel.

12.3.2 Developing Strategies for Enhancing Fire Resistance

When a bridge or tunnel is deemed to be critical from a fire hazard point of view (as determined in the first step), then strategies should be developed to mitigate fire hazard as part of the second step. The most effective and practical strategy to overcome the adverse effects of fire hazard on transportation structures (especially bridges) is through enhancing the fire resistance of structural members. The level of fire resistance possessed by a structural member varies with the type of construction material used. For instance, in the case of concrete structural members, the required fire resistance can be achieved through proper detailing, such as the provision of sufficient concrete cover thickness to rebars. Supplementary fire protection (fire insulation) is required for achieving fire resistance in conventional concrete members. In the case of structural members made of high-strength concrete, any fire-induced spalling in the member can be minimized through adding polypropylene or steel fiber to the concrete mix as discussed in Chap. 8. Similarly, in cases where FRP strengthening is added to concrete members in a bridge or a tunnel, the needed fire resistance can be achieved through supplementary fire insulation, which will limit temperature rise and loss of strength and bond properties in the FRP-strengthening system. In the case of tunnels, the most effective strategy for enhancing fire resistance is through adopting spalling mitigation measures for tunnel linings. This can be achieved through adding polypropylene fibers to the concrete mix as detailed in Chap. 8.

In the case of steel bridges, structural members are susceptible to damage and even failure under severe fire exposure. This is due to the high thermal conductivity and low specific heat of steel and because steel loses its strength and stiffness properties at a rapid pace (as compared to concrete). Fire resistance of such steel structural members can be enhanced through applying external fire insulation to the member. Also, the susceptibility of temperature-induced web buckling, which is one of the major drawbacks with slender sections used in bridge girders, can be minimized through the use of thicker sections. This strategy can enhance fire resistance and minimize post-fire damage. In the case of steel-concrete composite structural members, fire resistance can be enhanced by accounting for beneficial composite action between steel and concrete (i.e., through proper details of shear studs).

In the case of timber bridges, fire resistance can be enhanced through the provision of fire insulation to structural members. Alternatively, through overdesign of cross-sectional size of a wood member, the fire resistance can be enhanced. In order to identify the criticality of a bridge for fire hazard and then to develop the needed strategies for enhancing fire resistance, detailed calculations should be carried out, and this requires the application of structural engineering principles.

12.4 Approach for Evaluating Fire Resistance of Bridge Structures

Although fire represents a significant hazard to transportation infrastructure, it is of rare occurrence. In many fire incidents, fires burn out quickly or are extinguished through firefighting before the occurrence of significant damage to structural members. As a result, it may not be economical or practical to design all transportation infrastructure for fire hazard. Only those structures that are at a high risk of fire should be designed for fire safety. Fire hazard in bridges can be tackled, to a certain extent, through the provision of appropriate fire resistance to structural members.

The structural system in a bridge is composed of superstructural members (slabs and decks) and substructural members (girders, piers, and abutments). These structural members can be made of different materials and designed with specific characteristics. Thus, expected overall fire performance of a bridge is a function of fire resistance of both super- and substructural members. In order to quantify the fire risk and determine appropriate fire resistance measures, the fire resistance of each member is to be evaluated and checked. For identifying the vulnerability of a bridge to fire hazards, a fire-based importance factor can be applied to classify the fire risk associated with each structural member of the bridge (Kodur and Naser, 2013). Knowing the vulnerability of a bridge to fire, appropriate fire resistance strategies can be developed to overcome fire hazard in critical bridges.

12.4.1 General Procedure

The proposed approach for mitigating fire hazard in a bridge is composed of five steps. In Step 1, the magnitude of a fire hazard on a selected bridge is quantified empirically through the application of a fire-based importance factor. Such an importance factor for fire design of bridges can be derived along the same lines as that of the importance factor used in the design for wind, snow, and earthquake events (loading) (ASCE, 2010). However, unlike the aforementioned events, reliable statistical data on fires in bridges is not widely available. Moreover, the associated frequency and complex nature of fire characteristics add further challenges and complexities. Hence, a qualitative assessment based on rational engineering judgment seems to be the method of choice in the assessment of the risk of fire hazard in transportation structures. If the importance factor indicates that the fire risk to the bridge under consideration is high, then relevant strategies for mitigating fire hazard in that bridge are developed.

For this purpose, in Step 2, structural members of the selected bridge should be analyzed to evaluate inherent fire resistance. To do so, advanced analysis must be carried out, as there are no simplified or empirical approaches currently available specifically for transportation structures. Such advanced analysis involves evaluating the response of structural members under thermal and structural loading effects to determine inherent fire resistance. If the structural members (such as girders) of the bridge have low inherent fire resistance and cannot withstand the adverse effects of fire, then the configuration of bridge structural members should be modified through measures such as additional fire insulation to enhance fire resistance as part of Step 3. The modified structure with added fire provision is reanalyzed in Steps 4 and 5, and this iterative procedure continues until bridge structural members can withstand a design fire hazard. The analysis is carried out under various scenarios (fire, insulation thickness, etc.) to develop optimum strategies for enhancing fire resistance of structural members. The associated steps in this approach are illustrated through the flowchart shown in Fig. 12.1 and described in detail elsewhere (Kodur and Naser, 2013).

12.4.2 Evaluating Fire Risk (Step 1)

As part of Step 1, the magnitude of fire risk in a bridge should be quantified. This can be done by calculating a fire-based importance factor for a given bridge (or structural member) as proposed by Naser and Kodur (2015). For assessing the importance factor, relevant data on characteristics of the selected bridge and its criticality should be collected and analyzed. The importance factor for assessing fire risk in a bridge is a function of the vulnerability of structural members to resist fire and also resulting consequences of a fire incident on life safety, traffic flow, and property losses. The degree of vulnerability of super- and substructural components of a bridge to fire mainly depends on the fire resistance of various structural members of bridges. The impact of fire on a bridge is dependent on the critical nature of the bridge from the standpoint of traffic functionality.

The vulnerability of a bridge to fire hazard is mainly a function of key characteristics of structural members that influence their fire performance, and these factors include geometrical features, materials used in construction, loading and restraint conditions, and fire intensity. The critical nature of the bridge is the second main factor that determines the importance of a bridge, from a fire hazard point of view.

For deriving an importance factor of a bridge, a weightage factor approach can be applied (Kodur and Naser, 2013). As part of this approach, the key characteristics that influence fire hazard in bridges are grouped into five classes: geometrical features, material properties and design parameters (Class I), hazard

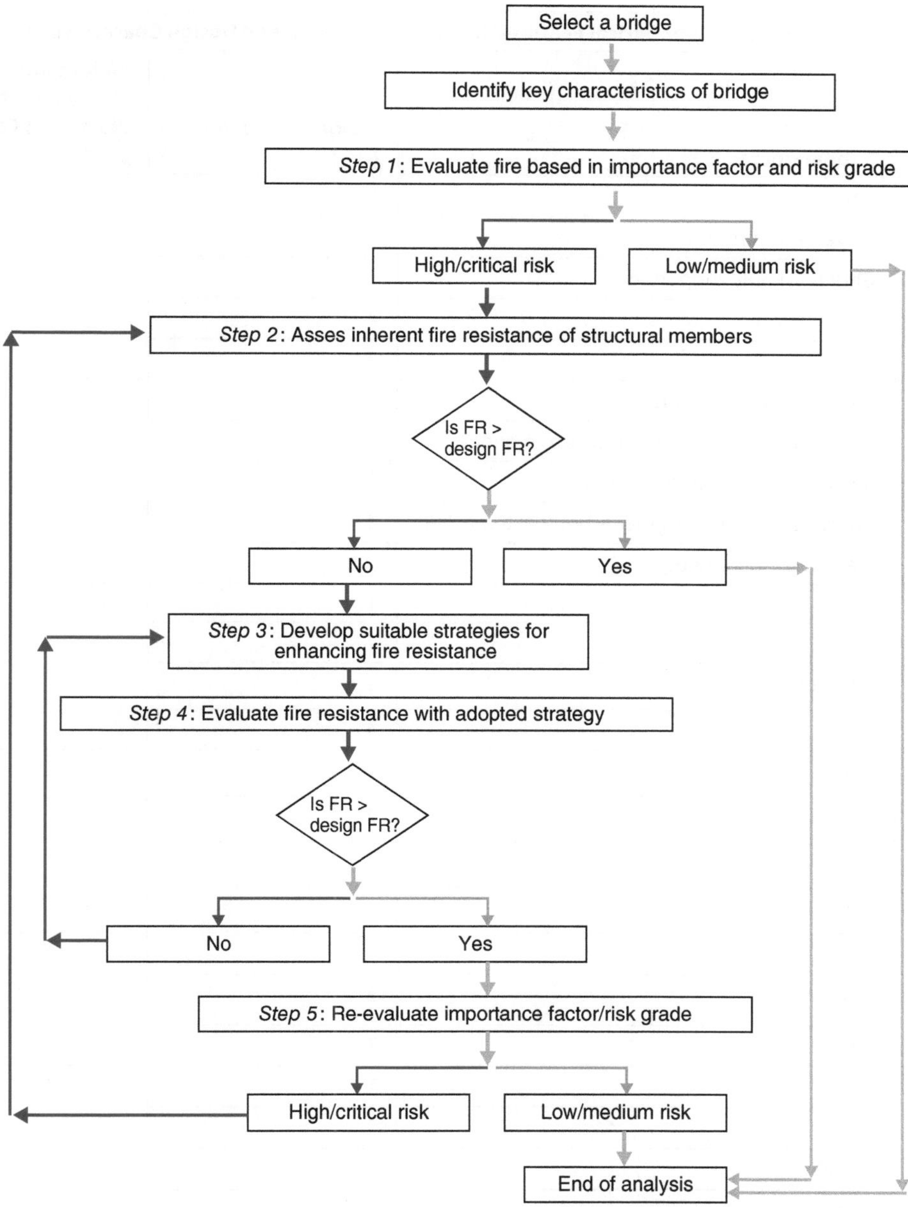

FIGURE 12.1 Flowchart of the proposed approach for mitigating fire hazard in bridges. (*With permission from Elsevier.*)

(fire) likelihood (Class II), traffic demands (Class III), economic impact (Class IV), and expected fire losses (Class V). Each class covers different parameters (characteristics) that contribute to the importance factor or the vulnerability of a bridge to fire. Within each parameter, there are various subparameters that determine the conditions of a specific structural member in a bridge. The five classes, with associated parameters and subparameters, are listed in Table 12.4. Based on engineering judgment and recommendations of previous studies, weightage factors should be assigned to different subparameters. These weightage factors ($\varphi_{i,x}$) are assigned on a scale of 1 to 5 based on the extent of influence of that subparameter to the fire performance of the structural member or impact to fire, where the largest value (5) indicates the highest risk (susceptibility) to fire hazard. Further information on the classes, parameters, subparameters, and rationale for assigning weightage factors can be found in a previous study (Kodur and Naser, 2013). The weightage factors for various subparameters (within each parameter) are tabulated in Table 12.4. As an example, under the material type subparameter in Class I, a bridge made of concrete members is assigned a weightage factor of 1, as it is more fire resistant than a steel-concrete composite bridge (with a weightage factor of 3) or a steel bridge (with a weightage factor of 5).

	Class I: Geometrical Features, Material Properties, and Design Characteristics ($\psi_g = 0.47$)			
Parameter	**Subparameter**	**Weightage Factor ($\varphi_{g,x}$)**	**Weightage Factor ($\varphi_{e,x}$) for Illustrated Case Study**	**Max. Weightage Factor ($\varphi_{g,x(max)}$)**
Structural system	Truss/Arch	1	3	5
	Girder—continuous	2		
	Girder—simply supported	3		
	Cable—stayed	4		
	Suspension	5		
Material type	Reinforced concrete bridge	1	3	5
	High-strength/(prestressed) concrete bridge	2		
	Steel-concrete composite bridge	3		
	Concrete bridge strengthened with external FRP	4		
	Steel and timber bridges	5		
Span (m)	<50	1	1	4
	50–200	2		
	200–500	3		
	>500	4		
Structural component	Deck	1	3	4
	Abutment	2		
	Girder	3		
	Pier	4		
No. of lanes	2	1	1	3
	2–4	2		
	>4	3		
Age (years)	<15	1	1	4
	15–29	2		
	30–50	3		
	>50	4		
Current rating	100	1	5	5
	60–80	2		
	40–60	3		
	20–40	4		
	<20	5		
Additional service features	1 deck	1	4	5
	2 decks + pedestrians	2		
	Accommodates railroad	3		
	Multilevel	4		
	Above water	5		

Table 12.4 Weightage Factors Based on the Different Features of a Bridge

	Class II: Hazard (Fire) Likelihood ($\psi_h = 0.22$)			
Parameter	**Subparameter**	**Weightage Factor ($\varphi_{h,x}$)**	**Weightage Factor ($\varphi_{e,x}$) for Illustrated Case Study**	**Max. Weightage Factor ($\varphi_{h,x\,(max)}$)**
Response time (min)	<5	1	1	5
	5–10	2		
	10–20	3		
	20–30	4		
	>30	5		
Historical/ architectural significance	Conventional (normal)	1	1	3
	Landmark	2		
	Highly prestigious	3		
Threat perception	None (low)	1	1	3
	Not available (medium)	2		
	Frequent (high)	3		
Fire scenario	Natural causes/malfunction of utilities control systems	1	1	5
	A collision and fire with other vehicles	2		
	A fuel tanker collision and fire with bridge substructure	3		
	Major fuel tanker collision and fire with multiple vehicles and against bridge substructure	4	4	
	Fire due to fuel freight ship collision with a bridge pier	5		
	Class III: Traffic Demand ($\psi_t = 0.11$)			
Parameter	**Subparameter**	**Weightage Factor ($\varphi_{t,x}$)**	**Weightage Factor ($\varphi_{e,x}$) for Illustrated Case Study**	**Max. Weightage Factor ($\varphi_{t,x\,(max)}$)**
ADT (vehicles/day)	<1000	1	3	5
	1000–5000	2		
	5000–15,000	3		
	15,000–50,000	4		
	>50,000	5		
Facility location	Rural	1	2	3
	Suburban	2		
	Urban	3		

Table 12.4 Weightage Factors Based on the Different Features of a Bridge (*Continued*)

Parameter	Subparameter	Weightage Factor ($\varphi_{e,x}$)	Weightage Factor ($\varphi_{e,x}$) for Illustrated Case Study	Max. Weightage Factor ($\varphi_{e,x(max)}$)
	Class IV: Economic Impact ($\psi_e = 0.12$)			
Closeness to alt. routes (km)	<10	1	1	3
	10–20	2		
	>20	3		
Time expected for repair (month)	<3	1	1	3
	3–9	2		
	>9	3		
Cost expected for repair	<1 million	1	2	3
	1–3 million	2		
	>3 million	3		

Parameter	Subparameter	Weightage Factor ($\varphi_{f,x}$)	Weightage Factor ($\varphi_{e,x}$) for Illustrated Case Study	Max. Weightage Factor ($\varphi_{f,x(max)}$)
	Class V: Expected Fire Losses ($\psi_f = 0.08$)			
Life/property losses	Minimum to no injuries	1	1	3
	Minimum casualties	2		
	Many casualties	3		
Env. damage	Minor damage	1	1	3
	Significant damage	2		
	Unacceptable damage	3		

Parameter	Subparameter	Weightage Factor ($\varphi_{e,x}$)	Max. Weightage Factor ($\varphi_{e,x(max)}$)
	Class VI: Fire Mitigation Strategies		
Security	Monitoring systems	1	4
	Guards	2	
	Restricted access zones	3	
	Fire detection systems	4	
Laws and regulations	Provide distinguished exits for large fuel tankers	1	4
	Limit operation timings	2	
	Limit vehicle speed	3	
	Limit transport size (20,000 L)	4	
Fire protection and insulation features	On-site firefighting equipment	1	5
	Use of flooding agents and/or foam deluge systems	2	
	1-hr insulation to main structural members	3	
	2-hr insulation to main structural members	4	
	Implementing structural fire design for bridge	5	

Table 12.4 Weightage Factors Based on the Different Features of a Bridge (*Continued*)

Similarly, under the structural components category in Class I, weightage factor is assigned in ascending order for deck ($\varphi_{g,sc} = 1$), abutment ($\varphi_{g,sc} = 2$), girder ($\varphi_{g,sc} = 3$), and pier ($\varphi_{g,sc} = 4$). The higher weightage factor to girders and piers is due to the fact that fire can be much more concentrated around piers and girders as compared to abutments, since fires tend to break out following a collision of a fuel tanker with a pier/other vehicle in the vicinity of a bridge. In addition, damage or collapse of a pier or girder can cause imminent collapse rather than damage/loss of deck.

Once the weightage factors are assigned, the approach for evaluating the importance factor comprises four main substeps. In the first substep, knowing the maximum weightage factor in each of the parameters of a bridge (listed in Table 12.4), a class factor (ψ_x) is to be calculated as

$$\psi_x = \frac{\sum \varphi_{x(\max)}}{\varphi_{\text{total}}} \tag{12.1}$$

where $\varphi_{x(\max)}$ = Maximum weightage factor of each parameter in class (x)
φ_{total} = Summation of maximum weightage factors of all parameters in all five classes

In the second substep, a class coefficient (Δ_x) is calculated as the ratio of the summation of the weightage factors of all subparameters in class (x) to the summation of the maximum weightage factor of all the parameters in the same class:

$$\Delta_x = \frac{\sum \varphi_{i,x}}{\sum \varphi_{x(\max)}} \tag{12.2}$$

where $\varphi_{i,x}$ = Weightage factor of subparameter (i) in class (x)
$\varphi_{x(\max)}$ = Maximum weightage factor of each parameter in class (x)

In the third substep, an overall class coefficient (λ) is evaluated as the summation of the product of class coefficient (Δ_x) and corresponding class factor (ψ_x):

$$\lambda = \sum \Delta_x \psi_x \tag{12.3}$$

Based on the overall class coefficient value (λ), a fire-based importance factor is assigned and the fire risk associated with bridges is grouped under four risk grades in the fourth substep. These risk grades are defined as low, medium, high, and critical. The limiting values for grouping a bridge under one of the risk grades is 0.8, 1.0, 1.2, and 1.5, respectively (see Table 12.5). If the bridge falls under a "low" or "medium" risk grade (e.g., a concrete bridge located in a rural area that serves a low volume of traffic), then the bridge is considered to be less susceptible to fire damage or collapse, and thus no additional measures may be needed to enhance fire safety of such a bridge. However, if the bridge falls under a "high" or "critical" risk grade, as in the case of a suspension steel bridge that serves a large volume of traffic and is located above a body of water, then the bridge is deemed to be somewhat or highly susceptible to fire-induced damage/collapse, and thus additional measures are needed to minimize the fire hazard on that bridge. It should be noted that about 5% of bridges fall under the "critical" risk category and about 10% to 15% fall under the "high" risk category. Thus, appropriate fire protection to structural members in "critical" and "high" bridges can minimize the adverse effects of fire hazard to a great extent.

In general, structural members in steel bridges that fall under the "high" or "critical" risk grade often have inherent fire resistance (failure time) of much less than 45 minutes under a typical bridge fire exposure. Thus, suitable strategies must be developed to bring down the fire risk of these bridges to a "low" or "medium" risk grade, so as to resist typical bridge fire exposure. This can be done through developing

Risk Grade	Overall Class Coefficient (λ)	Importance Factor (IF)
Critical	≥0.95	1.5
High	0.51–0.94	1.2
Medium	0.20–0.50	1.0
Low	<0.20	0.8

Table 12.5 Risk Grade and Associated Importance Factor for Fire Design of Bridges

relevant strategies to enhance fire resistance (FR) of main structural members of a bridge or adding additional fire mitigation strategies as discussed in Sec. 12.3.1. One such practical strategy is through the application of fire protection (insulation) to the main structural members of a steel bridge. The applied fire protection should provide 60 to 120 minutes of fire resistance to the main structural members, since previous bridge fire incidents have shown that steel girders in a bridge can fail in 20 to 30 minutes under intense fire exposure, and in many cases, open-air fires get controlled in 60 to 120 minutes through firefighting or attaining burnout conditions. Thus, enhancing the fire resistance of structural members to 60 to 120 minutes through the application of fire insulation to structural members can significantly lower the risk of collapse. Similarly, fire resistance of a concrete bridge with a "high" or "critical" risk grade can be minimized by providing thicker concrete cover to rebars so as to limit temperature rise in steel rebars. Full details on enhancing fire-based importance factors, including assessing weightage factors for various types of bridges, are given by Kodur and Naser (2013).

12.4.3 Fire Resistance Analysis (Step 2)

As part of Step 2, advanced analysis should be carried out to quantify the inherent fire resistance of structural members of the selected bridge. Of the different structural members in a bridge, girders tend to be more susceptible to fire-induced damage, since girders (or decks) in a bridge experience the impact of fire exposure to a much higher degree than piers or abutments. Therefore, fire safety of a given bridge can often be enhanced just by increasing the fire resistance of girders or decks (especially in steel bridges).

Advanced analysis in Step 2 can be carried out utilizing a three-dimensional nonlinear finite element model that is capable of handling thermomechanical problems and integrating nonlinear temperature-dependent material properties. The finite element model can be developed in ANSYS, ABAQUS, or other such software. In this type of model, a bridge girder can be discretized using three-dimensional elements. Two sets of discretization models are needed for undertaking thermal and structural analysis. For the thermal analysis (in ANSYS), the girder and slab can be discretized using PLANE55 elements, while SURF151 elements are used to simulate surface effect applications (thermal radiation and heat convection). For structural analysis, the girder can be discretized using SHELL181 elements for the girder and SOLID185 elements to model the (concrete) slab, while LINK8 elements are used to model the steel reinforcement in the concrete slab, assuming a full bond between steel reinforcement and concrete exists. Surface-to-surface fully bonded contact elements (CONTA174 and TARGE170) are typically used to account for the composite action between the concrete slab and the top flange of the steel girder.

The temperature-dependent thermal and mechanical properties of steel and concrete are to be input, and these properties can follow relations specified in Eurocodes 2 and 3 to other relevant codes. The presence of any fire insulation is to be incorporated explicitly in thermal analysis. The predicted temperature profile is input into the structural model as a thermal body load.

To discretize the boundary conditions present in the structure, support conditions of the bridge girder are applied on line nodes at the lower face of the bottom flange, while boundary conditions for symmetry are applied along girder cross-section nodes at the midspan of the girder. Also, if the continuity effect of the concrete slab is to be accounted for, the slab is to be restrained from lateral movement. To evaluate the realistic failure of bridge girders, strength and deflection limit states need to be applied (rather than critical temperature-limiting criterion). For instance, failure of a bridge occurs when the load-carrying capacity (moment, shear, axial) falls below the level of the applied loading effect or when the midspan deflection of a fire-weakened girder exceeds the specified limiting deflections. Full details on the discretization of structural members, temperature-dependent properties of material to be adopted, and failure modes to be considered in undertaking fire resistance analysis are given elsewhere (Kodur et al., 2017).

After evaluating inherent fire resistance, if the inherent fire resistance of a bridge (girder) is higher than the required (design) fire resistance (generally of 60 to 120 minutes), then no additional measures may be needed to enhance the fire safety of such a bridge. However, if the fire resistance of a bridge member is less than the required fire resistance, then suitable fire mitigation strategies should be developed to enhance fire resistance of such structural members.

12.4.4 Deriving Optimum Fire Protection Strategies (Steps 3, 4, and 5)

As discussed above, through the application of fire protection (insulation) to steel bridge girders, the fire resistance of main structural members can be enhanced. Similarly, in concrete bridges, the provision of higher cover thickness or other detailing requirements can improve fire resistance. Thus, through enhancing the fire resistance of structural members' vulnerability of a bridge to fire hazard can be minimized. The enhanced fire resistance of a reconfigured (insulated) bridge through adding fire insulation of other measures has to be confirmed by undertaking another set of finite element analysis as part of Step 4. If the applied fire insulation or other measures enhance the fire resistance of vulnerable members to meet design objectives (achieving 60 to 120 minutes of fire resistance), then the importance factor of the bridge structural member is reassessed in Step 5 (by applying a similar procedure as per Step 1, but taking into account the applied fire protection measures). If the "overall fire risk" grade of the structural member decreases to a "medium" or "low" risk grade, this indicates that the fire risk to the bridge has been minimized. At this point, the analysis is said to be complete.

If the fire resistance of the reconfigured structural members is still less than the needed fire resistance and/or if the bridge still falls under a "critical" or "high" fire risk grade, the fire protection measures are to be readjusted, such as by increasing the thickness of the fire insulation, by providing a higher concrete cover, or by increasing the cross section of the members, and the above steps (Steps 3 to 5) are to be repeated until the fire risk grade of the bridge decreases to the "medium" or "low" risk category. The step-by-step procedure for undertaking the above-discussed procedure is illustrated in Fig. 12.1, and its validation is detailed in Kodur and Naser (2013).

12.4.5 Illustration of Proposed Approach for Fire Resistance

The applicability of the above-discussed five-step approach in a practical situation is illustrated for a steel bridge that experienced major damage due to fire. This steel bridge was designed and built without any special consideration to fire safety (resistance) provisions. The response of this bridge was under ISO 834, and hydrocarbon fire exposure was simulated. A brief summary of results is presented here. Full details on the analysis including step-by-step calculations can be found elsewhere (Kodur et al., 2017).

Description of the Bridge and Fire Incident

On January 5, 2002, a tanker truck carrying 9900 gallons of diesel fuel collided with the pier of the I-65 overpass at the I-20, I-59, and I-65 junction near Birmingham, Alabama. The intense heat from the burning of flammable gasoline fuel reached about 1100°C, and resulting high temperatures in steel girders led to the degradation of capacity and stiffness of steel girders, causing the main span (steel girder number 7) to sag about 3 m. Following the fire, the bridge was shut down and commuters were detoured to alternate routes. The repairs costed about $1.325 million and the bridge was reopened for traffic after 54 days of necessary repairs.

Characteristics of the Bridge

This steel bridge was composed of three simply supported spans: the main span and two side-approach spans. The middle (main) span, over the highway, was of 36-m length, while the side-approach spans were 25 m long on each side of the main span. Each span consisted of seven girders, spaced equally at a transverse spacing of 2.15 m and supported laterally with cross "X" bracing.

The bridge was composed of steel girders, made up of built-up sections. The yield strength of steel used in the girder was 350 MPa. The steel girders were supporting a concrete slab of 170-mm thickness with an effective slab width of 2.15 m. The compressive strength of concrete used in slab was 40 MPa. The flange plates were of 457-mm width and 28-mm thickness, and the web plate had dimensions of 1344-mm depth and 12-mm thickness. The steel girders were stiffened with intermediate stiffeners at a spacing of 1.1 m along the span length and with a thickness of 12 mm, while bearing stiffeners of 25-mm thickness were placed at the support locations. The discretized bridge girder is shown in Fig. 12.2.

Evaluating Fire Risk Based on Importance Factor (Step 1)

The various geometric and material characteristics of this bridge and details of the above-described fire incident were compiled from the literature. Then, the above-discussed "fire-based importance factor" approach was applied to evaluate vulnerability of this bridge to fire. Since the unprotected steel girders generally exhibit much lower fire resistance than reinforced concrete piers or concrete decks, the probability of fire-induced failure in steel girders is much higher than that in concrete piers/concrete decks. Hence, the fire-based importance factor for the overall bridge is often dictated mainly by the fire performance of steel girders. This is due to the fact that temperature-induced degradation in strength and stiffness properties of steel occurs at a much faster pace than that of concrete. Further, concrete members exhibit higher fire resistance compared to steel members. Applying the outlined procedure (as per Step 1 in Fig. 12.1), the fire-based importance factor for the girders can be evaluated as follows.[2]

[2] See Kodur and Naser (2013) for detailed step-by-step calculations.

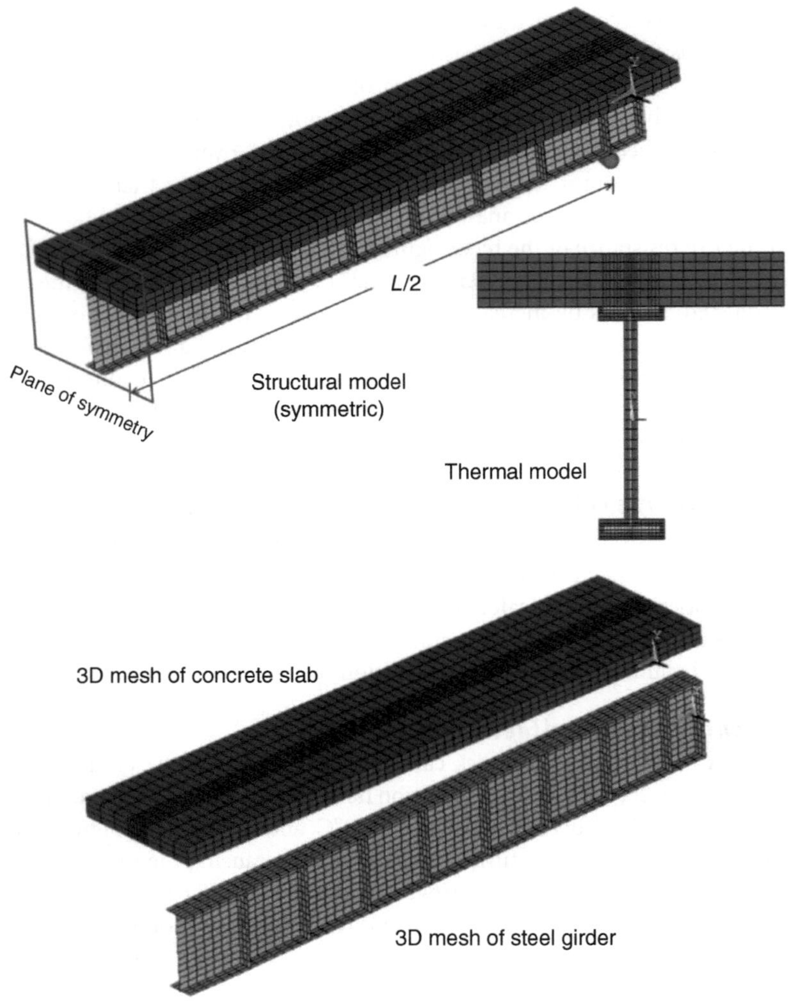

Figure 12.2 Discretization of a steel bridge girder for thermal and structural analysis. (*With permission from Elsevier.*)

Based on the compiled information from open literature, the weightage factors for parameters and subparameters are given below (and in italics in Table 12.4). First, the individual coefficients (Δ_x) for the different classes (from Table 12.4) are calculated by applying Eq. 12.2, as shown below.

Using weightage factors for various subparameters and the maximum weightage factor for each class as given in the second column from the right in Table 12.4, the class coefficient (Δ) is calculated as

$$\Delta_g = \frac{3+3+1+1+3+3+4+5}{5+5+4+3+4+4+5+5} = \frac{23}{35} = 0.66$$

$$\Delta_h = \frac{1+1+1+4}{5+3+3+5} = \frac{7}{16} = 0.44$$

$$\Delta_t = \frac{3+2}{5+3} = \frac{5}{8} = 0.63$$

$$\Delta_e = \frac{1+1+2}{3+3+3} = \frac{4}{9} = 0.44$$

$$\Delta_f = \frac{1+1}{3+3} = \frac{2}{6} = 0.33$$

Then, the overall class coefficient (λ) is calculated as per Eq. 12.3. (Please note that class factors ψ are given in Table 12.4).

$$\lambda = \Delta_g \times \psi_g + \Delta_h \times \psi_h + \Delta_t \times \psi_t + \Delta_e \times \psi_e + \Delta_f \times \psi_{f0}$$

$$\lambda = 0.66 \times 0.47 + 0.44 \times 0.22 + 0.63 \times 0.11 + 0.44 \times 0.12 + 0.033 \times 0.08 = 0.53$$

By correlating the overall class coefficient (λ) with range in Table 12.5, λ translates into an importance factor of 1.2, and thus the risk grade for the girders (bridge) is taken as "high."

Evaluating Inherent Fire Resistance of Steel Members (Step 2)

Since the I-65 bridge falls under the "high" risk category, this bridge is highly vulnerable to fire hazard. In order to develop appropriate strategies for overcoming the vulnerability of this bridge to fire hazard, the inherent fire resistance of structural members (i.e., girders) is to be evaluated. To evaluate the inherent fire resistance of the girder through numerical simulation, finite element analysis should be undertaken as part of Step 2. Full details for undertaking such analysis are given in Kodur et al. (2017). Steel girder number 7 of the main span (with a length of 36 m) of the I-65 bridge, which experienced the most damage in a previous fire incident, is selected for fire resistance analysis. This girder is analyzed independently, since the main span is simply supported and separated by expansion joints from the adjacent spans. The fire resistance analysis is carried out under the combined effect of fire exposure and applied service loading. Two sets of analysis were carried out, one under ISO 834 fire exposure and the second under hydrocarbon fire exposure.

Results from thermal analysis on the I-65 girder under both hydrocarbon and ISO 834 fires are plotted in Fig. 12.3 to illustrate temperature distribution in the composite girder as a function of fire exposure time. It can be seen that the temperature in the top flange is much lower than that in the bottom flange of the girder. This is mainly due to the "heat sink" effect of the concrete slab, which absorbs much of the heat from the top flange as well as the top portion of the web. Also, temperature rise in the web is more rapid when compared to that in the bottom flange, due to high slenderness of the web (lower thickness with a large exposed surface area). It should be noted that hydrocarbon fire exposure is much more severe than a standard fire scenario (ISO 834), since these fires are characterized by high-intensity burning with peak temperatures reaching as high as 1000°C within the first few minutes of fire.

The large variation in temperature between the steel girder and the slab generates a thermal gradient effect across the depth of the composite section. Such thermal gradients are influenced by the severity of fire exposure and result in different levels of thermal strain at the bottom flange (and web) of the steel girder as compared to that in concrete slab. This can lead to the development of a large curvature (thermal bowing) in the composite bridge girder.

The structural response of the I-65 bridge girder is illustrated in Fig. 12.4, where midspan deflection is plotted as a function of fire exposure time under ISO 834 and hydrocarbon fire scenarios. At the early stage of fire, midspan deflection increases linearly until yielding occurs in the steel girder. Most of the deflection that occurs at the early stage of fire exposure is mainly due to thermal bowing resulting from the development of thermal gradients. Therefore, steel girders (or other structural members) experience different levels of deflection depending on the extent of thermal gradients arising due to different fire severities generated under two fire scenarios (see Fig. 12.3). Hence, the girder in the case of hydrocarbon fire experiences much higher deflection at the early stage (i.e., first 5 minutes) of fire exposure as compared to that under ISO 834 fire exposure (i.e., first 15 minutes).

The deflection in this girder increases with fire exposure time due to deterioration in strength and stiffness properties of steel and spread of plasticity in the section resulting from temperature rise in steel. This increase is much more predominant in the case of hydrocarbon fire, as compared to ISO 834 fire. With increased fire exposure time, the web of the steel girder starts to develop some level of instability due to temperature-induced local buckling arising from rapid temperature rise in the slender web.

During the final stages of fire, the rate of deflection increases rapidly due to plastification of the bottom flange and buckling of the web which results from the rapid temperature-induced degradation in strength and stiffness properties of steel (as well as the associated effects of high-temperature creep). Finally, the steel girder experiences runaway failure due to loss of flexural capacity.

Results plotted in Fig. 12.4 show that fire resistance (failure time) of a steel girder is highly influenced by the level of fire severity. In the case where the I-65 bridge girder is exposed to hydrocarbon fire (as in the case of a tanker fire accident), the steel girder fails in just 9 minutes, while the same girder survives for 24 minutes when subjected to less severe ISO 834 fire exposure. This variation results from the rapid burning that occurs in the case of hydrocarbon fire as compared to that of ISO 834 fire exposure.

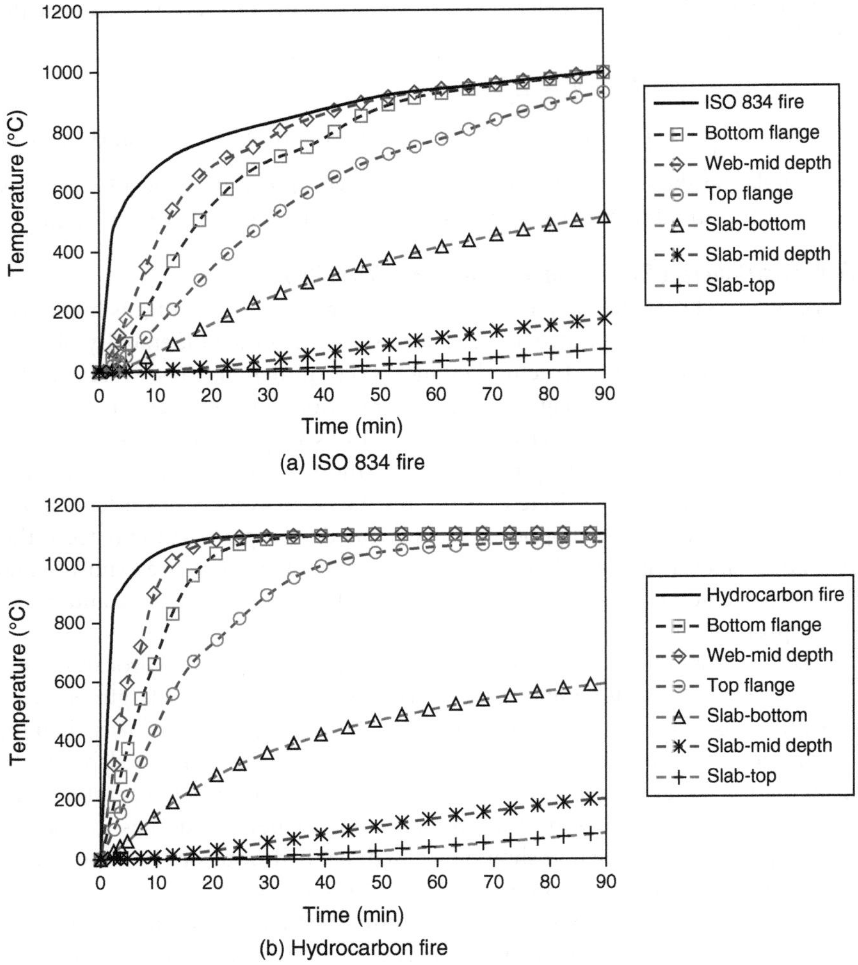

Figure 12.3 Cross-sectional temperature progression in a bare I-65 bridge girder. (*With permission from Elsevier.*)

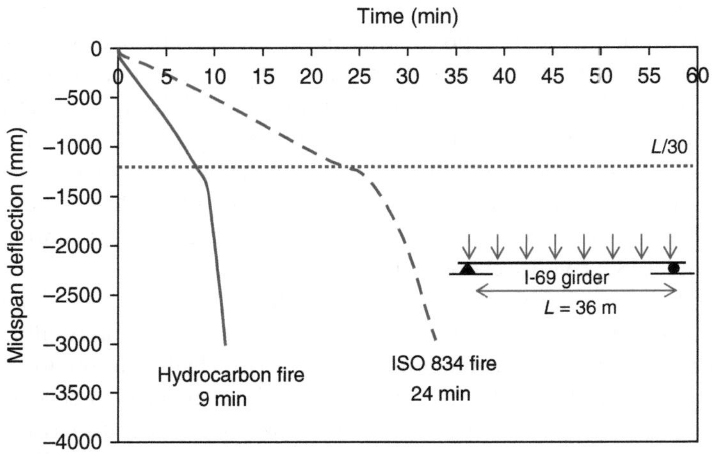

Figure 12.4 Midspan deflection in the uninsulated I-65 bridge girder under various fire scenarios. (*With permission from Elsevier.*)

Developing Strategies for Enhancing Fire Resistance and Evaluating Fire Resistance of Bridge (Steps 3 and 4)

The uninsulated steel girder in the above bridge experienced failure early into fire exposure, and thus the fire performance of this girder is deemed poor. Failure of a girder can lead to substantial damage or even complete collapse of the bridge in a fire incident (as occurred during the fire incidents on this bridge on January 5, 2002). To minimize the vulnerability of this steel bridge to fire hazard, relevant strategies must be developed to enhance fire resistance of the steel girder, as outlined in Step 3 of Fig. 12.1. One such strategy is the application of fire insulation to steel girders as done in the case of steel-framed buildings (see Sec. 12.3).

To illustrate this strategy, a gypsum-based spray-on fire insulation (SFRM, for spray-applied fire-resistive material) is hypothetically applied on the steel girder (Kodur et al., 2017). In order to arrive at an optimum fire protection thickness and insulation scheme, finite element analysis of the I-65 bridge girder is carried out in ANSYS utilizing various thicknesses of SFRM, including 7, 12.7, 14, 16, 18, 20, 25, and 30 mm. With each insulation thickness, the steel girder was analyzed under two fire scenarios (ISO 834 and hydrocarbon), and corresponding fire resistance was evaluated (as part of Step 4).

Figure 12.4 shows the predicted temperature progression in the insulated (reconfigured) I-65 steel girder corresponding to a 16-mm SFRM thickness on steel girders. The thermal response of the insulated steel girder follows a similar trend of that observed in the uninsulated girder, but with a much lower rise in sectional temperatures, as can be seen by comparing plots in Fig. 12.5. Similarly, midspan deflection in insulated girder increases at a much slower rate than that in the case of bare steel girder, which can be seen by comparing deflection trends in Figs. 12.4 and 12.6. The slower rise in steel temperatures and associated deflections is due to the effect of applied fire insulation on the steel girder.

Fire resistance of this girder, with 16-mm-thick fire insulation, increases to 113 minutes (from 24 minutes for a bare girder) under ISO 834 fire exposure, while the corresponding fire resistance of the girder increases to 76 minutes from 9 minutes under hydrocarbon fire conditions (from 9 minutes for a bare girder). Similarly, with 30-mm-thick insulation, fire resistance of this steel girder increases to 124 minutes

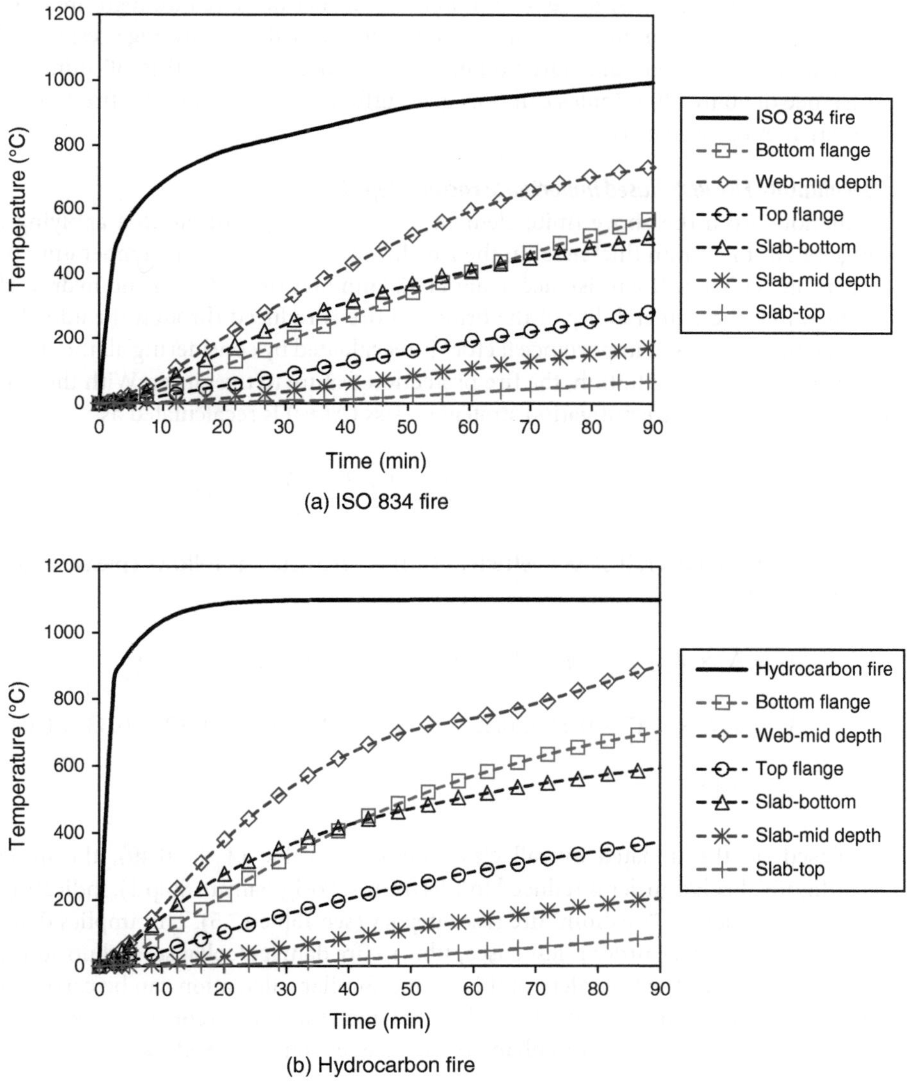

FIGURE 12.5 Cross-sectional temperature progression in an insulated (with 16 mm insulation thickness) I-65 bridge girder. (*With permission from Elsevier.*)

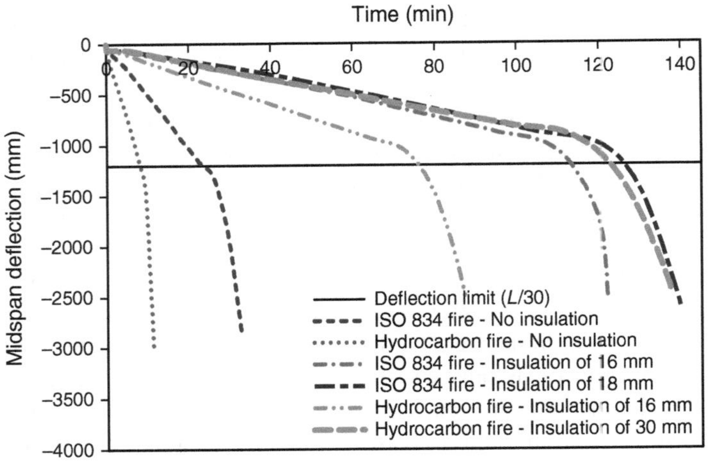

Figure 12.6 Midspan deflection in an insulated I-65 bridge girder under various fire scenarios. (*With permission from Elsevier.*)

under ISO 834 fire exposure, while the corresponding increase in fire resistance of this girder is 120 minutes under hydrocarbon fire conditions.

This enhanced fire resistance to 60 minutes or higher (achieved through fire insulation) can significantly lower the risk of fire-induced collapse or damage. Typically, the average response time for firefighters to arrive at a fire incident and start firefighting operations can be 20 to 30 minutes. The provision of fire resistance of 60 to 120 minutes can thus give sufficient time to fight the fire or for the fire to burn out on its own, in most scenarios.

Reevaluating Fire Risk-Based Importance Factor (Step 5)

Predictions from the above finite element analysis clearly indicate that applying fire protection (in the range of 18- to 30-mm thickness) in the I-65 bridge girder can delay temperature progression within steel girders and enhance fire resistance to about 120 minutes (for ISO 834 and hydrocarbon fire scenarios). To confirm that the vulnerability of the bridge to fire is reduced through the added fire protection on steel girders, the "fire-based importance factor" is reevaluated by considering all the characteristics of the bridge (as in Step 1), and in addition, the fire protection feature to the girder. With the added fire protection, the class coefficient for fire mitigation strategies class (Δfms) is recalculated as[3]

$$\Delta_{fms} = \frac{0+0+3}{4+4+5} = \frac{3}{13} = 0.23$$

Then the updated overall class coefficient (λu) is calculated as follows (please note that class factors ψ are given in Table 12.4).

$$\lambda_u = \Delta_g \times \psi_g + \Delta_h \times \psi_h + \Delta_t \times \psi_t + \Delta_e \times \psi_e + \Delta_f \times \psi_f - \Delta_{fms} \times \psi_{fms}$$

$$\lambda_u = 0.60 \times 0.47 + 0.44 \times 0.22 + 0.63 \times 0.11 + 0.44 \times 0.12 + 0.33 \times 0.08 - 0.23 \times 0.18$$

$$\lambda_u = 0.49$$

Based on the updated overall class coefficient value ($\lambda_u = 0.49$), the importance factor for the reconfigured bridge girder is reduced to 1.0 (from 1.2 originally in Step 1), indicating that this reconfigured bridge falls under the "medium" fire risk category (see Table 12.5). This implies that the bridge after adding insulation to the steel girder is less susceptible to fire damage/collapse, and no additional measures may be needed to enhance the fire safety of this bridge. Similar calculation can be carried out for other structural members (piers, deck, etc.) made of different material types (concrete, steel, wood, etc.), and optimal strategies can be developed for enhancing the fire resistance of bridges.

[3]See Kodur et al. (2017) for detailed step-by-step calculations.

12.5 Strategies for Mitigating Fire Hazard in Tunnels

Tunnel fires, as described in Sec. 12.2, are generally of high intensity and represent a very harsh environment not only for commuters and firefighters but also for structural members. Even seemingly minor events can quickly escalate to large fires in tunnels. In current practice, the main areas of concern during tunnel fires have been for the safe evacuation of commuters. For this reason, ventilation conditions, in particular smoke ventilation, are well addressed in the design of long tunnels. Other active fire protection systems are utilized to some extent in most long tunnels to overcome the problem of fire hazard. However, limited passive fire protection provisions are incorporated in the design of tunnels. This is mainly due to the lack of calculation methods and associated design guidance for fire resistance design in codes and standards. The provision of structural fire protection measures in tunnels can greatly enhance the protection of escape routes, avoiding major business interruption, and can even minimize complete or partial collapse of the tunnel structure during a fire incident. Therefore, both active and passive fire protection measures need to be incorporated to minimize the adverse impact of a fire in a tunnel.

The active fire protection systems currently used in tunnels include water mist and foam deluge systems, heat detectors, sprinklers, smoke extractors, and smoke alarm systems. These active fire protection measures help to slow down the spread and growth of fire and smoke during a tunnel fire to a great extent and thus reduce the heat release rate. Moreover, these systems also help limit the spread of fire to other vehicles by wetting the other combustible materials and cooling the ceiling gas temperature, thereby slowing down burning, which in turn can minimize damage to tunnel structures. For instance, little structural damage was observed during a fire incident in the Melbourne Burnley Tunnel (Schütz, 2014), due to the presence of water deluge system in the tunnel.

On another front, communication systems are being recommended for tunnels, and when installed, they can form the first line of defense during a fire incident. These systems include, for example, video smoke detection systems. These systems analyze data from video cameras to pinpoint the exact location of a heat spot and the direction and size of the fire. Though not compulsory for tunnels under 3000 m, the European Directive 2004/54/EC (European Directive 2004/54/EC, 2004) recommends installation of automatic incident detection (AID)[4] systems.

Currently, the passive protection measures are incorporated in tunnels to mitigate fire-induced spalling of concrete in tunnel linings. These measures serve as the last line of defense in case of fire, to maintain the structural integrity of the tunnel. In particular, spalling mitigation measures are needed in the case of newer concretes, since these concretes are highly susceptible to spalling due to their low permeability. Following are specific measures that can mitigate fire-induced spalling of concrete and enhance fire resistance of tunnel linings.

- In the case of new tunnels, polypropylene fibers can be added to concrete to mitigate fire-induced spalling in concrete (Kodur et al., 2003). These polypropylene fibers melt at about 160°C, thereby enhancing porosity in concrete. With increased porosity, steam generated within concrete during a fire event can escape. This will reduce the pore pressure buildup in concrete linings, which in turn reduces spalling concrete (Bilodeau et al., 2004).
- In the case of existing or new tunnels, coatings composed of vermiculite-cement-based materials can be applied in varying thicknesses depending on the expected fire intensity in tunnels and limiting temperature requirement at the interface of the coating and concrete of tunnel linings. These coatings serve as a shield during fire exposure and slow down the temperature rise in the reinforcement and concrete of tunnel linings, thereby improving the fire resistance. Alternatively, thermal barriers in the form of insulation boards and panels made of calcium silicate aluminate materials can be installed on tunnel linings. These insulation boards can limit the temperature rise of tunnel linings, thereby preventing fire-induced spalling in concrete, as well as minimizing strength loss in concrete and rebar.
- In the case of new or existing tunnels, detailed numerical simulations can be carried out to develop optimum fire protection measures for different structural members used in tunnels. Currently, there are very few calculation methods or validated numerical models available to carry out such calculations so as to develop optimum fire resistance strategies in tunnels.

[4] AID systems are typically video-based and use algorithms to analyze video images and detect deviations from "normal" patterns, automatically alerting operators of nonroutine situations.

12.6 Emerging Trends and Research Needs

Although there have been some research and development efforts in recent years to address fire hazard issues in transportation structures, there is still a lack of specific fire resistance provisions and reliable numerical tools for designing and developing optimum strategies for overcoming the adverse effects of fire. There is therefore an urgent need and opportunity for further research to development efforts to mitigate fire hazard in transportation structures.

12.6.1 Numerical Models for Evaluating Fire Resistance

There is a need for developing risk-based approaches to evaluate the vulnerability of a tunnel (or a bridge) to fire hazard considering all critical aspects that influence the fire performance of a structure. Further, test data on fire growth characteristics, temperature-dependent variation in material properties, and structural design features specific to those used in bridge and tunnel structures should be generated to validate numerical models. In addition, there is a need for incorporating required fire resistance provisions in codes and standards for specific type of bridges and tunnels.

12.6.2 Post-Fire Residual Capacity Assessment and Repair Strategies

Structural members in a bridge or a tunnel under severe fire exposure may experience significant deformations and structural damage resulting from temperature-induced phenomenon such as spalling in concrete girders/piers, web shear buckling in steel girders, or rupture of connections due to the development of thermally induced forces. Alternatively, exposure to moderate- or low-intensity fire may not result in noticeable deformations or damage in structural members, but localized damage can still occur, such as spalling in concrete linings (and girders) or local buckling in the web of steel girders. Nevertheless, following a fire, a weakened structure cannot be reopened to traffic, even after the fire is fully extinguished, until residual capacity of structural members is evaluated. Currently there is lack of data, provisions, techniques, and guidelines for the rapid assessment of fire-damaged transportation structures, whose characteristics and design features are significantly different from those of building structures. Thus, there is an urgent need for developing rapid assessment techniques (including nondestructive test methods) and calculation methods for evaluating the post-fire conditional assessment of members in transportation structures. Such residual capacity assessment techniques will be highly useful for developing appropriate retrofitting strategies for upgrading fire-damaged structural members. In addition, further research and development efforts are needed to develop guidance on appropriate repair and strengthening methods for retrofitting fire-damaged structural members in transportation structures.

12.7 References

ASCE. *Minimum Design Loads for Buildings and Other Structures. ASCE 7-10.* Reston, VA: ASCE, 2010.

ASHRAE. *Control of Fire and Smoke. ASHRAE 5.6.* Atlanta, GA: ASHRAE, 2019. https://tc0506.ashraetcs.org/index.php.

Aziz, Esam M., Venkatesh K. Kodur, Jonathan D. Glassman, and Maria E. Moreyra Garlock. "Behavior of Steel Bridge Girders under Fire Conditions." *Journal of Constructional Steel Research* 106 (March 2015): 11–22. https://www.sciencedirect.com/science/article/pii/S0143974X14003253.

Bai, Yong, William R. Burkett, and Phillip T. Nash. "Rapid Bridge Replacement under Emergency Situation: Case Study." *Journal of Bridge Engineering* 11, no. 3 (May 2006). https://doi.org/10.1061/(ASCE)1084-0702(2006)11:3(266).

Battelle. *Comparative Risks of Hazardous Materials and Non-Hazardous Materials Truck Shipment Accidents/Incidents.* Washington, DC: Federal Motor Carrier Safety Administration, 2004.

Bilodeau, A., V. K. R. Kodur, and G. C. Hoff. "Optimization of the Type and Amount of Polypropylene Fibres for Preventing the Spalling of Lightweight Concrete Subjected to Hydrocarbon Fire." *Cement and Concrete Composites* 26, no. 2 (2004): 163–174.

European Directive 2004/54/EC. 2004. *Directive 2004/54/EC of the European Parliament and of the Council of 29 April 2004 on Minimum Safety Requirements for Tunnels in the Trans-European Road Network.* Document 32004L0054. https://eur-lex.europa.eu/legal-content/EN/TXT/?uri=CELEX%3A32004L0054.

Garlock, Maria, Ignacio Paya-Zaforteza, Venkatesh Kodur, and Li Gu. "Fire Hazard in Bridges: Review, Assessment and Repair Strategies." *Engineering Structures* 35 (February 2012): 89–98. https://doi.org/10.1016/j.engstruct.2011.11.002.

IBC. 2018 International Building Code. IBC, 2018. https://codes.iccsafe.org/content/IBC2018?site_type=public.

Kodur, V. K., E. M. Aziz, and M. Z. Naser. "Strategies for Enhancing Fire Performance of Steel Bridges." *Engineering Structures* 131 (January 2017): 446–458. https://doi.org/10.1016/j.engstruct.2016.10.040.

Kodur, V. K. R., Fu-Ping Cheng, Tien-Chih Wang, and M. A. Sultan. "Effect of Strength and Fiber Reinforcement on Fire Resistance of High-Strength Concrete Columns." *Journal of Structural Engineering* 129, no. 2 (2003): 253–259.

Kodur, Venkatesh, Li Gu, and Maria E. Moreyra Garlock. "Review and Assessment of Fire Hazard in Bridges." *Transportation Research Record: Journal of the Transportation Research Board* 2172, no. 1 (January 2010): 23–29. https://doi.org/10.3141/2172-03.

Kodur, V. K. R., and M. Z. Naser. "Importance Factor for Design of Bridges against Fire Hazard." *Engineering Structures* 54 (2013): 207–220. https://doi.org/10.1016/j.engstruct.2013.03.048
Naser, M.Z., and V. K. R. Kodur "A Probabilistic Assessment for Classification of Bridges against Fire Hazard." *Fire Safety Journal* 76 (2015): 65–73. https://doi.org/10.1016/j.firesaf.2015.06.001.
Naser, M. Z., and V. K. R. Kodur. "Comparative Fire Behavior of Composite Girders under Flexural and Shear Loading." *Thin-Walled Structures* 116 (2017): 82–90. https://doi.org/10.1016/j.tws.2017.03.003.
NBCC. *National Building Code of Canada* 2015. https://www.nrc-cnrc.gc.ca/eng/publications/codes_centre/2015_national_building_code.html.
NFPA. 2017a. *NFPA 15: Standard for Water Spray Fixed Systems for Fire Protection*. https://www.nfpa.org/codes-and-standards/all-codes-and-standards/list-of-codes-and-standards/detail?code=15.
NFPA. 2017b. *NFPA 502: Standard for Road Tunnels, Bridges, and Other Limited Access Highways*. https://www.nfpa.org/codes-and-standards/all-codes-and-standards/list-of-codes-and-standards/detail?code=502.
NYDOT. *Bridge Fire Incidents in New York State*. Albany, NY: New York State Department of Transportation, 2008.
Peris-Sayol, G., I. Paya-Zaforteza, S. Balasch-Parisi, and J. Alós-Moya. 2017. "Detailed Analysis of the Causes of Bridge Fires and Their Associated Damage Levels." *Journal of Performance of Constructed Facilities* 31, no. 3 (June 2017). https://doi.org/10.1061/(ASCE)CF.1943-5509.0000977.
PIARC. Systèmes et Équipements Pour La Maîtrise Des Incendies et Des Fumées Dans Les Tunnels Routiers/Systems and Equipment for Fire and Smoke Control in Road Tunnels. Association mondiale de la route/World Road Association. 2007. https://www.piarc.org/en/order-library/5425-en-Systems and equipment for fire and smoke control in road tunnels.htm.
Schütz, D. "Fire Protection in Tunnels: Focus on Road and Rail Tunnels." SCOR, 2014.
United States Department of Transportation. "Highway Statistics, 1995–2010."

12.8 Nomenclature

Ψ	Class factor
Δ_g	Geometrical features, material properties, and design characteristics for class factor
Δ_h	Hazard (fire) likelihood class factor
Δ_t	Traffic demand class factor
Δ_e	Economic impact class factor
Δ_f	Expected fire losses class factor
Δ_{fms}	Fire mitigation strategies
$\varphi_{x(\max)}$	Maximum weightage factors of each parameter in class (x)
$\varphi_{i,x}$	Weightage factor of subparameter (i) in class (x)
Δ	Class coefficient
λ	Overall class coefficient
λu	Updated overall class coefficient
IF	Importance factor

12.9 Numerical Examples

12.1

Evaluate the fire-based importance factor for the Bill Williams River Bridge, Arizona, which was subjected to a fire incident on July 28, 2006. Refer to Kodur and Naser (2013) for details of this bridge and characteristics of the fire that occurred on this bridge.

Characteristics of the Bridge

The Bill Williams River Bridge was built in 1967 and was made of prestressed concrete girders and a composite concrete deck. The 14 precast AASHTO Type III concrete girders were of 23.3-m span with a 165-mm cast-in-place concrete deck on top.

Description of the Bridge Fire Incident

A fuel tanker carrying diesel fuel of about 29,000 L in the truck overturned and spilled fuel, leading to ignition and rapid burning. This fuel burned for two-and-a-half hours and affected three spans (#8, #9, and #10) of the bridge. After inspection and necessary repairs, the bridge was reopened to traffic. The total cost of retrofitting the bridge was estimated at $700,000.

Evaluating the Importance Factor

The approach discussed in Sec. 12.4 can be applied to evaluate the fire-based importance factor for this bridge. Based on the available data for the bridge, the various parameters and subparameters needed for assessing the importance factor are collected as per Table 12.1. Then, five substeps (as part of Step 1) discussed in Sec. 12.4 should be applied to evaluate the importance factor:

- *Substep 1.* Collect data and statistics on Bill Williams River Bridge (as discussed in Sec. 12.4.2).
- *Substep 2.* Assign weightage factors (φ) for different parameters [see Kodur and Naser (2013) for details].
- *Substep 3.* Calculate individual class coefficients (Δ_x).

The individual class coefficients (Δ) are computed as follows:

$$\Delta_x = \frac{\sum \varphi_{i,x}}{\sum \varphi_{x(\max)}}$$

where $\varphi_{i,x}$ = Weightage factor of subparameter (i) in class (x)
$\varphi_{x(\max)}$ = Maximum weightage factors of each parameter in class (x)

$$\Delta_g = \frac{2+4+2+3+1+3+5+2}{5+4+5+4+3+4+5+5} = \frac{22}{35} = 0.63$$

$$\Delta_h = \frac{3+1+1+4}{5+3+3+5} = \frac{9}{16} = 0.56$$

$$\Delta_t = \frac{1+1}{5+3} = \frac{2}{8} = 0.25$$

$$\Delta_e = \frac{1+1+1}{3+3+3} = \frac{3}{9} = 0.33$$

$$\Delta_f = \frac{3+3}{3+3} = \frac{6}{6} = 1.0$$

- *Substep 4.* Calculate overall class coefficient (λ).[5]

The overall class coefficient (λ) is given by

$$\lambda = \Delta_g \times \psi_g + \Delta_h \times \psi_h + \Delta_t \times \psi_t + \Delta_e \times \psi_e + \Delta_f \times \psi_f$$

$$\lambda = 0.63 \times 0.47 + 0.56 \times 0.22 + 0.25 \times 0.11 + 0.33 \times 0.12 + 1.0 \times 0.08 = 0.57$$

The "overall class coefficient" for this bridge was found to be 0.57, and thus the importance factor for this bridge works out to be 1.2 (as per Table 12.5), which places the bridge under the "high" risk grade. Since the bridge falls under this grade, implementing some of the fire mitigation strategies would enhance the fire performance of the bridge. For instance, implementing the use of monitoring systems, limiting the size of fuel/chemicals in tankers to 20,000 L, or using flooding agents and/or foam deluge systems can reduce the vulnerability of the bridge to fire hazard (alternatively, the fire response of the girder can be enhanced through installing fire protection on the girder). In this case, the above strategies would reduce the vulnerability of the bridge to future fire hazard. This can be verified through calculating the importance factor with added fire mitigation strategies.

- *Substep 5.* Calculate updated importance factor to account for updated fire mitigation strategies.

[5] Please note that class factors ψ are given in Table 12.4.

$$\Delta_{fms} = \frac{1+4+2}{4+4+5} = \frac{7}{13} = 0.54$$

$$\lambda_u = \lambda - \Delta_{fms}\psi_{fms}$$

$$\lambda_u = 0.57 - 0.54 \times 0.18 = 0.47$$

Hence, the updated risk grade and importance factor for the bridge with added provision are "medium" and 1.0, respectively.

12.2

Apply the advanced analysis method to evaluate the vulnerability of I-80/880 MacArthur Maze Bridge against a fire incident similar to that which occurred on April 29, 2007. The following information relates to the bridge characteristics and fire incident.

Description of the Bridge Fire Incident

The fire started when a tanker truck carrying 8600 gallons of gasoline overturned on a connector ramp of I-80/880, underneath the I-580 expressway. The firefighters responded to the incident within 14 minutes, but the intense fire resulted in temperatures reaching 1100°C. This intense heat led to the deterioration of strength and stiffness properties in the steel box support girder, leading to the collapse of two I-580 spans. The first span of the overpass collapsed in about 17 minutes following the start of the fire, and a second adjacent span also collapsed about 5 minutes after the first collapse. This incident cost $9 million to repair the bridge and an estimated economic loss of $6 million per day (for four weeks) from the closure of bridge.

Characteristics of the Bridge

The MacArthur Maze Bridge was composed of six steel girders (built-up section), supporting a concrete slab of 19.7-mm thickness and effective width of 2.0 m. The first span that collapsed under fire was 25.6 m long, while the second collapsed span was 23.5 m long. The dimensions of the top flange plate were of 304.8-mm width and 25.4-mm thickness, while the bottom flange had the same width as the top flange but a 50.8-mm thickness. The web plate had dimensions of 1080-mm depth and 12.7-mm thickness. The steel girders were stiffened with intermediate stiffeners at a spacing-to-depth ratio of 1.0 along the span and a thickness of 12 mm. The steel girders had a yield strength of 345 MPa, while the concrete used in the slab had a compressive strength of 24.8 MPa.

Evaluating the Fire Risk Based on Importance Factor (Step 1)

The various characteristics of the bridge and details of the fire incident were collected from the available data in literature (Peris-Sayol et al., 2017). Following the procedure outlined in Sec. 12.4 and as detailed in Kodur et al. (2017), the fire-based importance factor (IF) for this bridge (main girders) works out to be 1.2, which places this bridge under the "high" risk grade. Bridges in this IF category can experience some level of collapse/damage in main structural elements, resulting in a partial shutdown of traffic operations. This vulnerability can be minimized by protecting steel girders with fire insulation.

Evaluating the Inherent Fire Resistance of Steel Members (Step 2)

This bridge is analyzed to evaluate fire resistance through finite element analysis software (ANSYS) as discussed in Sec. 12.4. A typical girder of this bridge is analyzed with simply supported end conditions under two possible fire exposure scenarios: ISO 834 and hydrocarbon fire.

Results from ANSYS analysis are utilized to evaluate inherent fire resistance of the steel girder. As expected, the cross-sectional temperature profile in the girder follows a similar trend as the I-65 bridge (shown in Fig. 12.3) but increases at a slightly lower rate. This is due to thicker flanges and lower web slenderness in the I-80/880 bridge girder, as compared to the I-65 bridge girder (discussed in Sec. 12.4). The structural response of I-80/880 steel girders is illustrated in Fig. E12.1, where midspan deflection is plotted as a function of fire exposure time under ISO 834 and hydrocarbon fire scenarios. Results presented in this figure show that the I-80/880 steel girder fails without any fire protection in about 12 minutes under hydrocarbon fire and in 35 minutes under ISO 834 fire exposure. Also, it can be seen that the bare steel girder exhibits a higher fire resistance than the I-65 bridge girder. This is attributed to a lower web slenderness and shorter span of the I-80/880 bridge as compared to the I-65 bridge girder.

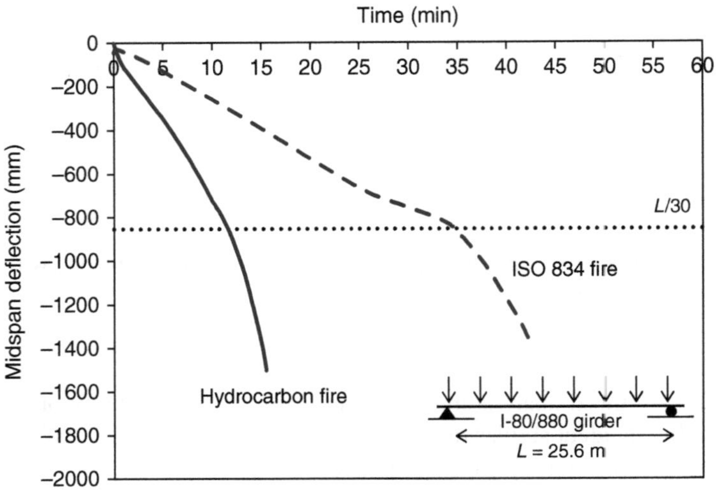

FIGURE E12.1 Midspan deflection in uninsulated I-80/880 bridge girder under various fire scenarios.

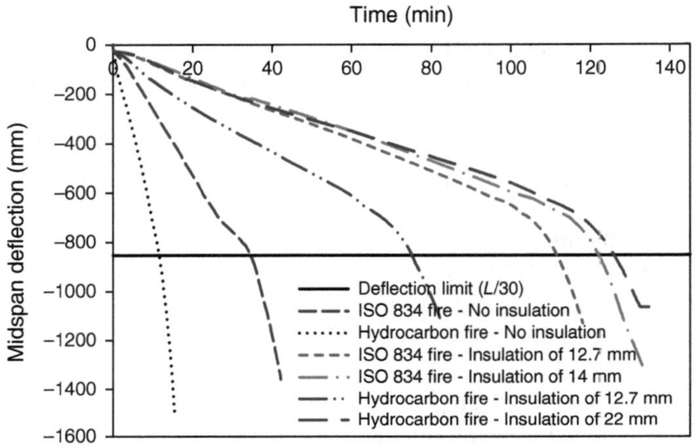

FIGURE E12.2 Midspan deflection in insulated I-80/880 bridge girder under various fire scenarios.

Developing Strategies for Enhancing Fire Resistance (Steps 3 and 4)

The above results clearly show that this bridge experienced substantial damage and collapse under fire due to the lower fire resistance of the girder. Also, it can be inferred that the bridge could have performed better if the fire resistance of girders were higher. In order to enhance fire resistance, the girders were provided with supplemental fire insulation. Further, to optimize the fire protection thickness and insulation scheme, a finite element analysis of the steel girder was carried out in ANSYS utilizing varying thicknesses of fire insulation including 7, 12.7, 14, 16, 20, 22, and 25 mm. For each insulation thickness, the steel girder was analyzed and the corresponding fire resistance was evaluated.

A summary of results from numerical analysis is shown below for various fire insulation thicknesses on girders. The structural response of this girder with an insulation thickness of 12.7 mm (for a 60-minute fire rating), and 14 and 22 mm (for a 120-minute fire rating for ISO 834 and hydrocarbon fire scenarios) is selected as the optimum insulation thickness to enhance fire resistance. It can be seen from Fig. E12.2 that midspan deflection decreases significantly due to supplemental fire insulation on the steel girder. This results in a slower progression of temperatures in the steel section. The use of 14-mm insulation thickness enhanced the fire resistance of the girder from 35 to 120 minutes under ISO 834 fire exposure, while the fire resistance of same girder increased from 12 to 122 minutes under hydrocarbon fire conditions with insulation of 22-mm thickness.

Reevaluating the Fire Risk-Based Importance Factor (Step 5)

To illustrate the lower fire risk with added fire insulation, after the fire protection is applied, the fire-based importance factor is reevaluated for the insulated girder. The fire-based importance factor of the insulated girder works out to be 1.0, and the modified girder configuration falls under the “medium” fire risk category, which implies the bridge is less susceptible to fire damage.

CHAPTER 13

Residual Capacity of Fire-Damaged Structures

13.1 Overview

This chapter presents steps and approaches that can be applied for the conditional assessment of fire-damaged structures. The residual properties of various construction materials after exposure to elevated temperatures are discussed. Simple and advanced calculation methods needed to carry out a residual capacity assessment of fire-damaged structures are reviewed. Emerging trends and future research needs in this particular research area are highlighted.

13.2 Importance of Residual Capacity

When fire occurs in a structure, structural members undergo increased deformations and loss of load-carrying capacity as a result of temperature-induced degradation in properties of constituent materials. However, statistical data clearly indicates that the complete collapse of a structural member in a fire event is extremely rare. In fact, the probability of such failure in reinforced or prestressed concrete structures, or in the case of properly insulated steel or composite structures, is very low (Beitel and Iwankiw, 2008). Nonetheless, structural members can experience some level of damage following fire exposure. In most of these cases, structures can be brought back to reoccupancy and normal use following a fire after proper post-fire assessment and associated repairs. Therefore, following a fire, it is imperative to assess the state of structural members before reopening for occupancy to ensure that the structure has sufficient residual capacity. Such an assessment is critical to ensure occupant safety and also to develop appropriate retrofitting and repair measures in fire-damaged structures.

The extent of damage a structure undergoes during fire exposure depends on numerous factors. Of these factors, peak fire temperature reached, duration of burning, severity of fire exposure, and excessive reserve capacity of the member have the highest influence on residual capacity. In the case of a severe fire, structural members might experience significant loss in their load-carrying capacity. Such loss of capacity in concrete or composite structures can result due to high rebar temperatures attained during a fire and also from any loss of cross section due to possible fire-induced spalling in concrete. The capacity of loss in steel or composite structures can occur due to temperature-induced loss of strength and buckling in steel, which could result from various issues, including a rapid temperature rise in the member resulting from the debonding of fire insulation. In the case of timber structures, significant loss of capacity can result from the charring process due to the fact that wood is a combustible material and such charring can lead to the loss of effective sectional area.

The current approach for estimating the residual capacity of fire-damaged structures is mostly through visual inspection and through applying rules of thumb wherein physical examination of a fire-damaged structure is undertaken mainly by looking for visual signs of excessive cracking, deformation, buckling, spalling, and so on. Such inspection is primarily qualitative and highly dependent on the expertise and experience of assessing engineers or experts. As such, visual inspection may not yield a reliable assessment of residual capacity.

Another approach currently used to evaluate residual capacity of fire-damaged structures is through testing cores or coupons taken from fire-damaged portions of structures. In these tests, cores of concrete or steel coupons or wood sections taken from fire-damaged structural members are tested to determine leftover (residual) strength in these cores. Results from strength tests are compared to the original (design) strength (prior to fire) to evaluate leftover capacity. This approach, although slightly better than visual inspection, does not yield an accurate assessment of residual capacity. Further, physical testing of cores requires significant effort and time, which is not practical, as the fire-exposed structures are to be opened for occupancy immediately following a fire. Alternatively, nondestructive tests or calculation methods can be applied for evaluating the residual capacity of fire-exposed structural members (Felicetti et al., 2009; Kodur and Agrawal, 2016). However, the applicability of nondestructive techniques and numerical approaches for evaluating the residual capacity under varying conditions of fire exposure, loading, and structural parameters of the fire-damaged structures is not well established. This is mainly due to the lack of validated approaches and data relating to residual properties of materials and fire exposure scenarios. Some of the recent advances with respect to residual capacity assessment are presented in the following sections.

13.3 Residual Material Properties after High-Temperature Exposure

In order to assess the residual capacity of a fire-damaged structure, residual properties of materials following high-temperature exposure are needed. As discussed in previous chapters, properties of construction materials degrade with temperature rise; however, most properties recover to some extent following the cooldown to room temperature. The extent of recovery in properties depends on the peak temperature reached in the material, the type and characteristics of material, the rate of heating and cooling, and the loading (stress) state during fire exposure. The available information on residual properties of different construction materials is summarized in this section.

13.3.1 Concrete

Material-level tests on temperature-exposed concrete cylinders by Schneider (1988) showed that concrete undergoes distinct thermal property changes during the heating and cooling cycles. Specific heat and thermal conductivity of concrete decrease during the cooling cycle, owing to the irreversible loss of moisture (water) and microcracking in concrete during the heating phase. The extent of residual thermal expansion or shrinkage that remains in concrete after cooling down to ambient conditions is dependent on the peak temperature experienced during a fire event. There is no broad guidance in codes and standards on the residual thermal properties of concrete. In most advanced analysis studies, for tracing the responses of a structure during the cooling or residual phase, researchers have used the same thermal properties as those during the heating phase.

In terms of the compressive strength of concrete, Eurocode 2 recommends the use of a 10% reduction in strength properties (from corresponding reduction factors at peak temperature during heating) to allow for the simple representation of residual strength in concrete. As an illustration, a concrete specimen (of $f_c' = 50$ MPa) when heated to 600°C peak temperature (during a fire) will have its strength reduced to 62% of room compressive strength. However, when this concrete specimen is cooled down to ambient (20°C) conditions, the concrete recovers some of its loss in strength and thus will have 56% of original strength (as shown in Fig. 13.1).

Based on residual strength tests, some researchers have suggested that the compressive strength of concrete following cooldown can decrease by 20% of its original value. Moreover, the extent of deterioration is influenced by the cooling method (air cooled or water quenched) as well. Chang et al. (2006) showed that the residual compressive strength of concrete decreases with an increase in peak temperature during the heating phase. This rate of decrease is found to be lower when peak heating temperature is below 200°C. In general, the residual compressive strength after exposure to a peak heating temperature of 200°C is about 90% of the original room temperature strength; however, the corresponding reduction after attaining peak temperatures of 400, 600, and 800°C and cooldown to ambient temperature is about 65%, 40%, and 15% of original strength values, respectively. Chang et al. (2006) proposed the following relation for residual normalized compressive strength (f_{cr}'/f_c') in terms of peak exposure temperature in concrete (T) ranging from 20 to 800°C, and this is expressed as

$$f_{cr}'/f_c' = 1.008 + T/450 \ln(T/5800) \tag{13.1}$$

Elastic modulus of concrete, taken as the secant modulus at 40% of the peak stress on the compressive stress-strain curve, decreases at a faster pace than that of compressive strength during the heating phase of concrete, and this is true for the residual modulus of concrete as well. The residual elastic moduli at 200, 400, and 600°C reduce to about 80%, 40%, and 6% of the original modulus, respectively. The normalized

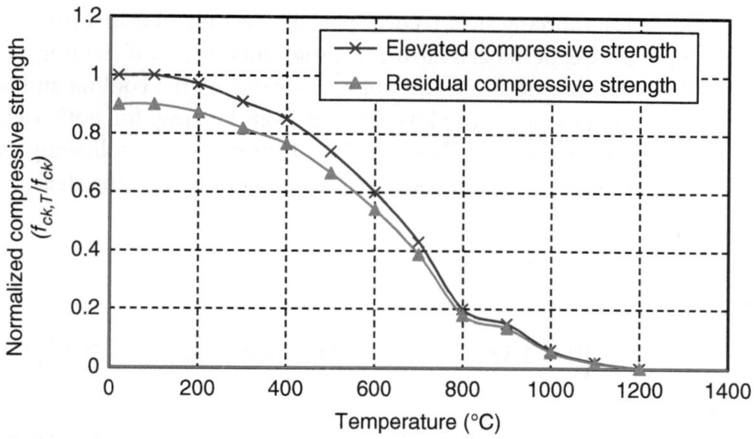

FIGURE 13.1 Normalized compressive strength of concrete with peak exposure temperature.

reduced elastic modulus (E_{cr}/E_c) in compression with temperature (T) can be evaluated through the following relations (Chang et al., 2006):

$$E_{cr}/E_c = -0.00165T + 1.033,\ 20°\text{C} < T < 125°\text{C} \tag{13.2}$$

$$E_{cr}/E_c = 1/1.2 + 18(0.0015T)^{4.5},\ 125°\text{C} < T < 800°\text{C} \tag{13.3}$$

where E_c and E_{cr} are the original and reduced elastic modulus in compression and T is the peak temperature attained during heating.

Tensile strength of concrete decreases with an increase in peak temperature exposure. For the same temperature rise and level, the reduction in residual tensile strength is greater than that in the compressive strength, especially for temperatures less than 400°C. The relations for normalized residual tensile strength, (f'_{tr}/f'_t), with peak temperature (T) is expressed as (Chang et al., 2006):

$$f'_{tr}/f'_t = 1.05 - 0.0025T,\ 20°\text{C} < T < 100°\text{C} \tag{13.4}$$

$$f'_{tr}/f'_t = 0.8,\ 100°\text{C} < T < 200°\text{C} \tag{13.5}$$

$$f'_{tr}/f'_t = 1.02 - 0.0011T,\ 200°\text{C} < T < 800°\text{C} \tag{13.6}$$

where f'_t and f'_{tr} are the original and reduced tensile strength and T is the peak temperature attained during heating.

The residual stress-strain response of concrete following cooldown is also affected by the recovery in its strength properties, as well as by the closure of preexisting cracks caused by the heating and cooling cycle, which lead to a pronounced concave-up of the stress-strain curve at the initial stages of loading. Therefore, the shape of the ascending stress-strain curve at a given temperature during the heating of concrete is different from that in the residual (cooldown) state, and the shape varies with the peak temperature reached during heating. Besides, with the increase in peak temperature, the descending branch of stress-strain response becomes flatter due to the softening of material. Thus, for advanced analysis, it is important to utilize the complete residual stress-strain response of concrete to evaluate the residual capacity of concrete structures. The residual compressive stress-strain response of concrete can be traced using the relation proposed by Chang et al. (2006).

Anderberg and Thelandersson (1976) reported that much of the transient creep deformations in concrete that occur during first-time heating of concrete under load do not recover following cooldown of concrete.

13.3.2 Structural Steel, Reinforcing Steel, and Prestressing Steel

Thermal and mechanical properties of steel, which degrade at elevated temperatures attained during fire exposure, recover to some extent upon cooldown to ambient temperature. Strength and stiffness properties of structural and reinforcing steel recover partially or fully to initial room temperature values, depending on peak temperature experienced during heating. Nonetheless, if yielding occurs in steel during the heating process, subsequent plastic strains do not recover during the cooling and residual state. Overall, residual properties of structural steel are less affected from heating, for both yield and ultimate strength, than reinforcing or prestressing steel. Tao et al. (2013) proposed the following expressions to evaluate residual yield ($f_{y,r,T}$) and ultimate ($f_{u,r,T}$) strength of structural and reinforcing steel.

For structural steel:

$$f_{y,r,T} = \begin{cases} f_y & T \leq 500°\text{C} \\ [1 - 2.33 \times 10^{-4}(T - 500) - 3.88 \times 10^{-7}(T - 500)^2]f_y & T > 500°\text{C} \end{cases} \tag{13.7}$$

$$f_{u,r,T} = \begin{cases} f_u & T \leq 500°\text{C} \\ [1 - 1.95 \times 10^{-4}(T - 500)]f_u & T > 500°\text{C} \end{cases} \tag{13.8}$$

For reinforcing steel:

$$f_{y,r,T} = \begin{cases} f_y & T \leq 500°\text{C} \\ [1 - 5.82 \times 10^{-4}(T - 500)]f_y & T > 500°\text{C} \end{cases} \tag{13.9}$$

$$f_{u,r,T} = \begin{cases} f_u & T \leq 500°\text{C} \\ [1 - 4.85 \times 10^{-4}(T - 500)]f_u & T > 500°\text{C} \end{cases} \tag{13.10}$$

Figure 13.2*b* shows the residual stress-strain relation for reinforcing steel after heating to certain peak temperature (*T*). These degradation trends are plotted based on data reported by Neves et al. (1996) for reinforcing steel having a yield strength of 420 MPa. In the absence or lack of specific data, this residual stress-strain response of reinforcing steel can be used for structural steel.

The residual elastic modulus of prestressing steel wire does not get affected if the steel is heated to below 720°C (Fan and Lu, 2002). The residual 0.2% offset yield strength and the ultimate tensile strength of various prestressing steels have been shown to decrease significantly between 400 and 600°C. The *PCI Manual* (PCI, 2011) provides some guidance for quantifying the residual strength of prestressing strands after exposure to a certain peak temperature. For peak temperatures less than 400°C, the strands retain 100% of their initial strength on cooling, but after reaching peak temperatures equal to 482 and 600°C, they retain only 70% and 50% of their initial strength, respectively.

Hertz (2004) proposed a generalized closed-form solution for predicting the residual strength of various building materials after being subjected to certain peak temperatures during heating. This relation is given as

$$\xi(T) = k + \frac{1 - k}{1 + \dfrac{T}{T_1} + \left(\dfrac{T}{T_2}\right)^2 + \left(\dfrac{T}{T_8}\right)^8 + \left(\dfrac{T}{T_{64}}\right)^{64}} \tag{13.11}$$

The parameter ξ is the ratio of material strength at peak temperature T (°C) to the same property at ambient upon cooldown. The parameter k represents the ratio between the minimum and maximum value of the material property. Parameters T_1 to T_{64} are determined by fitting the experimental data for specific material under consideration. MacLean (2007) lists values of parameters (coefficients) for the case of prestressed steel to be used in this equation. By applying the coefficients in Table 13.1, the residual strength of prestressing steel can be evaluated.

Special types of steel used in structural applications and bolts experience different levels of residual properties after exposure to a certain peak temperature. Limited experimental data exist on post-fire mechanical properties of structural bolts. Yu (2006) carried out residual shear strength tests on A325 bolts after exposure to various peak temperatures up to 800°C, and the results from these tests are shown in Fig. 13.3. Overall, bolts

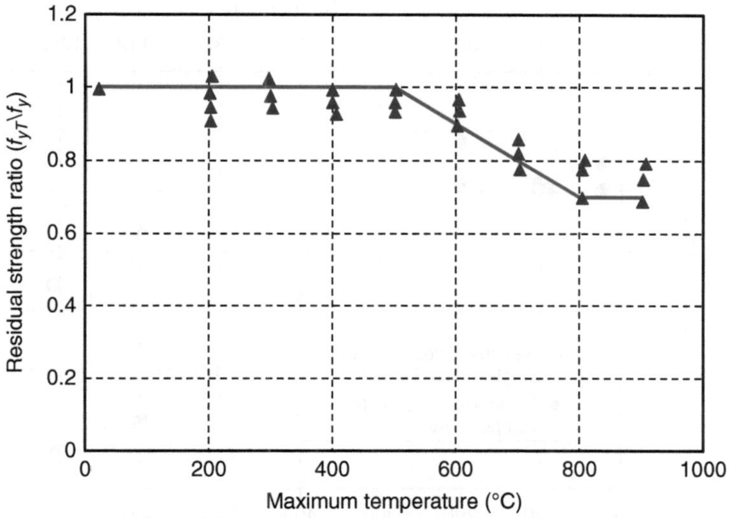

(a) Normalized residual strength of hot-rolled reinforcing steel

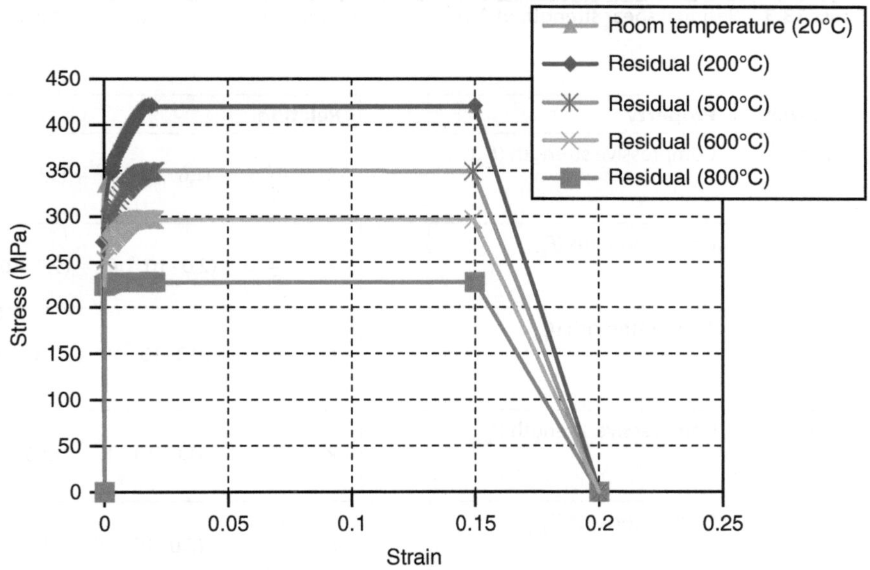

(b) Residual stress-strain curves for reinforcing steel having yield strength of 420 MPa

FIGURE 13.2 Residual properties of hot-rolled reinforcing steel.

Residual Stress	k	T_1	T_2	T_8	T_{64}
Yield	0.20	100,000	750	550	650
Ultimate	0.20	100,000	950	550	650

TABLE 13.1 Parameters to Be Used in Predicting Residual Strength in Prestressing Steel

as compared to structural steel retain much less residual strength, especially when the peak temperature attained during fire exposure exceeds 600°C.

13.3.3 Masonry

Masonry, similar to concrete, regains much of its strength upon cooldown from a peak temperature attained during heating. Limited experimental data exists on the residual mechanical properties of masonry. The published data suggests that cementitious masonry materials (e.g., mortars, concrete, or calcium silicate blocks) generally undergo a decrease in strength on cooldown following peak temperature exposure,

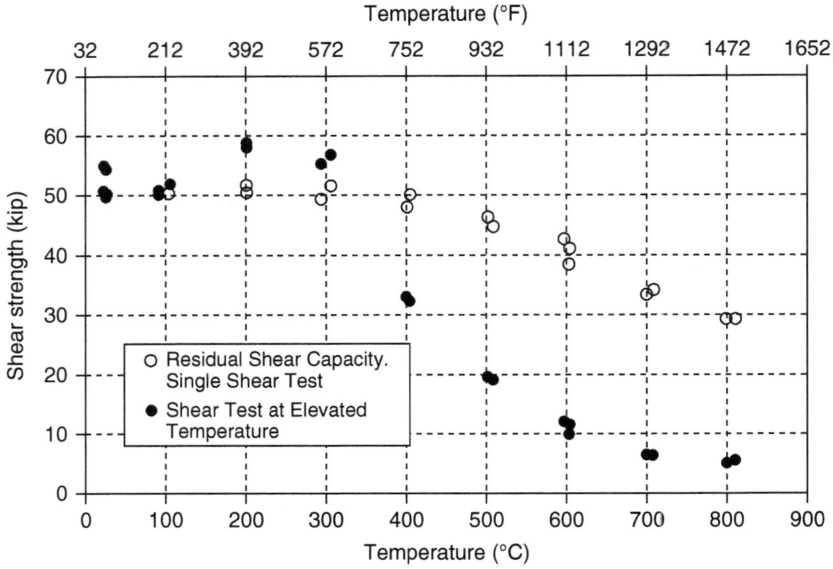

Figure 13.3 Residual shear strength of ASTM A325 bolts post fire exposure. (*With permission from ASCE.*)

Material	Property	Relation
Masonry	Compressive strength ($f_{c,\theta}$)	$k_\theta = \frac{f_{c,\theta}}{f_c} = -(1.0 \cdot 10^{-6})\theta^2 + (5.0 \cdot 10^{-4})\theta + 0.9898$
	Elastic modulus (E_θ)	$k_\theta = \frac{E_\theta}{E} = -(2.0 \cdot 10^{-6})\theta^2 + (9.0 \cdot 10^{-4})\theta + 0.9833$
	Tensile strength ($f_{t,\theta}$)	$k_\theta = \frac{f_{t,\theta}}{f_t} = -(0.3 \cdot 10^{-5})\theta^2 + (1.3 \cdot 10^{-3})\theta + 0.9762$
Brick	Compressive strength ($f_{bc,\theta}$)	$k_\theta = \frac{f_{bc,\theta}}{f_{bc}} = -(1.0 \cdot 10^{-6})\theta^2 + (3.0 \cdot 10^{-5})\theta + 0.9998$
	Elastic modulus ($E_{bc,\theta}$)	$k_\theta = \frac{E_{b,\theta}}{E_b} = -(2.0 \cdot 10^{-6})\theta^2 + (1.1 \cdot 10^{-3})\theta + 0.9789$
Mortar	Compressive strength ($f_{mc,\theta}$)	$k_\theta = \frac{f_{mc,\theta}}{f_{mc}} = -(5.0 \cdot 10^{-6})\theta^2 + (2.7 \cdot 10^{-3})\theta + 0.9486$
	Elastic modulus ($E_{m,\theta}$)	$k_\theta = \frac{E_{m,\theta}}{E_m} = -(2.0 \cdot 10^{-6})\theta^2 + (2.0 \cdot 10^{-4})\theta + 0.9964$

Table 13.2 Residual Properties of Masonry (*Adapted from Russo and Sciarretta, 2012*)

while clay bricks experience an increase in modulus (stiffness) and strength, and a decrease in strain capacity. Russo and Sciarretta (2012) proposed expressions for residual mechanical properties of brick, mortar, and masonry in terms of peak temperature attained during heating, as shown in Table 13.2.

13.3.4 Timber

The residual strength of timber upon cooldown to ambient temperature depends on a number of factors including residual moisture content (in fire-exposed wood), heating rate, burning period, peak temperature, and to some extent, species of timber. As per reported information in *Wood Handbook* (USDA, 2010), wood after 50 days of oven heating reaching to a peak temperature of 115 and 135°C can experience 10% and 38% reduction in the modulus of elasticity upon cooldown. Knudson and Schniewind (1975) and Schaffer (1973) developed graphs showing the effect of temperature exposure on residual tensile and

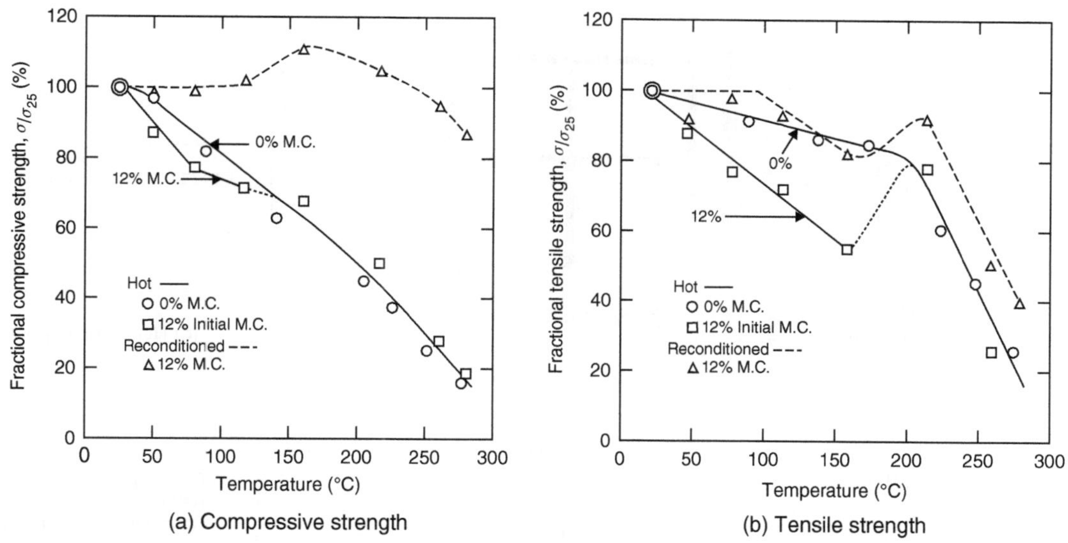

FIGURE 13.4 Compressive and tensile strength of timber as a function of peak temperature in the hot state as well as after cooling to ambient temperature. (*Adapted from Schaffer, 1973.*)

compressive strength of wood. This data, which is plotted in Fig. 13.4, shows the effect of temperature exposure on the residual strength properties of wood. In this figure, Schaffer (1973) has noted that compressive strength decreases rather uniformly with temperature, and independent of initial moisture content; however, much of residual strength is recovered after cooling and reconditioning of wood. On cooldown, when wood is reconditioned back to having 12% moisture content, it regains much of its original strength (as can be seen in Fig. 13.4). Further, the residual tensile strength reduces significantly only after peak temperature goes beyond 200°C, and this degradation depends on the amount of wood moisture. The significant reduction in strength when wood is heated beyond 200°C can be attributed to permanent physical changes in characteristics of the wood due to alteration in its microstructure.

13.4 Residual Behavior of Fire-Exposed Structures

The residual behavior of a structure during cooldown following fire exposure is a quite complex phenomenon and is influenced by a number of factors, including fire severity, level of applied loading present on the structure, material type and its properties, and restraint conditions of the structure. Of these factors, the material type (i.e., concrete, steel, wood, or masonry) the structure is made of has a significant influence on the residual capacity retained in a structure following fire exposure. The typical residual response is illustrated for the case of a concrete structure exposed to fire.

13.4.1 Typical Post-Fire Behavior of a Structural Member

Generally, concrete structural members retain much of their capacity following a fire incident. However, the extent of residual capacity in a structural member depends on both temperature history and structural parameters present during fire exposure. The typical residual behavior is illustrated in Fig. 13.5 through the tracing response of a typical RC beam in a building following a fire incident. As illustrated in Fig. 13.5, there are three distinct phases in the response of the beam during and after a fire: heating, cooling, and residual. The rise in fire temperature in the building leads to increased sectional temperatures in both concrete core and rebars in the RC beam. As shown in Fig. 13.5, the temperatures in rebar (and inner layers of concrete) gradually rise due to the low thermal conductivity and high heat capacity of concrete. Large temperature gradients develop between the fire-exposed surface and inner layers of concrete. These thermal gradients lead to a rise in deflection in the beam. As fire (exposure time) progresses, sectional temperatures within the RC beam stabilize and the effect of thermal gradients is less pronounced. Nonetheless, midspan deflection in the RC beam continues to increase due to temperature-induced degradation in strength and stiffness properties of concrete and rebar (see Fig. 13.5).

During the *heating phase*, once the fire attains flashover, cross-sectional temperatures rise in rebar and concrete, leading to a deterioration of load-carrying capacity of the beam, together with a rise in deflection, due to temperature-induced degradation in strength and modulus properties of concrete and rebar. Then, the fire enters the *cooling phase* and fire temperatures begin to decay. However, cross-sectional temperatures

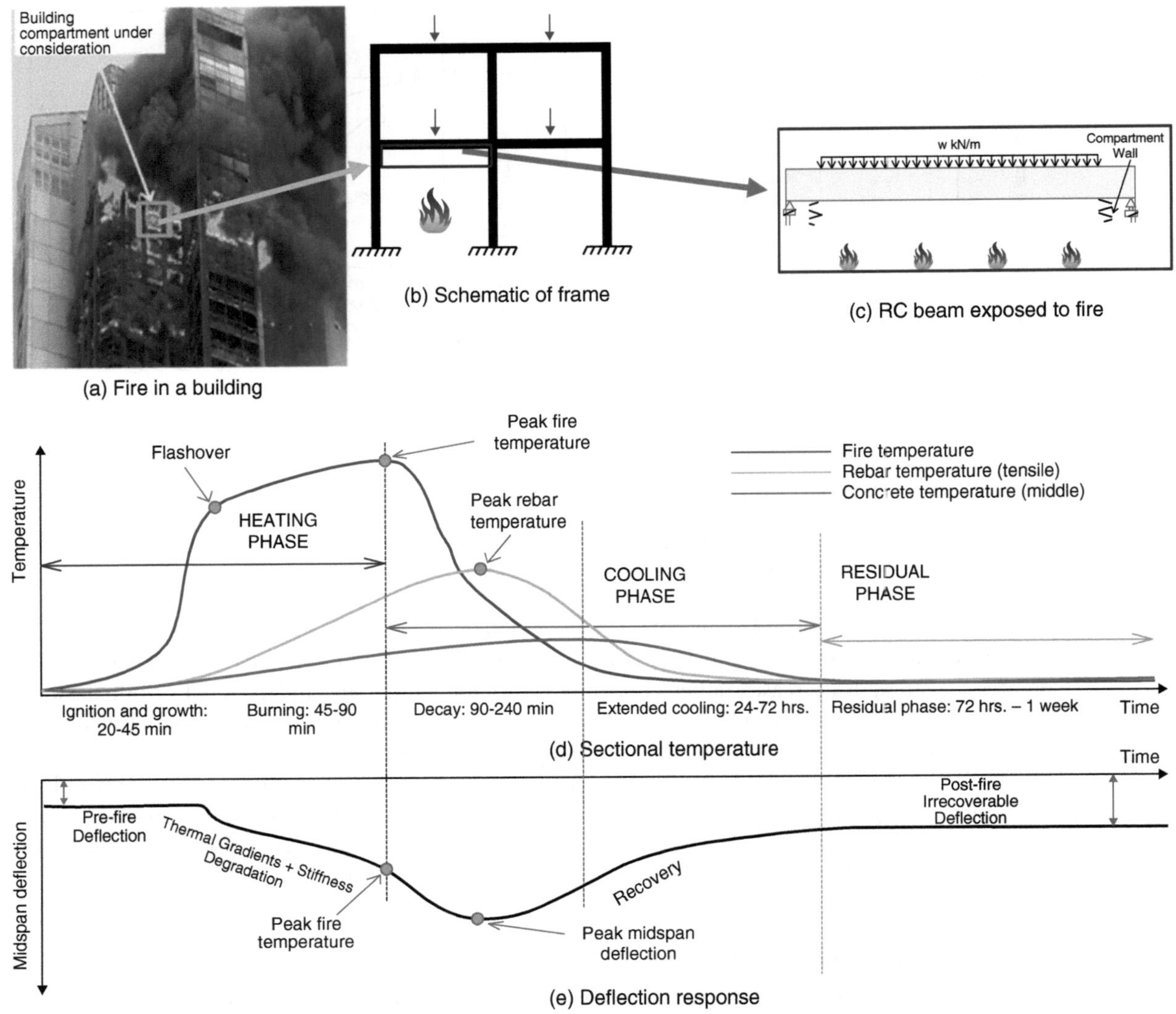

Figure 13.5 Response of a typical RC beam during and after fire exposure. (*With permission from Agrawal, 2020.*)

within the structural member remain significantly high for a prolonged duration following burnout conditions or after the fire is extinguished, owing to the high thermal inertia of concrete. In fact, temperatures within the structural member may not cool down fully to ambient conditions until 24 to 48 hours after the fire has been extinguished, depending on the size (thermal mass) of the member and fire severity. Then the beam enters the *residual phase*, and with gradual cooldown, the beam recovers part of its capacity and peak deflections attained during the heating phase. After complete cooldown to ambient conditions, an unrecoverable residual deflection can be present in the member. These residual deflections represent the state of structural damage in the beam resulting from fire exposure, and the magnitude of such residual deformation depends on the peak temperature reached, load level, boundary conditions, and temperature-induced degradation in material properties.

The structural response during three distinct phases is shown in Fig. 13.5 through progression (and recovery) of midspan deflection with fire exposure time. The various salient points during a fire incident (including cooldown), together with approximate onset times, are also shown in Fig. 13.5*c*,*d* for a typical fire scenario in a building.

The above-illustrated residual response is for typical concrete beams exposed to fire. The overall residual response of other structural members (such as columns, slabs, and walls) made of other materials (i.e., steel, composite, or wood) is similar to the one described above for RC beams. However, the extent of recovery in residual capacity and level of residual deformations is dependent on a number of factors, including fire exposure severity, load level, material-specific mechanical properties, restraint conditions, and additional temperature-induced phenomena that develop in different material types (steel, wood, etc.). These factors should be given due consideration in evaluating residual capacity of fire-damaged structures.

13.4.2 Factors Governing Residual Capacity

The extent of damage to a structure resulting from a fire incident, or the extent of remaining capacity following a fire, is influenced by a number of factors, including fire severity, type and properties of construction materials (steel, concrete, or wood), recovery time following fire exposure, temperature-induced effects (i.e., charring, spalling, buckling, bond slip, etc.), load level, and restraint conditions present during fire exposure. Many of these factors are interdependent and can vary widely under different fire exposure scenarios.

The severity of fire exposure can influence residual capacity of structures significantly. If the fire is severe, structural members might lose much of their capacity or even undergo collapse. However, in moderate to minor fire exposure scenarios, structural members might retain much of their capacity following a fire. Fire severity primarily depends on fuel load, ventilation characteristics, and geometrical parameters of the fire compartment. These factors may evolve with the growth of the fire; for instance, a sudden increase in ventilation can occur due to breakage of glass windows. Furthermore, the presence of active fire protection systems such as sprinklers and firefighter intervention influences the rate of fire temperature rise, peak fire temperature reached, and time taken for the structure to cool down. Similarly, the cooling phase of fire is influenced by the fire compartment characteristics as well as the actions of firefighting.

Properties of construction materials vary during the heating and cooling phases of fire exposure, as well as on cooldown to ambient conditions. For example, thermophysical, mechanical, and deformational properties of concrete change substantially during the heating phase of fire exposure primarily due to the breakdown of C-S-H (calcium silica hydrate) gel and the loss of moisture present in concrete. The properties of concrete change further during the cooling phase due to microcracking and chemical changes occurring during heating. Also, load-induced thermal strain (LITS) that develops in concrete during the first time heating under load does not recover upon cooldown of a concrete member. Similarly, temperature-induced effects, such as creep and buckling in structural and cold-formed steel structures, spalling in concrete structures, charring of timber in wood structures, and debonding of FRP in FRP-strengthened structural members, can influence the residual capacity retained in a fire-exposed structure.

Residual properties of constituent materials, upon cooldown of a structure following a fire, are different from the original (initial) properties prior to fire exposure or the reduced properties during the heating or cooling phase of fire. The extent of recovery of properties following a fire event is a function of time and can take days to weeks in fire-exposed concrete to regain much of the original (room temperature) compressive strength. However, in the case of structural steel or reinforcing steel, much of the properties recover to the original state quickly if the peak temperature attained in steel is below 500°C. If peak temperature is above 500°C, the residual strength retained is only part of its original room temperature strength. In concrete structures, the bond between concrete and rebars can be affected following exposure to elevated temperature due to temperature degradation in bond strength. The residual properties of structural (or reinforcing or prestressing) steel can recover much quicker than that of concrete and wood. In some cases, reconditioning of wood (via treating) can help in regaining much of lost structural properties.

In addition to fire severity and material properties, structural parameters, such as the extent of load and restraint conditions present on a structure during fire, also have a significant influence on the residual response and residual capacity of a member. To illustrate this, the response of a concrete beam (similar to the one shown in Fig. 13.5) exposed to the same fire exposure scenario (as illustrated in Sec. 13.4.1) but under varying load levels (of 30%, 40%, and 50% of original capacity) present during fire exposure is plotted in Fig. 13.6 (Agrawal and Kodur, 2019). The progression of midspan deflection with fire exposure time (including that under an extended cooling period) is plotted in Fig. 13.6*a*, while the load deflection response following the complete cooldown of the beam after fire exposure is plotted in Fig. 13.6*b*. It can be clearly seen in Fig. 13.6*a* that both the rate of rise in deflection and the maximum deflection occurring during fire exposure increase with increasing load level. Further, as can be seen in Fig. 13.6*b*, a higher load level produces higher residual deflections and results in a larger reduction in (residual) capacity, even when subjected to the same fire exposure (thermal) conditions.

13.5 Conditional Assessment of Fire-Damaged Structures

Assessment of fire-induced damage in a structure can be carried out through a combination of visual inspection, nondestructive testing, in situ testing, and application of simplified or advanced analysis approaches. Following a fire incident, the first step in assessing damage is to broadly classify the extent of

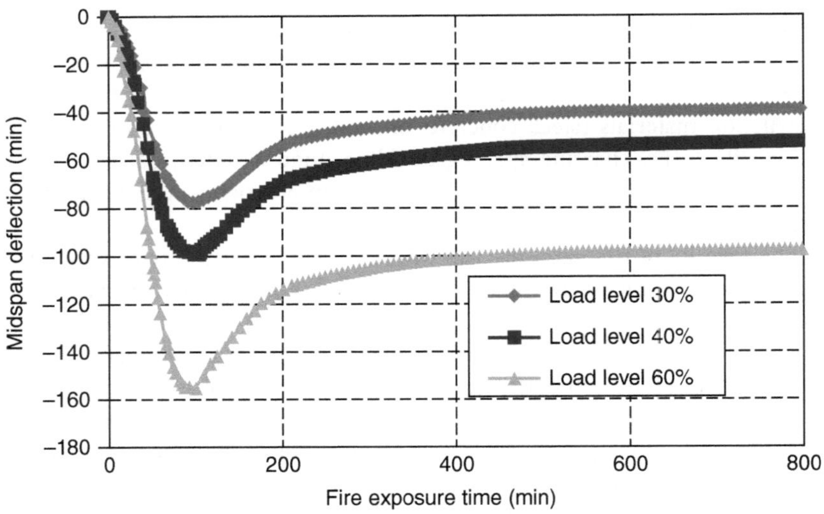

(a) Midspan deflection during heating and cooling phases of fire

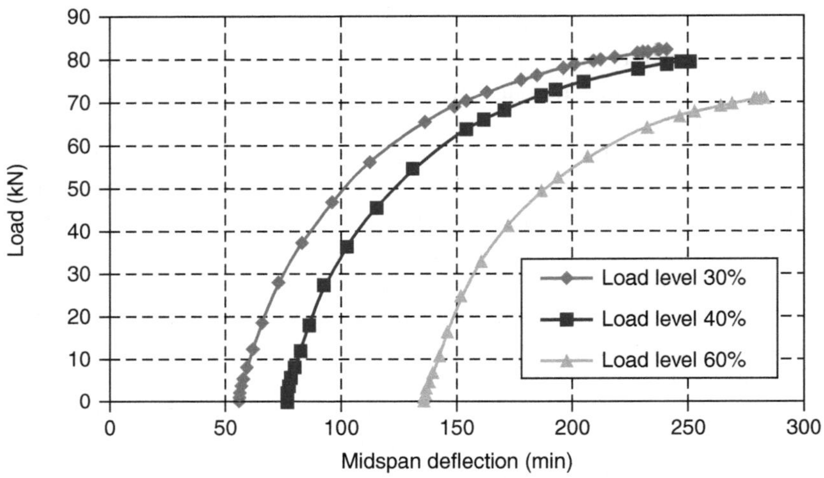

(b) Residual load-deflection response following cool down

FIGURE 13.6 Effect of load level on the response of an RC beam during and following fire exposure.

damage in a structure. Such damage classification helps in applying the right type of technique for assessing residual capacity of fire-damaged structures.

13.5.1 Damage Classification

Temperature-induced damage following fire exposure can be nonstructural (cosmetic or surficial) or structural. Nonstructural damage includes discoloration and surface changes (smoke, texture, appearance) resulting from smoke, soot, and so on, and there is no loss of structural capacity in this scenario. Structural damage, on the other hand, implies some level of reduction in capacity of a structural member resulting from temperature-induced degradation in mechanical properties of materials, as well as the loss of cross section or other associated factors (e.g., local buckling).

FiB (2007a) and a few other standards and guides provide limited recommendations for characterizing damage in fire-exposed structural members. According to FiB, the damage to a structure following fire can be grouped under five classes as listed in Table 13.3. These five classes represent cosmic (nonstructural) damage (Class 1), technical (nonstructural) damage (Class 2), surface (structural) damage (Class 3), sectional (structural) damage (Class 4), and member (structural) damage (Class 5). Visual inspection (assessment) techniques can be utilized as the first step to group the structural members affected in a fire to a damage class ranging from Class 1 to 5. This classification makes it possible to develop different repair strategies for the rehabilitation of fire-damaged structures. The description of damage together with the

Class	Characterization	Damage Description	Extent of Residual Capacity Loss	Residual Capacity Assessment
1	Cosmetic damage, surface	Characterized by soot deposits and discoloration. In most cases soot and color can be washed off. Uneven distribution of soot deposits may occur. Permanent discoloration of high-quality surfaces may cause their replacement. Odors are included in the class (they cannot be removed fully, but chemicals are available for their removal to some extent)	No loss of capacity	No need for assessment
2	Technical damage, surface	Characterized by damage on surface treatments and coatings. Limited extent of concrete spalling or corrosion of unprotected metals. Painted surfaces can be repaired. Plastic-coated surfaces need replacement or protection. Minor damages due to spalling may be left in place or may be replastered	None to very minimal loss of capacity	Simplified analysis
3	Structural damage, surface	Characterized by some concrete cracking and spalling, lightly charred timber surfaces, some deformation of metal surfaces, or moderate corrosion. This type of damage includes also Class 2 damages and can be repaired in similar ways	Minimal loss of capacity	
4	Structural damage, cross section	Characterized by major concrete cracking and spalling in the web of I-beams, deformed flanges, and partly charred cross sections in timber members, degraded plastics. Damages can be often repaired in the existing structure. Within the class are also (a) the large structural deformations that reduce the load-bearing capacity, and (b) the large dimensional alterations that prevent the proper fitting of the different substructures and systems into the building. This applies in particular to metallic constructions	Some (moderate) level of capacity loss	Simplified/ advanced analysis
5	Structural damage components	Characterized by severe damages to structural members and components, with local failures in the materials and large deformations. Concrete constructions are characterized by extensive spalling, exposed reinforcement, and damaged compression zones. In steel structures, extensive permanent deformations are due to diminished load-bearing capacity caused by high temperature. Timber structures may have almost fully charred cross sections. Mechanical decay in materials may occur as a consequence of the fire. Class 5 damages usually will cause the dismissal of the structure	Significant loss of capacity	Advanced analysis

Table 13.3 Classification of Damage in Fire-Exposed Structures as per FiB Recommendations (*Adapted from FiB, 2007a*)

extent of residual capacity lost and the type of analysis needed for residual capacity assessment is listed in Table 13.3.

Structural members grouped under Class 1 undergo very minor changes, such as very limited to no cracking, spalling, deflection rise. There may be only blackening of the surface of the member resulting from smoke, and this damage is grouped under nonstructural damage. In such cases, any further detailed assessment of the structure is generally not required, as the structure is likely to retain all of its original capacity. The structure can be reinstated to its original state with the undertaking of nonstructural (cosmetic) repairs such as smoke cleaning and repainting.

Structural damage due to fire exposure is said to be minimal to low, in the case of structural members, grouped under damage Classes 2 and 3. In these cases, structural members undergo only some level of cracking, minor spalling, discoloration of concrete, signs of limited local buckling, and a low level of deformations during fire exposure. In the case of Class 2 damage, any significant loss of load-bearing capacity in the structural member is unlikely. However, under Class 3, if temperature-induced structural damage is evident from visual inspection, the extent of loss of residual capacity needs to be further assessed. Residual capacity of a structural member in such a scenario (Class 3) can be reasonably assessed by applying simplified (empirical) analysis methods combined with the testing of coupons taken from fire-damaged structural members. Severe structural damage (damage Classes 4 and 5) in fire-exposed concrete members is said to have occurred in the case of a structural member undergoing large cracking in concrete members, moderate to high levels of concrete spalling with rebars partially or fully exposed in concrete members, high levels of local buckling or bulging in the case of steel or composite members, or high levels of charring in wood members combined with noticeable residual deformations (or deflections) in the member (concrete, or steel, or wood). In such cases, some level of loss of capacity is likely and advanced calculation methods need to be applied for evaluating the realistic load-bearing capacity of structural members. These are some well-established techniques and rules of thumb in the literature that can aid in the conditional assessment of fire-damaged structures. These rules and techniques are different for different material types and are covered under the concrete, steel, and wood subsections below.

13.5.2 Concrete and Masonry Structures

Visual observation of physiological changes in concrete and masonry structures can provide an initial estimate of damage following a fire exposure. Changes in concrete (surface) color, thermal cracking (crazing), extent of spalling, and rebar conditions provide insight into the extent of temperature-induced damage in a concrete or masonry structure, and these crucial observations can be used for damage classification as per the scheme listed in Table 13.3. This visual classification scheme is applicable to both axial and flexurally loaded concrete members.

Concrete undergoes distinct changes in color with heating as a function of temperature. This color change can be attributed to distinct chemical and physiological changes in cement, fine aggregate, and coarse aggregate. It should be noted that these color changes tend to be somewhat similar for both siliceous and calcareous aggregate concretes, with the only difference being that the development of a red (pink) color at different temperatures is not as prominent in calcareous aggregates as it is dependent on the presence of certain ferric compounds. This change in color is permanent (upon cooldown) and can be utilized to estimate the peak temperature reached in a fire-exposed concrete member. The color of siliceous aggregate concrete (mostly containing quartzite or granite) changes to pink or red in the 300 to 600°C range, whitish gray in the 600 to 900°C range, and buff in the 900 to 1000°C range, as illustrated in Fig. 13.8 (Hager, 2014). The pink/red discoloration results from the presence of iron compounds, in the fine or coarse aggregate, which dehydrate or oxidize in this temperature range. The mineralogical changes mortar and cement paste undergo at different temperature levels are also similar to that of concrete, as can be seen in Fig. 13.8. In the case of high-performance concrete (HPC), the presence of a high amount of silica fume has some influence on color changes, but may not be noticeable through the naked eye. Also, due to a lower amount (or lack) of aggregates in HPC, as well as a lower sand (quartz) content, the pink color may not be as obvious in fire-exposed HPC. It should be noted that while similar color graphs (as in Fig. 13.8) are not available for carbonate aggregate concretes, other tools such as graphs showing variation in hue levels at different temperature ranges can be used to identify peak temperature experienced in a carbonate or other aggregate concrete types (Short et al., 2001). In addition to the coloring scheme shown in Figs. 13.7 and 13.8, mineralogical and strength changes in concrete resulting at various temperatures are summarized in Table 13.4 and can be highly useful in gauging the extent of strength loss in conventional concrete types.

The above color change patterns can be used to correlate and get an estimate of the peak temperature experienced by concrete, which in turn can give an idea of the strength loss in concrete due to exposure to certain peak temperature (Hager, 2014; Short et al., 2001). In all aggregate types, the percentage loss of

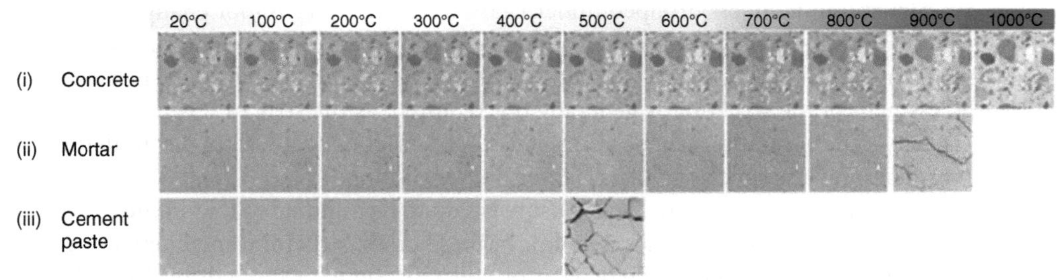

FIGURE 13.7 Color change in conventional siliceous aggregate concrete, heated to temperatures ranging from 100 to 1000°C. (*Hager, 2014.*)

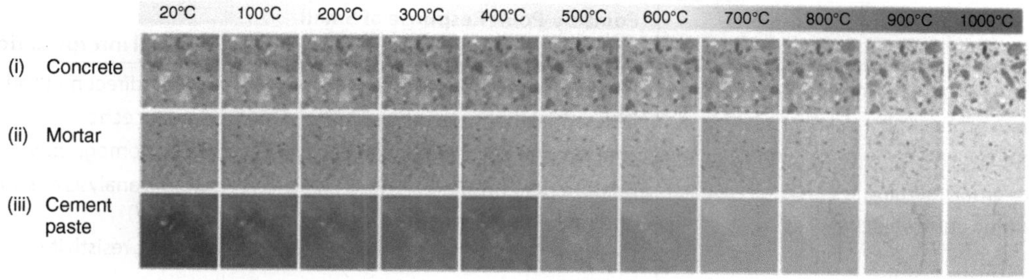

Figure 13.8 Color change in high-performance siliceous aggregate concrete, heated to temperatures ranging from 100 to 1000°C. (*Hager, 2014.*)

Heating Temperature	Changes Caused by Heating	
	Mineralogical Changes	Extent of Strength Loss
70–80°C	Dissociation of ettringite, causing its depletion in the cement matrix	None
105°C	Loss of physically bound water in aggregate and cement matrix commences, causing an increase in the capillary porosity and minor microcracking	None
120–163°C	Decomposition of gypsum, causing its depletion in the cement matrix	None
250–300°C	Pink/red discoloration of aggregate caused by oxidation of iron compounds commences at approximately 300°C. Loss of bound water in cement matrix and associated degradation becomes more prominent	Little (below 10%)
450–500°C	Dehydroxylation of portlandite, causing its depletion in the cement matrix Red discoloration of aggregate may deepen in color up to 600°C Flint aggregate calcines at 250–450°C and will eventually (often at higher temperatures) change color to white/gray. Normally isotropic cement matrix exhibits patchy yellow/beige color in cross-polarized-light, often completely birefringent by 500°C	Marginal (20%–30%)
573°C	Transition of α to β quartz, accompanied by an instantaneous increase in volume of quartz of about 5% in a radial cracking pattern around the quartz grains in the aggregate	Moderate (30%–40%)
600–800°C	Decarbonation of carbonates; depending on the content of carbonates in the concrete, e.g., if the aggregate used is calcareous, this may cause a considerable contraction of the concrete due to release of carbon dioxide; the volume contraction will cause severe microcracking of the cement matrix	Significant; more than 50% strength loss at 600°C and reaching to 80% at 800°C
800–1200°C	Complete disintegration of calcareous constituents of the aggregate and cement matrix due to both dissociation and extreme thermal stress, causing a whitish gray coloration of the concrete and severe microcracking. Limestone aggregate particles become white	All/most of the strength lost around 1000°C
1200°C	—	Concrete starts to melt, i.e., pulverizes (becomes powder)
1300–1400°C	—	Completely melted

Table 13.4 Mineralogical and Strength Changes to Concrete Caused by Heating (*Adapted from Gosain et al., 2008*)

strength was found to be independent of the original level of strength but influenced by the sequence of heating and loading (Hager, 2014).

The above-listed visual observations can be supported by testing coupons taken from a fire-damaged structure to get an indirect indication of the condition of the concrete, especially when the level of damage in the member is moderate to high (Classes 3 to 5). These tests include strength, rebound hammer,

Average Response of the Concrete Cover	Point-By-Point Response of Small Samples	Special Interpretation Techniques
Schmidt rebound hammer Windsor probe Capo test BRE internal fracture Ultrasonic pulse velocity	Small-scale mechanical tests Differential thermal analysis (DTA) Thermo-gravimetric analysis (TGA) Dilatometry (TMA) Thermoluminescence Porosimetry Colorimetry Microcrack-density analysis Chemical analysis	UPV indirect method Impact echo Sonic tomography Modal analysis of surface waves (MASW) Electric resistivity

Table 13.5 Nondestructive Assessment Techniques of Fire-Damaged Concrete Structures

ultrasonic pulse velocity (UPV), Windsor probe, BRE internal fracture, and thermoluminescence. While the tests on core concrete or steel coupons are valuable aids in assessment, they do not give a complete picture of the extent of deterioration. In fact, assessing the residual capacity of concrete structures exposed to fire is quite a difficult task, because the traditional destructive or nondestructive testing techniques are generally not suitable for strength assessment of such a highly heterogeneous layered material. The possible solutions to this problem generally involve a closer inspection of the average response of the concrete cover, a point-by-point analysis of small samples taken at different depths, or some special techniques aimed at interpreting the overall response of the fire-exposed concrete member. Table 13.5 lists some of the approaches recommended in FiB (2007b) for nondestructive assessment of fire-damaged concrete structures at different levels of damage.

When the damage classification is grouped under Class 3, 4, or 5 (as in Table 13.3), any idea of strength degradation in the damaged concrete and rebars is highly helpful to estimate the residual capacity of the member. The most direct method of estimating the strength of in situ concrete is by testing cores cut from the damaged structural member. Extracting cores from a fire-damaged structure is usually the first step for a number of test methods (colorimetry, chemical analysis, etc.). It should also be noted that cutting a core itself provides useful information on the damage depth, provided that a continuous recording of both the coring depth and the energy consumption is carried out.

The use of the ultrasonic pulse velocity technique for the assessment of thermally damaged concrete is well established due to the direct relation between pulse velocity and dynamic elastic modulus (Wang et al., 2013). This test requires a flat surface and is therefore appropriate only for the member surfaces with no spalling. The method has been found to be particularly suitable for walls and trough floors damaged due to a localized fire.

The Windsor probe test involves shooting a steel probe into the surface of the concrete (Malhotra and Carino, 2003). The measured length of the probe (left exposed) can be correlated with leftover compressive strength. This test has been found to be very quick and simple to get compressive strength and can be used even on surfaces subjected to some level of spalling, provided that these surfaces are reasonably flat. The BRE (Building Research Establishment) internal fracture test involves drilling a hole (6-mm diameter) in which a wedge anchor bolt with an expanding sleeve is placed 20 mm below the surface (Malhotra and Carino, 2003). The torque required to pull the anchor out gives an indication of strength from a calibration chart. It should be noted that this test may not provide as accurate results as those using the Windsor probe test.

Another technique, referred to as thermoluminescence testing, relies on the measurement of the residual thermoluminescence in small amounts of sand drilled from the concrete. Since a major loss of thermoluminescence occurs in the range of 300 to 500°C, that is, when concrete strength begins to degrade, this yields an estimate of the extent of strength loss and thus the degree of fire-induced damage in concrete. By quantifying the thermoluminescence, the temperature profiles experienced by the concrete layer can be determined from the heated surface. An advantage of this technique is that only small holes need to be drilled for sampling. However, performing this kind of test requires special equipment and requires a high level of expertise.

Since most of the thermally induced transformations in concrete are not reversible, reheating of the damaged concrete does not lead to any significant transformations until the maximum temperature experienced during the previous fire incident is exceeded. The evidence of an ongoing transformation during the second material heating in a laboratory furnace can be provided by thermodilatometric analysis (TMA), thermogravimetric analysis (TGA), and differential thermal analysis (DTA) measurements. Many of these analyses were carried out by Harmathy (1965).

Additional techniques, based on a point-by-point examination of small samples, can aid to assess the extent of defects in concrete, which can directly be related to material decay. One such technique is mercury-intrusion porosimetry, which is commonly used to evaluate the cumulative volume of pores within a wide range of sizes. Both the quantity and the average size of pores in concrete microstructure increase owing to high-temperature exposure. Modern digital image analysis is another technique that measures microcrack density (namely, the total length of cracks per unit surface).

It should be noted that many of the above-discussed techniques and approaches are applicable for evaluating the conditional assessment of masonry structures as well.

13.5.3 Steel Structures

When exposed to fire, steel structures experience thermally induced damage in the microstructure of steel that may alter its constituent properties. Assessing these altered properties requires a combined knowledge of metallurgical changes and structural response variations under high-temperature exposure.

Some of the above-discussed techniques and procedures such as nondestructive testing methods outlined for conditional assessment of concrete structures can also be applied to estimate fire damage in steel and composite structures. In addition, chemical composition analysis, metallographic observations, and testing of cores for evaluating physical properties (yield strength, ultimate strength, ductility, toughness, etc.) can be applied for estimating fire-induced damage in steel structures.

Well-documented visual observations on a steel structure following a fire incident can provide the most useful information concerning the peak fire temperature during heating and the corresponding peak temperature attained in steel. Peak fire temperature reached during a fire event can usually be estimated by examining the compartment contents surrounding a fire-damaged structure. For example, wood and paper ignite at 320°C, and most plastics melt or burn between 82 and 150°C. In the case of composite construction, changes in nearby concrete surface (color, cracking, etc.) may indicate the severity of fire experienced by steel members. For example, color changes and surface spalling can indicate the level of temperature rise in concrete members, as discussed in previous sections.

The condition and state of exposed steel surface can provide valuable insight into peak fire temperature reached during a fire incident. For example, a tightly adhering mill scale indicates a steel temperature considerably below 650°C. Structural steel members exposed to temperatures above 650°C will develop a coarse, eroded surface markedly different from the appearance produced by mill rolling (Dill, 1960). Although numerous types of coatings, markers, and paints are applied on structural steel, they usually are not designed to withstand elevated temperatures. At temperatures above 315°C, most coatings usually will be blistered, discolored, or even flaked off depending on the exposure to open flames. Any identifiable residue of coatings or paints is an indication that the steel has not experienced critical temperature (in excess of 500°C) during heating. A temperature rise above 500°C will cause an unrestrained member to increase in length or large forces to develop.

Tide (1998) rationalized that the above-described visual inspection points are important for establishing peak fire temperature experienced by steel during heating and also to get an idea of the residual capacity of the fire-damaged steel structure. However, assessing the geometry of the steel member after fire exposure is of foremost importance in order to evaluate the condition of the damaged steel member. For this purpose, Tide categorized fire-damaged steel members into three categories:

- *Category 1.* Straight members that appear unaffected by the fire.
- *Category 2.* Members noticeably deformed, as can be seen from local buckling, twisting, deflection, peeling of fire insulation, and so on, but could be heat-straightened if economically justified.
- *Category 3.* Members severely deformed, as measured through high levels of (local) buckling, large deflection (deformation), and so on, that only under extreme circumstances would repair be given any consideration.

During inspection of a whole steel structure, the condition of each fire-exposed individual member needs to be determined, and only after that, the safety of the whole structure (e.g., frame) can be established. A member inventory should be performed before any assessment for repair or replacement can begin.

Further, camber and sweep of each fire-exposed structural steel member should be determined using appropriate measurement techniques (plumb bob, stringline, laser device, etc.) (Tide, 1998).

In the case of slightly deformed Category 1 members—that is, those with deformations greater than rolling tolerances—such members need to be further analyzed to determine the needed repair level. Depending on individual circumstances, the analysis will determine if these members can be accepted unconditionally (as is), heat-straightened, stabilized with supplemental braces, or reinforced with supplementary plates and shapes. Category 2 steel members may require additional attention because a more critical decision (to repair or to replace) is to be made, and this is often a function of the nearby (adjoining) members' condition. In general, a steel beam is much easier to replace than a fire-damaged column supporting several floors. If a steel member, classified as Category 2, is heat-straightened, the change in metallurgical and structural properties is often minor. The decision of repairing or replacing Category 2 members is usually dependent on expediency, economics, or overcoming the human psychological rejection of what appears to be damaged steel (Tide, 1998).

In most cases, Category 3 members are often directly rejected since these members will have very little residual capacity left. In fact, salvaging a Category 3 member is only likely in a rare or critical situation where the removal of the damaged member is inappropriate or impossible. In such scenarios, Category 3 members are first heat-treated, repaired, and then strengthened as required.

Due to the fact that the connection behavior under elevated temperature exposure is different than the behavior of the main steel member (because of their relative compactness), inspecting connections is imperative in horizontal and vertical restrained members. The axial force developed in a restrained member can impose large restrained forces on end connections. Under such conditions, a steel beam (or column) may buckle or deform to accommodate the newly developed fire-induced forces. Under these conditions, connection distress can be easily identified. For example, when a fire-damaged Category 1 steel beam cools, if the connection has fractured, then the steel beam will pull away from the adjacent member, which usually reveals the extent of damage in the connection. In other cases, as the buckled beam cools and shortens, the connection materials, bolts or welds, would have fractured and thus indicate the need for immediate repair.

Other than visual inspection, there are also several nondestructive test procedures applicable to examine fire-exposed structural steel. One of the most common techniques is surface hardness testing. This test can be utilized to determine the occurrence of transformation in steel due to a rapid heating and accelerated cooling cycle. Measurements from this test provide an empirical determination of the approximate tensile strength of steel. Hardness readings on fire-exposed steel members indicate the strength loss based on estimated metallurgical degradation.

In destructive testing, a sample (steel coupon) is cut from the fire-damaged steel member and tested in the laboratory in order to determine specific physical property and residual stress information. Structural steel exposed to temperatures below 400°C during heating generally experiences a minor reduction of yield strength, tensile strength, and ductility. If the exposed peak temperature is higher than 400°C, the reduction level depends on temperature-induced changes in steel microstructure, original mill rolling conditions, and actual duration and range of temperature exposure. When sampling and testing for yield strength of steel, one must recognize that yield strength varies within the member cross section and along its length. Charpy V-notch (CVN) testing of removed samples is to be carried out when fatigue loading or a brittle fracture potential exists.

13.5.4 Timber Structures

Large-size (cross-section) wood members under short fire exposure may retain much of their capacity. In such cases, the residual capacity can be closely estimated by knowing the initial strength properties of the uncharred wood and the residual (uncharred) cross section of the member. The effective cross section of a fire-exposed wood member can be determined by measuring the extent of char depth—that is, by scrapping of charred layers.

Nondestructive evaluation (NDE) methods can also be applied to evaluate the post-fire residual capacity of timber structures. While extensive work has been done on the use of NDE techniques for assessing the residual capacity of wood affected by biological degradation, only limited NDE work has been done on wood exposed to elevated temperatures. Some of the NDE techniques that can be used are visual and density determination, screw withdrawal, microindentations and penetration, resistance to microdrilling technology, sound/stress wave, ultrasonic pulsing, chemical analysis, and near infrared testing (Malhotra and Carino, 2003).

One detrimental outcome of fire-induced thermal degradation in wood is the reduction in the density of fire-exposed wood when the heating temperature is in the range of 200 to 350°C. Taking into account both mass and volumetric changes, the change in density can equate up to 60% of original density at a temperature of 340°C. Thus, one option for NDE evaluation is to measure the density itself. A core sample from a fire-exposed wood member can be removed to examine the density gradient. It should be noted that the strength properties of wood following elevated temperature exposure are much more affected than the stiffness properties (Kretschmann, 2010).

13.6 Calculation Approaches for Evaluating Residual Capacity

When there are indications of structural damage, based on a conditional assessment as described in Table 13.4, the residual capacity of such structural members must be asserted before the structure is opened for occupancy. Although it is preferable to test cores or samples from the fire-damaged structure to arrive at the exact load-carrying capacity, this may not always be possible due to time constraints, as well as from a practical consideration. In such scenarios, calculation approaches are to be applied to evaluate post-fire residual capacity of structural members.

General Procedure

The extent of strength and stiffness degradation in a structural member following fire exposure is dependent on a number of factors, including fire severity, reduction in properties of constituent materials, load level, and boundary (support) conditions. Many of these factors are interdependent and can vary significantly in different scenarios. Thus, a residual capacity evaluation of fire-exposed members is quite complex and the computed capacity can vary widely depending on the assumptions made and input properties used in the analysis.

The calculation approaches for evaluating the residual capacity of structural members can be undertaken at two levels. The first level is based on simplified sectional analysis, wherein strength equations similar to that for room temperature capacity evaluation are utilized, with due consideration to strength reduction factors for material properties (resulting from high-temperature exposure), in calculating residual capacity. The strength reduction factors for materials are calculated based on peak temperature experienced by the material within the cross section of a member under fire exposure. The second level of analysis is through advanced analysis wherein the detailed residual response of a fire-exposed structure is evaluated with due consideration to all critical factors governing the response of the member in the entire range of fire exposure including the cooldown phase.

13.6.1 Simplified Approaches

When the fire-induced damage in a structural member is minor to moderate (damage Class 2 or 3 as per Table 13.3), simplified sectional analysis approaches can be applied to evaluate approximate residual capacity. In such simplified approaches, residual capacity of a member is evaluated by applying room temperature strength equations, but taking into account strength degradation occurring in materials due to elevated temperature exposure. In a few studies, such simplified approaches have been validated by comparing the calculated residual capacity prediction with the capacity measured through laboratory testing, and a good correlation was seen (Kodur et al., 2010). The application of a simplified approach for evaluating residual sectional capacity of fire-exposed concrete members is illustrated here. The same procedure can be applied for other types of structural members made of wood, steel, and so on.

The simplified approaches involve estimating peak fire temperature reached during a fire incident. The peak fire temperature can be estimated through reviewing first responders' fire incident reports, from visual observations, or by applying other indirect methods such as fire growth calculations (through the use of relevant information on compartment characteristics) as per the guidance provided in Sec. 13.5. As an example, peak fire temperature can be estimated by examining the color changes and/or microstructural changes on the exposed surfaces through a thermostructural analysis of concrete. Guidance is available in the literature to link concrete color to fire temperature (see Fig. 13.8). So if the concrete color is pink or red, it is an indication that peak concrete temperature reached during heating is in the range of 300 to 600°C. While the exact color of concrete is dependent on several factors (e.g., aggregate mineralogy), the color transformation of concrete can provide a rough correlation to the maximum temperature reached during a fire event. The temperature on the concrete surface, as indicated by color, can be assumed to be approximately 90% of the maximum temperature of the fire (in the compartment). Thus, an estimate of temperature reached in fire can be back-calculated (Kodur et al., 2013). Knowing the peak fire temperature or peak surface temperature, temperature at various depths within concrete members (including rebars) can be calculated by applying Wickström's equation (Wickström, 1986) or by the 500°C isotherm method as in Eurocode 2 (CEN, 2004), which were discussed in Chap. 8.

As an illustration, the peak temperature (T_{smax}) in a concrete layer of rebar of a concrete member subjected to a peak fire temperature of T_{fmax} can be evaluated using the following equation (Kodur et al., 2010):

$$T_{smax} = \lambda T_{fmax} \tag{13.12}$$

where T_{smax} is the maximum rebar (or concrete layer) temperature, T_{fmax} is the maximum fire temperature, and λ is a modification factor to account for specific fire and section (and materials) characteristics. λ depends on the axis distance (distance from the center of the rebar to the nearest exposed surface or to a specific concrete layer depth), the depth-to-width ratio of the concrete section, and the fire duration. λ can be calculated as per the equation below (Kodur et al., 2010):

$$\lambda = 1.45\left(t_h + \frac{t_c}{2}\right)^{0.2}\left(0.4 + 0.03\frac{H}{B}\right) - 5x \tag{13.13}$$

where t_h is the duration of the heating phase (hours), t_c is the duration of the cooling phase (hours), H is the maximum rebar temperature, B is the section width, and x is the axis distance to rebar (m). This equation is capable of predicting the peak rebar (or concrete layer) temperature of RC members with a reasonable level of accuracy. In lieu of the above equation, detailed thermal analysis, through finite element- or finite difference-based models, can be carried out to arrive at sectional temperatures in the member.

With the maximum temperature reached in the rebar or various concrete depths known, the residual strength in concrete and steel can be arrived at using residual strength-temperature relations (as per the discussion in Sec. 13.7). Then residual capacity of a fire-exposed concrete beam can be computed by modifying room temperature capacity equations specified in applicable design codes to account for residual properties of reinforcing steel and concrete. For example, the residual flexural capacity of an RC beam can be evaluated as

$$M_n = A_s f_{yT}\left(d^* - \frac{A_s f_{y,r,T}}{1.7b^* f'_{cT}}\right) \tag{13.14}$$

where M_n is the residual flexural capacity of the fire-exposed RC beam, A_s is the area of tension reinforcement (steel), $f_{y,r,T}$ is the residual strength of reinforcing steel, d^* is the effective depth of the fire-damaged concrete section, b^* is the width of the fire-damaged concrete section, and f'_{cT} is the reduced compressive strength of concrete. If concrete strength varies over the depth of the compression block, a layered approach can be applied to evaluate the reduced compressive strength of concrete at each layer and then summing up the strength of various layers to get the resultant residual compressive force.

The above approach was shown to give conservative predictions of residual capacity in fire-damaged RC beams (Kodur et al., 2010). However, this method assumes that concrete cover thickness is not severely reduced under fire conditions (due to the occurrence of fire-induced spalling). In cases where a significant portion of the concrete cover spalls off, leaving rebars partially or fully exposed to fire, the residual flexural capacity computed through this sectional approach may not be conservative.

In the case of fire-exposed concrete columns (compression member) wherein cover on concrete to rebars is fully spalled off, the concrete in the cover region is to be ignored in the calculation of residual capacity. Furthermore, any buckling in rebars arising from spalling or excessive deformation needs to be taken into consideration. To account for such local buckling, a buckling strength reduction coefficient, k_{st}, is to be applied. Jiang et al. (2007) proposed the following relation to calculate the buckling reduction coefficient k_{st} as

$$k_{st} = \left(1.136 - 111.6\frac{e}{s}\right) \times \left(0.972 + 0.004\frac{s}{d_s} - 0.003\left(\frac{s}{d_s}\right)^2\right) \tag{13.15}$$

where e is the initial bending displacement of steel rebar, s is the spacing of stirrups, and d_s is the diameter of steel rebar.

If, however, there is partial or no loss of concrete cover, the residual strength of the column can be calculated as a summation of contribution from the strength of rebars and concrete layers within the concrete core (beneath the rebar cage), but with a reduced strength of concrete and steel resulting from higher temperature exposure. Accordingly, the residual capacity of a fire-exposed RC column is given as

$$N_r = A_{\text{core}} f_{ct} + k_{st} A_s f_{yt} \tag{13.16}$$

where N_r is the residual strength of a fire-exposed RC column, A_{core} is the area of "core region with temperature less than 500°C," A_s is the area of steel rebars, k_{st} is the reduction coefficient due to buckling (only used when cover concrete to rebars spalls off), f_{ct} is the residual strength of concrete, and f_{yt} is the residual strength of steel rebar.

The above-listed empirical equations (Eqs. 13.5 and 13.6) have a number of limitations and can only be used for concentrically loaded non-slender columns with a square cross section, as these equations do not take into account the interaction of bending and compression. Also, the above approach is applicable only when the spalling in columns does not reach beyond the reinforcement cage. In the case of severe spalling conditions (beyond the reinforcement cage), advanced analysis should be applied to evaluate residual load-bearing capacity in fire-exposed columns.

13.6.2 Advanced Analysis

There are a number of limitations with the above-discussed simplified approaches. Further, due to large approximations, the calculated residual capacity may not be accurate. These limitations can be overcome through the application of advanced analysis. Recently, a three-stage calculation procedure to evaluate the detailed residual response of fire-damaged structures was proposed by various researchers (Agrawal and Kodur, 2019; Aziz and Kodur, 2013). The three-stage analysis for evaluating the residual response of a structural member comprises Stage 1—prior to fire exposure, Stage 2—during fire exposure (involving heating and cooling phase of fire), and Stage 3—post-fire exposure (after complete cooldown of the member). The response in each stage can be traced by applying any finite element-based software package, such as ANSYS or ABAQUS. The three-stage procedure for residual capacity evaluation is illustrated in a flowchart in Fig. 13.9.

As part of Stage 1 of the advanced analysis approach, the room temperature capacity of the fire-damaged structural member is to be evaluated through a detailed finite element analysis of the member. In the analysis, the load is gradually incremented in steps on the structural member until failure occurs. This room temperature capacity reflects the actual capacity of the member (not the design capacity) prior to fire exposure. This realistic capacity of the structure prior to fire exposure can be useful for comparison with the residual capacity of the member to be evaluated in Stage 3 to ascertain the extent of capacity degradation from fire damage in the member.

In Stage 2 of the analysis, the response of the structural member is evaluated under a specified fire exposure scenario, load level, and restraint conditions as present (or closely possible) during a fire event. The fire resistance analysis is carried out using finite element-based programs (such as ABAQUS, ANSYS, etc.) at various time increments until the failure occurs in the structural member under fire exposure, or until the fire-exposed member cools down to ambient conditions. The procedure for undertaking Stage 2 of the analysis is exactly the same as that laid out in Chaps. 7 to 11. Relevant fire exposure scenarios and other structural parameters (such as load, restraint, etc.) form the input to the thermal and structural components of the analysis. At the end of each time increment, response parameters generated from thermal and structural analyses are utilized to check the state (failure) of the structural member under different failure limit states. In this stage, temperature-dependent thermal and mechanical properties of construction materials that are distinct during heating and cooling phases of fire exposure can be input into a finite element analysis for evaluating realistic response.

Following the cooling down of the structural member to room temperature, and if there is no failure of the structural member in Stage 2, residual capacity analysis (Stage 3) should be carried out. The temperature-induced residual stresses and strains that exist in the structural member after fire exposure, as obtained from Stage 2 of the analysis in the form of residual deformations, form the initial condition that should be imposed on the member during Stage 3 of the analysis. These residual deformations result from accumulated damage in the structural member due to temperature-induced material degradation and plastic strains. This state of the structural member is the initial state for Stage 3 of the analysis. For this analysis, residual

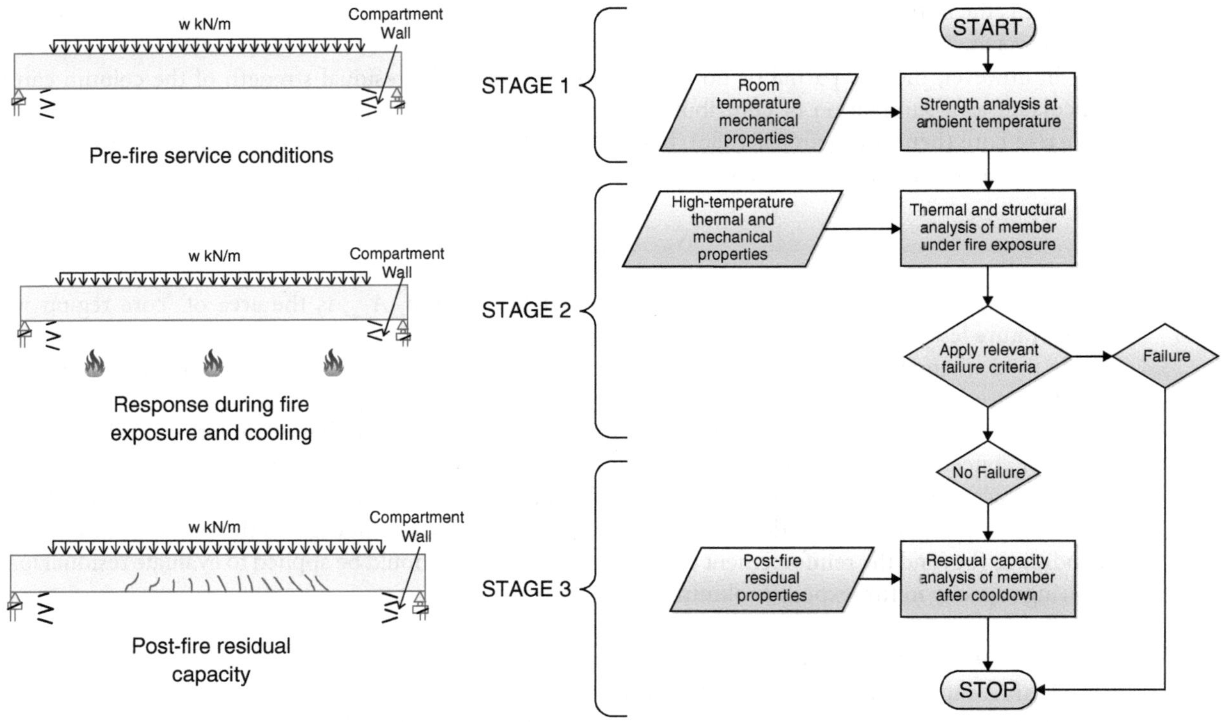

Figure 13.9 Flowchart illustrating the three stages involved in residual capacity analysis. (*With permission from Agrawal, 2020.*)

material properties of the fire-damaged structural member (as laid out in Sec. 13.3) should be input. As part of Stage 3 analysis, the fire-damaged structural member is loaded incrementally (as in Stage 1 analysis) until failure, and in this way the load-deflection response of the structural member is traced. The peak value on the load-deflection curve represents the residual capacity in the fire-damaged structural member. Through this three-stage approach, a realistic assessment of the extent of fire damage induced in a structural member and its residual capacity can be made.

In evaluating residual capacity, due consideration should be given to different failure criteria at each stage of the analysis. For example, in Stage 1 of the analysis (at ambient conditions), strength limit state generally governs the failure, and the peak load (or moment) at which failure occurs represents the realistic initial capacity of the structural member. In Stage 2, under fire exposure, the temperature-induced buckling (of the web) might dominate the failure limit state in steel structural members due to higher slenderness of the web as compared to the flanges. Also, significant deflections that develop under fire conditions can lead to a high level of fire-induced forces at the connections when such girders are restrained from thermal expansion. These aspects (web buckling in steel, spalling in concrete, etc.) should be considered in evaluating failure. Therefore, stability limit state criteria, in addition to strength criterion, should be considered in evaluating the failure in Stage 2 of the analysis. In Stage 3, the strength (or deflection) limit state generally governs the failure and should be applied to evaluate the residual capacity of the fire-exposed members. The above procedure can yield detailed residual response, including load, deflection, and so on of a structure throughout fire exposure and following cooldown of the member.

13.6.3 Illustration of Advanced Analysis

Recently, a number of researchers have applied advanced analysis approaches for assessing the capacity of fire-damaged concrete and steel structures (Wang et al., 2008; Aziz and Kodur, 2013; Kodur and Agrawal, 2016). The application of the advanced analysis approach for a residual capacity assessment of a fire-damaged reinforced concrete beam is illustrated here.

A three-stage approach for evaluating the residual response of fire-damaged reinforced concrete beams was applied by Kodur and Agrawal (2016). The analysis was carried out through a finite element model developed in ABAQUS as per the steps listed in Fig. 13.9. Full details of the analysis, including assumptions made, discretization of the beam, and validation of the approach, can be found elsewhere (Kodur and Agrawal, 2016).

Beams Analyzed

Two identical concrete beams, designated as beams B1 and B2 (shown in Fig. 13.10), are analyzed to illustrate a residual capacity evaluation following fire exposure composed of distinct heating and cooling phases. Key response parameters including deflections, capacity, and sectional temperatures (in Stage 2 only) were calculated during each stage of analysis to monitor the response of each beam.

Analysis Details

The constitutive material models for concrete and steel are defined within the framework of ABAQUS software. Two separate submodels, namely, structural and thermal models, are developed in ABAQUS to carry out the three stages of analysis. A structural model is needed to carry out strength analysis in Stages 1, 2, and 3, while a thermal model is needed for heat transfer (thermal model) calculations in Stage 2 only (to compute sectional temperatures in the structural member). The Stage 2 analysis of a fire-exposed RC beam is undertaken using the sequentially coupled thermomechanical calculation procedure. In this procedure, sectional temperatures generated in the heat transfer analysis are applied for subsequent structural analysis, but no reverse dependency exists.

In the case of a thermal model in a 3D space, concrete was discretized using DC3D8 elements (eight-noded linear brick element), while steel reinforcement was discretized using DC1D2 elements (two-noded link element) available in the ABAQUS library. The DC1D2 elements have nodal temperature as the only active degree of freedom. In the case of a structural model, the concrete beam is discretized using eight-noded continuum elements with reduced integration (C3D8R) and two-noded link elements (T3D2) for reinforcing steel.

Input Parameters for Analysis

Various parameters such as beam geometry, load distribution, boundary conditions, fire scenario, and material properties should be input to carry out the different stages of analysis. Details of the beams and the time-temperature curves of fire exposure are presented schematically in Fig. 13.10.

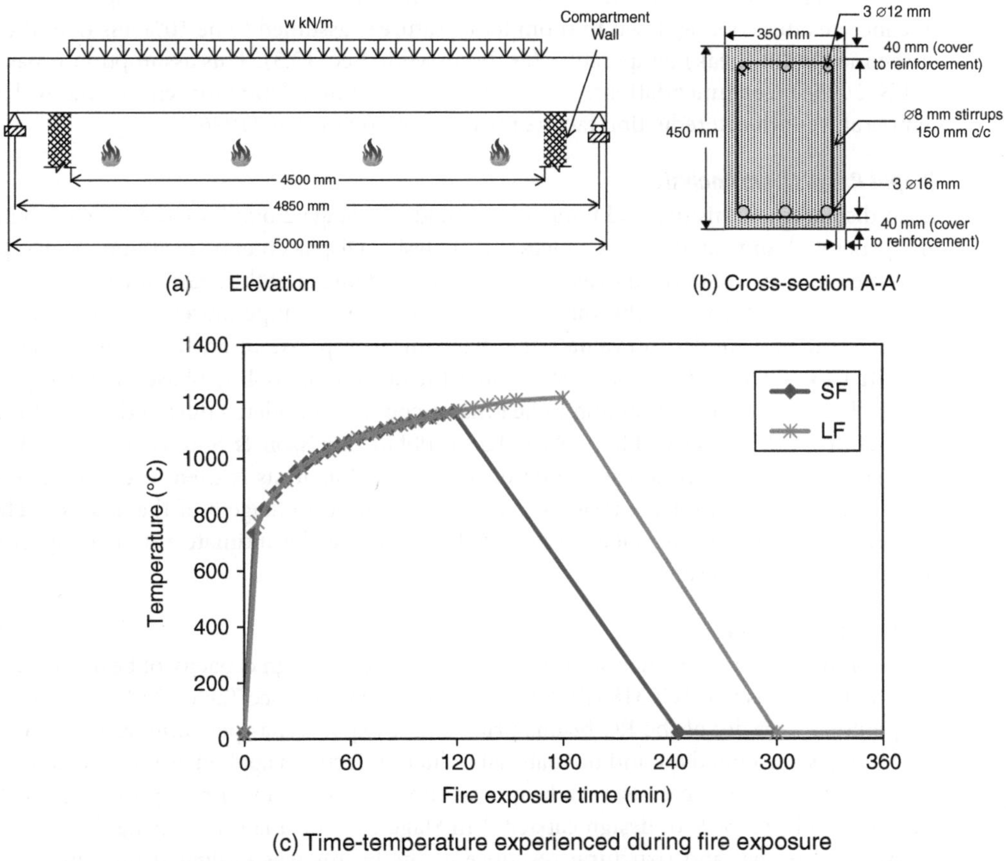

Figure 13.10 Dimensions, reinforcement details, loading, and fire scenarios assumed for a case study on RC beams.

Beam Designation	Fire Exposure	ACI Design Capacity at Room Temperature (kN·m)	Load Level Present During Fire Exposure (%)	Predicted Residual Capacity (kN·m)
B1	SF*	191	50	189
B2	LF**	191	50	164

*Short fire scenario with 2-hour heating phase.
**Severe fire scenario with 3-hour heating phase.

Table 13.6 Summary of Analysis Parameters Adopted and Results from Case Study

Both beams are of identical cross-sectional dimensions and reinforcement details. The steel of the main reinforcing bars and stirrups is of yield strength 450 and 280 MPa, respectively. The average compressive (cylinder) strength of concrete in the beam (prior to fire exposure, as tested from the cores taken from non-fire-exposed sections) is taken to be 50 MPa. The beams are assumed to be simply supported in all three stages of analysis. A uniformly distributed load was assumed to be acting on each beam, generating a bending moment of about 50% of the beam capacity according to ACI 318 (ACI, 2008), and this represents the load (stress) level typically present in buildings during pre-fire exposure conditions (see Table 13.6). The fire exposure scenarios adopted to simulate fire in the compartment (where beams are located) are calculated as per Eurocode 1 recommendations (CEN, 2002) and composed of a distinct heating and cooling phase. Varying the fuel load from 500 to 750 MJ/m^2 resulted in a heating phase duration of 2 hours for the Short Fire (SF) scenario and 3 hours for the Severe Fire (LF) scenario, respectively. Linear cooling (decay) of fire temperatures was adopted as per Eurocode 1 (CEN, 2002) recommendations to simulate cooling of the beams (see Fig. 13.10).

Besides these parameters, various material properties are to be input during different stages of analysis. For Stage 1 analysis, typical uniaxial stress-strain relations for concrete and reinforcing steel at room temperature (as per ACI 318 specifications) are adopted. In Stage 2 of analysis, during the heating phase, the temperature-dependent variations in thermal and mechanical properties of concrete and reinforcing steel are assumed to follow that of Eurocode 2 and 3 provisions (CEN, 2004, 2005a). In Stage 3 of the analysis, after cooling of the fire-exposed RC beam, the residual uniaxial compressive and tensile strength of concrete after cooling down to room temperature is assumed to be 10% less than the strength attained at the maximum (peak) temperature (as discussed in Sec. 13.3). This assumption is based on Eurocode 4 (CEN, 2005b) recommendations. The residual stress-strain relation for reinforcing steel is calculated using temperature-induced reduction factors provided by Neves et al. (1996).

Output Results from Analysis

Load displacement history and temperature fields (in Stage 2 only) are the primary output variables that are generated during different stages of analysis. Displacements resulting from applied loading at incremental steps are utilized to ascertain condition (failure) of the beam during Stage 1 of the analysis. In Stage 2, the output from the thermal analysis, namely nodal temperatures, is applied as a thermal body load on the structural model to evaluate the mechanical response of an RC beam under fire exposure. An identifier to ascertain if the material is under the heating or cooling phase, so as to apply the appropriate material properties corresponding to heating or cooling, is updated during the structural analysis using a subroutine, UFIELD, available in ABAQUS (Hibbitt, Karlsson & Sorensen, 2012). Also, the maximum temperature experienced at each node during thermal analysis is used to calculate residual mechanical properties to be used for the residual capacity evaluation in Stage 3 of the analysis. The load (moment) deflection response from Stage 3 is utilized to evaluate the ultimate residual capacity of fire-exposed reinforced concrete beams.

Results and Discussion

Based on the above set of analysis, the room temperature design capacity of beams B1 and B2 is calculated to be 191 kN·m as per ACI 318 (ACI 2008) design equations (see Table 13.6). However, the realistic room temperature capacity of the RC beams prior to fire exposure, as per finite element analysis, when strain hardening is accounted for and no material reduction factor is applied, was calculated to be 250 kN·m.

Following Stage 1 analysis, each beam was subjected to distinct fire exposure scenarios under constant load (stress) level (50% of design capacity) in Stage 2 of the analysis. During Stage 2 of the analysis (fire exposure), thermal and structural response of the beams was evaluated not only until the end of fire exposure but over an extending cooling period—specifically, the period from fire extinguishing time to

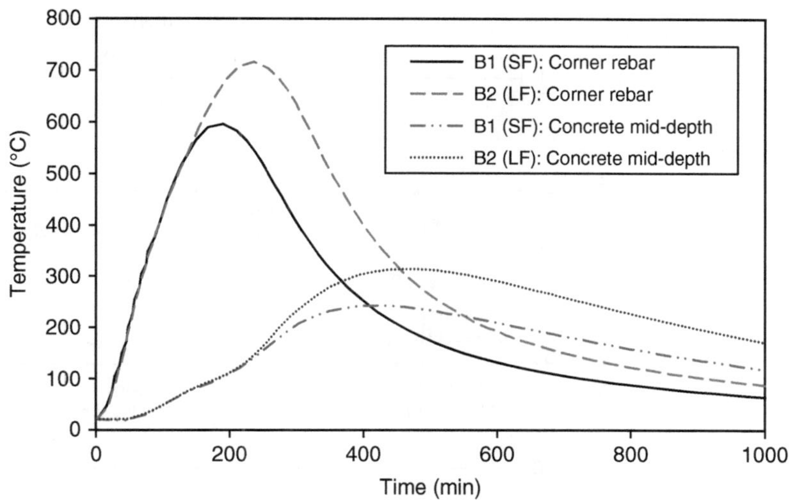

FIGURE 13.11 Predicted sectional temperatures in beams B1 and B2. (*With permission from Agrawal, 2020.*)

complete cooling down of the member, which included time taken for the entire cross section of the beam to revert back to room temperature. This was over 24 hours after the fire started.

The thermal response of the analyzed beams, B1 and B2, during fire exposure is shown in Fig. 13.11 by plotting rebar and concrete temperatures at different locations in the cross section, as a function of time. Unlike fire temperatures that rose rapidly in the first few minutes, the temperatures within the cross section of both beams started to rise only after 10 to 15 minutes into fire exposure, when fire (compartment) temperature was already in excess of 700°C. After that temperatures within the beam cross section of beams B1 and B2 rise monotonically during the heating phase, which lasted for 120 minutes and 180 minutes, respectively.

Furthermore, cross-sectional temperatures in both beams continue to increase even as fire temperatures decay, owing to the high thermal inertia of concrete. In fact, cross-sectional temperatures reach peak values during the cooling (decay) phase of fire exposure in both beams. Peak rebar temperatures in beams B1 and B2 are calculated to be 592 and 715°C at 170 and 240 minutes, respectively, well after the heating phase of fire exposure has ended. Similarly, peak temperatures at concrete mid-depth are calculated to be 240 and 315°C, respectively. Therefore, the relatively less severe fire, i.e., Short Fire (SF), exposure adopted for beam B1 leads to relatively lower cross-sectional temperatures as compared to those experienced in beam B2. Nonetheless, both beams do not fail during the heating or extended cooling phases after fire exposure.

The structural response of beams B1 and B2 during Stage 2 of analysis can be gauged through plotted midspan deflections as a function of time in Fig. 13.12. During the early stage of fire exposure, deflection rise is mainly governed by the level of applied loading and thermal gradients that develop within the beam cross section. As fire exposure progresses further, cross-sectional temperature within the section begins to rise and thermal gradients decrease along the depth. Midspan deflection in beams continues to increase, but at a relatively gradual rate, owing to the gradual reduction in mechanical properties, especially the elastic modulus of reinforcing steel.

As the rebar temperatures continue to increase steadily, peak midspan deflection reaches to about 80 and 120 mm in beams B1 and B2, respectively. The peak midspan deflection in beam B2 under more severe exposure increases significantly by almost 50% (see Fig. 13.12), and this is mainly due to the higher peak rebar temperature in beam B2 (700°C) as compared to beam B1 (600°C; see Fig. 13.11).

Both beams do not experience failure during the heating or cooling phases of fire exposure. Midspan deflections start to recover once the rebar and concrete temperatures revert to ambient temperatures due to the fire subsiding (cooldown). This can be primarily attributed to the recovery in strength and modulus properties in reinforcing steel, once the beam enters the extended cooling phase. The recovery in midspan deflection during the cooling phase of fire exposure is governed by cross-sectional temperatures, especially rebar temperatures.

Midspan deflections reach a steady state in both beams B1 and B2 as rebar temperatures cool down to below 150°C. However, noticeable residual deformations remain in the fire-exposed beams and the beams

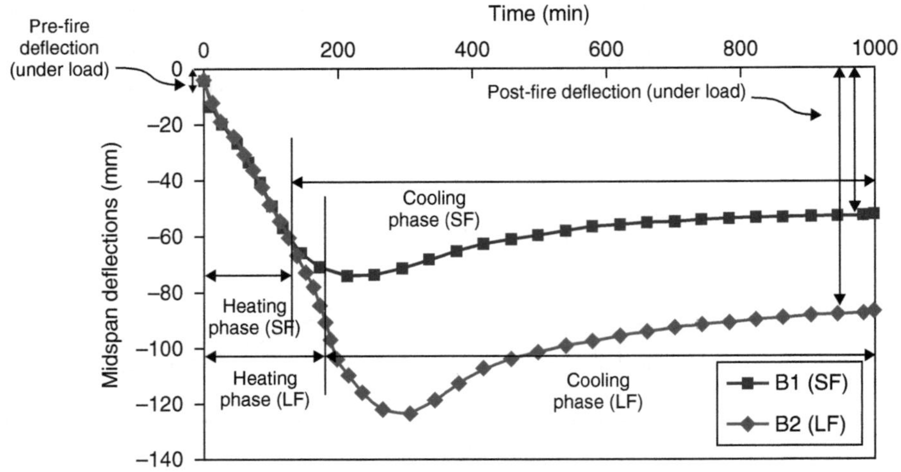

Figure 13.12 Predicted midspan deflections during fire exposure in beams B1 and B2. (*With permission from Agrawal, 2020.*)

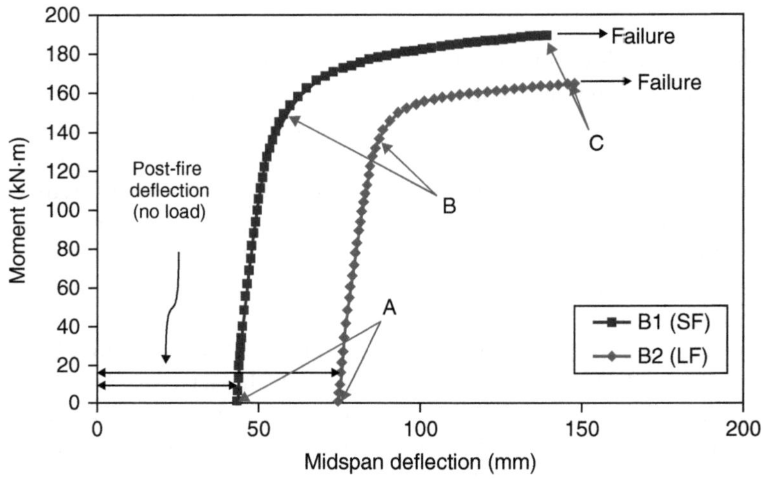

Figure 13.13 Predicted residual moment-deflection response for fire-damaged beams.

do not revert back to their pre-fire configuration (in terms of deflections) even after the applied loading is removed. This is primarily due to irreversible temperature-induced damage in concrete, which does not recover all of its strength and stiffness properties upon cooldown to ambient conditions, as well as residual plastic strains in steel reinforcement and concrete. Such residual deformations adversely affect future serviceability of the fire-damaged concrete structure, as these deformations can be significantly larger than pre-fire (room temperature) deflection in the member. The residual deformation was calculated to be 43 mm for beam B1 and 74 mm for beam B2 (see Fig. 13.12), and this was assumed to be the initial state for the residual strength analysis in Stage 3.

The residual moment-deflection response of beams B1 and B2 is evaluated in Stage 3 of the analysis and is plotted in Fig. 13.13. Both fire-damaged beams exhibit three key phases in deflection progression: linear response (marked as A-B in Fig. 13.13), onset of yielding in steel reinforcement (marked as B in Fig. 13.13), and plastic deformation until failure (marked as B-C in Fig. 13.13). In the first phase (see A-B in Fig. 13.13), moment-deflection response of fire- damaged beams follows a linear trend as seen in a cracked section, until the onset of yielding in steel reinforcement. This can be attributed to extensive tensile cracking and temperature-induced material degradation that occur in the beams during fire exposure. The yielding moment for fire-damaged beams, B1 and B2, was calculated to be 135 and 150 kN·m, respectively. Following yielding, the beams deform in a plastic fashion with further increase in moment-carrying capacity due to the strain hardening effect in steel reinforcement.

The peak moment-carrying capacity in fire-damaged beams B1 and B2 was calculated to be 189 and 164 kN·m, respectively (as can be seen in Fig. 13.13). This calculated residual capacity is comparable to room temperature design capacity computed as per ACI 318 (ACI, 2008) even after significant fire damage, and this is due to the fact that strain hardening of steel is conservatively not accounted for in ACI 318 (ACI, 2008) design equations for estimating capacity of beam in Stage 1 (Kodur et al., 2010). However, the residual capacity of beams (in B1 and B2) represents a recovery of 76% and 65%, respectively, with respect to realistic room temperature capacity of the beam calculated using finite element analysis. Thus, following a fire incident, the fire-damaged member may satisfy the design limit state from a strength consideration but needs to be retrofitted to provide a comparable level of safety (reserve capacity) as existed prior to the fire incident.

The above-illustrated advanced analysis procedure can be applied for other types of structural members made of steel, composite, or timber materials. However, due consideration should be given to appropriate material properties and failure limit states in those unique materials. The application of the advanced analysis procedure to evaluate the residual capacity of fire-exposed steel members is illustrated by Aziz and Kodur (2013).

13.7 Repair Strategies

Following conditional assessment and residual capacity analysis, appropriate repair strategies are to be applied to repair the fire-damaged structure. Knowing the extent of loss in structural capacity, appropriate repair and retrofitting techniques can be adapted to enhance the structural capacity to its original level (Torgal et al., 2017). This is to be achieved keeping in mind that the repaired structure should have at least the same level of load-bearing capacity as before the fire, and also that the repaired structure should meet the same fire safety level as that prior to the fire incident.

Depending on the damage class (as per Table 13.3), repair solutions can vary from simple actions such as cleaning, painting, and aesthetic renovation (for structures with Class 1 damage); repairing surface of members (Class 2 damage) only using approved materials (such as polymer-modified mortars, paints, and coatings) or replacing single elements (as in the case of steel or prefabricated-concrete members grouped under Class 3 or 4 damage); or adopting more complex retrofitting procedures (members with Class 4 or 5 damage) such as adding extra insulation layers reinforcement (i.e., FRP composites), fixing buckled steel sections,[1] or adding extra material to enhance sectional size. Selecting an appropriate repair material is a critical step in the repair process. Repair materials must be compatible with the base material, strength needs, available technical resource, and other specific requirements of the fire-damaged structure. In general, repair materials should be of similar configuration to that of the original construction material. These repair techniques and retrofitting solutions vary with the type of construction material the structure is made of and are covered below for concrete, steel, and wood structures.

13.7.1 Concrete Structures

In the case of fire-damaged concrete structures, fresh concrete can be used to fill in damaged (or spalled) sections. Alternatively, shotcrete, a cementitious repair material, can be used to repair fire-damaged RC members. In contrast to conventional concrete, the use of shotcrete omits the need for formwork and reduces the time for repairs. Shotcrete exhibits a short setting time and develops mechanical properties at an early age and hence facilitates quick repair. Ferrocement is another cementitious material and is composed of layers of wire meshes and mortar that can be applied on fire-damaged concrete structural members. While applying ferrocement is labor-intensive, the workmanship required to prepare this mix is simple and the bulk of its raw materials are locally available.

When structural capacity is to be significantly enhanced, additional steel reinforcement or steel plates should be installed in fire-damaged structures. In recent years, fiber-reinforced polymers (CFRP, GFRP, or AFRP) are often used to enhance structural capacity of defective RC structures.[2] FRP plates or strips can be attached to the surface of a fire-damaged concrete beam or slab after the concrete surface is prepared. Similarly, FRP sheets can be wrapped around the exterior face of fire-damaged RC columns as well. The

[1] In the case of concrete, masonry, and timber members, all the external layers that have been subjected to high temperatures (in excess of 300°C in concrete and steel and 100°C for wood) need to be removed through chiseling, sand blasting, and cleaning with high-pressure water jets.

[2] Details on the application of FRP in strengthening, including calculation methods for capacity evaluation, is covered in Ahmed and Kodur (2011) and Kodur and Yu (2013).

advantages of FRP materials include ease of application, high strength-to-weight ratio, excellent mechanical strength, and resistance to corrosion and chemical attack. These materials also exhibit low density, which is important because they add less weight to the repaired structure and can be conveniently applied without using heavy machinery. FRP is also suitable for structures of various geometry types or cross section. One of the main issues that arises with the use of FRP in retrofitting members is the low fire resistance of the repaired member. The retrofitted structural member should be checked to ensure it meets the required fire resistance (rating). If it does not, then the retrofitted member should be provided with appropriate fire insulation to meet the required resistance.

13.7.2 Steel Structures

The heat-straightening technique can be applied to "fix" fire-damaged steel and composite members. Prior to the undertaking of repair, the type and extent of damage are to be categorized based on conditional assessment as described in Sec. 13.5. *Federal Highway Administration Guide* (FHWA, 2008) groups the extent of damage in steel members under four categories:

- *Category S.* This type refers to damage as a result of bending about the "strong" or major axis. For rolled or built-up shapes, the web element is bent about its strong axis with one flange element in compression and the other in tension. In addition to plastic deformation, the compression flange and web will sometimes exhibit local buckling due to the high compressive stresses.
- *Category W.* This category refers to damage as a result of bending about the "weak" or minor axis. For rolled or built-up I-shapes, the neutral axis is usually within, or near, the web. Consequently, the web may not yield or deform into the inelastic range. If neither is laterally restrained, the flange elements are bent about their strong axes and usually exhibit a classical flexural yielding pattern.
- *Category T.* This type refers to damage as a result of torsion or twisting about the longitudinal axis of a member. For rolled or built-up shapes, if neither is laterally braced, the flange elements tend to exhibit flexural plastic deformation in opposite directions. The web is often stressed at levels below yield. If one flange is constrained (such as in the case of a composite bridge girder), then the unconstrained flange element is subjected to plastic deformation, and yielding may also occur in the web.
- *Category L.* This category includes damage that is localized in nature. Local flange or web buckling, web crippling and damage at bracing locations, and bends or crimps in plate elements of a cross section typify this behavior.

The main repair strategy for fixing fire-damaged steel members is through heat-based adjustment (i.e., heat straightening). When the member has undergone large distortions during fire, it has to be brought back to its original configuration through such heat-straightening techniques. After that, retrofitting materials such as steel plates or FRP layers (as discussed above for concrete members) can be installed to enhance the capacity of fire-damaged steel members (Schnerch et al., 2007). Following the repair, the retrofitted member should be provided with appropriate levels of fire insulation to meet fire resistance requirements as per applicable codes.

13.7.3 Timber Structures

In the case of wooded structural members, when there is nominal capacity loss during a fire incident, the first step in the repair process is to remove any charred layers of wood. Following this, the surfaces of the wood member should be clean-painted to remove odor and to fix cosmetic damage (coloring). King (2002) indicated that in most cases structural strengthening of fire-damaged wood framing members (in buildings) is not required if the char depth is less than 6 mm. In the case of partially fire-damaged wood (more than 6 mm char depth) where there is some level of capacity loss, repair strategies often consist of reinforcing the original damaged member by attaching a supplemental piece of wood to it. This action is referred to as *sistering*. To prevent future damage, epoxy-repaired timber should be protected by a thick outer wood layer or other protective material such as gypsum board. In addition, FRP and metallic strengthening systems, as in the case of concrete and steel structures, can also be added to enhance the capacity of fire-damaged timber structures. In the case of wood structures, where there is a significant loss of capacity, the members should be replaced. Following the fire, the retrofitted wood structure should be checked for fire resistance requirements, and if it does not possess same level of fire resistance as before, then appropriate fire insulation should be provided, as discussed in Chap. 10.

13.8 Emerging Trends and Future Needs

Although fire incidents in buildings and civil infrastructure have increased over the years, collapse of structures due to fire is rare. This is due to efficient and effective firefighting, as well as the presence of active and passive fire protection systems in buildings. However, when structural members experience some level of damage, the extent of residual capacity of such members is to be ascertained before opening for reoccupancy to ensure safety. Even to date there are very limited engineering approaches and guidance for ascertaining the residual capacity of fire-damaged structures. Therefore, there is an urgent need for research and development efforts in this area to advance the state of the art.

13.8.1 Reliable Approaches for Residual Capacity Assessment

There is scope for basic and advanced research aimed at developing advanced numerical models for evaluating the post-fire residual capacity of fire-weakened structural members. Such numerical models should account for variation in temperature-dependent properties of materials (during both heating and cooling phases), as well as for material-specific phenomena (spalling in concrete, buckling in steel, charring in wood, etc.) and structural parameters (restraint, load level, etc.). Such advanced models should be applied to carry out a series of parametric studies to quantify the influence of critical parameters on the residual capacity of structural members under different fire scenarios. A promising technique, referred to as *event reconstructing,* in which advanced analysis can be carried out to trace the demand-to-capacity ratio in a fire-damaged structure, was explored by Pfister et al. (2002). This technique can be further developed to identify retrofitting solutions to "fix" the fire-damaged members.

13.8.2 Sensing Techniques for Continual Assessment During or Following a Fire

Currently, there are some nondestructive evaluation (NDE) techniques for assessing the condition of a structure following a fire incident. However, most of these techniques are not validated for damage assessment following fire exposure. During fire exposure, a structure (material) undergoes microstructural changes, and this is different from that under other loading conditions. Therefore, there is a need for establishing the validity of currently used NDE techniques to assess structure (material) damage following fire exposure.

There is scope for extending advanced *thermography* techniques, which involve applying a heat source (or mechanical stimulation) as excitation input to identify defects to get an estimate of fire-induced damage and thus assess the residual capacity in a fire-damaged structure (Hiasa et al., 2018). This requires the development of suitable thermographic sensors for incorporation in unmanned vehicles (UVs), which can be deployed during a fire event to assess real-time damage (Moore, 2011). Unmanned aerial vehicles (UAVs or drones) can also collect color and infrared images to allow rapid inspection of structures during or after a fire incident. Further, UAVs can also be used to pinpoint severely damaged structural members to provide an immediate and accurate assessment of the level of damage, which would help authorities and first responders to make critical decisions such as quickly entering a building or a tunnel following a fire incident.

In recent years, the use of UAVs in disaster management can be explored due to the latest advances in UAV technology, as well as emerging technologies such as big data and IoT (internet of things). However, one area where there is not much progress is damage assessment during a fire event. This is mainly due to the lack of thermal sensors that are capable of monitoring response under elevated temperatures. Therefore, there is an urgent need for developing thermal sensors for incorporation in UAVs so as to monitor response parameters (deflection, stresses, color change) during a fire event.

13.9 References

ACI. *ACI 318-08: Building Code Requirements for Reinforced Concrete*. Vol. 552. Detroit, MI: ACI, 2008.

Agrawal, Ankit. "Residual Response of Reinforced Concrete Beams Following Fire Exposure." PhD Dissertation, Michigan State University, 2020.

Agrawal, A., and V. Kodur. "Residual Response of Fire-Damaged High-Strength Concrete Beams." *Fire and Materials* 43, no. 3 (2019): 310–322.

Ahmed, Aqeel, and Venkatesh Kodur. "The Experimental Behavior of FRP-Strengthened RC Beams Subjected to Design Fire Exposure." *Engineering Structures* 33, no. 7 (2011): 2201–2211. https://doi.org/10.1016/J.ENGSTRUCT.2011.03.010.

Anderberg, Yngve, and Sven Thelandersson. "Stress and Deformation Characteristics of Concrete at High Temperatures. 2. Experimental Investigation and Material Behaviour Model." *Bulletin of Division of Structural Mechanics and Concrete Construction* 54 (1976). https://lup.lub.lu.se/search/publication/1245737.

Aziz, Esam Mohammed. Response of fire exposed steel bridge girders. Michigan State University, 2015.

Aziz, Esam, and Venkatesh Kodur. "An Approach for Evaluating the Residual Strength of Fire Exposed Bridge Girders." *Journal of Constructional Steel Research* 88 (2013): 34–42. https://doi.org/10.1016/J.JCSR.2013.04.007.

Beitel, J., and Nestor Iwankiw. 2008. *Analysis of Needs and Existing Capabilities for Full-Scale Fire Resistance Testing.* NIST GCR 02-843-1. Baltimore, MD: Highes Associates.

CEN. *Eurocode 1: Actions on Structures. Part 1-2 : General Actions—Actions on Structures Exposed to Fire. EN 1991-1-2.* Brussels: European Committee for Standardization, 2002.

CEN. *Eurocode 2: Design of Concrete Structures. Part 1-1: General Rules and Rules for Buildings. Eurocódigo. EN 1992-1-1.* Brussels: European Committee for Standardization, 2004.

CEN. *Eurocode 3: Design of Steel Structures—Part 1-2: General Rules—Structural Fire Design. EN 1993-1-2.* Brussels: European Committee for Standardization, 2005a.

CEN. *Eurocode 4: Design of Composite Steel and Concrete Structures—Part 1-2: General Rules—Structural Fire Design. EN 1994-1-2.* Brussels: European Committee for Standardization, 2005b.

Chang, Y. F., Y. H. Chen, M. S. Sheu, and G. C. Yao. "Residual Stress-Strain Relationship for Concrete after Exposure to High Temperatures." *Cement and Concrete Research* 36, no. 10 (2006): 1999–2005. https://doi.org/10.1016/J.CEMCONRES.2006.05.029.

Dill, F. H. 1960. "Structural Steel after a Fire." *Proceedings of National Steel Construction Conference,* May 5–6, Denver, CO. Chicago, IL: American Institute of Steel Construction, 1960.

Fan, J., and Z. Lu. "Experimental Study on Materials Properties of Prestressed Steel Wire Post High Temperatures." *Industrial Construction* 32, no. 9 (2002): 30–31.

Felicetti, Roberto, Pietro G. Gambarova, and Alberto Meda. "Residual Behavior of Steel Rebars and R/C Sections after a Fire." *Construction and Building Materials* 23, no. 12 (2009): 3546–3555. https://doi.org/10.1016/J.CONBUILDMAT.2009.06.050.

FHWA. "Guide for Heat-Straightening of Damaged Steel Bridge Members Final Draft of the AASHTO Guidelines 5. Report Date." 2008. http://www.ntis.gov.

FiB. *FRP Reinforcement in RC Structures: Technical Report.* International Federation for Structural Concrete. 2007a. https://www.fib-international.org/publications/fib-bulletins/frp-reinforcement-in-rc-structures-115-detail.html.

FiB. *FRP Reinforcement in RC Structures, Bulletin No. 40.* 2007b. https://doi.org/10.1371/journal.pntd.0001792.

Gosain, N., F. Drexler, and Dilip Choudhuri.. "Evaluation and Repair of Fire-Damaged Buildings." *Structure Magazine* (September 2008): 18–22.

Hager, I. 2014. "Colour Change in Heated Concrete." *Fire Technology* 50, no. 4 (2014): 945–958.

Harmathy, T. "Effect of Moisture on the Fire Endurance of Building Elements." *Moisture in Materials in Relation to Fire (ASTM).* 1965. http://www.astm.org/DIGITAL_LIBRARY/STP/PAGES/STP48429S.htm.

Hertz, K. D. "Reinforcement Data for Fire Safety Design." *Magazine of Concrete Research.* 2004. http://orbit.dtu.dk/ws/files/144764856/Reinforcement_Data_for_Fire_Safety_Design_ver_2_Proofs.pdf.

Hiasa, Shuhei, Recep Birgul, Masato Matsumoto, and F. Necati Catbas. "Experimental and Numerical Studies for Suitable Infrared Thermography Implementation on Concrete Bridge Decks." *Measurement* 121 (June 2018): 144–159. https://doi.org/10.1016/J.MEASUREMENT.2018.02.019.

Hibbitt, Karlsson & Sorensen, Inc. "ABAQUS Standard User's Manual." Pawtucket (America). 2012.

Jiang, F., C. Zhu, and J. Xue. "Nonlinear Analysis of Load-Carrying Capacity for Bar Buckling Based on Arc-Length Method." *Building Science* 23, no. 11 (2007): 9–12.

King, M. *NICE Guidelines for Fire and Smoke Damage Repair.* The National Institute of Disaster Restoration, 2002.

Knudson, R. M, and A. P. Schniewind. "Performance of Structural Wood Members Exposed to Fire." *Forest Products Journal* (1975). http://agris.fao.org/agris-search/search.do?recordID=US7501067.

Kodur, V. K. R., and Ankit Agrawal. "An Approach for Evaluating Residual Capacity of Reinforced Concrete Beams Exposed to Fire." *Engineering Structures* 110 (March 2016): 293–306. https://doi.org/10.1016/J.ENGSTRUCT.2015.11.047.

Kodur, V. K. R., M. B. Dwaikat, and R. S. Fike. "An Approach for Evaluating the Residual Strength of Fire-Exposed RC Beams." *Magazine of Concrete Research* 62 (2010): 479–488. https://doi.org/10.1680/macr.2010.62.7.479.

Kodur, V. K. R., N. K. Raut, X. Y. Mao, and W. Khaliq. "Simplified Approach for Evaluating Residual Strength of Fire-Exposed Reinforced Concrete Columns." *Materials and Structures* 46, no. 12 (2013): 2059–2075. https://doi.org/10.1617/s11527-013-0036-2.

Kodur, V. K. R., and Baolin Yu. "Evaluating the Fire Response of Concrete Beams Strengthened with Near-Surface-Mounted FRP Reinforcement." *Journal of Composites for Construction* 17, no. 4 (2013): 517–529. https://doi.org/10.1061/(ASCE)CC.1943-5614.0000348.

Kretschmann, David. 2010. "Mechanical Properties of Wood." In *Wood Handbook: Wood as an Engineering Material.* Centennial, ed. General Technical Report FPL ; GTR-190. Madison, WI : U.S. Dept. of Agriculture, Forest Service, Forest Products Laboratory, 2010: pp. 5.1-5.46., 190:5.1-5.46; Chapter 5. https://www.fs.usda.gov/treesearch/pubs/all/37427.

MacLean, Kevin J. N. 2007. *Post-Fire Assessment of Unbonded Post-Tensioned Concrete Slabs: Strand Deterioration and Prestress Loss.* M.Sc. Engineering thesis. Ontario: Queen's University. https://qspace.library.queensu.ca/handle/1974/954.

Malhotra, V. M., and N. J. Carino. *Handbook on Nondestructive Testing of Concrete.* Boca Raton, FL: CRC Press, 2003.

Moore, J. 2011. "UAV Fire-Fighting System." Issued November 29, 2011. https://patents.google.com/patent/US20130134254A1/en.

Neves, I. Cabrita, João Paulo C. Rodrigues, and António de Pádua Loureiro. "Mechanical Properties of Reinforcing and Prestressing Steels after Heating." *Journal of Materials in Civil Engineering* 8, no. 4 (1996): 189–194. https://doi.org/10.1061/(ASCE)0899-1561(1996)8:4(189).

PCI. 2011. MNL 124, Design for Fire Resistance of Precast Prestressed Concrete. Chicago, IL: PCI.

Pfister, Christian, Rudolf Brázdil, and Mariano Barriendos. "Reconstructing Past Climate and Natural Disasters in Europe Using Documentary Evidence Historical Climatology." *Pages News* 10, no. 3 (December 2002). www.pages-igbp.org/products/newsletters/ref2002_3.html.

Russo, S., and F. Sciarretta. "Experimental and Theoretical Investigation on Masonry after High Temperature Exposure." *Experimental Mechanics* 52, no. 4 (2012): 341–359. https://doi.org/10.1007/s11340-011-9493-0.

Schaffer, E. L. "Effect of Pyrolytic Temperatures on the Longitudinal Strength of Dry Douglas-Fir." *Journal of Testing and Evaluation.* 1, no. 4 (1973):319–329. https://compass.astm.org/download/JTE10025J.11533.pdf.

Schneider, Ulrich. "Concrete at High Temperatures: A General Review." *Fire Safety Journal* 13, no. 1 (1988): 55–68. https://doi.org/10.1016/0379-7112(88)90033-1.

Schnerch, D., M. Dawood, S. Rizkalla, and E. Sumner. "Proposed Design Guidelines for Strengthening of Steel Bridges with FRP Materials." *Construction and Building Materials* 21, no. 5 (2007): 1001–1010. https://doi.org/10.1016/j.conbuildmat.2006.03.003.

Short, N. R., J. A. Purkiss, and S. E. Guise. "Assessment of Fire Damaged Concrete Using Colour Image Analysis." *Construction and Building Materials* 15, no. 1 (February 2001): 9–15. https://doi.org/10.1016/S0950-0618(00)00065-9.

Tao, Zhong, Xing-Qiang Wang, and Brian Uy. "Stress-Strain Curves of Structural and Reinforcing Steels after Exposure to Elevated Temperatures." *Journal of Materials in Civil Engineering* 25, no. 9 (2013): 1306–1316. https://doi.org/10.1061/(ASCE)MT.1943-5533.0000676.

Tide, R. H. R. "Integrity of Structural Steel After Exposure to Fire." *Engineering Journal—AISC* 35, (1998): 26–38. https://s3.amazonaws.com/academia.edu.documents/44439158/Integrity_of_Structural_Steel_After_Exposure_to_Fire.pdf?AWSAccessKeyId=AKIAIWOWYYGZ2Y53UL3A&Expires=1544130297&Signature=BybAQYAFfNLbTFMLCO0cfF6NoKw%3D&response-content-disposition=inline%3B filename%3DIntegrity_of_Structural_Steel_After_Expo.pdf.

Torgal, Fernando Pacheco, R. E. Melchers, Nele de Belie, Xianming Shi, Kim Van Tittelboom, and Andrés Sáez Perez. *Eco-Efficient Repair and Rehabilitation of Concrete Infrastructures.* Cambridge: Woodhead Publishing, 2017.

USDA. *Wood Handbook: Wood as an Engineering Material.* Madison, WI: U.S. Department of Agriculture, Forest Service, Forest Products Laboratory, 2010. https://doi.org/General Technical Report FPL-GTR-190.

Wang, Song-Yung, Te-Hsin Yang, Li-Ting Lin, Cheng-Jung Lin, and Ming-Jer Tsai. "Fire-Retardant-Treated LoSw-Formaldehyde-Emission Particleboard Made from Recycled Wood-Waste." *Bioresource Technology* 99, no. 6 (2008): 2072–2077. https://doi.org/10.1016/J.BIORTECH.2007.03.047.

Wang, Wei-yong, Bing Liu, and Venkatesh Kodur. "Effect of Temperature on Strength and Elastic Modulus of High-Strength Steel." *Journal of Materials in Civil Engineering* 25, no. 2 (February 2013). https://doi.org/10.1061/(ASCE)MT.1943-5533.0000600.

Wickström, U. 1986. *A Very Simple Method for Estimating Temperaturein Fire Exposed Concrete Structures.* http://www.diva-portal.org/smash/record.jsf?pid=diva2%3A961620&dswid=881.

Yu, L. 2006. "Behavior of Bolted Connections during and After a Fire." https://repositories.lib.utexas.edu/bitstream/handle/2152/2959/yud72766.pdf.

13.10 Nomenclature

T	Temperature (°C)
f'_{cr}	Residual compressive strength
f'_{c}	Compressive strength of concrete
E_{cr}	Residual modulus
E_{c}	Modulus
f'_{tr}	Residual tensile strength
f_{t}	Tensile strength of concrete
E_{o}	Modulus property at ambient conditions
$f_{y,r,T}$	Residual yield strength
$f_{u,r,T}$	Residual ultimate strength
ξ	Ratio of material strength at peak temperature T (°C) to the same property at ambient upon cooldown
$f_{c,\theta}$	Compressive strength of masonry
E_{θ}	Modulus of masonry
$f_{t,\theta}$	Tensile strength of masonry
$f_{bc,\theta}$	Compressive strength of brick
$E_{bc,\theta}$	Modulus of brick
$f_{mc,\theta}$	Compressive strength of mortar
$E_{m,\theta}$	Modulus of mortar
T_{smax}	Maximum rebar temperature
T_{fmax}	Maximum fire temperature

λ	Modification factor to account for specific fire and section (and materials) characteristics
t_h	Duration of the heating phase (hours)
t_c	Duration of the cooling phase (hours)
H	Maximum rebar temperature
B	Section width
x	Axis distance to rebar (m)
A_s	Area of tension reinforcement
$f_{y,r,T}$	Residual strength of reinforcing steel
d^*	Effective depth of the fire-damaged concrete section
b^*	Width of the fire-damaged concrete section
f'_{cT}	Reduced compressive strength of concrete
k_{st}	Buckling reduction coefficient
N_r	Residual strength of fire-exposed RC column
A_{core}	Area of core region with temperature less than 500°C

13.11 Numerical Examples

13.1

Determine the residual capacity of an RC column exposed to fire in a building. The square cross-sectioned (305 mm in size) column is 3 m in length and is made of 42 MPa concrete. The column has four 20-mm diameter longitudinal bars and 10-mm diameter stirrups spaced at 305 mm, and the clear concrete cover is 50 mm. The yield strength of the longitudinal bars is 444 MPa and that of stirrups is 427 MPa. The typical service load present on the column during fire exposure is 2400 kN. Assume that this column was exposed to a fire composed of 2 hours of heating as per ISO 834 exposure followed by 2 hours of cooling. Use the approach presented in Sec. 13.6.1.

Solution

Ultimate load-bearing capacity of the column at room temperature is

$$N = 0.8 \times (A_c f_c + A_s f_y) = 0.8 \times (305 \times 305 \times 42 + 1256 \times 444) = 3571771\,\text{N} = 3572\,\text{kN}$$

Step 1. Estimation of maximum fire temperature and fire duration.

Although we can make a coarse estimation of the maximum fire temperature from the color change of the exposed surface of the concrete, here we take the temperature of the fire (at 120 minutes) to be according to ISO 834 standard fire curve, which is given by

$$T_f = 20 + 345 \times \log_{10}(8 \times t + 1) = 20 + 345 \times \log_{10}(8 \times 120 + 1) = 1049°\text{C}$$

Step 2. Calculation of maximum rebar temperature experienced during fire.

Distance from the column surface to the center of rebar is 50 + 20/2 = 60 mm, so $x = 0.06$ m and $y = 0.06$ m. Knowing that $t_h = 120$ minutes (2 hours) and $T_f = 1049°\text{C}$, then η_x, η_y, η_w, and T_{bar} can be calculated as

$$\eta_x = \eta_y = 0.18 \ln\left(\frac{t_h}{x^2}\right) - 0.81 = 0.18 \ln\left(\frac{2.0}{0.06^2}\right) - 0.81 = 0.3276$$

$$\eta_w = 1 - 0.0616 t_h^{-0.88} = 1 - 0.0616 \times 2.0^{-0.88} = 0.9665$$

$$T_{bar} = [\eta_w(\eta_x + \eta_y - 2\eta_x\eta_y) + \eta_x\eta_y]\,T_f$$

$$= [0.9665 \times (0.3276 + 0.3276 - 2 \times 0.3276 \times 0.3276) + 0.3276 \times 0.3276] \times 1049 = 559°\text{C}$$

With the above equations, the location of the 500°C isotherms in concrete core can be calculated. This location (through a trial-and-error approach) is found to be at 84 mm. The effective area of concrete with temperature less than 500°C = (305 − 84)×(305 − 84) = 48,841 mm².

Step 3. Determination of residual strength in rebar and concrete.

The reduction coefficient for evaluating residual strength rebar as per Eurocode 3 is

$$k_{y,T} = 0.61 - 0.5(T - 500)/200$$

$$k_{y,T} = 0.61 - 0.5(559 - 500)/200 = 0.47$$

Thus, the residual strength of rebar after fire exposure is 440 × 0.47 = 203.5 MPa. The reduction coefficient for effective concrete in the "core region" (with a temperature less than 500°C) is assumed to be 1.0. Hence, there is no reduction in residual strength of concrete in the core region (from 42 MPa).[3]

Step 4. Calculation of residual capacity of the fire-exposed RC column.

The residual capacity of RC column can be determined by the following equation:

$$N = A_{\text{core}} f_{ct} + k_{st} A_s f_{yt} = 48841 \times 42 + 1 \times 1256 \times 203.5 = 2306.9 \text{ kN}$$

Here, stability factor k_{st} is taken to be 1.0, since no spalling has been reported.

Summary. Applying the approach proposed by Kodur et al. (2013), the residual capacity of fire-exposed RC column is found to be 2306.9 kN. The residual capacity of this column is lower than the service load, which is 2400 kN. Thus, this fire-damaged RC column needed to be strengthened to enhance the load-bearing capacity to its original (pre-fire) strength level.

13.2

Using advanced analysis, evaluate the post-fire residual response of a simply supported steel bridge girder shown in Figure E13.1. The steel bridge comprises five hot-rolled steel girders of W33×141 supporting a reinforced concrete slab of 200-mm thickness. The steel girder is assumed to be in full composite action with the slab and to be laterally supported by transverse diaphragms at the midspan, as well as at both ends, to prevent lateral movement. The bridge girder is 13.2 m in span length and has two expansion joints at its ends with a width of 36 mm. The girders are assumed to be fabricated from grade 50 steel (yield strength of 350 MPa), while the concrete used in the slab is of compressive strength of 30 MPa. Assume the bridge is exposed to a severe hydrocarbon fire.

Solution (for advanced analysis procedure refer to Aziz and Kodur, 2013)

Step 1. Develop the geometric model of the girder and mesh it using thermal elements: SOLID70, and structural elements: SOLID65, SOLID46, SOLID45, LINK8, and INTER205, respectively. Discretize the girder as shown in Fig. E13.2.

Step 2. Input temperature-dependent material properties of concrete and reinforcing steel as per Eurocode 2 provisions. Also input hydrocarbon fire temperature curve and corresponding convection coefficient of $\alpha_c = 50$ W/(m²·°C) as recommended in Eurocode 1.

Step 3. Carry out a heat transfer analysis to get temperature rise across the depth of the girder. Results from thermal analysis, sectional temperature at various depths, are plotted in Fig. E13.3.

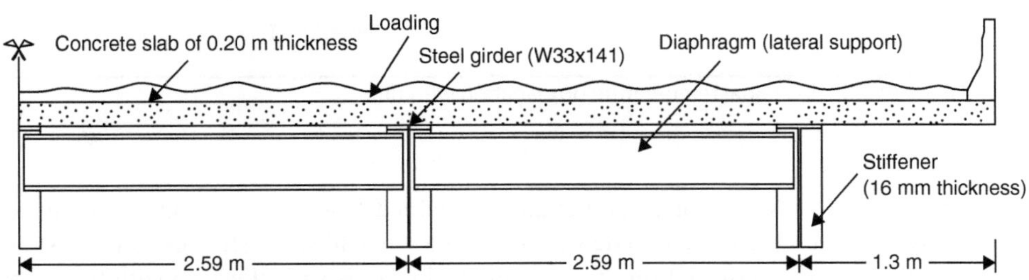

FIGURE E13.1 Bridge girder exposed to fire. (With permission from Mohammed, 2015.)

[3]These reduction factors can be evaluated using residual strength temperature relations in applicable codes/guidance documents or as per guidance given in Sec. 13.3.

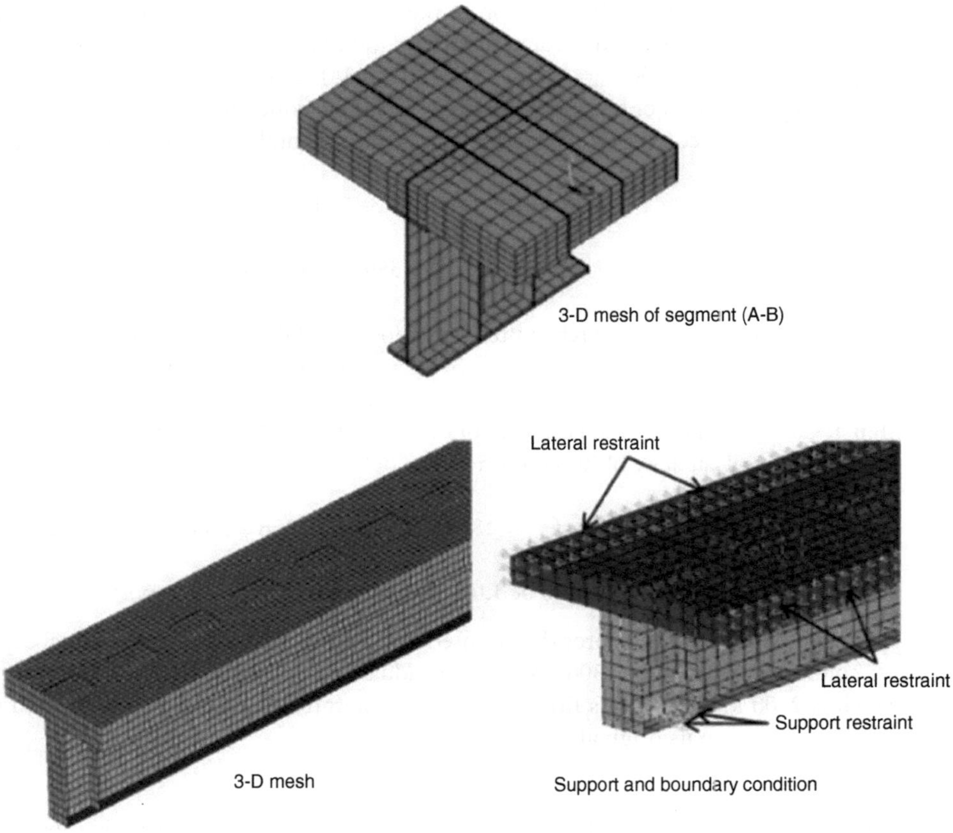

FIGURE E13.2 Discretization of a girder for fire resistance analysis.

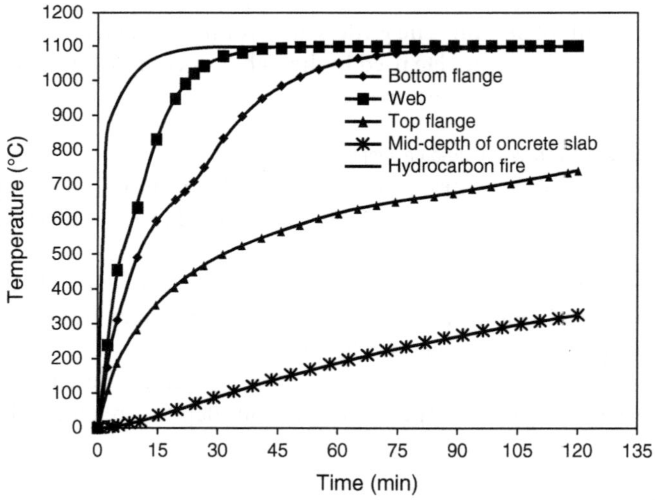

FIGURE E13.3 Sectional temperature propagation with time.

Step 4. The cross-sectional temperatures obtained from Step 3 form the input to the second stage of analysis: structural analysis. The structural analysis will show stresses as well as midspan deflection as a function of fire exposure time. A variation of this midspan deflection with time is shown in Fig. E13.4 as predicted by Aziz and Kodur (2013). Much of the deflection recovers following cooldown of the girder, with only a very small amount of residual deflections. In addition to hydrocarbon fire, the deflections are also generated for two other scenarios, external design fire as per Eurocode 1 and a moderate design fire.

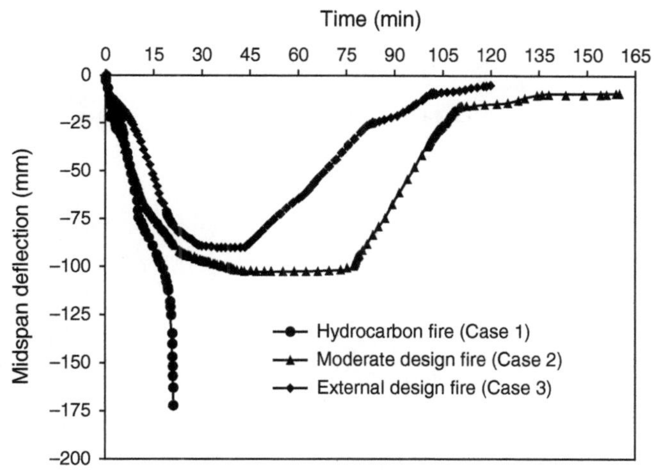

Figure E13.4 Evolution of deflection with fire exposure time.

Case	Fire Scenario	Max. fire Temperature	Max. Steel Temperature	Room Temperature Capacity (kN·m)	Residual Capacity (kN·m)	% of Original Capacity
Case 1	Hydrocarbon fire	1100°C	1000°C	5104	Failure under fire	–
Case 2	Moderate fire	800°C	795°C	5104	2915	57%
Case 3	External fire	680°C	670°C	5104	3665	72%

Table E13.1 Results from Residual Strength Analysis on Fire-Exposed Bridge Girder

Step 5. The maximum temperature reached in steel is utilized to compute the residual yield strength of steel, while the residual compressive strength of concrete is evaluated based on EC4 strength-temperature relations. Air cooling is assumed during the decay phase, since the material properties in Stage 3 of the numerical analysis was on the same basis. Finally, the residual moment capacity of the steel girder is evaluated by applying appropriate strength equations but using residual strength of steel and concrete. Results of residual capacity prediction are listed in Table E13.1.

Index

Note: Page numbers followed by *f* refer to figures; by *t,* tables; by *n,* footnotes.

D

E

9 781260 128581

Printed by Libri Plureos GmbH in Hamburg,
Germany